Thomas Liddell Ainsley

A Guide Book to the Local Marine Board Examination

The Ordinary Examination. 39th Edition

Thomas Liddell Ainsley

A Guide Book to the Local Marine Board Examination
The Ordinary Examination. 39th Edition

ISBN/EAN: 9783337417635

Printed in Europe, USA, Canada, Australia, Japan

Cover: Foto ©Lupo / pixelio.de

More available books at **www.hansebooks.com**

A

GUIDE BOOK

TO THE

LOCAL MARINE BOARD

EXAMINATION.

THE ORDINARY EXAMINATION.

BY THOMAS L. AINSLEY,
TEACHER OF NAVIGATION.

Thirty-ninth Edition.

SOUTH SHIELDS:
PRINTED AND PUBLISHED BY THOMAS L. AINSLEY, 16, MARKET PLACE,
AND 86, BUTE STREET, CARDIFF.
LONDON: SIMPKIN, MARSHALL, & Co., STATIONERS' HALL COURT;
HAMILTON, ADAMS, & Co., PATERNOSTER ROW;
SAMPSON, LOW, MARSTON, & Co., FLEET STREET;
R. H. LAURIE, 53, FLEET STREET; CHARLES WILSON
(LATE NORIE AND WILSON), 157, LEADENHALL STREET; J. IMRAY AND SON,
89 AND 102, MINORIES; J. D. POTTER, 31, POULTRY.
LIVERPOOL: G. PHILIP AND SON, CAXTON BUILDINGS.
EDINBURGH: J. MENZIES & Co.
SYDNEY & MELBOURNE: GEORGE ROBERTSON.

FRODSHAM AND KEEN,
CHRONOMETER,
Watch & Nautical Instrument Makers,
9, St. George's Crescent, LIVERPOOL.

SOUTH SHIELDS:
THOMAS L. AINSLEY, PRINTER, 16, MARKET PLACE.

PREFACE TO THE FIRST EDITION.

This Work is intended as a Guide to the Officers of all grades of the Merchant Service, in the examinations they are required to undergo before the Local Marine Board. It will be issued in Two parts:—Part I containing what is termed the Ordinary Examination, and Part II containing the Extra Examination.

The present volume, which relates to the Ordinary Examination, contains model solutions of examples in the various problems required of Candidates when under Examination, with numerous Exercises to each Problem, together with a variety of Examination Papers. It also contains all requisite information respecting the Deviation of the Compass; Lights of the English, St. George's, and Bristol Channels, &c.; Stowage of Cargoes; Invoice; Charter Party; Bottomry Bonds, &c.

In the preparation of the articles on Seamanship, the following works have been consulted:—"The Kedge Anchor," by W. Brady, U.S.N.; "The Seaman's Friend," by R. H. Dana, jun; "The Sheet Anchor," by Darcy Lever, Esq.; while my obligations to other works have been duly acknowledged. The works of ABBOTT, LEES, STEELE, and M'CULLOCH, &c., are the authorities that have been consulted on the subjects of Charter Party, Bills of Lading, &c.

T. L. A.

South Shields, July 10th, 1856.

ADVERTISEMENT

TO

THE THIRTY-EIGHTH EDITION.

In this Edition of the "GUIDE BOOK," such alterations and additions have been made in the work as were necessary to adapt it to the present requirements of the Examinations—considerable alterations in the Examination Papers having recently come into operation.

T. L. A.

South Shields,
May 30th, 1880.

ERRATA ET CORRIGENDA.

Page 345, Ex. 9, for index correction $+ 2° 30''$ read $+ 2' 30''$.
" 349, Ex. 9, for time by chronometer $8^h 7^m 37^s$ read $8^h 7^m 27^s$.
" 352, Ex. 12, for Feb. $28^d 12^h 25^m 35^s$ read Feb. $28^d 12^h 26^m 30^s$.

CONTENTS.

	PAGE
Notices of Examinations of Masters and Mates	1
Examination in Colour Blindness	10
Places and Days of Examinations	12
Explanation of Signs used	13
Principles and Practice of Arithmetic	14
Decimal Fractions	41
On Logarithms	58
Multiplication by Logarithms	73
Division by Logarithms	76
Trigonometrical Tables	81
Natural Sines, &c.	81
Trigonometrical Ratios	84
Logarithmic Sines, &c.	85
Navigation—Definitions of, &c.	97
Preliminary Rules in Navigation—To Find Difference of Latitude	106
To Find Meridional Difference of Latitude	107
To Find the Latitude in	108
To Find Middle Latitude	108
To Find the Difference of Longitude	109
To Find the Longitude in	110
The Compass	112
Correcting Courses—Leeway	120
Variation	122
Deviation of the Compass	130
Methods of Finding the amount of the Deviation	150
Correction of Compass Bearings	157
Napier's Diagram	161
Fundamental Formulæ of Navigation	182
On the Traverse Table	183
Traverse Sailing	189
Parallel Sailing	194
Middle Latitude Sailing	197
Mercator's Sailing	200
The Day's Work	206
Preliminary Rules in Nautical Astronomy—The Conversion of Civil into Astronomical Time	223
The Conversion of Astronomical into Civil Time	224
The Conversion of Longitude into Time	225
The Conversion of Time into Longitude	225
On Finding the Greenwich Date	227
To Reduce the Sun's Declination	230
To Find Polar Distance	237
To Find the Equation of Time	238
Correction of Sun's Observed Altitude	242

CONTENTS.

	PAGE
To Find the Latitude by Sun's Meridian Altitude	244
Variation by an Amplitude	250
On Finding the Time of High Water, by Admiralty Tide Tables	258
To find the Rate of a Chronometer	268
Greenwich Date by Chronometer	271
To Find the Hour-angle	275
On Finding the Longitude by Chronometer	277
On Sumner's Method	287
On Finding the Variation by an Azimuth	304
On Finding the Latitude by Reduction to Meridian	313
On Finding the Latitude by a Meridian Altitude of a Fixed Star	321
Examination Papers	324
Quadrant and Sextant	379
Adjustments of the Quadrant, Sextant, &c.	384
How to Find the Course to Steer in a Known Current	389
Soundings	389
On the Chart	394
Mercator's Chart	394
Practical Examination in the use of the Chart	398
Answers	409

EXAMINATION OF MASTERS AND MATES

FOR

CERTIFICATES OF COMPETENCY

Under "*The Merchant Shipping Act, 1854*,"

AND

VOLUNTARY EXAMINATION IN STEAM.

---

1. UNDER the provisions of "The Merchant Shipping Act, 1854," no "Foreign-going Ship"* or "Home Trade Passenger Ship"* can obtain a clearance or transire, or legally proceed to sea, from any port in the United Kingdom, unless the Master thereof, and in the case of a Foreign-going Ship, the First and Second Mates or Only Mate (as the case may be), and in the case of a "Home Trade Passenger Ship" the First or only Mate (as the case may be), have obtained and possess valid Certificates, either of competency or Service, appropriate to their several stations in such ship, or of a higher grade; and no such ship, if of *one hundred tons burden or upwards*, can legally proceed to sea unless at *least one officer besides the Master* has obtained and possesses a valid Certificate, appropriate to the grade of Only Mate therein, or to a higher grade; and every person who having been engaged to serve as Master, or as First or Second or Only Mate of any "Foreign-going Ship," or as Master or First or Only Mate of a "Home Trade Passenger Ship," goes to sea as such Master or Mate without being at the time entitled to and possessed of such a Certificate as the Act requires, or who employs any person as Master, or First, Second, or Only Mate of any "Foreign-going Ship," or as Master or First or Only Mate of any "Home Trade Passenger Ship," without ascertaining that he is at the time entitled to and possessed of such Certificate, *for each offence incurs a penalty not exceeding fifty pounds.*

2. Every Certificate of *Competency* for a "Foreign-going Ship" is to be deemed to be of a higher grade than the corresponding Certificate for a "Home Trade Passenger Ship," and entitles the lawful holder to go to sea in the corresponding grade in such last-mentioned Ship; but *no Certificate for a "Home Trade Passenger Ship" entitles the holder to go to sea as Master or Mate of a "Foreign-going Ship."*

3. *Certificates of Competency* will be granted to those persons who pass the requisite examinations, and otherwise comply with the requisite conditions. For this purpose examiners have been appointed, and arrangements have been made for holding examinations at the ports and upon the days mentioned in the Table marked A. appended hereto. The days for examination are so arranged for general convenience, that a candidate wishing to proceed to sea, and missing the day at his own port, may proceed to another port where an examination is coming forward.

4. Candidates for examination must give in their names to the Local Marine Board if the place where they intend to be examined is a port where there is a Local Marine Board, on or before the day of examination (except in the case of London† and Liverpool), and

---

* By a "Foreign-going Ship" is meant one which is bound to some place out of the United Kingdom beyond the limits included between the River Elbe and Brest; and by a "Home Trade Passenger Ship" is meant any Home Trade ship employed in carrying Passengers; and it is to be observed that *Foreign Steam Ships when employed in carrying Passengers between places in the United Kingdom are subject to all the Provisions of the Act, as regards Certificates of Masters, Mates, and Engineers, to which British Steam Ships are subject*: s. 291 of the Merchant Shipping Act, 1854, and s. 5 of the Merchant Shipping Acts, &c., Amendment Act, 1862.

† At London applications for examination must be made on Fridays from 10 till 4, and on Saturdays from 10 till 3.

At Liverpool applications for examination must be made on Tuesdays, Wednesdays, Thursdays, and Saturdays, during office hours.

must conform to any regulations in this respect which may be laid down by the Local Marine Board; and if this be not done, delay may be occasioned.

5. Testimonials of character, and of sobriety, experience, ability, and good conduct on board ship will be required of all applicants, and without producing them no person will be examined. As such testimonials may have to be forwarded to the office of the Registrar-General of Seamen in London for verification before any certificates can be granted, it is desirable that candidates should lodge them as early as possible. The testimonials of servitude of Foreigners and of British Seamen serving in foreign vessels, which cannot be verified by the Registrar-General of Seamen, must be confirmed either by the Consul of the country to which the ship in which the candidate served belonged or by some other recognized official authority of that country, or by the testimony of some credible person on the spot having personal knowledge of the facts required to be established. Upon application to the Superintendent of the Mercantile Marine Office candidates will be supplied with a form (Exn. 2), which they will be required to fill up and lodge with their testimonials in the hands of the examiners.

6. Services which cannot be verified by proper Entries in the Articles of the Ships in which the Candidates have served cannot be counted. Thus,—for instance, A Man will state his Service to have been as Second or Only Mate, and to support his assertion will produce a Certificate of Discharge or of Employment by the Master stating that he served as Mate, when on reference to the Articles it appears that he has actually been rated as Boatswain; the service in such a case will not be regarded as having been in the capacity of Mate.

Whenever a Man has, from any cause, been regularly promoted on a vacancy in the course of the Voyage from the rank for which he first shipped, and such promotion, with the ground on which it has been made, is properly entered in the Articles and in the Official Log Book, he will of course receive credit for his service in the higher grade for the period subsequent to his promotion.

7. The examinations will commence early in the forenoon on the days mentioned in Table A, appended hereto, and will be continued from day to day until all the candidates whose names appear upon the Superintendent's list on the day of examination are examined.

8. Where the Local Marine Board are in every respect satisfied with the testimonials of a candidate, service in the coasting trade may be allowed to count as service, in order to qualify him for examination for a Certificate of Competency for Foreign-going Ships as a Mate, and two years' service as Mate in the coasting trade may be allowed to count as service for a Master's Certificate, provided the candidate's name has been entered as Mate on the Coasting Articles, and provided he has already passed an examination.

QUALIFICATIONS FOR CERTIFICATES OF COMPETENCY FOR A "FOREIGN-GOING SHIP."

The qualifications required for the several ranks undermentioned are as follow:—

9. A SECOND MATE must be seventeen years of age, and must have been four years at sea.

In Navigation.—He must write a legible hand, and understand the first five rules of arithmetic, and the use of logarithms. He must be able to work a day's work complete, including the bearings and distance of the port he is bound to, by Mercator's method; to correct the sun's declination for longitude, and find his latitude by meridian altitude of the sun; and to work such other easy problems of a like nature as may be put to him. He must understand the use of the sextant, and be able to observe with it, and read off the arc. (See List A, page 10).

In Seamanship.—He must give satisfactory answers as to the rigging and unrigging of ships, stowing of holds, &c.; must understand the measurement of the log-line, glass, and lead-line; be conversant with the rule of the road, as regards both steamers and sailing vessels, and the lights and fog signals carried by them, and will also be examined as to his acquaintance with "the Commercial Code of Signals for the use of all Nations."

10. An ONLY MATE must be nineteen years of age, and have been five years at sea.

In Navigation.—In addition to the qualification required for a Second Mate, an Only Mate must be able to observe and calculate the amplitude of the sun, and deduce the variation of the compass therefrom, and be able to find the longitude by chronometer by the usual methods. He must know how to lay off the place of the ship on the chart, both by bearings of known objects, and by latitude and longitude. He must be able to determine the error of a sextant, and to adjust it; also to find the time of high water from the known time at full and change. (See List A, page 10).

In Seamanship.—In addition to what is required for a Second Mate, he must know how to moor and unmoor, and to keep a clear anchor; to carry out an anchor; to stow a hold; and to make the requisite entries in the ship's log. He will also be questioned as to his knowledge of the use and management of the mortar and rocket lines in the case of the stranding of a vessel, as explained in the official log-book.

11. A FIRST MATE must be nineteen years of age, and have served five years at sea, of which one year must have been as either Second or Only Mate, or as both.*

In Navigation.—In addition to the qualification required for an Only Mate, he must be able to observe azimuths and compute the variation; to compare chronometers and keep their rates, and find the longitude by them from an observation of the sun; to work the latitude by single altitude of the sun off the meridian; and be able to use and adjust the sextant by the sun.

In Seamanship.—In addition to the qualification required for an Only Mate, a more extensive knowledge of seamanship will be required as to shifting large spars and sails, managing a ship in stormy weather, taking in and making sail, shifting yards and masts, &c., and getting heavy weights, anchors, &c., in and out; casting a ship on a lee-shore; and securing the masts in the event of accident to the bowsprit.

12. A MASTER must be twenty-one years of age, and have been six years at sea, of which at least one year must have been as First or Only Mate, and one year as Second Mate.

In addition to the qualification for a First Mate, he must be able to find the latitude by a star, &c. He will be asked questions as to the nature of the attraction of the ship's iron upon the compass, and as to the method of determining it. He will be examined in so much of the laws of the tides as is necessary to enable him to shape a course, and to compare his soundings with the depths marked on the charts. He will be examined as to his competency to construct jury rudders and rafts; and as to his resources for the preservation of the ship's crew in the event of wreck. He must possess a sufficient knowledge of what he is required to do by law, as to entry and discharge, and the management of his crew, and as to penalties and entries to be made in the official log; and a knowledge of the measures for preventing and checking the outbreak of scurvy on board ship. He will be questioned as to his knowledge of invoices, charter-party, Lloyd's agent, and as to the nature of bottomry, and he must be acquainted with the leading lights of the channel he has been accustomed to navigate, or which he is going to use. (See List B, page 10).

In cases where an applicant for a certificate as Master Ordinary has only served in a fore and aft rigged vessel, and is ignorant of the management of a square-rigged vessel, he may obtain a certificate on which the words "*fore and aft rigged vessel*" will be written. This certificate does not entitle him to command a square-rigged ship. This is not, however, to apply to Mates, who, being younger men, are expected for the future to learn their business completely. (See also page 9).

13. An EXTRA MASTER'S EXAMINATION is voluntary and intended for such persons as wish to prove their superior qualifications, and are desirious of having certificates for the highest grade granted by the Board of Trade.

In Navigation.—As the vessels which such Masters will command frequently make long voyages, to the East Indies, the Pacific, &c., the candidate will be required to work a lunar observation by both sun and star, to determine the latitude by the moon, by Polar star off the meridian, and also by double altitude of the sun, and to verify the result by Sumner's method. He must be able to calculate the altitudes of the sun or star when they cannot be

---

* Service in a superior capacity is in all cases to be equivalent to service in an inferior capacity.

observed for the purposes of lunars,—to find the error of a watch by the method of equal altitudes,—and to correct the altitudes observed with an artificial horizon.

He must understand how to observe and apply the deviation of the compass; and to deduce the set and rate of the current from the D. R. and observation. He will be required to explain the nature of great circle sailing, and know how to apply practically that knowledge, but he will not be required to go into the calculations. He must be acquainted with the law of storms, so far as to know how he may probably best escape those tempests common to the East and West Indies, and known as hurricanes.

In Seamanship.—The extra examination will consist of an inquiry into the competency of the applicant to heave a ship down, in case of accident befalling her abroad; to get lower masts in and out; and to perform such other operations of a like nature as the Examiner may consider it proper to examine him upon.

QUALIFICATIONS FOR CERTIFICATES OF COMPETENCY FOR A "HOME TRADE PASSENGER SHIP."

14. A MATE must write a legible hand, and understand the first four rules of arithmetic. He must know and understand the rule of the road, and describe and show that he understands the Admiralty regulation as to lights. He must be able to take a bearing by compass, and prick off the ship's course on a chart. He must know the marks in the lead line, and be able to work and heave the log.

15. A MASTER must have served one year as a Mate in the Foreign or Home Trade. In addition to the qualifications required for a mate, he must show that he is capable of navigating a ship along any coast, for which purpose he will be required to draw upon a chart produced by the Examiner the courses and distances he would run along shore from headland to headland, and to give in writing the courses and distances corrected for variation, and the bearings of the headlands and lights, and to show when the courses should be altered either to clear any danger, or to adapt it to the coast. He must understand how to make his soundings according to the state of the tide. He will also be questioned as to his knowledge of the use and management of the mortar and rocket lines in the case of the stranding of a vessel, as explained in the Official Log Book.

A first-class Pilot may be examined for a Master's Certificate of Competency for Home Trade Passenger Ships, notwithstanding that he may not have served in the capacity of Mate.

GENERAL RULES AS TO EXAMINATIONS AND FEES.

16. The candidates will be allowed to work out the various problems according to the method and the tables they have been accustomed to use, and will be allowed five hours to perform the work; at the expiration of which time, if they have not finished, they will be declared to have failed, unless the Local Marine Board see fit to extend the time.

17. The fee for examination must be paid to the Superintendent of the Mercantile Marine Office (Shipping Master). If a candidate fail in his examination, half the fee he has paid will be returned to him by the Superintendent of the Mercantile Marine Office on his producing the Form Exn. 17, late HHI, which will be given him by the Examiner. The fees are as follow:—

FOR "FOREIGN-GOING SHIPS."

	£	s.	d.
Second Mate	1	0	0
First and Only Mate, if previously possessing an inferior certificate	0	10	0
If not	1	0	0
Master, whether Extra or Ordinary	2	0	0
Master, if previously in possession of a certificate for "fore and aft rigged vessels"	1	0	0

N.B.—*Any person having a Master's Certificate of Competency for Foreign-going Ships may go up for an extra examination without payment of any Fee, but if he fails in his first examination, half a Master's Fee will be charged for each subsequent examination.*

FOR "HOME TRADE PASSENGER SHIPS."

Mate £0 10 0
Master 1 0 0

18. If the applicant passes he will receive the Form Exn. 16, late GG, from the Examiner, which will entitle him to receive his Certificate of Competency from the Superintendent of the Mercantile Marine Office, at the port to which he has directed it to be forwarded. If his testimonials have been sent to the Registrar to be verified, they will be returned with his certificate.

19. If an applicant is examined for a higher rank, and fails, but passes an examination of a lower grade, he may receive a certificate accordingly, but no part of the fee will be returned.

20. In every case the Examination, whether for Only Mate, First Mate, or Master, is to commence with the problems for Second Mate.

21. In all cases of failure the candidate must be re-examined *de novo*. If a candidate fails in *Seamanship* he will not be re-examined *until after a lapse* of Six Months, to give him time to gain experience. If he fails three times in *Navigation* he will not be re-examined until after a lapse of Three Months.

22. As the examinations of Masters and Mates are made compulsory, the qualifications have been kept as low as possible; but it must be distinctly understood that it is the intention of the Board of Trade to raise the standard from time to time, whenever, as will no doubt be the case, the general attainments of officers in the merchant service shall render it possible to do so without inconvenience; and officers are strongly urged to employ their leisure hours, when in port, in the acquirement of the knowledge necessary to enable them to pass their examinations; and Masters will do well to permit apprentices and junior officers to attend schools of instruction, and to afford them as much time for this purpose as possible.

EXAMINATION OF MASTERS AND MATES WITH REFERENCE TO THE COMMERCIAL CODE OF SIGNALS FOR THE USE OF ALL NATIONS.—INSTRUCTIONS TO EXAMINERS.

23. In transmitting the accompanying copy of the latest edition of the Commercial Code of Signals for the use of the Examiners, the Board of Trade desire to direct attention to the principal points connected with this Code as to which Candidates for Examination should be questioned.

24. At the same time, as the subject is probably new to some of the Examiners themselves, the Board recommend to them a perusal of the *Report of the Signal Committee* of 1855 (which will be found at the commencement of the Signal Book), and also the *first few pages of the Book*. The information therein given will be found sufficient to make the Examiners theoretically acquainted with the characteristics of the New Code, and the advantages it claims to possess over other Codes, and will enable them to appreciate and urge upon Candidates for Examination the facilities which the new System of Signalling affords for easy and rapid communication.

25. The "comprehensiveness" and "distinctness" of the Commercial Code are its chief recommendations.

26. The form of the Hoist generally indicates the nature of the Signal made, so that an observer can at sight understand the character of the Signal he sees flying.

27. The Examinations should tend to elicit a knowledge of the distinctive features of the Code above alluded to.

With this object the Examiners should make the 2, 3, and 4 Flag Signals on the Frame board which is furnished for the purpose (*always taking care first to show the Ensign and the Code Pennant at the Gaff*),* questioning the Candidates as to the distinguishing Forms of the respective Hoists, which will be indicated according as a Burgee, or a Pennant, or a Square Flag, is uppermost.

* The object of this is, of course, to distinguish the Signals from those of another Code.

28. The Candidate ought to know how to find in the Signal Book the communication or the inquiry he desires to make, and how to make the Signal. The Signal to be made should *invariably* be sought for by the Candidate in the Vocabulary and Index, Part II, and never in Part I.

29. The Candidate ought to know how to interpret a Signal.

The Examiner should place a Signal on the Frame board, and vary the Signal by showing a 2 or 3 Flag Signal, or a "Geographical" or a "Vocabulary" Signal, or the name of a Merchant Ship or a Ship of War.

The two latter Signals would not of course be found in the Signal Book. The candidate ought to point them out in the *Code List of Ships*.

30. A candidate ought to be able to read off a Signal at sight, so far as to name the Flags composing the Hoist.

31. He ought to know the use of the Code Pennant, and of the Pennants C and D, "Yes" and "No."

32. The candidate should be practised in the use of the Spelling Table, by being made to spell his own name, or some word not in the Vocabulary of the Code.

33. As Ships of War use a different set of Code Flags, the candidate ought to be aware of the fact, and should know that a plate of the Admiralty Flags is to be found in the Signal Book, as well as plates of the Code Flags which Foreign Ships of War will use in signalling to Merchant Vessels. He should also know that every official Log Book contains plates of these Code Flags.

34. A knowledge of the Distant Signals should be required of the Candidate, their object, and the mode of signalling by the Distant Code, which will be found at the end of the Signal Book.

For this purpose two Black Balls, 2 Black Square Flags, and 2 Black Pennants will be furnished with the Frame board, and the candidate should be required to make one or two Distant Signals, and to read off one or two made by the Examiners.

The Ball being the distinguishing symbol of the Distant Signal, any Pennants or Flags of the Code may be employed in conjunction with it, irrespective of colour. The Black Pennants and Flags are merely sent as showing best in the light background of the Frame board.

SEMAPHORES.

35. We have as yet no Semaphores on our coasts. The French, however, have upwards 110 such stations established on their coasts, at which the Commercial Code of Signals *only* is used.

36. A plate at the end of the Signal Book explains the method by which the arms of the Semaphore are made to represent by their position (up, down, or horizontal,) the three symbols used for distant signalling, viz., a Flag, a Ball, or a Pennant. Before making Signals with the Semaphore, the Black Disc with the White rim should be placed on the top of the Semaphore Mast, as it properly forms a part of the Mast itself.

37. The Board of Trade think it of consequence to observe that as the Commercial Code has (in its integrity) been translated into French, and as copies of the Signal Book are furnished to all French Vessels of War and Semaphore stations, any Englishman can now by this Code make his wants known to them.

Other nations are now negotiating for the adoption of the Commercial Code, and from the favour with which Foreigners seem to have accepted the Code wherever it has been presented to their notice, there is every reason to believe that in a short time the Mercantile Marine of all nations will have the advantage of being able to communicate by an "Universal Language of Signals."

38. Her Majesty's Government have done all in their power to promote the use of the Commercial Code, and the Government of India and nearly all the Colonial Governments have adopted it, and a large number of Signal Books and Code Lists have already been circulated in the British Possessions abroad.

The Examiners are to insert in the Report of Examinations (under the heading of "Remarks") the words "passed (or failed) in Commercial Code of Signals," as the case may be.

MASTERS' AND MATES' VOLUNTARY EXAMINATIONS IN STEAM.

39. Arrangements have been made for giving to those Masters and First or Only Mates who are possessed of or entitled to certificates of competency, an opportunity of undergoing a voluntary examination as to their practical knowledge of the use and working of the steam engine. These examinations are conducted on the premises, and under the superintendence of the Local Marine Boards, at such times as they may appoint for the purpose; and the Examiners are selected by the Board of Trade, from the Engineer Surveyors appointed under the fourth part of "The Merchant Shipping Act, 1854."

40. Any Master or Mate desiring to be examined in Steam must deliver to the Superintendent of the Mercantile Marine Office a statement in writing to that effect, upon the Form of Application (Exn. 2, late EE); if the applicant has a Certificate of Competency, such certificate must be delivered to the Shipping Master along with his statement. If he is about to pass an examination for a Certificate of Competency at the same time, the applications should be sent in together.

41. A fee of one pound must be paid by the applicant for the examination *in Steam*, and the Superintendent of the Mercantile Marine Office will thereupon inform him of the time and place at which he is to attend to be examined, and the examination will then and there proceed in the same manner as the other examinations. If the applicant fails, and has given in his certificate, it will be at once returned to him, but *no part of the fee he has paid will be returned*.

42. If he passes, the Report (Exn. 14, late FF) will be sent to the Board of Trade, and the Certificate of Competency with the Form (Exn. 2, late EE) to the Registrar General of Seamen; the words "*Passed in Steam*," with the date and place of examination, will then be entered on the certificate and its counterpart, and the certificate will be sent to the Superintendent of the Mercantile Marine Office of the port named in the Application (Exn. 2, late EE) to be delivered to the applicant in the usual manner.

43. The examination is *vivâ voce*, and extends to a general knowledge of the practical use and working of the steam-engine, and of the various valves, fittings, and pieces of machinery connected with it. Intricate theoretical questions on calculations of horse-power or areas of cylinders and valves, or any of the more difficult questions which appertain to steam engines and boilers, will not be asked. The examination will in fact be confined to what a master of a steam-vessel may be called upon to perform in the case of the death, incapacity, or delinquency of the engineer.

44. If the applicant fails to answer some few of the questions, and yet, in the opinion of the Examiner, possesses such a competent knowledge of the parts of the engine generally, and such other practical knowledge of the subject as will enable him to effect the object in view, the Examiner will exercise his discretion as to whether a sufficiently high standard of knowledge has been attained, and pass him or not accordingly.

45. The Examiner will provide drawings and working sections, on a sufficiently large scale, of the various parts of the steam-engine, and of the valves and slides, &c., as may be necessary, and will require the applicant to make use of them in giving his answers to the various questions put to him; and, if an opportunity offer, the applicant will be permitted, under the guidance of the Examiner, to start and stop the engine of some vessel which may have her steam up.

CERTIFICATES OF SERVICE.

46. A Certificate of *Service* entitles an Officer who had served as either Master or Mate in a British Foreign-going Ship before the 1st January, 1851, or as Master or Mate in a Home Trade Passenger Ship before the 1st January, 1854, to serve in these capacities again; and it also entitles an Officer who has attained or attains the rank of Lieutenant, Master,

Passed Mate, or Second Master, or any higher rank in the service of Her Majesty or of the late East India Company, to serve as Master of a British Merchant Ship, and may be had by application to the Registrar-General of Seamen, Adelaide Place, London Bridge, London, or to any Superintendent of a Mercantile Marine Office in the Outports, on the transmission and verification of the necessary certificates and testimonials.

LOCAL MARINE BOARD EXAMINATIONS—NOTICE TO CANDIDATES—OFFICIAL NOTICE.

1. Candidates are required to appear at the examination room punctually at the time appointed.

2. Candidates are prohibited from bringing into the examination room books or papers of any kind whatever. The slightest infringement of this regulation will subject the offender to all the penalties of a failure.

3. In the event of any Candidate being detected in defacing, blotting, writing in, or otherwise injuring any book or books belonging to the Board, the papers of such Candidate will be detained until the book or books so defaced be replaced by him. He will not, however, be at liberty to remove the damaged book, which will still remain the property of the Board.

4. In the event of any Candidate being discovered copying from another, or affording any assistance or giving any information to another, or communicating in any way with another, during the time of examination, he will subject himself to a failure and its consequences.

5. No Candidate will be allowed to work out his problem on a slate or on waste paper.

6. No Candidate will be permitted to leave the room until he has given up the paper on which he is engaged.

7. Candidates will be allowed to work out the various problems by the method and tables they have been accustomed to use, and will be allowed five hours to perform the work. At the expiration of the five hours they will, if they have not finished, be declared to have failed, unless the Local Marine Board or Examiner see fit to lengthen the period in any special case. If however, the period is lengthened in any case, the special circumstances of that case, and the reasons for lengthening the period must be reported to the Board of Trade by the Examiners at the time they send in the report on Form FF.

8. Candidates will find it more convenient, both here and at sea, to correct the declination and other elements from the Nautical Almanac by the "hourly differences" which have been given in that work in order to facilitate such calculations; they will thereby render themselves independent of any proportional or logarithmic table for such purpose.

9. The corrections by inspection from tables given in many works on navigation will not be allowed (see Tables IX, XI, and XXI, in Norie's Epitome; Tables 21 and 38 in Raper's Navigation; every correction must appear on the papers of the Candidates. The First-class and Extra Master are referred to page 519 of the Nautical Almanac, 1867, for further information on this subject.

10. Candidates are expected to bring their answers to all problems within, or not to exceed, a margin of one mile of position from a correct result.

11. In finding the longitude by chronometer the logarithms used in finding the hour-angle should be taken out for seconds of arc.

12. In all other problems the logarithms to the nearest minute will be sufficiently correct for all grades, except Extra Master, from whom a degree of precision will be required, both in the work and in the results, beyond what is demanded from the inferior grades.

THOMAS GRAY, Assistant Secretary.

Board of Trade, Marine Department, *January* 1st, 1869.

INSTRUCTIONS TO EXAMINERS.

Service as Second Mate to be subject to certain conditions equivalent to service as First Mate for the purpose of qualifying for Examination for a Master's Certificate.

Board of Trade, Marine Department.

August, 1878.

On and after the 1st August, 1878, the following regulations shall be substituted for the regulations now in force relating to the amount and class of sea service required of a Candidate for an Ordinary Master's Certificate, viz. :—

A Master must be twenty-one years of age, and have been either six years at sea, of which one year must have been as First or Only, and one year as Second Mate, or he must have been six and a half years at sea, of which two and a half years must have been as Second Mate, during the last twelve months of which he must have been in possession of a First Mate's Certificate.

T. H. FARRER, Secretary.

THOMAS GRAY, Assistant Secretary.

EXAMINATION IN SUMNER'S METHOD BY PROJECTION.

The Board of Trade have decided that on and after the 1st January, 1878, candidates for examination for the grades of First Mate and Master shall be examined in Sumner's Method by Projection.

This subject shall be considered as forming part of the Navigation Examination.

PARTICULARS OF EXAMINATION.

Candidates will be required to ascertain their longitude by chronometer worked with two assumed latitudes, one greater and one less than the latitude by dead reckoning.

They are to mark off the two positions so ascertained on the chart, and are then to connect them with a straight line, which will show the bearing of any land it may intersect, and draw a line at right-angles to this, in the direction of the sun, showing the sun's true bearing.

With reference to a second observation, the candidates will not be for the present obliged to perform the calculations. The longitudes corresponding to the two latitudes are to be furnished to them by the Examiner, together with the course and distance made good by the ship between the two observations. The candidates will then be required to correct the first line of equal altitude for the ship's change of station, in the interval between the two observations, to project the line of equal altitude corresponding to the second observation on the chart, showing by its intersection with the first line of equal altitude, as corrected for change of station, the position of the ship at the time of the second observation. Outline charts, extending from 46° to 49° and from 49° to 52° of latitude respectively, will be furnished by the Board of Trade to the different Examiners for this purpose.

NOTICE TO OFFICERS AND SEAMEN IN THE MERCANTILE MARINE.

COLOUR BLINDNESS.

The Board of Trade have decided that on and after the 15th March, 1880, the following arrangements shall be made in respect to the Examination of persons as to their ability to distinguish colours.

1. Examinations in Colour shall be open to any person serving or about to serve in the Mercantile Marine.

2. Any person desirous of being examined must make application to a Superintendent of a Mercantile Marine Office on Form Exn. 2ᵃ, and pay a fee of One Shilling.

3. He must on the appointed day attend for examination at the Examiner's Office; and if he passes he will receive a Certificate to that effect.

4. In future the examination of a Candidate for a Master's or Mate's Certificate, who does not, at the time of making application, hold a Certificate of Competency of any grade, will commence with the Colour test, and if the Candidate fails in that test he will not be allowed to present himself for examination in Navigation and Seamanship. The fee he has paid for Examination for a Certificate of Competency will include the fee for the Colour test, and, with the exception of One Shilling, will be returned to him.

5. A Candidate who has obtained a Certificate before these regulations came into force, and who on presenting himself for Examination for a Certificate of a higher grade is unable to pass the Colour test, will notwithstanding be permitted to proceed in the Examination in Navigation and Seamanship for the Certificate of the higher grade; but

(1.) Should he pass this Examination, the following statement will be written on the face of the higher Certificate which may be granted to him, viz.: "This Officer has failed to pass the Examination in Colours."

(2.) Should he fail to pass the Examination in Navigation and Seamanship, a like statement, relating to his being Colour blind, will be made on his inferior Certificate before it is returned to him.

Information as to places and hours of Examination may be obtained from a Superintendent of a Mercantile Marine Office.

THOMAS GRAY,
Assistant Secretary.

Board of Trade, *March,* 1880.

PORTS AT WHICH EXAMINATIONS IN COLOUR TAKE PLACE.

ABERDEEN.—Examiner of Masters and Mates. Wednesday and Thursday in each week.

BELFAST.—Examiner of Masters and Mates. Any day.

BRISTOL.—Examiner of Masters and Mates. Second and fourth Tuesday and following Wednesday in each month.

CARDIFF.—Principal Officer of the Board of Trade. Any day.

CORK.—Examiner of Masters and Mates. Any day.

DUBLIN.—Examiner of Masters and Mates. Any day.

DUNDEE.—Examiner of Masters and Mates. Thursday and Friday in each week.

GLASGOW.—Examiner of Masters and Mates. Tuesdays and Wednesdays in examination weeks for Masters and Mates.

GREENOCK.—Examiner of Masters and Mates. Tuesday and Wednesdays in examination weeks for Masters and Mates.

HULL.—Examiner of Masters and Mates. Second and fourth Tuesday and following Wednesday in each month. Principal Officer of the Board of Trade. Any day.

LEITH.—Examiner of Masters and Mates. Any day.

LIVERPOOL.—Examiner of Masters and Mates. Wednesday and Saturday in each week.

LONDON.—Examiner of Masters and Mates. Monday and Tuesday in each week.

NORTH SHIELDS.—Principal Officer of Board of Trade. Any day.

PLYMOUTH.—Examiner of Masters and Mates. Any day.

SOUTHAMPTON.—Superintendent of the Mercantile Marine Office. Any day.

SOUTH SHIELDS.—Examiner of Masters and Mates. Thursdays and Fridays in examination weeks for Masters and Mates.

SUNDERLAND.—Examiner of Masters and Mates. Thursdays and Fridays in examination weeks for Masters and Mates.

SWANSEA.—Board of Trade Officer. Any day.

Appendix A.

EXAMINATION DAYS

AT

PLACES.	DAYS. For Masters and Mates.
Aberdeen*	Wednesday in each week.
Belfast*	First and third Tuesday in each month.
Bristol*	Tuesday in each week.
Cork	Second and fourth Monday in each month.
Dublin*	Wednesday in each week.
Dundee	Thursday in each week.
Glasgow*	Thursday in each week.
Greenock*	Tuesday in each week.
Hull	Second and fourth Tuesday in each month.
Leith*	Tuesday in each week.
Liverpool*	Every week—Monday and Tuesday "Foreign Trade;" Thursday and Friday "Home Trade Passenger" and "Foreign Trade."
London*	The examination in Navigation commences every Monday, and the examination in Seamanship takes place as soon as the Navigation examination is finished. Master's voluntary examination in Steam held on Friday in each week.
Shields, South*	Thursday in each week alternately with Sunderland.
Southampton*	Nil.
Sunderland*	Thursday in each week alternately with South Shields.
Plymouth	Tuesday in each week.

N.B.—The examination days are liable to occasional alteration, and Candidates are therefore advised to ascertain the actual date of examination from the Superintendent of the Mercantile Marine Office.

* At these places Masters' Extra Examinations are held.

EXPLANATION OF SIGNS USED.

The following signs are made use of both in arithmetic and algebra:—

SIGNS OF OPERATION.

1. The sign $+$, called *plus* (which is the Latin for *more*), signifies *additive*, or to be *added*, and shows that the number before which it stands is to be *added*; thus $3 + 4$ (read *three plus four*) means that 4 is to be added to 3, making 7.

2. The sign $-$, called *minus* (which is the Latin for *less*), signifies *subtractive*, or to be *subtracted*, and shows that the number before which it stands is to be *subtracted*; thus $13 - 5$ (read as *thirteen minus five*) means that 5 is to be subtracted from 13, leaving 8.

3. The sign $\times$ (into), signifies *multiplied by*, and shows that the numbers between which it stands are to be *multiplied*; thus 3×4 (read *three into four*, or *three multiplied by four*), means that 3 is to be multiplied into 4, making 12. Sometimes a full stop at the bottom of the figure is used for this; thus, 2×7, or 2.7, are both used to express twice seven.

4. The sign $\div$, signifies *divided by*, and means that the number which stands before it is to be divided by the one which follows it; thus $14 \div 2$ (read *fourteen by two*), means that 14 is to be divided by 2, making 7. The operation of division is also indicated by writing the divisor under the dividend with a line between them; thus 14 by 2 is also frequently denoted thus $\frac{14}{2}$.

5. The sign $=$, signifying *equal to* (or amounting to), means that the numbers between which it stands are equal to each other; that is, have the same arithmetical value, each taken as a whole.

Examples of the preceding, with the results in each case, will stand thus:—

    1. 14 and 3 equals 17, or $14 + 3 = 17$.   3. $7 \times 5 = 35$.
    2. $10 - 3 = 7$.   4. $14 \div 2 = 7$, or $\frac{14}{2} = 7$.

6. The signs : *is to*, :: *so is*, are the signs of *proportion*: as $2 : 4 :: 8 : 16$; that is, as 2 is to 4 so is 8 to 16.

7. The signs $\overline{14 - 4} + 10 = 20$, show that the difference between 4 and 14 added to 10 is equal to 20. The line drawn over 14 and 4 is called a *vinculum*.

8. The signs $10 - \overline{2 + 5} = 3$ signify that the sum of 2 and 5 taken from 10 is equal to 3.

9. 8^2 is read 8 *squared*, and means that the 8 is to be multiplied by itself; thus, $8 \times 8 = 64$; hence 64 is called the *square* of 8.

10. The sign $\sqrt{}$, prefixed to any number, signifies that the *Square Root* of that number is required; thus,

$\sqrt{64}$ is read *the square root of* 64, and means that number which when multiplied by itself gives 64; hence 64 is called the square of 8.

THE PRINCIPLES AND PRACTICE OF ARITHMETIC.

CHAPTER I.
DEFINITIONS, PRELIMINARY NOTIONS, NUMERATION, AND FUNDAMENTAL OPERATIONS.

ARTICLE I.—DEFINITION I.

1. **Arithmetic** is the science which treats of *numbers*—of the mode of expressing them—of the manner of computing by them—and of the various uses to which they are applied in the practical business of life.

2. A **Unit** or **Unity** is the name given to that quantity which is to be reckoned as *one* when other quantities of the same kind are to be measured. Thus, each of the terms, *a* man, *a* house, *a* pound, &c., denotes one individual of its kind, being the same as *one* man, *one* house, *one* pound, &c., respectively: and these are the basis or elements by means of which *several* men, *several* houses, *several* pounds, &c., may be computed.

3. **Number** is the relation of a quantity to its unit, the notion of number being suggested by successive repetitions of the individual unit; or number is the name by which we signify how many objects or things are considered whether *one* or *more*. When, for instance, we speak of one ship, two steamers, three masts, or four yards, the number of things referred to will be one, two, three, or four, according to the case; and so one, two, three, four, and the rest, are called numbers. Numbers thus viewed are termed *whole numbers* or *integers;* and for the sake of uniformity, the unit is considered the first or least integer.

4. Numbers used to express one or more individuals of *specified* kinds are called *concrete* numbers; whereas *two, three, four,* &c., by themselves, not particularizing the *kinds* of individuals, are termed *abstract* numbers.

NOTATION.

5. **Def. 1.**—*Notation* is the method of expressing any proposed number by certain symbols or characters appropriated for that purpose.

6. **Def. 2.**—The *symbol* or *representative* of unit or unity is 1; but instead of other numbers being expressed by assemblages or multitudes of units *placed* together, which would soon become embarrassing, other characters or symbols have been invented, by means of which every number, however great, may be expressed; and instead of a different symbol being adopted for every different number, which would soon become equally inconvenient, *all* numbers are expressed by means of the following ten symbols, or as they

are usually termed, *figures*, and sometimes *digits*, which have their names respectively annexed:

1	2	3	4	5	6	7	8	9	0
one	two	three	four	five	six	seven	eight	nine	zero

the first *nine* of which are called by their names; and the *last*, which is variously denominated *nought*, *cypher*, or *zero*, when standing by itself has no signification, or at most, denotes the absence of number, and is to be regarded merely as an *auxiliary* digit, for the purposes to be hereafter explained.

7. **Def. 3.**—Whenever a figure is placed on the *right* of the same or any figure, it has, by *universal agreement*, the effect of increasing the value of the last-mentioned figure *tenfold*, at the same time that it retains its own value.

Thus, beginning with the auxiliary digit 0, we have the following numbers and their representations:

10	11	12	13	14	&c.
ten	eleven	twelve	thirteen	fourteen	&c.
20	21	22		&c.	
twenty	twenty-one	twenty-two		&c.	

and it is obvious that by means of *two* figures, this kind of notation may be continued till we arrive at ninety-nine, whose symbol is 99.

8. **Def. 4.**—Beyond this number, the use of *two*, either of the *same* or *different* figures, will not enable us to go, but a repetition of the contrivance in the last Article will, by means of *more* figures, supply the defect.

Thus, supposing the effect of any figures being placed on the right of symbols formed as above, to be to increase all their values *tenfold*, we shall have

100	101	102	&c.
one hundred	one hundred and one	one hundred and two	&c.

so likewise of succeeding numbers; thus, we have

234   587
two hundred and thirty-four   five hundred and eighty-seven

and again 999 will be *nine hundred and ninety-nine*, which is the largest number capable of being expressed by *three* figures.

Here, the *first* figure on the right hand is said to occupy the *unit's place*, the *second* the place of *tens*, and the *third* that of *hundreds*.

Of the auxiliary digit 0, the sole use is in the effect specified in the last two articles; and all figures to the *right* of it will therefore be unaffected by it.*

9. **Def. 5.**—In estimating numerical magnitudes, we proceed in order from *hundreds*, to *thousands*, *tens of thousands*, and *hundreds of thousands*, *millions*.

---

* The word *cypher* is from the Arabic word *Tsaphara*, blank or void. At the end or in the middle of any number the cypher is of use to keep the significant digits in their proper rank, when the units or the hundreds or any other denomination may be wanting, *e.g.*, 60 means 6 tens followed by no units, 606 means 6 hundreds with no tens, but 6 units. At the beginning of a number cyphers would be useless; if so placed they could only indicate the absence of any higher class, *e.g.*, 096 means only 9 tens and 6 units, the cypher showing that there are no hundreds, which is equally intelligible if the cypher be omitted.

tens of millions, and *hundreds of millions*, in precisely the same manner as we have done above from *units* to *tens*, and from *tens* to *hundreds*.

10. Agreeable to the principle of Art. 7, it is *assumed* that "any figure placed on the right of one or more others, has the effect of increasing every one of them tenfold, without altering its own value," and this enables us to express with facility any number whatever.

Thus—

1. 1000 will represent one *thousand*.
2. 5493 will represent five *thousands*, four *hundreds*, and ninety-three.
3. 23456 will represent twenty-three *thousands*, four *hundreds*, and fifty-six.
4. 729054 will represent seven hundred and twenty-nine *thousands* and fifty-four.
5. 1803205 will represent one *million*, eight hundred and three *thousands*, two *hundreds*, and five.
6. 32754081 will represent thirty-two *millions*, seven hundred and fifty-four *thousands*, and eighty-one.
7. 473025004 will represent four hundred and seventy-three *millions*, twenty-five *thousands*, and four.

11. If the first three figures, beginning from the right hand, be denominated so many *units*, tens of *units*, and hundreds of *units*, it follows that the next three figures taken the same way will be *thousands*, tens of *thousands*, and hundreds of *thousands*; the next three in order will be *millions*, tens of *millions*, and hundreds of *millions*, and so on.

Whence to express in figures any number proposed, we have only to consider in which of these divisions each part of it ought to be found, observing that *three* figures from the right must be taken to make each division complete, before we proceed to the next.

EXAMPLES.

Ex. 1. Express, by means of figures, thirty-five thousand eight hundred and nineteen.
Here, eight hundred and nineteen belongs to the *first* division on the right, and is written 819; also, thirty-five thousand must be found in the *second* division from the right, and is 35; whence the proposed number may be expressed by 35819.

Ex. 2. Express in figures the number five million twenty-five thousand six hundred and seven.
In this case the first division on the right will be 607; the second will be 025, the digit o being affixed to the left of the others without altering their value, to make up the required number of *three*; and the third is 5, so that the expression will be 5025607.

Ex. 3. Express by figures the following number, five hundred and seventy million two hundred and six thousand and fifty-four.
Here, the first division is 054, the o altering only the value of the figures in the *subsequent* divisions; the second division is 206; and the third is 570, whence the number proposed is correctly expressed by 570206054.

EXAMPLES FOR PRACTICE.

1. One hundred. 2. One hundred and one. 3. One hundred and ten. 4. Nine thousand and nine. 5. Nine thousand and ninety. 6. Nine thousand nine hundred and nine. 7. Five thousand and seventy-four. 8. Ten thousand seven hundred. 9. Ninety thousand and ninety. 10. Three hundred and five thousand. 11. Nine hundred thousand nine hundred. 12. Five hundred and five thousand five hundred and fifty. 13. One million

three thousand and eight. 14. Five million thirty thousand and forty-nine. 15. Nine million nine hundred thousand and six. 16. Fifty-eight million and nine. 17. Seventy million three hundred and two thousand four hundred and forty-one. 18. Two hundred and twenty-two millions and thirty-five. 19. Six hundred and four million sixty thousand and five. 20. Eight hundred million three thousand and thirty-three. 21. Nine hundred million nine hundred thousand and nine hundred. 22. Seven hundred million and seven. 23. One hundred and eighty million. 24. Five hundred million. 25. Five hundred and eighty million two hundred and forty-five thousand one hundred and ninety-two. 26. Seven hundred and seven million seven thousand and seventy-seven.

12. This method of notation can never present any difficulty, provided it be carefully remembered that every division of figures as we proceed from the right hand towards the left must be *completed*, as far as it is possible, and, by a little practice, we shall be enabled to write down any number by beginning at the *left* hand.

EXAMPLE.

Ex. 1. To write down six hundred and thirteen millions five hundred and nineteen, we observe that the division of millions will be 613; that of thousands will be 000, and that of units 519; so that the number is expressed in arithmetical symbols by 613000519.

13. It will be observed, from what has been said, that each of the nine figures or digits, 1, 2, 3, 4, 5, 6, 7, 8, 9, has an *absolute* value of itself, whereas the auxiliary digit 0 has no such value; and on this account the former are termed *significant* figures, in contra-distinction to the last. It will, moreover, have occurred to the reader that every one of these significant digits, in addition to its *absolute* value, which is fixed and certain, possesses also a *local* value dependent upon the situation in which it is placed; thus, in the expression of the number four thousand three hundred and twenty-one, which will be 4321, the 1 in the first place on the right hand retains its absolute value; the second figure 2, in its situation, denotes two *tens*, or *twenty*; the third is three *hundred*, and the fourth is four *thousand*; so that the local values of 2, 3, and 4, are respectively, *ten* times, a *hundred* times, and a *thousand* times, as great as their absolute values; and it is the circumstance of assigning to each of the significant figures a *local* as well as an absolute value, which confers upon the system the immense power which it possesses.

NUMERATION.

14. **Def. 6.**—*Numeration* is the art of reading or estimating the value of a number expressed by figures, and is, therefore, the *reverse* of Notation.

15. From the circumstance of every figure possessing a local as well as an absolute value, it follows that the value of each figure must be estimated by the place which it occupies; hence, a figure standing by itself expresses so many *units*; a figure in the second place so many *tens*; a figure in the third place so many *hundreds*, and so on; consequently, if we suppose any numerical expression to be divided into *periods* or portions, each consisting of three figures as far as they go, the figures of the period on the right will be

units and tens and hundreds of *units;* those of the next will be units, tens and hundreds of *thousands;* those of the third will be units, tens and hundreds of *millions*, and so on.

Thus,

1. 25 is twenty-five.
2. 304 is three hundred and four.
3. 5287 is five thousand two hundred and eighty-seven.
4. 70639 is seventy thousand six hundred and thirty-nine.
5. 306583 is three hundred and six thousand five hundred and eighty three.
6. 1648305 is one million six hundred and forty-eight thousand three hundred and five.
7. 53024367 is fifty-three million twenty-four thousand three hundred and sixty-seven.
8. 257008005 is two hundred and fifty-seven million eight thousand and five.

In each of these instances we conceive the expression to be separated into periods of three figures each as far as they go, beginning at the right hand, as in 257008005, we observe that 005 is the first period, 008 the second, and the third period is 257, denotes two hundred and fifty-seven millions 008 eight thousands and 005 five units.

16. The last article will be rendered more clear by the following scheme, called the *Numeration Table*:—

Hundreds of Millions.	Tens of Millions.	Millions.	Hundreds of Thousands.	Tens of Thousands.	Thousands.	Hundreds.	Tens.	Units.
9	8	7	6	5	4	3	2	1
	9	8	7	6	5	4	3	2
		9	8	7	6	5	4	3
			9	8	7	6	5	4
				9	8	7	6	5
					9	8	7	6
						9	8	7
							9	8
								9

wherein the local value of every figure in each of the horizontal rows is pointed out by the name written *upwards* at the top of the whole; thus, in the *third* horizontal line from the bottom the figures will be read *nine hundred and eighty-seven,* and in the second line from the top, *ninety-eight millions, seven hundred and sixty-five thousand, four hundred and thirty-two.*

Examples for Practice.

Express in words:—

1.	43	9.	505	17.	87054	25.	1000001	33.	20084216	41.	202202200
2.	60	10.	550	18.	70707	26.	8047328	34.	5001860	42.	100100101
3.	12	11.	1000	19.	60880	27.	4090300	35.	8080808	43.	275008005
4.	21	12.	2020	20.	99404	28.	5210007	36.	55700005	44.	100010001
5.	100	13.	3303	21.	903756	29.	6030405	37.	76014059	45.	79030184
6.	101	14.	4004	22.	202202	30.	9009900	38.	6006606	46.	408076032
7.	110	15.	7707	23.	400400	31.	41041014	39.	56700505	47.	401400056
8.	500	16.	8880	24.	550550	32.	3000006	40.	120015015	48.	908500060

ADDITION.

17. Addition is the collecting together of two or more numbers, and the amount of all of them is termed their *sum*. The sign $+$ *(plus)* is employed to indicate addition, as $7 + 2$ signifies that 2 is to be added to 7. Also, the sign $=$ *(equal)* signifies that the numbers between which it is placed are equal: thus, $8 + 1 = 9$.

The process of addition depends upon the principle that the sum of two numbers is equal to the sum of their respective parts. Thus,

Let it be required to find the sum of two numbers, 1724 and 4638, and explain the process.
$1724 = 1$ thousand $+ 7$ hundreds $+ 2$ tens $+ 4$ units,
$4638 = 4$ thousands $+ 6$ hundreds $+ 3$ tens $+ 8$ units,
and as the sum of these two numbers is equal to the sums of their respective parts, that sum is
5 thousands $+ 13$ hundreds $+ 5$ tens $+ 12$ units.

To each of the four parts into which the first number is separated add the part of the second which is under it, beginning at the units. Thus, 8 units and 4 units are 12 units, that is, 1 ten and 2 units; again, 3 tens and 2 tens are 5 tens; 6 hundreds and 7 hundreds are 13 hundreds, or 1 thousand and 3 hundreds. Lastly, 4 thousands and 1 thousand are 5 thousands, hence the sum is either 5 thousands 13 hundreds 5 tens and 12 units, or 6 thousands 3 hundreds 6 tens and two units $= 6362$.

Hence,

18. The rule for simple addition is as follows:

RULE I.

Write the numbers to be added together in vertical columns so that the units of all the numbers may be in one column, the tens in the second, the hundreds in the third, and so on. Draw a line under the last number, and, beginning with the column of units, add *successively the* numbers *contained in each column; if the sum does not exceed nine, write it down under the line, but if it contains tens reserve them to be* added *to the* next *column, writing down only the* units *of each column, and under the* last *column put the entire sum, whatever it may be. If the* sum *of any column be an* exact number *of tens, write* 0 *for the units and* carry *the tens to the next column.*

EXAMPLES.

Ex. 1. Let it be required to find the sum of 26389, 38127, 2815, 6497, 835, and 3745.

Write the numbers as at the side, so that the figures of the same class shall be in the same vertical column; then taking the sum of each class, we find there are 38 units, 27 tens, 31 hundreds, 25 thousands, and 5 tens of thousands. Now 38 units are 3 tens and 8 units, then writing 8 below the units column, carry the 3 tens to the 27 tens, which together make 30 tens, or 3 hundreds and 0 tens. Write 0 below the column of tens and reserve the 3 hundreds to be added to the 31 hundreds; this gives 34 hundreds, or 3 thousands and 4 hundreds, and writing 4 below the column of hundreds, carry the 3 thousands to the 25 thousands, and we get 28 thousands, or 2 tens of thousands and 8 thousands. Writing the 8 below the column of thousands, carry the 2 tens of thousands, making the entire sum $= 78408$.

```
26389
38127
 2815
 6497
  835
 3745
-----
78408
```

19. Verification of Addition.—The usual verification is to add both upwards and downwards and see if the sums agree. This is generally sufficient. If more is required, or if the student cannot get a long column to cast the same way both up and down, he can cut it up and add each portion separately; then add the sums.

EXERCISES IN SIMPLE ADDITION.

(1)	(2)	(3)	(4)	(5)	(6)	(7)	(8)
311413	543123	536123	123456	761284	657890	692387	876578
452734	234512	453215	234561	612874	278679	4956	495
130421	713145	1234	345612	8719	5798	87658	54939
3718	104234	4231	456223	46759	67843	769378	8797
24561	36142	51234	561234	587999	488567	5790	358428
341323	3451	613254	612345	987678	37429	87958	768453

(9)	(10)	(11)	(12)	(13)	(14)	(15)	(16)
662593	846914	516398	425396	567453	169964	145673	197794
395266	415327	854627	674958	654359	435434	366535	543543
841923	723456	735829	827694	531769	744315	679654	765976
356627	674216	916355	731045	765453	476757	341345	415161
725983	328427	827146	556677	147954	496059	569765	954131
346783	736259	633289	889900	645679	695969	694313	643167

(17)	(18)	(19)	(20)	(21)	(22)	(23)	(24)
987825	916427	695024	986257	985626	372519	586372	148537
736349	625736	538426	427385	796842	463726	477754	697296
856925	346831	827836	514986	915638	298534	638831	526438
734316	857936	735985	726326	809274	851372	951490	723649
827842	735784	216515	915817	444444	319628	479291	859698
936736	426467	859827	734482	913258	738543	863748	852619
842625	849753	910756	386012	872364	497791	376546	419648
759519	358358	683625	219863	410698	345345	356633	777777
846325	647846	745841	391285	742367	679567	459681	999999
987846	386921	526606	842163	946208	161514	453148	555555
333445	666777	888999	615827	807609	131549	567963	724483
335445	666777	888999	736846	915827	761346	313499	952637

25. Add together the addends (1) under exercises (1), (9), and (17); (2) under (2), (10), and (18); (3) under (3), (11), and (19); (4) under (4), (12), and (20); (5) under (5), (13), and (21); (6) under (6), (14), and (22); (7) under (7), (15), and (23); and (8) under (8), (16), and (24).

26. Add together three hundred and nine million, four hundred and seventeen thousand, and eighty-seven; six hundred and seventy-five thousand, and forty-nine; seven thousand and ninety-seven million, eight hundred and fourteen thousand, three hundred and five; seventy-nine million, five hundred and four thousand, and forty-nine; six thousand and seventy-eight million, four hundred and thirty-nine thousand, six hundred and forty-seven; seven thousand million, eight hundred and seventy-six thousand, four hundred and twenty-nine.

SUBTRACTION.

20. The process of finding a number which shall be equal to the difference of two numbers is called *subtraction*. It is customary to call the quantity from which the subtraction is made, the **minuend**; the quantity to be subtracted, the **subtrahend**; and the result of the subtraction, the **difference**. Thus, then, we have, minuend − subtrahend = difference.

We may also write this as

$$\text{minuend} = \text{subtrahend} + \text{difference},$$

which shows the connection between subtraction and addition.

The operation of subtraction is indicated or expressed by the sign —, which is read *minus* or *less by*, with the use of the sign =; thus, the excess of 7 above 3 will be expressed in the form $7 - 3 = 4$, which is read 7 minus 3 equals 4; where the sign — between 7 and 3 denotes the subtraction of the latter from the former, and the sign = between 3 and 4 shows the *equality* of the excess to 4.

21. The process of subtraction involves two principles; the one is the equal augmentation or diminution of the numbers. In either way, the difference of the two numbers will not be altered; for if the greater number be either increased or diminished by 7, for example, and the less be increased by 7, the numbers themselves will be altered.

The other principle is this: since 12 exceeds 7 by 5, and 8 exceeds 6 by 2, then 12 and 8 together, or 20 exceeds 7 and 6 together, or 12 by 5 and 2 together or 7.

Let it be required to take 231 from 574.

Write the numbers as in the margin, units under units, tens under tens, and hundreds under hundreds; then 4 units exceed 1 unit by 3 units, 7 tens exceed 3 tens by 4 tens, 5 hundreds exceed 2 hundreds by 3 hundreds. Therefore, by the second principle, all the first column together exceeds all the second column by all the third column together, that is, by 3 hundreds 4 tens 3 units, or 343 which is the difference between 574 and 231.

Hundreds.	Tens.	Units.
5	7	4
2	3	1
3	4	3

Again, let it be required to subtract 23957 from 802126.

	Hundreds of Thousands.	Tens of Thousands.	Thousands.	Hundreds.	Tens.	Units.
802126 =	8	0	2	1	2	6
23957 =		2	3	9	5	7

Now here a difficulty immediately arises since 7 is greater than 6, and cannot be taken from it, neither can 5 be taken from 2, 9 from 1, 3 from 2, nor 2 from 0. To obviate this we must have recourse to the first principle, and add the same quantity to both these numbers, which will not alter their difference. Add ten to the first number, making 16 units, and add ten also to the second number, but, instead of adding ten to the number of units, add one to the number of tens, making 6 tens. Again, add ten tens to the first number, and one hundred to the second, then add ten hundred to the first, and one thousand to the second, and so on, adding equal quantities to each. In this way the numbers will be changed into the following :—

Hundreds of Thousands.	Tens of Thousands.	Thousands.	Hundreds.	Tens.	Units.
8	10	12	11	12	16
1	3	4	10	6	7
7	7	8	1	6	9

and the difference 778169 is obtained in the usual manner.

Hence, *when the upper figure is the less we must augment it by ten, and retain one to be added to the lower figure immediately to the left.*

22. The rule for simple subtraction is as follows :—

RULE II.

1°. *Put the smaller number under the greater, taking care, as in addition, that units shall be under units, tens under tens, hundreds under hundreds, and so on.*

2°. *Beginning at the* units, *take each figure in the* lower *line from the figure above it, if the lower figure be not the greater of the* two, *setting down the remainder below it.* (See the operation in Ex. 1, page 17).

3°. *But if any figure in the* lower *line be greater than that above it, add* 10 *to the* upper *one, and then take the* lower *figure from that* sum, *setting down the remainder, and carrying one (i.e. adding 1) to the* next lower *figure, and with which proceed as before, and so on, till the whole is finished.* (See Ex. 2, page 17).

The following examples will illustrate this rule.

EXAMPLES.

Ex. 1. Let it be required to subtract 42572 from 76594.
In the units column, 2 from 4 leaves 2, set down 2.
In the tens column, 7 from 9 leaves 2, set down 2.
In the hundreds column, 5 from 5 leaves 0, set down 0.
In the thousands column, 2 from 6 leaves 4, set down 4.
In the ten thousands column, 4 from 7 leaves 3, set down 3.

From 76594 *minuend.*
Subt. 42572 *subtrahend.*
Rem. 34022 *difference.*

Ex. 2. Subtract 7495 from 9263.
In the process adopted in practice the figures in the minuend are not, as in the second example No. 21, page 16, actually altered: and perhaps we might more simply explain the practical process as follows:

9263
7495
———
1768

To subtract 5 from 3 is impossible, so separate 1 ten from the 6 tens, and adding it to the 3 units, say 5 from 13 leaves 8. Now we are supposed to have separated 1 ten from the 6 tens, but as the figure really remains 6, we still have to take 1 from it; also we have to take from it the 9 in the lower line, so instead of taking away first 1 and then 9 more, take away 10 at once; but 10 from 6 being impossible, separate 1 from the place of hundreds, and adding it as 10 tens to the 6 tens, say 10 from 16 leaves 6. As we have not really taken 1 from the 2 hundreds, we have still 1 to take from it; also we have to take the 4 in the lower line; instead of taking first 1 and then 4, take away 5 at once; but 5 from 2 being impossible separate 1 from the place of thousands and add it as 10 hundreds to the 2 hundreds, and say 5 from 12 leaves 7. As we have not really diminished the figure 9 in the place of thousands, we have still 1 to take from it, and likewise we have to take away the 7 in the lower line, so taking away 8 at once from 9 we have 1 left in the place of thousands, and the entire difference is 1768.

Ex. 3. Let it be required to subtract 27385 from 64927.
Then placing the former number under the latter (as in the margin), we proceed thus:

64927 *minuend.*
27385 *subtrahend.*
———
37542 *difference.*

In the units column, 5 from 7 leaves 2, set down 2.
In the tens column, 8 from (not 2) but 12 leaves 4, set down 4 and carry 1.
In the hundreds column, (3 + 1, *i.e.* 3 + 1 carried) from 9 = 4 from 9 leaves 5, set down 5.
In the thousands column, 7 from (not 4) but 14 leaves 7, set down 7 and carry 1.
In the ten thousands column, (2 + 1, *i.e.* 2 + 1 carried) from 6 = 3 from 6 leaves 3, set down 3.

Ex. 4. As another example, let 86025704 be subtracted from 130741392.
Then, having arranged the numbers as in the margin, we proceed thus: 4 from 12, 8, carry 1; 1 and 0, 1, 1 from 9, 8; 7 from 13, 6, carry 1; 1 and 5, 6, 6 from 11, 5, carry 1; 1 and 2, 3, 3 from 4, 1; 0 from 7, 7; 6 from 10, 4, carry 1; 1 and 8, 9, 9 from 13, 4, carry 1; and 1 carried being taken from 1, leaves 0, therefore, the remainder is 44715688.

From 130741392 *minuend.*
Subt. 86025704 *subtrahend.*
———
Rem. 44715688 *difference.*

23. **Verification of Subtraction.**—The best verification is to add the subtrahend and difference. This ought to give back the minuend, or original quantity from which the subtraction was made.

EXERCISES IN SIMPLE SUBTRACTION.

(1)	(2)	(3)	(4)	(5)	(6)	(7)	(8)
706105	804601	980001	600501	702001	601002	501001	602004
84694	265061	980000	600492	26000	46003	20106	11906

(9)	(10)	(11)	(12)	(13)	(14)	(15)	(16)
701628	508000	403000	393436	321288	345876	206011	8500000
20449	119	26901	219050	213788	123457	48605	90909

(17)	(18)	(19)	(20)	(21)
10000001000	10100011101011	10101160330	10110011010110	479863217896
7077070077	1011100110110	80209337	10011101011	241826424862

Take each subtrahend 12 times from its minuend in the following examples:—

(22)	(23)	(24)	(25)	(26)	(27)
7432326	6677298	7213545	4362579	6002109	8100630
157689	67527	57636	9873	45108	6156

28. What number remains when one million four hundred thousand six hundred and nineteen has been taken from one hundred millions and two?

29. What is the difference between sixteen thousand and eighteen, and one million?

30. If eighty-nine be added to fourteen thousand six hundred and forty-three, how many must be added to the sum to make it ten millions?

31. How many must be added to the sum of 99, 416305, and 2108, that it may exceed the difference between 19104 and 605 by 1143200?

32. What must be added to 7543 to make it 16000?

33. The sum of two numbers is 84207, the less is 12327, what is the greater?

24. By help of the plus (+) and minus (—) signs, we can easily connect together in a single row a set of numbers, of which some are to be added and others subtracted; thus $4 + 6 - 3 - 2$ means that 4 and 6 are to be *added*, and 3 and 2 are to be *subtracted*, so that $4 + 6 - 3 - 2 = 5$. Instead of subtracting first 3 and then 2, we may, of course, subtract 5 at once; so that the above is the same as $10 - 5 = 5$. In some cases where additive and subtractive quantities are indicated, there is a difficulty which may be stated in a simple form by means of an example.

$$2 - 7 + 8 - 1$$

If we begin from the left-hand end, our first operation is to subtract 7 from 2, which cannot be done directly; we have recourse, therefore, to a different arrangement, namely, to take all the additions first, and all the subtractions last. We write it, therefore, as

$$2 + 8 - 7 - 1$$

After adding the 2 and the 8, we may either subtract the 7 and the 1 in succession, or we may add the 7 to the 1 and subtract their sum from the sum of the 2 and the 8. The simplest way, therefore, of getting at the result of a set of mixed additions and subtractions is to add the additive quantities, and then (separately) add the subtractive quantities, and subtract the sums. Thus, if we have

$$125 + 427 - 684 + 237 - 15$$

125	684	789
427	15	699
237	–––	–––
–––	699	90 result.
789		

EXAMPLES FOR PRACTICE.

1. $361 + 483 - 246 - 179 = 419.$
2. $573 - 184 + 602 - 67 = 924.$
3. $12064 + 700628 - 109641 + 637 - 2604 = 601086.$
4. $23596 - 625 + 72311075 - 13758 - 3506185 + 6879 = 68820880.$
5. $1 - 2 + 4 - 8 + 16 - 32 + 64 - 128 + 256 - 512 + 1024 - 2048 + 4096 - 8192 + 16384 - 32768 + 65536 - 131072 + 262144 - 524288 + 1048576 - 2097152 + 4194304 = 2796203.$

MULTIPLICATION.

25. **Multiplication** is a short method of finding the amount of a number repeated any number of times; thus, when 3 is multiplied by 4, the number produced by the multiplication is the sum of 3 repeated 4 times, which sum is equal to $3 + 3 + 3 + 3$ or 12.

26. The number which is repeated, or, in other words, is to be multiplied, is called the **multiplicand**; the number denoting the repetitions—*i.e.*, the number by which it is to be multiplied—is called the **multiplier**; and the amount, or the number which is found by multiplying the former by the latter is called the **product**.

The operation of MULTIPLICATION is expressed by the sign ×, which is read *into*, or *multiplied by*; thus, $5 \times 7 = 35$. Sometimes a full stop at the bottom of the figure is used for this; thus, 2×7 or 2.7, are both used to express twice seven. So again, $4 \times 5 \times 13 = 260$, expresses the continued product of 4, 5, and 13.

Multiplier × Multiplicand = Product. Multiplicand × Multiplier = Product.

27. If more than two numbers are multiplied together, the result is called the **continued product**.

28. The multiplicand and multiplier are termed **factors** of the product, because they are factors or makers of the product.

29. To perform the operation of multiplication readily, a *table*, called the *multiplication table*, must first be learnt, and the result which arises from multiplying one number by another, provided neither be greater than 12, must be committed to memory; it is one of the few operations in arithmetic where the memory of rules is indispensible.

THE MULTIPLICATION TABLE.

1	2	3	4	5	6	7	8	9	10	11	12
2	4	6	8	10	12	14	16	18	20	22	24
3	6	9	12	15	18	21	24	27	30	33	36
4	8	12	16	20	24	28	32	36	40	44	48
5	10	15	20	25	30	35	40	45	50	55	60
6	12	18	24	30	36	42	48	54	60	66	72
7	14	21	28	35	42	49	56	63	70	77	84
8	16	24	32	40	48	56	64	72	80	88	96
9	18	27	36	45	54	63	72	81	90	99	108
10	20	30	40	50	60	70	80	90	100	110	120
11	22	33	44	55	66	77	88	99	110	121	132
12	24	36	48	60	72	84	96	108	120	132	144

In this table the first horizontal line consists of the first twelve numbers in order; the second consists of the products when multiplied by 2; the third contains their products when by 3; the fourth when multiplied by 4, and so on; and the table is repeated in the following manner.

Thus, to make use of the *second* line of figures, we say—

twice 1 are 2	twice 5 are 10	twice 9 are 18
twice 2 are 4	twice 6 are 12	twice 10 are 20
twice 3 are 6	twice 7 are 14	twice 11 are 22
twice 4 are 8	twice 8 are 16	twice 12 are 24

NOTE.—It should be noticed that the product of any two numbers is the same whichever of them is taken as the multiplier—*e.g.*, 7 times 3 = 3 times 7. This is illustrated in the above Table by a double row of dark figures forming the diagonal of the square.

30. When two numbers are to be multiplied together, it is a matter of indifference, so far as the product is concerned, which of them be taken as the multiplicand or multiplier; in other words, the product of the first multiplied by the second, will be the same as the product of the second multiplied

by the first. Thus, 4×5 and 5×4 express the same thing, namely, 20. This is best seen as follows :—

```
  *   *   *   *   *
  *   *   *   *   *
  *   *   *   *   *
  *   *   *   *   *
```

Here there are 20 stars, or asterisks. It is evident that the total number —20—is not altered by the way in which we choose to group them, in order to count them. Thus we may either say there are 5 in each line, and 4 lines, or that there are 4 in each column, and 5 columns. Now one of these ways of counting takes the 20 as 4×5, the other as 5×4. Hence these are simply two ways of arriving at the same result, or product.

32. The fundamental principles upon which the process of multiplication depends are these :—

(1) If we separate any multiplicand into any number of parts, and multiply each part severally by any number and add the results, the *whole* multiplicand is thus multiplied, *e.g.*, 15, which may be separated into 8 and 7 is multiplied by 9, if 8 and 7 be each multiplied by 9, and the results added together.

(2) If the multiplier be separated into any number of parts, and the multiplicand be multiplied severally by each of those parts and the results added together, this is equivalent to multiplication by the *whole* multiplier, *e.g.*, if it be required to multiply 17 by 12, and we multiply 17 by 4, and 17 by 8, and add these results, we have then taken 17 exactly, $4 + 8$ times or 12 times.

Let it be required to multiply 6739 by the single figure 8.
Since the product 6739 by 8 is evidently equal to the sum of the products of all its parts, and
$$6739 = 6000 + 700 + 30 + 9$$
we must, therefore, multiply each of these parts by 8, and add together the results; we have, therefore, the following operation :—

Ten thousands.	Thousands.	Hundreds.	Tens.	Units.				
	6	7	3	9	6739			
					8			
			7	2	72 =	product of	9 by 8	(*a*)
		2	4		240 =	,,	30 ,, 8	(*b*)
	5	6			5600 =	,,	700 ,, 8	(*c*)
4	8				48000 =	,,	6000 ,, 8	(*d*)
5	3	9	1	2	53912 =	,,	6739 ,, 8	

Now 9 units multiplied by 8 gives 72 units (*a*), 3 tens multiplied by 8 gives 24 tens (*b*), 7 hundreds multiplied by 8 gives 56 hundreds (*c*), and 6 thousands multiplied by 8 gives 48 thousands (*d*). Writing these results in the ordinary way, and adding them together, as above, the *sum*, which is the required *product*, is 53912.

In practice the partial products 72, 240, 5600, and 48000, are not written down, but combined mentally into one sum; thus, we say 8 times 9 units are 72 units, that is to say, 7 tens and 2 units, we accordingly write down 2 in the units' place, and carry 7 to the place of tens; 8 times 3 tens are 24 tens, and the 7 tens carried make 31 tens, or 3 hundreds and 1 ten. Put the 1 in the tens' place and carry the 3 hundreds; 8 times 7 hundreds are 56 hundreds, and the 3 hundreds carried make 59 hundreds, or 5 thousands 9 hundreds; write the 9 in the place of hundreds and carry 5 to the thousands. We have next, 8 times 6 thousands are 48 thousands, and the 5 thousands carried are 53 thousands, which is the entire number of thousands, the whole product being 53912.

$$\begin{array}{r} 6739 \\ 8 \\ \hline 53912 \end{array}$$

The reason of the following rule will now be evident.

When the multiplier is not greater than 12.

RULE III.

1°. *Place the multiplier under the multiplicand, units under units, and draw a line under the multiplier to separate it from the product.*

2°. *Commencing at the* unit's figure *multiply each figure of the* multiplicand *by the* multiplier. *If the* product *of the* multiplier *and* any figure *of the* multiplicand *is less than ten, set it down under that digit; but set down the right-hand figure only of the* product *when it is a number of more than* one figure, *and carry, as in addition. Set down the last product in full.*

EXAMPLES.

Ex. 1. Multiply 6209748 by 7.

To work out this sum, commence by placing the multiplier under the multiplicand, units under units. Beginning at the right hand, and multiplying each figure in the multiplicand separately, the work is as follows:—

7 times 8, 56, write down the 6 and carry 5; 7 times 4, 28, which, with 5 carried, make 33, set down 3 and carry 3; 7 times 7, 49, and 3 carried, are 52, set down 2 and carry 5; 7 times 9, 63, and 5 are 68, set down 8 and carry 6; 7 times 0, 0 and 6 are 6, set down 6; 7 times 2 are 14, set down 4 and carry 1; 7 times 6 are 42, and 1 43, set down 43.

$$\begin{array}{r} 6209748 \\ 7 \\ \hline 43468236 \end{array}$$

The following, worked like the above example, require no further explanation.

Multiply	73816	1073142	460052	531462
By	8	9	11	12
Product	590608	8165278	5060572	6377544

EXAMPLES FOR PRACTICE.

1. 34264896×2
2. 65431987×3
3. 37654198×4
4. 37986578×5
5. 91823740526×6
6. 6521734782×7
7. 48586873×8
8. 57324178×9
9. 987654321×10
10. 891237654×11
11. 647853291×12
12. 919273654×12

13. Multiply 745234 six times successively by the following multipliers, 2, 3, 4, 5, 6, 7, 8, 9, 11, and 12.

32. To multiply any number by 10, we have only to remove each of the figures of the multiplicand one place to the left and their value will be increased ten times, or, in other words, any number is multiplied by 10 by annexing *one* cypher, by 100 by annexing *two* cyphers, by 1000 by annexing *three* cyphers, &c.; *e.g.*, $85 \times 10 = 850$, for, by annexing the cypher, the 5

units have become 5 *tens*, and the 8 *tens* have become 8 *hundreds*, i.e., *the several parts* of the multiplicand have each received a tenfold increase, and, therefore, the whole number has been multiplied by 10. Again, $2376 \times 100 = 237600$, where the value of each figure is increased a hundred times by writing to the right of the multiplicand as many cyphers as there are in the multiplier.

(a) When the significant figure of the multiplier is not a unit, as for example 30, 400, or 700. Since these multipliers are the same, as 10 times 3, 100 times 4, or 1000 times 7; the multiplicand is first multiplied by the significant figure 3, 4, or 7, (by Rule III), afterwards the product is multiplied by 10, 100, or 1000 as in Art. 32) by writing one, two, or three cyphers to the right of the product. Thus, to multiply 468 by 700, we have the operation in the margin.

```
    468
    700
  327600
```

When the multiplier consists of many figures, the process is as follows:—

EXAMPLE.

Ex. 1. Multiply 4786 by 2783. That is, to take 4786 2783 times and add them all together; or, to take it 2000 times, 700 times, 80 times, and 3 times, and the sums together; or, to multiply it by 2000, by 700, by 80, and by 3, and add the products together.

Ordinary form (B)

(A)
```
4786 ×    3 =    14358
4786 ×   80 =   382880
4786 ×  700 =  3350200
4786 × 2000 =  9572000
And the sum is 13319438
∴ 4786 × 2783 = 13319438
```

Having placed the multiplier under the multiplicand, units under units, tens under tens, &c., proceed thus:—

(1) $4786 \times 3 = 14358$
(2) $4786 \times 8 = 38288$
  ∴ 4786×8 *tens* = 38288 *tens*
(3) $4786 \times 7 = 33502$
  ∴ 4786×7 *hunds.* = 33502 *hunds.*
(4) $4786 \times 2 = 9572$
  ∴ 4786×2 *thous.* = 9572 *thous.*
Now, by simple addition, the sum of these products
∴ $4786 \times 2783 = 13319438$

	Thousands	Hundreds	Tens	Units
	4	7	8	6
	2	7	8	3
	1	4	3	5	8
	3	8	2	8	8
	3	3	5	0	2
	9	5	7	2	
=	1	3	3	1	9	4	3	8

If the ordinary method of performing the operation (see under B above) be compared with the detailed form (under A), it will be observed that by arranging the figures in the second line of multiplication one place to the left of those in the first, those in the third one place to the left of those in the second, and so on, we retain the figures in each line in their proper places without the addition of the cyphers at the end of each line.

The reason of the following rule will now be evident.

33. When the multiplier is greater than 12, we proceed as follows:—

RULE IV.

1°. *Place the multiplier under the multiplicand so that the units of the former may be under those of the latter, the tens under tens, &c., and draw a line under the whole.*

2°. *Write down the* product *of the* whole multiplicand *by the* unit's digit *of the* multiplier, *as in rule. In like manner write down the* product *of the* multiplicand *by each of the remaining figures of the* multiplier, *observing to place the right-hand figure of each line in the column* under *the* figure *of the* multiplier *from which it came* (see Ex. 1).

(a) *If the* multiplicand *contains a* cypher, *treat it as if it were a* number, recollecting that $0 \times 1 = 0$, $0 \times 2 = 0$, *and so on.*

(b) *If one or more of the* figures (not final figures) *of the* multiplier *be* 0, *the corresponding* partial product, *or products, will be* 0 *or cyphers, and the lines may be* entirely omitted, *recollecting to give its* proper value *to the* product, *arising from* multiplying *by the* next figure (see Ex. 2).

3°. *Then* add *all these partial products together, and their* sum *will be the* entire product *of the* factors.

EXAMPLES.

Ex. 1. Multiply 821436 by 672576.

```
      821436
      672576
      ------
      4928616
     5750052
     4107180
     1642872
    5750052
   4928616
   --------
   552478139136
```

Here the first figure (6) of the first partial product is set below the figure 6 in the multiplier, the first figure (2) of the second partial product is set below 7, the multiplying figure, the first figure (2) of the third partial product is set below 5, the figure multiplied by, and so on; the first or right-hand figure on each row standing directly under the multiplying figure producing it.

Ex. 2. Multiply 32407 by 6005.

```
      32407
       6005
      -----
     162035
    194442
    ---------
    194604035
```

Here the first figure (5) of the first partial product is placed immediately below 5, the figure multiplied by, and since the next two figures of the multiplier being cyphers, the corresponding products will be 0 or cyphers, and are entirely omitted, and the multiplicand is then multiplied by 6 and the right-hand figure of the product is set in the column directly below the multiplier 6.

Ex. 3. Multiply 729 by 63817 = 63817 × 729.

```
      63817
        729
      ------
     574353
    127634
   446719
   --------
   46512593
```

Ex. 4. 700013 × 54836 = 54836 × 700013

```
      54836
     700013
     ------
     164508
     54836
   383852
   -----------
   38385912868
```

The following example will illustrate a point of some utility connected with the position of the first figure obtained by multiplication.

Ex. 5. Multiply 3421975 by 11912.

In this example it will be observed that though there are five figures to multiply by, we have only three separate lines of products. The reason is that in the first place we multiply by 12, next by 9, and then by 11, which shortens the sum by two whole lines—a considerable saving. It is necessary, however, to be careful to *place the first figure* in each case *obtained by multiplication directly* under *the figure by which you are* multiplying.

```
    3421975
      11912
    --------
   40063700
   30797775
   37641725
   ----------
   40761566200
```

Examples for Practice.

1. 78954236 × 34
2. 98765240 × 57
3. 93876129 × 95
4. 67869578 × 903
5. 23589647 × 673
6. 1001001 × 999
7. 815085 × 20048
8. 6437063 × 5006701
9. 958866 × 804002
10. 378421896 × 5928578
11. 28814412 × 12345678
12. 987654321 × 1234567890
13. 4771213 × 602059999
14. 461761 × 930937 × 972744 × 708421
15. 9090909 × 1111111 × 999999 × 1010101
16. 9999 × 9998oool × 999700029999
17. 9999 × 9998 × 1000 × 10001 × 10002
18. 9999 × 9999 × 9999 × 9999 × 9999 × 9999

34. If the multiplier to any proposed multiplicand consists of any one or more of the nine digits, followed by a cypher, or any number of cyphers, then multiply according to the following

RULE V.

1°. *Place the* multiplier *under the* multiplicand, *so that the* significant *figure of the* multiplier *shall stand* under *the* unit's figure *of the* multiplicand, *and* multiply *the* successive *figures of the* multiplicand *by the* significant figure *of the* multiplier, according to Rule IV.

2°. *Then, to the* product *thus obtained, place to the* right *the* same number *of* cyphers *as are* contained *in the* multiplier.

Example.

Multiply 123456789 by 80 and 800000.

Multiplicand   123456789
Multiplier   80
─────────────────────────
Product   9876543120

Multiplicand   123456789
Multiplier   800000
─────────────────────────
Product   98765431200000

In the first of these examples we multiply first by 8, according to Rule III, then annex (*i.e., join to*) to the product *one* cypher, because the multiplier contains *one* cypher, in order to preserve the product in its proper place, as the product of 8 *tens*. In the second example the same rule is followed, but five cyphers are annexed, because the multiplier contains *five* cyphers, in order to preserve the product in its proper place as the product of 8 *hundred of thousands*.

35. If the multiplier and multiplicand both end with cyphers, we may omit them in the working, and proceed according to the following

RULE VI.

Multiply the significant *figures of the factors, as directed in* Rule IV. *Then, to the* product, affix *as many* cyphers *as have been* omitted from the end *of both the* multiplier and multiplicand.

The principle of this annexation has been already explained (No. 32).

EXAMPLES.

Thus, if 570 be multiplied by 3200, and 4076800 by 307000.

```
      (1)                    (2)
      570                 4076800
     3200                  307000
     ----                 -------
      114                  285376
      171                  122304
     ----                 -------
   1824000              1251577600000
```

In the first case, the 7 multiplied by 2 is the same as 70 multiplied by 200; and 70 multiplied by 200 gives 14000. In the second the product of the significant figures is 40768 × 307 = 12515776, to this *five* cyphers must be annexed, because 100 × 1000 = 100000; and 12515776 × 100000 = 1251577600000.

EXAMPLES FOR PRACTICE.

1. 80437050 × 90
2. 98311950 × 110
3. 4381792800 × 5600
4. 41508473 × 7002800
5. 42456008000 × 5400
6. 765321 × 760
7. 325758 × 72000
8. 567423 × 4080
9. 49864023 × 708600470
10. 3000100030 × 400100010000

36. It is sometimes advantageous to split up a multiplier which is the product of two or more numbers, and multiply by its factors; thus, if we have to multiply by 36, it is easier to multiply in this case by 6 and 6 (6 × 6 = 36), or by 4 and 9 (4 × 9 = 36), than to multiply by long multiplication, that is, by 3 tens and 6. In any case we have two rows of multiplication, but in the last case we have an addition into the bargain.

EXAMPLE.

Multiply 57894362 by 48.

Since 48 = 6 × 8; or, 4 × 12 = 48, then the answer is found as below.

```
         57894362                            57894362
                6                                   4
         --------                            --------
6 times the multiplicand = 347366172   4 times the multiplicand = 231577448
                                 8                                       12
                           ---------                              ---------
                           2778929376 Ans.                        2778929376 Ans.
```

∴ (6 times 8 times =) 48 times the multiplicand.   ∴ (4 times 12 times =) 48 times the multiplicand.

EXAMPLES FOR PRACTICE.

1. 685732 × 15
2. 356628 × 36
3. 434560 × 56
4. 279819 × 72
5. 356718 × 81
6. 2936298 × 84
7. 108143145 × 121
8. 117974340 × 132
9. 128699280 × 144

37. **Verification of Multiplication.**—I. *By casting out nines.*—Add together the figures of the multiplicand, multiplier, and product separately, not counting any 9 that may occur, rejecting also 9 whenever, in adding up, the sum amounts to 9 or more; note each result. Multiply the first two

remainders, *i.e.*, the remainder arising from casting out nines in the multiplicand and multiplier, retaining, as before, only what is left after the rejection of all the nines from this product, if the sum of the digits exceed nine; then, if the remainder which thus arises is the same as that from the product of the two factors, the operation is likely to be correct, unless there be some compensation of errors, or some figures misplaced.*

Thus, in the annexed example, we say (omitting the 9) 3 and 7 are 10; then 1 and 6 are 7, which write down. Again, 2 and 8 are 10; then 1 and 3 are 4, which is also put down near the multiplier. Lastly, the product of 4 and 7 are 28, and 2 and 8 are 10, which is one above 9. Write, then, 1 near the product, and cast the nines out of the product thus, 1 and 8 are 9; 8 and 2 are 10; 1 and 5 are 6 and 3 are 9; 2 and 8 are 10; which being 1 above 9 shows that the operation most probably is correct.

```
Multiply 90376....7
          2083....4
         ——————
         271128
         723008
        1807520
        ————————
       188253208....1
```

The usual way of noting the four results is to make a cross, to put the first in the left-hand opening, the second in the opposite opening, the third above, and the fourth below. If the upper and lower results are the same, the work, as just observed, is *most likely* correct, but otherwise it is wrong.

In this example both the multiplier and multiplicand leave no remainder when the nines are cast out, *i.e.*, both numbers are divisible by 9, without a remainder.

Then 0 × 0 = 0, place this 0 at *b*; next, casting out the nines from the product, there is no remainder. Hence, the operation, most probably, is correct.

```
  21800475   (b)
      1089    0
  —————————  0 × 0
  196104275   0
```

The method may even be used in detail to verify each step of the sum, as follows:—

```
   349751    remainder    2
    28637        „         8
   ————                   ——
   2448257       „         5
   1049253       „         6
   2098506       „         3
   2798008       „         7
    699502       „         4
   ————————      „        ——
  10015819387    „         7
```

Thus checking every separate multiplication and the addition as well as the complete result.

II. The truth of all results in multiplication may be proved by using the multiplicand as multiplier, and the multiplier as multiplicand; if the product thus obtained be the same as the product found at first, the results are in all probability true.

---

* It is plain, that if any of the figures in the product were made to exchange places, the agreement of the third and fourth results would remain, though the product would be wrong; as would also be the case if one figure were increased and another diminished by the same number; all, therefore, that we can safely infer is, that the agreement spoken of must have place if the work be correct, so that if it fail the work is wrong. Suppose, for instance, that we had made 73084163 × 7854 = 554270392192, then, applying the test, we get from the first factor the result 5, for the second the result 6, and from the product of these the above-stated product of the two numbers the result is 4. This product, therefore, is incorrect, and, upon revising the multiplier, we find that the 3, after the nought, should have been 2.

$$\begin{array}{c} 3 \\ 5 \times 6 \\ 4 \end{array}$$

DIVISION.

38. The object of division is to find how many times one number is contained in another. The quantity to be divided is called the dividend, the quantity by which we divide is the divisor, the *number of times* is the quotient, and what remains over (if any such there be) is called the remainder. Dividend = divisor × quotient + remainder.

The operation of Division is expressed by the sign ÷, which is read *by* or *divide by*; thus, $42 \div 7 = 6$, implies that the result of the division of 42 by 7 is 6. The number 42 which is divided is called the dividend, that which divides, i.e., 7 the divisor and the result 6, the quotient.

39. The first idea of obtaining the result is to use subtraction and count the times we have to use it.

Thus, to find how many times 8 is contained in 34.

$$\begin{array}{r} 34 \\ 8 \end{array} \quad (1)$$
$$\begin{array}{r} 26 \\ 8 \end{array} \quad (2)$$
$$\begin{array}{r} 18 \\ 8 \end{array} \quad (3)$$
$$\begin{array}{r} 10 \\ 8 \end{array} \quad (4)$$
$$2$$

We see we can take 8 away from 34 *four* times in succession, and then we leave 2. But if we had helped ourselves by the multiplication table (of *eight* times) we might have done it more shortly. For since $5 \times 8 = 40$, 8 will go 5 times into 40 exactly; therefore, 8 will not go 5 times into 34. Again, $4 \times 8 = 32$, and thus 8 will go 4 times into 34 and leave something over. This "something over" is evidently $34 - 32$, or 2.

So long as the quotient is a small number the process of continued subtraction may be employed, but when the dividend contains the divisor a large number of times, it would be necessary to abridge the operation by taking away as many times the divisor at once as we please, provided the number of times is marked at each step. For example, to divide 115 by 12 we may take away 8 times 12 at once, and afterwards take away 12; therefore, 12 may be subtracted 9 times, and the remainder is 7.

$$\begin{array}{r} 115 \\ \underline{96} = 8 \text{ times } 12 \\ 19 \\ \underline{12} = \text{once } 12 \\ 7 = \text{remainder.} \end{array}$$

In order to avoid the labour of repeated subtractions, we may lay down the following principle, viz., that if we separate any dividend into any number of parts, and find how often the divisor may be subtracted from these parts (or how often the divisor is *contained*, as it is called, in each of these), we shall, by adding these results, obtain the correct quotient of the whole dividend divided by that divisor, because it is evident that the *whole* dividend will contain the divisor as many times as its *several parts together* contain it.

Let it be required to divide 3168 by 27. Here the quotient will consist of three digits, and therefore there will be at least 3 separate subtractions Now the figure in the hundreds' place cannot be more than 1, and if the product 27 hundreds, or 2700, be subtracted from the total product 3168, the remainder 468 must contain the products of the tens and units of the quotient multiplied by the divisor 27. We now enquire how often 27 is contained ten times in 468, and this is found to be only once ten times; then subtracting the partial product 27 tens, or 270 from 468, the remainder is 198. Lastly, we have to divide 198 by 27, which gives 7 for a quotient and a remainder 9; and, therefore, 3168 contains 27, 100 + 10 + 7, or 117 times, leaving 9 for the remainder.

```
3168
2700 = 100 times 27
----
468
270 = 10 times 27
---
198
189 = 7 times 27
---
9
```

It will be seen that as often as 27 is contained in 31, so many hundred times will it be contained in 3100, or in 3168; and as often as 27 is contained in 46, so many ten times it will be contained in 460, or 468, and in this manner any quotient figure is just as readily obtained as the last or unit's figure of it.

40. The preceding articles contain the principles of division, and all that remains is to apply them in the most economical manner.

EXAMPLE.

Suppose we have to divide 2987618 by 3605.

Operation with cyphers in full.
```
3605)2987618(800 + 20 + 8,
     2884000       or 828
     -------
      103618
       72100
     -------
       31518
       28840
       -----
        2678
```

Operation without annexing cyphers.
```
3605)2987618(828
     28840
     -----
     10361
      7210
     -----
      31518
      28840
      -----
       2678
```

41. If the divisor be not greater than 12.

RULE VII.

1°. *Set down the dividend with a line under it to separate it from the future quotient; and write the divisor on the left of the dividend with a curved line between them.*

2°. *By the multiplication table find the* greatest *number of times the* divisor *is* contained *in the* first *figure of the* dividend, *or, if necessary, the* first two *or* first three *figures of the* dividend, *and place the figure denoting the times* directly *under the* figure *divided, or under the* last *figure if more than one have been taken, and* carry *what is over, that is, regard the figure which is* over *to be* prefixed *to the following figure of the dividend.*

3°. *Divide this number by the divisor, set down the result as the next figure of the quotient, carry the remainder to the next figure of the dividend; and if the divisor is* not *contained in any figure of the dividend, place a* cypher *in the* quotient *and* prefix *this figure to the next one of the dividend as if it were a remainder, and proceed in the same manner till all the figures of the dividend are exhausted.* The number thus found is the quotient.

EXAMPLES.

Ex. 1. Divide 25602 by 3.

Placing the dividend and divisor (3) as in the margin, we proceed thus:— 3 is contained in 2, no times, so that nothing is to be placed under the 2; 3 is contained in 25, *i.e.*, 25 thousands, 8 times, (*i.e.*, 8 thousands) and 1 over, 8 and carry 1; this 1 regarded as *prefixed* to the 6, gives the number 16, *i.e.*, 16 hundreds, 3 is in 16, 5 times and 1 over; 3 is in 10 (*i.e.*, 1 hundred) 3 times and 1 over; 3 is in 12, 4 times. Therefore, the quotient is 8534; and this is the complete quotient as there is no *remainder*.

3)25602
$\overline{}$
8534

Ex. 2. Divide 7804623 by 5.

We say, 5 in 7, 1 and 2 over; 5 in 28, 5 and 3 over; 5 in 30, 6; 5 in 4, 0 and 4 over; 5 in 46, 9 and 1 over; 5 in 12, 2 and 2 over; 5 in 23, 4 and 3 over. As there is here a remainder, we annex it, with the divisor 5 under it, to the figures of the quotient, and call 1560924⅗ the complete quotient.

5)7804623
$\overline{}$
1560924⅗

Ex. 3. Divide 84111648 by 12.

We say, 12 in 8, no times, so that nothing is to be placed under the 8; 12 in 84 goes 7 times and 0 over; 12 in 1, no times and 1 over, put down 0 under the 1; 12 in 11 0 times and 11 over; this 11 regarded as prefixed to the next figure of the dividend gives the number 111; therefore say, 12 in 111, 9 times and 3 over; 12 in 36, 3 times and 0 over; 12 in 4, 0 times and 4 over; 12 in 48, 4 times.

12)84111648
$\overline{}$
7009304

EXAMPLES FOR PRACTICE.

1. $135792695 \div 2$
2. $584697386 \div 3$
3. $399345884 \div 4$
4. $298244760 \div 5$
5. $400678493 \div 6$
6. $276586437 \div 7$
7. $6947421006 \div 8$
8. $2470263075 \div 9$
9. $254096146 \div 10$
10. $1101182267 \div 11$
11. $1095137170 \div 12$
12. $59437055312 \div 12$

Divide each of the following dividends six times successively by each of the following divisors—2, 3, 4, 5, 6, 7, 8, 9, 11, 12.

1. 6154778
2. 7000000
3. 4086791
4. 8821000
5. 2000707
6. 9370005
7. 9914060
8. 4706009

42. **If the divisor be greater than 12.**

RULE VIII.

1°. *Place the divisor and dividend in the same line, separated by a small curve line, and on the right of the dividend draw another line of the same kind;* thus,—

divisor) dividend (quotient

2°. *Mark off from the* left-hand *side of the* dividend *a number of figures* equal *in number to those of the* divisor, *or one more, if necessary, and find the* greatest *number of times the* divisor *is contained in this number; place the figure which denotes this number on the* right *as the first figure of the quotient.*

3°. *Multiply the divisor by this number and place the product under the number marked off at the left of the dividend.* Subtract *the said* product *from that part of the* dividend *under which it stands.*

4°. *Bring down the* next *figure of the* dividend *and place it to the right of the remainder, and if the* number *thus formed be* greater *than the* divisor, *find the* greatest *number of times the* divisor *is contained in it, and write this number as the* second *figure of the quotient; but if this number be* less *than the* divisor *bring down the* next *figure of the dividend, or more, until a number not* less *than the* divisor, *is formed*, remembering *to place* a cypher *in the* quotient *for every* figure *brought down*, except the last; *find how often the divisor* is contained in this number; *then multiply, subtract, and bring down, &c., as before, till all the figures of the dividend are exhausted.*

The number thus obtained is the quotient required.

Note.—When the divisor is large, the learner will find assistance in determining the quotient figure, by finding how many times the first one or two figures on the left hand of the divisor is contained in the first one or more of those of the dividend. This will give pretty nearly the right figure. Some allowance must, however, be made for carrying from the product of the other figures of the divisor, to the product of the first into the quotient figure. If any product be greater than the number which stands above it, the last figure in the quotient must be changed for one of smaller value; but if any remainder be greater than the divisor, or equal to it, the last figure of the quotient must be changed for a greater.

Examples.

Ex. 1. Let it be required to divide 256434 by 346.

Looking at the leading figures of the divisor, and also at that of the dividend, with the view of seeing whether the latter contains the former, which it does not, 3 being greater than 2; we therefore commence with the number 25, formed by the first *two* figures of the dividend, and seeing that 3 is contained in 25 8 times, we should put 8 for the first quotient figure; but bearing in mind that when the *whole* divisor is multiplied by this 8 we must attend to the *carryings;* we perceive that 8 is too great, we therefore try 7, and find 7 times 346 to be 2422, a number less than 2564 above it, so that we can *subtract;* the remainder is 142, which, when the next figure of the dividend is brought down, becomes 1423. We now take this as a dividend, and looking at the *leading figures* in this new dividend and the divisor, we see that that latter *will go* 4 times, we therefore put 4 for the second quotient figure, and multiplying and subtracting we get 39 for the second remainder, and, by bringing down another figure we get 394 for a new dividend; the divisor goes into this *once*, so that the quotient is 741, and the final remainder 48; this remainder must be annexed with the divisor underneath to the quotient figures, so that the complete quotient is 741 $\frac{48}{346}$, which is the 346th part of 256434.

```
346)256434(741 quotient.
    2422
    ────
    1423
    1384
    ────
     394
     346
     ───
      48
```

Principles and Practice of Arithmetic. 37

Ex. 2. Divide 10841971621.4 by 5783.

```
                          5783 ) 10841971621.4 ( 18748005 quotient.
Quotient figure (1)......  5783
                          ─────
                           50589 ............... 1st remainder with next figure.
              (8)......   46264
                          ─────
                           43257 ............... 2nd       ,,           ,,
              (7).......  40481
                          ─────
                           27761 ............... 3rd       ,,           ,,
              (4).......  23132
                          ─────
                           46296 ............... 4th       ,,           ,,
              (8)........ 46264
                          ─────
                             322 ............... 5th       ,,           ,,
    ☞    (0)...........
                          
                            3221 ............... 6th       ,,           ,,
    ☞    (0)...........
                          
                           32214 ............... 7th       ,,           ,,
              (5).......... 28915
                          ─────
                            3299  final remainder.
```

It must be noticed that if any *dividend* formed by a remainder and a figure brought down should be less than the divisor, that the divisor will go *no times* in that dividend; so that a o will be the corresponding quotient figure; and that, then, a second figure must be brought down as in the operation annexed. The steps marked ☞ are inserted merely to show the principle. In practice we simply put down the two noughts in the quotient, and go at once to 32214 for the divisor.

Ex. 3. Divide 6421284 by 642.

```
                                                              642)6421284(10002
```

642 goes once into 642, and leaves no remainder. Bring down the next figure (1) of the dividend, then 642 is in 1 no times, put o in the quotient after the 1. The next figure of the dividend (2) being brought down to the right of the 1 forms the dividend 12, then 642

```
                                                                  642
                                                                  ───
                                                                  1284
                                                                  1284
```

is in 12 no times, put a cypher as the next quotient figure; bring down the next figure of the dividend (8), then 642 in 128 goes no times, write o as the corresponding quotient figure; but the next figure brought down (4) makes the quantity brought down 1284, which contains the divisor twice, and gives no remainder.

EXAMPLES FOR PRACTICE.

1. 983296 ÷ 13	4. 9500864 ÷ 43	7. 56703264 ÷ 123
2. 800062 ÷ 23	5. 9943000 ÷ 78	8. 571414204 ÷ 809
3. 20067690 ÷ 37	6. 8904030 ÷ 89	9. 94010610 ÷ 987
10. 14710962989869 ÷ 1709	12. 1395243584 ÷ 5678	14. 940870015 ÷ 8764
11. 2107791630 ÷ 3654	13. 3855999705 ÷ 6789	15. 1000000000 ÷ 9999
16. 1000000000 ÷ 1001	21. 668c943744279021 ÷ 95400621	
17. 100000000000000 ÷ 11111	22. 72193263111263526.9 ÷ 987654321	
18. 4842315713782 ÷ 570634	23. 200000001876.0631 ÷ 31622777	
19. 815240906170 ÷ 763054	24. 79222833122805843200 ÷ 879510067	
21. 490002800004 ÷ 900702	25. 2962296118145871914806.56 ÷ 972744	

26. Divide 1000........ (with as many noughts added as may be necessary to give ten places of figures in the quotient) by 2302585093.

43. When the divisor is a composite number, that is, can be separated into two or more factors, the division can be effected by the following rule:—

RULE IX.

1°. *Divide by one factor, setting down the quotient and remainder.*

2°. *Divide the quotient thus obtained by another factor, setting down the quotient and remainder, and so on, till all the factors are employed.* The last quotient will be the answer required.

3°. *The proper remainder is found, when the divisor is resolved into but two factors, by multiplying the second remainder by the first divisor, and to the product adding the first remainder,* but when more than two factors are employed *by multiplying the remainders after every line by all the divisors except their own, and adding the results.*

EXAMPLES.

Ex. 1. Divide 569736869 by 15.

Here, since 15 is the product of 3 and 5, it is obvious that the quotient may be obtained from successive divisions by 3 and 5.

$$15\begin{cases} 3 & | \ 569736869 \\ 5 & | \ 189912289\ldots 2 \\ & \ \ \ 37982457\ldots 4 \end{cases}$$

Here, the remainder 2 in the first quotient is 2 units of the upper line; but the remainder 4 in the second line consists of 4 units of the second line; and as each unit in the second line is *three times* as great as each unit in the upper line, the remainder 4 is equal to 3×4 units of the upper line, *i.e.*, is equal to 12 ordinary units, hence the whole remainder is $2 + 12$, or is 14.

Ex. 2. Divide 8327965 by 72.

Here, since 72 is the product of 8 and 9, the quotient is obtained by successive divisions by 8 and by 9; otherwise, since $6 \times 12 = 72$, we may first divide by 6 and then divide the resulting quotient by 12.

$$72\begin{cases} 8 & | \ 8327965 \\ 9 & | \ 1040995\ldots 5 \\ & \ \ \ 115666\ldots 1 \end{cases}$$

Ex. 3. Divide 8327965 by 99.

In this instance 99 is the product of 9 and 11; we divide by 9 and the quotient thus arising we divide by 11.

$$99\begin{cases} 9 & | \ 8327965 \\ 11 & | \ 925329\ldots 4 \\ & \ \ \ 84120\ldots 9 \end{cases}$$

To deduce the remainders which would have been left had the divisions been performed by 72 and 99 in the usual way, we may observe that the first partial remainder 5 (Ex. 2) must be *units;* but the second remainder must be regarded as so many collections of 8 units each, and that the first partial remainder 4 (Ex. 3) must be *units;* but the second dividend being as so many collections of 9 units each, the second remainder must, in this case, be regarded as so many collections of 9 units each; hence the true remainders in these examples are respectively

$$1 \times 8 + 5 = 13, \text{ and } 9 \times 9 + 4 = 85.$$

Ex. 4. Divide 2071998 by 192.

The factors for 192 are $4 \times 6 \times 8 = 192$.

$$192\begin{cases} 4 & | \ 2071998 \\ 6 & | \ 667999\ldots 2 \\ 8 & | \ 111333\ldots 1 \times 4 + 2 = 6 \\ & \ \ \ 13916\ldots 5 \times 6 \times 4 + 6 = 126 \end{cases}$$

Dividing by 4 the remainder is 2, dividing by 6 the remainder is 1, dividing by 8 the remainder is 5, which gives a total remainder 126, found thus: the remainder 2 is two units, the second remainder is one ₁ and 2, making 6, the third remainder 5 is five times 24 (4×6) or 120, because having divided by $4 \times 6 = 24$, the quotient 11133⅓ are twenty-fours, so that any remainder must consist of twenty-fours; 120 added to the previous remainder 6, gives 126. Thus the answer is—quotient 13916, remainder 126.

Ex. 5. Divide 24533279 by 432.

Since $6 \times 8 \times 9 = 432$, we may divide successively by these numbers.

$$432\begin{cases} 6 & | \ 24533279 \\ 8 & | \ 4088879\ldots 5 \\ 9 & | \ 511109\ldots 7 \times 6 + 5 = 47 \\ & \ \ \ 56789\ldots 8 \times 8 \times 6 + 47 \\ & \ \ \ \ \ \ \ \ \ \ \ \ \ = \text{remainder } 431. \end{cases}$$

The first remainder is 5, the second remainder is 7, then 7 times the first divisor ($= 6$) $+ 5$ (the first remainder) $= 47$. The third remainder is 8, then 8 times 8 (the second divisor) $= 64$, and 64 times the first divisor $6 = 384 +$ remainder 47, gives the final remainder 431.

EXAMPLES FOR PRACTICE.

1. $6489175432689467 \div 14$
2. $598432789648320758 \div 22$
3. $56983475689268 \div 36$
4. $59864326859168 96 \div 63$
5. $987654321012345 \div 66$
6. $543210123456789 \div 121$
7. $35913213916219 11 \div 132$
8. $4902550716552769 \div 144$

44. Division may also be abridged where the divisor is terminated by a cypher or cyphers; we proceed as follows:—

RULE X.

1°. *Cut off the cyphers from the divisor, and as many figures from the right-hand of the dividend as there are cyphers so cut off, at the right-hand end of the divisor, then proceed with the remaining figures in the usual manner* (Rule VII or VIII), *and if there are anything remaining after the division annex those figures which are cut off from the dividend; otherwise, the figures cut off will be the remainder.*

NOTE.—The same rule applies when the divisor and dividend both terminate with cyphers.

EXAMPLES.

Ex. 1. Divide 3704196 by 20.

$$2{,}0)370419{,}6$$
$$\overline{185209\tfrac{16}{20}}$$

In the first of these examples you mark off with a comma the cypher or 0 in the divisor, and the first figure 6 to the right in the dividend; this is equivalent to dividing both divisor and dividend by 10. You next divide the remaining figures 370419, to the left in the dividend, by the divisor 2, according to Rule VII; thus is obtained the quotient 185209, and remainder 1; to this remainder you annex the figure 6, which was cut off, and you have the complete remainder 16. The quotient may now be correctly represented thus, $185209\tfrac{16}{20}$.

Ex. 3. Divide 271830 by 30.

$$3{,}0)27183{,}0$$
$$\overline{9061 \; Ans.}$$

1st. Cut off a cypher from the divisor, and also one from the dividend.
2nd. Divide 27183, the remainder of the dividend, by 3, the remainder of the divisor: the quotient 9061 is the answer.

Ex. 2. Divide 31086901 by 7100.

$$71{,}00)310869{,}01(4378\tfrac{3101}{7100}$$
$$284$$
$$\overline{268}$$
$$213$$
$$\overline{556}$$
$$497$$
$$\overline{599}$$
$$568$$
$$\overline{31}$$

In the second example you follow the same rule; that is, you cut off two cyphers in the divisor and two figures in the dividend, and obtain the quotient in the usual way, which is 4378, and remainder 31; to this 31 annex the two figures cut off from the dividend, and you have the complete remainder 3101.

EXAMPLES FOR PRACTICE.

1. $9357864837986496 \div 50$
2. $67400869473 8425 \div 700$
3. $987654321670 \div 3000$
4. $483795864973106789 \div 120$
5. $6550000280034 \div 6300$
6. $26799534687 \div 7890000$

45. **Verification of Division.**—(1.) Multiply the quotient by the divisor, or the divisor by the quotient, and to the product add the remainder, if there be one. The result ought to be the same as the dividend; because we are only adding the divisor the same number of times, as it was subtracted in the operation of division.

(2.) *Subtract the remainder, if any, from the dividend, and divide the difference so obtained by the quotient.* The result should be equal to the divisor, if the working be correct.

MISCELLANEOUS EXAMPLES.

1. Express in figures: Ten thousand and four. Four thousand and four. Forty-four thousand and four.
2. $29483 + 7648 + 32479 + 586 + 298364 + 98765 + 897 + 789 + 5678 + 99$.
3. From 6794006897 take 3985160534, and from 100010 take 10011.
4. Multiply 94785830 by 78060, and 879510067 by 90076096000.
5. Divide 5688208152 by 594, and 100000000000000 by 967.
6. Express in figures: One hundred million one hundred thousand and one hundred.
7. Add together 90473, 9456, 263, 59, 45694, 5437, 87668497, 2837, 9865, 3652, 999, and 8888.
8. Find the difference between 100000000000 and 87649786. Take one hundred thousand two hundred and twelve from four million one hundred and one.
9. Multiply 326904678 by 3060900.
10. Divide 236487698743 by 85409, and $5754054870008880 \div 8009909$.
11. Express in figures: One hundred and three million eighty thousand two hundred and seven.
12. Add together 69074, 6745, 723, 29, 931648, 9005, 76245, 54267, 47096, and 7777.
13. From 7860007000 take 6974208506.
14. Multiply 167409678 by 768900. How much greater is 678×76 than $54675 \div 9$?
15. Divide 600000700649008805 by 98706543.
16. Express in words and in figures how much greater the value of one 5 is than the other in the number 658457.
17. Multiply 129847 by 468. If, in the process, you shift all the figures resulting from the multiplication of the multiplicand by 4 two places further to the left and then add, of what two numbers will the result be the product?
18. What number subtracted from 850967 will leave 3946? The 365th part of a number is 101001, what is the number?
19. The digits in the units' and millions' places of a number are 4 and 6 respectively, what will be the digits in the same places when 99999 is added to the number?
20. What number must be added to sixty-nine thousand four hundred and twenty seven to produce three hundred and twenty-five millions seven thousand and twenty-one?
21. Find the sum, difference, and product of 12345678 and 288144412. Find the sum, difference, and product of 1234567 and 4321089.
22. 15996 tons of coal are exported in 43 ships: how many tons does each ship on the average carry?
23. How many years of 365 days each in 46355 days?
24. How often can you subtract 6 from 47112? What number subtracted three times from 78467 will leave 41426?
25. How many ships, each carrying 673 men, can transport an army of 22882 men?
26. By what number must you divide 7460020 in order that the quotient may be 52907 and the remainder 133?
27. 2036809 divided by a certain number gives a quotient 2031 with a remainder of 1474: find the dividing number?
28. A ream of paper contains 20 quires of 24 sheets each; on each page there is room for 34 lines of writing: how many may be written in the ream?
29. What is the number of holes in a sheet of perforated zinc, containing 1519 square inches, if there be 85 in the square inch?
30. What will remain after subtracting 213 as often as possible from 83216?
31. The product of two numbers is 1270374 and half of one of them is 3129: what is the other number?
32. Find the sum, difference, product, and quotient of 1653125 and 13225.

33. Find the sum, difference, product, and quotient of 9765625 and 78125.
34. What number subtracted three times from 78467 will leave 41416?
35. A number increased by 13 times itself amounts to twenty millions five thousand and six: find the number.
36. If the sum of 250 and 173 be multiplied by their difference, and the product divided by 33: find the result.
37. Find the difference between the sum of 4715 added to itself 398 times, and the sum of 2017 added to itself 408 times.
38. A man died in 1873, aged 94; his son died in 1827, aged 17: how old was the father when the son was born?
39. A gentleman being asked how old his son is, replied "In 23 years he will be as old as I was when he was born, and I shall be 58:" how old is the son?
40. The Mariner's Compass was invented in Europe in the year 1302; how long was that before the discovery of America by Columbus, which happened in 1472?
41. The distance of one of Jupiter's moons from that planet is 490140 miles, and of another 652650: find how much further one is from Jupiter than the other.
42. From the centre of Saturn to the outside of his inner ring is 71870 miles, to the inside of the same ring is 54560: find the breadth of the ring.

DECIMAL FRACTIONS.

46. **Arithmetical** operations become lengthy and troublesome if they involve many vulgar fractions of different denominations; it becomes necessary, therefore, to devise a method of expressing fractions in such a manner that they may be easily reduced to the same denomination. To effect this all fractions are reduced to others having for denominators 10, 100, 1000, &c. Such fractions are called **Decimal Fractions**.

47. Decimals occur so frequently in all computations relating to Nautical Astronomy, that it becomes absolutely necessary to have a knowledge of their application and their relation to Vulgar Fractions.

48. In the Notation of Integers or common numbers, the actual value of each figure depends upon its position with respect to the place of units, its value in any one position being one-tenth of what it would be if it stood one place further to the left; thus the number 1111 denotes one thousand, one hundred, one ten, and one unit, or 1000 + 100 + 10 + 1, where the second unit beginning with the right-hand one is ten times the first, the third is ten times the second, the fourth ten times the third, and so on; or beginning with the first on the left, the second is the tenth part of the first, the third the tenth part of the second, and so on, till we come down to the last unit, which is merely one; or in other words, the figures decrease in a tenfold ratio from left to right.

49. Now we may evidently extend this principle still further, and on the same plan may represent one-tenth of *one*, one-tenth of *this*, or one-hundredth of *one*, one-thousandth of *one*, and so on, by simply putting some mark of separation between the *integers* and these *fractions*. The mark actually used

is a *dot* or *full stop*, and is called the *decimal point*, thus 1111·1111.* The unit (or 1) next the dot, on the left, is 1; the unit one place from this on the left is 10; the next is 100; the next 1000, and so on. In like manner, the unit next the decimal point, on the right, is $\frac{1}{10}$, the next $\frac{1}{100}$, the next $\frac{1}{1000}$, and so on. In other words, any figure one place to the right of the unit's place will be one-tenth of what it would be if it were in the unit's place, and will thus really denote a decimal fraction; any figure two places to the right of the unit's place will be one-hundredth of what its value would be if it were in the unit's place, and so on for any number of figures, as in the following table, which may be regarded as an extension of the numeration table.

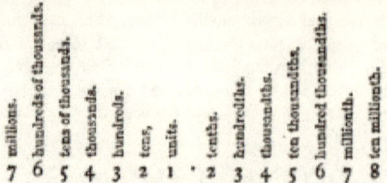

50. This being agreed upon, it follows that a decimal may either be considered as the sum of as many fractions as it contains digits, or as a single fraction; thus :—

·567 = $\frac{5}{10} + \frac{6}{100} + \frac{7}{1000} = \frac{567}{1000}$.

·0305 = $\frac{0}{10} + \frac{3}{100} + \frac{0}{1000} + \frac{5}{10000} = \frac{0305}{10000}$.

18·204 = 18 + $\frac{2}{10} + \frac{0}{100} + \frac{4}{1000} = \frac{18204}{1000}$.

51. Hence, *a decimal is always equivalent to the vulgar fraction whose numerator is the decimal considered as integral, that is, the number itself, when the decimal point is suppressed, and whose denominator is 1 followed by as many cyphers as there are decimal places in it.*

52. We generally speak of any figure in a decimal as being in *such a place of decimals*; thus, for instance, in 3·14159, we should say that the 5 is in the fourth place of decimals, the 9 in the fifth place, and so on, reckoning from left to right.

53. The figures 1, 2, 3, 4, 5, 6, 7, 8, 9, in a decimal are sometimes called *significant* figures or digits; thus, in such a decimal as ·0002345, we should say that 2 is a significant digit, because it is the first figure which indicates a number, the cyphers only serving to fix the place in which the 2 occurs.

54. Numbers made up of whole numbers and fractions, either vulgar or decimal, are called mixed numbers; for instance, 368·414 is a mixed number, the figures which precede the decimal point (the 3, the 6, and the 8) are whole numbers or integers, while those which follow the point (·414) are decimals.

---

* The decimal point should be put at the top of the line of figures, thus—5·7, because 5.7 with a stop at the bottom is used in most works to mean 5 × 7 = 35.

55. To read off, or express in words, decimal fractions, *read the decimal figures as if* whole *numbers, and to the last figure* add *the* name *of the order determined by the place it occupies;* thus, ·734 is read *seven hundred and thirty-four* thousandths; 58·64327 is read fifty-eight, *together with sixty-four thousand three hundred and twenty-seven* hundred-thousandths; ·080905 is read *eighty thousand nine hundred and five* millionths.

In reading decimals as well as whole numbers, the *unit's* place should always be made the *starting* point. It is advisable for the learner to apply to every figure the name of its order, or the place which it occupies, before attempting to read them. Beginning at the unit's place he should proceed towards the right, thus—*units, tenths, hundredths, thousandths,* &c., pointing to each figure as he pronounces the name of its order. In this way he will be able to read decimals with as much ease as he can whole numbers.

56. The value of the decimal figures depending entirely on the place they occupy with respect to the point which separates the units from the tenths, any number of cyphers on their right may be annexed or effaced, without altering the value of the *significant* figures. For instance, 0·7 is the same as 0·70, because the number that expresses the decimal fraction becomes ten times greater while its parts become hundredths, and are therefore diminished ten times,

$$\text{thus } \tfrac{7}{10} = \tfrac{70}{100} = \tfrac{700}{1000}, \&c.,$$

and hence it is evident that annexing cyphers to the right hand of decimals does not change their value, for we only multiply both numerator and denominator by 10, 100, &c., and consequently does not alter their value at all. Again, take a decimal such as ·56, which, as already explained, means 5 tenths 6 hundredths, it will follow that ·560 means 5 tenths, 6 hundredths, *no* thousandths; whence the addition of the cypher to the right-hand has made no alteration in the value of the decimal. In fact,

$$·56 = \tfrac{56}{100} \text{ and } ·560 = \tfrac{560}{1000} = \tfrac{56}{100}.$$

Similarly ·23, ·230, and ·2300, are all of equal value, for expressed as fractions they are respectively $\tfrac{23}{100}$, $\tfrac{230}{1000}$, and $\tfrac{2300}{10000}$.

57. But placing cyphers between the decimal point and the other decimal figures does alter the value of the decimal, because this alters the place of the significant digits, the value being diminished ten times for each cypher that is prefixed,

$$\text{thus } ·7 = \tfrac{7}{10}, \ ·07 = \tfrac{7}{100}, \ ·007 = \tfrac{7}{1000}, \text{ and so on.}$$

We infer from this, that as the value of a decimal is *decreased* ten-fold for every cypher added to the left-hand, we do in fact *divide* a decimal by 10, by 100, by 1000, &c., as we shift the decimal point one, two, three, &c., places to the *left*; and that conversely by shifting the decimal point one, two, three, &c., places to the *right*, we multiply the decimal by 10, by 100, by 1000, &c. For instance, the expression 56·789 is divided by 10 if written 5·6789, is divided by 100 if written ·5678, and is divided by 1000 if written ·056789; whereas the expression ·00723 is multiplied by 10 if written ·0723, is multiplied by 100 if ·723, and is multiplied by 1000 if written 7·23.

Examples for Practice.

Express as decimals—

1. $\frac{7}{10}$, $\frac{7}{100}$, $\frac{7}{1000}$, and $\frac{73}{100}$; also $\frac{1}{10}$, $\frac{111}{100}$, $\frac{11}{100}$, and $\frac{7233}{1000}$.

2. $\frac{1}{100}$, $\frac{11}{1000}$, $\frac{111}{10000}$, $\frac{1000001}{1000000}$, $\frac{1}{10}$, $\frac{1}{100}$, $\frac{1}{1000}$, $\frac{1}{10000}$, and $\frac{131}{100000}$.

3. $\frac{93}{100}$, $\frac{9001}{1000}$, $\frac{900011}{100000}$, $\frac{90000101}{10000000}$, $\frac{111}{\text{millionths}}$, $\frac{121}{\text{thousandths}}$, $\frac{321}{\text{tenths}}$, $\frac{101}{\text{thousandths}}$.

4. $\frac{2222}{1000}$, $\frac{213}{100}$, $\frac{11111}{10000}$, $\frac{1}{10000}$, $\frac{12}{10000}$, $\frac{2122}{100}$, $\frac{123}{10}$.

5. $\frac{11111}{10000}$, $\frac{122212}{100000}$, $\frac{12222222}{10000000}$, $\frac{21232222}{10000000}$.

6. In the following mixed numbers write the fractional parts in decimals:—

   7 $\frac{1}{1000}$, 43 $\frac{1111}{10000}$, 9 $\frac{1000000}{10000000}$, 1 $\frac{1}{1000000}$, and 35 $\frac{111111}{1000000}$.

7. Express as decimal fractions the following:—

   Seventy-three *thousandths*; one hundred and ninety-seven *ten thousandths*; one *millionth*; two hundred and sixty-one *hundred thousandths*; one thousand and one *ten millionths*.

8. Express as decimals the following:—

   One, and fifty-four *hundredths*; twenty-four, and seventy-nine *thousandths*; three hundred and fifteen, eight *thousandths*, and fifty *millionths*; eleven hundred *millionths*; nine *thousandths*, and three hundred *thousandths*.

9. One tenth; three hundredths; five thousandths; one hundred and five thousandths; two millionths; sixty millionths; forty-one and eight hundredths; one thousand and one thousandth; thirty and six millionths; one hundred thousandth; two thousand three hundred and seventy-five hundred millionths.

10. Express in words the following decimals and mixed numbers:—

    ·283, ·5321, ·74895, ·82·056, 27·8354, 34·0009, 43·101007, 23·75, 2·375, ·2375, ·00002375.

11. Express in words the following:—

    ·6, ·92, ·5498, 7·07, 26·405, ·000001, ·00037, 11·101101, ·0440308, ·82344, ·13236.

12. Write in words 9·0457; 4004·0000345; 3·400; 524000634·0008034; ·000003705; ·000024056; 7005·000000674; 100000·0000001; 10·001; 9·000028; 1·0006003.

13. 1·000001; ·1000001; ·00000001; 1·113004; 9·203167; 4·3008004; 27·4627350.

14. What is the effect of moving the decimal point backwards or forwards?

ADDITION.

58. Decimals, or integers and decimals mixed, may be added together precisely as in whole numbers, care being taken so to arrange the figures that all the decimal points fall exactly under one another. This will ensure that *tenths* fall under *tenths*, *hundredths* under *hundredths*, &c. The reason of this arrangement will appear from the following consideration: if this rule were not observed, tenths would fall under hundredths, or hundredths under thousandths, as the case might be; and we should be attempting to add together fractions which had not common denominators. But if we arrange the decimal points all exactly beneath one another, tenths fall under tenths, hundredths under hundredths, &c., in other words, by so arranging them we at once bring the several fractions to a common denominator, and can proceed to add them together. The decimal point, in the answer, will fall exactly beneath the decimal points in the quantities to be added. When the sum of any figures exceeds 10, 20, &c., *carrying* to the next denominator will be performed exactly as in whole numbers, whether the given quantities are all decimals, or are mixed integers or decimals. For as the value of each figure decreases tenfold as we proceed from left to right, the rules of ordinary addition are immediately applicable.

We have, therefore, the following rule for addition:—

RULE XI.

1°. *Place the quantities so that their decimal points shall be in the same vertical line;* for then the quantities of the same denomination will stand together.

2°. *Then proceed as in addition of whole numbers.*

EXAMPLES.

Ex. 1. For instance, let it be required to add together ·8, ·78, and ·678.

```
 ·8
 ·78
 ·678
------
2·258
```

Where we see that after writing in the answer 8 in the place of *thousandths,* that 7 *hundredths* and 8 *hundredths* added together make 15 *hundredths;* but 15 *hundredths* are 1 *tenth* and 5 *hundredths,* writing 5 in the place of *hundredths,* and carrying one to the place of *tenths,* we obtain 22 *tenths;* but 22 *tenths* are properly written as 2 *integers* and 2 *tenths.*

Ex. 2. Again, where integers and decimals are mixed.

```
 3·007
42·6
  ·3975
46·0045
```

Where writing 5 in the place of *ten thousandths,* the sum of 7 *thousandths* and 7 *thousandths* is 14 *thousandths;* writing 4 in the place of *thousandths,* and carrying 1 to the place of *hundredths,* we obtain 10 as the sum in the *hundredths* place; but 10 *hundredths* are 1 *tenth,* carrying 1 to the place of *tenths,* we have 10 *tenths;* but as 10 *tenths* are 1 *unit,* we carry 1 to the place of integers, and write 6 in the place of *units,* and 4 in the place of *tens.*

Ex. 3. Add together 0·35, 47·4, and 9·12.

```
 0·35
47·4
 9·12
------
56·87
```

Ex. 4. Add together 23·628, 4·1056, ·0137, and ·0042.

```
23·628
 4·1056
  ·0137
  ·0042
-------
27·7515
```

Ex. 5. Add together 1234·6789, 13, 170, ·0054, ·5, and 87·142.

```
1234·6789
  13
 170
    ·0054
    ·5
  87·142
---------
1505·3263
```

Ex. 6. Add together 66199·3226, ·301, 54·5, ·00632, 1000, ·07, and 32745·80008.

```
66199·3226
    ·301
   54·5
    ·00632
 1000
    ·07
32745·80008
-----------
100000
```

EXAMPLES FOR PRACTICE.

Find the value of

1. ·225, 3·086, 12·17, ·0051, and 729·54; 2·63, ·263, ·0263, and ·000263.
2. 8·1, 40·652, 98·51, 695·7, and 43·9706; 69·75, 0·97, 0·059, 673·5, 4·8, and 931·6.
3. 897·4, 63·18, 400·03, 7·9, 63·9, and 5·0079; ·00162, ·1701, 325, 2·7031, and 3·000701.
4. 3608·26, 360·826, 36·0826, 3·60826, and ·22314; 467·3004, 28·78249, 1·29468, and 3·78241.
5. 36·053, ·0079, ·000952, 417, 85·5803, ·0000501.
6. 87·1 + 0·376 + ·0056 + 49 + 3·009 + ·709; 293·0072, 89·00301, 29·84567, 924·00369, and 71·39602.
7. ·8 + ·046 + 9·1 + 3·09 + 8·6409 + 32; 1·721341, 8·620047, 51·720345, 2·684, and 62·304607.
8. 1 + ·1 + ·01 + ·001 + ·0001; 4·07 + ·6201 + ·936 + 29·08 + 1·0101 + 7.
9. 1 + ·2 + ·03 + ·004 + ·0005; ·7, 50·08, 312·907, ·4093, 494·5, and 87·003.

SUBTRACTION.

59. In subtraction of decimals, or of integers and decimals mixed, for reasons precisely similar, the decimal points must be arranged to fall exactly beneath one another, and then the smaller quantity can be subtracted from the larger in the same manner as in whole numbers, *thousandths* being taken from *thousandths*, *hundredths* from *hundredths*, and *tenths* from *tenths*. The decimal point in the answer will fall exactly beneath the decimal points in the subtrahend and minuend, cyphers may be added (or supposed to be added) to the *right* of the decimal figures in the minuend, as this will not alter the value (see page 42), and the subtraction may proceed as in whole numbers.

We have, therefore, the following rule for subtraction:—

RULE XII.

1°. *Place the quantities so that their decimal points shall be in the same vertical line.*

2°. *Next proceed as in subtraction of whole numbers.*

EXAMPLES.

Ex. 1. Subtract ·756 from ·897.

$$·897$$
$$·756$$
$$\overline{·141}$$

Here the difference between 6 *thousandths* and 7 *thousandths* is 1 *thousandth*, between 9 *hundredths* and 5 *hundredths* is 4 *hundredths*, between 8 *tenths* and 7 *tenths* is 1 *tenth*.

Ex. 2. From 37·6 take ·907.

In this instance 37·6 may be written 37·600.

$$37·600$$
$$·907$$
$$\overline{36·693}$$

Ex. 3. From 98765·4321 take 99·99.

$$98765·4321$$
$$99·99$$
$$\overline{98665·4421}$$

Ex. 4. Subtract ·97658 from 5·1394.

$$5·1394$$
$$·97658$$
$$\overline{4·16282}$$

In subtracting 8, 0 is supposed to occupy the place above it as 5·13940 = 5·1394.

Ex. 5. Subtract ·0000999 from ·01.

$$·01$$
$$·0000999$$
$$\overline{·0099001}$$

Ex. 6. From 1 take ·47712.

$$1$$
$$·47712$$
$$\overline{·52288}$$

In examples of this kind (Ex. 5, 6, and 8) when the number of decimal figures in the lower line exceeds the number of figures in the upper, it is advisable to mentally supply cyphers to make up the deficiency in the upper line. This may be done without altering the value of the upper line.

Ex. 7. Subtract 247·258746 from 347·258745.

$$347·258745$$
$$247·258746$$
$$\overline{99·999999}$$

Ex. 8. From 1 take ·000001.

$$1$$
$$·000001$$
$$\overline{·999999}$$

Examples for Practice.

Subtract

1. 3·07 from 6·501; ·79999 from 9; 2·9989 from 3; ·999999 from 9.
2. ·0090806 from 39·857; ·00032 from 32; ·876534 from 8·21314; 364·3123 from 456·0546.
3. ·99 from 1; ·00000099 from 99; ·000001 from 10; 3·29 from 999; 25·6050 from 567·392.
4. ·9682347 from 65·00001; ·79999 from 9; 9·163 from 81·6823401.
5. ·000001 from ·0001; ·000004 from ·0004; ·00032 from 32; ·87623 from 24681.
6. From 700 take 7 hundredths; from ·0001 take ·0000001.
7. From 42 hundredths take 42 thousandths; 154 millionths from 6231 hundred thousandths.
8. From 96 thousandths take 909 ten thousandths; 92 thousandths from 29 thousand.

MULTIPLICATION.

60. We have stated that for every place we shift the decimal point to the *right* we increase the value of the decimal *ten-fold*, for every place we shift it to the *left* we decrease it *ten-fold*. Now, in multiplying two decimals together, since the law of *local value* hold with regard to the digits comprising the decimals, the process of multiplication will be performed exactly as in ordinary whole numbers; the only matter requiring consideration will be the proper position of the decimal point.

Suppose we have to multiply 4·935 by 6·28, and let us suppose the decimal point in each case removed to the extreme right. Then (Art. 57, page 43) we have multiplied the number 4·935 by 1000, and the number 6·28 by 100, and we have obtained the numbers 4935· and 628· respectively. Now, 4935 × 628 = 3099180, but as we increased our original numbers one thousand and one hundred fold respectively, it is evident our product is increased 1000 × 100, or one hundred thousand fold. Dividing, therefore, the above result, 3099180 by 1000000, or what is the same thing (Art. 57, page 43), writing it 3099180· and removing the decimal point 5 places to the left, we get for the product of the numbers 4·935 and 6·28, the result 30·99180. It will be seen that the number of decimal places on the product, namely, 5, is the sum of the numbers of the decimal figures in the two given numbers.

We have, therefore, the following rule for multiplication:—

RULE XIII.

Multiply the numbers together, as whole numbers, and point off as many decimal places in the product (beginning at the right) as there are decimal places in the multiplier and multiplicand together.

When the decimal places to be pointed off are more in number than the figures of the product, make up the proper number by prefixing cyphers to the product.

Examples.

Ex. 1. Multiply 34·11 by 3·72.

```
    34·11
     3·72
   ------
     6822
    23877
    10233
   ------
  126·8892
```

In 34·11 are two decimals; in 3·72 are two; therefore four decimal places are pointed off.

Ex. 2. Multiply 236000 by ·48.

```
    236000
       ·48
   -------
      1888
       944
   -------
  113280·00
```

The product of 236 by ·48 is 11328; in 236000 are no decimals; in ·48 are two decimals; therefore two places are pointed off in the product.

Ex. 3. Multiply 56·3 by ·08.

$$56·3$$
$$0·08$$
$$\overline{4·504}$$

In 56·3 is one decimal; in ·08 are two; therefore three places are pointed off in the product.

Ex. 5. Multiply ·0048 by ·000012.

$$·0048$$
$$·000012$$
$$\overline{·0000000576}$$

The product of 48 by 12 is 576; in ·0048 are 4 decimals; in ·000012 are six decimals; therefore the product must contain ten decimals (four and six) and seven cyphers are prefixed to 576, whence the product is ·0000000576, as above.

Ex. 4. Multiply 5·63 by ·00005.

$$5·63$$
$$0·00005$$
$$\overline{0·0002815}$$

In 5·63 are two decimals; in ·00005 are five; therefore three cyphers must be prefixed to the product 2815, and seven decimals marked off.

Ex. 6. Find the value of 1·005 × ·005 × ·0064.

$$1·005$$
$$·005$$
$$\overline{·005025}$$
$$·0064$$
$$\overline{\begin{array}{r}20100\\300150\end{array}}$$
$$\overline{·0000321600}$$

EXAMPLES FOR PRACTICE.

Find the value of

1. 2·5 × 4; ·25 × 40; 2·5 × 476; 2·5 × 4·76; ·0025 × 4·76; ·025 × ·0476.
2. ·0002 × ·00101; 90·01 × 0·034; ·0008 × ·00014; and ·6005 × ·0035.
3. 21·56 × ·0035; 24·35 × ·074; 35·85 × 2·09; and ·004716 × ·21240656.
4. 100·0008 × ·000306; 7535060 × 62·3906; and 31·50301 × 17·0352.
5. 25067823 × ·0000001; 394·2003 × ·00000003; and ·834567834 × ·00000008.
6. 47·83 by 10, 100, 1000, $\tfrac{1}{10}$, $\tfrac{1}{100}$, $\tfrac{1}{1000}$; ·5 × 1000; ·75 × 100000.
7. 22·5 × ·0241 × ·0024; ·0003 × ·01 × 500000; ·006 × ·00012.
8. 2·7 × ·27 × ·027 × 270; ·2 × ·04 × ·008 × 64000; 8·004 × ·004.
9. 1·1 × ·011 × 1·01 × ·0101; ·013 × 1·6 × ·007 × 3·05; 1003 × 6·12.

DIVISION.

61. Let it be required to divide 37·015 by 6·73.

By shifting the decimal point to the right of the dividend and divisor so as to turn both into whole numbers, we increase the dividend 1000-fold, and the divisor 100-fold. The former of these alterations will have the same effect as multiplying the quotient by 1000, the latter the same as dividing by 100; so that the quotient will be 10 times too great, and must be further divided by 10, i.e., one decimal place must be pointed off to give a correct result.

Had it been required to divide 370·15 by 6·73 where there is the *same* number of decimal places in both dividend and divisor, by shifting the decimal points so as to make both whole numbers, we should increase the dividend 100-fold, and the divisor 100-fold; this would not affect the value of the result, and the quotient would be a whole number requiring no decimal point at all.

If the given quantities had been 370·15 by ·673, so that there had been fewer decimals places in the dividend than in the divisor, by converting both into whole numbers we should have increased the dividend 100-fold and the divisor 1000-fold. This would have decreased the divisor 10-fold, and to obtain the correct result we should have had to multiply the quotient by 10.

We can hence determine the following practical rule for the division of decimals:—

RULE XIV.

Divide as in whole numbers. The rule for placing the decimal point is, *that the* quotient *must have as many* decimal figures *as the* decimal places *in the* dividend *exceed those in the* divisor; *that is, the* quotient *and* divisor *together must contain as many* decimals *as the* dividend.

EXAMPLES.

Ex 1. Divide 640·87458 by 64·5.

We here proceed as in ordinary division, except that a cypher is annexed to the dividend for carrying in the division. The dividend 640·874580 contains six decimal places, the divisor 64·5 contains one, therefore the quotient 993604 must contain (6 − 1) *five* decimal places; then five decimal places counted from the right to the left gives 9·93604, the quotient required.

```
64·5)640·87458(993604
     5805
     ————
      6037
      5805
      ————
       2324
       1935
       ————
        3895
        3870
        ————
         2580
         2580
         ————
```

Ex. 2. Divide ·239316 by 3256.

```
·3256)·239316(·735
      22792
      —————
       11396
        9768
       —————
        16280
        16280
       —————
```

Here the number of decimal places in the dividend ·2393160 (including the cypher supplied to carry on the division) is *seven*; the divisor (·3256) contains *five* decimal places, and *four* from *seven* leaves *three*; then three places counted off from right to left in the quotient gives ·735 as the quotient required.

Ex. 3. Divide 762·151 by ·00325.

```
325)762·151(234508
    650
    ———
    112 1
     975
    ———
    1465
    1300
    ———
     1651
     1625
     ———
      2600
      2600
      ———
```

In this instance the number of decimal places (including cyphers supplied to carry on the division) in the dividend is *five*; the number in the divisor is also *five*, and the difference is nothing (0); therefore, since there are no places to mark off, the required quotient is 234508, an integer.

(*a*) *When the* number of decimal places *in the* dividend *is less than the* number *in the* divisor; *annex as many cyphers (and in the case where the dividend is a* whole number, cyphers preceded *by the* decimal point*) to the dividend as will make the number equal to the number in the* divisor, *and then proceed as in whole numbers;* the quotient will be an integer. *If there be a remainder and the division be carried on further by affixing cyphers to the successive remainders, all the quotient figures thus obtained will be* decimal.

H

Ex. 4. Divide 14·4 by ·0012.

The divisor contains *four* decimal places and the dividend only *one*; annex *three* cyphers to the dividend so as to make the number of decimal places in both divisor and dividend equal, and then divide by 12, and the work will stand thus :—

·0012)14·4000

12000

As the number of decimal places in the dividend is equal to that of the divisor, we have none to cut off. The answer is therefore 12000, an integer.

Ex. 5. Divide 3028·5 by ·673.

In this instance the divisor contains *three* decimal places, and the dividend only *one*; therefore *two cyphers* are annexed to make the number of decimal places in each equal.

·673)3028·500(4500
     2692
     ————
     3365
     3365
     ————
       00

The number of decimal figures in the dividend and divisor being equal, we have none to cut off in the quotient. The answer 4500 is therefore an integer.

(b) *When the number of figures in the quotient is not sufficient to make up the required number of decimals, cyphers must be prefixed.*

Ex. 6. Divide ·10304 by 9200.

9200)·10304(112
      9200
      ————
      11040
       9200
      ————
       18400
       18400
       ————

The number of decimal places in the dividend (counting the two cyphers supplied to carry on the division) is *seven*; the number of decimal places in the divisor is 0; the difference (7 − 0 =) *seven*, the number of decimal places the quotient must have; hence *four* cyphers are prefixed to 112, and the decimal point being placed before them gives ·0000112, the quotient required.

Ex. 7. Divide ·00000000676542672 by ·834567834.

·834567834)·00000000676542672
·834567834)676542672(8
           676542672

In the dividend are *seventeen* decimal places; in the divisor *nine*; the difference of these, or (17 − 9) *eight*, is the number the quotient must contain, hence *seven* cyphers must be prefixed to the 8, and the decimal point placed before them, making ·00000008, the quotient required.

(c) When the number of decimal places in the *dividend* is *less* than the number in the *divisor*, annex as many cyphers to the dividend as will make the number equal to the number in the divisor, and then proceed as in *whole numbers*.

Ex. 8. Divide 570 by ·005.

The divisor contains *three* decimal places and the dividend none, annex *three* cyphers preceded by a decimal point to the dividend, so as to make the number of decimal places in divisor and dividend equal, and then divide by 5, and the work will stand thus :—

·005)570·000

114000

The number of decimal places in both dividend and divisor being equal, there are none to cut off. The answer 114000 is therefore an integer.

Ex. 9. Divide 2107206 by 42·06.

Annexing *two* cyphers, preceded by a decimal point, to the dividend, and thereby making the number of decimal places in both dividend and divisor equal, and then dividing by 42·06, the work stands thus :—

42·06)2107206·00(50100
      21030
      ————
       4206
       4206
       ————
        00

The answer 50100 is evidently an integer, since there are no decimals to cut off, because the number of decimal places in both dividend and divisor are equal.

Reduction of Decimals.

The division may always be carried to any degree of accuracy by annexing cyphers to the dividend, as seen in Example 6 (a).

62. The decimal point may be removed altogether from both the divisor and dividend, by continually multiplying each by 10; for the quotient will thus remain unaltered. The first decimal in the quotient will then appear only with the first cypher annexed to carry on the division.

EXAMPLE.—Divide 55·80 by 0·04. Multiplied by 10 they become 558 and 0·4 multiplied again they become 5580 and 4, the quotient of which is 1395.

This easy process furnishes a complete security against wrongly placing the decimal point in the quotient.

EXAMPLES FOR PRACTICE.

1. Divide 99·12105 by 5, 7, 9, 36, and 88.
2. ,, ·6052158 by 3, 6, 9, 24, and 6400.
3. ,, 236·041 by 1·75; 67234 by 85; 60·0001 by 1·01; 300·402 by 12·1.
4. ,, 325·67543 by 20·02; 684234·6 by 2682; 73·8243 by ·061; ·00006 by ·003.
5. ,, 34·82725 by 39·5275; 897·25 by ·98725; 42·9365 by 387·25; and 3·1415926 by 180, true to seven decimal places.
6. ,, 17·084522 by ·024; 1237·0519 by ·5425; 762·151 by ·00325.
7. ,, ·00033 by ·011; 140·02564 by 1·871; 4·32067 by ·001; ·59 by 80000; 167342 by ·002; ·001024 by 3051757·8125; 1001 by 16384.
8. ,, 992 by ·37; ·1599 by 4100; ·019 by 190; 1 by ·152; ·2 by 23·2.
9. ,, 1·95 by ·00013; 9·614 by ·000019; ·25 by 31·25; 8·92 by 237·6567.
10. ,, ·03679 by 2·83; 165·434 by 36·2; ·027472 by 3·434; 61000 by ·825.
11. ,, 17·171717 by 343·4; 1255 by 1·004; 12·55 by 1004; 7·231068 by ·12.
12. ,, ·012550 by 1004000; 12·55 by ·01004; 1255 by 10·04; 4 by ·00001.
13. ,, ·001255 by 1004; ·01 by 1000; 6821091·97627 by 88·03.
14. ,, ·6 by 6; 6 by ·6; ·06 by 60; ·006 by ·6; 600 by ·6; 600 by ·06; and ·006 by 600.
15. ,, ·00636056 by ·86; 6100 by ·825; 23 by ·000579; 37·69416 by ·156; ·0016 by ·000008; 6 by ·0000001; ·8 by ·0000002; ·000054 by 9; 4000 by ·000125.
16. ,, 36 by 10000; 9·3 by 10; 51·306 by 100; 8 by 10000; 2·0076364 by 1000000.
17. ,, 1⅐ to 10 decimal places; 2⅔⅗ to 12 decimal places; 10⅗⅓ to 25 decimal places.

REDUCTION.

63. The great convenience of decimals makes it often desirable to reduce vulgar fractions to a decimal form.

To reduce a vulgar fraction to a decimal.

We have to change the fraction to another equivalent fraction whose denominator is of the form 10, 100, 1000, &c. To do this we multiply the numerator and denominator of the fraction by 10, 100, 1000, &c., as may be necessary, *i.e.*, we add a certain number of cyphers (the same to each); we then divide the numerator and denominator by the original denominator. These operations will not alter the value of the fraction. If the numerator by the addition of the cyphers becomes divisible by the denominator, without remainder, the required decimal is found; if not, a circulating or recurring decimal is produced as is shown in the following examples:—

Ex. 1. To reduce ⅝ to a decimal.

$$\tfrac{5}{8} = \tfrac{5000}{8000} \div 8 = \tfrac{625}{1000} = ·625.$$

Here we multiply numerator and denominator by 1000 and divide them by 8. The resulting fraction $\tfrac{625}{1000}$ represented as a decimal is ·625.

Ex. 2. To reduce $\frac{123}{625}$ to a decimal.

$$\frac{123}{625} = \frac{1230000}{6250000} = \frac{1968}{10000} = \cdot 1968.$$

Here we multiply numerator and denominator by 10000, and then divide them by 625. The resulting fraction $\frac{1968}{10000}$ represented as a decimal is ·1968.

```
625)1230000(·1968
    625
    ———
    6050
    5625
    ————
    4250
    3750
    ————
    5000
    5000
```

The work is shortened thus:—We put down the numerator 123 as dividend, and denominator 625 as a divisor, and adding cyphers as often as required, we obtain as a quotient the significant digits of the decimals; and the number of cyphers added to the dividend will be the number of places to be marked off in the question.

Hence to convert vulgar fractions into decimals we proceed by the following rule.

RULE XV.

Annex a cypher to the numerator, and then divide by the denominator; if there be a remainder, annex another cypher, and continue the division, still annexing a cypher, either till the division terminates without a remainder, or till as many decimals as are considered necessary are obtained: the quotient, with a decimal point before it, will be the value of the fraction in decimals.

EXAMPLES.

Ex. 1. Reduce $\frac{1}{5}$ to a decimal fraction.

```
5)1·0
  ———
  0·2
```

Dividing 10 by 5 (the cypher being added) we find that $\frac{1}{5} = 0·2$. That $\frac{1}{5} = 0·2$ is easily proved, for $\frac{1}{5} = \frac{10}{50}$; consequently, by dividing both the numerator and denominator by 5, we have $\frac{1}{5} = \frac{2}{10} = ·2$.

Ex. 2. Convert $\frac{3}{8}$ into a decimal fraction.

```
8)3·000
  —————
  ·375
```

That $\frac{3}{8} = ·375$ is proved thus, $\frac{3}{8} = \frac{3000}{8000}$; consequently, dividing the numerator and denominator by 8, (the denominator of the fraction), we have $\frac{3}{8} = \frac{375}{1000} = ·375$.

Ex. 3. Reduce $\frac{1}{3}$ to a decimal fraction.

```
3)1·0000
  ——————
  ·3333, &c.
```

Dividing 10 by 3 gives 3, the next cypher added gives another 3, and so on continually.

Ex. 4. Reduce $\frac{6}{11}$ to a decimal.

```
11)6·0
   ————
   ·5454
```

It is plain from the remainder that 54 would recur continually, so that $\frac{6}{11}$ is equal to a *recurring* decimal; 54 being the *recurring period*.

Ex. 5. Reduce $\frac{25}{36}$ to a decimal.

$$36 = 4 \times 9 \quad 36 \begin{cases} 4)25\cdot00 \\ \overline{9)\ 6\cdot25} \end{cases}$$

·694444, &c.

Hence $\frac{25}{36} = ·694$ which is called a *mixed recurring* or *circulating* decimal, consisting of the non-recurring part 69, and the recurring part 444, and usually written with a point or dot above the figure which is repeated.

Ex. 6. Reduce $\frac{3}{14}$ to a decimal.

$$14 = 2 \times 7 \quad 14 \begin{cases} 2)3\cdot000 \\ \overline{7)1\cdot500} \end{cases}$$

·2142857

Hence $\frac{3}{14} = ·2142857$; the recurring part 142857 having a point above its first and last figures being called its period. If the whole decimals recurs, it is called a *pure circulator.*

Ex. 7. Convert $\frac{8}{113}$ into a decimal.
113)8·00(0·07079, &c.
   791

   900
   791

   1090
   1017
   ---
     73 &c.

When the 0 is annexed to the 8, the divisor 113 will go *no time;* therefore the first decimal place is to be occupied with a cypher. Annexing now a second 0, the next decimal figure is 7, and the work proceeds as above. The quotient shows that $\frac{8}{113} = ·07079$, &c.; the decimals may be carried out to any extent.

Ex. 8. Reduce $\frac{1}{128}$ to a decimal.
128)1·000(·0078125
      896

      1040
      1024

       160
       128

       320
       256

       640
       640

Ex. 9. Reduce $\frac{101}{8103}$ to a decimal.
$\frac{101}{8103} = ·0123291015625$.

Ex. 10. Reduce $\frac{1}{512}$ to a decimal.
$\frac{1}{512} = ·001953125$.

Ex. 11. Reduce $\frac{7}{512}$ to a decimal.
$\frac{7}{512} = ·013671875$.

Ex. 12. Reduce $\frac{19}{43}$ to a decimal.
$\frac{19}{43} = ·04512067156348373557187827911857291275970619$.

EXAMPLES FOR PRACTICE.

1. Change into equivalent decimals $\frac{7}{10}$, $\frac{11}{14}$, $\frac{23}{50}$, $\frac{11}{32}$, $\frac{7}{10}$, $\frac{4}{50}$, $\frac{9}{512}$, and $\frac{11}{44}$.

2. Reduce to decimals $\frac{7}{19}$, $\frac{11}{14}$, $\frac{13}{16}$, $\frac{1}{17}$, $\frac{223}{360}$, $\frac{112}{115}$, $\frac{91}{8}$, and $\frac{9}{512}$, carrying interminants to seven decimal places.

64. It becomes important to observe what fractions will produce terminating decimals. Suppose a fraction in its lowest terms; then in reducing it to a decimal we multiply the numerator by 10, 100, 1000, &c. Now the numerator contains no factor common to the denominator, and by this multiplication we introduce the factors 2 and 5 as often as we please and no others. Unless, then, the denominator contains no other factors except twos and fives, this multiplication cannot render the numerator divisible by it. Hence the only fractions which will produce terminating decimals are those whose denominators contain only 2 and 5 as prime factors. All other fractions will produce circulating decimals, though in many cases the period is so long that it would be tedious to find it.

65. Decimals are most frequently used to make calculations on numbers that have been obtained by observations of some kind, by measuring, for instance, or weighing; and it is very seldom indeed that the accuracy of these observations can be relied on to within one five-thousandth part of the unit employed. Now if we cannot rely on the measurement beyond three decimal places, it is needless to carry the result derived from it any farther. In all operations with decimals, then, whether terminating or repeating, we may usually stop at the third or fourth place, and need very rarely go beyond the fifth or sixth. We may, however, attain any degree of exactness that may be required, by carrying the decimals far enough.

66. With respect to repeating decimals, if perfect accuracy be necessary, they must in most cases be reduced to vulgar fractions before they are added, subtracted, multiplied, or divided. In almost all the applications of decimals, however, an approach to accuracy is sufficient, and this is attained by carrying the decimal only to a moderate number of digits, and omitting the rest. If, in converting a vulgar fraction into a decimal, we stop after the third digit, for instance, adding unity to that digit, if the next be 5 or upwards, it does not differ from its exact value by more than one five-thousandth part of the unit employed. Thus, ·172 differs from $17\frac{2}{43}\frac{7}{7}$ by ·000$4\frac{3}{7}\frac{7}{2}$, which is less than ·0005. Similarly, ·983 differs from ·98276 by ·000$2\frac{3}{17}$, which is also less than ·0005.

67. To reduce any quantity or fraction of one denomination to the decimal of another denomination.

RULE XVI.

Reduce the number of the lowest *denomination to a decimal of the next* higher *denomination,* prefix *to this decimal the number of its denomination given in the question, if any, and reduce this also to a decimal of the next* higher *order, and so on till all the numbers of the given denomination are exhausted, and the decimal of the required denomination has been obtained :* the last result will be the answer.

EXAMPLES.

Ex. 1. Let it be required to express 17s. 5¼d. as a decimal of £1.

The process will be first to express the *fractional* part of a penny as a *decimal* of a penny; placing the 5 as a whole number before this decimal, to divide that result by 12, in order to reduce it to the decimal of a shilling; placing the 17 as a whole number before this decimal, to divide that result by 20 in order to reduce it to the decimal of a pound. This will be written as follows:—

4) 1·

12) 5·25 pence

20)17·4375 shillings

·871875 of a pound.

It will be seen from this, that whatever we should divide by in whole numbers in or £sr to bring pence into shillings, or shillings into pounds, that we must likewise divide by in this case, only marking off correctly the decimal results.

Ex. 3. Find what decimal of an hour is 40^m.

There are 60 minutes in an hour; hence 1 minute is $\frac{1}{60}$ of an hour, and 40 minutes is $\frac{40}{60}$ of 1 hour, which gives 0·66 of 1 hour.

60)40·00

0·66 of an hour.

Ex. 2. Express as decimals of a degree 27° 18′ 35″.

60 | 35″
60 | 18·5833
   | 27·30971⁄2

Here for convenience of arrangement we write the 35″ uppermost, and the 18′ and 27° directly under it, and draw a vertical line to the left of the line opposite these numbers; write for divisors the number of that denomination which makes one of the next higher —namely, 60 opposite the seconds, since 60″ = 1′, and 60 opposite the minutes, because 60′ = 1°. Then dividing 35 by 60 we get ·58⅓, which we write after the minutes, which gives 18′·58⅓; this again divided by 60, the number of minutes in a degree, gives the quotient ·3097½, which being annexed to the degrees, 27°, gives the answer 27°·3097½.

Ex. 4. Find what decimal of an hour is 15^m.

Here 1 minute is $\frac{1}{60}$ of 1 hour, 15 minutes is $\frac{15}{60}$ of 1 hour; hence $\frac{15}{60}$ gives 0·25 of 1 hour.

60)15·00

0·25 of an hour.

Ex. 5. Find what decimal of 1 degree is 8° 37″.

37″ are $\frac{37}{60}$ of 1′, or 0·61 of 1′; then 1′ is $\frac{1}{60}$ of 1°; hence 8°·61 are $\frac{8·61}{60}$ of 1°, or 0°·143.

$$60)37″$$
$$60)8·616$$
$$0·143 \text{ of a degree.}$$

Ex. 7. Find what decimal of 1 mile (nautical) is 700 feet.

There are 6080 feet nearly in a nautical mile; hence 1 foot is $\frac{1}{6080}$ of a mile, and 700 feet are $\frac{700}{6080}$ of 1 mile, which gives 0·115 of 1 mile nearly.

$$6080)7000·0(0·115$$
$$6080$$
$$\overline{9200}$$
$$6080$$
$$\overline{31200}$$
$$30400$$

Ex. 6. Find what decimal of 1 day is 3ʰ 42ᵐ.

42ᵐ are $\frac{42}{60}$ of 1 hour, or 0ʰ·7; and 1ʰ is $\frac{1}{24}$ of 1 day; hence 3ʰ·7 is $\frac{3·7}{24}$ of 1 day, or 0ᵈ·154166, &c.

$$60)42ᵐ$$
$$24)3·7ʰ$$
$$\overline{0·154166}$$

Ex. 8. Find what decimal of 1 foot is 8¾ inches.

First, ¾ is 0·75 of 1 inch; hence 8¾ inches are 8·75 inches. Then, 1 inch is $\frac{1}{12}$ of 1 foot; hence 8·75 inches are $\frac{8·75}{12}$, or 0·729 of 1 foot.

$$12)8·75$$
$$\overline{0·729}$$

68. *Or, reduce the given quantities to the lowest denominations when there are more than one, and also the integer to which it is referred, to the same denomination: then divide the given quantity by the integer thus reduced.*

Ex. 1. (Ex. 7 above.) The given quantity 700 feet, being all of one denominator requires no further reduction. The integer 1 mile, reduced to the same denomination, is 6080 feet; then 700 divided by 6080 gives 0·115.

Ex. 2. (Ex. 8 above.) 8 inches and 3 quarters are 35 quarters, and 1 foot reduced to the same denomination is 48 quarters; then 35 divided by 48 gives 0·729.

EXAMPLES FOR PRACTICE.

1. Express as decimals of an hour 17ᵐ; 29ᵐ; 42ᵐ; 35ᵐ; 48ᵐ; and 58ᵐ.
2. Reduce 5ᵈ 12ʰ 25ᵐ 39ˢ·92 to decimals of a week; 183¼ᵈ to decimals of a year.
3. Express as decimals of a day the following quantities:—12ᵈ 14ʰ 13ᵐ 12ˢ; 29ᵈ 17ʰ 11ᵐ 45ˢ; 15ᵈ 17ʰ 48ᵐ 54ˢ; and 119ᵈ 5ʰ 19ᵐ 15ˢ.
4. Express as decimals of a degree the following quantities:—8° 11′ 15″; 19° 40′ 45″; 104° 16′ 7½″; and 82° 19′ 30″.
5. What decimal is 3 qrs. 21 lbs.; 12 cwt. 1 qr. 16 lbs.; 14 cwt. 1 qr.; 10 cwt. 1 qr. 14 lbs.; 13 tons 6 cwt. 7 lbs.; 1 ton 13 cwt.; 17 tons 18 cwt. 3½ lbs. of a ton.?
6. 6s. 4d.; 8s. 8½d.; 4s. 7½d.; 15s. 11¼d.; 19s. 10¼d.; £1 13s. 11½d.; ⅔ of £1.

69. To find the value in a lower denomination of any decimal of a higher denomination.

RULE XVII.

1°. *Note the number of parts which the unit or integer of the given quantity contains of the next inferior denomination, and multiply the given decimal by this number;* the product is the given quantity expressed in that denomination.

2°. *If this product has a decimal part, multiply this decimal by the number of parts which the unit of the present denomination contains of the next inferior denomination to that just before employed;* this product is the quantity which the given decimal contains of the *next denomination.*

3°. *Proceed (if there still be decimals), in like manner, to the lowest denomination in which the decimal is required to be expressed.*

EXAMPLES.

Ex. 1. Find the number of feet in 0·115 of a mile.

The next inferior denomination to that of miles is here feet, of which the number in one mile is ....

$$\begin{array}{r} 0\cdot 115 \\ \times\ 6082 \\ \hline 230 \\ 920 \\ 6900 \\ \hline \end{array}$$

Ans. (in the lowest denomination required) 699·430 feet.

Ex. 2. Find the number of seconds in 0·7 of a minute.

The next inferior denomination to that of minutes is seconds of which the number in a minute is ....

$$\begin{array}{r} 0\cdot 7 \\ \times\ 60 \\ \hline \end{array}$$

Ans. 42·0 seconds.

Ex. 3. Find the number of inches and eighths in 0·48 of a foot.

The next inferior denomination to that of feet is inches, of which the number in a foot is.......................

$$\begin{array}{r} 0\cdot 48 \\ \times\ 12 \\ \hline 5\cdot 76\ \text{inches.} \end{array}$$

The next proposed inferior denomination to inches is eighths, of which the number in an inch is ....

$$\begin{array}{r} \times\ 8 \\ \hline \end{array}$$

Ans. 6·08 eighths.

Ex. 4. What is the value of ·625 of a cwt.?

The next inferior denomination to that of a cwt. is qrs., of which the number in a cwt. is ....

$$\begin{array}{r} \cdot 625 \\ \times\ 4 \\ \hline 2500\ \text{qrs.} \end{array}$$

The next inferior denomination to qrs. is lbs., of which the number in a quarter is ....

$$\times\ 28$$

Ans. 14·0 lbs.

EXAMPLES FOR PRACTICE.

Find the value of the following expressions:—

1. ·25 cwt.; ·375 qrs.; ·625 lbs.; ·975 ton; ·7768 ton; ·24956 ton; ·0675 cwt.
2. ·491 day; ·343 week; ·534 year; ·53058875 day.
3. ·2957795 degree; 7·85425 degrees; 64·3815 degrees; 10·8725 degrees.
4. ·487936324 ton of sea water in cubic feet and inches (35 ft. = 1 ton).
5. Express the same in gallons of 277·274 cubic inches.

MISCELLANEOUS EXAMPLES.

1. The height of the highest mountain is about 28000 feet: what decimal is that of the earth's diameter, which is 8000 miles?
2. The parallax of the star α Centauri is given as 0·9187, or $\frac{19}{22}$ of a second: show by how much the vulgar fraction differs from the decimal fraction.
3. Reduce 29^d 12^h 44^m 21^s·82 to decimals of a day.

Reduction of Decimals.

4. Add together 2·095 hours, ·07 days, ·05 weeks, and express the same as the decimal of 365·25 days.

5. A nautical mile is 6082·66 feet, and an imperial mile 5280 feet; express each of these miles as decimals of the other. Also find how near the results are to the decimal values of $\frac{23}{20}$ and $\frac{20}{23}$.

6. A sidereal day is $23^h\ 55^m\ 4^s\cdot09$; express this as a decimal of a common day—that is, of 24^h—and give the result to nine decimal places.

7. If 90 degrees correspond to 100 French grades, how many degrees are there in the sum of 41·45 degrees and 41·45 grades.

8. A mètre is 39·37079 English inches, a kilomètre is 1000 mètres; express as decimals of each other a kilomètre and an English mile.

9. If the length of a degree of latitude is 365000 feet, and a mètre one ten-millionth of 90 degrees: find its length in feet.

10. Express in figures: Thirty-four and two thousandths, and by it divide 28255662. What alteration must be made in the quotient if the decimal part in the dividend be moved eight places to the left?

11. The sidereal year being $365^d\ 6^h\ 9^m\ 9^s\cdot6$, and the tropical year $365^d\ 5^h\ 48^m\ 49^s\cdot7$: reduce their difference to the decimal of a tropical year.

12. Supposing the velocity of electricity be 288000 miles per second, and the earth's circumference to be 25000 miles: calculate to seven places of decimals the time of transmission of an electric telegraph to the antipodes.

13. A French mètre is 39·37 inches nearly: show that a foot is equal to ·304 mètre, nearly.

14. Find the sum, difference, and product of 86·25 and 39·625, and divide the sum of the three results by 6·25.

15. The circumference of a circle is 3·14159 times its diameter. Find the circumference of circles whose diameters measure 13·7 feet, 1·96 yards, and 28·342 miles, respectively.

16. What is the difference between the fifteenth and sixteenth parts of 297·9832?

17. A penny is ·08975 inches thick: find the height of a pile of 1000 pennies.

18. A cubic inch of water weighs 252·458 grains, and the weight of an imperial gallon of water is 10 lbs. avoirdupois: find (to three places of decimals) the number of cubic inches in an imperial gallon, there being 7000 grains in 1 lb. avoirdupois.

19. A cubic foot weighs 445 lbs.: what does a pound measure in inches?

20. A gallon of water weighs 10 lbs., and measures 277·274 cubic inches: what is the weight of a cubic foot and the measure of a ton?

21. It having been calculated that the air-pump of a marine engine can lift 8616·96 tons of salt water in 6 hours, by considering that a cubic foot of salt water weighs 64 lbs.: what is the error and actual weight of the salt water, when the true weight of a cubic foot is 64·16875 lbs.?

22. If a sovereign weighs 123·274 grains: how many sovereigns will weigh 10 lbs. 8 oz. 8 dwts. 5 grs.?

23. Divide 2·021 by 1000, 20·21 by ·001, 23·0142 by 121, 23014200 by ·0121, and 2301·420 by 0·0012100.

24. If the mean diameter of the earth be 504979200 inches in length, express its length in feet, yards, poles, furlongs, and miles.

25. A cubic foot of gold is extended by hammering, so as to cover an area of 6 acres: find the thickness of the gold in decimals of an inch, correct to the first two significant figures.

ON LOGARITHMS.

70. Logarithms are numbers arranged in Tables for the purpose of facilitating arithmetical computations. They are adapted to the natural numbers, 1, 2, 3, ... in such a manner that by means of them

    the operation of Multiplication is changed into that of Addition;

Division	Subtraction;
Involution	Multiplication;
Evolution	Division;*

71. Take any whole numbers, as 18, 813, 6489; the first consists of two, the second of three, and the third of four figures or *digits*. Again, in the mixed number 739·815, the whole number or integral part (739) consists of three *digits*.

72. By multiplying a number by itself, *one, two, three*, &c., times successively, we obtain the *second, third, fourth,* &c., powers of that number; hence, a *power of a number is the number arising from successive multiplication by itself*. Thus, $3 \times 3 = 9$ is the square or second power of 3; and $5 \times 5 \times 5 = 125$, the cube or third power of 5; and so on.

These operations are denoted by means of *Indices*, or small figures placed on the right of the numbers, a little above the line; thus, $2^2 = 2 \times 2 = 4$, $3^3 = 3 \times 3 \times 3 = 27$, and $2^5 = 2 \times 2 \times 2 \times 2 \times 2 = 32$, where the *Index* or *exponent* denotes the number of factors employed.

73. When there are a series of numbers, such that each is found from the previous one by the addition or subtraction of the same number, they are said to be in arithmetical progression. 1, 3, 5, 7, 9, 11, &c., are in arithmetical or equi-different progression, since each number is found by adding 2 to the immediately preceding.

---

* No proof can here be offered that numbers must exist possessing the properties under which we call them logarithms; neither can any account be here given of the methods of computing such logarithms. The reader will accept the statements that if such numbers exist, bearing the properties aforesaid, they are called logarithms. He must also accept the tables which are published, recording logarithms for the several numbers to which they profess to belong, though he cannot at present verify the computations of these several logarithms; and he will be informed how he may use these tables to effect with comparative ease many calculations which would otherwise be most laborious.

The truth is, though it requires for its demonstration higher algebra than this work presupposes the reader to be acquainted with, that not only has every number a logarithm, but it has an infinite variety of logarithms, constructed, as the term is, on different scales or bases. The base of any system of logarithms is defined by the fact that in that system *unity* is its logarithm.

Any number *might* be used as a base; but in fact there are only two numbers which are ever really used.

The one is an unterminating decimal, 2·7182818 . . . ., denoted generally by the letter *e*. This is the base of what is called the natural or Naperian system; and the advantage of it consists in the ease with which logs. are computed, to this base; but which we cannot here explain.

The other is 10, which is the base in ordinary use, and with this base log. 10 = 1. Logarithms to this base are the only ones which will now be considered in their practical use.

74. Again, the numbers 3, 6, 12, 24, &c., are in geometrical progression, for each number is formed from the one immediately preceding by multiplying by 2. If we take the following series of powers, $3^1, 3^2, 3^3, 3^4, 3^5$, &c., we find that the exponents proceed in arithmetical progression, and the quantities themselves in geometrical progression.

75. **Def.**—Logarithms are a series of numbers in *arithmetical* progression answering to another series in *geometrical* progression, so taken that 0 in the former corresponds with 1 in the latter.

Thus, 0, 1, 2, 3, 4, 5, 6, &c., are the logarithms or *arithmetical* series,
and 1, 2, 4, 8, 16, 32, 64, &c., are the numbers or geometrical series,
answering thereto—the latter being called the *natural number*.
Or, 0, 1, 2, 3, 4, 5, the logarithms,
and 1, 5, 25, 125, 425, 5125, the corresponding numbers.
Or, 0, 1, 2, 3, 4, 5, the logarithms,
and 1, 10, 100, 1000, 10000, 100000, the corresponding numbers.

In which it will be seen, that by altering the common ratio of the geometrical series, the same arithmetical series may be made to serve as logarithms of any series of numbers. As above, when the common ratio of the geometrical series are 2, 5, and 10 respectively.

76. The common ratio in the geometrical series corresponding to the common difference of 1 in the arithmetical series is called the **base of the system**. Thus, the base of the first specimen exhibited is 2, the base of the second is 5, and the base of the third is 10.

In the specimens just exhibited we have, in each, taken two ascending progressions, but they might equally well have been two descending progressions, or the one descending and the other ascending. Logarithms, however, as now used in practice, are limited to the case of two progressions, either both ascending or both descending—the former giving the logarithms of integers, the latter of fractional numbers.

But a better way of considering logarithms is as follows:—

77. **Def.**—The logarithm of a number to a given base is the index of the power to which the base must be raised to give the number.

For instance, if the base of a system of logarithms be 2, 3 is the logarithm of 8, because $8 = 2^3 = 2 \times 2 \times 2$.

And if the base be 5, then 3 is the logarithm of the number 125, because $125 = 5^3 = 5 \times 5 \times 5$.

There may be thus as many different systems of logs. as we please; but, for practical use, it is necessary to select and adhere to one. That usually employed now is called *Briggs' system*.

78. We now proceed to describe what is called the **common system of logarithms**. In the common system of logarithms unity is assumed to be the logarithm of 10; that is, 10 is the constant base. All the logarithms registered

in the Tables commonly used, are indices of the radix or base 10; a Table of logarithms of numbers is in fact nothing more than a Table of the exponents of 10 placed against the several numbers themselves. Accordingly—

0 is the log. of	1,	because	$1 = 10^0$
1	,,	10,	,,	$10 = 10^1$
2	,,	100,	,,	$100 = 10^2$
3	,,	1000,	,,	$1000 = 10^3$
4	,,	10000,	,,	$10000 = 10^4$
		&c.		&c.

Now, if the above Tables were amplified by the insertion of the logarithms of all the numbers between 1 and 10, between 10 and 100, &c., we should have a Table of logarithms of all numbers from 1 to 10000; and whatever may be the difficulty of determining the intermediate logarithms, it is at once easily seen that the logarithms of all numbers between 1 and 10, *i.e.*, between 10^0 and 10^1 must lie between 0 and 1, and will be 0 + a fraction, that is, a decimal less than 1; of all numbers between 10 and 100, *i.e.*, between 10^1 and 10^2 must lie between 1 and 2, and will be 1 + a fraction, or a decimal between 1 and 2; of all between 100 and 1000 will be 2 + a fraction, and so forth; or the integral part of each intermediate logarithm will be *one less* than the number of integral figures in the quantity of which it is the logarithm. Thus, the logarithms of 2, 3, 4, &c., to 9, have 0 as the integral part; those of 10, 11, 12, &c., to 99, have 1 as the integer; those of 100, 101, 102, &c., to 999, have 2 as the integer; and so forth. Hence Tables of logarithms usually supply only the fractional or decimal part; the integral part is always known from the number of integers in the value whose logarithm is wanted. Very few logs. can be expressed in terminating decimals, but this causes little inconvenience, since a log. carried to six or seven decimal places is sufficiently exact for all common purposes.

79. The integers 1, 2, 3, 4, &c., which are the logarithms of 10 and its powers (see 78), are chief indices, and the logarithms intermediate to these, as for instance 1·778151 (which is the logarithm of 60) cosisting of an integer and a decimal fraction, though they are also indices, are usually referred to as consisting of an *index** and *mantissa*†, the integral part being specially termed the *index* or *characteristic*, because it indicates, by being *one less*, how many integral places are in the corresponding natural number, and the annexed decimal being called the *mantissa*.

EXAMPLE.—In the log. 4·616339, the figure (4) standing to the left of the decimal point is the *characteristic* or *index*, and the remaining portion (·616339) is the *mantissa* or *decimal part*.

* In order to avoid confusion from the use of the word "index" to signify two things, we shall throughout this work employ the term *characteristic* when speaking of logarithms, and *index* when speaking of roots or powers.

† *Mantissa*, a Latin word signifying an additional handful; something over and above an exact quantity.

80. To find the characteristic of the logarithm of any number greater than unity we have, therefore, the following rule:—

RULE XVIII.

The characteristic of the logarithm of a number greater than unity, i.e., of a whole or mixed number, is one less than the number of the digits of its integer part

Thus: the characteristic of the logarithm of 849 is 2; for the number 849 is an integer consisting of three digits (that is the number between 100 and 1000) and 1 less than 3 is 2. Also, the index of the log. of 264·96 (which is a mixed number) is 2, since the integral part of the number, namely 264, is a number between 100 and 1000, or consists of 3 digits, and one less than 4 is 3. Again, 3 is the characteristic of the logarithm of 3847·216, since this number has 4 integral digits; while 0 is the characteristic of the logarithm of 3·847216, since this number has one integral digit.

Again, the characteristic of the log. of a number of *one* place of integers (such as 5 or 5·08, or 5·0801) is 0. Again, every number with *two* places of integers (such as 50, or 50·8, or 50·813) is 1. Again, every number with *three* places of integers (such as 508, or 508·2, or 508·25) has for its characteristic 2, and so on.

EXAMPLES FOR PRACTICE.

Write down the characteristics of the logarithm of the following numbers:—

1.	365	6.	69710	11.	474000	16.	473·908
2.	4·8	7.	45·82	12.	4256·45	17.	54793000
3.	643·75	8.	8640	13.	3·9	18.	21256·8
4.	28·9	9.	75	14.	8	19.	2·14006
5.	6	10.	7·265	15.	18	20.	50·7406

81. It has been shown that in the common system of logarithms (Briggs') the log. of 1 is 0; consequently, if we wish to extend the application of logs. to fractions, we must establish a convention by which the logs. of numbers wholly decimal, *i.e.*, less than unity, may be represented by numbers less than zero, *i.e.*, by *negative numbers*.

Extending, therefore, the above principles to negative exponents, since

$10^0 = 1 = 1$ 0
$10^{-1} = \frac{1}{10} = 0·1$ $\bar{1}$ is the logarithm of ·1 in this system.
$10^{-2} = \frac{1}{100} = 0·01$ $\bar{2}$,,   ·01   ,,
$10^{-3} = \frac{1}{1000} = 0·001$ $\bar{3}$,,   ·001   ,,
$10^{-4} = \frac{1}{10000} = 0·0001$ $\bar{4}$,,   ·0001   ,,
&c.   &c.

It follows from this, that when the number is a decimal with all its digits significant, in value between 1 and $\frac{1}{10}$, its log. is negative, yet not so small as the log. of $\frac{1}{10}$, which is — 1. Its log. therefore will be something between 0 and — 1, or — 1 with some positive decimal added. Hence — 1 is its characteristic. When the number is a decimal with zero as its first digit, in value therefore below $\frac{1}{10}$ but not so low as $\frac{1}{100}$, its log. is less than — 1, but not so small as —2, and so will be —2 with some positive decimal attached. Thus $\bar{2}$ is the characteristic. The log. of a decimal between ·01 and ·001 is some number between —2 and —3, and its characteristic is —3; of a number between ·001 and ·0001 its log. is between —3 and —4, and its characteristic is —4; and generally, following this reasoning, it will appear that the characteristic of a decimal fraction is negative, and may be known from its denoting the place of the first significant figure of the decimal, as being the 1st, 2nd, 3rd, &c., place after the point; hence,

82. To find the characteristic of any number less than unity, *i.e.*, of a decimal, we have the following

RULE XIX.

The characteristic of the logarithm of a number less than unity, and reduced to the decimal form, is negative and one more than the number of cyphers following the decimal point.

A negative characteristic is denoted by writing over it the negative sign (—), thus $\bar{1}, \bar{2}, \bar{3}$, &c.*

Thus the characteristic of the logarithm of ·00521 is $\bar{3}$, since the number of cyphers following the decimal point increased by 1 is 3.

Similarly the index of log. of ·156   is $\bar{1}$
,,   ,,   ·0156   is $\bar{2}$
,,   ,,   ·00046   is $\bar{4}$
,,   ,,   ·000000721 is $\bar{7}$

83. But in order to avoid the confusion that might arise by the addition and subtraction of *negative* indices, the following rule is frequently used.

RULE XX.

Add 1 to the number of cyphers between the decimal point and the first significant figure, and subtract from 10; the remainder is the index required.

Thus the characteristic of the log. of ·04 is $\bar{2}$, or 8, since 1 added to the number of cyphers following the decimal point is 2, then 2 from 10 is 8.

Similarly the index of log. of ·140   is 9.
,,   ,,   ·0149   is 8.
,,   ,,   ·00064   is 6.
,,   ,,   ·000000721 is 3.

(a) If the characteristic of a vulgar fraction is required, it must first be reduced to an equivalent decimal fraction, and then the index is found by the rule.

Thus, the index of log. $\frac{1}{8}$, or of log. ·125 is $\bar{1}$ or 9.
,,   of log. $\frac{1}{25}$, or of log. ·04 is $\bar{2}$ or 8.
,,   of log. 24$\frac{2}{5}$, or   24·4   1.

EXAMPLES FOR PRACTICE.

Write down the characteristics of the logarithms of the following decimal fractions :—

1. ·045   6. ·000000t   11. ·4537   16. ·037299
2. ·9   7. ·01   12. ·009   17. ·00000052018
3. ·0004   8. ·0003127   13. ·0000008 18. ·000000105379
4. ·6798   9. ·02803   14. ·000064   19. ·5
5. ·0062   10. ·7007   15. ·000485   20. ·000000000382

---

* The negative sign (—) is written *above* the characteristic, thus $\bar{2}$, instead of before it, to show that it affects only the characteristic and not the mantissa, which remains positive. If it were written in front of the complete logarithm it would signify that the entire logarithm was negative, but such logarithms are never employed in the operations connected with navigation.

84. The characteristic may also be found as follows:—

RULE XXI.

Place your pen between the first and second figure, (not cypher), and count one for each figure or cypher, until you come to the decimal point; the number thus given will be the characteristic: but observe that if you count to the left *you must subtract the number found from* 10, *and consider the* remainder *as the characteristic.*

Thus, in finding the log. of 4·6017, if you place the pen between the first figure (4) and second (6), it falls on the decimal point; in this case the characteristic is 0. Next, in the case of log. of 4601·7, place your pen between 4 and 6, and count $\overset{4}{\vert}\overset{601\cdot7}{123}$; the characteristic is 3. Next, in the case 4601700, here the decimal point falls behind the last cypher (No. 5). Hence, counting as before, we have $\overset{4}{\vert}\overset{601700}{123456}$ and the characteristic is 6.

Again, in the case of log. ·00046017 the first significant figure is 4. Hence, counting, we have $\overset{\cdot 0}{\vert}\overset{\vert 0046017}{4\vert 321}$, but here we count to the left, so that the characteristic is negative, or $\bar{4}$, which taken from 10 is 6. Again, in the case of log. of ·46017, we have $\overset{\cdot}{\vert}\overset{4\vert 6017}{1}$, and the characteristic is $\bar{1}$, or 9.

85. The mantissa of the logarithm depends entirely on the relative value of the figures composing the quantity whose logarithm it is, and not at all upon the *numerical* value of that quantity: thus, the mantissa of the log. of 13 is ·113943, which is also the mantissa of 1·3, or 130, or 1300, for in each case the 1 and the 3 have the same relative value. So the mantissa of a logarithm is always the same, if the significant figures remain the same, and is not altered by the addition of cyphers to the right or left of these figures, or what is equivalent, by the multiplication or division of the quantity by 10, or any power of 10; it is only the characteristic which alters its value by an alteration in the position of the decimal point, 1 being added to the characteristic for every place the decimal point is removed to the right, that is, for every 10 by which the quantity is *multiplied;* or, 1 is *subtracted* from the characteristic for every place the decimal point is removed to the left, that is, divided by 10.

The logarithm of	745800	being	5·872622
that of	74580	is	4·872622
,,	7458	,,	3·872622
,,	745·8	,,	2·872622
,,	74·58	,,	1·872622
,,	7·458	,,	0·872622
,,	·7458	,,	$\bar{1}$·872622 or 9·872622
,,	·07458	,,	$\bar{2}$·872622 or 8·872622
,,	·007458	,,	$\bar{3}$·872622 or 7·872622
,,	·0007458	,,	$\bar{4}$·872622 or 6·872622

ON TABLES OF LOGARITHMS OF NUMBERS.

86. In RAPER's, NORIE's, and the collection of nautical tables intended to accompany this work, the Tables of the Logarithms of Numbers are arranged so as to give the mantissæ of the natural numbers from 1 to 10000. If the reader will open an ordinary table of logarithms, such as is contained in the above-mentioned works, he will find a short table of logs. from 1 to 100 immediately preceding the general table, and giving the entire logarithm; following which is the general table, on opening which he will find a vertical column on the left side of the page containing three digits and ten columns of logarithms headed by the digits 0, 1, 2, ... 9. These last are fourth digits to be attached to the former three, so that the table thus embraces numbers from 1000 up to 9999. Opposite to every such number is a number with six places of figures. This is a decimal, though to save printing the decimal point is not printed, and it is the decimal part, or mantissa, of the logarithm of the number to which it corresponds. The characteristics are never printed, but are prefixed according to the Rules XVIII, XIX, and XX. Hence from such a table we can take out the logarithm of any number with any four significant digits.

In the margin of all tables of logarithms the difference of the successive logarithms in that part of the table is set down. In some tables, also, there are little tables in the margin called *tables of proportional parts*. These are placed under every successive difference, and contains for that difference the number to be added in respect of each unit's digit, so as to form the logarithm of numbers of six digits. These numbers are found as directed by the Rule XXV, page 66.

87. If the number be given, its logarithm may be found as follows:—

To find the logarithm of a number consisting of not more than two digits, *i.e.*, which does not exceed 100, using the short table of logarithms, from 1 to 100, preceding the general table.

RULE XXII.

Seek for the number proposed, considered as a whole number, in the column at the top of which is No., and the logarithm will be found opposite to it in the next column to the right hand. Prefix the proper characteristic (by changing it if necessary) to the mantissa (see Rules XVIII and XIX, pages 61 and 62). The result is the logarithm sought.

NOTE.—It may be observed here, once for all, that the proposed number must be considered a whole number, and in case a decimal point occurs in the given number, no notice is taken of it till we come to the insertion of the characteristic; and should cyphers occur *between the decimal point and the first significant figure*, these also, are disregarded in entering the table, being only taken into account when determining the characteristic.

Ex. 1. Required the logarithm of 21, 2·1, ·21, and ·021.

In the first page of the Table, and in one of the vertical columns marked No., we find 21, against which stands 1·322219, the logarithm sought. Since the mantissa of the logarithm of any number consisting of the same figures is the same whether the number be integral, fractional, or mixed, the logarithms of the numbers 2·1, ·21, and ·021 will have the same decimal part as 21, the characteristic only being changed, consequently the logarithm of 2·1 is 0·322219, the logarithm of ·21 is 9·322219, and the logarithm of ·021 is 8·322219.

Ex. 2. To find the logarithm of 52, 5·2, ·52, and ·00052:—

In the Tables we find the log. of 52 is 1·716003, and, therefore, simply changing the index, the log. of 5·2 is 0·716003, ·52 is 9·716003, and the log. of ·00052 is $\bar{4}$·716003 or 6·716003.

No.	Nat. No.	No.	Nat. No.	No.	Nat. No.	No.	Nat. No.
1.	5	7.	4½	13.	94	19.	·0091
2.	9	8.	·004	14.	½ or ·5	20.	25·0
3.	·009	9.	2·4	15.	¾ or ·75	21.	·024
4.	·01	10.	24	16.	2·5	22.	·000035
5.	·0001	11.	·24	17.	¼ or ·25	23.	·000057
6.	14	12.	·0021	18.	·09	24.	ruvu

88. To find the logarithm of a number consisting of not more than three places of figures (from 100 to 1000).

RULE XXIII.

Find the given number in the left-hand column of the Table, and opposite it, in the next column, will stand the mantissa or decimal part of the logarithm. Prefix the characteristic according to Rules XVIII and XIX. The result is the logarithm sought.

NOTE.—When we say "three figures" we mean independently of cyphers either to the right or to the left. Thus we should include 6340, 73200, and ·00265 under the head of this problem.

NOTE.—*If the number is less than three figures* make up three by placing cyphers, if not already present (or by supposing them placed), on the right of the number, cyphers so added being regarded as decimal; then proceed as directed in the above Rule XXIII. Thus the logarithm of 75 is the same as that of 75·0; the logarithm of 8 is the same as of 8·00; and that of ·035 is the same as of ·0350.

EXAMPLE.

Ex. 1. Required the log. of 476, 4·76, and ·00476.

We seek in the left-hand column of the Table for 476, against which in the column marked 0 at the top, stands the mantissa corresponding thereto; and this part by the rule is the same for each of the above numbers. Now prefixing the index according to the number of integral figures in the natural number, we find the log. of 476 is 2·677607; of 4·76 is 0·677607; and of ·00476 is 7·677607.

Again, the logarithm of 576 is 2·760422; that of 39·4 is 1·595496; that of ·0253 is 2·403121.

No.	Nat. No.	No.	Nat. No.	No.	Nat. No.	No.	Nat. No.
1.	100	5.	673	9.	8·96	13.	·0147
2.	145	6.	794	10.	1·47	14.	434
3.	2·94	7.	982	11.	·147	15.	·0000448
4.	361	8.	4·83	12.	·901	16.	448000

89. If the number contains four places of figures, exclusive of final cyphers, or cyphers included between the decimal point and the first significant figure.

RULE XXIV.

Find the first three figures in the vertical column on the left marked No., *and the fourth in the horizontal column at the top of the page.* Under this last, and opposite the three figures, will be found the mantissa of the logarithm sought. Prefix the index according to Rules XVIII and XIX. The result is the logarithm sought.

EXAMPLE.

Ex. 1. Required the logarithms of 4587 and of 0·0004587.

The first three figures (viz., 458) being found in the column to the left marked No., and the fourth (7) in the line of digits at the top of the page, the decimal part of logarithm (·661529) is found in the same horizontal line as the three first figures of the given number, and in the same column as the fourth. The characteristic is 3, being one less than the number of *integers* in the whole number; whence the completed logarithm is 3·661529. The logarithm of ·0004587 is $\bar{4}$·661529, the characteristic being negative, and one more than the number of prefixed cyphers.

Again, the logarithm of 3470 is 3·541330; that of 3·492 is 0·543074; and that of 0·3468 is $\bar{1}$·540079; that of 74·39 is 1·871515; that of 325600 is 5·512648, in which case the mantissa of 3256 is taken out, since it is the same as the mantissa of 3256000.

No.	Nat. No.	No.	Nat. No.	No.	Nat. No.	No.	Nat. No.
1.	1000	4.	5432	7.	·01012	10.	987·6
2.	1234	5.	26·06	8.	94·87	11.	·06843
3.	25·65	6.	2·606	9.	7·777	12.	·002784

NOTE.—The foregoing rule may be used not only in the case of numbers consisting of four places of figures, but may be made to include all numbers consisting of less than four significant digits, and so enable us to dispense with the Rules XXII and XXIII. Thus, if the number consists of less than four figures, make up four by placing cyphers, if not already placed (or by supposing them placed), on the right of the number; cyphers so added being regarded as decimals. Then proceed to find the mantissa of the log. by the foregoing rule.

Thus, the log. 75 is the same as the log. of 75·00; the log. of 8 is the same as the log. of 8·000; and that of ·035 the same as that of ·03500 (3500 in the tables).

90. Although the tables in RAPER, NORIE, and the "Nautical Tables" accompanying this work are constructed so that the mantissæ corresponding to more than four figures cannot be taken out directly, yet the mantissæ of numbers containing five or six figures can be found from them without much trouble by means of the tabular difference taken out of the extreme right hand column of the page (see 86).

If the number consists of more than four figures other than final cyphers, or if the number be a decimal fraction, cyphers immediately following the decimal point, we use

RULE XXV.

1°. *Cut off the first four figures and consider the rest as a decimal.*

2°. *Find the mantissa corresponding to the first four figures* (Rule XXIV).

3°. *Multiply the tabular difference by the decimal cut off*, i.e., by the remaining figures of the given number, *and cut off from the right-hand as many figures as there are in the* multiplier, *but at the same time adding unity if the highest figure thus cut off* is not less than 5.

4°. *Add the integer part of this product to the figures of the mantissa just found.*

These proportional parts are thus compiled on the supposition that the difference between *the numbers (nearly equal to each other) is proportional to the difference between their logarithms.* This proportion can be shown to be approximately true.

The result is the mantissa of the required logarithm.

The characteristic or index is found by Rules XVIII and XIX, pages 61 and 62.

EXAMPLES.

Ex. 1. Required the logarithm of 28434.

Tab. diff.
153
× 4
―――
61,2 or 61

Mantissa of 2843 = ·453777
Tab. diff. 153 × 4 = 61·2 = + 61
Characteristic = 4
―――――――――――――
The log. of 28434 = 4·453838

We seek in the left-hand column of the Table for 284 (the first three digits) and also at the top of the page in one of the horizontal columns we find 3 (the fourth figure), then in a line with the former and in the column with the latter at the top we have 453777, which is the mantissa of 2843. In a line with the quantity in the right-hand column marked Diff., stands tab. diff. 153; which multiplied by 4, the remaining digit of the given number, produces 612; then cutting off one digit from this (since we have multiplied by only one digit) it becomes 61, which being added to 453777 (the mantissa of 2843) makes 453838, and, with the characteristic, 4·453838, the required logarithm.

The logarithm of 284·34 is 2·453838, and the log of ·028434 is $\bar{2}$·453838 or 8·453838.

Ex. 2. Required the logarithm of 12806.

Tab. diff.
338
× 6
―――
202,8 or 203

Mantissa of 1280 = ·107210
Tab. diff. 338 × 6 = 202·8 = + 203
Characteristic = 4
―――――――――――――
The log. of 12806 = 4·107413

Ex. 3. Find the logarithm of 873457.

Tab. diff.
50
+ 57
―――
28,50 or 29

Mantissa of 8734 = ·941213
Tab. diff. 50 × 57 = 28·50 = + 29
Characteristic = 5
―――――――――――――
The log. of 873457 = 5·941242

The mantissa of the first four figures is found thus:—Opposite the 873 and under 4 stands 941213; then in the right-hand column in a line with this stands the diff. 50, which being multiplied by 57, the remaining digits of the given number, makes 2850; from this we cut off *two* digits to the right (since we have multiplied by *two* digits), when it becomes 28; but as the highest digit cut off is 5, we add unity, which makes 29. Then 5·941213 (the logarithm of 8734) + 29 = 5·941242 is the required logarithm.

Ex. 4. Required the logarithm of 628007.

Mantissa of 6280 = ·797960
Tab. diff. 69 × 07 = 4·83 = + 5
Characteristic = 5
―――――――――――――
The log. of 628007 = 5·797965

Tab. diff.
69
× 07
―――
4,83 or 5

The log. of 628·067 is 2·798006, and the log. of ·0062S067 is $\bar{3}$·798006 or 7·798006. The mantissa of the log. of each of these numbers being the same, the index only being varied. (See Rules XVIII and XIX.)

1.	38475	7.	435·60	13.	200000	19.	365152
2.	384·75	8.	78·624	14.	·056214	20.	997·1370
3.	12345	9.	2·2055	15.	·0098563	21.	32·1928
4.	543·21	10.	0·78362	16.	643786	22.	1·032764
5.	66666	11.	10000	17.	1129·06	23.	1000½
6.	9244·8	12.	·000800073	18.	·998095	24.	596·423

91. **To find the natural number corresponding to a given logarithm.**—If the logarithm be given, the number which corresponds to it may be found by the following rules, which are the converse of those last given for finding the logarithm when the number is given.

Since the characteristic denotes how many places the first significant figure stands to the right or left of the unit's place; conversely, therefore, if logs. be given having for characteristics 1, 2, 3, ... $\bar{1}$, $\bar{2}$, $\bar{3}$, ... there are in the integral parts of the number to which these logs. belong, 2, 3, 4, ... 0, $\bar{1}$, $\bar{2}$, ... digits respectively. In illustration of these remarks take the following:—

Log. 4·589950 (in which characteristic 4) gives			38900
3·589950	············	3 ..	3890
2·589950	············	2 ..	389
1·589950	············	1 ..	38·9
0·589950	············	0 ..	3·89
$\bar{1}$·589950 or 9·589950 }	············	$\bar{1}$ or 9 ..	·389
$\bar{2}$·589950 or 8·589950 }	············	$\bar{2}$ or 8 ..	·0389
&c.		&c.	

In which it will be observed that the first answer must consist of five integers, because the index of the given logarithm is 4; that the second answer must contain four integers, because the index of the given logarithm is 3; that the third answer must contain three integers, because the index of the logarithm is 2, &c., &c.; and that the sixth answer must be a decimal fraction having the first significant figure in the place of tenths, because the logarithmic index is $\bar{1}$; and lastly, that the seventh answer must be a decimal fraction having the first significant figure in the place of hundredths, because the logarithmic index is $\bar{2}$.

92. From the foregoing it is evident that when the figures of the natural number have been found, we must place the decimal point so that the number of integral figures may be one more than the characteristic denotes. Cyphers must be supplied to the right, if necessary, to make up the number, hence

RULE XXVI.

Add 1 to the characteristic of the given logarithm, and mark off to the left the number of figures for whole numbers; the rest (if any) will be decimals.

If the characteristic is negative place the decimal point to the left of the natural number found, along with as many cyphers as may remove the first significant figure to that place of decimals which the index expresses; that is, one cypher fewer than the number denoted by the characteristic, whence, to find the place of the decimal, we have the following

RULE XXVII.

The number corresponding to a logarithm with a negative index is wholly decimal, and the number of cyphers following the decimal point is one less than the characteristic of the logarithm.

But instead of the negative characteristic its *arithmetical* complement is sometimes used, in which case we proceed by

RULE XXVIII.

Add 1 to the index, and subtract the number thus found from 10; the remainder is the number of cyphers to be prefixed to the figures taken out of the Tables. Place the dot before the first cypher.

93. To find the natural number corresponding to any given logarithm.

When the mantissa or decimal part of the logarithm can be found exactly in the Table, we proceed by

RULE XXIX.

1°. *Seek out the mantissa, and take from the column No. the three figures in the same horizontal row.*

2°. *From the head of the column take the fourth figure.*

3°. *From the characteristic find by the rules already given the position of the decimal point*, and so adjust the local value of the figures. (Rules XXVI, XXVII, and XXVIII, No. 92, page 68).

(*a*) *When the characteristic of the given logarithm requires a greater number of digits to the left of the decimal point than there are in the number found by the above rule, the deficiency is made up by adding a sufficient number of cyphers to the right.*

(*b*) *If the natural number is a decimal fraction, and the final figure or figures are cyphers, they need not be written down.*

EXAMPLES.

Ex. 1. Given the logarithm 2·698970 to find the natural number.

Entering the Table with the decimal part ·698970, we find the natural number corresponding to it to be 5, or 50, or 500, or 5000, &c., but as the index of the logarithm is 2, the natural number must contain three integral figures. Hence the natural number of 2·698970 is 500.

Ex. 2. Given the logarithm $\bar{3}$·539954 or 7·539954: find the number.

Entering the Table with the decimal part, we find the corresponding number is 3467; to this we prefix two cyphers, since the index is $\bar{3}$; or adding 1 to 7 (8), and subtract 8 from 10, we have 2, the number of cyphers to be prefixed, and then the decimal point; hence the number corresponding to 7·539954 is ·0034567.

Ex. 3. What number corresponds to the logarithm 4·214314.

The decimal part of the log. being found opposite 163 and under the figure 8 at the top of the page: therefore the digits of the required number are 1638. But as the characteristic is 4, there must be in it 5 places of integers. A cypher is annexed (see Rule XXIX, (*a*)). Hence the required number is 16380.

Ex. 4. Required the natural numbers corresponding to logs. 0·176091 and $\bar{4}$·176091.

(1) The mantissa ·176091 stands in the Table opposite 150, and the column with 0 at the top; and the characteristic 0 shows that one of these is integral, whence the number sought is 1·500 or 1·5 (see Rule XXVI, page 68).

(2) The mantissa of second log. being the same as that of the first, the corresponding number will consist of the same significant figures, but the characteristic $\bar{4}$ shows that the first significant figure (1) must occupy the fourth place to the right of the decimal point, whence the number sought is ·00015. (See Rule XXVII or XXVIII, pages 68 and 69.)

Ex. 5. Required the natural number whose logarithms are respectively 1·813514, 0·303412, 4·996993, $\bar{2}$·299943 or 8·299943, 4·000000, $\bar{4}$·000000, 7·816109, we shall find them to be as follows:—

$$
\begin{array}{rcl}
1\cdot813514 &=& \log. \text{ of } 65\cdot09 \\
\cdot303412 &=& \text{,,} \quad 2\cdot011 \\
4\cdot996993 &=& \text{,,} \quad 99310 \\
\left.\begin{array}{l}\bar{2}\cdot299943 \\ \text{or } 8\cdot299943\end{array}\right\} &=& \text{,,} \quad \cdot01995 \\
4\cdot000000 &=& \text{,,} \quad 10000 \\
\bar{4}\cdot000000 &=& \text{,,} \quad \cdot0001 \\
7\cdot816109 &=& \text{,,} \quad 65480000
\end{array}
$$

Where it will be observed that the first answer must contain only two integers, as the index of the given logarithm is 1; that the second must contain only one integer as the characteristic is 0; that the third must consist of five integers, because the index of the given logarithm is 4, and therefore to 9931, the number found in the Table, a cypher is annexed, (see Rule XXIX, (a); and that the fourth answer must be a decimal, having the first significant figure two places to the right of the decimal point because the characteristic is $\bar{2}$; the fifth answer must consist of five integral figures (a cypher being annexed to make up the number) since the characteristic is 4; the mantissa of the sixth log., or ·000000, gives the corresponding natural number 1000, but adjusting the decimal punctuation, or the local value of the figures, the characteristic $\bar{4}$ denotes that the first significant figure (1) must stand in the fourth decimal place, and, therefore, three cyphers must be prefixed, and the natural number will be ·0001—the three final cyphers not being written down. Finally, the mantissa of last log. being found in the table gives the natural number corresponding as 6548, to which annex four cyphers; the characteristic 7 determines the number to consist of 8 integral figures.

No.	Log.	No.	Log.	No.	Log.	No.	Log.
1.	0·477121	9.	3·898506	17.	2·990561	25.	$\bar{7}$·991093
2.	0·041393	10.	2·538574	18.	4·541579	26.	7·903524
3.	0·973128	11.	1·394977	19.	$\bar{1}$·744058	27.	$\bar{2}$·621488
4.	1·161368	12.	3·845098	20.	$\bar{1}$·501196	28.	$\bar{9}$·901349
5.	0·812245	13.	7·000000	21.	$\bar{7}$·875061	29.	$\bar{3}$·662758
6.	2·767898	14.	5·825426	22.	6·602060	30.	$\bar{4}$·851258
7.	0·394452	15.	4·698970	23.	$\bar{8}$·845098	31.	6·913761
8.	1·478422	16.	5·000000	24.	3·605197	32.	5·868527

94. When, as usually happens, the mantissa cannot be found exactly in the Tables, but lies between two successive records in the Tables, and it is proposed to find the corresponding number correct to six places of figures, other than final cyphers immediately following a decimal point, the number is to be found by the method of proportional parts, on the supposition that, between two successive records in the table, the number advances in proportion to the increase of the logarithm.

95. To find the natural number corresponding to a given logarithm, when more than four figures are required. We proceed by

RULE XXX.

1°. *Having found the next lower mantissa in the Tables, note the four figures which correspond to it.*

2°. *From the given logarithm subtract that taken out of the Tables, divide the remainder (annexing as many cyphers as there are digits required above four) by the tabular difference, and reduce the quotient to the form of a decimal.*

3°. To the four figures already found, add this decimal, and shift the decimal point to suit the characteristic of proposed logarithm.

The result will be the required number.

NOTE.—It is needless to annex many cyphers to the dividend. We cannot with safety carry the natural number to more than six figures when the tabular difference contains three, or to more than five when the tabular difference contains only two.

EXAMPLES.

Ex. 1. Given the logarithm 3·543027 to find the natural number.

```
    Given logarithm              3·543027
    Mantissa next lower in Table  ·542950 which corresponds to 3491.
                 Tab. diff. = 124)·7700(·62
                                 744
                                 ───
                                 260
                                 248
```

Attaching this (·62) to the four figures, we have 349162, &c. The decimal punctuation or local value of the figures of the number can now be adjusted, and as the index is 3, we obtain, by pointing off four figures to the left, 3491·62, the natural number sought.

Ex. 2. Given the logarithms 5·654329 and 2̄·654273 to find the natural numbers.

```
    Given logarithm              5·654329
    Mantissa next less in Table  ·754273 which corresponds to 4511.
                 Tab. diff. = 96)5600(58
                                 480
                                 ───
                                 800
                                 768
                                 ───
         Ans. 451158.             32
```

654273, which corresponds with the natural number 4511, is the logarithm next less than the given one; therefore the first *four* digits of the required number are 4511. Adding two cyphers to 56, the difference between 654273 and the given logarithm, it becomes 5600, which being divided by 96, the *tabular difference* corresponding with 4511, gives 58 as quotient and 32 as remainder. The *integers* of the required number (one more than 5, the characteristic) are, therefore, 451158. The mantissa of the second log. being the same as the first one, the natural number will contain the same significant figures, viz., 451158, but the characteristic 2̄ shows that the first significant figure of the nat. no. (4) must stand in the second place to the right of the decimal point; therefore, the nat. no. corresponding to 2̄·654273 is ·0451158.

Ex. 3. Let it be required to find the number of which the logarithm is 3·104831.

```
    Given logarithm              3·104831
    Mantissa next lower in Table  ·104828 which corresponds to 1273.
                 Tab. diff. = 341)3000(·008
                                 2728
                                 ────
                                  272
```

Therefore, 3·104831 = log. of 1273·009 nearly. In dividing by tab. diff. we take remainder 3 and a cypher, then 341 in 30 goes no times, which we place down in the quotient, then taking another cypher we have 300, which contains 341 no times, lastly, 341 goes into 3000 eight times with 272 for remainder. The remainder 272 being more than half the quotient the last figure of the quotient (8) is increased by 1 or unity.

No.	Log.	No.	Log.	No.	Log.	No.	Log.
1.	2·931214	10.	4·994603	19.	0·230449	25.	7·246631
2.	3·625343	11.	4·925936	20.	1·217845	26.	1·998813
3.	4·851906	12.	5·091512	21.	3·98467·1		or 9·998813
4.	4·361730	13.	2·535224		or 7·984671	27.	1·895090
5.	1·725364	14.	3·744726	22.	4·463726		or 9·895090
6.	1·972521	15.	5·831835		or 6·463726	28.	4·931847
7.	5·659707	16.	2·415671	23.	2·241877		or 6·931847
8.	5·734968	17.	4·841989		or 8·241877	29.	3·565942
9.	5·823904	18.	4·092561	24.	6·371000		or 5·565942

96. **Def.**—The arithmetical complement of a number is the number by which it falls short of the units of the next higher denomination. (If x is any number whatever, then the arithmetical complement of $x = 10 - x$.) It is abbreviated into *ar. co.*

To find the arithmetical complement* of the logarithm of a number.

RULE XXXI.

Begin from the left; subtract every figure from 9 up to the lowest significant figure, which subtract from 10. Repeat the cyphers at the end, if any.

(a) When the characteristic is negative it must be added to 9.

EXAMPLES.

Ex. 1. Find the ar. co of 1·79043.

(a) Ar co. log. required 8·02957.

(a) Thus we say, beginning at the left hand, 1 from 0, 9 from 9, 7 from 9, 0 from 9, 4 from 9, and 3 from 10.

The ar. co. of 3·607218 is 6·392782
· · · 0·714000 „ 9·286000
· · · 5·631642 „ 4·368358

Ex. 2. Find the ar. co. of ·9085640.

(b) Ar. co. log. required 0·913360.

(b) In this example we proceed as before; thus 9 from 9, 0 from 9, 8 from 9, 0 from 9, 5 from 0, and as the next figure, 4, is the *lowest significant figure* (see Rule), we take it from 10, which leaves 6; lastly, the cypher at the end is repeated.

The ar. co. of 2·170630 is 7·829370
· · · 1·217034 „ 10·782966
· · · 3·178680 „ 12·821320

97. A subtractive quantity is, by this means, made additive. The process is equivalent to subtracting the number from 10, and the reason of it is evident on considering that to add 3 and subtract 10 is the same as to subtract 7. In like manner, instead of subtracting 42ᵐ 10ˢ for example, we may add 17ᵐ 50ˢ (the complement of 60ᵐ), provided we subtract 1ʰ (or 60ᵐ); and thus any number of quantities, of which some are additive and some subtractive, may be rendered all additive, provided that the larger numbers which are employed in taking the complements be themselves subtractive.

MISCELLANEOUS.

98. We here insert a collection of numbers, the logarithms of which are to be taken out of the Tables.

1.	3	11.	63·5	21.	844·4	31.	93·7654	41.	10000000
2.	0·1	12.	6390	22.	·92096	32.	5279·9	42.	·000000362
3.	4·9	13.	·1463	23.	·0899	33.	50000	43.	30000·9
4.	38	14.	3·874	24.	10000	34.	700090	44.	10000·9
5.	380	15.	6754	25.	4800	35.	264000	45.	594500
6.	100	16.	·0876	26.	9080·8	36.	404207	46.	88590200
7.	·0001	17.	·3467	27.	·00058	37.	500909	47.	287642
8.	25·6	18.	1·083	28.	·035872	38.	48·617	48.	0·003564
9.	3·88	19.	0·125	29.	·000448	39.	93·514	49.	·000856436
10.	900	20.	0·0009	30.	4480000	40.	·032764	50.	63480000

* A very curious and valuable artifice, discovered by GUNTER about 1614.

99. Required the natural number of the following logarithms:—

1.	2·309630	10.	0·565021	19.	$\bar{3}$·954243	28.	5·606389	37.	$\bar{7}$·883030
2.	3·676968	11.	0·778441	20.	$\bar{3}$·959041	29.	5·000000	38.	3·625343
3.	0·954243	12.	2·769504	21.	4·705864	30.	$\bar{2}$·881955	39.	1·725364
4.	1·698970	13.	5·774152	22.	0·415974	31.	$\bar{1}$·167317	40.	$\bar{5}$·627407
5.	0·000000	14.	5·421604	23.	$\bar{1}$·000000	32.	7·875061	41.	$\bar{3}$·686216
6.	2·000000	15.	3·000000	24.	$\bar{3}$·954243	33.	0·000186	42.	0·400573
7.	2·564494	16.	6·394452	25.	$\bar{2}$·716003	34.	6·947385	43.	5·002559
8.	3·563362	17.	1·415674	26.	5·654243	35.	$\bar{2}$·961081	44.	4·321547
9.	2·621754	18.	1·188591	27.	0·434294	36.	0·763947	45.	0·875061

100. Finally, we recommend the student to commit to memory the following table of logarithms to two places:—

No.	Log.	No.	Log.	No.	Log.
1.	∞	4.	60	7.	85
2.	30	5.	70	8.	90
3.	48	6.	78	9.	95

MULTIPLICATION BY LOGARITHMS.

101. In multiplication we proceed by

RULE XXXII.

1°. *Find the logarithms of the numbers, the product of which is required.* (For the method of taking out the log. of a number see pages 64 to 67.)

NOTE.—If any of the quantities is a decimal, either the negative characteristic of that quantity or its arithmetical complement is to be used (see Rules XIX and XX, page 62.)

2°. *Add these together, the sum will be the logarithm of the product.*

3°. *Find from the Tables the corresponding number.* (For the method of finding the corresponding number to a log., see pages 68 to 71.)

This will be required product.

NOTE 1.—When the characteristics are negative and subtracted from 10 (see Rule XX, page 62), if the sum of such characteristic exceeds the sum of tens borrowed, the product will be a whole number; otherwise it will be a decimal. (See Ex. 15, page 75.)

NOTE 2.—When the characteristics of the logarithms to be added are all positive, it is evident that their sum will be positive.

NOTE 3.—*If the characteristics are all negative,* their sum diminished by the figure—if any—carried from the sum of the mantissæ or positive decimal parts will be negative. (Ex. 9.)

NOTE 4.—*If some characteristics are positive and the others negative,* find the sum of the positive characteristics together with any figure which may be carried from the decimal part of the logarithm; also add the negative characteristics together; subtract the less of these quantities from the greater and prefix to the difference the sign belonging to the greater. But if a positive and a negative characteristic are exactly equal to each other, cancel both; this is done in practice by simply drawing the pen through them. (Ex. 13.)

EXAMPLES.*

1. Multiply 77 by 100. The log. of 77 and 100 being taken from the table, we have

    77 log. 1·886491
    100 log. 2·000000

    7700 log. 3·886491

We have here added the logs. of the given factors, and having sought in the Table for the mantissa ·886491, we have found the figures of the nat. no. corresponding to be 7700; the index 3 determines *four* of these to be integral; hence the product is 7700 (Rule XXVI, page 68).

2. Multiply 97 by 83. The log. of 97 and 83 being taken from the Table, we have

    97 log. 1·986772
    83 log. 1·919078

    8051 log. 3·905850

We add the logs. of the given factors, and then seek in the Table for the mantissa ·905850, which corresponds to the natural number 8051; the index 3 determines *four* of these to be integral; hence the product is 8051 (Rule XXVI, page 68).

* In these examples, and for several of the subjoined Exercises, the logarithmic is more tedious than the ordinary method of calculation; the purpose here intended being simply to make the student familiar with the process of finding products logarithmically. It must be remembered too, that by the logarithmic process, we generally obtain only an approximate value of the required result.

3. Multiply 378 by 50.

$$378 \text{ log. } 2\cdot577492$$
$$50 \text{ log. } 1\cdot698970$$
$$18900 \text{ log. } 4\cdot276462$$

The mantissa of log., viz., ·276462, is found *exactly* in the Table in a line with 189, and under 0; but as the characteristic 4 requires 5 digits in the integer part, we therefore add a cypher (0), which gives 18900 as the nat. no. corresponding to the proposed log. This is according to Rule XXIX (a), page 69.

5. Multiply 963 by 48·9 by common logarithms. The log. of 963 and 48·9 being taken from the Table, we have

$$963 \text{ log. } 2\cdot983626$$
$$48\cdot9 \text{ log. } 1\cdot689309$$

$$\text{log. } 4\cdot672935$$
(next lower in Table) ·672929 gives 4709

Product 47090·7   92)6·00(06
                      552

                       48

We have here added the logs. of the given factors together, and having sought for the given mantissa ·672935, which is not to be exactly found in the Tables, we obtain the next less mantissa ·672929, which we subtract from the given mantissa; the difference is 6, to which two cyphers are annexed, and then we divide by the tabular difference 92, whence we obtain 07 nearly; the remainder, 48, being more than half the divisor, 1 is added to the last figure in the quotient (6); attaching these to the four figures obtained previously, we have 470907; the characteristic 4 determines *five* of these to be integral; hence the product is 47090·7 (Rule XXVI, page 68). The multiplier containing *one* decimal place, the product is worked out to one place of decimals.

8. Multiply 29·42 by 8·6 by common logarithms.

$$29\cdot42 \text{ log. } 1\cdot468643$$
$$8\cdot6 \text{ log. } 0\cdot934498$$

$$2\cdot403141$$
(next lower in Table) 403120* gives 2530

Product 253·012   171)2100(12 +

In this instance the characteristic of the log. of the product is 2, hence the integral part of the natural number must contain 3 figures; but since there are decimals in both factors, there must be decimals in the product—as many decimal places as there are in both the multiplier and multiplicand together. In 29·42 are two decimal places, and in 8·6 one; hence in the product three decimal places are required, making, with the three integral figures, in all six places. Now the next lower mantissa found in the table gives the four corresponding figures 253*, leaving *two* figures to be found. (See Rule XXX, page 70.)

* This log. is taken from NORIE, and is incorrect in the last decimal figure, which ought to be 1, as given in RAPER's table; the true log. being ·40312052.

4. Multiply 3456 by 500.

$$3456 \text{ log. } 3\cdot538574$$
$$500 \text{ log. } 2\cdot698970$$
$$1728000 \text{ log. } 6\cdot237544$$

The characteristic 6 requires 7 digits in the integer part of product, we therefore annex 3 cyphers which gives 1728000 as the nat. no. required. (See Rule XXIX (a), page 69.)

6. Multiply 734 by 23.

$$734 \text{ log. } 2\cdot865696$$
$$23 \text{ log. } 1\cdot361728$$

log. 4·227424
(next lower mantissa) 227372 corresponds
                       ------ to 1688.
Product 16882   258)520(2
                            516

7. Multiply 498 by 376.

$$498 \text{ log. } 2\cdot697229$$
$$376 \text{ log. } 2\cdot575188$$

$$187248 \text{ log. } 5\cdot272417$$
$$306$$

Diff. 232)11100(48 nearly

9. Multiply ·0567 by ·00339.

Both multiplier and multiplicand being decimals, the characteristics of those factors will be negative, but instead we use their arithmetical complements thus:—

$$\cdot0567 \text{ log. } 8\cdot753583$$
$$\cdot00339 \text{ log. } 7\cdot530200$$

$$\cdot0001922 \text{ log. } 6\cdot283783$$

Here 10 is borrowed to find the characteristic both of the multiplicand ·0567, and the multiplier ·00339 (see Rule XX, page 62). The sum of the characteristics, including the 1 carried from the decimal part of the log., amounts to 16; reject 10 and write down 6 for the index of the log. of product. Then, seeking in the Table for the decimal part, viz., ·283783, the natural number corresponding to it is found to be 1922; and since the sum of the indices 16 is 4 *less* than the 20 borrowed, (see Rule XXXII, Note 1, page 73) the product is a decimal fraction, and the *first significant digit* must stand in the fourth decimal place; hence the product is ·0001922.

Or thus—using negative indices:—

$$\cdot0567 \text{ log. } \bar{2}\cdot753583$$
$$\cdot00339 \text{ log. } \bar{3}\cdot530200$$

$$\cdot0001922 \text{ log. } \bar{4}\cdot283783$$

In adding, when we come to the place of tenths, the process is 5 and 7 are 12, 2 to put down and 1 to carry, and since the characteristics are both negative ($\bar{2}$) and ($\bar{3}$), we diminish their sum ($\bar{5}$) by the number carried (1), which leaves $\bar{4}$ for the index (see Rule XXXII, Note 3, page 73). We prefix 3 cyphers because the index being 4 the first significant figure of product must stand in the fourth place from the decimal point.

10. Multiply 99·9 by 8·63.

    99·9 log. 1·999565
    8·63 log. 0·936011

  862·136 log. 2·935576
                558

         5,0)180,0
              36

11. Multiply 436 by 19·7.

    436 log. 2·639486
    19·7 log. 1·294466

  8589·2 log. 3·933952
             43

     5 l)90(2 nearly

12. Find the product of ·073 by ·00028 by logarithms.

  ·073 log. 8·863323
  ·00028 log. 6·447158

  ·00002044 log. 5·310481

Or, using the negative characteristic, thus :—

  ·073 log. $\bar{2}$·863323
  ·00028 log. $\bar{4}$·447158

  ·00002044 log. $\bar{5}$·310481

In adding, when we come to the place of tenths, the process is 5 and 8 are 13, 3 to put down and 1 to carry; and this 1 being a positive quantity. Hence in the above, +1, −2, and −4 are to be algebraically added together to form the new characteristic. The sum of the two characteristics (both negative) viz.,— 2 and −4 is −6, which diminished by + 1 leaves − 5 for the new characteristic. We prefix *four* cyphers, because the characteristic being 5 and this with the first significant figure must stand in the fifth decimal place. (Rule XXVII, page 68.)

13. Multiply 24000 by ·000783.

  24000   log. 4·380211
  ·000783 log. 6·893762

  18·7919 + log. 1·273973

Here 10 was borrowed in determining the index of the log. of ·000783, and since the sum of the indices (including 1 carried from the decimal part of log.) is *eleven*, we reject or *pay back* the 10 borrowed, which leaves 1 for the index, and the nat. number corresponding is found to be 187919, and we mark off to the right two figures (1 more than the characteristic) whence the answer is 18·7919+.

Or thus—using negative indices: Ex. 13.

  24000   log. 4·380211
  ·000783 log. $\bar{4}$·893762

  18·7919 + log. 1·273973

Here the 1 which is carried after adding 1, 8, and 3 (in the place of tenths), instead of increasing the 4, leaves 5. This is according to Rule XXXII, Note 4, page 73.

14. Multiply ·0172 by ·00214.

  ·0172 log. 8·235528
  ·00214 log. 7·330414

  ·00003680 8 log. 5·565942

In this instance 10 is borrowed, in finding the index of the log. both of the multiplier and multiplicand, and 10 is rejected from the sum, which sum (15) being 5 *less* than the amount borrowed (20), indicates that the product must be a decimal fraction, and the *first* significant digit stands in the *fifth* decimal place; hence the product is ·00003680 8. This is according to Rule XXXII, Note 1, page 73.

Or thus—using negative indices: Ex. 14.

  ·0172 log. $\bar{2}$·235528
  ·00214 log. $\bar{3}$·330414

  ·00003680 8 log. $\bar{5}$·565942

The characteristics of both logs. being negative, the sum of them is taken, and this, with the negative sign over it, is put down as the characteristic of the log. of product. We prefix four cyphers to the number taken out of the Table, since the characteristic being 5, the first significant figure of the product must stand in the fifth place from the decimal point.

15. Required the product of 17·25, 0·82, and 0·065.

  17·25 log. 1·236789
  ·82 log. 9·913814
  ·065 log. 8·812913

  0·919425 log. 9·963516

Here the sum of the characteristics of log. of both of the second and third factors, and subtracting the sum of the indices, to from 20 leaves 1; the sum being *less* than the number borrowed, the product is a decimal, and hence the first significant figure must occupy the first place to the right of the decimal point. (See Rule XXXII, Note 1, page 73.)

Or thus :—

  17·25 log. 1·236789
  ·82 log. $\bar{1}$·913814
  ·065 log. $\bar{2}$·812913

  0·919425 log. $\bar{1}$·963516

Here we have 1 to carry from the mantissa, which adds 1 to the positive characteristic 1 (characteristic of log. 17·25, as above) makes positive 2. Now the sum of the negative indices is $\bar{3}$ (negative 3), and, therefore, since where one is positive and the other is negative, the difference is the characteristic; we have + 2 from $\bar{3}$ leaves $\bar{1}$ for the characteristic, (see Rule XXXII, Note 4, page 73) and the first significant figure of the quotient must occupy the first place to the right of the decimal point. (Rule XXVI, page 68.)

Examples for Practice.

1. Multiply by logs. 85 by 70; 39 by 27; 100 by 10; and 369 by 9.
2.   „   538 by 1·74; 601 by 18; 250 by 12·5; and 3964 by 7.
3.   „   20·42 by 0·5; 3·646 by 0·75; 2·745 by 0·24; and 5·792 by 6·5.
4.   „   5671 by 4·7; 517 by 6596 60·609 by 72; 1·955 by 10·04; and 758875 by 8.
5.   „   127 by 304; 476 by 100; 80·08 by 5·98; 5760 by 30; and 970 by 630.
6.   „   37·6 by 249; 44·4 by 22·2; 182·7 by 250; 2807 by 200; and 63·055 by 84.
7.   „   280054 by 50; 30967 by 90; 23716 by 350; and 45670 by 690.
8.   „   81·33 by 15·3; 47·6 by 6·82; 10000 by 10; and 4·02674 by ·0123456.
9.   „   78960 by 400; 756·875 by 8; 94·055 by 74; and 1975 by 10·76.
10.   „   732 by 543; 58·7 by 66·4; 3000 by 100·14; and 60060 by 700.
11.   „   543·29 by 3800·62; 90·43 by 712·2; 87·305 by 4·09; and 1·20936 by 46.
12.   „   348·25 by 7·125; 498·256 by 41·2467; 56·3426 by ·023579; ·123456 by 26813·9.
13.   „   ·0001468 by ·000395; 0·0006 by 10·0004; ·605 by ·00000091; and 35·691 by ·0048.
14.   „   ·00146 by ·039; 5900 by ·00071; 4·189 by ·00071; and 247·55 by 56·72.
15.   „   527·45 by 1·6938; 10·5526 by 317·145; ·007461 by ·3351767; and ·0700397 by ·0086752.
16.   „   ·1 by ·1; ·0001 by ·00001; ·011 by 1·01 and ·00101; and 1000 by 100.

DIVISION BY LOGARITHMS.

102. In division we proceed by

RULE XXXIII.

1°. *Find the logarithms of the numbers the quotient of which is required.*

NOTE.—If the dividend or divisor, or both, are decimals, the negative characteristic of that quantity, or its arithmetical complement, is to be used.

2°. *Subtract the logarithm of the divisor from that of the dividend, (adding 10 to the characteristic of this last, if required); the difference will be the logarithm of the quotient.*

3°. *Find from the Tables the corresponding number.*

This will be the required quotient.

NOTE 1.—When the divisor is greater than the dividend, the characteristic of the logarithm of the quotient will come out negative—the quotient itself being, evidently, a decimal; but if we wish to avoid the use of negative characteristics it will be necessary to add 10 to the characteristic of the dividend when subtracting the logarithm of the divisor, and the characteristic of the remainder is the arithmetical complement of the negative characteristic of the quotient. (See Ex. 4, 5.)

NOTE 2.—If, for the sake of convenience, the line containing the quantity to be subtracted, when the quantities have been written down one under the other, is called the take line and the quantity from which it is to be subtracted the *from* line, then subtracting in the usual way until we come to the characteristics; if their signs are alike take the difference of them, and if the *from* line is the *greater*, prefix to the remainder the *given sign*; but if the *take* line is the *greater* prefix the *contrary* of the given sign. If the signs are different, take the *sum* of the characteristics and prefix the sign of the *from* line. The figure borrowed when subtracting the decimal part of the logarithm, when carried to the characteristic, is always to be added, and therefore make a negative characteristic less, thus 2 carried to 5 makes it 3.

NOTE 3.—Otherwise, if one or both of the given terms are decimals, remove the decimal points till the factors contain whole numbers, and the dividend the greatest; then if the dividend be more places removed than the divisor, remove the decimal point of the quotient as many places to the *left hand*, but if the divisor be more places removed, then remove the decimal point of the quotient as many places to the *right* hand. If the dividend and divisor be equally removed, the quotient is not to be altered.

1. Divide 3192 by 76.

The log. of 3192 is taken out according to Rule XXIV, page 65, and the log. of 76 by Rule XXII, page 64.

$$3192 \text{ log. } 3\cdot504063$$
$$76 \text{ log. } 1\cdot880814$$

Quotient 42·0 log. $1\cdot623249$

3. Divide 579416 by 4324.

Log. of 5794 = 7·62978   Tab. diff. 75
Parts for 16 + 12   × 16
Log. of 579416 = 5·762990   450
                                    75
                                  12,00

579416 log. 5·762990
4324 log. 3·635886

134·0 log. 2·127104

4. Divide 34 by 582.

34 log. 1·531479
582 log. 2·764923

Quotient ·05842 log. 8·766556

In this instance 10 is *added* to the characteristic of the dividend to enable the subtraction of the log. of divisor to be made, and to avoid negative characteristics; the divisor is greater than the dividend, the quotient therefore is a decimal. (See Note 1, p 76.)

Or thus, using negative characteristics :—

34 log. 1·531479
582 log. 2·764923

·05842 log. $\bar{2}$·766556

In this example, when carrying 1 to the characteristic 2, we have to subtract 3 from 1 which gives —2 (negative 2); or according to Note 2, the characteristics having like signs (+) their difference is taken, and the *take* line being the greater, prefix to the remainder a contrary sign (—) to the given one.

6. Find the quotient of ·09983 ÷ ·67.

·09983 log. 8·999261
·67 log. 9·826075

·149 log. 9·173186

Before subtracting the log. of divisor from that of the dividend 10 must be added to the characteristic of the dividend; the quotient is therefore a decimal. Or using negative characteristics the work will stand thus :—

·09983 log. $\bar{2}$·999261
·67 log. $\bar{1}$·826075

·149 log. $\bar{1}$·173186

In this instance both characteristics are of the same sign (—), the *from* line the greater; the characteristic of log. of quotient is marked with sign (—).

2. Divide 830772 by 982.

The log. of 830772 is taken out by Rule XXV, page 65. We seek in the left-hand column of the Table (No.) for 830 (the first three digits), and also at the top of the page in one of the horizontal columns for the fourth figure 7, then in a line with the first and under the latter we have 919444. In a line with this quantity and in the right-hand column marked *Diff.* stands 52, which multiplied by the remaining figures of the nat. number, viz. 72, produces 3744; then cutting off two digits from these (since we multiplied by *two* digits) it becomes 37, which being added to 919444, the mantissa of 8307, makes 919481, and with the characteristic 5, is 5·919481. The work will stand thus :—

Log.   8307 = 919444   Tab diff. 52
Diff. for 72 +   37   × 72
                919481   104
                                      364
                                     37,44

830772 log. 5·919481
982 log. 2·992111

Quotient 846·0 log. 2·927370

5. Divide 3672 by 51000.

3672 log. 3·564903
51000 log. 4·707570

Quotient ·072 log. 8·857333

Here 10 is added to the characteristic of the dividend before subtracting log. of divisor. The divisor being greater than the dividend the quotient is evidently a decimal. (See Note 1, page 76.)

Or thus, using negative characteristics :—

3672 log. 3·564903
51000 log. 4·707570

·072 log. $\bar{2}$·857333

In this instance, when carrying 1 to the characteristic 4, we have to subtract 5 from 3, which gives —2 (negative 2); or by Note 2.— Take the difference of characteristics, as they are of the same sign (+), and prefix a contrary sign (—) to the remainder, the *take* line being the greater.

7. Divide ·01958 by ·4828.

·01958 log. 8·291813
·4828 log. 9·683767

·04056 log. 8·608046

Here 10 has to be borrowed in subtracting the log. of divisor from that of dividend; or using negative characteristics the work will stand thus :—

·01958 log. $\bar{2}$·291813
·4828 log. $\bar{1}$·683767

·04056 log. $\bar{2}$·608046

To obtain the characteristic of the quotient ($\bar{2}$) the 1 that is carried, and which is positive, is added to the $\bar{1}$ producing 0, which has then to be subtracted from $\bar{2}$, leaving $\bar{2}$.

8. Divide 18·792 by ·000783.

$$\begin{array}{lrlrlr}\text{Log.} & 1879 & = & 273927 & \text{Diff.} & 232 \\ \text{Diff. for} & 2 & = & +\ 46 & & 2 \\ \hline \text{Log.} & 18\text{·}792 & = & 1\text{·}273973 & & 46,4 \end{array}$$

Log. 18·792 = 1·273973
Log. ·000783 = 6·891762

Log. 24000 = 4·380211

The divisor here is less than the dividend, the former (18·792) being a mixed number, whilst the latter is a decimal (·000783); the quotient, therefore, is an integer.

Or, using the negative characteristic of divisor.

Log. 18·792 = 1·273973
Log. ·000783 = $\bar{4}$·893762

Log. 24000 = 4·380211

In the subtraction it will be seen that carrying 1 to the $\bar{4}$ we say 1 and $\bar{4}$ make $\bar{3}$, and $\bar{3}$ taken from 1 leaves + 4. The characteristics being of different names, + and −, their sum is taken, and the remainder takes the same sign as the *from* line—in this case it is positive (+). In writing down the result the + is left out.

9. Divide 26843 by ·03010.

$$\begin{array}{lrlrlr}\text{Log.} & 2684 & = & 428782^* & \text{Tab. diff.} & 162 \\ \text{Diff. for} & 3 & = & +\ 49 & & \times\ 3 \\ \hline \text{Log.} & 26843 & = & 428831 & & 48,6 \\ & & & & & \text{or 49} \end{array}$$

26843 log. 4·428831
·03010 log. 8·478566

891794 log. 5·950265

* Raper 3.

$$\begin{array}{r}219 \\ \hline 49)4600(94 \text{ nearly.}\end{array}$$

Or thus:—

26843 log. 4·428831
·03010 log. $\bar{2}$·478566

891794 log. 5·950265

Here the characteristic of the second log. is $\bar{2}$, but following the rule, we have changed it to positive 2. The characteristic of first log. being positive 4, the two are added, the sum being positive 6, but having borrowed 1, the correct characteristic is 5, and being positive, the quotient will contain 6 integral figures.

10. Divide ·8 by ·0000002.

·8 log. 9·903090
·0000002 log. 3·301030

4000000 log. 6·602060

The divisor being *less* than the dividend, the quotient is evidently an integer, and the characteristic denotes that it is a whole number consisting of seven places of figures; cyphers are therefore annexed to make up the required number.

Or, Log. ·8 = $\bar{1}$·903090
Log. ·0000002 = $\bar{7}$·301030

Log. 4000000 = 6·602060

The characteristics are both negative, take their difference, and prefix to the remainder the contrary sign to the given one, as the take one is the greater.

11. Divide ·00815 by ·000275.

0·00815 log. 7·911158
0·000275 log. 6·439333

Quotient 29·6364 log. 1·471825

Or, ·00815 log. $\bar{3}$·911158
·000275 log. $\bar{4}$·439333

29·6364 log. 1·471825

The index of the divisor $\bar{4}$ being supposed changed to positive 4, the difference between which and $\bar{3}$ leaves positive 1 for index of quotient. Or, proceeding according to Note 2—Since the characteristics have *like* signs, take their difference; the remainder takes a positive sign, or a contrary sign to the *take* line, which is the greater.

12. Divide 469·76 by 0·937.

469·76 log. 2·671877
0·937 log. 9·971740

Quotient 501·345 log. 2·700137

Or, 469·76 log. 2·671877
0·937 log. $\bar{1}$·971740

501·345 log. 2·700137

13. Divide 6 by ·0000001.

6 log. 0·778151
·0000001 log. 3·000000

6000000 log. 7·778151

The divisor is *less* than the dividend, the quotient, therefore, is a whole number, and the characteristic 7 indicates that it consists of 8 places of figures; annex cyphers to make up the number.

Or, 6 log. 0·778151
·0000001 log. $\bar{7}$·

6000000 log. 7·778151

14. Divide ·012550 by 1004000.

·012550 log. 8·098644
1004000 log. 6·001734

·0000000125 log. 2·096910

The divisor being greater than the dividend, the quotient is a decimal.

Or, ·012550 log. 2̄·098644
1004000 log. 6·001734

·0000000125 log. 3̄·096910

The characteristic of the dividend is 7̄, that of the divisor positive 6; then according to Note 2, the signs being *unlike*, take the *sum* of the characteristics, prefixing the sign of the *from* line (−).

15. Divide ·027472 by 3·434.

Log. 2747 = ·438859   Diff. 158
Diff. for   2 = +   3 :   × 2

Log. 2747 = ·438891   31,6

Log. ·02747 = 8̄·438891
Log. 3·434 = 0·535800

Log. ·008000 = 7̄·903091
Or, using negative characteristics, thus:—
Log. ·02747 = 1̄·438891
Log. 3·434 = 0·535800

Log.   ·008 = 3̄·903091

Examples for Practice.

1. Divide 6391 by 77; 21636 by 36; 6384 by 76; and 93750 by 750.
2. ,, 9504000 by 98; 45000 by 9; 6071000 by 8; and 58469 by 981.
3. ,, 382·746 by 593; 218432 by 495; 300360 by 100·12; and 365·55 by 5·5.
4. ,, 783254 by 250689; ·79632 by ·019354; ·0092852 by ·0003461; and ·654831 by ·474586.
5. ,, ·0008464 by ·0002852; ·05826 by ·95381; ·019354 by ·79632; ·0003461 by ·0092852; ·00005 by 2·5, by 25, and by ·0000025.
6. ,, 77000000 by 9999; 680300 by 681500; 100·002 by 1·0012; and 75759·6 by 13·062.
7. ,, 1·32704 by ·0358; ·7156 by 2·68878; 87·641 by ·000368; and ·563426 by ·023574.
8. ,, 999999 by 10101; 57634·1 by 276·4; 69·7565 by ·97564; and 352740 by 56780.
9. ,, 40048000 by 800; 11123100 by 340; 1869210 by 90; and 1875000 by 15000.
10. ,, 75·2484 by 8·59; 147392 by 440; 1962810 by 10·04; and 888888 by 88000.
11. ,, 248·25 by 364·87; ·235316 by 293·864; 5·6949 by 53·058; 3876000 by 1200; and 42 by ·00007.
12. ,, ·06314 by ·0007241; ·004728 by 0·2382; 36·49 by 192·24; ·048869 by ·0071698.
13. 19 ÷ 72; 19 ÷ 72; ·19 ÷ 72; 19 ÷ ·0072; 6 ÷ ·0000003; and 9 ÷ ·0000003.
14. ·01237 ÷ ·10846; 28·7642 ÷ ·083456; ·010011 ÷ 0993; and ·048869 ÷ ·0071698.
15. ·1 ÷ ·0004572; 1 ÷ ·0011636; 11·2221 ÷ 111; 4000 ÷ ·000125; and ·562625 ÷ 52643.
16. ·0001 ÷ ·0001; 1000000 ÷ ·0000001; 10 ÷ 100; ·00020001 ÷ ·000001; 1000 ÷ $\frac{1}{10}$.

103. When it is proposed to find the value of an expression in which both multiplication and division are signified, the sum of the logarithms of the factors of the dividend, diminished by the sum of the logarithms of the factors of the divisor, will be the logarithm of the value required.

Example.

Thus: to find the value of $\dfrac{209 \times 573 \times 63}{287 \times 2101}$

287 log. 2·457882
2101 log. 3·322426

5·780308

209 log. 2·320146
573 log. 2·758155
63 log. 1·799341

6·877642
5·780308

Ans.: 12·5122 log. 1·097334

104. It is very often expedient to transform the logarithm of a divisor into that of a multiplier, and it is customary, in such calculations, to avoid not only negative logarithms, but negative indices also, by substituting for a subtraction logarithm its arithmetical complement *(See No. 96, page 72)*. This makes the operation consist of a single addition, only we must diminish the result by subtracting 10 for every arithmetical complement that has been used. By this means the process of division is less open to error from mistakes when logarithms with negative characteristics would be subtracted.*

To apply this method to the example above:—Having found in the Table the log. of the divisor 287, we may at once transform it into the addition logarithm 7·542118, and similarly, for the log. of 2101 we may write 6·677574, and then the calculation will proceed continuously as follows:—

$$\begin{array}{r}209 \text{ log. } 2\cdot320146\\573 \text{ log. } 2\cdot758155\\63 \text{ log. } 1\cdot799341\\287 \text{ ar. co. } 7\cdot542118 - 10\\2101 \text{ ar. co. } 6\cdot677574 - 10\\\hline 1\cdot097334 \text{ } Ans.: \text{ } 12\cdot5122\end{array}$$

EXAMPLES FOR PRACTICE.

1. $\dfrac{7 \times 8\cdot73}{\cdot54963}$ $\qquad$ $\dfrac{84 \times \cdot00769 \times 683}{598 \times \cdot0030146 \times \cdot039}$ $\qquad$ $\dfrac{8\cdot4 \times \cdot0769 \times \cdot00673}{59\cdot8 \times \cdot0000146 \times \cdot0039}$

2. $\dfrac{67\cdot038 \times \cdot010705 \times 4\cdot1525}{7854 \times 3\cdot1416 \times \cdot086725}$ $\qquad$ $\dfrac{28\cdot045 \times 1\cdot3564 \times \cdot0942537}{48\cdot375 \times 2\cdot71828 \times 52359}$

3. Divide ·06314 × ·7438 × ·102367 by ·007241 × 12·9476 × ·496523, and compare the result with the product of 8·71979 × ·057447 × ·0206168.

105. **Degree of Dependence.**—The number of places of figures which may be obtained in a result derived from any table of logarithms is the same, usually rejecting prefixed cyphers, as the number of decimals to which the logarithms are carried. But towards the end of the Table the last place thus obtained cannot always be depended upon within a unit, that is, provided the mantissa of log. is greater than ·9388. Thus, for instance, the log. 3·7575 corresponds to the no. 5721 and the log. 3·7576 to 5722, nearly. It will moreover be noticed that the log. tables are, in fact, useless for dealing with numbers consisting of more than eight places of figures. Thus, for instance, we should not find any difference between the log. of 23·47832 and the log. of 23·4783297.

This remark should be kept in mind, because it is mere waste of time to employ more figures than are required to insure a certain degree of precision in the result.

* To divide by any number n is the same in effect as to multiply by its reciprocal $\frac{1}{n}$ (that is, the quotient of unity divided by that number, and is so called from an exchange of places between a numerator and denominator: thus, the reciprocal of $\frac{2}{3}$ is $\frac{3}{2}$, that of 6, or $\frac{6}{1}$, is $\frac{1}{6}$). Therefore to subtract log. n is the same in effect as to add log. $\frac{1}{n} = 0 - \log. n$.

TRIGONOMETRICAL TABLES.

106. There are two kinds of trigonometrical tables; the first, called the *Table of Natural Sines, Cosines, &c.*, contains the numerical values of the sines, cosines, tangents, &c., that is, of the trigonometrical ratios for each given value of the angle; the second, called the *Table of Logarithmic Sines, &c.*, contains the logarithms of the numbers in the first Table.*

TABLE OF NATURAL SINES, &c.

107. The trigonometrical functions† or ratios are numbers which are capable of being calculated from geometrical principles, and accordingly certain series have been investigated, and certain algebraic expedients devised for the general purpose of determining the trigonometrical ratios. With such aid the sines, cosines, &c., of all angles from 0° to 90° (*i.e.*, for all values of A, from A = 0 up to A = 90) have been computed to several places of decimals and arranged in tables called *Tables of Natural Sines, Cosines, &c.* In some tables the angles succeed each other at intervals of 1″, in others at intervals of 10″; but in ordinary tables (as Table XXVI, NORIE) at intervals of 1′, and to the last mentioned we shall refer.

108. The statement of the method by which such tables are constructed is unsuitable to the pages of the present work. The mode of using them in computation we shall now proceed to explain.

109. The arrangement of this table will be understood from a simple inspection. It contains the sines, cosines, &c., of angles between zero and 90°, generally for every minute, and the fluctuations of angles containing a number of degrees, minutes, and seconds, have to be found by interpolation similar in their nature to those that are required to be used in tables of logarithms of numbers. This interpolation is based upon the supposition that the differences of the sines, &c., are proportional to the differences of the angles, and this proportion, though theoretically inexact, gives, in general, a sufficient approximation, provided the difference of the angles of the table are sufficiently small.

110. Referring to the Tables (Table XXVI, NORIE) it will be seen that the degrees are given at the *top* of the column, and the minutes down the *left* hand side of the page for the sines.

And, for the cosines, the degrees are given at the *bottom* of the page, and the minutes up the *right* hand side of the page.

---

* The usual trigonometrical tables are given in conjunction with tables of logarithms, and they more frequently give logarithms only than sines, cosines, &c., themselves. When logarithms were invented they were called *artificial* numbers, and the originals for which logarithms were computed, were accordingly called *natural* numbers. Thus, in speaking of a table of sines, to express that it is not the logarithms of the sines which are given, but sines themselves, that table would be called a table of *natural* sines, and the logarithms of these would be called not *logarithms of sines* but *logarithmic sines, &c.*

† By the *functions of angles* (sometimes called their *trigonometrical* or *geometrical functions*) are meant their sines, tangents, secants, versed sines, and chords; the word *function* signifying any quantity that is dependent on another changing as it changes.

The difference of the trigonometrical ratios for 100″ are given at the foot of each column.

111. In using these Tables, we have either to find the sine, cosine, &c., of an angle whose value is given in degrees (°), minutes (′), and seconds (″); or to find the corresponding angle in degrees, minutes, and seconds.

112. If the value of the angle be given in degrees and minutes only, the sine, cosine, &c., is found directly from the Tables, in which are registered the values of the trigonometrical ratios.

All the numbers contained in such Tables as Norie's Table XXVI must be understood as decimals.

Thus, nat sine 7° 7′ = ·123890
„ sine 59 40 = ·863102
„ cosine 15 30 = ·963630
„ cosine 71 12 = ·322266

113. As the sines, cosines, &c., pass through all their possible *numerical* values while the angle varies from 0° to 90°, the tables are not extended beyond 90°; such computations would be superfluous, for the sine or cosine of an angle between one and two right angles, viz., of an angle greater than 90° is the same in numerical value as the sine, cosine, &c., of an angle as much below 90°, and is known from the recorded sine or cosine of its supplement.*

Whence also

Nat. sine 156° 42′ = sine 23° 18′ = ·395546
„ cosine 108 48 = cosine 71 12 = ·322266
„ sine 140 16 = sine 39 44 = ·639215
„ cosine 140 16 = cosine 39 44 = ·769028

114. If the angle contains seconds, we must proceed by the *method of proportional parts*, as in the following examples:—

RULE XXXIV.

1°. *Find from the Table the nat. sine, cosine, &c., which corresponds to the degrees and minutes.* (Norie, Table XXVI.)

2°. *Multiply the difference by the seconds, and divide by* 100.

NOTE.—To divide by 100 we have merely to *cut off the two* right-hand figures.

3°. *If the required quantity be a nat. sine, tangent, or secant, add the result to the last figures obtained in* 1°; *if it be a cosine, cotangent, or cosecant, subtract.* The result will be the required sine, cosine, &c.

NOTE 1.—The reason of this rule is founded on the principle that for a small interval, such as one minute, the increase of the sine is proportional to the increase of the angle.

NOTE 2.—It is necessary to bear in mind that the sine, tangent, and secant (under 90°) for which the tables are constructed increase as the arc increases, whilst the cosine, cotangent, and cosecant decrease as the arc increases. This will require the corrections connected with a sine, a tangent, or a secant to be added, and those connected with a cosine, a cotangent, or a cosecant to be subtracted whether arcs or their functions be sought from the tables.

---

* *Def.*—The supplement of an angle is the result when the angle is subtracted from 180°. In other words, an angle and its supplement together make 180°, or two right angles, thus, 23° 19′ is the supplement of 156° 41′, and 156° 41′ is the supplement of 23° 19′.

EXAMPLES.

Ex. 1. Find the nat. sine of $12° \ 44' \ 27''$.
 Nat. sine $12° \ 44' = \cdot 220414$
 Tab. diff. $\dfrac{437 \times 27}{100} = + \ 128$

 $\overline{}$
 $\cdot 220542$
Ans.: Nat. sine $12° \ 44' \ 27'' = \cdot 220542$

To obtain the parts for the second we multiply the tabular difference by the number of seconds and divide by 100, thus:—

 Tab. diff. 473
 No. of seconds × 27
 $\overline{}$
 3311
 946
 $\overline{}$
 127,71
 128 nearly.

Ex. 2. Find the nat. cosine of $31° \ 28' \ 42''$.
 Nat. cosine $31° \ 28' = \cdot 852944$
 Tab. diff. $\dfrac{253 \times 42}{100} = - \ 106$

 $\overline{}$
 $\cdot 852838$
Ans.: Nat. cosine $31° \ 28' \ 42'' = \cdot 852838$

 Tab. diff. 253
 Seconds × 42
 $\overline{}$
 506
 1012
 $\overline{}$
 106,26
 106

EXAMPLES FOR PRACTICE.

To find the nat. sine of

1. $34° \ 48' \ 15''$ 3. $71° \ 20' \ 43''$ 5. $46° \ 22' \ 37''$ 7. $53° \ 7' \ 49''$
2. $60 \ \ 7 \ 18$ 4. $21 \ 44 \ 21$ 6. $76 \ 57 \ 49$ 8. $86 \ \ 3 \ 17$

To find the nat. cosine of

1. $14° \ 15' \ 3''$ 3. $80° \ 22' \ 22''$ 5. $46° \ 31' \ 41''$ 7. $38° \ 31' \ 10''$
2. $70 \ 47 \ 40$ 4. $5 \ 22 \ 10$ 6. $29 \ 40 \ 48$ 8. $8 \ 19 \ 17$

115. If the value of the sine, cosine, &c., be given, and it is required to find the angle, we use the following rule:—

RULE XXXV.

1°. *Find in the Tables the next lower nat. sine, nat. cosine, &c., and note the corresponding degrees and minutes.*

2°. *Subtract this from the given sine, cosine, &c., multiplying the difference by 100; divide by the tabular difference, and consider the result as seconds.*

3°. *If the given value be that of a sine, tangent, or secant, add these seconds to the degrees and minutes found in* 1°*; if it be that of a cosine, cotangent, &c., subtract. The result will be the required angle.*

NOTE.—In taking out the angle for a natural cosine we may take out the *next greater natural cosine*, and subtract the given natural cosine from it; and having found the seconds (''), as above, they are *additive*. The trigonometrical ratio corresponding to the *next less angle* being written down in every case, confusion will be avoided, as the additional seconds will always be *additive*.

EXAMPLES.

Ex. 1. Given the natural sine = $0\cdot 732156$: find the angle.

 Given nat. sine 732156
 Sine $47° \ 4' = \underline{732147}$ next lower in Table XXVI, NORIE.

 Tab. diff. = 317   327)900(3'' nearly (additional seconds for nat. sine.)
     981   Ans.: $47° \ 4' \ 3''$.

The log. 732156 is sought for in Table XXVI, Norie, but as it cannot be found exactly, the next less is taken which corresponds to 47° 4′. The difference of the logs. is then found, two cyphers added (which is equivalent to multiplying by 100), and the product divided by the tabular difference; the quotient is the additional seconds.

Ex. 2. Given the natural cosine 853267: find the angle.

Given nat. cosine 853267
Cosine 31° 16′ = 853248 next lower in Table XXVI. Norie.

Tab. diff. = 253   253)1900(7″ (to be *subtracted*).
                      1771
                      ———
                       129   Ans.: 31° 25′ 53″.

$$\begin{array}{r} 31°\ 16'\ 0'' \\ -\ \ \ \ \ \ 7 \\ \hline 31\ 25\ 33 \end{array}$$

Ex. 3. Find the angle whose natural cosine is 728713.
Proceeding according to Note, page 83.

Here nat. cosine of required angle = ·728713
Nat. cosine of next less angle, or 43° 13′ = ·728769

Tab. diff. = 334   334)5600(17″ nearly, to be *added*.
                      334
                      ———
                      2260
                      2338

∴ angle required = 43° 13′ 17″.

EXAMPLES FOR PRACTICE.

Given the nat. sines, to find the angle.

1. ·898002 3. ·8   5. ·444   7. ·740912   9. ·75214
2. ·370383 4. ·920411 6. ·20389 8. ·46? or ·529221 10. ·96

Given the nat. cosine, to find the angle.

1. ·448807 3. ·726998 5. ·514841 7. ·769388 9. ·817726   11. ·999000
2. ·948397 4. ·702017 6. ·914237 8. ·974822 10. ·215515 12. ·6

TABLES OF LOGARITHMS OF TRIGONOMETRICAL RATIOS.

116. **The Trigonometrical Ratios** being numbers, have logarithms that correspond to them. In practice the logarithmic are generally far more useful than the natural sines, &c., though the latter are often necessary, or in some simple kinds of calculation, preferable.

117. As the sines and cosines of all angles, and the tangents of angles less than 45°, are less than radius or unity, being proper fractions, the logarithms of the value of these quantities, properly, have negative characteristics. In order to avoid the inconvenience of printing negative logarithms, and for other reasons, 10 *is added to* the characteristic before it is registered in the table of logarithmic sines, &c., so that we find the characteristic 9 instead of $\bar{1}$, 8 instead of $\bar{2}$, &c.

Thus, on referring to the Table of Natural Sines (Table XXVI, Norie), we find natural sine of 16° = ·275637. If we calculate the logarithm of ·275637, we find its value is $\bar{1}$·440338; if to this 10 is added we find that

Log. sine 16° = 9·440338.

To preserve uniformity, the characteristics of the logarithms of all the other ratios, namely, of the log. tangents, cotangents, secants, and cosecants are increased by 10. In trigonometrical operations this is convenient, but principally because the extraction of roots very seldom occurs.

It may be observed here that the uniform addition of 10 to the characteristic gives the logarithm of 10000 *million* times the natural number.

Thus, 9·599317 is the log. of 3979486000, and this latter number is the natural sine corresponding to a radius of 10000 *millions*, instead of a radius of unity.

118. **Usual arrangement of Tables of Logarithmic Sines, Cosines, &c.**—The table of logarithmic sines, cosines, tangents, cotangents, secants, and cosecants, contain all arcs from zero (0°) through all magnitudes up to 90°, the log. of radius, as just stated, being 10. At the top of the page is placed the number of degrees, and in the left-hand column each minute of the degree, opposite to which are arranged the numerical values of the log. sine, cosine, &c., of the corresponding angle in those columns, at the *top* of which those terms are placed. The headings of the columns run along the *top*, thus, as far as 44°. The degrees from 45° to 90° are placed at the bottom of the page, and the minutes of the degree arranged in a right-hand column, so that the angles read off on the right-hand side are complemental to those read off at the points exactly opposite on the left-hand side, the values of the sines, cosines, tangents, &c., being found in the columns at the *bottom* of which those terms are found. This arrangement is rendered practicable by the circumstance of every angle between 45° and 90° being the complement of another between 45° and 0°, every sine of an angle less than 45° is the cosine of another greater than 45°, every tangent is a cotangent, &c.; the sines, tangents, &c., of angles being respectively equal to the cosines, cotangents, &c., of the complements of the same angle.

The following shows the usual arrangement of such tables:—

M	Sin.	D.	Cosec.	Tan.	D.	Cot.	Sec.	D.	Cos.	M
M	Cos.	D.	Sec.	Cot.	D.	Tan.	Cosec.	D.	Sin.	M

Besides the columns headed "sine, tangent," &c., are three smaller columns headed "Diff." They contain, in most tables, the differences between the values of the consecutive logarithms in the contiguous columns on either side, but corresponding to a change of 100″ in the arc (not the difference corresponding to 60″ of arc or angle); and it must be kept in mind that the same difference is common to the sine and cosecant, to the tangent and cotangent, and to the secant and cosine. They are inserted for the convenience of finding the values of the sines and cosines, &c., of angles which are expressed in degrees, minutes, and seconds.

119. The above, as just stated, is the usual arrangement of most tables, but in the earlier editions of NORIE and some other works the arrangement is somewhat different.

The columns are arranged thus:—

M	Sine.	Diff.	Cosine.	Diff.	Tangent.	Diff.	Cotangent	Secant.	Cosecant.	M
M	Cosine.	Diff.	Sine.	Diff.	Cotangent	Diff.	Tangent.	Cosecant.	Secant.	M

Since the same difference is common to the sine and cosecant, to the tangent and cotangent, in this arrangement, then, it must be particularly borne in mind, that the first "Diff." column (from the left) belongs to the first column of logarithms on the left hand of the page, and is also the "Diff." for the first column on the right of the page; that the second column of "Diff." (from the left) belongs to the second column of logarithms from either the right or left of the page; and that the third column of "Diff." belongs to the third column from either the right or the left, which may be otherwise expressed, thus:—A cosecant takes a sine "diff."; a secant takes a cosine "diff."; and cotangent takes a tangent "diff."

120. **In the use of these Tables**, as in that of the natural sines, two questions present themselves:—First, having given the angle in degrees, minutes, and seconds, required the log. sine, log. cosine, &c. Second, having given the log. sine, log. cosine, &c., required the value of the angle in degrees, minutes, and seconds.

121. When an angle is presented **in degrees and minutes only**, the tabular logarithm of its sine, tangent, &c., will be found (Table XXV, NORIE, or Table 68, RAPER) simply by inspection, according to the following:—

RULE XXXVI.

1°. **If the angle or arc is less than** 45°. *Find the* page *having the given degrees at the* top, *and in the left-hand marginal column find the* minutes, *then opposite the minutes, and in the column which is marked at the* top *with the name of the ratio, will be found the logarithm sought.*

2°. **If the angle be greater than** 45°. *Look for the* page *having the given degrees at the* bottom, *and find the* minutes *in the right-hand column; the logarithm of the proposed function of the angle will be found* opposite *the* minutes *in the column* marked *at the* foot *with the* name *of the* ratio *whose logarithm is sought.*

EXAMPLES.

Ex. 1. Find the log. sine of 37° 47′.

As the arc is *less* than 45°, by looking at the *top* of the table for the degrees (37°), and in the *first* column on the left for the minutes (47′), we find in the column having at its *top* the word sine, the figures 9·787232, which is the log. sine of the arc required.

Ex. 2. Find the log. tang. of 75° 34′.

Here, as the arc is *greater* than 45°, looking at the *bottom* of the table for the degrees (75°), and in the *last* or right-hand column for the minutes (34′), we find in the column having tang. at the *bottom* 10·589431, which is the log. tangent of 75° 34′.

Log. sine of	40° 4′ is 9·808669	Log. sine of	57° 5′ is 9·924001
Log. cosine of	21 38 ,, 9·968278	Log. cosine of	79 51 ,, 9·246069
Log. tangent of	84 13 ,, 10·994466	Log. tangent of	21 50 ,, 9·602761
Log. cotangent of	55 58 ,, 9·829532	Log. cotangent of	27 45 ,, 10·278911
Log. secant of	70 20 ,, 10·472954	Log. secant of	44 59 ,, 10·150389
Log. cosecant of	8 35 ,, 10·826092	Log. cosecant of	69 54 ,, 10·027291

Examples for Practice.

Take out the logarithms of the following trigonometrical ratios.

1. Log. sine	9° 10'	7. Log. cos.	53° 28'	13. Log. sine	11° 20'
2. Log. cosec.	40 40	8. Log. sine	51 49	14. Log. cosec.	35 41
3. Log. cosine	12 48	9. Log. sec.	60 34	15. Log. cosine	23 14
4. Log. tang.	37 26	10. Log. cotang.	79 19	16. Log. sec.	47 54
5. Log. cotang.	8 25	11. Log. cosec.	45 45	17. Log. cotang.	70 39
6. Log. sec.	43 1	12. Log. sine	53 56	18. Log. sine	57 12

122. If the value of the angle be given in **degrees, minutes, and seconds**, we proceed by

RULE XXXVII.

1°. *Find from the table the sine, tangent, secant, cosine, &c., which corresponds to the degrees and minutes;* also *take out the number in the contiguous column headed* "Diff." *on the same line* (See Nos. 118 and 119, page 85.)

2°. *Multiply the tabular difference* ("Diff.") *by the seconds, reject the last* two *figures (always two) of the product for the division by 100, and the remaining figures will furnish the proper correction for seconds.*

NOTE 1.—If the value of the two figures cut off is not less than fifty, one must be added to the first right-hand figure left.

3°. *If the required quantity be a sine, tangent, or secant, add the result to the last figures obtained in 1°; if it be a cosine, cotangent, or cosecant, subtract.**

The result will be the required sine, tangent, secant, cosine, &c.

NOTE 2.—The process above is sufficiently accurate, unless for the sines and tangents of very small angles, and for the tangents and secants of angles very near 90°. When an angle of degrees, minutes, and seconds, and of less magnitude than 3°, occurs in calculation, neither the logarithmic sine nor the logarithmic tangent will be found very accurately from the ordinary Tables. In some books, as HUTTON's "Mathematical Tables," a special Table is given, containing the logarithmic sines and tangents to every second in the first two degrees of the quadrant. By that Table we should find the correct log. tang. of 1° 25' 45" to be $\bar{2}$·3970503, whereas, by using the tab. diff. for 1° 25' and 1° 26' in the ordinary Table, we should get the less accurate result, $\bar{2}$·3970448, because for such small angles, the successive tabular differences for one minute shows too rapidly a wide departure from equality. When an angle of degrees, minutes, and seconds, and within less than 3° of 90° occurs in calculation, we cannot, for the reason just stated, obtain very accurately from the ordinary Tables either the logarithmic or the natural tangent. Thus, the true log. tang. of 88° 4' 15" is 1·6029497; but by the ordinary Tables we would get for the last three figures 552. NORIE gives the log. sin. and log. tang. to every *ten* seconds of the first two degrees of the quadrant, and RAPER gives the log. sines to every second up to 1° 30', and to every ten seconds up to 4° 30'.

---

* In some Tables these differences are those due to 1 minute, or 60 seconds, and are got by simply subtracting the greater of the logarithms from the less. The difference d, due to any smaller number (a) of seconds is found from such Tables by the proportion 60 : a :: D : d, so that $d = \frac{Da}{60}$. But as before observed the differences usually given in the Tables are those due not to 60 seconds but to 100 seconds, so that in these Tables, $d = \frac{Da}{100}$; and thus d is found somewhat more readily.

EXAMPLES.

Ex. 1. Find the log. sine of 6° 36′ 27″.

Here the given number of degrees (6°) being less than 45°, look for them in the head line at the top of the page, turning over the leaves till the proper page is found, then in that page look in the second line for the name of the column wanted, viz., the sine; and in the left-hand vertical column marked M at the top, find the number of minutes (36′); having found the minutes, then in the same line and under sine is found 9·060460, which is the log. sine corresponding to 6° 36′. Now this log. being found in the first column on the left, the tabular difference must be taken out of the first "diff." column from the left. It will be noticed that there is no diff. exactly opposite to 36′, but between 36′ and 37′ will be found the diff. 1817, which multiplied by the seconds (27″) gives 49059, and rejecting the two last figures from this product (for the division by 100) gives quotient 490, which being increased by 1, since the figures cut off exceed 50 (see Note 1, page 87) gives 491 as the correction of the logarithm for the seconds. The work will stand thus:—

$$\begin{array}{ll} \text{Log. sine } 6° 36' = 9·060460 & \text{Tab. diff. } 1817 \\ 27'' \text{ gives } + 491 & \phantom{\text{Tab. diff. }} \times 27 \\ \hline \phantom{\text{Log. sine } 6° 36' = } 9·060951 & \phantom{\text{Tab. diff. }} 12719 \\ & \phantom{\text{Tab. diff. }} 3634 \\ & \phantom{\text{Tab. diff. }} \overline{} \\ & \phantom{\text{Tab. diff. }} 490,59 = 491 \end{array}$$

Ans.: Log. sine 6° 36′ 27″ = 9·060951.

Ex. 2. Find the log. cosine of 13° 5′ 32″.

The log. cosine of 13° 5′ is 9·988578, and the tabular difference corresponding to the log. cosine of the given degrees and minutes is 50; this being multiplied by 32 (the given number of seconds), and pointing off *two figures* to the right, is 16 to be *subtracted*, because the cosine is a *decreasing* log.; therefore—

$$\begin{array}{ll} \text{Log. cosine } 13° 5' = 9·988578 & \text{Tab. diff. } 50 \\ 32' \text{ gives } - 16 & \phantom{\text{Tab. diff. }} \times 32 \\ \hline \phantom{\text{Log. cosine } 13° 5' = } 9·988562 & \phantom{\text{Tab. diff. }} 100 \\ & \phantom{\text{Tab. diff. }} 150 \\ & \phantom{\text{Tab. diff. }} \overline{} \\ & \phantom{\text{Tab. diff. }} 16,00 \\ & \phantom{\text{Tab. diff. }} \text{or } 16 \end{array}$$

Ans.: Log. cosine 13° 5′ 32″ = 9·988562

The parts for the seconds are subtracted in this instance, *being a colog*. (See Rule XXXVII, 3°.

Ex. 3. Find the log. tangent of 72° 59′ 8″.

The log. tangent of 72° 59′ is 10·514209, and the tab. diff. corresponding to the given degrees and minutes is 753; this being multiplied by 8 (the number of seconds), and pointing off *two figures* to the right is 60, which is additive; thus:—

$$\begin{array}{ll} \text{Log. tang. } 72° 59' 0'' = 10·514209 & \text{Tab. diff. } 753 \\ \text{Parts for } 8'' = + 60 & \phantom{\text{Tab. diff. }} 8 \\ \hline \text{Log. tang. } 72° 59' 8'' = 10·514269 & \phantom{\text{Tab. diff. }} 60,24 \end{array}$$

Ex. 4. Find the log. cotangent of 73° 21′ 7″.

The log. cotangent of 73° 21′ is 9·475763, and the tab. diff. corresponding to the cotangent of the given degrees and minutes is 767: this being multiplied by 7 (the given number of seconds), and pointing off *two figures* to the right is 54, which is to be *subtracted* in this instance, being a colog.

$$\begin{array}{ll} \text{Log. cotang. } 73° 21' 0'' = 9·475763 & \dfrac{(\text{Tab. diff. } 767) \times 7}{100} = 53,69 \\ \text{Parts for } 7'' = - 54 & \text{or } 54 \\ \hline \text{Log. cotang. } 73° 21' 7'' = 9·475709 \end{array}$$

The parts for the seconds are subtracted in this instance, *being a colog*. (See Rule XXXVII, 3°.

Ex. 5. Take out log. sine 1° 5′ 34″.

Here the angle whose log. sine is sought being less than 2°, it must, therefore, be taken out of the special part of the Table (see Table XXV, page 107, NORIE). The next less angle to be found in the Table is 1° 5′ 30″, the log. sine of which 8·279941, and the corresponding tabular "Diff." (for 10″ in this part of the Table) is 1104, which multiplied by 4, the seconds over 30, gives 4416, and cutting off one figure from the right, for the division by 10, gives the correction 442, to be added to the logarithm taken out of the Table; thus the work stands as follows:—

Log. sine	1° 5′ 30″ = 8·279941		Tab. diff.	1104
Parts for	4 = + 442			4
Log. sine	1 5 34 = 8·280383			441,6
				or 442 nearly.

Ex. 6. Required the cosecant of 3° 7′ 21″.

Log. cosecant	3° 7′ 0″ = 11·264646		Tab. diff.	3857
Parts for	21 = − 810			21
Log. cosecant	3 7 21 = 11·263836			3857
				7714
				809,97

Ex. 7. Take out the cosine of 88° 20′ 46″. (See Table XXV, NORIE, page 108). In the special part of the Table, at the bottom part, we get 88° 20′ 40″ (the next less angle), and the cosine opposite is 8·460761, and the corresponding tabular "Diff." (for 10″ in this part of the Table) is 729, which multiplied by 6, the seconds over 40′, gives 4374, and cutting off one figure from the right, for the division by 10, gives the correction 437 to be subtracted from the logarithm taken out of the Table; thus the work stands as follows:—

Log. cosine	88° 20′ 40″ = 8·460761		Tab. diff.	729
Parts for	6 = − 437			6
Log. cosine	88 20 46 = 8·460324			437,4
				or 437

123. For the functions of an angle between 90° and 180° we may take the same functions of its supplement; hence,

To find the logarithm of a trigonometrical ratio of an angle greater than 90°, we have the following

RULE XXXVIII.

Subtract the angle from 180° and look for the remainder, which is called its supplement in the Tables.

EXAMPLES.

Ex. 1. Find the log. sine of 110° 24′. Subtract it from 180°.

From	180° 00′	
Subtract	110 24	
Remainder	69 36	(Supplement).

Look for the log. sine of remainder (namely 69° 36′), which is 9·971870; or log. sine 110° 24′ = 9·971870.

Ex. 2. Find the log. secant of 95° 43′; also the log. cosecant of the same.

Subtracting 95° 43′ from 180° 0′ gives remainder 84° 17′, and look for the log. secant of 84° 17′, which is 11·001701; ∴ log. secant of 95° 43′ is 11·001701.

Again, look for the cosecant of 84° 17′, which is 10·002165; ∴ log. cosecant of 95° 43′ is 10·002165.

Ex. 3. Find the log. tangent of 128° 55′ 47″.

From 180° 0′ 0″ Subtract 128° 55′ 47″ (Remainder) = 51° 4′ 15″ (Supplement).

∴ Supplement of the given angle = 51° 4′ 13″.

Log. tangent 51° 4′ 0″ = 10·092664		Tab. diff. 431
Parts for 13 = + 56		× 13
Log. tangent 51 4 13 = 10·092720		1293
		431
		56,03
		or 56

∴ Log. tangent 128° 55′ 47″ = 10·092720. *Ans.*

124. But a *readier* way, and the better practical method, is to proceed as follows :—

RULE XXXIX.

Diminish the given angle by 90°, *and look out the remainder in the tables, observing that if the trigonometrical ratio have* "co" *prefixed to it drop the* "co," *but if it have not* "co," *prefix it, then find the logarithm corresponding to the new ratio and angle. Or,*

If A *denote any angle less than* 90°, *then*

For sine ....(90 + A) take out........ cosine A
„ tangent......(90 + A) ——— ......cotangent A
„ secant ......(90 + A) ——— ...... cosecant A
„ cosine ......(90 + A) ——— ...... sine A
„ cosecant ....(90 + A) ——— .... secant A
„ cotangent ....(90 + A) ——— ...... tangent A

Obs.—This rule may easily be remembered by observing that to the sine, tangent, and secant, *co* is prefixed, while from the cosine, cosecant, and cotangent, the *co* is dropped, and in each case the excess of 90° of the angle is used.

EXAMPLES.

Ex. 1. Find the log. cosine of 110°.

To find the log. cosine of 110°, or log. cosine (90 + 20), take out the log. sine 20°, which is 9·534052.

Ex. 2. To find the log. secant of 160° 12′, take out the cosecant 70° 12′, which is 10·026465.

Log. cosine of	143° 24′	= Log. sine	53° 24′	is	9·904617
Log. cosecant of	99 37	= Log. secant	9 37	„	10·006146
Log. sine of	109 2	= Log. cosine	19 2	„	9·975583

Ex. 3. Find the log. cosecant of 131° 45' 19".

Subtracting 90° from 131° 45' 19" = 41° 45' 19".

Log. cosecant 131° 45' 19" = log. secant 41° 45' 19".

Log. secant	41° 45' 0" = 10·127228		Tab. diff.	188
Parts for	+ 19 = + 36		×	19
Log. secant	41 45 19 = 10·127264			1692
				188
				35,72

In this instance "co" is prefixed to the given trigonometrical ratio, then, according to rule, "co" is dropped, and the log. corresponding to the new ratio is taken out for the remainder resulting from the given angle when diminished by 90°.

Ex. 4. Required the log. tangent 99° 32' 58".

Log. tangent 99° 32' 58" = cotangent 9° 32' 58".

Log. cotangent	9° 32' 0" = 10·774844		Tab. diff.	1288
Parts for	+ 58 = − 747			58
Log. cotangent	9 32 58 = 10·774097			10304
				6440
				747,04

125. In RAPER the required logs. are given to every half minute, and, therefore, the required log. is to be taken out to the nearest less arc to that given; and adjoining it will be seen a column of "Parts," from which the correction for the remaining seconds is to be taken, and this correction is to be *added* if the log. taken out be a sine, tangent, or secant, but *subtracted* if it be a cosine, cosecant, or cotangent (that is, if the log. have *co* prefixed). See Note 2, page 82.

In most cases the parts are given in the adjoining column for every second up to 30", but when the angle is small, or large, some are given for each second up to 10". In this case, if the parts for a larger number of odd seconds than 10 are required, take them out in instalments. When the angle is very small, or very large, the parts are not calculated at all, but the difference for a half minute is given opposite to each logarithm. In this case multiply the given difference by the number of the odd seconds and divide by 30. The result will be the parts required.

EXAMPLES.

Ex. 1. Find log. sine 37° 19' 51".

Log. sine	37° 19' 30" =	9·782713
Diff. for	21 =	+ 58
Log. sine	37 19 51 =	9·782771

Ex. 2. Find log. cotang. 64° 53' 39".

Log. cotang.	64° 53' 30" =	9·670813
Diff. for	9 =	− 49
Log. cotang.	64 53 39 =	9·670764

Ex. 3. Find log. tang. 8° 32' 18".

Log. tang.	8° 32' 0" =	9·176224
Diff. for	10 =	143
	8 =	115
Log. tang.	8 32 18 =	9·176482

Ex. 4. Find log. cosine 83° 41′ 57″.

Log. cosine	83° 41′ 30″ =	9·040915	
Diff. for	27 =	− 515	
Log. cosine	83 41 57 =	9·040399	

Diff. for 10″ = 191
Diff. for 10 = 191
Diff. for 7 = 133
─────────────
Diff. for 27 = 515

Ex. 5. Find log. cosec. 3° 7′ 21″.

Log. cosec.	3° 7′ 0″ =	11·264646
Diff. for	21 =	812
Log. cosec.	3 7 21 =	11·263834

Diff. for 30″ = 1160
21
─────
1160
2320
─────
3,0)24360
─────
Diff. for 21″ = 812

126. But for the purpose of lessening the labour of finding the log. sines and cosines in the case where the parts are not given, two other tables have been constructed (Tables 66 and 67).

Table 66 gives the logarithms of sines of small angles from 0° to 1° 30′, and the logarithms of cosines of large angles from 88° 30′ to 90° to each second. Table 67 gives the logarithms of sines of angles from 1° 30′ to 4° 31′ and the log. cosines of angles from 85° 29′ to 88° 30′ to every ten seconds. In Table 67 there are columns of "parts" by which the log. of the sine or cosine of an angle containing odd seconds may be found, and conversely. Each page of this Table is divided into six spaces by horizontal lines, and there are two columns of parts in each space. The left-hand column of parts is to be used when the angle is in the upper half of the space, and the right-hand column when it is in the lower.

EXAMPLES.

Ex. 1. Take out log. sine 1° 5′ 34″.

Log. sine 1° 5′ 34″ = 8·280383. *Ans.*

Ex. 2. Take out log. cosine 88° 43′ 57″.

Log. cosine 88° 43′ 57″ = 8·344790. *Ans.*

Ex. 3. Find log. sine 4° 26′ 18″.

Log. sine 4° 26′ 10″ =	8·888446	
Diff. for	8 =	216
Log. sine 4 26 18 =	8·888662	

In this instance the parts are taken from the right-hand column, because the angle, 4° 26′ 10″, is situated in the lower half of the space.

EXAMPLES FOR PRACTICE.

Required the log. sine, tangent, secant, cosine, cotangent, and cosecant corresponding to the following arcs:—

1. 6° 53′ 56″
2. 29 9 30
3. 37 49 14
4. 56° 54′ 17″
5. 10 10 6
6. 70 47 40
7. 1° 49′ 47″
8. 87 28 45
9. 1 0 40
10. 115° 34′ 41″
11. 119 40 48
12. 101 40 19

Take out of the Table the following:—

13.	Log. sine	2° 40′ 10″	22.	Log. cosec.	127° 30′ 40″
14.	Log. sine	170 30 39	23.	Log. cosec.	141 16 51
15.	Log. sine	1 49 47	24.	Log. sine	2 0 53*
16.	Log. cosine	89 59 19	25.	Log. cotang.	89 23 37
17.	Log. cosine	88 40 56	26.	Log. cosine	87 23 27
18.	Log. cosine	108 40 60	27.	Log. cosine	88 50 29
19.	Log. tang.	1 8 7	28.	Log. tang.	1 2 18
20.	Log. cotang.	3 7 8	29.	Log. sec.	101 8 7
21.	Log. tang.	114 9 30	30.	Log. sine	110 11 18

127. If the value of the log. sine, log. cosine, &c., *i.e.*, the logarithm of a trigonometrical ratio, be given, and it is required to find the corresponding angle in degrees and minutes, we use

RULE XL.

Look for the logarithm in the several columns of the table marked at the top or bottom with the name of the given trigonometrical ratio, which being found exactly or the nearest, whether larger or smaller, to it, will give the degrees and minutes answering to the given logarithm, being careful to observe that when the name of the given ratio is found at the top of the table, then the degrees are to be taken from the top and the minutes from the left-hand marginal column; but if the name of the ratio is found at the bottom of the table, take the degrees from the bottom and the minutes from the right-hand side of the page.

NOTE.—In using the Table *inversely*, for example, in searching for the angle which has 9·611294 for the logarithm of its sine, the student must not distinguish sine from cosine, nor tangent from cotangent, but must consider *sines* and *cosines* as one table, *tangents* and *cotangents* as one table, and must cast an eye on both, and get to 9·611294 as fast as he can. For want of this caution some beginners will turn over page after page until they come to 45°, and then back again to the very page that was first opened.

EXAMPLE.

Ex. 1. Required the angle corresponding to the log. sine 9·729223.

In page 142, Table XXV, No. 12, under the word "Sine," and opposite 25′ in left-hand marginal column, are the exact figures, the degree (being sought at the head of the page, because the column in which the figures are found is named at the head) is 32°; therefore, the angle is 32° 25′.

If the angles for the cosine of the same logarithm be required, the degrees are found at the bottom, and the minutes in the right-hand column, and is 57° 35′ accordingly.

Log. sine	9·731009 = 32° 34′		Log. cosine	9·995555 = 8° 11′
Log. sine	9·871073 = 48 0		Log. cosec.	10·030580 = 68 45
Log. tang.	9·787036 = 31 29		Log. cotang.	10·508820 = 17 13
Log. tang.	10·047850 = 48 9		Log. cosine	9·718497 = 58 28
Log. sec.	10·043673 = 25 16		Log. cosec.	10·307885 = 29 29
Log. sec.	10·566325 = 74 15		Log. cotang.	11·197235 = 3 38

* When the tabular difference is considerable, as in this instance, the log. is easier reduced from the log. of the nearest minute.

128. If the value of the log. sine, log. cosine, &c., **be given**, and it is required to find the corresponding angle, in degrees, minutes, and seconds, we use

RULE XLI.

1°. *Find in the Tables* (XXV, NORIE) *the next lower log. sine, log. cosine, &c., and note the corresponding degrees and minutes; also, take the number from the corresponding part of the adjoining column of " Diff."*

2°. *Subtract this from the given log. sine, log. cosine, &c., multiply the difference by* 100, i.e., *annex two cyphers, divide by the tabular difference, and consider the result as seconds.*

3°. *If the given value be that of a log. sine, log. tangent, or log. secant,* add *these seconds to the degrees and minutes found in* 1°; *if it be that of a log. cosine, log. tangent, or log. cosecant,* i.e., *if the log. have* co *prefixed,* subtract.

The result will be the required angle.

NOTE.—If the given log. be a *cosine, cosecant,* or *cotangent,* we may seek out the next *greater* to the given log.: then proceed by 2° to find the seconds, which *add* to the degrees and minutes as found by 1°.*

EXAMPLES.

Ex. 1. Given log. sine = 9·422195 (or 1̄·422195): find the angle.

We take out 9·421857, the log. sine of 15° 19', as it is the logarithm *next less* than the given one, which we take, as the logarithms in the columns increase with the angle. The difference of these logarithms is 338, and if two cyphers be affixed to the difference, and the number then divided by 768, taken from the column of Diff. in the Table, we have 44 for the number of seconds to be *added* to the degrees and minutes before taken out. The work will stand thus:—

           Given log. sine   9·422195
           Tab. log. sine next less 9·421857 = log. sine 15° 19'

       Tab. diff. for 100" = 768)·33800(44" additional seconds.
                                   3072
                                   ─────
                                   3080
                                   3072

           Therefore 9·422195 = log. sine of 15° 19' 44".

Ex. 2. Given log. cosine = 9·873242 (or 1̄·873242): find the angle.

Here we take out 9·873223, the log. cosine of 41° 41', as it is the log. cosine in the Table next less than 9·873242. The difference between these two logarithms is 19; and if two cyphers be affixed to the difference we get 1900; whence 1900 divided by 187, the number from the column of " Diff." gives 10 for the number of seconds to be subtracted. Hence the required angle is 41° 40' 50". The work will stand thus:—

    Given log. cosine   9·873242
    Tab. log. cosine next less 9·873223 = log. cosine 41° 40'   19 × 100
                               ────────   ──────── = 10" *subtractive.*
                                   19   187
           Tab. diff. for 100" = 187.
           Therefore, 9·873242 = log. cosine of 41° 40' 50"

───────────────────────

* By writing down the log. corresponding to the *next less angle in every case* the seconds are always *additive,* thus avoiding confusion as to their mode of application. (See Note, Rule XXXV, page 83.)

Proceeding according to Note, Rule XXXV, page 83.

    Given log. cosine   9·873242
    Tab. log. cosine next greater 9·873335 = 41° 40' (next less angle).

        Tab. diff. for 100" = 187)9300(50", nearly, *additive*.

          ∴ angle required = 41° 40' 50".

EXAMPLES FOR PRACTICE.

Required the Angles (to the nearest second), the Log. Sine of which is:—

1.	9·741279	4.	8·600700	7.	9·500000	10.	7·456430	13.	9·900000
2.	9·926100	5.	9·518317	8.	9·800000	11.	9·999631	14.	8·846217
3.	8·707654	6.	9·929638	9.	9·909176	12.	9·974538	15.	8·462167

Find the Arc to the Log. Cosine of

1.	9·787140	4.	9·995637	7.	9·517232	10.	9·932338	13.	7·799520
2.	9·750333	5.	9·179726	8.	9·212036	11.	9·998970	14.	9·000000
3.	8·134758	6.	9·273216	9.	8·361861	12.	8·281485	15.	9·013628

Find the Arc to the Log. Secant of

1.	10·013839	3.	10·000765	5.	10·746129	7.	10·022719	9.	10·315400
2.	10·205665	4.	10·048398	6.	11·005231	8.	11·642535	10.	11·200000

Find the Arc to the Log. Cosecant of

1.	10·347194	4.	10·974476	7.	11·000873	10.	10·070362	13.	10·009000
2.	10·252108	5.	10·121000	8.	11·467931	11.	10·900000	14.	10·061462
3.	11·005231	6.	11·442539	9.	11·166007	12.	11·079003	15.	11·290123

Find the Arc to the Log. Tangent of

1.	10·636863	4.	10·827204	7.	11·276400	10.	9·642876	13.	10·060431
2.	10·000100	5.	10·150328	8.	8·297036	11.	9·846175	14.	8·668612
3.	10·287342	6.	8·961007	9.	9·716135	12.	11·281456	15.	8·258262

Find the Arc to the Log. Cotangent of

1.	9·742961	4.	10·060431	7.	8·327691	10.	9·100100	13.	8·460000
2.	10·876432	5.	10·710880	8.	10·010101	11.	10·825001	14.	9·374611
3.	10·287632	6.	11·197568	9.	8·781464	12.	8·272775	15.	12·069844

MISCELLANEOUS.

1. If sine A = ·432651, find log. sine A.
2. If tang. A = 3, find log. tang. A.
3. Given log. cos. A = 9·236713, find nat. cos. A.
4. Given log. tang. 35° 20' = 9·850593, find log. cotang. 35° 20' without using any tables at all.
5. Find the log. cosec. 68° 45' 24" from the table of natural sines only.
6. Given log. sec. A = 11·024680, find nat. cos. A.
7. Given log. cosine A = 9·450981, find A (1) from a table of log. cosines, and (2) from a table of nat. cosines.
8. Given nat. sec. A = 2·005263, find A (1) from a table of nat. sines and cosines, and (2) secant from a table of log. secants.
9. Sine 36° × tang. 54° = ·654.
10. Find by the tables the angle whose sine is $\sqrt{\tfrac{1}{15}}$.
11. Given log. cot. A = 11·015627, find nat. cot. A.
12. Find nat. cot. 45° 18' 17" from the table of cotangents.
13. Find to the nearest second the angle whose sine is $\tfrac{1}{4}$, $\tfrac{1}{3}$, $\tfrac{1}{2}$, and $\tfrac{192}{193}$.
14.   ,,   ,,   tang. is $\tfrac{23}{24}$, $\tfrac{115}{117}$, and $\tfrac{116}{117}$.

129. It is also necessary to have a distinct conception of the limits to which the Trigonometrical Ratios tend when the angles become right-angles. The following are the Trigonometrical Ratios for the angles 0° and 90°:—

$$
\begin{aligned}
\text{Sin.} \quad 0° &= 0 & \text{Sin.} \quad 90° &= 1 \\
\text{Cos.} \quad 0° &= 1 & \text{Cos.} \quad 90° &= 0 \\
\text{Tang.} \quad 0° &= 0 & \text{Tang.} \quad 90° &= \infty \\
\text{Cot.} \quad 0° &= \infty \; * & \text{Cot.} \quad 90° &= 0 \\
\text{Sec.} \quad 0° &= 1 & \text{Sec.} \quad 90° &= \infty \\
\text{Cosec.} \quad 0° &= \infty & \text{Cosec.} \quad 90° &= 1
\end{aligned}
$$

And the following, therefore, are the Logarithms of their Trigonometrical Ratios:—

$$
\begin{aligned}
\text{Log. sin.} \quad 0° &= -\infty & \text{Log. sin.} \quad 90° &= 0 \\
\text{Log. cos.} \quad 0° &= 0 & \text{Log. cos.} \quad 90° &= -\infty \\
\text{Log. tang.} \quad 0° &= -\infty & \text{Log. tang.} \quad 90° &= \infty \\
\text{Log. cot.} \quad 0° &= \infty & \text{Log. cot.} \quad 90° &= -\infty \\
\text{Log. sec.} \quad 0° &= 0 & \text{Log. sec.} \quad 90° &= \infty \\
\text{Log. cosec.} \quad 0° &= \infty & \text{Log. cosec.} \quad 90° &= 0
\end{aligned}
$$

130. When these values occur amongst others requiring to be added to or subtracted from them, the learner must be careful to remember *that the addition to* or *subtraction from them of finite numbers cannot alter them*. Hence the explanation of the results in the following:

EXAMPLES.

Ex. 1. Add together log. cot. 0° and log. sine 20°.

$$
\begin{aligned}
\text{Log. cot.} \quad 0° &= \infty \\
\text{Log. sine} \quad 20° &= 9\cdot534032 \\
\hline
\text{Ans.} \quad &\infty
\end{aligned}
$$

Ex. 2. Add together log. cos. 90° and log. tang. 45°.

$$
\begin{aligned}
\text{Log. cos.} \quad 90° &= -\infty \\
\text{Log. tang.} \quad 45° &= 10\cdot000000 \\
\hline
\text{Ans.} \quad &\infty
\end{aligned}
$$

Ex. 3. From log. cos. 0° take log. sine 62° 48'.

$$
\begin{aligned}
\text{Log. cos.} \quad 0° &= -\infty \\
\text{Log. sine} \quad 62° 48' &= 9\cdot949105 \\
\hline
\text{Ans.} \quad &-\infty
\end{aligned}
$$

Ex. 4. From log. tang. 21° 48' 30" take log. cot. 90°.

$$
\begin{aligned}
\text{Log. tang.} \quad 21° 48' 30'' &= 9\cdot602212 \\
\text{Log. cot.} \quad 90° &= -\infty \\
\hline
\text{Ans.} \quad &\infty
\end{aligned}
$$

131. In the event of a bad or obliterated figure in the table, it may be convenient to know that the tangents are found by subtracting the cosines from the sines, adding always 10, or the radius; the cotangents are found by subtracting the tangents from 20, or the double radius, and the secants are found by subtracting the cosines from 20, the diameter of a circle whose radius is 10.

---

* This mathematical symbol is called *infinity*.

NAVIGATION.

DEFINITIONS.

132. Navigation is a general term denoting that science which treats of the determination of the place of a ship on the sea, and which furnishes the knowledge requisite for taking a vessel from one place to another. The two fundamental problems of navigation are, therefore, the finding at sea the present position of the ship, and the determining the future course.

133. The place of a ship is determined by either of two methods, which are independent of each other:—1st. By referring it to some other place, as a fixed point of land, or a previous defined place of the ship herself. 2nd. By astronomical observations.

134. It has been customary to employ the term NAVIGATION in a restricted sense to the first of these methods; the second is usually treated of under the head of NAUTICAL ASTRONOMY.

Navigation and Nautical Astronomy are the two great co-ordinate divisions of the "*Art of Sailing on the Sea*," as the old writers quaintly worded it. The first branch of the art is accomplished by means of the Mariner's Compass, which shows the *direction* of the ship's track; the Log, which, with the help of sand-glasses for measuring small intervals of time, gives the velocity or the rate of sailing, and thence the distance run in any interval; and also a Chart of appropriate construction; in short, this branch of the art relates to the directing the ship's course under the varying forces of winds and currents, and the estimation of her change of place. The second division is that branch of practical astronomy by which the situation of the observer on the globe is ascertained by a *comparison of the position of his Zenith with relation to the heavens with the known position of the Zenith of a known place* at the same moment. The principal instruments are the sextant for measuring the altitudes and taking the distances of heavenly bodies; and a chronometer to tell us the difference in time between the meridian of the ship and the first meridian; also a pre-calculated astronomical register, such as the Nautical Almanac, the Connaissance de Temps of France, &c. The solution of problems in nautical astronomy requires the use of spherical trigonometry, which is therefore characteristic of this method of navigation.

135. **A Sphere** is a solid body bounded by a surface, every point of which is equally distant from a fixed point within it; this fixed point is called the **centre**; the constant distance is called the **radius**.

Every section of a sphere by a plane is a circle.

136. **A Great Circle** of a sphere is a section of the surface by a plane which passes through its centre. **A Small Circle** of a sphere is a section of the surface by a plane which does not pass through its centre.

Or, a *great* circle is the circle of a sphere having for its centre the centre of a sphere, thus dividing the sphere into two equal parts; no greater circle can be traced upon its surface. All other circles are called *small* circles.

All great circles of a sphere have the same radius. All great circles bisect each other.

137. The **Axis** of any circle of a sphere is that diameter of the sphere which is perpendicular to the plane of the circle. The extremities of the axis are called the **poles** of the circle.

o

138. The extremities of that diameter of a sphere which is perpendicular to the plane of a circle are called the *poles* of that circle. In the case of a small circle, the poles are distinguished as the *adjacent* and *remote* pole.

All parallel circles have the same poles. The distance of every point in the circumference of a circle from either of its poles is the same. The poles of a great circle are 90° distant from every point of the circle.

139. Regarding any great circle as a *primary* circle, all great circles which pass through its poles are called its *secondaries*.

All secondaries cut their primary at right-angles.

The arc of a great circle is measured by the angle subtended by it at the centre of the sphere, which is also the same as the angle of inclination, at its pole, of two secondaries drawn through its extremities.

140. The earth is nearly a globe or sphere.

The ordinary proofs of this are of the following nature:—1st. When a vessel is seen at a considerable distance on the sea, in any part of the world, the hull is entirely or partly concealed by the water, though the masts are visible. 2nd. Ships have actually and repeatedly made the circuit of the globe; that is, by sailing from a port in a westerly direction they have returned to it in an easterly direction. 3rd. When we travel a considerable distance from north to south, a number of new stars appear, successively, in the heavens, in the quarter to which we are advancing, and many of those in the opposite quarter gradually disappear, which would not happen if the earth were a plane in that direction. 4th. In an eclipse of the moon, which is caused by the intervention of the body of the earth between the sun and moon, the shadow of the earth thrown on the moon is found in all cases, and in every position of the earth, to be a circular figure; the earth, therefore, which casts that shadow, must be a round body.

141. The earth, however, is not a perfect sphere, but of the figure of an oblate spheroid very nearly, that is, a figure traced out by an ellipse revolving round its shortest axis, being flattened in at the poles, and bulging out in a corresponding degree at the equatorial regions—the curvature being less as we recede from the equator to the poles; such a figure, in fact, as would be produced if a hoop were slightly flattened by pressure, and then made to revolve about the shortest diameter thus produced.

The shortest diameter (that which joins the poles) being 7899 statute miles, and that of the fullest parts (about the equator) being nearly 26½ more.

We can, of course, in a work like this, give no intelligible account of the refined mathematical processes by which the most probable values of the flattening in, and of the absolute dimensions have been obtained. It is sufficient to say that from a combination of the measurements of ten arcs of the meridian, BESSEL has deduced the following results:—*

Greater, or equatorial diameter . . . . . . 41,847,192 feet = 7925·604 miles.
Lesser, or polar diameter . . . . 41,707,324 „   = 7899·114 „
Difference of diameter, or polar compression . 139,768 „   =   26·471 „
Proportion of diameters, as 299·15 to 298·15.

And from the result it follows that the polar diameter is shorter than the equatorial by about $\frac{1}{300}$ (one three hundredth) part. This quantity is technically called the *compression*.†

---

* Astronomische Nachrichten, No. 438.

† The best values for its dimensions, however, appear to be those given by Capt. CLARKE.

Equatorial diameter . . . . . . 41847662 feet = 12754937 metres.
Polar axis . . . . 41707536 „   = 12712227 „

142. **The Axis of the Earth** is that diameter about which it is supposed to turn round once in twenty-four hours. The direction of this rotation is from west to east, thus causing all the heavenly bodies to have an apparent motion from east to west.

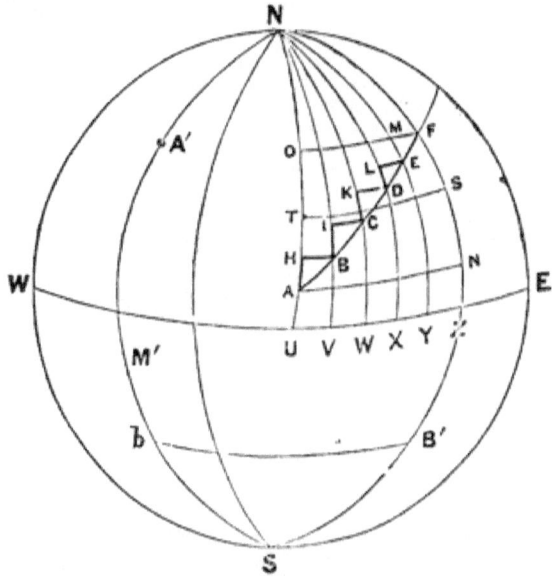

143. **Poles.**—The two extremities of the axis of the earth are called the *poles* of the earth, distinguished respectively as the *North Pole* and *South Pole* —N S (see Fig.) The former being that to which we in Europe are nearest. As they are the extremities of a diameter they are 180° apart.

144. **Equator** (from Latin *æquare*, to divide into equal parts), called also by seamen the Line, is a great circle circumscribing the earth, every point of which is equally distant from the poles, being 90° from each, as W M'E ; and dividing the globe into two equal parts called hemispheres ; that towards the north pole is called the northern hemisphere, as N W E, and the other the southern hemisphere, as S W E. (See Figure above.)

If a plane be supposed to pass through the centre of the earth at right-angles to its axis, it will intersect its surface in a great circle called the EQUATOR.

<small>The equator is chosen as the primary circle for co-ordinates. At all places on this circle the sun rises at 6ʰ A.M., and sets at 6ʰ P.M., all the year round ; the days and nights are therefore equal, being 12ʰ each.</small>

145. **The Meridian** of any place is a semi-circle passing through that place and the poles, and therefore cutting the equator at right-angles, as N M'S, N W S, N Z S. (See Figure.) The other half of the circle is called the *opposite meridian*. Every point on the surface of the earth may be conceived

to have a meridian passing through it; hence there may be as many meridians as there are points in the equator. Of all these innumerable meridians one is always selected as the *Initial Circle of Longitude*, or, as it is commonly called, the *First Meridian*. It is a matter of arbitrary choice amongst different nations; thus the first meridian with us is that of Greenwich, whilst the French refer to Paris, &c.

Meridians (*L. Meridies*, from *medius dies*, mid-day) are so called because they mark all places which mark noon at the same instant, for when any one of the meridians is exactly opposite the sun it is mid-day with all places situated on that meridian; and with the places situated on the opposite meridian it is consequently midnight. They are secondaries to the Equator, and on them Latitudes are reckoned North and South from their primitive. They also mark out all places which have the same longitude, and are hence called "*Circles of Longitude*."

Every portion of the meridian lies north and south; and places lying north and south of each other are said to be on the same meridian.

The direction of the meridian towards the north pole is called *north*, and marked N.; the opposite direction is called *south*, marked S. Directions at right-angles to the meridians are called *east* and *west*; the right hand looking to the north *east*, the left hand *west*: they are marked E. and W.

146. **Latitude** is the distance from the equator, measured in degrees (°), minutes ('), and seconds ("),* on the meridian of the place, or its angular distance from the equator measured by the arc of the meridian intercepted (cut off), between the place and the equator, or by the corresponding angle at the centre of the sphere: it is marked *north* (N.), or *south* (S.), according as the place is to the north or south of the equator. Thus the arc A' M' (Fig., page 99), is the latitude of a place A' (supposed Greenwich), and is marked N., because A' is to the north of W M' E; and the latitude of B' is M' B', and marked S., because the place B' is to the south of the equator, whilst O U, or its equal F Z, is the latitude of O, or of F.

As the latitude begins at the equator (lat. 0°), and is reckoned thence to the poles (lat. 90°), where it terminates, therefore the greatest latitude a place can have is 90°, and all other places must have their latitude intermediate between 0° and 90°.

147. **Parallels of Latitude** are small circles of the sphere parallel to the equator, that is, equidistant from it in every point, and hence all the places of the same latitude being at the same distance from the equator, are said to be on the same parallel; thus (Fig., page 99) A N, T S, O F, and *b* B' are portions of parallels of latitude, and all places on O F, and *b* B', &c., have the same latitude, being on the same parallel.

148. **Co-Latitude** is the complement of the latitude to 90°; thus the co-latitude of A' (Fig., page 99) is A N, of B' is B' S.

149. The **Difference of Latitude** (abbreviated *diff. lat.*) between two places, or of the parallels O F and T S, or of any places on these parallels, is the arc of a meridian included between their parallels of latitude, showing how far one of them is to the northward or southward of the other; thus

---

* All circles, great or small, are supposed to be divided into 360 equal parts called degrees (°) 60' (minutes) make one degree, and 60" (seconds) make one minute.

(Fig., page 99) A' b is the difference of latitude of the two places A' and B'; F S between the places F and T, or O T. The difference of latitude between two places can never exceed 180°.

The **difference of latitude of the ship** is therefore the distance made good in a north or south direction. This is also called her "*northing*" or "*southing*," these names being indicated by the initials N. and S.

150. It is evident that when two places are on the *same* side of the equator, their diff. lat. is found by subtracting the less latitude from the greater; and that when they are on *opposite* sides of the equator, that is, when one place is in north latitude and the other in south latitude, the *sum* of their latitudes is the *diff.* lat. Thus the diff. of lat. of A' and B', which is A' b, is the sum of the north lat. A' M', and of the south lat. Z B', or M' B'.

151. **Meridional Parts.**—At the equator a degree of longitude is equal to a degree of latitude; but as we approach the poles, while (upon the supposition that the earth is a sphere) the degrees of latitude remain the same, the degrees of longitude become less and less. In the chart, on Mercator's projection, the degrees of longitude are made everywhere of the same length, and, therefore, to preserve the proportion that exists at every part of the earth's surface between the degrees of latitude and the degrees of longitude, the former must be increased from their natural lengths, more and more as we recede from the equator. The lengths of small portions of the meridian thus increased, expressed in minutes of the equator, are called meridional parts; and the *meridional parts for any latitude* is the line, expressed in minutes (of the equator), into which the latitude is thus expanded. The meridional parts computed for every minute of latitude from 0° to 90°, from the *Table of Meridional Parts*, which is chiefly used for finding the meridional difference of latitude in solving problems in Mercator's sailing, and for constructing charts on Mercator's projection.

152. The **Meridional difference of Latitude** is the quantity which bears the same ratio to the difference of latitude that the difference of longitude bears to the departure. It is the projection of the difference of latitude on the Mercator's chart, and takes its name from the meridional parts, by the use of a table of which parts it is found.

153. **Middle Latitude.**—When the two places are situated on the *same* side of the equator, the middle latitude is the latitude of the parallel passing midway between them; its value is therefore half the sum of the latitudes of the two places. When the two places are situated on *opposite* sides of the equator the simple "middle latitude" is replaced by the two half latitudes of each of the places. (See RAPER's Navigation, page 98.)

154. **Longitude** is the arc of the equator intercepted between the first meridian and the meridian of the place, and is, therefore, the measure of the angle between the two meridians; thus, (Fig., page 99) take N A' M' b S as the meridian of Greenwich, then, the longitude of A', or of any place on the meridian N A' M' b S is O, and taking N U as the meridian of T, then the arc of the equator M' U, reckoned in degrees (°), minutes ('), and seconds("), or the angle M' N U, which M' U measures, is the longitude of T from M', the

meridian of Greenwich; the arc M′Z, or the angle M′NZ is the longitude of the points Z, N, S, and F, or of any place on the meridian NZ; the arc WM′, or the angle WNM′, which WM measures, is the longitude of the meridian NWS, or of any place on that meridian.

Longitude is reckoned from the first meridian, both eastward and westward, till it meets at the opposite point of the equator, therefore the longitude can never exceed 180°.

It will be evident that the latitude alone will be insufficient for the determination of the position of a place. If we state that a certain place is 45° north of the equator, it will be impossible to ascertain certainly the place in question, inasmuch as there is a circle of points on the earth, all of which are 45° north of the equator. If we suppose a circle drawn round the surface of the northern hemisphere parallel to the equator, at the distance from the equator of 45°, every point of such circle will be equally characterised by the latitude of 45°. But if we state its latitude and longitude, we can fix at once and unequivocally, the position of the place. Thus, let us suppose that its latitude is 50° north, its longitude 30° east of Greenwich; its position will be found by imagining a circle parallel to the equator, drawn upon the northern hemisphere at a distance of 50° from the equator; then supposing a meridian drawn through Greenwich intersecting this parallel, and another drawn so as to cross the equator at a point 30° east of the former; the place in question will be upon the line parallel to the equator first drawn, inasmuch as it will be 50° north of the equator, and it will also be in the meridian last drawn, inasmuch as it will be 30° east of Greenwich. Since, then, it will be at the same time upon both these lines, it will necessarily be at the point where they cross each other at the east of the standard meridian of Greenwich. The place of a ship on the apparently indefinable and trackless face of the ocean can, in this manner, be as accurately marked down and discussed as any known and visible spot on the stable land.

155. **Difference of Longitude** between two places is the arc or portion of the equator included between their meridians, or, which is the same thing, the corresponding angle at the pole. To measure, therefore, the diff. of longitude of two places, we must follow down their meridians to the equator, and then take the included portion of the equator itself. It is named East or West, according to the direction in which the ship is proceeding; thus, if we take A and F (Fig., page 99) to represent two places on the surface of the globe, the arc UZ, or the angle UNF, is diff. of long. between A and F, and is East, the arc UZ being the difference of HK, and HO is the difference of the longitudes of PK and PO, or of any two places on those meridians, and WZ, the sum of WM′, and M′Z is the difference of the longitudes of the meridians NWS and NZB′S.

156. **Horizon.**—The remote bounding circle which, to an eye elevated above the surface of the ocean, appears to unite sea and sky, is called the *visible* or *sea-horizon*. A plane conceived to touch the surface of the earth at any place, and to be extended to the heavens, is called the *sensible horizon* of that place. And a plane parallel to this, but passing through the centre of the earth, is called the *rational horizon* of that place.

157. When a ship, in sailing from one place to another, preserves the same angle with the meridians, as she crosses them in succession, she is said to sail on a **Rhumb Line**. Thus, a ship in sailing from A to F (Fig., page 99) is supposed to describe on the sea a curve AF, which cuts the meridians NA, NB, NC, &c., at the same angle; that is, the angles NAF, NBF, NCF, are supposed to be equal. The rhumb line coincides with the meridian when

the course is due N. or S., or with a parallel of latitude when the course is due E. or W. On any other course but these the rhumb line is a spiral, approaching nearer and nearer to one of the poles at every convolution, but never reaching it.

Such a curve is appropriately called the *Equiangular Spiral*, and the *Loxodromic Curve*; and also because, in sailing on it, we keep on the same rhumb or point of the compass, it is called the rhumb curve. That such a curve may be drawn through any two given points will appear from this consideration,—that from one of the points an infinite number of these curves can be drawn, making different angles with the meridian, and on some one of these the second point must lie. It is evident also that only one of these curves can pass through the two points. It is the track used ordinarily in navigation, for when out of sight of land the compass determines the ship's track, and hence the selection of that track which makes a constant angle with the meridian, the advantages of such a selection being that the seaman is not required to alter his course. It would seem desirable to take the shortest route on the voyage, and this is the arc of a great circle; but the great circle drawn between two places—except it happens to be on the equator, or a meridian itself—cuts successive meridians at different angles, as a little consideration will show. When in sight of his port the compass is no longer needed, and the rhumb line is given up, and the port is made for on the great circle. When accurately following the compass course, we are, in strictness, only approximating, though very closely approximating, to a rhumb line, on account of the continuous change in the variation, due to the magnetic pole and the pole of the earth not being coincident.

158. **The Course**, from one point of the earth's surface to another, is the constant angle which the rhumb curve, joining the two points, makes with the meridians, or it is the direction in which a ship sails from one place to another, this direction being referred to the meridian, which lies truly north or south, or to the north or south line of the compass by which the ship is steered. The former is distinguished as the *True Course*, the latter as the *Compass Course*.

The course *steered* is the angle between the meridian and the ship's head.

The course *made good* is the angle between the meridian and the ship's real track on the surface of the sphere.

The course is reckoned from the north towards the east or west, when the ship's head is less than eight points from the north; and similarly from the south point.

The course is measured in *points* of $11° 15'$ each, or in degrees and minutes.

159. **The Distance** between two places is the arc of the rhumb line joining them, expressed in nautical miles of 60 to the degree of latitude. Thus (Fig., page 99) the length of line A F, expressed in minutes of a great circle of the earth, is called the *distance*.*

It must never be lost sight of that the distance is not necessarily nor generally the *shortest* distance between the two places, that is, the distance as the "crow flies." On a Mercator's Chart the rhumb curve is represented by a straight line, but it must be borne in mind, that equal parts of any such line do not represent equal distances on the earth.

The Meridian Distance between two places is the arc of a parallel of latitude between them.

---

* Minutes of a great circle are usually called *nautical miles*, or simply miles.

160. **Departure** is the sum of all the intermediate meridian distances made in going from one place to another, computed on the supposition that the distance is divided into indefinitely small equal parts. It is the distance, in nautical miles, made good towards the east or west, and such departure is expressed in *miles*, and not like the longitude, in *arc*. When the two places are on the same parallel, the departure is identical with the distance. When the places do not differ much in latitude, and are on the same side of the equator, an approximation to the departure is found in the arc of the parallel of middle latitude included between the meridians of the two places.

If the subjoined right-angled plane triangles be taken to illustrate the terms defined above (Nos. 158 to 160), A B will represent the distance sailed, that is, the length of A F on the globe (see Fig. 1); A C drawn N. and S., or in the meridian, shows the angle C A B, the *course;* A C will represent A O (Fig. 1), while B C drawn E. and W. will represent the sum of the small departures H B, I C, K D, &c., from the successive meridians which it crosses.

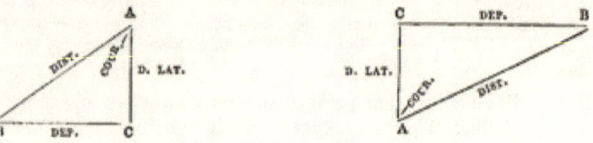

161. If a ship's course be due north or south, she sails on a meridian, and therefore makes no departure; hence the distance sailed will be equal to the difference of latitude.

If a ship sails either due east or due west, she sails on a parallel of latitude; in which case she makes *no difference of latitude*, and the departure is identical with the distance.

When the course is 4 points, or 45 degrees, the difference of latitude and departure are equal.

When the course is *less* than 4 points, or 45 degrees, the difference of latitude exceeds the departure; but when it is *more* than 4 points, or 45 degrees, the departure exceeds the difference of latitude.

162. **Magnetic Course** is the angle which the ship's track makes with the magnetic meridian; such an angle can only be shown by a compass not affected with deviation; but since the compasses of all iron ships have more or less deviation, and any course steered by such compass is magnetic in a certain sense, it has been deemed necessary to distinguish these when *corrected for deviation* as *correct magnetic courses.*

163. **Compass Course** is the angle which the track of the ship makes with the north and south line of the compass card; such a course is affected with deviation and variation; applying the deviation the result is the correct magnetic course; applying both the deviation and the variation it becomes the true course.

164. **The True Bearing*** of an object or place is the angle contained between the meridian and the direction of the object, and is the same thing as the course towards it. It is thus qualified to distinguish it from the "Compass" and "Correct Magnetic Bearing."

165. **Correct Magnetic Bearing.**—The "correct magnetic bearing" of an object is the angle which its direction makes with the magnetic meridian; such an angle can only be found by a compass not affected with deviation. It is the bearing observed with the azimuth compass after being corrected for local deviation.

A magnetic bearing, as given in "sailing directions" and on charts, is the *correct* magnetic bearing—in respect to a compass affected with deviation.

166. **Compass Bearing.**—The bearing of an object as taken by the compass. It is the angle between the direction of the needle of the standard compass on board the ship of the observer and the direction of the object; it is therefore affected by the deviation and variation of the compass; but the deviation to be applied in this case is that due to the azimuth (direction) of the ship's head, *not that on the point of bearing:* when this correction is applied it becomes the *correct* magnetic bearing, and if, further, the correction for variation be applied, the true bearing or azimuth is deduced; E. deviation and variation to the right, W. deviation and variation to the left.

Taking a bearing of an object is called setting it.

The bearings of two objects taken from the same place constitute **Cross Bearings**, the lines of direction of the two objects intersecting or crossing each other at the place of the observer.

167. The **Tropics** of **Cancer** and **Capricorn** are the parallels of latitude 23° 28′ N. and S. The Sun is vertical at noon twice in the year to every place between the tropics, and never to any place outside of them.

The space between the tropics is called the *Torrid Zone*.

168. The parallel of latitude which is 23° 28′ from the north pole is called the *Arctic Circle;* and that which is at the same distance from the South pole is called the *Antarctic Circle*. Within these circles the sun does not set during part of the summer, nor rise during part of the winter.

The spaces within these circles are called the *Frigid Zones*.

The spaces between the tropics and the polar circles are called the *Temperate Zones*.

---

* Before the introduction of iron in such large quantities into the construction and equipment of steamers and iron sailing ships, bearings and courses were deemed to be sufficiently well defined when spoken of as true and magnetic—the latter qualifying term being used simply to indicate *the direction by compass as affected by variation only*, according to the locality. But on board an iron ship compass bearings and compass courses though magnetic—inasmuch as they are the indications of a magnetic needle—are no longer such in the old sense of the term, since they are affected by deviation. Under these circumstances it has been found necessary, especially in respect to this class of ships, to adopt a modification of the old nomenclature, and so "bearing" or "course" admits of an additional qualifying term not previously recognised, viz., *correct* magnetic.

PRELIMINARY RULES IN NAVIGATION.

169. DEF.—The latitude and longitude of the place left are called the *latitude from* and *longitude from;* the latitude and longitude of the place arrived at are called the *latitude in* and *longitude in.*

170. Given the latitude from and latitude in or to, *to find the true difference of latitude.*

To find the difference of latitude. (For definition, &c., see Nos. 149 and 150, pages 100 and 101.)

RULE XLII.

1°. When the latitudes have *like* names—*Subtract the less latitude from the greater, and multiply the degrees in the remainder by* 60, *adding in the minutes.* The result is the true difference of latitude.

2°. When the latitudes have *unlike* names—*Take the sum of the two latitudes, reduce it to minutes.* The result is the true difference of latitude.

3°. To name the diff. lat.—*If the* latitude to *is* North *of the* latitude from, *mark the diff. of latitude* North (N.); *but if* latitude to *is* South *of* latitude from, *mark diff. latitude* South (S.)

Latitudes are reckoned north and south of the equator. If these different directions are considered the one positive and the other negative, the difference of latitude of two places is always found by taking the *algebraic* difference of their latitudes.

EXAMPLES.

Ex. 1. Find the diff. of lat. between Tynemouth Light, in lat. 55° 1′ N., and the Naze of Norway, in lat. 57° 58′ N.

```
Lat. Tynemouth   55°  1′ N.
Lat. Naze        57  58  N.
                 ─────────
                  2  57
                     60
                 ─────────
                    177 N.
```

The lat. *from* (Tynemouth) and lat. *to* (Naze) being of the *same name*, that is, both North, the *difference* of them is taken for the diff. lat., and since we have to pass from the lower North lat. to a higher, the diff. lat. is marked North (N.)

Ex. 2. Required the diff. of lat. between Cape Formosa, in lat. 4° 15′ N., and St. Helena, in lat. 15° 55′ S.

```
Lat. C. Formosa    4° 15′ N.
Lat. St. Helena   15  55  S.
                 ─────────
                  20  10
                      60
                 ─────────
D. lat.            1210 S.
```

The lat. *from* (C. Formosa) is *North*, and the lat. *to* (St. Helena) is *South*, it is evident that the ship must sail South in order to pass from North lat. into South; whence we put South (S.) to the diff. of lat.

Ex. 3. A ship from lat. 32° 40′ N., sails to lat. 20° 47′ N.: what is the diff. of lat. made?

```
Lat. from   32° 40′ N.
Lat. to     20  47  N.
           ─────────
            11  53
                60
           ─────────
D. lat.       713 S.
```

The ship here passes from a *higher* N. lat. to a *lower* N. lat., and to do so must evidently sail S.; whence we mark diff. lat. S.

Ex. 4. Required the diff. of lat. between Port Natal, in lat. 29° 53′ S., and Akyab, in lat. 20° 8′ N.

```
Lat. Port Natal   29° 53′ S.
Lat. Akyab        20   8  N.
                 ─────────
                  50   1
                      60
                 ─────────
D. lat.            3001 N.
```

As the ship (Port Natal) is in S. hemisphere and Akyab is in the N. hemisphere, to pass from the former into the latter the ship must sail N.

Ex. 5. A ship from lat. 50° S. arrives in lat. 45° 29′ S.: what is the diff. of lat. ?

Lat. from 50° 0′ S.
Lat. in   45 29 S.

D. lat.   4 31 = 271 N.

Here the ship passes from a *higher* to a *lower* S. lat., and to do so must evidently sail N.; whence the diff. lat. is marked N.

Ex. 6. A ship from lat. 13° 45′ S., arrives in lat. 26° 15′ S.: required diff. lat.

Lat. from 13° 45′ S.
Lat. in   26 15 S.

       12 30 = 750 S.

Here the ship passes from a *lower* to a *higher* S. lat., and to do so must evidently sail S.: whence S. is marked against the diff. of lat.

(*a*) When one of the places has no latitude, or is on the Equator, the latitude of the other place is equal to the difference of latitude.

Ex. 7. A ship from a place A, lat. 0, is bound to a place B, lat. 25° S.: required the diff. lat.

Since lat. is reckoned from the Equator lat. 0° (N. or S.), to pass from 0° to 25° S., the ship must evidently sail S.; whence the lat. of B (25°) is diff. of latitude, and is marked S.

Ex. 8. A ship from a place A, in lat. 10° N., arrives at a place B, in lat. 0°: required the diff. lat. made.

One place being on the Equator, and the other in 10° N., the diff. of lat. is evidently 10° or 600′, and is named S., because it is evident the ship must sail South to pass from 10° N. to 0° N.

EXAMPLES FOR PRACTICE.

Required the difference of latitude between the place A and the place B in each of the following examples:—

1. Lat. A 55° 0′ N.   2. Lat. A 50° 38′ N.   3. Lat. A 58° 24′ S.
      B 58 23 N.   B 42 48 N.   B 63 17 S.
4. Lat. A 3 42 S.   5. Lat. A 13 15 S.   6. Lat. A 0 0
      B 1 48 N.   B 0 0   B 2 37 S.
7. Lat. A 10 10 N.   8. Lat. A 49 52 S.   9. Lat. A 0 17 S.
      B 0 0   B 42 13 S.   B 1 17 N.

171. To find the meridional difference of latitude, having given the latitude from and latitude in. (For definition, see page 101, Nos. 151 and 152).

RULE XLIII.

Take the meridional parts for the two latitudes from the Table of meridional parts; take the difference *if the latitudes are of the* same name, *but their* sum *if the names are* unlike. *The result is the meridional difference of latitude.*

EXAMPLES.

Ex. 1. Lat. A 49° 10′ N., lat. B. 27° 40′ N.: find the mer. diff. of lat.

Lat. A 49° 10′ N. M. parts 3397
    B 27 40 N.   „   1729

        Mer. d. lat. 1668

Ex. 2. Lat. left 49° 58′ S., and lat. bound to 32° 42′ S.: find mer. diff. of lat.

Lat. left 49° 58′ S. M. parts 3471
Lat. to   32 42 S.   „   2078

        Mer. d. lat. 1393

Ex. 3. Lat. left 29° 53′ S., and lat. to 20° 8′ N.: required mer. diff. of lat.

Lat. left 29° 53′ S. M. parts 1880
Lat. to   20 8 N.   „   1234

        Mer. d. lat. 3114

Ex. 4. Lat. from 46° 40′ N., and lat. to 34° 22′ S.: find the mer. diff. of lat.

Lat. left 46° 40′ N. M. parts 3173
Lat. to   34 22 S.   „   2198

        Mer. d. lat. 5371

EXAMPLES FOR PRACTICE.

Find the meridional difference of latitude in each of the following examples:—

1. Lat. from 34° 40′ N.   Lat. in 33° 20′ N.   4. Lat. from 15° 44′ N.   Lat. in 4° 20′ S.
2.   „   24 12 S.   „   15 18 N.   5.   „   60 20 S.   „   67 10 S.
3.   „   49 10 S.   „   52 47 S.   6.   „   0 0   „   4 20 N.

172. To find the latitude in, having given the latitude from and true difference of latitude.

RULE XLIV.

1°. When the latitude from and true difference of latitude have a *like* name—*To the latitude from* add *the true difference of latitude (turned into degrees, minutes, and seconds, if necessary)*: *the sum will be the* latitude in, *of the* same name *as the* latitude from.

2°. When the latitude from and true difference of latitude have *unlike* names—*Under the latitude* from, *put the true difference of latitude (in degrees and minutes, if necessary)*: *the* remainder *marked with the name of the* greater *is the* latitude in.

EXAMPLES.

Ex. 1. A ship from lat. 59° 27′ S., sails South, until the diff. lat. is 374 miles: required the lat. come to.

$$6,0)37,4$$
$$\overline{6° 14′ S.}$$

Lat. from 59° 27′ S.
D. lat.   6 14 S.
Lat. in   65 41 S.

Ex. 2. A ship from lat. 2° 25′ N. sails South, 180 miles: what lat. is she in?

$$6,0)18,0$$
$$\overline{3° 0′}$$

Lat. from 2° 25′ N.
D. lat.   3   0 S.
Lat. in   0 35 S.

In this example it is evident that as the diff. lat. is more than the lat. left, the ship must have crossed the Equator, and consequently has come into South lat.

Ex. 3. A ship from lat. 55° 1′ N. sails North, 94 miles: find the lat. in.

$$6,0)9,4$$
$$\overline{1° 34′}$$

Lat. from 55° 1′ N.
D. lat.   1 34 N.
Lat. in   56 35 N.

Ex. 4. A ship from lat. 28° 39′ N. sails South, 131 miles: required the lat. in.

$$6,0)13,1$$
$$\overline{2° 11′}$$

Lat. from 28° 39′ N.
D. lat.   2 11 S.
Lat. in   26 28 N.

Ex. 5. A ship from lat. 0° 49′ S. sails North, 83 miles: required the lat. in.

$$6,0)8,3$$
$$\overline{1° 23′}$$

Lat. from 0° 49′ S.
D. lat.   1 23 N.
Lat. in   0 34 N.

Ex. 6. A ship from lat. 3° 12′ N. sails South, 192 miles: required the lat. arrived at.

$$6,0)19,2$$
$$\overline{3° 12′}$$

Lat. from 3° 12′ N.
D. lat.   3 12 S.
On the Equator 0 0

EXAMPLES FOR PRACTICE.

Find the latitude in, in each of the following examples:—

1. Lat. from 31° 10′ N.   D. lat. 172′ N.
2.   „   29 38 N.   „   104 S.
3.   „   3   2 S.   „   190 N.
4.   „   2 56 S.   „   357 N.
5.   „   0   0   „   168 S.
6. Lat. from 0° 8′ N.   D. lat. 182′ S.
7.   „   0 39 N.   „   59 S.
8.   „   3 58 N.   „   238 S.
9.   „   4 48 S.   „   288 N.
10.   „   35 25 S.   „   229 S.

173. To find the middle latitude, having given the latitude from and latitude in. (For definition see No. 153, page 101.)

RULE XLV.

The name being supposed *alike*, that is, both *North* or both *South*—*Add together the true latitudes, and take half the sum; the result is the middle latitude.*

NOTE.—When the names are unlike, the middle latitude (which is seldom required but for obtaining the departure) should be found by means of a table; but in this case it may perhaps be as well to avoid the use of the middle latitude in any of the common problems of navigation.

EXAMPLES.

Ex. 1. Find the mid. lat., having given the lat. from 50° 25' N., and lat. in 47° 12' N.

  Lat. from 50° 25' N.
  Lat. in   47 12 N.
      2)97 37
  Mid. lat. 48 48

Ex. 2. Lat. from 6° 28' S., lat. in 14° 50' S.: required the mid. lat.

  Lat. from 6° 28' S.
  Lat. in   14 50 S.
      2)21 18
  Mid. lat. 10 39

EXAMPLES FOR PRACTICE.

Required the middle latitude in each of the following examples:—

1. Lat. from 16° 10' S.   D. lat. 138' S.   4. Lat. A 63° 53' S.   Lat. B 59° 10' S.
2.   „   1 40 S.   „   61 S.   5. „   56 10 N.   „   50 15 N.
3.   „   36 22 N.   „   90 S.   6. „   67 20 S.   „   61 42 S.

174. To find the difference of longitude, having given the longitude from and longitude to. (For definition see No. 155, page 102.)

RULE XLVI.

1°. When the longitudes are of the *same* name—*Take their difference and reduce the same to minutes, place* E. *or* W. *against the remainder, according as the longitude* to *is East or West of longitude* from.

2°. When the longitudes are of *contrary* names—*Take the sum of the two longs., which sum, if less than* 180°, *is the diff. of long., and attach* E. *or* W., *according as the long.* to *is East or West of long.* from; *but when the sum exceeds* 180° *subtract it from* 360°, *for the diff. of long., and reduce the remainder thus found to minutes, attaching to it the* contrary name *to that found in the usual way.*

Longitudes are reckoned East or West of the first meridian. If these different directions are considered one positive and the other negative, the difference of longitude of two places is always found by taking the *algebraic* difference of their longitudes.

EXAMPLES.

Ex. 1. Find the diff. of long., having given the long. from 89° 42' W., and long. in 79° 42' W.

  Long. from 89° 42' W.
  Long. in   79 42 W.
       10 0
       60
  D. long.   600 E.

The ship here passes from a *high* W. long. to a *lower*, and diff. long. must be E. to do so.

Ex. 2. Required the diff. of long., having given the long. from 12° 20' E., and long. in 2° 45' W.

  Long. from 12° 20' E.
  Long. in   2 45 W.
       15 5
       60
  D. long.   905 W.

The ship here passes from E. long. to W. long., and in order to do so diff. long. must be W.

Ex. 3. A ship from Cape Bajoli, long. 3° 48' E., is bound to Cape Sicie, in long. 5° 51' E.: required the diff. of long.

  Long. Cape Bajoli 3° 48' E.
  Long. Cape Sicie 5 51 E.
         2 3
         60
  D. long. 123 E.

The long. to Cape Sicie is E. of long. from Cape Bajoli, therefore, diff. of long. is marked E. The ship must evidently sail E.

Ex. 4. A ship from Tynemouth, in long. 1° 25' W., is bound to long. 7° 12' E.: required the diff. of long.

  Long. from 1° 25' W.
  Long. to   7 12 E.
       8 37
       60
  D. log.   517 E.

The ship here is about to cross the meridian of Greenwich (long 0°) and pass from W. long. to E. long., whence the diff. of long. must be E. to do so.

Ex. 5. Find the diff. long. between Acapulco, long. 99° 54' W., and Pellew Island, long. 134° 21' E.

Long. Acapulco	99° 54' W.
Long. Pellew Island	134 21 E.
Being greater than 180° it is subtracted from	234 15 E. 360 0
Diff. of long. is	125 45 W. 60
D. long.	7545 W.

By going E. and W. from Greenwich, the two places in this example will be found to be 234° 15' asunder, but as both places are for our purpose upon one circle, the smaller arc of the circle must be taken to find how far apart the places Acapulco and Pellew Island are separated; so that the sum 234° 15' is subtracted from 360°, the whole circumference of a circle, for the required answer.

Ex. 6. A ship from long. 177° 50' E. arrives in long. 178° 10' W.: what diff. of long. has she made?

Long. left	177° 50' E.
Long. in	178 10 W.
Being greater than 180° it is subtracted from	356 0 W. 360 0
Diff. of long. is	4 0 E. 60
D. long.	240 E.

Ex. 7. A ship from long. 5° 12' W. is bound to a port in long. 90° W.: what diff. of long. must she make?

Long. from	5° 12' W.
Long. to	90 0 W.
	84 48 60
D. long.	5088 W.

The ship here passes from a *less* to a *greater* W. long.; and therefore the diff. of long. must be W. to do so.

Ex. 8. A ship from long. 165° E. is bound to a place in long. 72° 12' E.: what diff. of long. must she make?

Long. left	165° 0' E.
Long. to	72 12 E.
	92 48 60
D. long.	5568 W.

The ship in this example sails from a *greater* to a *less* long. (E. long.), the diff. long. is therefore of a different name to the long. left.

EXAMPLES FOR PRACTICE.

Required the difference of longitude between a place A and a place B in each of the following examples:—

1.	Long. A 9° 29' W.	Long. B 4° 29' W.	7.	Long. A 0° 55' E.	Long. B 7° 3' E.
2.	,, 1 25 W.	,, 7 2 E.	8.	,, 40 10 E.	,, 33 10 E.
3.	,, 6 11 E.	,, 5 45 W.	9.	,, 178 30 W.	,, 178 30 E.
4.	,, 0 0	,, 4 20 W.	10.	,, 176 34 E.	,, 176 34 W.
5.	,, 4 20 W.	,, 0 10 E.	11.	,, 38 32 W.	,, 8 43 E.
6.	,, 7 2 E.	,, 0 0	12.	,, 5 12 W.	,, 25 12 W.

175. To find the longitude in, having given the longitude from and the difference of longitude.

RULE XLVII.

1°. When the longitude from and the difference of longitude have *like* names—*To the longitude from* add *difference of longitude (turned into degrees, if necessary)*: the sum, *if not* more than 180°, *will be* the longitude in, *of the same name as the* longitude from; *but if the* sum exceed 180°, subtract *it from 360°, and the remainder is the* long. in and *of a contrary* name to long. from.

2°. When the longitude left and difference of longitude have *unlike* names —*Under* longitude from, *put difference of longitude (in degrees and minutes, if necessary)*; *take the* less *from the* greater; *the* remainder, *marked with the name of the* greater, *is the* longitude in.

EXAMPLES.

Ex. 1. A ship from long. 5° 12' W. makes diff. long. 113' W: required the long. in.

```
                Long. from  5° 12' W.
6,0)11,3        D. long.       1 53  W.
                Long. in       7  5  W.
  1° 53' W.
```

Ex. 2. A ship from long. 1° 25' W. sails E. until her diff. of long. is 177': required her long. in.

```
                Long. from  1° 25' W.
6,0)17,7        D. long.       2 57  E.
                
  2° 57' E.     Long. in, 1 32 E.
```

Ex. 3. A ship from long. 0° 57' E. sails W. until her diff. of long. is 201': find the long. in.

```
                Long. from  0° 57' E.
6,0)20,1        D. long.       3 21  W.
                Long. in       2 24  W.
  3° 21' W.
```

Ex. 4. Let the long. left be 174° 4' W., and the diff. of long. 797' W.: required the long. in.

```
6,0)79,7        Long. from  174°  4' W.
                D. long.     13  17  W.
  13° 17' W.
Being greater than 180°      187 21  W.
subtract from                360  0
                Long. in     172 39 E.
```

Ex. 5. Long. from 3° 40' W., diff. of long. 220' E.: required the long. in.

```
                Long. from  3° 40' W.
6,0)22,0        D. long.       3 40  E.
                Long. in       0  0
  3° 40' E.
On the meridian of Greenwich.
```

Ex. 6. A ship from long. 177° 40' W. makes 140' diff. of long. to the W.: required the long. arrived at.

```
                Long. from  177° 40' W.
6,0)14,0        D. long.       2 20  W.
                
  2° 20' W.     Long. in    180   0  W.
                    or,     180   0  E.
```

EXAMPLES FOR PRACTICE.

Required the longitude in, or arrived at, in each of the following examples:

1. Long. from 5° 48' W. D. long. 110' W.
2.   „   0 59 W.   „   137 E.
3.   „   29 10 E.   „   114 E.
4.   „   3 10 E.   „   220 W.
5.   „   2 47 W.   „   242 E.
6.   „   3 12 E.   „   237 W.
7. Long. from 41° 29' W. D. long. 139' E.
8.   „   94 4 E.   „   115 W.
9.   „   98 54 E.   „   302 E.
10. „   178 13 E.   „   201 E.
11. „   177 6 W.   „   237 W.
12. „   179 59 W.   „   2 W.

13. Define meridian of the earth, equator, parallel of latitude. Which of these are great circles, and why?

THE COMPASS.

176. **The Compass*** is simply an instrument which utilises the directive power of the magnet. A magnetised bar of steel, apart from disturbing forces and free to move, points in a definite direction, and to this direction a l others may be referred, and a ship guided on any desired course.

There are various adaptations of the instrument, according to the use it is specially intended for. The compass intended for use on board ship is called the "**Mariner's Compass**," and according to the purpose it is intended for it is named the **Steering Compass**, the **Standard Compass**, and the **Azimuth Compass**.

177. The **Mariner's Compass** consists of a circular card, which represents the horizon of the observer; the circumference or edge of the card being divided according to two systems of notation into points and degrees.

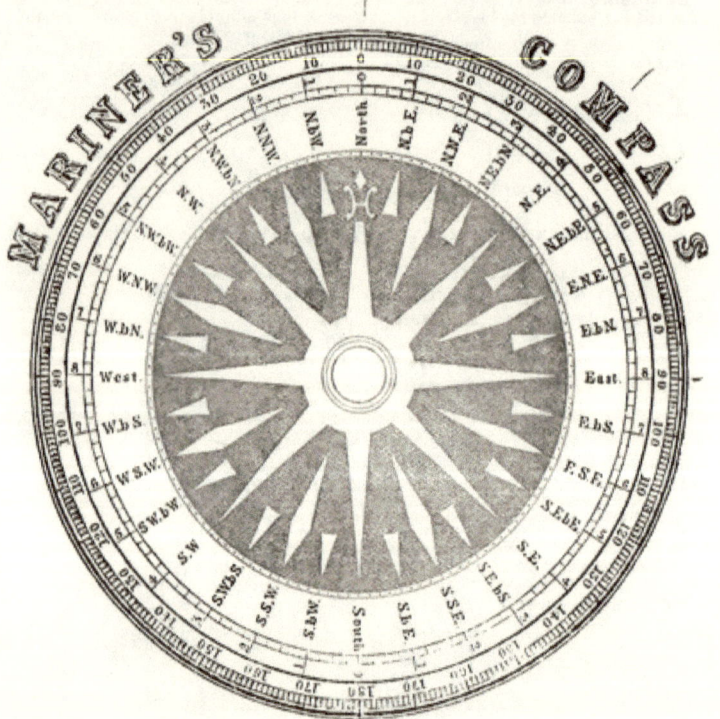

* The origin of the compass is very obscure. The ancients were aware that the loadstone attracted iron, but were ignorant of its directing property. The instrument came into use in Europe sometime in the course of the thirteenth century.

(1). **By Points.**—There are 32 *points;* and each of those divisions is again sub-divided into four parts called *quarter points*. A point of the compass being therefore the 32nd part of the circumference of a circle is equal to $11° 15'$. The four principal points, or, as they are called, the *cardinal points*, are the North (represented by N.), South (S.), East (E.), West (W.), the East being to the right, and West to the left, when facing the North.

All the points of the compass are called by names composed of these four terms.

Thus, the points half-way between the cardinal points are called after the two adjacent cardinal points; hence the point midway between the North and East is called North-east, and represented by N.E.; so midway between South and East is called South-east (written S.E.); in like manner we get South-west (written S.W.), and North-west (written N.W.)*

A point half-way between one of these last and a cardinal point is called, in like manner, by a name composed of the nearest cardinal point and the adjacent points, N.E., N.W., S.E., and S.W. Thus, the point half-way between N. and N.E. is called North-north-east (written N.N.E); the point between E. and N.E. is called East-north-east (written E.N.E.); and so we have E.S.E., S.S.E., S.S.W., W.S.W., W.N.W., and N.N.W. The points next the eight principal points, namely, N., N.E., E., S.E., S., S.W., and N.W., are named by placing *by* between the letter representing the point to which it is adjacent and the next cardinal point in the same direction. Thus, the point next to N., on the east side, is called North by East, *i.e.*, North in the direction towards East (written N. by E.); that next N.E., towards the North, is called North-east by North (N.E. by N.), *i.e.*, North-east in the direction towards North; and so we have N.E. by E., E. by N., E. by S., S.E. by E., S.E. by S., S. by E., S. by W., S.W. by S., S.W. by W., W. by S., W. by N., N.W. by W., N.W. by N., N. by W.; in this manner we get other sixteen points. We have thus got names to all the thirty-two points of the compass.

Each point is again sub-divided into half points and quarter points.

A *half point*, which is the middle division between two points, is called after that one of its adjacent points which is either a cardinal point or is the nearest to a cardinal point. Thus, the middle division between N. and N. by E. is called North-*half*-east (written N. ½ E.) Half points near N.E., N.W., S.E., and S.W., take their name from these points. Thus we say N.E. ½ N., N.E. by E. ½ E.†

---

* These new directions also give names to the four quarters of the compass, as, when we say that "the wind is in the S.W. quarter," meaning thereby not exactly S.W., but somewhere between S. and W.

† In naming the half and quarter points it is advisable in some cases to sacrifice system to simplicity. Thus, for example, seamen commonly say N.N.E. ½ E. instead of N.E. by N. ½ N.; we do not, however, say E.N.E. ½ E., though this is simpler than E. by N. ½ N., since it is at once seen to be 6½ points. It would of course be more systematic, as a matter of geometry, to reckon the half points always from N. or S., because the ship's course is reckoned from the meridian; but on the other hand, as a matter of names, regard will be had to the whole points between which it falls, and to the order in which these are taken.

The same holds for a quarter and for three-quarters as for a half point, all of which are named upon the same principle as the subordinate points.

In chosing the name to use we must be guided by circumstances. In some problems it is convenient always to reckon uniformly from North or South, but generally the simpler name will be the preferable one; and similarly for quarters and three-quarters of a point.

(2). **By Degrees.**—The whole circumference is divided into three hundred and sixty degrees (360°), each degree into sixty minutes (60′). This furnishes a notation for the compass more minute than points, half points, and quarter points. We still reckon from the cardinal points: thus, to indicate a division which has 72° 48′ to the East of North we write N. 72° 48′ E.

178. The name of the opposite point to any proposed point is known at once without referring to the compass, by simply reversing the name or the letters which compose it—thus, the opposite of N. being S. and of E. being W., the opposite point of N.E. by N. is at once known to be S.W. by S., the opposite of W. $\frac{3}{4}$ S. is E. $\frac{3}{4}$ N., and so on.

179. Repeating the points in any order is called *boxing the compass*; to do this is, of course, one of the first things a seamen learns.

180. As the ship's course, which is sometimes expressed in points and sometimes in degrees, is always reckoned from the North or South point, the seaman has to refer at once, in using the Tables, to the *number of points* or *degrees* in any course given by *name*. The following table, which exhibits the degrees, minutes, and seconds in each quarter point of the compass, will be convenient for reference.

A TABLE OF THE ANGLES,
which every Point and Quarter Point of the Compass makes with the Meridian.

NORTH		Points	°	′	″	Points	SOUTH	
		0	2	48	45	0 $\frac{1}{4}$		
		0 $\frac{1}{2}$	5	37	30	0 $\frac{1}{2}$		
		0 $\frac{3}{4}$	8	26	15	0 $\frac{3}{4}$		
N. by E.	N. by W.	1	11	15	0	1	S. by E.	S. by W.
		1 $\frac{1}{4}$	14	3	45	1 $\frac{1}{4}$		
		1 $\frac{1}{2}$	16	52	30	1 $\frac{1}{2}$		
		1 $\frac{3}{4}$	19	41	15	1 $\frac{3}{4}$		
N.N.E.	N.N.W.	2	22	30	0	2	S.S.E.	S.S.W.
		2 $\frac{1}{4}$	25	18	45	2 $\frac{1}{4}$		
		2 $\frac{1}{2}$	28	7	30	2 $\frac{1}{2}$		
		2 $\frac{3}{4}$	30	56	15	2 $\frac{3}{4}$		
N.E. by N.	N.W. by N.	3	33	45	0	3	S.E. by S.	S.W. by S.
		3 $\frac{1}{4}$	36	33	45	3 $\frac{1}{4}$		
		3 $\frac{1}{2}$	39	22	30	3 $\frac{1}{2}$		
		3 $\frac{3}{4}$	42	11	15	3 $\frac{3}{4}$		
N.E.	N.W.	4	45	0	0	4	S.E.	S.W.
		4 $\frac{1}{4}$	47	48	45	4 $\frac{1}{4}$		
		4 $\frac{1}{2}$	50	37	30	4 $\frac{1}{2}$		
		4 $\frac{3}{4}$	53	26	15	4 $\frac{3}{4}$		
N.E. by E.	N.W. by W.	5	56	15	0	5	S.E. by E.	S.W. by W.
		5 $\frac{1}{4}$	59	3	45	5 $\frac{1}{4}$		
		5 $\frac{1}{2}$	61	52	30	5 $\frac{1}{2}$		
		5 $\frac{3}{4}$	64	41	15	5 $\frac{3}{4}$		
E.N.E.	W.N.W.	6	67	30	0	6	E.S.E.	W.S.W.
		6 $\frac{1}{4}$	70	18	45	6 $\frac{1}{4}$		
		6 $\frac{1}{2}$	73	7	30	6 $\frac{1}{2}$		
		6 $\frac{3}{4}$	75	56	15	6 $\frac{3}{4}$		
E. by N.	W. by N.	7	78	45	0	7	E. by S.	W. by S.
		7 $\frac{1}{4}$	81	33	45	7 $\frac{1}{4}$		
		7 $\frac{1}{2}$	84	22	30	7 $\frac{1}{2}$		
		7 $\frac{3}{4}$	87	11	15	7 $\frac{3}{4}$		
East.	West.	8	90	0	0	8	East.	West.

181. The card for practical use is generally made of mica covered with paper, so as to be as light as possible. Two or more magnetic needles,* which are small steel bars magnetised, are fixed below the circular compass card, but parallel with its meridional line, so that the N. ends of the needles shall coincide (in direction) with the N. end of that line, and the S. ends of the needles with the S. end of the same line. An inverted conical brass socket, called a cap, with a hard stone in its centre, is passed through a hole in the centre of the card, and the whole is then accurately balanced on a sharp centre or pivot rising from the middle of a brass or copper bowl, and sufficiently large to admit of the card moving freely within it: the cover of the bowl is glass, which, while protecting the card from wind and weather, admits of its indications being distinctly seen. There is also a vertical line drawn inside the bowl which is called the *lubber's line*. The bowl, having a weight fixed to it below, is placed in *gimbals*, which are brass hoops or rings so arranged as to admit of motion about two horizontal axis at right-angles to one another, *i.e.*, each turning upon two pivots at opposite points of the hoop next greater in size; by this means the loaded bowl remains nearly horizontal during the confused and irregular motion of the ship.

To the deck, in front of the helmsman's position, a stand called a *Binnacle* is firmly fixed, which may be of any shape—octagonal, square, or pillar-like—sometimes of wood, sometimes of brass: within it are supports or bearings into which the pivots or outer rings of the compass bowl fit, and its movable top or cover is fitted with a glass front and a lamp or lamps to cast a light on the compass card by night. This constitutes the *Steering Compass*.

182. The helmsman steers the ship so that a line parallel to the keel passes over the centre of the card, and the point prescribed as the course. Care is taken to place the box so that the *lubber's point* in the bowl and the centre of the card are in a line fore-and-aft, or parallel to the keel; but as the lubber's point deviates a little from its proper position when the ship is heeled over, seamen do not implicitly depend upon it, as, indeed, the name implies.

183. **The Azimuth Compass** is a compass of superior construction, particularly adapted to observe bearings. It is mounted on a stand, and is fitted with two small frames carrying vertical wires, called *sight-vanes*, for the purpose of observing objects elevated above the horizon. In one of these vanes there is a long and very narrow slit, and in the other is an opening of the same kind, but wider, and having a wire up and down the middle of it, exactly opposite the slit.

184. In the best modern instruments, a horizontal ring is expressly provided to carry the vertical wire frame, and instead of having a wire next the eye, a glass prism, acting by internal reflection, is placed there, so arranged that one-half of the pupil of the eye can observe the wire on the further side of the horizontal ring and the distant object, and the other half of the pupil can see the graduations of the compass card by internal reflection in the

---

* The object of using several magnets is to increase the magnetic moment of a given weight of steel.

prism. This prism is a solid piece of glass, whose sides are parallelograms and ends triangles. The compass card is very carefully and minutely graduated; besides the points and quarter points being marked, the circumference over which the prism passes is graduated in degrees, and usually cut to every 20′, and this graduation is arranged so that we may read off the bearing at once, and is reckoned in more ways than one, for facilitating taking bearings from different cardinal points. The card can be brought to rest by a stop. There is also a contrivance for throwing the card off its centre when the instrument is not in use, to prevent the fine pivot being worn, and the sensibility of the compass impaired. This instrument is known as the **Prismatic Azimuth Compass.**

185. In observing bearings on board ship the card should never be stopped, but two or more bearings being read off as quickly as convenient, the mean should be used; for, as the vessel, and consequently the compass card, have always some motion, the card may not therefore be stopped exactly in the middle of its vibration, which, as it may be supposed to vibrate equally on both sides of the line of direction of the object, is essential to the true result.

186. **The Standard Compass** on board ship is the one placed in a particular spot on deck, or above it. It should be placed in the middle of the ship, and fixed on a permanent and secure pillar or support, raised at such a height (not less than 5 feet) as to permit amplitudes of the sun and bearings of the land to be conveniently observed by it. In the Royal Navy it is used as an azimuth compass, being fitted with an azimuth circle, which is graduated so as to show the angle between the ship's head and any heavenly body, as measured on the horizon (thus acting as a *dumb card*); the sight vanes and reading prism being fitted to the azimuth circle in such a way as to turn freely in azimuth without moving the compass or disturbing the card. The card of the azimuth compass should not exceed 7 to 7½ inches diameter.

It should also be in a position as far as possible removed from any considerable mass of iron—at least 5 feet from iron deck beams—and should not be within 10 feet of the extremity of any elongated iron mass, especially if vertical, such as funnels, stanchions, or the spindle of the wheel; and it should be received as a general rule that no iron, subject to occasional removal, is to be placed within 15 feet of this compass, either on the same deck, or that below it.

But in the mercantile marine the practice prevails of taking these bearings by means of a dumb card, many of which are in use, and answer the purpose intended.* They can be placed in any part for observing, and the true bearing of the ship's head is determined, so that by comparing it with the bearing shown by the Standard Compass the deviation of the latter can be ascertained. Under these circumstances we recommend that the only consideration for determining the selection of a place for the Standard Compass should be favourable conditions connected with its compensation.

* Perhaps the best and most useful of these is the one known as BAIN & AINSLEY'S "Compass Corrector."

187. **Steering Compasses** being placed according to the requirements of the ship, the moderate and uniform amount of deviation generally attainable at the Standard Compass by selection of position, cannot always be secured. Still we should do the best we can, for if, as frequently happens, the steering wheel is placed near an iron stern-post or rudder-head, and further fitted with an iron spindle—near which, of necessity, the steering compass is fitted—then large and perplexing deviations may be expected, defying even approximate correction by magnets, causing much inconvenience to the helmsman, and possibly a total loss of the services of the compass on the ship proceeding into southern latitudes.*

The following rules to avoid the inconvenience and even danger just pointed out, have been recommended in selecting a place for steering compasses:—"Not to be within half the width of the ship from the stern-post or rudder-head; the spindle of the steering wheel and the foremost support on which the wheel works *not* to be of iron; avoid vertical iron." The needle should be at least 3 ft. 6 in. from iron deck beams, and as much higher as can be made convenient to the helmsman.

In addition to the rules already given for the guidance of seamen, the following (given by Capt. EVANS, Superintendent of the Admiralty Compass

---

* **Bridge Compass.**—Those who arrange for the construction and equipment of iron steamers ought not to lose sight of the fact that the bridge compass is the most important of all. It is on the bridge where the officer of the watch is stationed on all occasions when caution is required, and if the compass before him is a reliable one he is better enabled to navigate the ship safely than by a compass situated in any other part. Yet many errors are committed in placing this compass. Take an instance, "where the bridge compass was rendered untrustworthy from the ventilator of the engine-room being *close* in front of it. Two reasons were assigned for this. The one was that the bell-mouthed top required to be turned round in order to regulate the amount of ventilation, it was convenient to have it within reach of the bridge. The other was, that in placing the bridge farther from the ventilator a greater evil would be incurred, since it would require that the compass should either be brought nearer to the iron mast or nearer to the funnel. It was not understood that this ventilator, small as it was compared with the mast or the funnel, produced five times as much error as these two combined, if placed equally near. The head of the ventilator was a blue pole by induction, and was on a level with the compass, and when the bell-mouth, or cowl, was turned towards the stern, not two feet from the compass on the bridge, it caused two points of deviation, but when the cowl was turned towards the bows only a half point of error was produced, and when turned to the starboard or port side other complications resulted rendering the compass useless, because the red pole of the ventilator was so far distant as to render no appreciable amount of repulsion to compensate the attraction of the upper pole. But the blue and red poles of the mast and funnel were situated one above the compass and the other below, so as nearly, if not altogether, to compensate each other. Thus, being ignorant of the laws which regulate deviation, in attempting to avoid an imagined evil, a serious real error was committed. This compass being found unworthy of reliance was removed, and the captain in ascending the Gulf of St. Lawrence had continually to refer to the Standard Compass on the deck, which in hazy weather rendered the navigation more difficult than it would have been had there been a trustworthy bridge compass." Again, it not unfrequently happens that the bridge on which the compass stands is placed directly over the engine and close against the central iron stanchion which supports the rail running along the bridge, while according to another arrangement the compass stands on a semicircular piece of planking projected from the bridge forward with the railing carried round it, with perhaps, as in some instances, five and in others four iron stanchions in close proximity to the compass, while in other cases where the circular space is just large enough for the binnacle there may be only three iron stanchions, but in all these cases these iron stanchions are only a very few inches from the card, and the card itself within six inches of the plane of the top of the stanchions. It cannot be a matter of surprise that no dependence can be placed on compasses so placed.

Department) are worthy the attention of the Naval Architect and those superintending the equipment of the ship:—

(1). In all designs for the construction of iron ships, a place to be prepared for the Standard Compass, and to be shown in the plan.

(2). The Standard Compass not to be within half the breadth of the ship from the rudder-head and stern-post or iron-cased screw well, not to be nearer an iron deck or iron deck beams than five feet.

(3). In ships built near *North*, the Standard Compass to be as far *forward* as the requirements of the ship will permit. In ships built head near *South*, to be as far aft as the requirements of the ship will permit, subject to Rule 2. In ships built nearly *East* or *West*, the Standard Compass not to be near either extremity of the vessel.

(4). To be as far as possible from transverse iron bulkheads.

(5). As far as possible, no masses of iron—as boilers, engines, bulkheads, or stanchions—should be placed below the compass, or within $55°$ of the vertical line through the centre, the angle being drawn from the compass as centro to the centre of the mass in question.*

(6). To the above we would add, not to be nearer the break of the poop, either before or aft, than half the breadth of the vessel.

NOTE.—"**Comparative Merits of Large and Small Compasses.**—Of late years much diversity in practice has prevailed as to the size of compasses for use on board ship. The Admiralty Standard Card, for example, is fitted with needles, the maximum lengths of which are $7\frac{1}{2}$ inches, while in large passenger steam vessels the needles are frequently 12 to 15 inches, and even longer. The chief object in the employment of large compasses is to enable the helmsman to steer to degrees, and a more accurate course is presumed to be preserved.

"With reference to this increased size it must be observed that competent authorities limit the length of efficient compass needles to 5 or 6 inches; beyond this limit an increase of length is alone accompanied by an increase of directive power in the same proportion, and if the thickness of the needle be preserved, the weight, and consequently the friction, increases in the same ratio. No advantage of directive power is therefore gained by increase of length, but with the increased weight of the card and appendages, the increase of friction probably far exceeds the increase of directive force; sluggishness is the result, which is further exaggerated by the extreme slowness of oscillation of long needles compared with short ones.

"Large cards, however convenient in practice, are therefore not without danger, for the course steered may deceive the seaman by seeming right to the fraction of a degree, but which avails little if the card itself is wrong half a point, and the ship in consequence hazarded. In the opinion of the writer, the present Admiralty Standard Card is as large as should be used for the purposes of *navigation*; and that as regards safety in the long, steady,

---

* Investigation has shown that the effect of a sphere of iron within this cone is prejudicial by diminishing the directive force and increasing the heeling error to windward—when without the cone it would be beneficial in both respects. Hence the recommendation.

With reference to the magnetic character of boilers, or tanks, it has been stated that the effect is the same as if they were solid bodies, on the assumption that magnetism exists entirely on the surface of iron masses. This is not the case; it is, however, true that the effect of hollow masses of iron increases very rapidly with the increase of the thickness of the iron, so that the limit of thickness is speedily reached when the effect of the body is sensibly the same as if it were solid; for example, in a tank 4 feet in diameter and 1-10th of an inch thick, the effect is about $\frac{1}{2}$ of a solid mass of the same size; in a similar sized tank $\frac{2}{3}$ of an inch thick, the effect would be about half that of a solid mass. See a valuable investigation by Mr. ARCHIBALD SMITH, in the Phil. Trans. for 1865, pages 304—318.

and fast ship, the choice is really between the Admiralty Card and a smaller one. In short, the question may be thus stated: the smaller a card the more correctly it points, the larger a card the more accurately it is read."—*Manual of the Deviation of the Compass, by Capt. Evans, R.N.*

188. There is no advantage in having a large number of compasses in a ship: since, unlike the mean results of a number of chronometers, for example, the mean results of any number of compasses need not necessarily be near the truth, as they may all be largely in error, and that error may be all in one direction. Hence the necessity of depending upon one compass alone, but that compass should be in the best position in the ship, of the best manufacture, and the constant attention of the navigator should be devoted to ascertain its errors.

ADJUSTMENTS OF THE COMPASS.

189. (1). The direction of the magnetism of the needle or the "magnetic axis" should be in a line along *the middle of the needle itself*, otherwise the needle will not point with exactness to the magnetic North and South. To examine whether this is the case reverse the needle on the card. If after this reversion the N. and S. points of the card are also found to be reversed, the adjustment is good.

As this error obviously affects all points of the compass alike, it may be included in the total variation of the particular compass as found by observation, and therefore need not be made the subject of special examination.

(2). *The pivot must be in the centre of the graduated circumference of the card.* If it is not, the difference of bearing of two objects will not be the same when measured on different parts of the edge. This adjustment is generally good.

(3). *The line joining the eye-vane and the object-vane*, called the *"line of sight" of the Azimuth Compass, must pass directly over the pivot.* This condition is examined by noting carefully the bearing of a distant object, and then turning the compass half round so as to reverse the vane and the slit, and then repeating the observation with an object eight points from the first. The bearings taken directly should be identical with those taken by reversion. The effects of this error, if any, may be eliminated by taking the mean of the direct and reversed bearings every time the instrument is used.

(4). *The sight-vanes must be vertical,* i.e., *the eye-vane and the object-vane must* each be vertical.

This can be examined only on shore, by observing whether the wires coincide through their length with a plumb line, or any vertical edge. When this adjustment is not perfect, or when the bowl is not maintained in a strictly horizontal position, bearings are most correctly obtained when the object is low.

CORRECTING COURSES.

190. The corrections of the compass are those quantities which must be applied to the indications of the instrument to obtain the reading that would be given if the north point of the compass card always corresponded to the north point of the horizon. Three corrections are sometimes necessary to be applied to the course steered by compass, to reduce it to the true course; and the converse. These are called

1. The Leeway.
2. The Variation of the Compass.
3. The Deviation of the Compass.

1. LEEWAY.

191. The angle included between the direction of the fore-and-aft line or keel of a ship, and that in which she moves through the water, as indicated by her wake, is called the **Leeway**.

A ship is said to be **on the port tack** when the wind is on her port side, that is, on the left-hand side of a person looking forward; and **on the starboard tack** when the wind is on her starboard side, that is, on the right-hand side of a person looking forward.*

When the ship is not going before the wind, she will not only be forced forward in the direction of her head, but in consequence of the wind pressing against her sideways, her actual course will be to "*leeward*" of the apparent course she is lying. The amount of leeway **differs** in different ships; depending on their construction, on the sails sets, the velocity forward, and other circumstances. Experience and observation are required to judge what amount of leeway to allow in each case. The correction for *leeway* is necessary to deduce the course made good from the course steered, and it is one of the corrections to be applied in reducing the compass course to the true course in the day's work; the correction being allowed according to

RULE XLVIII.

When the ship is on the port tack, allow the leeway to the right of the course steered; but when on the starboard tack, allow it to the left, the observer looking from the centre of the compass towards the point the ship is sailing upon.

EXAMPLES.

Ex. 1. The course steered is N.W. by W., the wind N. by E., leeway 1½ points.

The ship has the starboard tacks aboard; therefore, the leeway (1½ points) allowed to the left of N.W. by W., gives corrected *Course* W. by N. ¾ N.

Ex. 2. Course by Compass S. by E., wind E. by S., leeway 2¾ points.

The ship is on the port tack, then 2¾ points allowed to the right of S. by E., is S. by W. ¾ W., the Course corrected for leeway.

* A ship is said to be *on the tack* of the side from which the wind comes, even if it be on the quarter.

Ex. 3. Course N.E. by N., the wind N.W. by N., the leeway 1 point.

The ship being on the *port* tack, 1 point to the *right* of N.E. by N is N.E., the corrected *Course.*

Ex. 4. Course steered West, the wind N.W. by N., leeway 3½ points.

The ship is on the *starboard* tack, 3½ points to the *left* of West is S.W. ½ W., the compass course made good.

192. The points of the compass are frequently spoken of in calculation with reference to their position to the right or **left** of the **cardinal** point towards which the spectator, who is supposed placed in the centre of the compass, is looking.

Supposing the given point of the compass have *North* in it, then looking from the centre of the card over the cardinal point North, he has the quadrant from North to East on his right hand, and the quadrant from North to West on his left hand.

Hence any point between North and East is said to be to the **right** of North, and every point between North and West is said to be to the **left** of North; thus, N.N.E. is said to be "two points" to the right of North (for shortness usually written 2 pts. R. of N.), and W.N.W. "six points" left of North (written 6 pts. L. of N.)

Again, suppose the given point of the compass to have *South* in it, then the observer, looking from the centre of the card and facing *South*, has the quadrant from South to East on the **left** hand, and the quadrant from South to West on the **right** hand.

Hence any point between the S. and E. is said to be to the **left** of South, and any point between S. and W. is said to be to the **right** of South; thus S.E. by S. is said to be "three points" to the **left** of S. (for shortness written 3 pts. L. of S.), and W. by S. is "seven points" **right** of S. (usually written 7 points R. of S.)

Adopting this notation the work in the above examples will stand thus:—

Examples.

Ex. 1.

Course steered N.W. by W. is	5 pts.	L of N
Leeway carries ship	1½ ,,	L of N
Sum is corrected course	6½ ,,	L of N
	or W. by N. ½ N.	

Ex. 2.

Course steered S. by E. is	1 pt.	L of S
Leeway carries ship	2¾ ,,	R of S
The difference is	1¾ ,,	R of S
	S. by W. ¾ W.	

Ex. 3.

Course steered N.E. by N. is	3 pts.	R of N
Leeway carries ship	1 ,,	R of N
Sum is corrected course	4 ,,	R of N
	or N.E.	

Ex. 4.

Course steered West is	8 pts.	R of S
Leeway carries ship	3½ ,,	L of S
The diff. is corrected course	4½ ,,	R of S
	or S.W. ½ W.	

Examples for Practice.

Correct the following courses for leeway.

	Course Steered.	Wind.	Leeway.		Course Steered.	Wind.	Leeway.
1.	S.S.W.	S.E.	1¼	5.	E. ¾ N.	N. by E.	1¾
2.	S.W. ½ W.	W.N.W.	2¼	6.	N.W. ¾ N.	N.E. by E.	1¾
3.	N. by E.	E. by N.	¾	7.	S.W. by W.	S. by E.	2¾
4.	N.N.E. ½ E	N.W. ½ N.	2	8.	N.E. ¼ E.	N. by W.	1½

(a) *When the ship is* hove-to, *take the* middle point *between that to which she comes up and that to which she* falls off *for the* compass course, *and correct this for leeway.*

EXAMPLES.

Ex. 1. A ship lying-to under her mainsail, with her starboard tacks aboard, comes up E. by S., and falls off to N.E. by E., making 5 points leeway. What compass course does she make good?

The middle point between E. by S. and N.E. by E. is E. by N., then 5 points to the *left* hand gives N.N.E., the compass course made good.

Ex. 3. A ship lying-to comes up S. by E. and falls off to S.E. by E., the wind being S.W., making 5 points leeway: required the compass course.

The middle point between S. by E. and S.E. by E. is S.E. by S., then 5 points to the *left* hand (the ship having starboard tacks on board) is East, the compass course made good.

Ex. 2. A ship lying-to under a close-reefed main topsail, with her port (larboard) tacks on board, comes up to S.S.W. and falls off to S.W. by W., making 2½ points leeway. What compass course does she make?

The middle point between S.S.W. and S.W. by W. is S.W. ½ S., then 2½ points to the *right* hand is W.S.W.

Ex. 4. A ship lying-to with port tacks on board, comes up W. by S. and falls off N.W. by W., making 5 points leeway. What course does she make good?

The middle point between W. by S. and N.W. by W. is W. by N., then 5 points to the *right* hand is N.N.W., the compass course made good.

2. THE VARIATION OF THE COMPASS.

193. The needle points to the magnetic North, which in few parts of the world agrees with the true North; the difference between them is called the *Variation of the Compass.**

The variation is said to be *easterly* when the North end of the needle is drawn to the eastward, and *westerly* when drawn to the westward of the true North; thus, when the North end of the needle points to that part of the horizon which is true N.N.W. ½ W., the variation is said to be 2¼ points West; but when it points to the N. by E. part of the horizon, the *variation* is said to be 1 point East.

194 The variation is different in different places,† and it is also subject to a slow change in the same place, and becomes alternately East and West.‡

---

* This is the term commonly employed by nautical men; but among men of science the term "Magnetic Declination" is usually substituted for "Magnetic Variation."

† At Greenwich, at the present time, the variation is 20° W., or the North end of the magnetic needle does not point exactly North, but 20° W. of North. In the West Indies the variation is 0; at Cape Farewell, 53° W.; at Cape Horn, 23° E.; at Hobart Town, 10° E.; at Canton 1° E.; and Cape of Good Hope, 29½° W. Generally in Europe, Africa, and the Atlantic, the variation is westerly, while in America and the Pacific it is easterly.

‡ "The system of Magnetic Meridians has undergone considerable changes in the times of modern accurate science. The southern point of Africa received from the Portuguese voyagers in the fifteenth century the name of L'Agulhas (the needle), because the direction of the compass needle, or the Local Magnetic Meridian, coincided with the Geographical Meridian: it now makes with it an angle of about 30° W. In the sixteenth century, the compass needle in Britain pointed East of North: it points from 20° to 30° (in different parts of the British Isles) West of North. At the present time, a change of the opposite character is going on: in 1819 the westerly declination at Greenwich was about 24° 23', which was probably its maximum; in the last 30 years it has diminished from 23½° to 20° nearly. It is believed that the magnetic poles are rotating round the geographical poles from East to West."—*A Treatise on Magnetism,* designed for the use of Students in the University. By GEORGE BIDDELL AIRY, M.A., L.L.D., D.C.L.

It also changes slightly at different times of the day.* Its value for each locality is indicated on charts, and always to be found by easy methods.

195. **Variation** is one of the "corrections" in deducing the **true course and bearing** from the **course and bearing** observed with the compass. It is given on the charts used in navigation.

The method of correcting Compass Courses or Bearings for Variation will be readily understood by means of an example.

Suppose the variation of the compass is found to be two points East—that is, the needle is directed two points to the right of the North point of the heavens—that is, points N.N.E. instead of N.; then the N.N.W. point of the compass card will evidently point to the true north, and every other point on the card will be shifted round two points. If, therefore, a ship is sailing *by compass* N.N.W., or, as it is usually expressed, her compass course is N.N.W., her true course will be North; that is, *two points to the right of the compass course*. In a similar manner it may be shown that when the variation is two points westerly, the true course will be *two points to the left of the compass course*.

196. To find the **true course**, the **compass course** being given.

RULE XLIX.

Allow easterly *variation to the* right *of the* compass course.
  ,,   westerly   ,,   left   ,,
looking from the centre of the card over the point to be corrected.†

EXAMPLES.

Taking the courses between North and South round by E.

Ex. 1. Course steered N.E. by E., variation 2¾ points *West*, to find the true course.

Here the compass course is N. 5 points E., and the variation is westerly, and hence must be applied to the *left*, thereby bringing it 2¾ points nearer to the North (N. 5 E. − 2¾ = N. 2¼ E.), that is, within 2¼ points of North; the true course is therefore N.N.E. ¼ E.

Ex. 2. Course steered the same, viz., N.E. by E., variation 1¾ points East.

Here the compass course is N. 5 points E., and the variation easterly, and hence must be applied to the *right*, thereby carrying the course *away from the North towards the East*, that is, 6¾ points to the eastward of North (N. 5 E. + 1¾ E. = N. 6¾ E.); the true course is therefore E. by N. ½ N.

---

* Besides the gradual changes which occur in terrestrial magnetism, both as regards direction and intensity of force, in the course of long periods of time, there are minute fluctuations continually traceable. To a certain extent these are dependent on the varying positions of the sun, and, to a much smaller extent, of the moon, with respect to the place of observation; but over and above all regular and periodic changes, there is a large amount of irregular fluctuations, which occasionally become so great as to constitute what is called a *magnetic storm*. These variations occur with great rapidity, causing deflections to the right and left comparable in their rate or period of alternation with ordinary telegraphic signalling; accidental variations of 70′ have been observed. "Magnetic Storms" are not connected with thunder storms, or any other known disturbance of the atmosphere, but are invariably connected with exhibitions of aurora borealis, and with spontaneous galvanic currents in the telegraph wires, and this connection is found to be so certain, that upon remarking the display of one of the three classes of phenomena, we can at once assert that the other two are observable (the aurora borealis sometimes not visible here, but certainly visible in a more northern latitude).

† The learner must be careful to remember when **correcting his courses** that he is to suppose himself *looking from the centre of the card over the point to be corrected.* When he places the compass card before him, mistakes very frequently occur in the application of the variation between the *East* and *West* points round by *South*; thus, taking the compass with the North point placed before or from the observer, while an error could scarcely arise when correcting courses in the N.E. and N.W. quadrants, it would be different with the S.E. and S.W. quadrants, unless he bore in mind that in the latter instance the compass card should be placed before him, as if he were facing the South. From what has been said it will be seen that in correcting courses, the significance of RIGHT on the face of a compass card, is *as the hands of a watch move over the dial*, and LEFT *the contrary direction*.

Ex. 3. Course by compass N.N.E., variation $2\frac{1}{4}$ points *West*, the *true course* $2\frac{1}{4}$ points to the left hand of N.N.E., or N. $\frac{1}{4}$ W.

Ex. 4. Compass course S. by E. variation $2\frac{1}{4}$ *East*, $2\frac{1}{4}$ points allowed to right of S. by E. is S. by W. $\frac{1}{4}$ W., or S. $1\frac{1}{4}$ W.

Ex. 5. Compass course S.E., variation $1\frac{1}{4}$ points *East*, then the *true course* (allowing the variation to the *right*) will be S.S.E. $\frac{3}{4}$ E., or S. $2\frac{3}{4}$ points E.

Ex. 6. But compass course S.E., variation $2\frac{1}{2}$ points *West*, then the *true course* (allowing the variation to the *left*) will be E. by S. $\frac{1}{2}$ S., or S. $6\frac{1}{2}$ points E.

Ex. 7. Compass course East, variation 2 points *West*, then allowing 2 points to the left gives *true course* E.N.E.

Ex. 8. Compass course E., variation $2\frac{3}{4}$ points *East*, then the *true course* (allowing the variation to the right hand) is S.E. by E. $\frac{1}{4}$ E.

Now proceeding to the courses between North and South round by West.

Ex. 9. Course by compass N.W. $\frac{1}{2}$ W., variation 2 points *West*, then the *true course* (allowing the variation to the *left*) will be W. by N. $\frac{1}{2}$ N., or N. $6\frac{1}{2}$ points W.

Ex. 10. Taking the same compass course, viz, N.W. $\frac{1}{2}$ W., when the variation is $1\frac{1}{2}$ points *East*, the *true course* (allowing the variation to the *right*) will be N.W. by N., or N. 3 points W.

Ex. 11. Again, compass course S.W. by S., variation $2\frac{3}{4}$ points *West*, the *true course* (allowing variation to the *left*) will be S. $\frac{1}{4}$ W.

Ex. 12. Compass course S.W. by S. (as before) variation $1\frac{3}{4}$ *East*, the *true course* (allowing the variation to the *right*) will be S.W. $\frac{3}{4}$ W., or S. $4\frac{3}{4}$ points W.

Ex. 13. Compass course S.S.W., variation $3\frac{1}{2}$ *West*, then allowing $3\frac{1}{2}$ W. to the left of S.S.W. gives S. by E. $\frac{1}{2}$ E., or $1\frac{1}{2}$ points E.

Ex. 14. Compass course W., variation $2\frac{1}{2}$ E., then the *true course* (allowing $2\frac{1}{2}$ points to the *right*) is N.W. by W. $\frac{1}{2}$ W., or N. $5\frac{1}{2}$ points W.

Ex. 15. But with compass course West, and variation $3\frac{1}{4}$ *West*, then allowing $3\frac{1}{4}$ points to the left of W., the *true course* is S.W. $\frac{3}{4}$ W., or S. $4\frac{3}{4}$ points W.

Ex. 16. Compass course N.N.W. $\frac{1}{2}$ W., variation $3\frac{1}{4}$ points *East*, then $3\frac{1}{4}$ points to the right of N.N.W. $\frac{1}{2}$ W., is N. $\frac{3}{4}$ E.

197. The learner should so familiarise himself with the compass card as to be able entirely to dispense with its use in correcting courses, and when he has acquired such knowledge, he will find the following rule serviceable, in which the points of the compass are treated numerically.

RULE I.

1°. *Put down the* points *and* quarter points *which the* compass course *is to the* right *or* left *of* North *or* South, *marking them* R. *or* L. *accordingly.*

2°. *Underneath put the* variation, *marking it also* R. *or* L., *accordingly as it is* E. *or* W.

3°. *If the names are* alike, *take the* sum, *with* that name, *for the true course.*

(a) *When the* sum amounts *to* 8 points, *it is either* E. *or* W.

(b) *When the* sum exceeds 8 points, *take it from* 16 points; *the* remainder *is the true course to be reckoned from the* opposite point to that which the compass course is reckoned from.

That is, it is to be reckoned from the North if it had previously been reckoned from S., but marked S. if previously marked N.; also, if marked L (left) change to R (right); but if marked R change to L.

4°. *If the names are* unlike, *take the* difference, *and mark it the* same name *as the* greater.

Correcting Courses.

(c) *If the variation, being* subtractive, *exceeds the amount from which it is to be subtracted, take the points of the course* from *the* variation, *and name it the course towards* West *if it had previously been* Easterly, *but towards the* East *if it had been* Westerly.

(d) *Also bear in mind that* o *points is either* North *or* South *as the case may be.*

The following are examples of this method of applying the variation, and the numbers and letters in brackets refer to the rule as given above:—

1. Compass Courses:—S.S.W.; N. by E. ½ E.; W.S.W.; and E. by N. Variation 3½ points Easterly. Required the True Courses.

S.S.W.	N. by E. ½ E.	W.S.W.	E. by N.
S.S.W. = 2 R. of S.	1½ R. of N.	6 R. of S.	7 R. of N.
Var. 3½ R. [3°]	3½ R. [3°]	3½ R. [b]	3½ R. [b]
Sum 5½ R. of S.	5	— 9½ R. of S.	— 10½ R. of N.
		16	16
S.W. by W. ½ W.	N.E. by E.		
Here the *sum* is taken for true course, the names being alike.	Here the names being alike the *sum* is taken. (See No. 3°).	6½ L. of N.	5½ L. of S.
		W. by N. ½ N.	S.E. by E. ½ E.

2. Compass Courses:—N.N.W.; S. by E.; W. ½ N.; and E. by S. Variation 2½ W.

N.N.W.	S. by E.	W. ½ N.	E. by S.
2 L. of N.	1 L. of S.	7½ L. of N.	7 L. of S.
2½ L. [3°]	2½ L. [3°]	2½ L. [b]	2½ L. [b]
4½ L. of N.	3½ L. of S.	— 10 L. of N.	— 9½ L. of S.
N.W. ½ W.	S.E. ½ S.	16	16
		6 R. of S.	6½ R. of N.
		W.S.W.	E. by N. ½ N.

3. Compass Courses:—N.E. ½ E.; S.W. ¾ W.; N. by E.; and S. by W. Variation 2¼ points West.

N.E. ½ E.	S.W. ¾ W.	N. by E.	S. by W.
4½ R. of N.	4¾ R. of S.	1 R. of N.	1 R. of S.
2¼ L. [4°]	2¼ L. [4°]	2¼ L. [c]	2¼ L. [c]
2¼ R. of N.	2½ R. of S.	1¼ L. of N.	1¼ L. of S.
N.N.E. ¼ E.	S.S.W. ½ W.	N. by W. ¼ W.	S. by E. ¼ E.

4. Compass Courses:—N.W. by W.; S.E. by E.; N. by W. ½ W.; and S. by E. Variation 3¼ points East.

N.W. by W.	S.E. by E.	N. by W. ½ W.	S. by E.
5 L. of N.	5 L. of S.	1½ L. of N.	1 L. of S.
3¼ R. [4°]	3¼ R. [4°]	3¼ R. [c]	3¼ R. [c]
1¾ L. of N.	1¾ L. of S.	1¾ R. of N.	2¼ R. of S.
N. by W. ¾ W.	S. by E. ¾ E.	N. by E. ¾ E.	S.S.W. ¼ W.

5. N.N.E., Variation 2 points W.; S. by E., Variation 1 point E.; W. by S., Variation 1 point E.; and E.S.E., Variation 2 points W.

N.N.E., Var. 2 W.	S. by E., Var. 1 E.	W. by S., Var. 1 E.	E.S.E., Var. 2 W.
2 R. of N.	1 L. of S.	7 R. of S.	6 L. of S.
2 L.	1 R.	1 R.	2 L.
0 [d]	0 [d]	8 R. of S. [a]	8 L. of S. [a]
N.	S.	W.	E.

6. North, Variation 2 points E.; South, Variation 2 points W.; West, Variation 2 points W.; and East, Variation 2 points E.

N., Var. 2 E.	S., Var. 2 W.	W., Var. 2 W.	E., Var. 2 E.
0 = N.	0 = S.	8 R. of S.	8 L. of S.
2 R. of N.	2 L. of S.	2 L.	2 R.
2 R. of N. [3°]	2 L. of S. [3°]	6 R. of S. [4°]	6 L. of S. [4°]
N.N.E.	S.S.E.	W.S.W.	E.S.E.

198. If the learner has carefully gone through the preceding examples, he will have noticed that *Easterly* variation in its application to Compass Courses *increases* them in the *N.E.* and *S.W.* quarters of the compass, and *decreases* them in the *N.W.* and *S.E.* quarters. *Westerly* variation *decreases* the courses in the *N.E.* and *S.W.* quadrants, and *increases* it in the *N.W.* and *S.E.;* we have, therefore,

RULE LI.

Westerly variation is — from all points between N. and E. ....S. and W.
Easterly variation is + to all points between N. and E. ....S. and W.
Westerly variation is + to all points between N. and W.....S. and E.
Easterly variation is — from all points between N. and W.....S. and E.

We shall now proceed to illustrate the foregoing rule, which is very generally used in the correcting of courses.

1. Compass Courses:—N.N.E.; S. by W. ½ W.; W.N.W.; S.E. ½ E. Var. 3¼ E.

N. 2 E.	S. 1½ W.	N. 6 W.	S. 4½ E.
+ 3¼ E.	+ 3¼ E.	− 3¼ E.	− 3¼ E.
N. 5¼ E.	S. 4¾ W.	N. 2¾ W.	S. 1¼ E.
N.E. by E. ¼ E.	S.W. ¾ W.	N.N.W. ¾ W.	S. by E. ¼ E.

2. Compass Courses:—E.N.E.; W. by S.; N.N.W.; S. by E. Var. 3¼ E.

N. 6 E.	S. 7 W.	N. 2 W.	S. 1 E.
+ 3¼ E.	+ 3¼ E.	− 3¼ E.	− 3¼ E.
N. 9¼ E.	S. 10¼ W.	N. 1¼ E.	S. 2¼ W.
16	16	N. by E. ¼ E.	S.S.W. ¼ W.
S. 6¾ E.	N. 5¾ W.		
E. by S. ¾ S.	N.W. by W. ¾ W.		

3. Compass Courses:—N.E.; S.W. ½ S.; N.W. ½ N.; S.E. ¼ S. Var. 2¼ W.

N. 4 E.	S. 3½ W.	N. 3½ W.	S. 3¾ E.
− 2¼ W.	− 2¼ W.	+ 2¼ W.	+ 2¼ W.
N. 1¾ E.	S. 1¼ W.	N. 5¾ W.	S. 6 E.
N. by E. ¾ E.	S. by W. ¼ W.		

Correcting Courses.

4. Compass Courses:—N. by E.; S. by W. ¼ W.; W. ½ N.; E. by S. Var. 2¼ W.

```
    N. 1  E.          S. 1¼ W.         N. 7½ W.         S. 7  E.
   − 2¼ W.           − 2¼ W.          + 2¼ W.          + 2¼ W.
   ─────             ─────            ─────            ─────
    N. 1¼ W.          S. 1  E.         N. 9¾ W.         S. 9¼ E.
                                         16               16
                                       ─────            ─────
                                       S. 6¼ W.         N. 6¾ E.
```

5. Compass Courses:—N.N.W. ¾ W.; S.S.E. ¾ E.; N.E. by E. ¼ E.; S.W. by W. ¼ W. Variation 2⅜ E.

```
    N. 2¾ W.         S. 2¾ E.         N. 5¼ E.         S. 5¼ W.
   − 2⅜ E.           − 2⅜ E.          + 2¾ E.          + 2¾ E.
   ─────             ─────            ─────            ─────
      0                 0              N. 8  E.         S. 8  W.
    ─────             ─────
    North.            South.            East.           West.
```

199. Sometimes it may be desirable to express the Variation in degrees, in which case we proceed as follows:—

RULE LII.

1°. *Correct the compass courses for leeway as before directed, and convert the number of points thus found into degrees, marking them* R *or* L, *according as they are right or left of* N. *or* S.

2°. *Underneath write the variation, marking it* R *or* L *according as it is* E. *or* W. *Take the* sum *with the common name, if the names are* alike, *and the* difference *with the name of the* greater, *if the names are* unlike. The result will be the number of degrees the *true course* is from N. or S. *according as the course, corrected for leeway, is reckoned from the* N. *or* S.

(a) *If, in taking the sum the number of degrees* exceed 90° *take the* supplement to 180°, *and reckon the* true course *from the* opposite *point to that from which the* course corrected for leeway *is reckoned; also change the letter* R *or* L.

EXAMPLES.

Compass Course.	Winds.	Leeway.	Var.	True Course.
S.W. ½ S.	W. by N. ½ N.	¾	23° W.	S. 8° W.
N. by E.	E. by N.	3¼	20° E.	N. 5° W.
W. ¾ N.	S.W. by S.	1	25° W.	S. 85° W.

```
S.W. ½ S. = 3½ R. of S.      N. by E. = 1 pt. R. of N.       W. ¾ N. = 7¼ L. of N.

            3½ R. of S.                   1  R. of N.                   7¼ L. of N.
Leeway       ¾ L.                        3¼ L.                           1  R.
            ─────                        ─────                          ─────
            2¾ R.                        2¼ L.                           6¼

         or 31° R. of S.              or 25° L. of N.                 or 70° L.
Var.        23  L.                       20                              25  L.
            ─────                        ─────                          ─────
             8 R. of S.                   5 L. of N.                    −95 L. of N.
            ─────                        ─────                          180
            S. 8° W.                     N. 5° W.                       ─────
                                                                         85 R. of S.
                                                                        ─────
                                                                        S. 85° W.
```

RULE LIII.

200. *To find the compass course*, the true course and variation being given.

Easterly variation is allowed to the left.
Westerly " " right.

Examples.

Taking the courses between North and South round by East.

Ex. 1. Let the true course be N.E. by E., where the variation is 1½ points *West*, the *compass* course (allowing westerly variation to the *right*) will be E.N.E. ½ E.

Ex. 2. Taking the same course, viz. N.E. by E., where the variation is 1½ points *East*, and then the *compass* course (allowing easterly variation to the *left*) will be N.E. ½ N.

Ex. 3. Suppose the *true* course to be S.E. by E., where the variation is 2¼ points *West*, the *compass* course (allowing variation to the *right*) will be S.S.E. ¾ E.

Ex. 4. But the same course, viz., S.E. by S., where the variation is 2¾ points *easterly*, will give the *compass* course (allowing easterly variation to the *left*) S.E. by E. ¾ E.

Taking the courses between North and South round by West.

Ex. 5. Let the *true* course be N.W. by W., where the variation is 2½ points *West*, then the *compass* course (allowing westerly variation to the *right*) will be N.N.W. ½ W.

Ex. 6. Suppose the course to be the same, viz., N.W. by W., where the variation is 2½ points *easterly*, the *compass* course (allowing easterly variation to the *left*) will be W. ½ N.

Ex. 7. With the *true* course *West*, and the variation 2 points *West*, then the *compass* course (allowing 2 points to the *right*) will be W.N.W.

Ex. 8. Taking the same course *West*, suppose the variation to be 2 points *East*, then the *compass* course (allowing 2 points to the *left*) is W.S.W.

Ex. 9. With the *true* course S.W. ½ S., where the variation is 2¼ points *West*, then the *compass* course (allowing westerly variation to the *right*) will be S.W. by W. ¾ W.

Ex. 10. But with the same course, viz., S.W. ½ S., where the variation is 1½ points *East*, the *compass* course (allowing easterly variation to the *left*) will be S.S.W.

201. Treating the points of the compass numerically, we proceed according to the following

RULE LIV.

Proceed according to Rule L, page 124, *in every particular, except that the variation is to be* allowed the **opposite way** *to that of correcting compass courses,* viz., WESTERLY *variation is to be allowed to the* **right** *and marked* R, *and* EASTERLY *variation is to be allowed to the* **left** *and marked* L.

Examples.

1. True Courses:—N.N.E.; S. by W. ½ W.; E. by N. ½ N.; W. by S. Var. 2½ W.

N.N.E.	S. by W. ½ W.	E. by N. ½ N.	W. by S.
2 R. of N.	1½ R. of S.	6½ R. of N.	7 R. of S.
2½ R.	2½ R.	2½ R.	2½ R.
4½ R. of N.	4 R. of S.	9 R. of N.	9½ R. of S.
N.E. ½ E.	S.W.	16	16
		7 L. of S.	6½ L. of N.
		E. by S.	W. by N. ½ N.

2. True Courses:—N.E. by E. ¼ E.; S.W. by W.; E. by S. ¼ S.; N.W. by W. Variation 3¼ E.

N.E. by E. ¼ E.	S.W. by W.	E. by S. ¼ S.	N.W. by W.
5¼ R. of N.	5 R. of S.	6¾ L. of S.	5 L. of N.
3¼ L.	3¼ L.	3¼ L.	3¼ L.
2 R. of N.	1¾ R. of S.	10 L. of S.	8¼ L. of N.
		16	16
N.N.E.	S. by W. ¾ W.	6 R. of N.	7¾ R. of S.
		E.N.E.	W. ¼ S.

3. True Courses:—N. by W. ½ W. and S. by E.; Variation 3¼ W. S. by W. and N. by E.; Variation 3¾ E.

N. by W. ½ W.	S. by E.	S. by W.	N. by E.
1½ L. of N.	1 L. of S.	1 R. of S.	1 R. of N.
3¼ R.	3¼ R.	3¾ L.	3¾ L.
1¾ R. of N.	2¼ R. of S.	2¾ L. of S.	2¾ L. of N.
N. by E. ¾ E.	S.S.W. ¼ W.	S.S.E. ¾ E.	N.N.W. ¾ W.

202. To convert true course into compass course, we may proceed according to the following

RULE LV.

Westerly variation is + to all points between N. and E. ....S. and W.
Easterly variation is — from all points between N. and E. ....S. and W.
Westerly variation is — from all points between N. and W.....S. and E.
Easterly variation is + to all points between N. and W.....S. and E.

DEVIATION OF THE COMPASS.

GENERAL STATEMENT OF FACTS AND LAWS OF MAGNETISM.

203. **Magnets, Natural and Artificial.**—Natural magnets, or *loadstones*, are exceedingly rare, although a closely allied ore of iron, capable of being strongly acted upon by magnetic forces, and hence called *magnetic iron ore*, is found in large quantities in Sweden and elsewhere. Artificial magnets are usually pieces of steel which have been permanently endowed with magnetism by the action of other magnets. The needle, or bar of steel, in the Mariner's Compass is an artificial magnet.

204. **Poles, Neutral Lines, and Axis.**—The property of attracting iron is very unequally manifested at different points of the surface of a magnet. If, for example, an ordinary bar-magnet be plunged into iron filings, these become arranged round the ends of the bar in feathery tufts, which decrease towards the middle of the bar where there are none. The name *poles* is used, in a somewhat loose sense, to denote the two terminal portions of a magnet, or to denote two points, not very accurately defined, situated in these portions. The middle position to which these filings refuse to adhere is called the *neutral line*. Every magnet, whether natural or artificial, has two poles and a neutral line. The shortest line joining the two poles is termed the *axis* of the magnet.

205. **Magnetic Meridian** at any station is best defined as the direction of the declination needle.

206. **The Magnetic Equator** or *aclinic line* is the line which joins all those places of the earth where the needle remains quite horizontal, or where there is no dip. This line does not coincide with the geographical equator, nor is it a great circle, but a somewhat irregular curve crossing the geographical equator at two points almost exactly opposite each other, one near the West coast of Africa, in the Atlantic, and the other in the middle of the Pacific Ocean, and never receding from it further than $12°$; the position of the two being nearly coincident in that part of the Pacific where there are few islands, and most divergent when traversing the African and American continents.

207. **Magnetic Poles.**—At two points, or rather small linear spaces on the earth's surface, the needle assumes a position perpendicular to the horizon, or the dip is $90°$. These two spots are called *Magnetic Poles*. At the North magnetic pole, the North pole of the needle dips; at the South magnetic pole, the south pole of the needle dips. The terrestrial magnetic poles do not coincide with the geographical ones, nor are these points diametrically opposite. The position of these poles are latitude $70°$ N., long. $97°$ W., and lat. $73\frac{1}{2}°$ S., long. $147°$ E.

The line of no variation passes through these poles, and the lines of equal variation converge towards them.

208. Magnetic Elements.—A knowledge of terrestrial magnetism implies a knowledge of (1) Declination or Variation, (2) Inclination, (3) Intensity. These are called the magnetic elements of the place at which they are observed.

209. Magnetic Needle.—Magnetic Declination or Variation.—Any magnet freely suspended near its centre is usually called a *magnetic* needle, or more properly a *magnetised* needle. When a magnetised needle is so suspended or mounted that it can vibrate in the horizontal plane, it will take a definite direction, to which it always comes back after displacement. In this position of stable equilibrium, one of its ends points to the direction called magnetic north, and the other magnetic south, which differ, in general, by several degrees from geographical (or true) north and south. This is the principle on which compasses are constructed. The angle between the magnetic meridian and the geographical meridian is called the **Variation of the Compass.**

Imaginary lines on the surface of the earth, passing through all points where the needle points due north and south, are called **Lines of no Variation**, and lines passing through all points where the needle is deflected from the geographical meridian are called **Lines of Equal Variation**. These are extremely irregular curves, and form two closed systems surrounding two points, which may be called **Centres of Variation**. One of these points is in Eastern Siberia, the other in the Pacific Ocean, in the vicinity of the Marquesas.

210. Dip, or Inclination.—When a needle is prepared in the unmagnetised state for mounting in a compass, with its centre of gravity very little below its point of support, and is adjusted to horizontality, on being magnetised it will place itself in a particular vertical plane called the magnetic meridian, and will take a particular direction in that plane. This direction is not horizontal, except at the equatorial regions of the earth, but inclined generally at a considerable angle to the horizon, and this angle is called *dip*, or *inclination*. Its value at Greenwich, at present, is about 67°, the end which points to the north pointing at the same time downwards.

In the northern hemisphere, generally, it is the north end of the needle which dips, and in the southern hemisphere it is the end which points south.

The value of the dip, like that of the variation, differs in different localities. It is greatest in the polar regions, and decreases with the latitude to the equator, where it is approximately zero.

Dip, like the variation, varies greatly, not only from place to place, but also from time to time. In 1843 the dip at Greenwich was about 69° 1', it has diminished, with a rate continually accelerating, till in 1868 it was 67° 56'. It is also subject to slight annual and diurnal variations, being about 15' greater in summer than in winter.

Intermediate to the poles and equator lines are drawn through all points where the needle makes the same angle with the horizon. These are called *Lines of Equal Inclination or Dip*.

To help the seaman to understand the above remarks, let him proceed as follows:—Having provided a little unspun silk, by means of a bit of wax, or otherwise, attach the silk fibre to the magnetic needle by a single point at its middle. Place a magnet on the table, and hold the needle over the equator of the magnet. The needle sets horizontal. Move it towards the north end of the magnet, the south end of the needle dips, the dip augmenting as the north pole is approached, over which the needle, if free to move, will set itself exactly vertical. Move it back to the centre, it resumes horizontality; pass it towards the south pole, its north end now dips, and directly over the south pole the needle becomes vertical, its north end being now turned downward. Thus we learn that on one side of the magnetic equator the north end of the needle dips; on the other side the south end dips, the dip varying from nothing to ninety degrees. If we go to the equatorial regions of the earth with a suitably suspended needle, we shall find the position of the needle horizontal. If we sail north, one end of the needle dips; if we sail south, the opposite end dips; and over the north or south terrestrial magnetic pole the needle sets vertical. In this manner we establish a complete parallelism between the action of the earth and that of an ordinary magnet.

NOTE.—The dip is of importance to the navigator, as it appears to regulate the local deviation of the compass. It also renders necessary an adjustment to secure the horizontality of the compass card.

211. The horizontal position of the needle and card is preserved by a sliding brass weight fitted for the purpose, or by dropping sealing wax on one end of the needle. This adjustment will often require to be repeated after a considerable change of place.

212. **Mutual Action of Poles.**—On presenting one end of a magnet to one end of a needle thus balanced, we obtain either repulsion or attraction, according as the pole which is presented is similar or dissimilar to that to which it is presented. *Poles of contrary names attract one another; poles of the same name repel one another.*

This property furnishes the means of distinguishing a body which is merely magnetic (that is, capable of temporary magnetization) from a permanent magnet. The former, a piece of soft iron, for example, is always attracted by either pole of a permanent magnet, while a body which has received permanent magnetization has, in ordinary cases, two poles, of which one is attracted where the other is repelled. Magnetic attractions and repulsions are exerted without modification through any body which may be interposed, provided it be not magnetic.

213. **Names of Poles.**—The phenomena of variation and dip above described evidently require us to regard the earth, in a broad sense, as a magnet, having one pole in the northern and the other in the southern hemisphere. Now, since poles which attract one another are dissimilar, it follows that the magnetic pole of the earth, which is situated in the northern hemisphere, is *dissimilar* to that end of a magnetised needle which points to the north. Hence, great confusion of nomenclature has arisen, the usage of the best

writers being opposite to that which generally prevails. Popular usage in this country, however, calls that end or pole of a needle which points to the north the *north pole*, and that which points to the south the *south pole*.*

214. **Magnetic Induction.**—When a piece of iron is in contact with a magnet, or even when a magnet is simply brought near it, it becomes itself, for the time, a magnet with two poles and a neutral portion between them. If we scatter filings over the iron they will adhere to its ends, as shown (204). If we take away the influencing magnet the filings will fall off, and the iron will retain either no traces at all, or only very faint ones of its magnetization. If we apply similar treatment to a piece of steel, we obtain a result similar in some respects, but with very important differences in degree. The steel, while under the influence of the magnet, exhibits much weaker effects than the iron; it is much more difficult to magnetise than iron, and does not admit of being so powerfully magnetised; but, on the other hand, it retains its magnetization after the influencing magnet has been withdrawn. This property of retaining magnetism, when once imparted, has been named *coercive force*. Steel, especially when very hard, possesses great coercive force; iron, especially when very pure and soft, scarcely any.

In magnetization by influence, which is also called *magnetic induction*, it will be found on examination that the pole which is next the inducing pole is of contrary name to it; and it is on account of the mutual attraction of dissimilar poles that the iron is attracted by the magnet. The iron can in its turn support a second piece of iron, this again can support a third, and so on through many steps. A magnetic chain can thus be formed, having two poles. An action of this kind takes place in the clusters of filings which attach themselves to one end of a magnetised bar, these clusters being composed of numerous chains of filings.

215. **Magnetization by the action of the Earth.**—The action of the earth on magnetic substances resembles that of a huge permanent magnet, and hence the terrestrial magnetism will induce magnetism precisely as explained in 214. All soft or cast iron rods or bars, or other elongated forms of soft or cast iron, unless the position of their length is at a right-angle to the line of the direction of the earth's magnetic force, are immediately rendered magnetic by induction from the earth, and the nearer the iron is in direction to the line of force or dip the greater will be the amount of induction. When a bar of soft iron is held on the magnetic meridian and parallel to the dip, it becomes immediately endowed with feeble magnetic polarity. The lower

---

* Sir W. THOMSON calls the north-seeking pole the *south* pole, and the other the *north* pole, because the former is similar to the south and the latter to the north pole of the earth. In like manner most French writers call the north-seeking pole of a needle the *austral*, and the other the *boreal* pole. FARADAY, to avoid the ambiguity which has attached itself to the names north and south pole, calls the north-seeking end the *marked*, and the other the *unmarked* pole. AIRY, for a similar reason, employs in his recent *Treatise on Magnetism*, the distinctive names *red* and *blue* to denote respectively the north-seeking and south-seeking ends; these names, as well as those employed by FARADAY, being purely conventional, and founded on the custom of marking the north-seeking end of a magnet with a transverse notch or a spot of red paint. MAXWELL and JENKIN, in a report to the British Association, call the south-seeking pole of a needle *positive*, and the north-seeking pole *negative*.

extremity is a north pole, and if the north pole of a small magnetic needle be approached, it will be repelled. If the bar is held vertically the lower end will still be a north pole, but of less intensity; the upper end a south pole, also of less intensity. If the bar is held horizontally north and south, the north end will be a north pole, but of still lesser intensity; the south end a south pole, also of lesser intensity. If we now turn the bar in the same horizontal plane its magnetism will diminish, and if placed in an east and west direction, it will lose its polarity, and if we turn it still further until its position is reversed, the magnetic poles of the bar will be reversed.

While the bar is held with its length in the direction of the dip, if it be struck repeatedly with an iron hammer, it will be found, on removing it, to be a true magnet, the end which was lowest being charged with north magnetism, and this magnetism is not transient like the induced magnetism of soft iron, changing its place in the bar with every change in the position of the bar, but is constant like that of a steel bar, retaining the same magnetism whatever the position of the bar. By reversing the position of the bar and striking it a few blows with the hammer, its magnetism is reversed. The magnetism of the bar so struck resembles that of a steel magnet in all respects but this, that while, perhaps, no change can be remarked in hours or days, it infallibly diminishes in a long time. To express this partially permanent character, the term Subpermanent Magnetism has been adopted.

216. A sphere of soft iron will be magnetised in the same way, however held. The diameter in the line of dip will be the axis of magnetism, and the lower and north half of the surface will be north, the upper and south half south.

In bodies of any other shape the effects will be similar.

In an iron ship on the stocks, intense magnetism is developed by the process of hammering; N. magnetism being developed in the part of a ship which is below and towards the north, and S. magnetism in the part which is above and towards the south.

217. In the northern hemisphere all vertical or upright bars, such as stanchions and angle-irons composing the frames of ships, are magnetised by induction, their lower ends being north poles, the upper ends south poles, the upper end attracting the north pole of the needle held near them. On the other hand, in the southern hemisphere, these conditions are reversed; the upper ends of vertical iron are north poles, repelling the north pole of a compass needle and attracting the south pole. On the magnetic equator, where there is no dip, vertical soft iron has no polarity, because its position is at right-angles to the earth's line of force or dip. It is different with horizontal pieces of soft iron; they exert the same influence on a compass needle in both hemispheres, and in all latitudes.

218. The hull of an iron ship acts as a permanent magnet on compasses placed outside the vessel as well as those placed inside; an iron ship must therefore be viewed in its effect on a properly placed magnet rather as one great magnet, than as an aggregation of smaller magnets.

Keeping in view that the inductive effect from the earth's magnetism is greatest in the line of the dip, and the existence of a neutral equatorial plane at right-angles to the line of dip in spherical bodies, we are prepared to see that each iron ship must have a distinct distribution of magnetism depending on the place of building, and the direction of the head and keel while building; the ship's polar axis and equatorial plane conforming more or less to the line of dip of the earth at the place where built, and a plane at right-angles to that line; abundant observation and experiment have proved this important general principle.

219. To illustrate this principle: let us suppose, as in the following Figs. 3, 4, 5, and 6, that four iron ships, or four composite-built ships, with ribs, beams, stanchions, and deck girders of iron, are building on the cardinal points of the compass, in a port in England where the line of the earth's total magnetic force is inclined 70° to the earth's horizontal magnetic force, or in other words, where the dip of the needle is 70°.

Fig. 3 shows the magnetic state of a ship built head North magnetic. The line marked Dip passes through the centre of the ship; it shows the direction of the line of the earth's magnetic force. The line marked Equatorial or Neutral line is the line of no deviation, and runs at right-angles to the Dip. The after body of the ship, or the portion which is shaded, has

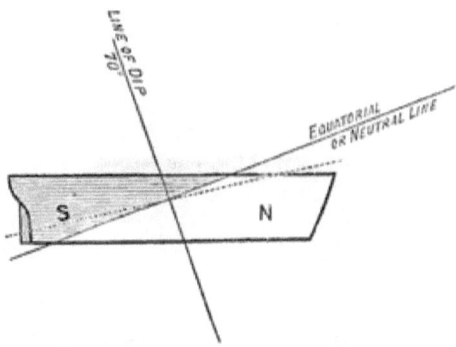

Fig. 3. Head North while building.

S. (*blue*) polarity, and the fore body, or white portion of the figure, N. (*red*) polarity; the upper part of the stern would have the S. (*blue*) polarity developed in a high degree; the lower part of the bows would have the N. (*red*) polarity equally developed. At the stern the north end of a compass needle would be strongly attracted; at the bow the south end of the needle would be strongly attracted; while a compass placed outside of the ship's topsides, above the line of no deviation, the north end of the needle will be attracted; if it be placed below that line the north end of the needle will be repelled and the south end attracted, in accordance with the law of magnetism. (No. 205.)

Fig. 4. Head South while building.

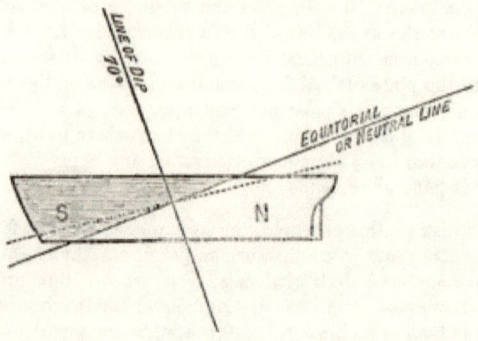

Fig. 4 represents the magnetic condition of a ship built head south. It will be seen by comparing Fig. 4 with Fig. 3 that the conditions are reversed; in Fig. 3 the magnetism of the after body of the ship is south (*blue*), while in Fig. 4 the after part of the ship possesses north (*red*) polarity; now the fore body of the ship has S. (*blue*) polarity, while in Fig. 3 it has N. (*red*) polarity; the upper part of the bow has S. (*blue*) polarity developed in a high degree, and the lower part of the stern N. (*red*) polarity equally developed. The N. (*red*) polarity of the stern repels the north end of the compass needle, and the S. (*blue*) polartiy of the bow attracts it. The dotted line crossing the equatorial line in Figs. 3 and 4 shows the probable position of the neutral line after the ship has been some time afloat, with her head in an opposite direction to that in which she was built, or after she had made a voyage.

The place of little or no deviation in a ship built head north is towards the bow, but in a ship built head south, towards the stern.

Fig. 5. Head East while building.

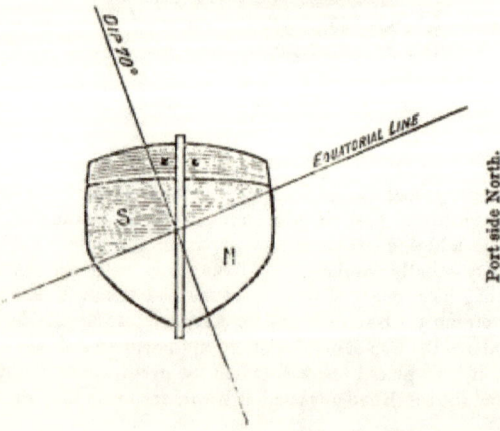

Fig. 5 is intended to show the magnetic state of a ship built head East. The whole of the upper part of the ship has S. (*blue*) polarity; the whole of the lower part has N. (*red*) polarity; but the S. (*blue*) polarity predominates on the starboard side, and the north end of a compass needle, if carried at the usual height of a compass along the amidship line of the upper deck from end to end, is attracted to the starboard side.

Fig. 6. Head West while building.

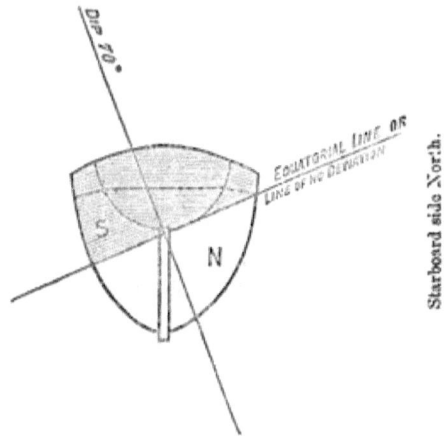

In Fig. 6, ship built head West, the magnetic conditions of Fig. 5, head East, are reversed; the whole of the upper part of the ship has still S. (*blue*) polarity, and the lower N. (*red*) polarity; but the magnetism of the port side of the upper works is developed in a higher degree than the starboard side, and the N. end of a compass needle, if carried along the upper deck from end to end, would be attracted to the port side. In other words, in these ships the whole of their decks have a S. (*blue*) polarity, yet in that part which was North while the ship was being built, this S. (*blue*) polarity is developed in a less degree than on the opposite side, consequently, the N. point of the compass is drawn towards that part of the ship in which the S. (*blue*) polarity is developed in the highest degree.

The deviation in both cases is rarely large, but less regular than in ships built head South.

T

Theoretically, there should be no spot of no deviation on the deck of ships built East or West.*

Ex. 7. Head North at Australia.

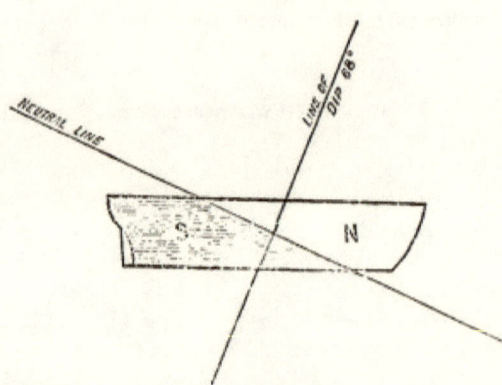

Fig. 7 represents an iron ship built head North in Australia, with a dip of about 68° South. In this ship the shaded part showing S. polarity lies below the equatorial line. It will be useful to compare this figure with Fig. 3, and mark the difference in the magnetic state of the two ships.

220. A little attention to the above diagrams will give the seaman a rough idea of the distribution of magnetism in iron ships; but it must be borne in mind that all large *detached* pieces of iron in a ship, such as iron masts, funnels, cylinders, and other masses of vertical iron, are independent magnets; in north magnetic latitude, their lower ends being north poles, their upper ends south poles.

221. The compasses of composite ships with iron frames and iron deck beams, are affected in the same way as those of ships built wholly of iron.

---

* From the special magnetic properties developed in a ship according to her position when building, it follows that a compass *aft*, in the usual place of the steering binnacle, the *character* of the deviation—*though not the amount*—may be approximately represented in a tabular form, as follows:—

Approximate magnetic direction of ship's head while building.	Approximate easterly deviation occurs when ship's head by compass is near	Maximum westerly deviation when ship's head by compass is near
N.	W.	E.
N.E.	N.W.	S.E.
E.	N.	S.
S.E.	N.E.	S.W.
S.	E.	W.
S.W.	S.E.	N.W.
W.	S.	N.
N.W.	S.W.	N.E.

DEVIATION OF THE COMPASS.

222. **The Deviation of the Compass** is the angle through which the magnetic needle is deflected from its natural position by the disturbing force of iron near it, that is, the angle included between the magnetic meridian and a plane passing through the poles of a compass needle.

The deviation is named East or West according as the north point of the compass so disturbed is to the east or west of its natural position.

Deviation consists of two principal parts, the Semicircular and the Quadrantal, following different laws, and requiring two different kinds of compensation; there is sometimes a third part of small amount called the Constant.

223. In the case of iron ships, as in that of iron bars (215), percussion and vibration, by hammering in rivetting, render the iron of which the vessel is constructed more susceptible to the inductive force of the earth, and causes the magnetism, which the iron of the ship thus acquires, to partake more of the character of permanent magnetism. Still this subpermanent magnetism undergoes a considerable diminution by being submitted to percussion, with the ship's head in a different position to that in which it was when she was being built, and especially if in a contrary direction. But the iron of which a ship is constructed always retains a large amount of this subpermanent magnetism as long as it remains in the form of a ship. The deviation arising from subpermanent magnetism is greater than that which is the result of transient induced magnetism. The polarity of the ship's magnetism, while she remains on the stocks, takes the direction of the earth's line of force or dip, and its effects on compasses will evidently depend on the direction the ship's head was whilst being built. Taking the case of a ship built head north (Fig. 3, page 135), the fore part of the ship has acquired north magnetism, and its action will be precisely the same as that of the north pole of a magnet; hence, on northerly courses, the north end of the compass needle will be repelled, and the directive power of the needle will be diminished. On southerly courses the north end of the needle points towards the stern, which has acquired subpermanent south magnetism, then the directive power of the needle is increased. On easterly and westerly courses the effects on the compass are greatest, since the force acts at right-angles to the needle; and on all intermediate positions of the ship's head the disturbances due to such positions are intermediate. As the ship's head is brought east of north, repulsion of the north end of the needle takes place, and westerly deviation is the result, and it reaches its maximum value when the fore-and-aft line of the ship is at right-angles to the needle; beyond that position the fore part of the ship attracts the south end of the needle, and westerly deviation is still the result. This attraction continues until the ship's head reaches south, when the line of action of the ship lies in the same direction as the needle, and no disturbance occurs, but the directive power of the needle is greater. On bringing the ship's head round west of south, the south pole of the needle still continues to be attracted, which causes easterly deviation, and it again attains its maximum when the fore-

and-aft line of the ship is at right-angles to the disturbed needle; this must occur to the north of west. After that point has been reached by the ship's head, the fore part of the ship repels the north end of the needle, easterly deviation still being the result until the ship's head is again at north. Thus we find that in an iron ship the disturbance of the compass is little or nothing when her head is on or near the points to which her head or stern were directed while building, and is greatest when the ship's head is directed to the points of the compass that were abeam while on the building slip; and, moreover, that easterly deviation is caused when the ship's head is in one half of the compass, and westerly deviation in the other. The deviation caused by subpermanent magnetism, and the effects of magnetism induced in vertical iron, has received the name of **Semicircular Deviation**, from producing opposite effects when the ship's head is on opposite semicircles of the compass, as the ship's head moves round a complete circle of azimuth. This error is caused by the subpermanent magnetism acquired in building, and the magnetism induced in vertical iron. The part due to subpermanent magnetism remains the same in kind, though different in amount, in all latitudes, unless the ship be subjected to strains or other mechanical violence. The part caused by the magnetism induced in vertical iron changes with a change of geographical position, or more correctly, as the dip changes, and is of contrary names on opposite sides of the magnetic equator, that is, if westerly deviation be produced on one side, easterly will be produced on the other. At the magnetic equator the earth's magnetism acts horizontally, and vertical soft iron will have no magnetism, and the semicircular deviation arising therefrom will disappear.

As a general rule the magnetism producing semicircular deviation, in a ship built in north magnetic latitude, attracts the north end of a compass needle to that part of a ship which was south from the compass while building; hence, the semicircular deviation in iron ships is generally represented by the effect of a magnet at the part of the ship which was south in building, with its south end towards the compass. Thus, in a ship built head north, the north end of the needle is drawn towards the stern. The following table will show the part of a ship towards which the north end of a needle is generally drawn, that is, the position of the permanent south pole developed in the process of construction.

Ship's head while building.	The north end of the compass needle on the poop or quarter deck is usually drawn.
North	towards the stern.
N.E.	,, starboard quarter.
East	,, starboard side.
S.E.	,, starboard bow.
South	,, bows or right ahead.
S.W.	,, port bow.
West	,, port side.
N.W.	,, port quarter.

Fig. 8.

Magnetic.

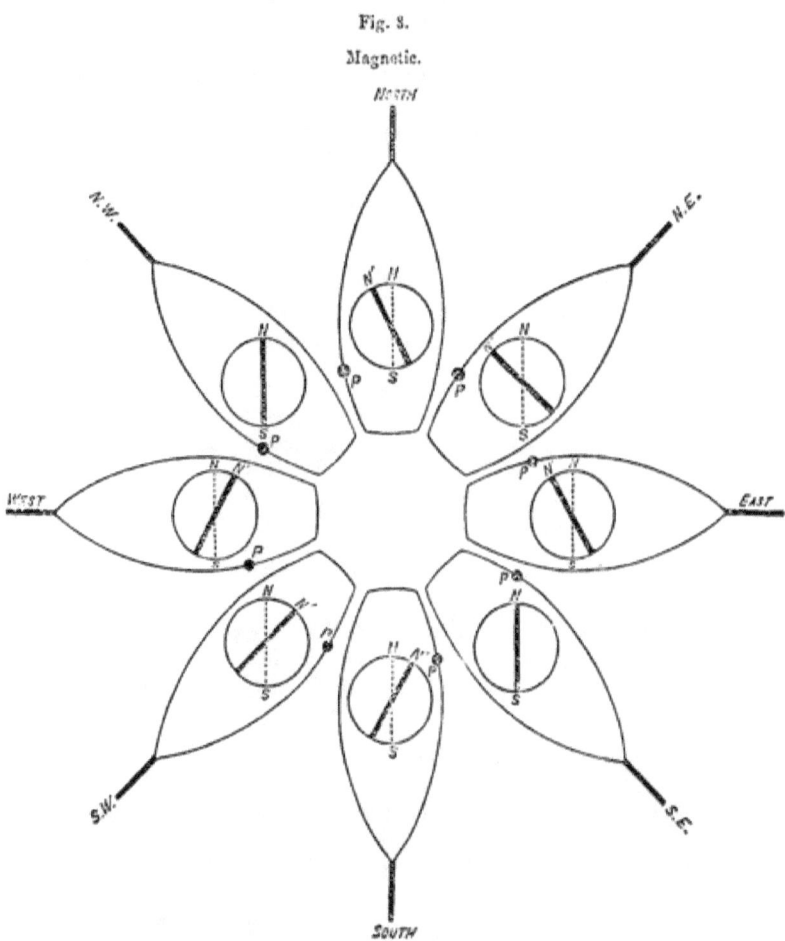

Fig. 8 will further illustrate the way in which the permanent magnetism, and the inductive magnetism of vertical iron acts upon the compass to produce semicircular deviation. Let it be supposed that the whole of the south polarity or attractive power of the above magnetism is concentrated in the point P on the port quarter of a ship built with her head near N.W. The ship is supposed to be swung round the compass, beginning at the N.W. point. The small circles represent the compass, the thick lines N'S the compass needle, the dotted lines the magnetic meridian or the direction of the needle when free from deviation. Beginning at N.W., and noting the position of the point P, it will be observed that there can be no semicircular deviation with ship's head in that direction, because the attractive force of the ship's magnetism at the point P is in a line with the compass needle N S. As the ship's head swings round towards the west, the relative positions of the point P and the compass needle will alter, and P will exert a pulling force upon the north end of the needle, causing it to deviate to the right from N to N', shown in the figure at West. The easterly deviation will increase until the ship's head swings to S.W., where it attains its greatest or maximum amount. After passing S.W. it gradually decreases past South until the ship's head reaches S.E., the opposite direction to that in which her head was built, where it is again zero or nothing. The point P is now on the opposite side of the compass to what it was when her head was at N.W., but it will be observed that it is in a line with the needle, and can exert no deviating influence over it.

As the ship swings with her head towards the East, the needle will gradually be drawn to the left hand until the westerly deviation attains its maximum at N.E. After passing N.E. the westerly deviation will decrease past North until the ship's head again reaches N.W., at which point there is no deviation. A very slight inspection of the figure will show that in the semicircle from N.W. round by the West to S.E., the deviation is easterly; while in the semicircle, or half the compass, from S.E. round by the East, the deviation is westerly. The above is merely given for the sake of illustration, but it must be remembered that no two ships are alike in their influence on the compass, nor will the ship's magnetism have the same effect on two compasses placed on different parts of the deck.

224. **Quadrantal Deviation** is so named from its being easterly and westerly, alternately, in the four quadrants as the ship moves round a complete circle of azimuth. It is caused by the transient or inductive magnetism of horizontal soft iron, such as iron deck beams, the iron spindle of the wheel, &c. It is zero or nothing when a ship's head is near the North, South, East, or West points, and greatest on the quadrantal points. It is generally easterly in the N.E. and S.W. quadrants, and westerly in the N.W. and S.E. quadrants of the compass. Quadrantal deviation remains unchanged in all magnetic latitudes, and provided that the iron in the ship be of good quality, the quadrantal deviation will be little, if at all, altered by the lapse of time.

To illustrate the way in which horizontal soft iron produces quadrantal deviation, let us suppose the whole of the induced magnetism in a ship to be represented by the soft iron bar B in Fig. 9. This cannot be so in actual practice, because the athwartship horizontal iron produces quadrantal deviation as well as the fore-and-aft iron, but we may suppose it may for the sake of clearness. The small circles represent the compass, the thick lines within the small circles the compass needle, the dotted lines within the compass the magnetic meridian. Beginning at north, it will be observed that the bar B is parallel with the magnetic meridian, and will therefore be an inductive magnet while it is in or near that position (215), its after end, marked S, being a south pole; but as the bar B is in a line with the compass needle N, it cannot exert any deviating power upon the needle, either to the right or left. As the ship's head swings towards the N.W., the relative positions of

Deviation of the Compass. 143

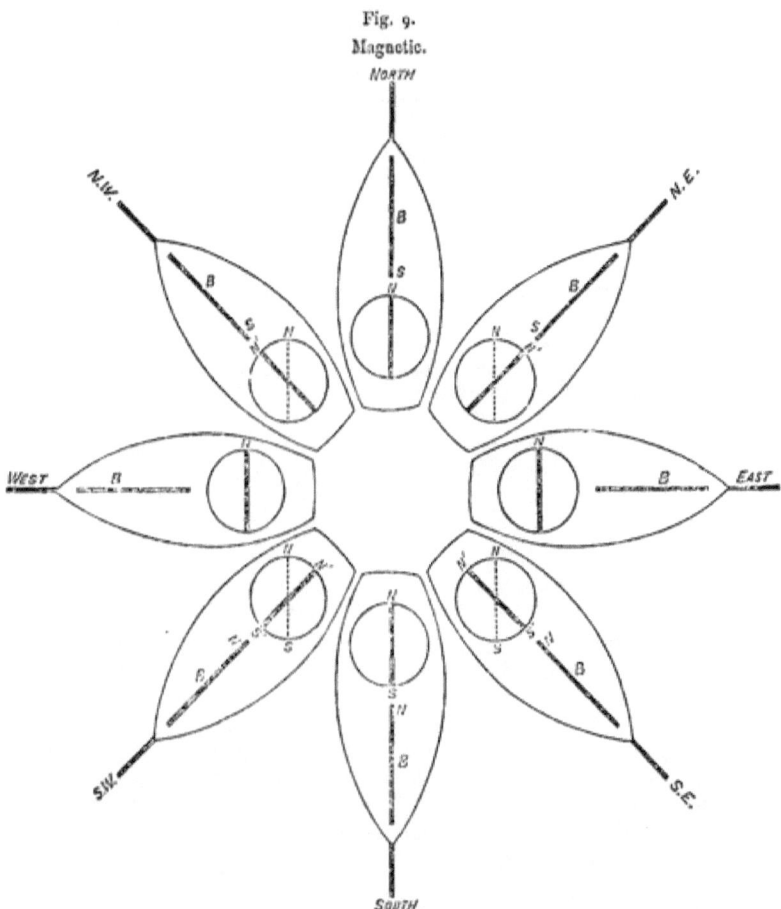

Fig. 9.

the bar B and the needle N are altered, and the south end of the bar draws the north end of the needle to the left from N to N'. As the ship's head approaches the west, the bar B loses its polarity, and at west it is at right-angles to the magnetic meridian, and ceases to exert any influence on the compass. The ship's head now swings towards the S.W., and the bar B, as it turns towards the south pole, again becomes an inductive magnet; its after end being a *north pole*, and drawing the south end of the compass needle from S to S'. When the ship's head reaches south there is no quadrantal deviation, because the bar B is in a line with the compass needle. As her head swings towards the S.E., the needle is drawn from S to S', causing westerly deviation. At east there is no deviation, for the same reason that there was none at west. After passing east, the after end of the bar B becomes a *south pole*, and draws the north end of the needle to the right-hand in the N.E. quadrant. As the ship's head approaches the north, the quadrantal deviation gradually decreases until it becomes nothing at north. The reader will observe that the bar B in this case produces easterly deviation in the N.E. and S.W. quadrants, and westerly deviation in the N.W. and S.E. quadrants. Cases may arise where the deviation is westerly in the N.E. and S.W. quadrants, but they are very rare.

225. **The constant part of the deviation** is generally very small, and is the same for every point of the compass, it often arises from defects in the compass itself. An error in the correct magnetic bearing of a distant object used to ascertain the deviation, will give an *apparent* constant deviation: for example, if the correct magnetic bearing of a lighthouse be S. 46° E., and the observer assumes it to be S. 44° E., and finds the deviation by it, there will be an error of 2° in the deviation thus found on every point of the compass; or, in other words, the westerly deviation will be 2° less, and the easterly deviation will be 2° more than it ought to be. When a ship is swung hurriedly, and her head is not allowed to remain for a minute or two on any point before observations are made, there is a temporary constant deviation produced; and this temporary deviation is easterly when the ship is swung to the left, as from East to North, and is westerly when the ship is swung to the right, as from North to East.

226. **Mechanical Compensation or Correction of the Compass by means of Magnets and Soft iron.**—These adjustments were first proposed by Mr. AIRY, the Astronomer Royal, and are now universal in the merchant service.

227. **Correction of the Semicircular Deviation.**—As this error is caused by magnetism induced in vertical iron, and by subpermanent magnetism acquired in building, it is very difficult in practice to separate the semicircular deviation caused by vertical iron from that caused by subpermanent magnetism. Could this be easily effected, a natural inference would be to compensate the former by vertical soft iron, and the latter by a magnet, each compensator placed so as to produce opposite effects to those of the ship.

An arrangement of this kind was adapted to several ships by Mr. RUNDELL, the Secretary of the Liverpool Compass Committee, who placed a vertical iron bar before the compass, leading from the keel (when possible) to a point some height above the level of the card. The distance from the compass and height to which it must be carried to be determined by experiment, but a difficulty has been experienced in adjusting this method so as to ensure success.

The order of proceeding for compensation is as follows:—

The ship must be upright, or on an even beam, with all her *iron* stores on board, in the positions which they are intended to occupy while at sea.

The position of the binnacle being decided on, draw a line upon the deck, fore-and-aft, through the centre of the place where the binnacle is to stand. Draw another line across the deck, at right-angles to the former, through the same centre.

Provide two or more powerful magnets from 18 inches to 2 feet in length.*

Let the ship's head be swung to the North or South, *correct* magnetic—either of these points will do. When the ship's head is steady at one of these points, observe whether there is any deviation; if there is any, lay one of the magnets on the deck athwartship, with its centre exactly on the fore-and-aft line drawn on the deck at some distance from the binnacle; move it gradually (not hurriedly) to or from the foot of the binnacle until the compass points correctly. The magnet may be placed either before or abaft the binnacle, whichever is most convenient, but its centre must always be over the fore-and-aft line drawn on the deck, and it must be kept at right-angles to the ship's keel. If the compass needle deviate to the left, the north end of the magnet must be placed to the left, and conversely.

After the compass has been made to point correctly at either the north or south points, swing her head round to the east or west correct magnetic (either will do), and steady her head on one of these points.

If there be any deviation, place the other magnet fore-and-aft, either on the port or starboard side of the binnacle, with its centre on the athwartship line drawn on the deck; move it to or from the foot of the binnacle until the compass points correctly.

If the ship was built with her head nearly north or south, two magnets may be required. These may either be placed one on each side of the compass, or both on the same side, as may be convenient; if placed on the same side of the binnacle, lay them an inch apart, and under all circumstances parallel, but always similar ends (N. or S.) directed to the same part of the ship.

The adjuster should be careful to see that the centre of the magnet is kept on the fore-and-aft line, so that one of the poles of the magnet be no nearer the binnacle than the other.

It must be remembered that this correction will only hold good for a small range of latitude, and while the ship's magnetism continues in the same state as when the correction was made.

228. **Correction of the Quadrantal Deviation.**—As the quadrantal deviation is caused by the action of horizontal soft iron, a natural inference is, that soft iron should be used for compensating this error, so placed as to cause opposite effects to those of the ship. The compensations for semicircular deviation being complete, the ship's head should be swung to one of the intercardinal points, N.E., N.W., S.E., or S.W., correct magnetic; and the binnacle being fitted with two small brass boxes, one on each side of and on a level with the compass; if there is any deviation place a quantity of small iron chain in the boxes until the compass points correctly. For greater certainty swing the ship to each of the other quadrantal points.

---

* Compensating magnets should be from 10 to 18 or 24 inches in length, their breadth one-tenth of their length, and their thickness one-fourth their breadth.

For the correction of large quadrantal deviations, cast iron correctors from 9 to 12 inches long, and 3 to 3¼ inches diameter, are preferred by the Liverpool Compass Committee.

The ship's head is to be steadied on one of the quadrantal points, and the correctors, one on each side of the compass, and on the same level as the needle, are to be moved to or from the compass until the quadrantal deviation is corrected, but not nearer than 1½ times the length of the needles from the centre of the card. It is only in very *rare* instances that the deviation is westerly on the N.E. and S.W. points, or easterly on the S.E. and N.W.; but if so, the correctors or chain-boxes are required to be placed on the fore-and-aft ends of the binnacle. The adjustment for the quadrantal deviation should always be made, as it tends to reduce the heeling error.*

In some cases there is a small amount of quadrantal deviation produced by horizontal soft iron running from the quarters to the opposite bows; iron in this position produces a quadrantal deviation, which is greatest when the ship's head is at N., S., E., and W., and least with the ship's head at N.E., S.E., S.W., and N.W.; it is, however, generally of so small amount that it may, in ordinary cases, be disregarded.

Although the correctors will compensate for a greater amount of deviation than the chain-boxes, they have in many instances been found to be insufficient, especially in cases of ships with iron decks.

Heeling Error.—Although a ship's compasses may be corrected by the above methods, they can only be depended upon so long as she remains upright. Besides the ordinary deviation of the compass there is a deviation caused by the heeling of iron ships, which may increase or decrease the deviation observed when the ship is upright. Cases have been observed in which the deviation from heeling has amounted to as much as two degrees for each degree of heel of the ship, that is, without altering the real direction of the ship's head, the apparent alteration in direction has amounted to 40° by heeling the ship from 10° to starboard to 10° to port. The effect is very serious in those parts where the wind is steady, and the ship inclined in the same direction for many days or weeks in succession.†

229. **To ascertain the amount of Heel.**—The instrument specially adapted to indicate the amount of heel is the *clinometer*. It consists of a brass semicircle, graduated at the edge to degrees, beginning at the middle of the arc and continued both ways; and to the centre a plumb line is attached. The instrument is fixed at right-angles to a fore-and-aft section of the ship, as a beam, or athwartship bulkhead, with the diameter placed upwards, and parallel to the deck. When the index points to 0, the vessel is upright, but

---

* These correctors are too frequently absent, and it should be remembered that they very essentially improve the action of the compass—not only diminishing the deviation, but increasing the directive force.

† "Usually, in an iron ship, when her head is placed north or south, the ship's inclination through an angle of n degrees disturbs the compass through an angle of n degrees; but in some particular instances it has been known to disturb the compass as much as $2n$ degrees. —*A Treatise on Navigation*, by G. B. AIRY, M.A., LL.D., D.L., page 182.

when she heels either way, the plumb line being free to move on its centre is always vertical, and the point at which it cuts the graduated edge shows the number of degrees that the vessel deviates from the perpendicular, that is, the heel of the ship. A compass card with the needle detached will answer the purpose, and an index may be made with a thread and plummet depending from the end.

230. **How the Deviation from Heeling is caused.**—The heeling error depends partly on vertical induction in transverse iron, and partly on vertical force arising from subpermanent magnetism in the ship, combined with that from vertical induction in vertical soft iron. The fore-and-aft iron is not disturbed from its horizontal position by heeling, consequently the athwartship beams then produce their full influence in disturbing the compass. When an iron ship heels over, forces, which before acted vertically, and did not disturb the horizontal compass needle, now act to one side and produce deviation; while transverse iron which was previously horizontal, becoming inclined, acquires magnetism by induction (215). In north magnetic latitude the upper or weather ends of athwartship beams, for example, become south poles, and the lower ends north poles; hence, from both these causes, the north end of the needle is drawn to windward. But, if the iron does not extend entirely across, as when a sky-light or hatchway is fitted, the opposite effects are produced; for then the end of the iron nearest the compass on the weather side is a north pole, and that nearest it on the lee side a south pole; and under these conditions the north end of the needle is drawn to leeward.

In vertical iron the force acting on the needle is no longer directly under it, but is shifted to the weather side of the ship, and thus in north magnetic latitude, as a general rule, the tendency of both horizontal and vertical iron is to draw the north end of the needle to windward. The vertical action of subpermanent magnetism modifies the result of these causes, and may either cause an increase or a diminution of the error so produced. If a ship has acquired subpermanent magnetism by having been built with her head north, there is a strong vertical force acting *downwards* (see Fig. 3, page 135) from the whole after body of the ship having south magnetism or polarity; this would conspire with the vertical induction in transverse iron, in attracting the north end of the needle to the weather side, as the ship heels over, and thereby increasing the change of deviation from other causes. On the other hand, if a ship be built with her head south, the vertical force acts *upwards* (see Fig. 4, page 136), the after part of the ship has acquired north magnetism, or polarity, and the north end of the needle, as the ship heels over, is repelled by it to the lee side, the vertical force acting in antagonism, in this case, to the transverse force, thus decreasing the error caused by soft iron. Thus is shown why in England the deviation of ships built there, with their heads northerly, are most affected by heeling.

In the ordinary position of the compass on the quarter-deck, we may, in most cases, if we know the direction in which the ship's head was built, anticipate the direction of the heeling error, and form an approximate estimate of its amount. Ships built with their heads from about S.W. to

S.E. by way of north, the upper parts have south polarity; and in those of this group built with their heads from N.W. to N.E., this south polarity is strongly developed near the position of the compass. In all these ships the north end of the compass needle will be drawn to windward, and forcibly so in the last named group. In the ships built with their heads from about S.W. to S.E. by way of south, their upper parts near the position of the compass have N. polarity, and hence the heeling error may be to leeward or to windward—and in either case small in amount—according as the vertical force, or force from transverse iron predominates.

231. **Position of ship's head for greatest and least change of Deviation from Heeling.**—There appears to be no deviation from heeling when the ship's head by compass is east or west, but it increases as the ship's head is moved from these points, and is greatest when the ship's head by disturbed compass is north or south. When the ship's head by compass is either east or west, the disturbing force, from the ship's heeling, acting at right-angles to the fore-and-aft midship line, tends to bring the needle into the magnetic meridian, and consequently no change in deviation can be produced from heeling. On the other hand, when the ship's head by compass is either north or south, the disturbing force acts at right-angles to the needle; hence the greatest change of deviation resulting from a vessel's heeling takes place when her fore-and-aft line is in the magnetic meridian.

232. In north latitude, in ships built with their heads to the north, with their compasses in the usual position, the deviation from heeling is much larger than in ships built with head to the south. In north latitude the north end of the needle is generally drawn towards the weather side of a ship, yet a small deviation to leeward has also been observed in north latitude, in some ships which were built in a southerly direction. In high south latitudes, where the dip is south, the north end of the needle has been observed to deviate to leeward. Compasses which are least affected by heeling in the northern hemisphere have **generally the greatest amount when south of the equator, and vice versa.***

---

* This should be particularly considered by masters of iron ships about to proceed to a port south of the equator.

The heeling error not only causes deviation, but also unsteadiness, and in some cases when a ship has rolled with her head north or south by compass, the card has spun, and this has rendered it impossible to keep the ship on any required point. It is not, however, always found that compasses having the largest heeling error are most productive of this inconvenience. "This unsteadiness depends on two conditions: first, the amount of the change of the deviation arising from any number of degrees of heeling; and second, the existence or non-existence of isonchronism between any number of the periods of the vibrations of the card and any number of the ship's rolling. Taking this latter condition into consideration when inconvenience has been experienced, arising from the unsteadiness of the compass, the card has been changed, since either a more sluggish or more active needle would be more or less in unison with the ship's roll." With a compensated card, while the ship remains in the same latitude, if we choose a card which is steady when the ship rolls with her head north or south, it will also be steady when her head is in any other direction. Still cases have arisen in which a compass that is steady in one latitude has vibrated, and even spun, when by changing the ship's place the earth's horizontal force has changed. To obviate this difficulty, washers, made to rest on the card with the cap passing through the middle, have been employed with advantage.

233. **Effects of Heeling.**—The effect of the heeling error, when the north end of the needle is drawn to windward, is to throw the ship to windward of her supposed position when steering on northerly courses; and to throw her to leeward when steering on southerly courses. Therefore, to make a *straight* course, when heeling, a ship should be kept away by compass on either tack on northerly courses; and she should be luffed up on either tack on southerly courses. The effect in the few cases in which the compass needle is drawn to leeward is the reverse, and in the southern hemisphere, also, the reverse of these rules holds good; but this is a point which can only be ascertained for each ship.

The heeling error may be expressed in terms of the deviation when upright, and the following are the results:—

On *Northerly Courses* :—

    Starboard tack—E. dev. is increased, W. dev. is decreased.
    Port tack—W. dev. is increased, E. dev. is decreased.

On *Southerly Courses* :—

    Starboard tack—W. dev. is increased, E. dev. is decreased.
    Port tack—W. dev. is decreased, E. dev. is increased.

And when the deviation when the ship is upright is small in amount and decreases by heeling, it may become reversed in name.

In the *few cases* in which the *North end of the compass needle is drawn to leeward*, the *rule* above is of course *reversed*.

234. **Correction of the Heeling Error.**—The mechanical correction of the heeling error is made by a magnet placed in a vertical position immediately below the centre of the compass card. The ship's head is to be placed north and south, correct magnetic; she is then heeled over to port and to starboard, and the magnet raised or lowered until the compass points correctly. In most cases the north end of the vertical magnet should be uppermost.

The deviation arising from a ship heeling being semicircular, this correction holds good only while a ship continues near the magnetic latitude where the adjustment was made; hence, arrangements must be made for sliding the magnet along as different latitudes are reached, and for removing it, and even reversing it in high latitudes of opposite names.

235. After all the compensations have been accurately made, there will still remain small residual errors; for these the ship must be swung, and a table of deviations made for use. When an arrangement of magnets is employed to neutralise those large deviations occasionally found, and caused by the iron ship's magnetism, the compass so corrected can never be considered as entirely compensated, and the deviation must be expected to change on change of latitude, and from other causes. It will thus be seen that the seaman can have no absolutely safe guide, except in the system of actual and **unceasing observation**.

METHODS OF FINDING THE AMOUNT OF THE DEVIATION.

236. When in port, there are two principal methods in general use for finding the deviation, viz.:—Method I, **by the known correct magnetic bearing of a distant object**, and Method II, **by reciprocal simultaneous bearings**, *i.e.*, with a compass on board and a compass on shore.

237. METHOD I.—**By the known bearing of a distant object.**—The requisite warps being prepared, the ship is to be gradually swung round so as to bring her head successively upon each of the 32 points of the Standard Compass; and when the ship and the compass card are perfectly steady, and her head exactly on any one point, the direct bearing of some well-defined object is to be observed with the Standard Compass, and registered. The ship's head is to be gently warped round in the same manner to the next point, and when duly stopped and steadied there, the bearing of the same object is to be again set, and again recorded; and so on, point after point, till the exact bearing of the one object has been ascertained with the ship's head on every separate point of the compass.

238. The object selected for this purpose should be at such a distance that the diameter of the space through which the ship revolves shall make no sensible difference in its real bearing, and should not exceed the one-hundredth part of the distance of the object. The distance must depend on the range the ship takes when swinging; if she be at anchor, in a tide-way, from 6 to 8 miles is not too much; brought up by the middle (in a dock) 2 miles will suffice.

239. The next step is to determine the *correct magnetic* bearing of the selected object from the ship; or in other words, the compass bearing it would have from on board if it were not disturbed by the attraction of the iron in the ship. This is effected by taking the compass to some place on shore (avoiding local influence) from which the part of the ship where the compass stood and the object of which the bearings had been observed shall be in one with the observer's eye, or else in the exactly opposite direction. The bearing of the object from that spot will evidently be the *correct* magnetic bearing from the ship by the compass. The difference between the *correct* magnetic bearing of the object and the successive bearings which were observed with the compass on board, when the ship's head was on the several points, will show the error of each of these points which was caused by the ship's iron; or, in other words, the Deviation of the Standard Compass according to the direction in which the ship's head was placed.

(b) The **correct magnetic bearing** of the distant object will be the mean value of all the observed bearings, if observed on equidistant points; or of four or more compass bearings, if taken also, on equidistant compass points.

240. **II.—By reciprocal bearings.**—Should there be no suitable object visible from the ship, and at the requisite distance as stated above, the deviations must be ascertained by the process of reciprocal bearings. A second compass is placed on shore where it will be entirely beyond the influence of iron of any description and where it can be distinctly

seen from the Standard Compass on board. Then take, simultaneously (known by pre-concerted signal), the bearing from each other of the compass on shore and the compass in the binnacle, as the ship is warped round so as to bring her head successively upon each of the thirty-two points of the Standard Compass on board, or on each alternate point.

To ensure the success of this operation, the compass on shore should not be more distant from the ship than is consistent with the most distinct visibility with the naked eye, of both compasses from each other. The observations should be made as strictly simultaneous as possible, the time at which each bearing is taken being noted both on shore and on board. It will be found convenient in practice, for the shore observer to chalk each observation on a black board, to be read at once from the ship, in order that the observation may be repeated if any apparent inconsistency presents itself.

Before this process is complete, the Standard Compass should be carried on shore, in order to be compared with the compass used there, by means of the bearing of some distant object, and the difference, if any, is to be recorded; and in all cases, when compasses are compared, the caps, pivots, &c., should be first carefully examined. The shore compass gives correct magnetic bearings.

The **difference** between the **correct magnetic bearing** of the Standard Compass as **observed from the shore**, and the **bearing** of the shore compass as **observed from the ship**, with her head in any particular point, reversed, *i.e.*, with 180° added or subtracted, will show the error on that point which was caused by the ship's iron; in other words the deviation of the Standard Compass according to the direction in which the ship's head was placed.

241. III.—**By Marks on the Dock Wall.**—This is a very convenient method where it can be practised. At Liverpool the correct magnetic bearings of the Vauxhall chimney, from various points of the dock walls, are painted in large figures on the walls, so that the bearing of the same chimney may be observed as the ship swings with the wind and tide; and at the same time that bearing marked on the wall, which is on a line between the Standard Compass and the chimney, is noted.

The **difference** between those bearings is **the deviation** for the point on which the ship's head is at the time.

In a similar manner, at Cronstadt, the correct magnetic bearings of a conspicuous point on a public building are painted on the mole.

242. If, during the operation of swinging, a haze obscures the shore compass, while the sun at the time is shining brightly, a number of points may be secured by time-azimuths, which otherwise might be lost. Time-azimuths are also advantageous where the second of the above methods cannot be used for want of an assistant observer for the shore compass; and when the first of the above methods is not available owing to the length of the ship and the scope of the moorings, combined with the most distant objects in sight, not being sufficiently far off to render the difference of their bearings insensible as the ship swings round to the tide. In such cases *Godfray's*

Azimuth Diagram, as also Azimuth or Sun's True Bearing Tables, computed for intervals of four minutes, by Staff-Commander J. BURDWOOD, R.N., published by the Admiralty, will be found useful as superseding the calculation for the determination of the True Azimuth.

243. Commander WALKER, R.N., has shown* that the deviation may be ascertained with sufficient accuracy by selecting a distant object, as before, "and as the ship swings by wind or tide from one point to another, write down the compass bearings of the distant object opposite the direction of the ship's head. As the ship swings round there will be two nearly opposite points of the compass on which the bearings of the distant object agree, and this should be the correct magnetic bearing of the object." The deviation is then found as in the first method.

244. **The Dumb Card.**—The difficulty of finding the correct magnetic bearing of the ship's head may be obviated, however, by using the *dumb-card*, *i.e.*, a compass card without the needle, slung in gimbals, with its centre over a fore-and-aft line of the vessel, and as near to its middle as possible. The card is fitted with sight vanes, similar to an azimuth compass. Having obtained the correct magnetic bearing of a distant object, place the card so that it shall point out that direction, and screw the sight vanes to the card, so as to cut the object with the thread. Then, as the ship is swung, the card must still be kept pointing out the correct magnetic bearing of the object by means of the sight vanes, and where the fore-and-aft line meets the edge of the card, must then be the correct magnetic bearing of the ship's head.

245. **To name the Deviation.**—Rule.—*When the reading by the shore compass* (reversed), *or the* correct *magnetic bearing of the distant object, is to the* right *of the reading by the compass on board, the deviation is* easterly; *when to the* left, westerly.

Thus, suppose the *correct* magnetic bearing from the shore compass, with ship's head at N.W., is N. 15° E., and the bearing of shore compass from the ship is S. 11° W.; to find deviation proceed thus:—

Reverse of the bearing by shore compass or correct magnetic bearing	S. 15° W.
Bearing from ship	S. 11 W.
Deviation	4 E.

When the ship's head lies N.N.E., let the binnacle compass bearing of the shore object or compass be N. 19° 30′ E., and the bearing of the binnacle compass from the shore compass be S. 27° 0′ W.: required the *deviation*.

The opposite point to S. 27° 0′ W. is N. 27° 0′ E, which is 7° 30′ to the right of N. 19° 30′ E. Hence the *deviation* is 7° 30′ E.

246. The directions of the ship's head having been taken by the compass in the ship, are therefore affected by the local attraction, and the apparent *compass bearing* of the ship's head differs from the correct magnetic bearing by the amount of the local deviation due to the position of the ship. For

---

* Magnetism of Ships and the Mariner's Compass.

instance, when the ship is apparently lying with her head east, it is not the true magnetic east; but supposing the local deviation to be one point easterly, the east point of the compass card will be drawn to E. by S., and the true magnetic direction of the ship's head will be E. by S.

The observations and tabulated results are incomplete until the *correct magnetic* bearing of the ship's head at each observation is found.

247. The following shows the arrangement of tabular forms for finding the deviation by the several processes described.

I. By bearing of a distant object.

Correct magnetic bearing of distant object from ship N. 63° 0′ W., distant 11 miles.

Ship's Head by the Standard Compass.	Bearing of Distant Object by the Standard Compass.	Deviation of Standard Compass.
East	N. 83° 20′ W.	20° 20′ E.
E. by S.	N. 82 15 W.	19 15 E.
E.S.E.	N. 81 5 W.	18 5 E.
S.E. by E.	N. 72 30 W.	16 30 E.
S.E.	N. 77 40 W.	14 40 E.

And similarly at all points of the compass.

II. By reciprocal bearings.

Time.*	Ship's Head by the Standard Compass.	SIMULTANEOUS BEARINGS.		Deviation of Standard Compass.
		From Standard Compass on board.	From the shore Compass.	
9ʰ 10ᵐ A.M.	North	S. 37° 50′ E.	N. 41° 0′ W.	3° 10′ W.
9 14 ,,	N. by E.	S. 45 0 E.	N. 42 25 W.	2 35 E.
9 17 ,,	N.N.E.	S. 51 40 E.	N. 43 30 W.	8 10 E.
9 21 ,,	N.E. by N.	S. 57 20 E.	N. 44 10 W.	13 10 E.
9 26 ,,	N.E.	S. 61 50 E.	N. 45 0 W.	16 50 E.
9 32 ,,	N.E. by E.	S. 65 30 E.	N. 46 0 W.	19 30 E.

And so on through all the points of the compass.

248. The seaman must remember that the corrections thus obtained belong to the compass by which the observations are made, and to that compass while it is in its proper place, and that these corrections will furnish no guide whatever to the effects of the iron on a compass placed in any other part of the ship; but if, while swinging, the direction of the ship's head by the other compasses is noted and tabulated, the deviation of all the compasses can be found.

---

* The time—as taken by compared watches—may be omitted if the shore observations can be clearly made out by being chalked on a black board.

The following is a Table of Deviations to which reference is to be made in working the following examples.

TABLE OF DEVIATIONS.

SHIP'S HEAD.	DEVIATION.	SHIP'S HEAD.	DEVIATION.
North	0° 22' W.	South	0° 16' E.
N. by E.	1 46 E.	S. by W.	1 50 W.
N.N.E.	3 20 E.	S.S.W.	3 16 W.
N.E. by N.	5 14 E.	S.W. by S.	4 48 W.
N.E.	7 14 E.	S.W.	6 16 W.
N.E. by E.	8 54 E.	S.W. by W.	7 40 W.
E.N.E.	10 44 E.	W.S.W.	9 18 W.
E. by N.	11 40 E.	W. by S.	10 34 W.
East	10 44 E.	West	11 50 W.
E. by S.	9 54 E.	W. by N.	11 10 W.
E.S.E.	9 8 E.	W.N.W.	10 16 W.
S.E. by E.	7 20 E.	N.W. by W.	9 18 W.
S.E.	6 18 E.	N.W.	7 52 W.
S.E. by S.	5 0 E.	N.W. by N.	6 18 W.
S.S.E.	3 24 E.	N.N.W.	5 2 W.
S. by E.	1 42 E.	N. by W.	3 10 W.

249. The purposes for which a Table of Deviations so formed are:—

1st.—To correct the course steered by the compass, in order that the correct magnetic course actually made good may be used in the calculation of the ship's reckoning, or to lay it down on the chart.

2nd.—If one or more bearings of the land are taken, to correct these bearings by the amount of deviation due to the direction of the ship's head at the time.

3rd.—If we wish to shape a course for a port, and having, either by calculation, or as taken from the chart, the correct magnetic course to be made good, so to apply the deviation as to obtain the compass course to be steered.

RULE LVI.

To find the correct magnetic course, having given the compass course and deviation.

Express the compass course in degrees, &c.; look in the Table of Deviations *for the* deviation opposite *the given course,* then,

    Easterly deviation allow to the right.
    Westerly ,, ,, left.

EXAMPLES.

Correct the following compass courses for deviation, as given in Table above:—

1. E.S.E. = 6 points L. of S.
6 points L. of S. = 67° 30' L. of S.
Deviation (Table) 9 8 R.

Cor. mag. course 58 22 L. of S.
  or S. 58 22 E.

In this instance, the deviation being Easterly, allow to the right.

2. N.N.W. = 2 points L. of N.
2 points L. of N. = 22° 30' L. of N.
Deviation (Table) 5 2 L.

Cor. mag. course 27 32 L. of N.
  or N. 27 32 W.

The deviation in this instance being Westerly, allow to the left.

3. S.W. = 4 points R. of S.

4 points R. of S. =	45° 0′	R. of S.
Deviation (Table)	6 16	L.
Cor. *mag. course*	38 44	R. of S.
	or S. 38 44	W.

4. W. = 8 points R. of S.

8 points R. of S. =	90° 0′	R. of S.
Deviation (Table)	11 50	L.
Cor. *mag. course*	78 10	R. of S.
	or S. 78 10	W.

5. W. ½ N. = 7½ pts. L. of N. = 87° 11′ L. of N.

Deviation =	11 40	L.
	98 51	L. of N.
	180 0	
Cor. *mag. course*	81 9	R. of S.
	or S. 81 9	W.

West	= 11° 50′	W.
W. by N.	= 11 10	W.
	4)0 40	
Dev. for ½ pt.	= 0 10	
Dev. for W.	= 11 50	
Dev. for W. ½ N.	= 11 40	W.

6. N.W. by W. ¾ W. = 5¾ L. of N. = 61° 41′ L. of N.

Deviation =	10 2	L.
	N. 71 43	W.

N.W. by W. =	9° 18′	W.
W.N.W.	= 10 16	W.
Change for 1 pt. =	58	
	4)58	
Change of dev. for ¼ pt. =	14	
Dev. for W.N.W.	= 10 16	
Dev. for N.W. by W. ¾ W. =	10 2	W.

The deviation (see Table) for N.W. by W. and W.N.W. is found, and the difference of these quantities is the correction for 1 point, which divided by 4 gives the correction for ¼ point = 0° 14′. Now compass course is ¼ point from W.N.W., therefore, apply correction 0° 14′ to the deviation for W.N.W., the result is the deviation for N.W. by W. ¾ W., and it is to be subtracted, because the deviation for N.W. by W. is less than for W.N.W.

7. N. ½ E. = ½ pt. R. of N. = 5° 38′ R. of N.

Deviation =	0 42	R.
Cor. *mag. course*	6 20	R. of N.
	or N. 6 20	E.

Deviation at N. =	0° 22′	W.
N. by E.	1 46	E.
	2)1 24	
	0 42	E.

One deviation being W., and the other E., half the difference of the two is taken for the deviation.

8. N. ½ W. = ½ pt. L. of N. = 5° 38′ L. of N.

Deviation =	1 46	L.
Cor. *mag. course*	N. 7 24	W.

Deviation at N. =	0° 22′	W.
N. by W.	3 10	W.
	2)3 32	
Dev. on ½ pt.	= 1 46	W.

Half the *sum* of the deviations for N. and N. by W. is taken for deviation on N. ½ W., both deviations being of the same name. Or proceed thus—take the deviations from Table for North and for N. by W.; take the difference and half it, apply this to the deviation for North, *adding* because the deviation is greater for N. by W. than for North.

Examples for Practice.

Correct the following courses steered for deviation, as given in the Table, page 154.

1. N.E. by N.	8. S.E. by E.	15. S.W. by S.	22. West.
2. North	9. N. by E. ½ E.	16. S. ½ E.	23. N.W. ¼ W.
3. N. by W. ½ W.	10. South	17. W. ½ S.	24. S. ½ W.
4. S.W. ½ W.	11. W. ½ N.	18. N.N.E. ½ E.	25. W. by S. ½ S.
5. W. ½ N.	12. S.E. ¼ S.	19. N. ¼ E.	26. E. ½ S.
6. S.S.W.	13. E. ¾ N.	20. W. by N.	27. East
7. E. by N.	14. N.W. ½ W.	21. S. by E.	28. W.N.W.

250. Proceeding to correct the courses for the deviations given in Table I, a second Table, arranged like the following, may be made for all the points of the compass.

TABLE II.

Courses steered by Compass.	Deviations.	Correct Magnetic Courses.
North........ or 0°	0°22′ W.	N. 0°22′ W. or North.
N. by E. .... „ N. 11 15′ E.	1 46 E.	N. 13 1 E. „ N. by E. ⅛ E.
N.N.E. ...... „ N. 22 30 E.	3 20 E.	N. 25 50 E. „ N.N.E. ¼ E.
N.E. by N. .. „ N. 33 45 E.	5 14 E.	N. 38 59 E. „ N.E. ½ N.
N.E. .... „ N. 45 0 E.	7 14 E.	N. 52 14 E. „ N.E. ⅝ E.
N.E. by E. .. „ N. 56 15 E.	8 54 E.	N. 65 9 E. „ N.E. by E. ¾ E.
E.N.E. ...... „ N. 67 30 E.	10 44 E.	N. 78 14 E. „ E. by N., nearly.
E. by N. .... „ N. 78 45 E.	11 40 E.	S. 89 35 E. „ East.
East .... „ 90	10 44 E.	S. 79 16 E. „ E. by S.
E. by S. .... „ S. 78 45 E.	9 54 E.	S. 68 51 E. „ E.S.E. ⅛ E.
E.S.E. ...... „ S. 67 30 E.	9 8 E.	S. 58 22 E. „ S.E. by E. ⅕ E.
S.E. by E. .. „ S. 56 15 E.	7 20 E.	S. 48 55 E. „ S.E. ⅜ E.
S.E. .... „ S. 45 0 E.	6 18 E.	S. 38 44 E. „ S.E. ½ S.
S.E. by S. .. „ S. 33 45 E.	5 0 E.	S. 28 45 E. „ S.S.E. ½ E.
S.S.E. ...... „ S. 22 30 E.	3 24 E.	S. 19 6 E. „ S. by E. ¾ E.
S. by E. .... „ S. 11 15 E.	1 42 E.	S. 9 33 E. „ S. ⅞ E.
South ...... „ 0	0 16 E.	S. 0 16 W. „ South.
S. by W. .... „ S. 11 15 W.	1 50 W.	S. 9 25 W. „ S. ⅞ W.
S.S.W. ...... „ S. 22 30 W.	3 16 W.	S. 19 14 W. „ S. by W. ¾ W.
S.W. by S. .. „ S. 33 45 W.	4 48 W.	S. 28 57 W. „ S.S.W. ½ W.
S.W. .... „ S. 45 0 W.	6 16 W.	S. 38 44 W. „ S.W. ½ S.
S.W. by W... „ S. 56 15 W.	7 40 W.	S. 48 35 W. „ S.W. ⅜ W.
W.S.W. .... „ S. 67 30 W.	9 18 W.	S. 58 12 W. „ S.W. by W. ⅛ W.
W. by S. .... „ S. 78 45 W.	10 34 W.	S. 68 11 W. „ W.S.W. ⅛ W.
West .... „ 90	11 50 W.	S. 78 10 W. „ W. by S.
W. by N. .... „ N. 78 45 W.	11 10 W.	N. 89 55 W. „ West.
W.N.W. .... „ N. 67 30 W.	10 16 W.	N. 77 46 W. „ W.N.W. ⅞ W.
N.W. by W. . „ N. 56 15 W.	9 18 W.	N. 65 33 W. „ N.W. by W. ⅞ W.
N.W. ...... „ N. 45 0 W.	7 52 W.	N. 52 52 W. „ N.W. ⅝ W.
N.W. by N... „ N. 33 45 W.	6 18 W.	N. 40 3 W. „ N.W. by N. ⅝ W.
N.N.W. .... „ N. 22 30 W.	5 2 W.	N. 27 32 W. „ N.N.W. ½ W.
N. by W. .... „ N. 11 15 W.	3 10 W.	N. 14 25 W. „ N. by W. ¼ W.

To obtain from the Table above the correct magnetic course of the ship from the course shown by the Standard Compass, look in the first column of the Table for the latter; the second column gives the deviation when her head is on that point; and in the third column (the deviation having been applied as directed in Rule LVI) the seaman will find the correct magnetic course given there, by inspection.

This Table will be found more useful than the common Table of Deviations, as it shortens the calculation, when it is required to fractions of a point, as a half, a quarter, &c., and when steering on a whole point, the *correct* magnetic course is known at sight.

The following examples will show the use of the Table.

EXAMPLES.

EX. 1. The ship's head by Standard Compass is N.E. ½ E.: what is the correct magnetic course? (Using the Table above.)

Corr. mag. course for N.E. = N. 52° 14′ E.
„ „ N.E. by E. = N. 65 9 E.

2)117 23

Corr. mag. co. for N.E. ½ E. = N. 58 41 E.

Here the courses taken from the Table are of the *same* name, therefore, half the sum is evidently the correct magnetic course corresponding to N.E. ½ E.

EX. 2. Find correct magnetic course when Standard Compass course is N. ½ E. (using the above Table).

Corr. mag. course for North = 0° 22′ W.
„ „ N. by E. = 13 1 E.

2)12 39

Corr. mag. course for N. ½ E. = 6 19 E.

Here the courses corresponding to North and N. by E. are of *contrary* names; hence, for a Standard Compass course, midway between N. and N. by E., we use half the difference of the *correct* magnetic courses corresponding to these points.

Ex. 3. Ship's head by Standard Compass is N. ¼ E.; find, by means of the Table, the corresponding correct magnetic course.

Corr. mag. course for North = 0° 22′ W.
„   „   N. ¼ E. = 13 1 E.
                        ─────────
                        4)13 23
Difference for ¼ point   =   3 21
Corr. mag. course for North   =   0 22 W.

Corr. mag. course for N. ¼ E. =   2 59 E.

Ex. 4. Required the correct magnetic course when ship's head by Standard Compass is W. ¾ N.

Corr. mag. course for West = S. 78° 10′ W.
„   „   W. by N. }
is N. 89° 55′ W. or S. 90° 5′ W. } S. 90 5 W.
                                    11 55
                                         3
                                    ─────────
                                    4)35 45
Difference for ¾ point   +   8 56
Corr. mag. course for West = S. 78 10 W.

Corr. mag. course for W.¾N.= S. 87 6 W.

251. **Correction of Compass Bearings.**—In order to correct the bearing of the object as taken by the Standard Compass, note the direction of the ship's head by that compass while taking the observation, then enter the first column of either Table I or II with that, and in the second column will be found the deviation to be applied to the bearing of the object. (See 164-166, page 105). Easterly deviation to be allowed to the right, and westerly to the left, as in Rule LVI.

CAUTION.—Be careful to remember that the deviation to be applied is that due to the compass course, *not that on the point of bearing;* and the consequence of a misapplication of the deviation, by applying that for the point of bearing instead of the deviation for the compass course, may lead into danger, if not loss.

EXAMPLES.

Ex. 1. The bearing by Standard Compass of the South Foreland is N.N.W., the course by the same is E.N.E.: required the correct magnetic bearing.

Taking out the deviation from the Table for the direction the ship's head was on at the moment the bearing was taken, we have

Bearing by Standard Compass of South Foreland   N. 22° 30′ W. or 22° 30′ L. of N.
Deviation by Table for E.N.E (applied to *right*)   10 44 E. „ 10 44 R.

Correct magnetic bearing   11 46 W. „ 11 46 L. of N.
                              or N. 11° 46′ W.

Ex. 2. Ship's head E.S.E. by compass, the bearing by the same compass of the Start Point is N. 20° W.: required the correct magnetic bearing.

Bearing of Start Point by Standard Compass   N. 20° W. or 20° L. of N.
Deviation by Table (applied to the *left*)   14 W. „ 14 L.

Correct magnetic bearing   N. 34 W. „ 34 L. of N.
                                or N. 34° W.

Ex. 3. Two islands bear S.E. and W.S.W.; the ship's head is N.E.: required the correct magnetic bearing of each.

Bearing by Standard Compass S.E.   = S. 45° E. or 45° L. of S.
Deviation by Table (applied to the *right*)   =   7 E. „ 7 R.

Correct magnetic bearing   = S. 38 E. „ 38 L. of S.
                                   or S. 38° E.

Bearing by Standard Compass W.S.W.   = S. 67° 30′ W. or 67° 30′ R. of S.
Deviation by Table (applied to the *right*)   =   7 14 E. „ 7 14 R.

Correct magnetic bearing   = S. 74 44 W. „ 74 44 R. of S.
                                   or S. 75° W.

Examples for Practice.

In the following examples the ship's compass course and the bearing of the object by compass are both given, and it is required to find the magnetic bearings of the objects, using the same deviation table (Table of Deviations, page 154).

No.	Ship's Head by Compass.	Compass Bearing.	No.	Ship's Head by Compass.	Compass Bearing.
1	West	East.	7	E. ¾ N.	N. ¼ W.
2	S.S.E.	E. by S. ¼ S.	8	N.E. ¼ E.	W. ¾ S.
3	N.E. by N.	North.	9	E. ¾ S.	W. by S. ¼ S.
4	W.N.W.	South.	10	N. ¾ E.	S. ¼ E.
5	W. by N. ½ N...	E. ¼ S.	11	S. ¼ E.	W. by N. ½ N.
6	S. by W. ½ W...	S. by W. ¼ W.	12	E. ¼ N.	N. by W. ¼ W.

252. Given a correct magnetic course by the chart between two points of land, to find the course that must be steered by compass.

RULE LVII.

Easterly deviation is allowed to the left,
Westerly ,, ,, right,
taking care that the deviation applied is that of the *correct* magnetic course.

Note.—In this case it is important to remember, not only is the general rule of applying the deviation *reversed*, but the correction to be applied is the deviation due to the given magnetic course, not that due to a compass course, as in Rule LVI; that is, to the correct magnetic course as found from the chart, or by calculation, the deviation, as due to that course, must be applied as directed above, in order to find the course to be *steered* by compass approximately. It will be observed that on those courses near which the deviation is considerable, and rapidly changing, the deviation on a given magnetic course is considerably different from that on the compass course of the same name. In such cases it will be necessary to again enter the table with the *approximate* course and get the corresponding deviation and *apply it* to the correct magnetic course; the result will be the compass course to be steered to make good the given correct magnetic course.

Example.

Ex. 1. Required the Compass Course that shall make correct magnetic W. by S.

Entering the first column of Table I, or II, with W. by S., the deviation on that point is found to be 10½° W., which allowed to the *right* would be about West; and since the deviation for this last does not differ from the deviation used, it may be considered that to make correct magnetic W. by S., the course to be steered is about West.

253. By comparing the first and third columns of Table II, the seaman may also by inspection, or a single interpolation, determine what course he will have to steer by the Standard Compass, in order to take up any given correct magnetic course. For example, let the given correct magnetic course be N.E., or N. 45° E.; on referring to column 3, it will be found that N. 45° E. lies nearly midway between N. 38° 59′ E. and N. 52° 14′ E., the

Standard Compass courses corresponding to which are N.E. by N. and N.E.; the course to be steered is consequently N.E. ½ N. If great accuracy be required, it will be necessary to find the exact proportion between the actual changes of the ship's head with reference to the horizon. Referring to the same example, he will find that the ship's head by Standard Compass between N.E. by N. and N.E., the actual angular change, is 13° 15′, represented by 11° 15′ of the compass. In shaping a course, therefore, between these points, the value of the half point is about 6° 37′, the quarter is 3° 18′, and similarly for smaller divisions of the rhumb.

To prevent, however, the possibility of error in such an important operation as that of shaping a course, a separate Table, may be advantageously constructed expressly for that object. See Table III, where the desired course being sought in the first column is immediately followed by the course to be steered by the Standard Compass, and given in degrees and minutes, as well as points and fractional parts.

TABLE III.

Correct Magnetic Course proposed to be steered.	Course that must be steered by the Standard Compass in order to make the Correct Magnetic Course.		
North	North	or nearly	North.
N. by E.	N. 10° E.	,,	N. ⅞ E.
N.N.E.	N. 19 E.	,,	N. by E. ¾ E.
N.E. by N.	N. 29 E.	,,	N.N.E. ⅝ E.
N.E.	N. 38½ E.	,,	N.E. ½ N.
N.E. by E.	N. 48½ E.	,,	N.E. ¼ E.
E.N.E.	N. 59 E.	,,	N.E. by E. ¼ E.
E. by N.	N. 68 E.	,,	E.N.E.
East	N. 79 E.	,,	E. by N.
E. by S.	East	,,	East.
E.S.E.	S. 78 E.	,,	E. by S.
S.E. by E.	S. 65 E.	,,	S.E. by E. ¾ E.
S.E.	S. 52½ E.	,,	S.E. ¾ E.
S.E. by S.	S. 39½ E.	,,	S.E. ½ S.
S.S.E.	S. 26½ E.	,,	S.S.E. ⅜ E.
S. by E.	S. 13 E.	,,	S. by E. ⅛ E.
South	South	,,	South.
S. by W.	S. 12½ W.	,,	S. by W. ⅛ W.
S.S.W.	S. 25½ W.	,,	S.S.W. ¼ W.
S.W. by S.	S. 39 W.	,,	S.W. ½ S.
S.W.	S. 52½ W.	,,	S.W. ½ W.
S.W. by W.	S. 65½ W.	,,	S.W. by W. ¾ W.
W.S.W.	S. 78 W.	,,	W. by S. ¼ S.
W. by S.	N. 89 W.	,,	West.
West	N. 78½ W.	,,	W. by N.
W. by N.	N. 68 W.	,,	W. by N. ¾ N.
W.N.W.	N. 58 W.	,,	N.W. by W. ¼ W.
N.W. by W.	N. 48 W.	,,	N.W. ¾ W.
N.W.	N. 38 W.	,,	N.W. ½ N.
N.W. by N.	N. 28 W.	,,	N.N.W. ½ W.
N.N.W.	N. 18 W.	,,	N. by W. ¾ W.
N. by W.	N. 8½ W.	,,	N. ¾ W.
North	North.		North.

Deviation of the Compass.

254. The following examples are designed to show the method of correcting courses for leeway, variation, and deviation.

EXAMPLES.

Ex. 1. Course steered E.N.E.; wind S.E.; leeway 2½ points; variation 17° E.; deviation 21° E.: required the true course.

Here the course by compass is E.N.E., or	6 points right of North.
The ship being on the starboard tack the leeway is applied to the left, and hence	2½ „ left.
The difference is the course corrected for leeway ..	3½ „ right of North.
Which expressed in degrees, &c., is	39° 22' right of North.
The variation and deviation are of *same* name, their sum, viz., (21° E. + 17° E.) 38° E., is	38 0 right.
True course	77 22 right of North.
	Or N. 77° 22' E.

Ex. 2. Course by compass N.N.W.; wind N.E.: leeway 2½ points; variation 45° W.; deviation 16° 52' E.; find the true course.

Here the ship's course is N.N.W.	2 points left of North.
The ship being on the starboard tack the leeway is applied to the left, and hence is	2½ „ left.
Therefore the sum is the course corrected for leeway	4½ „ left.
Which expressed in degrees, &c., is	50° 38' left of North.
The variation and deviation are of *contrary* names, their difference, viz., (45° W. − 16° 52' E.) 28° 8' W., is	28 8 left.
Sum	78 46 left of North.
True Course N. 78° 46' W., or W. by N.	

In this example the compass course and leeway are given in points, we therefore take the course and allow the leeway from the wind, which gives the course corrected for leeway, viz., 4½ points left of North, which expressed in degrees, &c., is 50° 38' L. of N.; then the difference of the variation and deviation is taken, as they are of *contrary* names, the remainder—which takes the name of the *greater*—is then applied; the result is the true course.

Ex. 3. Course by compass W. by S.; wind S. by W.; leeway 1¾ points; variation 36° 34' E.; deviation 13° 50' W.: find the true course.

Compass course W. by S. or	7 points right of South.
Leeway, port tack allowed to the right	1¾ „ right.
Sum exceeds 8 points	8¾ „ right of South.
Subtract from	16
	7¼ „ left of North.
or 81° 34' left of North.	
Variation and deviation are of contrary names, the difference, viz., (36° 34' E. − 13° 50' W.) = 22° 44' E.	22 44 right.
True course	58 50 left of North.
or N. 58° 50' W.	

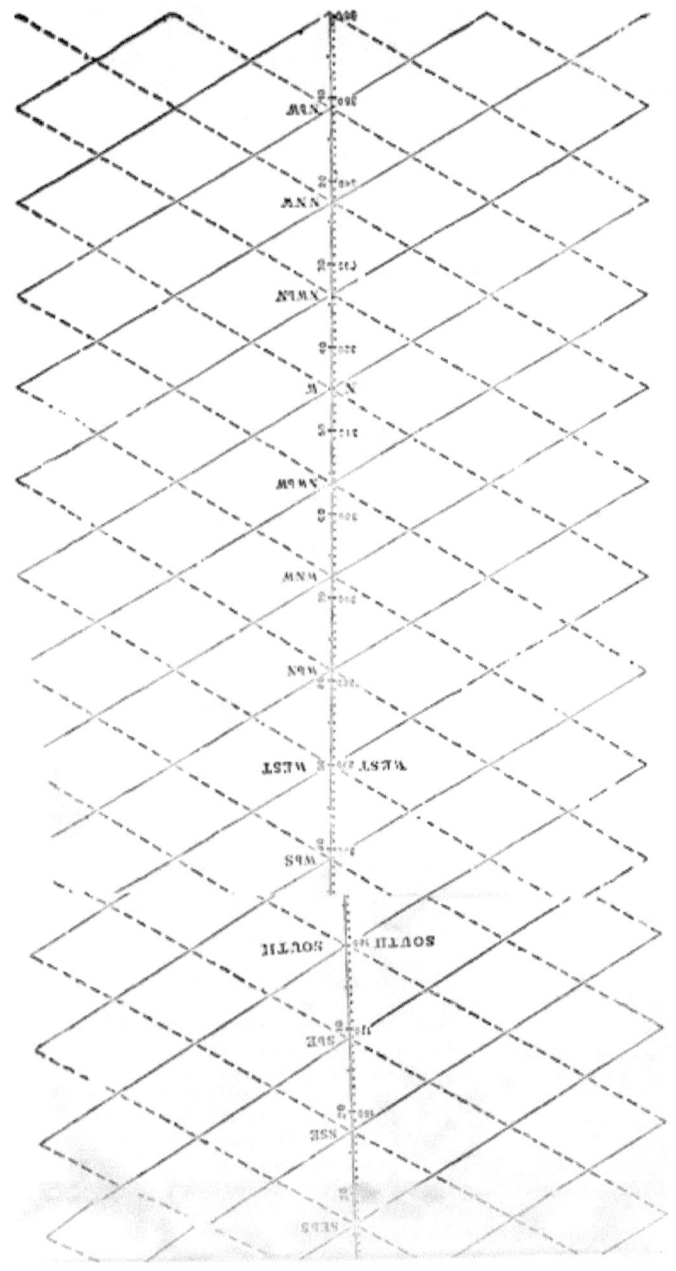

EXAMPLES FOR PRACTICE.

From the following Compass Courses find the True Courses:—

No.	Compass Courses	Winds.	Leeway.	Variation.	Deviation.
1.	N.E. by E.	N. by W.	2¾ pts.	2 pts. W.	3½ pts. E.
2.	North	E.N.E.	2 ,,	2 ,, W.	1 ,, E.
3.	N.N.W.	N.E.	3¼ ,,	2¾ ,, E.	1½ ,, E.
4.	West	S.S.W.	2¼ ,,	2¼ ,, E.	¾ ,, E.
5.	S.S.E. ¼ E.	S.W. ½ S.	3½ ,,	2¼ ,, E.	1¼ ,, W.
6.	E. ¾ S.	N.E. by N.	3½ ,,	1½ ,, W.	1½ ,, E.
7.	South	W.S.W.	2¼ ,,	16° 52′ E.	14° 4′ E.
8.	E. ¾ S.	S. by E.	1¾ ,,	42 0 E.	15 0 E.
9.	W.N.W.	North	3 ,,	42 0 E.	18 30 W.
10.	N.E. by E.	N. by W.	¾ ,,	14 0 E.	19 0 E.
11.	W. by S.	S. by W.	½ ,,	10 30 E.	19 0 W.
12.	South	E.S.E.	½ ,,	17 0 W.	3 0 E.
13.	West	N.N.W.	1½ ,,	18 30 E.	21 0 W.
14.	S.S.W. ¼ W.	S.E. by S.	2½ ,,	17 0 W.	5 0 W.
15.	N.W. by W.	N. by E.	3 ,,	25 0 W.	7½ ,, W.
16.	E. by N.	S.E. by S.	1¾ ,,	32 0 E.	12 0 E.
17.	W. by S. ½ S.	S. by W.	2 ,,	15 0 E.	15 0 W.
18.	E. ½ S.	N.N.E. ½ E.	2½ ,,	21 0 W.	4 0 W.
19.	S.W. by S.	W. by N.	1½ ,,	25 0 E.	10 0 W.
20.	South	E.S.E.	1½ ,,	52 0 W.	2 0 E.
21.	S.W. ½ S.	S.S.E. ½ E.	1½ ,,	52 0 W.	13 0 W.
22.	E. ½ S.	S. by E. ½ E.	¾ ,,	52 0 W.	24 0 E.
23.	East	S.S.E.	2¼ ,,	8 30 E.	15 35 E.
24.	W. ¼ N.	N.N.W.	1¾ ,,	8 30 E.	21 30 W.
25.	N. ¼ W.	W. by N.	¾ ,,	15 45 E.	6 0 W.
26.	E. ¼ N.	N.N.E.	2½ ,,	13 0 W.	20 0 E.
27.	N. by W. ¼ W.	N.E. ½ E.	2¼ ,,	53 0 W.	8½ ,, W.
28.	up S.E. } off E. by N. }	South	4½ ,,	1¼ pts. E.	2 pts. W. / 1½ ,, W.
29.	up S.W. ½ W. } off W. by N. }	S. ½ E.	5¼ ,,	1¼ ,, E.	2½ ,, E. / 1½ ,, E.
30.	up N.W. ¼ N. } off W. by N. }	N. by E.	4½ ,,	1¼ ,, E.	¾ ,, E. / 1½ ,, E.

NAPIER'S DIAGRAM.

255. It is often of the utmost importance in various branches of physical science to represent tables of *related numbers* by means of curve lines, or other figures that show *to the eye* the nature of the relations or laws expressed, or rather concealed, within the mass of figures constituting the tables. Not only does such a mode of representation at once manifest these laws—almost rendering them palpable—but it further points out in what cases natural laws are not represented, and therefore what the cases are that require a greater amount of observation. These modes of representation are commonly known as *Graphic Methods*.

Various "graphic methods" of delineating the deviation have been devised;* but the method introduced here is due to J. R. NAPIER, Esq., F.R.S., and is one peculiarly adapted for this purpose, as it is equally appli-

---

* Graphic methods for correcting the ship's course for the Deviation of the Compass have also been designed by Rear Admiral RYDER, Mr. ARCHIBALD SMITH, F.R.S., and Mr. W. W. RUNDELL. Admiral RYDER's which is an extension of NAPIER's diagram, is published by the Admiralty. Mr. SMITH's, known as the straight line method, is published by the Board of Trade, and also furnished to H.M. ships for fleet tactics, for which it is well adapted. Mr. RUNDELL's is known as the circular method. They are all useful in practice.

NAPIER'S DIAGRAM,

Showing Curve of Deviation.

Ship _____

West Deviation.   East Deviation.

EXAMPLES FOR PRACTICE.

From the following Compass Courses find the True Courses:—

No.	Compass Courses	Winds.	Leeway.	Variation.	Deviation.
1.	N.E. by E.	N. by W.	2¾ pts.	2 pts. W.	3½ pts. E.
2.	North	E.N.E.	2 ,,	2 ,, W.	1 ,, E.
3.	N.N.W.	N.E.	3¼ ,,	2¾ ,, E.	1½ ,, E.
4.	West	S.S.W.	2½ ,,	2½ ,, E.	¾ ,, E.
5.	S.S.E. ½ E.	S.W. ½ S.	3¾ ,,	2¼ ,, E.	1¼ ,, W.
6.	E. ¾ S.	N.E. by N.	3¼ ,,	1½ ,, W.	1½ ,, E.
7.	South	W.S.W.	2¼ ,,	16° 52′ E.	14° 4′ E.
8.	E. ¾ S.	S. by E.	1¾ ,,	42 0 E.	15 0 E.
9.	W.N.W.	North	3 ,,	42 0 E.	18 30 W.
10.	N.E. by E.	N. by W.	½ ,,	14 0 E.	19 0 E.
11.	W. by S.	S. by W.	¾ ,,	10 30 E.	19 0 W.
12.	South	E.S.E.	¼ ,,	17 0 W.	3 0 E.
13.	West	N.N.W	1¼ ,,	18 30 E.	21 0 W.
14.	S.S.W. ¼ W.	S.E. by S.	2½ ,,	17 0 W.	5 0 W.
15.	N.W. by W.	N. by E.	3 ,,	25 0 W.	7½ W.
16.	E. by N.	S.E. by S.	1¾ ,,	32 0 E.	12 0 E.
17.	W. by S. ½ S.	S. by W.	2 ,,	15 0 E.	15 0 W.
18.	E. ½ S.	N.N.E. ½ E.	2½ ,,	21 0 W.	4 0 W.
19.	S.W. by S.	W. by N.	1½ ,,	25 0 E.	10 0 W.
20.	South	E.S.E.	1¼ ,,	52 0 W.	2 0 E.
21.	S.W. ½ S.	S.S.E. ½ E.	1½ ,,	52 0 W.	13 0 W.
22.	E. ½ S.	S. by E. ½ E.	¾ ,,	52 0 W.	24 0 E.
23.	East	S.S.E.	2¼ ,,	8 30 E.	15 35 E.
24.	W. ¼ N.	N.N.W.	1¾ ,,	8 30 E.	21 30 W.
25.	N. ½ W.	W. by N.	½ ,,	15 45 E.	6 0 W.
26.	E. ½ N.	N.N.E.	2¼ ,,	13 0 W.	20 0 E.
27.	N. by W. ½ W.	N.E. ½ E.	2¼ ,,	53 0 W.	8½ W.
28.	up S.E. off E. by N.	South	4½ ,,	1¼ pts. E.	2 pts. W. 1½ ,, W.
29.	up S.W. ½ W. off W. by N.	S. ½ E.	5¼ ,,	1¼ ,, E.	2¼ ,, E. 1½ ,, E.
30.	up N.W. ½ N. off W. by N.	N. by E.	4½ ,,	1¼ ,, E.	¾ ,, E. 1¼ ,, E.

NAPIER'S DIAGRAM.

255. It is often of the utmost importance in various branches of physical science to represent tables of *related numbers* by means of curve lines, or other figures that show *to the eye* the nature of the relations or laws expressed, or rather concealed, within the mass of figures constituting the tables. Not only does such a mode of representation at once manifest these laws—almost rendering them palpable—but it further points out in what cases natural laws are not represented, and therefore what the cases are that require a greater amount of observation. These modes of representation are commonly known as *Graphic Methods*.

Various "graphic methods" of delineating the deviation have been devised;* but the method introduced here is due to J. R. NAPIER, Esq., F.R.S., and is one peculiarly adapted for this purpose, as it is equally appli-

* Graphic methods for correcting the ship's course for the Deviation of the Compass have also been designed by Rear Admiral RYDER, Mr. ARCHIBALD SMITH, F.R.S., and Mr. W. W. RUNDELL. Admiral RYDER's which is an extension of NAPIER's diagram, is published by the Admiralty. Mr. SMITH's, known as the straight line method, is published by the Board of Trade, and also furnished to H.M. ships for fleet tactics, for which it is well adapted. Mr. RUNDELL's is known as the circular method. They are all useful in practice.

cable whether the points on which the observations have been made are or are not precisely equidistant. It requires no calculation, and only a moderate degree of neat-handedness.

The method consists of two parts, the diagram and the curve. The diagram is the same for all vessels.

256. **Construction of the Diagram.**—In this method the diagram consists of a central or vertical line of convenient length—say 18 inches—which may be considered as representing the margin of the compass card cut at the north point, and straightened and extended in the following way :—

N   E   S   W   N

This line, which may be taken to represent no deviation, is divided into 32 equal parts, representing the 32 points of the compass, commencing at the top with North, and ranging in the order of N. by E., N.N.E., &c., and ending with North at the bottom. The central line is then intersected at each of the 32 points by two straight lines, one a *plain* line and the other a *dotted* line. The plain and dotted lines make an angle of 60° with the central line and with each other, and so forming a set of equilateral triangles with the central line, and converting the diagram into a simple addition and subtraction table. On the right side of the central line the *dotted* lines incline *downwards*, and the *plain* lines *upwards*. The reverse is the case on the left. The central line is further divided into 360 equal parts, representing degrees, and these divisions are numbered from 0° at the top to 360° at the bottom. They are also numbered, according to the usual mode of dividing the compass card, from 0° at North and South, up to 90° at East and West. This division of the central line into degrees serves also as a scale by which the deviations are laid off.

257. **Requisite Observations to be made.**—The least number of observed deviations available for obtaining a complete curve are the deviations on 4 points distributed equally, or nearly so, round the compass; but, if possible, the deviations should be observed on 8 or more points. If the observations are observed on 4 points only, these should be at or near N.E., S.E., S.W., and N.W., and from these it is possible to form a fairly approximate curve. The points next in importance are North, East, South, and West. If the deviations have been observed at or near the eight principal points, a curve can be drawn which will give the deviation on every point of the compass within very small limits of error.[*]

258. Cases may also occur in which by the ship swinging round at her anchors in a tide-way or to the wind, or by the aid of a steam-tug, the deviation may be observed on various directions of the ship's head, not being necessarily exact points of the compass; or similarly, whilst under steam or sail at sea, a number of azimuths of the sun may be observed, and hence the deviation obtained.

---

[*] The examination of a collection of curves made from actual observations, as in the report of the Liverpool Compass Committee, &c., will show that there is so much regularity that these interpolated deviations may generally be relied upon, although certain cases, such as the U.S.S. *Roanoke* (Report of Nautical Academy of Sciences, 1863), the irregularities are considerable.

In these cases the Graphic method here described furnishes a ready and effectual mode of obtaining a result on which the errors of individual observations are as far as possible compensated and any egregious errors eliminated.

259. **Construction of the Curve of Deviations.**—Easterly deviations are laid down to the right of the central line, westerly deviations to the left.*

The amount of the deviation is taken from the scale of degrees on the central line; then, if the deviation has been determined with the ship's head on an exact compass point, lay off the amount of the deviation on the *dotted*† line which passes through that point; but if not observed on the exact point, then on a line parallel to the dotted line, the compass course or direction of the ship's head being still taken from the central line, and mark the point so determined with a cross, or dot encircled in ink. Perform the same operation for each observed deviation. Then with a pencil and a light hand draw a flowing curve, passing as nearly as possible through all the crosses, or dots encircled; and when satisfied that the curve is good, draw it in ink. This is the curve of deviations.

If any of the pencil marks be out of the fair curve, it may be assumed that an error has been made in the observation for that point.

NOTE.—When the curve alters its form considerably near the North point as in several CURVES, EXAMPLES FOR PRACTICE, it will be advisable to bring the North points of the diagram together in the form of a drum and then draw the curve.

The process will be best understood by explaining the projection corresponding to the observations as given in the following table:—

Ship's Head by Standard Compass.	Deviation.	Ship's Head by Standard Compass.	Deviation.
North	6° 30' W.	South	5° 30' E.
N.E.	13 0 W.	S.W.	28 35 E.
East	22 15 W.	West	19 15 E.
S.E.	23 30 W.	N.W.	3 0 E.

Deviation Curve O in Diagram.

1. The first compass course on which an observation has been made is North, and the observed deviation is 6° 30' W. With a pair of dividers take from the central line a distance equal to deviation 6½°, and from North on the central line lay off the deviation on the dotted line which passes through that point towards the left—the deviation being W.; at the extremity of the distance make a dot or cross.

---

* In Rear Admiral RYDER's plan, the central line is the diagonal of a square and the other lines make angles of 45° with it, and at right-angles to each other and to the sides of the square, which sides are divided into 360°; the top and bottom representing correct magnetic courses, the sides compass courses. By this method the correct magnetic course corresponding to a given compass course, or the compass course corresponding to a given correct magnetic course, is found as by a table of double entry. The two methods, it will be seen, are the same in principle. Mr. NAPIER's will perhaps be found more convenient in construction by the expert; Admiral RYDER's more simple in use by the inexpert.

† If the table of deviations are given for the correct magnetic courses and not the compass courses or direction of the ship's head, the same process is gone through, except that the deviations are in that case laid off on the plain lines. It is, however, now generally understood that this procedure is contrary to practice and may lead to error.

2. The second compass course on which an observation has been made is N.E., and the observed deviation is 13° 0' W. With dividers take from the central line a distance equal to deviation 13°, and from N.E. on the central line lay it off on the dotted to the left—deviation being W.; and the point so determined mark with a cross or dot.

3. The third compass course on which an observation has been made is East, and the observed deviation is 22° 15' W. Take from central line 22¼°, and from E. on central line and on the dotted line passing through it, lay off the observed deviation to the left—deviation being W.; and mark the point so determined with a dot or cross.

4. Compass course S.E., observed deviation 23° 30' W. Take from the central line a distance equal to deviation 23½°, and from S.E. on the central line lay off on the dotted line passing through the same point the amount of deviation to the left—the deviation being W.; make a dot.

5. Compass course South, deviation 5° 30' E. Measure on the central line a distance equal to deviation 5½°, and having found compass course South on the central line, lay off the amount of deviation on the dotted line which passes through it towards the right—deviation being E.; and make a dot or cross.

6. Compass course S.W., deviation 28° 35' E. From the central line take a distance equal to observed deviation 28½°, and having found S.W. on the central line, lay off on the dotted line passing through that point the amount of deviation to the right—deviation being E.; make a dot or cross.

7. Compass course West, observed deviation 19° 15' E. Measure on the central line a distance equal to deviation 19¼°, and from West lay off on the dotted line passing through that course the amount of deviation towards the right—deviation E.; and make a dot or cross.

8. Compass course N.W., deviation 3° 0' E. Take from central line a distance equal to 3°, and lay off on the dotted line passing through N.W., the amount of deviation (3°) towards the right—deviation E.; and make a dot.

9. Then, with a pencil and a light hand, draw a flowing curve, passing as nearly as possible through all the crosses or dots, and if satisfied with the curve in pencil, draw it in ink.

NOTE.—The learner should take a pair of dividers and go through the above process on the diagrams here given (see Plate I). He should then take the blank diagram (see Plate II), and make the curve on it.

Ex. 2. Construct a curve of deviations, using for the purpose the following observations:— (See Deviation Curve A in Diagram.)

Ship's Head by Standard Compass.	Deviation.	Ship's Head by Standard Compass.	Deviation.
North	1° 15' W.	South	1° 50' E.
N.E.	22 30 E.	S.W.	15 0 W.
East	26 50 E.	West	26 0 W.
S.E.	17 0 E.	N.W.	27 0 W.

The following describes the process of construction:—

1. With a pair of dividers take from the central line a distance equal to deviation 1° 15', or 1¼°, and from North on the central line, lay the distance off on the dotted line passing through that point and towards the left—being W.; at the extremity of the distance make a dot or cross.

2. Take from the central line a distance equal to 22½° (22° 30'), and lay it off on the dotted line, from N.E. towards the right—being E.; make a dot or cross.

3. Take from the central line a distance equal to $26\frac{3}{4}°$ ($26°$ $50'$), and lay it off on the dotted line, from East towards the right—being E.; make a dot or cross.

4. Take from the central line a distance equal to $17°$ ($17°$ $0'$), and lay it off on the dotted line, from S.E. towards the right—being E.; make a dot or cross.

5. From the central line take a distance equal to $1\frac{3}{4}°$ ($1°$ $50'$), and lay it off on the dotted line, from South towards the right—the deviation being E.; make a dot or cross.

6. From the central line take a distance equal to $15°$, and lay it off on the dotted line, from S.W. towards the left—the deviation being W.; make a dot.

7. From the central line take a distance equal to $26°$, and lay it off on the dotted line, from N.W. towards the left—deviation being W.; make a dot.

8. From the central line take a distance equal to $27°$, and lay it off on the dotted line, from W. towards the left—deviation being W.; make a dot.

9. Repeat the admeasurement first made, from North, at the lower end of the central line.

10. Then, with a pencil and a light hand, draw a flowing curve, passing as nearly as possible through all the dots or crosses; when satisfied that the curve is good, draw it in ink. This is the curve of deviation.

Ex. 3. Construct a curve of deviations, using for the purpose the following observations:— (See Deviation Curve B in Diagram).

Ship's Head by Standard Compass.	Deviation.	Ship's Head by Standard Compass.	Deviation.
North	0° 22' W.	South	0° 16' W.
N.E.	7 14 E.	S.W.	6 16 W.
East	10 44 E.	West............	11 50 W.
S.E.	6 18 E.	N.W.	7 52 W.

Selecting those dotted lines which pass through the points representing the different directions of the ship's head, the deviations are laid off: thus, at North, we mark off $0\frac{1}{3}°$ ($0°$ $22'$) to the left on the dotted line passing through North—because deviation is West; at N.E., we take $7\frac{1}{4}°$ from the central line, and lay it off to the *right* (deviation being E.) on the dotted line passing through N.E.; at East, $10\frac{3}{4}$ is taken from the central line and laid off to the right on the dotted line passing through East; and so with the others, being careful to remember that the known deviations must be laid down on the *dotted* lines, *easterly* to the *right* and *westerly* to the *left* of the central line. The curve is then drawn neatly through all the points so laid down, and it will be found that the deviation for any other point taken from the curve corresponds with that taken from the Table of Deviations given at page 154; and the curve thus drawn can be used instead of the Table.

260. **How the Curve is used.**—The curve of deviations having been completed, the diagram affords a ready and convenient method of applying the deviation to the ship's course. This correction may be required as follows:— 1st, from the compass course which has been steered, it may be required to find the **correct magnetic course** to be laid down on the chart; 2nd, from the **correct magnetic course** given by the chart, it may be required to find the **compass course** on which the ship's head ought to be kept; 3rd, if one or more bearings of the land are taken, to **correct these bearings** by the amount of the deviation *due to the direction of the ship's head at the time*. The corrections are given by the following rules:—

261. To find the **Deviation** on any **Compass Course.**

RULE LVIII.

On the central line find the given course; then, with a pair of dividers, measure the distance from that point to where the curve cuts the dotted line passing through the course; but if no dotted line proceeds from the course, then measure from the course on the central line to the curve in a direction exactly parallel *to the nearest* dotted lines: *that distance measured on any part of the central line will give the deviation in degrees.*

EXAMPLES.

Ex. 1. What is the deviation on Compass Course N.E. by N. (*a*) for the deviation curve B; (*b*) using the deviation curve A?

(*a*) Having found the given course on the central line, with a pair of dividers measure the distance from N.E. by N. to where the curve cuts the dotted line proceeding from that point; this distance taken to the central line gives 6° E.

(*b*) Measuring with a pair of dividers the distance from N.E. by N. to where the curve cuts the dotted line proceeding from that point, the deviation measured on the central line is found to be 19° E.

Ex. 2. Required deviation in compass course W.S.W., using deviation A.

Find W.S.W. on the central line, measure the distance from that point to where the curve cuts the dotted line proceeding from it; this distance taken to the central line gives deviation $21\frac{1}{2}°$ W.

Ex. 3. What is the deviation for compass course N.E. $\frac{1}{2}$ N., using the curve C?

Place one leg of a pair of dividers at N.E. $\frac{1}{2}$ N. on the central line, and from thence measure the distance to the curve *in a direction exactly parallel* to the nearest dotted lines; this distance taken to the central line gives the deviation $12\frac{1}{2}°$ W., the deviation for N.E. $\frac{1}{2}$ N.

Ex. 4. Find the deviation for Standard Compass Course N. 84° W., using deviation curve A.

Find N. 84° W. on the central line, and placing one foot of a pair of dividers on that point, from thence measure the distance in the direction of an imaginary line drawn parallel to the nearest dotted lines; apply this distance to the central line, which shows the deviation for the ship's course is $26\frac{1}{2}°$ W.

EXAMPLES FOR PRACTICE.

Required the deviation for each of the following compass courses, using (*a*) the curve A, and (*b*) the curve C:—

S.S.E.; N.E. by N.; N.W.; N.W. by N.; S.W. by W.; W.N.W.; South; N. 48° E.; S. 52° W.; E. by S. $\frac{1}{4}$ S.; N. 41° W.; and S. 38° E.

NOTE.—Persons may differ one or two degrees in their estimate of what constitutes a fair curve; it is therefore quite likely that students may find their answers differ a degree or two from those given in this work.

For further exercises in this matter the learner may take the Table of Deviations given on page 154, and using the curve B, find the deviations for each of the 32 points of the compass; the results ought to agree pretty nearly with those given in the Table.

262. We have now the following easily applied solution of the two following problems:—

Problem I.—From a **Compass Course,** to find the corresponding **Correct Magnetic Course.**

RULE LIX.

On the central line find the given Standard Compass Course, and move on the dotted line drawn from it, or in a direction parallel to the *dotted lines till you reach the curve, and then move on a* plain line, *or in a direction parallel to the* plain *lines, till you get back to the central line.* The point on the central line at which you arrive is the correct magnetic course required.*

NOTE.—The directions in the above rule are easiest done by means of a pair of dividers. To move on the *dotted* line, or in a direction parallel to it, place one leg of a pair of dividers on the course, and the other leg at that point on the curve which is intersected by the *dotted* line proceeding from the course, or a point on the curve where a line included between the leg of the dividers on the central line and the leg on the curve shall be *exactly parallel* to the nearest dotted lines, then to return to the central line—keep the first leg of the dividers fixed and lift the other off the curve, move in a direction *parallel to the plain lines* until you reach the central line, the point where the dividers cut the central line shows the correct magnetic course; or keep the leg of the dividers which is on the curve fixed, and move the other leg off the central line parallel to the plain line until it again cuts the central line, which indicates the correct magnetic course required.

EXAMPLES.

Ex. 1. The course steered by standard compass is N.N.E.; what is the correct magnetic course to lay down on the chart (using the curve C in the diagram, Plate I.)

Find the given compass course N.N.E. on the central line, then take a pair of dividers, put one leg of the dividers on N.N.E., from which extend the other leg along the *dotted* line passing through the point till the curve is reached, then keeping the leg on the central line fixed, move the one off the curve, and then return to the central line in a direction parallel to the *plain* line; it will be found to intersect it at N. $13\frac{1}{2}°$ E., or N. by E. $\frac{1}{4}$ E., nearly, the required correct magnetic course.

Ex. 2. The course steered by compass is N.E. by N.: required the correct magnetic course (using Curve A, Plate I).

Follow the *dotted* line extending from N.E. by N. to where the curve cuts it, by placing one leg of a pair of dividers on the course found on the central line and the other at the point where the *dotted* passing through N.E. by N. cuts the curve, then keeping the leg on the central line fixed, lift the other from the curve and move in the direction of the nearest *plain* lines till the central line is reached; then the correct magnetic course will be found to be N. $52°$ E., or N.E. $\frac{3}{4}$ E., nearly.

Ex. 3. The course steered is S.S.W.: required the correct magnetic course (using curve C, Plate I).

Place one leg of the dividers on the compass course S.S.W. on the central line and the other on the place where the *dotted* line proceeding from it is intersected by the curve, then keeping the leg on the central line fixed, return with the other leg to the central line in a direction parallel to the *plain* line, it will be seen that the *correct* magnetic course is S.W.$\frac{3}{4}$ W

Ex. 4. The course steered by compass is S.E. $\frac{1}{2}$ E.: required the corresponding correct magnetic course (using curve C, Plate I.)

Place one leg of the dividers on S.E. $\frac{1}{2}$ E. on the central line, and the other leg on the curve, being careful to keep the two points of the dividers exactly parallel to the nearest *dotted* lines, then lift the leg off the curve and return to the central line, the place where this last line intersects the central line, shows the correct magnetic course to be S. $75\frac{1}{2}°$ E., or E. by S. $\frac{1}{4}$ S.

---

* It will be observed that this is merely the addition or subtraction, as the curve is to the right or left of the central line, of the deviation on the course, since the three sides of the triangle ABC passed over by the pencil or leg of dividers are all equal.

Ex. 5. Given the standard compass courses N. 38° E. and S. 49° W.: required the correct magnetic courses (using the A deviation curve, Plate I.)

Place one leg of the dividers on the standard compass course N. 38° E. on the central line, and move the other leg out in a direction parallel to the nearest dotted line until it meets the curve; then, keeping the leg which is on the central line fixed, move the other leg in the direction of the *plain* lines until it returns to the central line. The point arrived at shows the *correct* magnetic course is N. 58° E., nearly, or N.E. by E. ¼ E.

In a similar manner the correct magnetic course is found to be S. 33° W.

Examples for Practice.

In each of the following examples the compass course is given to find the corresponding *correct* magnetic course, using curve A and curve C.

Curve A.—N. 41° W.; N. 65° 30′ E.; S. 38° E.; S. 79° W.; West; N.E.; S.E. ¼ E.

Curve C.—N. 39° W.; N. 48° E.; S. 50° E.; N. 78° W.; N. 70° W.; N. 31½° E.; S. 76° E.; N. 70½° W.; N. 26° W.; and S. 47° W.

Problem II.—From a given **Correct Magnetic Course** to find the corresponding **Compass Course**.

RULE LX.

On the central line take the given correct magnetic course, and move on the plain *line drawn from that point or in a direction parallel to the* plain *lines till you arrive at the curve; and then move on a* dotted *line, or in a direction parallel to the* dotted *lines till you get back to the central line.* The point on the central line at which you arrive is the compass course required.*

See Note to Rule LIX, page 167.

Examples.

Ex. 1. Given correct magnetic course N.N.E., to find the corresponding compass course (using curve C).

Place one leg of the dividers on N.N.E. on the central line, extend the other leg to the spot where the *plain* line proceeding from N.N.E. meets the curve; then keeping the leg on the central line fixed, lift the one on the curve and return to the central line in a direction parallel to the *dotted* line, it will be found to intersect at N.E. by N., the required course by standard compass.

Ex. 2. What compass course will make correct magnetic S.E. (using curve A)?

Find S.E. on the central line; place one leg of the dividers on the spot and the other leg on the curve where the *plain* line that passes through S.E. cuts the curve; then keep the leg that is on the central line fixed, lift the other leg off the curve and move it in the direction of the nearest dotted line till it again touches the central line; the compass course that makes *correct* magnetic S.E. is shown on the central line to be S. 67½° E., or E.S.E.

Ex. 3. It is found from the chart that the correct magnetic course from the ship's position at noon to the Start Point is N. 86° E. What course must be steered by Standard Compass (using curve C)?

Find the correct magnetic course N. 86° E. on the vertical line; place one foot of the dividers on the spot, then follow thence with the other leg in a direction parallel to the nearest *plain* line until it meets the curve, and then return with the other leg to the central line in a direction parallel to the *dotted* line; the compass course required is S. 69½° E.

---

* To assist the memory the following rhyme is given in the Admiralty Manual:—

I.
"From compass course magnetic course to gain,
Depart by dotted and return by plain."

II.
"But if you wish to steer a course allotted,
Take plain from chart and keep her head on dotted."

Ex. 4. Required the compass course to make correct magnetic W. by N. (using curve A).

Place one leg of the dividers on W. by N. and the other on the curve where the *plain* line that passes through W. by N. cuts it; then keep the leg that is on the central line fixed, but lift the other off the curve and move it in the direction of the *dotted* line till it again touches the central line; it will then be seen that the compass course that makes *correct magnetic* W. by N. is N. 51° W., or N.W. ½ W.

Ex. 5. Given the correct magnetic courses N. 64° E. and N. 85° W., to find the Standard Compass course (using the C deviation curve, Plate I).

Find N. 64° E. on the central line; place one leg of the dividers on the point and the other on the curve, being careful to keep both legs of the dividers exactly parallel to the nearest *plain* line; keep the leg of the dividers that is on the central line fixed and move the other in the direction of the nearest *dotted* line till it meets the central line; the point of intersection in this instance is N. 85½° E.

In a similar manner the Standard Compass Course corresponding to correct magnetic course N. 85° W. is found to be S. 70° W.

EXAMPLES FOR PRACTICE.

In each of the following examples the correct magnetic course is given, to find the compass course (using curve A and curve C).

Curve A.—S. 73° 30′ W.; N. 42° 15′ E.; S. 15° 30′ W.; N. 14° 15′ E.; N. 62° 45′ E.; E. 15° S.; W. 45° S.; N.E. by N.; W.S.W.

Curve C.—S. 44° W.; N. 58½° E.; S. 5° W.; N. 24° E.; N. 83° E.; S. 50° E.; S. 22° W.; N.E. ¼ E.; and S.W. ½ S.

263. We shall now proceed to show the application of the foregoing rules to Questions 7, 8, 9, and 10, of List B, which contains the questions on the Deviation of the Compass required of Candidates for Certificates as Master Ordinary.

QUESTION 7. LIST B.

264. Given the **bearings** of a **distant object** *by Standard Compass* on eight *equidistant** points, to find the Correct Magnetic Bearing of the distant object† and thence the Deviation.

RULE LXI.

1°. **If the Compass Bearings are all of the same name,** i.e., *if they are all reckoned from* N. *or* S. *towards* E. *or* W.:

Take the sum of the Bearings in each column, then add these sums together, and divide by 8; *the result is the* Correct Magnetic Bearing of the distant object *of the same name as the* Standard Compass Bearings.

NOTE.—A bearing *North* or *South* must be expressed as 0°, whilst *East* and *West* bearing must be expressed as 90°, reckoned either from North or South (as most convenient) towards East or West; thus East may be written as N. 90° E., or as S. 90° E.; similarly West may be written either as N. 90° W. or S. 90° W.

---

* The 8 equidistant points are the 4 cardinal points and the four quadrantal points, viz., N.W., S.W., S.E., and N.E.

† The Correct Magnetic Bearing thus found is not strictly accurate; it will differ from the correct quantity by what is called the co-efficient A. The co-efficient A is found by adding the Deviations (algebraically),—that is, add together the Westerly deviations, also add the Easterly deviations together, and take the less from the greater, and mark the difference of the same name as the greater,—and divide by 8. This, however, is not required to be understood by Masters Ordinary, although it is required for Masters Extra.

z

2°. **If the Compass Bearing be of different names:**

(a) *If some of the* Bearings *are reckoned from* North *and the* others *from* South :

Take either *set from* 180°, *and they will all be reckoned from the* same *point* North *or* South ; the name *as to* East *or* West *remains unaltered ;* add *all the bearings together and divide by* 8, *the result is the* correct magnetic bearing *of the* same name *as the* unaltered bearings.

(b) *If some of the* Bearings *are towards the* East *and* others *towards the* West :

Find *the* sum *of those which are reckoned towards* East ; *and also the* sum *of those which are towards the* West ; *then take the* less *from the* greater, *and mark the* difference *of the* same name *as the* greater ; *the result* divided *by* 8 *gives the* Correct Magnetic Bearing *of the distant object, which is of the* same name *as the* difference.

NOTE.—On the form given at the Marine Board Examinations there is not sufficient space to perform the last mentioned addition and division : there is only room to find the first two sums ; the rest, however, can be finished in the margin.

3°. **To find the Deviation for each of the given Courses.**—(*a*) If the Correct Magnetic *and* Compass *bearings are of* like *names :*

Take their difference.

(*b*) *If* one *is reckoned from* North *and the other from* South, first *take the Correct Magnetic Bearing from* 180°, *and the remainder will have the* same name *as the* compass bearing, *then take the* difference *between the* Correct Magnetic *and* Compass Bearings.

(*c*) *If* both *bearings are from* North *or* both *from* South, *but* one *is* towards East *and the* other towards West, *take their* sum :

The difference *or* sum will be the deviation.

4°. **To name the Deviations.**—*If the* Correct Magnetic Bearing *is to the* right *of the* Compass Bearing *the deviation* is East, but if to *the* left *it is* West.

EXAMPLE I.

In the following Table give the correct magnetic bearing of the distant object, and thence the Deviation.

Ship's Head by Standard Compass.	Bearing of Distant Object by Standard Compass.	Deviation Required.	Ship's Head by Standard Compass.	Bearing of Distant Object by Standard Compass.	Deviation Required.
North	S. 41° W.		South	S. 46° W.	
N.E.	S. 59 W.		S.W.	S. 25 W.	
East	S. 64 W.		West	S. 18 W.	
S.E.	S. 60 W.		N.W.	S. 23 W.	

NOTE.—The above is the form in which the Table is given at the Local Marine Board Examinations. The Table is printed, excepting the columns of Bearings, which the Examiner fills up in writing. The following is the above example worked :—

Deviation of the Compass.

Ship's Head by Standard Compass.	Bearing of Distant Object by Standard Compass.	Deviation Required.	Ship's Head by Standard Compass.	Bearing of Distant Object by Standard Compass.	Deviation Required.
North	S. 41° W.	1° E.	South	S. 46° W.	4° W.
N.E.	S. 59 W.	17 W.	S.W.	S. 25 W.	17 E.
East	S. 64 W.	22 W.	West	S. 18 W.	24 E.
S.E.	S. 60 W.	18 W.	N.W.	S. 23 W.	19 E.

```
           S. 224° W.
           S. 112 W.
           ─────
        8) 336
           ─────
           S. 42° W. = Correct magnetic bearing of distant object.
```

Here we first add together the eight given bearings of the distant object, making 336°, and divide the sum by 8, giving as the result S. 42° W., the correct magnetic bearing of the distant object.

We next take the difference between the correct magnetic bearing thus obtained and each bearing given in the table; thus, with the ship's head at North, the bearing by compass is S. 41° W., and the difference between this and the correct magnetic bearing S. 42° W. is 1°, and because the *correct magnetic* bearing is to the *right* hand of the standard compass bearing the deviation is East. Again, with the ship's head at N.E., the compass bearing is S. 59° W., and the difference between this and the *correct magnetic* bearing S. 42° W. is 17°, but now the correct magnetic bearing is to the *left* of the compass bearing, hence the deviation is West, and so on with the remaining bearings, and the work will stand as follows:—

Ship's Head.	N.	N.E.	East.	S.E.
Correct mag. bear.	S. 42° W.	S. 42° W.	S. 42° W.	S. 42° W.
Standard com. bear.	S. 41 W.	S. 59 W.	S. 64 W.	S. 60 W.
Deviation	1 E.	17 W.	22 W.	18 W.

Ship's Head.	S.	S.W.	West.	N.W.
Correct mag. bear.	S. 42° W.	S. 42° W.	S. 42° W.	S. 42° W.
Standard com. bear.	S. 46 W.	S. 25 W.	S. 18 W.	S. 23 W.
Deviation	4 W.	17 E.	24 E.	19 E.

NOTE.—In the above work, the deviations are obtained by subtracting the less bearing from the greater, because they are all of the same name. (See LXI, 3° and 4°, page 170.)

QUESTION 8.   LIST C.   (See Rule LX, Problem II, page 168.)

From the above table construct a NAPIER's Curve, and give the courses you would steer by standard compass to make the following courses correct magnetic:—

(1.) S.E.   (2.) N.E. ¼ E.   (3.) S. 10° W.   (4.) E. ¼ N.

To Construct the Curve from the Table.—With a pair of dividers take from the central line 1°, the deviation for ship's head North; and lay it off from North, on the central line along the dotted line passing through the given point, and towards the right—being East; at the extremity of the distance make a dot or cross. Next take 17° from the central line, and with one foot of the dividers on N.E. on the central line, lay off this distance on the dotted line passing through the given point and to the left, because the deviation is West. Proceed in like manner with the deviation 22° W. at East; with 18° W. the deviation at S.E.; and with 4° W. the deviation at South. The deviation at S.W., West., and N.W. being easterly, must be applied to the right of the central line along the *dotted* lines proceeding from those points. Now draw with a pencil a curve passing as nearly as possible through the points found, and when satisfied with its uniformity, draw it in ink.

To find the Standard Compass Course.—(1.) Place one foot of a pair of dividers on S.E. on the central line, and the other foot on the point where the *plain* line extending from S.E. is cut by the curve; then keeping the first foot fixed and lift that on the curve, moving it in a direction parallel to the nearest *dotted* line, towards the central line; the compass course that makes *correct* magnetic S.E. is shown on the central line to be S. $29°$ E., or S.S.E. $\frac{3}{4}$ E., (easterly).

(2.) Take N.E. $\frac{1}{4}$ E. on the central line, then placing one foot of the dividers on that spot extend the other foot from this point and parallel to the nearest *plain* line until it cuts the curve; then keeping the first foot fixed, move the one on the curve parallel to the nearest *dotted* line and towards the central line; the compass course that makes *correct* magnetic N.E. $\frac{1}{4}$ E. is shown on the central line to be E. by N. $\frac{3}{4}$ N.

(3.) Place one foot of the dividers on S. $10°$ W. on the central line, and extend the other foot parallel to the nearest *plain* line, and to the *right* until it meets the curve; then keeping the foot on the central line fixed, move the foot on the curve thence parallel to the nearest *dotted* line until it arrives at the central line, which shows that the compass course to be steered is S. $9°$ W., nearly, in order to make S. $10°$ W. correct magnetic.

(4.) One foot of the dividers being placed on E. $\frac{1}{4}$ N., move the other foot to the left from that spot and parallel to the nearest *plain* line until it is cut by the curve; from thence, keeping the first foot fixed on the central line, move parallel to the nearest *dotted* line till the central line is reached; the compass course that makes correct magnetic E. $\frac{1}{4}$ N. is shown on the central line to be S. $71\frac{1}{2}°$ E.

QUESTION 9.   LIST B.

Suppose you steer the following courses by the standard compass, find the correct magnetic courses from the curve drawn:— (See Problem I, Rule LIX, page 167.)

(1.) S.S.E. $\frac{3}{4}$ E.   (2.) S. $\frac{1}{4}$ W.   (3.) E. by N. $\frac{3}{4}$ N.   (4.) N. $\frac{1}{4}$ W.

(1.) **To find the Correct Magnetic Course.**—With one foot of the dividers on S.S.E. $\frac{3}{4}$ E. on the central line, move the other foot on a line to the *left* and parallel to the nearest *dotted* line until it cuts the curve; then keeping the first foot fixed, move the foot on the curve from thence parallel to the nearest *plain* one and towards the central line; the *correct* magnetic course is at once seen to be S. $45\frac{1}{2}°$ E.

(2.) Placing one foot of the dividers on S. $\frac{1}{4}$ W. on the central line, move the other foot on a line to the right, parallel to the nearest *dotted* line and cutting the curve; from thence, keeping the first foot fixed on the central line, move in a direction parallel to the nearest *plain* line until the central line is reached; the *correct* magnetic course is thus shown to be S. $10°$ W.

(3.) Take the point on the central line representing E. by N. $\frac{3}{4}$ N., place a leg of the dividers on that point and move the other leg in a direction parallel to the *dotted* lines, and after meeting the curve, return parallel to the *plain* lines, until the central line is again reached; the correct magnetic course is thus found to be N. $48\frac{1}{2}°$ E.

(4.) One leg of the dividers being fixed on N. $\frac{1}{4}$ W. found on the central line, move the other leg in a direction parallel to the *dotted* lines till the curve is reached; and from thence returning to the central line in a direction parallel to the *plain* lines, we find *correct* magnetic course is North.

QUESTION 10.   LIST B.

265. Given the Bearings of two (or more) distant objects by the Standard Compass and also the Azimuth of the Ship's Head: required the correct magnetic bearing of these objects.

RULE LXII.

1°. *Find the* Deviation *corresponding to the* direction of the ship's head, *taking it from the Napier's Deviation Curve with a pair of dividers.*

2°. *Apply the* Deviation *thus found* (keeping the legs of the dividers the same distance apart) *to the Bearing of distant object by Standard Compass,* thus—*Place one leg of the dividers on the part of the central line which represents the Standard Compass Bearing,* and lay the other leg on the central line.

Upwards if deviation for ship's head is W.
Downwards if deviation for ship's head is E.

The number of degrees *there indicated will be the* correct magnetic bearing *from* N. *or* S. *towards the* E. *or* W.

The following bearings of distant objects have been taken by the Standard Compass as above; with the ship's head as given, find the correct magnetic bearing.

	Ship's Head.	Compass Bearing.		Ship's Head.	Compass Bearing.
1.	West	East.	3.	E. ¼ N.	N. ¾ W.
2.	S.S.E.	E. by S. ¼ S.	4.	N.E. ¼ E.	W. ½ S.

On the central line find the given course, West, and with a pair of dividers measure the distance from the point to where the curve cuts the *dotted* line proceeding from the course; this distance taken to the central line gives deviation 24° E. for ship's head at West.

We next apply the deviation to East—the bearing of distant object by standard compass—by placing one leg of dividers on central line at East and laying the other leg on the central line *downwards* (the deviation being East), the number of degrees there indicated; whence the correct magnetic bearing of distant object is S. 66° E.

Again, ship's course S.S.E. is found on the central line; with dividers measure from thence to where the *dotted* line proceeding from given course is cut by the deviation curve; this distance, taken from the central line, gives deviation 13° W.

The standard compass bearing of distant object taken when ship's head was at S.S.E. is E. by S. ¼ S., or S. 76° E., to which apply deviation, found as above, by placing one leg of dividers on S. 76° E. on the central line, and applying the other leg *upwards* on the central line (deviation being West); the result is the correct magnetic bearing S. 89° E.

The deviation due to E. ¼ N.—the direction of the ship's head—is found on deviation curve to be 22½° W.

Apply this deviation to compass bearing N. ¾ W., or N. 8° W., to *upwards* on the central line (deviation being West), which gives correct magnetic bearing of distant object N. 30½° W.

N.E. ¼ E. is the next given direction of the ship's head, and applied to the central line, the deviation for which on curve is found to be 18° W.

The deviation thus found being applied *upwards* from W. ¼ S., or S. 87° W., gives the correct magnetic bearing of distant object S. 69° W.

EXAMPLE II.

I. In the following Table give the correct magnetic bearing of the distant object, and thence the Deviation:—

(Correct Magnetic Bearing of distant object = N. 89° 4' W.)

Ship's Head by Standard Compass.	Bearing of Distant Object by Standard Compass.	Deviation Required.	Ship's Head by Standard Compass.	Bearing of Distant Object by Standard Compass.	Deviation Required.
North	S. 87° W.	3° E.	South	N. 87° W.	2° W.
N.E.	S. 70 W.	21 E.	S.W.	N. 75 W.	14 W.
East	S. 71 W.	20 E.	West	N. 68 W.	21 W.
S.E.	S. 81 W.	10 E.	N.W.	N. 72 W.	17 W.

```
        S. 310 W.                              N. 302 W.
        S. 418 W.             (180° × 4) = 720
        ───────                              ─────────
        8)728                                 S. 418 W.
        ───────
        S. 91 W. = Correct Magnetic Bearing.
          180
        ─────
    or N. 89 W.
```

In this example the bearings given in the left hand column are reckoned from S. towards W., while the bearings in the right hand column are reckoned from N. towards W.; therefore, before adding up the latter column, each bearing must be subtracted from 180°, and the remainder, in each case, is the bearing reckoned from S. towards W.

Subtracting N. 87° W. from 180° = S. 93° W. which put in the place of N. 87° W.
          N. 75 W. ,, 180 = S. 105 W. ,, ,, N. 75 W.
          N. 68 W. ,, 180 = S. 112 W. ,, ,, N. 68 W.
          N. 72 W. ,, 180 = S. 108 W. ,, ,, N. 72 W.

We may, however, proceed as above, viz.:—add up the Bearings reckoned from N. towards W.; and since there are four bearings so reckoned (from N. towards W.); take the sum from 720 (180 × 4); the remainder is the sum of the bearings to be reckoned from S. towards W. It is evident that to subtract the sum of the *four* bearings from 720° is the same thing as to subtract each bearing from 180° and to add the remainders.

Ship's Head.	N.	N.E.	East.	S.E.
Correct magnetic bearing	S. 91° W.	S. 91° W.	S. 91° W.	S. 91° W.
Compass bearing	S. 88 W.	S. 70 W.	S. 71 W.	S. 81 W.
Deviation	3 E.	21 E.	20 E.	10 E.

Ship's Head.	South.	S.W.	West.	N.W.
Correct magnetic bearing	N. 89° W.	N. 89° W.	N. 89° W.	N. 89° W.
Compass bearing	N. 87 W.	N. 75 W.	N. 68 W.	N. 72 W.
Deviation	2 W.	14 W.	21 W.	17 W.

II. From the above Table construct a NAPIER's curve, and give the courses you would steer by standard compass to make the following courses correct magnetic:—

    (1.) W. by S. ¾ S.   (2.) N. ½ E.   (3.) E. ¾ N.   (4.) S.E. ¾ S.
    Answers:—(1.) West. (2.) N. 1° E. (3.) N. 60° E. (4.) S. 49° E.

NOTE.—For the method of constructing the curve, see No. 259, page 163, and for Rule to find standard compass course to steer, see Rule LX, page 168.

III. Suppose you steer the following courses by standard compass, find the correct magnetic courses from the curve drawn :—

(1.) North. (2.) S.S.W. ¼ W. (3.) E. by S. ½ S. (4). N.E. ½ E.

Answers :—(1). N. 3° E. (2.) S. 15° W. (3.) S. 59° E. (4.) N. 71° E.

IV. You have taken the following bearings of a distant object by your standard compass as above; with the ship's head as given, find the correct magnetic bearing.

Ship's Head.	Compass Bearings.	Ship's Head.	Compass Bearings.
S. ¾ W.	South.	N.N.E. ¾ E. ..	S.W. ¼ S.
N.W. by W. ..	E. ¾ N.	S.E.	E. ¼ S.

By the curve the deviation for ship's head S. ¾ W. is 5½° W.; then placing one leg of the dividers on compass bearing South on central line, and laying the other leg *upwards* (deviation being West) the correct magnetic bearing of distant object is found to be S. 5½° E.

The deviation by the curve for ship's head N.W. by W. is 17° W.; then placing one leg of the dividers on the part of central line representing compass bearing E. ¾ N., the other leg applied *upwards* on the central line intersects it at N. 63° E.: what is the correct magnetic bearing?

For position of ship's head N.N.E. ¾ E. the deviation by the curve is 18½° E.; placing one leg of the dividers on the compass bearing S.W. ¼ S., and laying the other leg *downwards* (deviation being East) it shows the correct magnetic bearing to be S. 60½° W.

Ship's head S.E. gives on curve deviation 10½° E.; one leg of dividers placed E. ¼ S. and the other leg applied *downwards* (deviation being East) gives on central line correct magnetic bearing S. 76½° E.

EXAMPLE III.

Question 7. List B.

I. From the following Table find the correct magnetic bearing of the distant object, and thence the deviation :—

Ship's Head by Standard Compass.	Bearing of Distant Object by Standard Compass.	Deviation Required.	Ship's Head by Standard Compass.	Bearing of Distant Object by Standard Compass.	Deviation Required.
North	N. 1° E.		South	N. 3° W.	
N.E.	N. 13 W.		S.W.	N. 1 E.	
East......	N. 15 W.		West	N. 9 E.	
S.E.	N. 8 W.		N.W.	N. 12 E.	

In this example some of the bearings are towards East and the others towards West; we therefore write down in one column those which are reckoned towards East, and add these up; we likewise write down in another column those which are reckoned towards West, and add up this column; the less sum being subtracted from the greater gives as a remainder N. 16° 0' W., which divided by 8 gives N. 2° W. for the correct magnetic bearing: the work will stand thus:—

Ship's Head.	Compass Bearings.	Ship's Head.	Compass Bearings.
North	N. 1° E.	N.E.	N. 13° W.
S.W.	N. 1 E.	East	N. 15 W.
West	N. 9 E.	S.E.	N. 8 W.
N.W.	N. 12 E.	South	N. 3 W.
	N. 23 E.		N. 39 W.
			N. 23 E.
			8) N. 16 W.

Correct magnetic bearing is N. 2 W.

The deviation for each position of the ship's head is next to be found, and we proceed as follows:—

Ship's Head.	North.	N.E.	East.	S.E.
Correct magnetic bearing	N. 2° W.	N. 2° W.	N. 2° W.	N. 2° W.
Compass bearing	N. 1 E.	N. 13 W.	N. 15 W.	N. 8 W.
Deviation	3 W.	11 E.	13 E.	6 E.

Ship's Head.	South.	S.W.	West.	N.W.
Correct magnetic bearing	N. 2° W.	N. 2° W.	N. 2° W.	N. 2° W.
Compass bearing	N. 3 W.	N. 1 E.	N. 9 E.	N. 12 E.
Deviation	1 E.	3 W.	11 W.	14 W.

QUESTION 8. LIST B.

From the above Table construct a NAPIER'S curve, and give the courses you will steer by standard compass to make the following courses, correct magnetic:—

1. N.E. ½ E.   2. W.S.W.   3. N.N.W.   4. S.S.E.
5. N.N.E.   6. S. 18¼° W.   7. N. 4° E.   8. S. 62½° E.

NOTE.—For the method of constructing the curve, see No. 259, page 163, and for Rule for finding standard compass courses to steer, see Problem II, Rule LX, page 168.

Answers.—1. N. 40° E.   2. S. 76° W.   3. N. 14½° W.   4. S. 26° E.
5. N. 18 E.   6. S. 19 W.   7. N. 5 E.   8. S. 73 E.

QUESTION 9. LIST B.

Suppose you steer the following courses by the standard compass, find the correct magnetic courses from the curve drawn:—

1. E. by S. ¼ S.   2. E. ¾ N.   3. S. 85° W.   4. N. 1° E.

For the Rule for working this Question, see Problem I, Rule LIX, page 167.

Answer.—1. S. 65½ E.   2. N. 84½ E.   3. N. 74½ W.   4. N. 1½ W.

QUESTION 10. LIST B.

You have taken the following bearings of distant objects by your standard compass as above, with the ship's head as given, find the correct magnetic bearings. (See Rule LVIII, page 166).

Ship's Head.	Compass Bearing.	Ship's Head.	Compass Bearing.
S.E. by E.	N. 68° E.	S.S.W. ¼ W.	N. 4° E.
W. ¼ N.	S. 54 W.	N. ¾ E.	South.

The deviation by curve for ship's head S.E. by E. is 7½° E., and the correct magnetic bearing corresponding to N. 68° E. is found to be N. 75½° E.

Ship's head W. ¼ N. gives on curve the deviation 11° W., whence the correct magnetic bearing corresponding to compass bearing S. 54° W. is found to be S. 43° W.

Ship's head S.S.W. ½ W. gives deviation 1½° W. on the curve, and the correct magnetic bearing corresponding to compass bearing N. 4° E. is found to be N. 2½° E.

Ship's head N. ¾ E. gives on curve the deviation 0, whence the correct magnetic bearing is South—the same as the compass bearing.

Examples for Practice.

Example I.

I. From the following Table find the correct magnetic bearing of the distant object, and thence the deviation:—

(S.S. B————. Heeling 10° to Port.)

Ship's Head by Standard Compass.	Bearing of Distant Object by Standard Compass.	Deviation Required.	Ship's Head by Standard Compass.	Bearing of Distant Object by Standard Compass.	Deviation Required.
North	N. 77° W.		South	S. 76° W.	
N.E.	W. 1 S.		S.W.	W. 12 N.	
East	W. 19 S.		West	W. 19 N.	
S.E.	S. 62 W.		N.W.	N. 72 W.	

Answer:—Correct magnetic bearing, West, or S. 90° W.

In such a case as this, where the bearings are near the East or West points, and are given from different points, they must all be expressed from either the North or South point by taking their equivalents as follows, where they are all expressed for the North point.

The first bearing is expressed from the North towards the West. The second bearing is expressed from the West towards the South, but by adding 90° we get the equivalent bearing N. 91° W.; similarly, the third bearing becomes N. 109° W. The fourth bearing S. 62° W. is taken from 180°, and this becomes N. 118° W., and the fifth bearing is also taken from 180°, and the equivalent bearing is thus found to be N. 104° W. The sixth and seventh bearings being reckoned from W. to N., subtracting each from 90°, whence we get N. 78° W. and N. 71° W. as the respective equivalent bearings. The last bearing being N. 72° W., is written down as it stands, unchanged.

Answer:—Deviations—1. 13° W.   2. 1° E.   3. 19° E.   4. 28° E.   5. 14° E.   6. 12° W.   7. 19° W.   8. 18° W.

II. From the above Table construct a Napier's curve, and give the courses you would steer by the standard compass to make the following courses correct magnetic:—

1. N.E. ¼ N.   2. E. by S. ½ S.   3. S.S.W. ¼ W.   4. N.N.W. ¼ W.

Answer:—1. N. 42° E.   2. N. 87° E.   3. S. 28° W.   4. N. 10° W.

III. Given the following courses steered by standard compass to find the correct magnetic courses from curve drawn:—

1. N. by E. ½ E.   2. S.S.E. ½ E.   3. W. ¼ S.   4. W. by N. ¼ N.

Answer:—1. N. 6½° E.   2. S. 2° E.   3. S. 69° W.   4. N. 89° W.

IV. You have taken the following bearings of distant objects by your standard compass with the ship's head S.E. by E. ½ E.: find the correct magnetic bearing:—

Standard compass bearings:—W. by N. ½ N. and N. ¾ W.

Answer.—Deviations:—1. 26° E.   2. 20° W.

Correct magnetic bearings:—1. N. 51° W.   2. N. 18° W.

Example II.

I. From the following table find the correct magnetic bearing of the distant object and thence the deviation :—

Ship's Head by Standard Compass.	Bearing of Distant Object by Standard Compass.	Deviation Required.	Ship's Head by Standard Compass.	Bearing of Distant Object by Standard Compass.	Deviation Required.
North	N. 42° W.		South	N. 44° W.	
N.E.	N. 17 W.		S.W.	N. 68 W.	
East......	N. 9 W.		West	N. 76 W.	
S.E.......	N. 17 W.		N.W.	N. 64 W.	

Answer :—Correct magnetic bearing N. 42° W.

Deviation 0°; 25° W.; 33° W.; 25° W.; 2° E.; 26° E.; 34° E.; 22° E.

II. From the above table construct a Napier's curve and give the courses you would steer by standard compass to make the following courses correct magnetic :—

1. N.W. ½ W. 2. S.E. ½ S. 3. E. ¼ S. 4. W. ¾ S.

Answer :—1. N. 84° W. 2. S. 24° E. 3. S. 58° E. 4. S. 53½° W.

III. Suppose you steer the following courses by standard compass, find the correct magnetic courses from the curve drawn :—

1. W. by N. ¾ N. 2. South. 3. E. by N. ½ N. 4. N. by E. ¾ E.

Answer :—1. N. 39° W. 2. S. 1¾° W. 3. N. 43½° E. 4. N. 7° E.

IV. You have taken the following bearings of distant objects by your standard compass as above; with the ship's head as given, find the correct magnetic bearing :—

Ship's Head.	Bearing of Distant Object by Standard Compass.	Ship's Head.	Bearing of Distant Object by Standard Compass.
S.E.	N. by E. ½ E.	W. by S. ¾ S. ..	E. ½ N.
E. by S.	N. by W. ¾ W.	N.E. ½ E.	S.W. ¾ S.

Answer :—Deviations for S.E. is 25° W.; for E. by S. is 32° W.; for W. by S. ¾ S. is 32° E.; N.E. ½ E. is 26½° W.

Correct magnetic bearings N. 80° W.; N. 52° W.; S. 64° E.; S. 10° W.

Example III.

I. From the following table find the correct magnetic bearing of the distant object, and thence the deviation :—

Ship's Head by Standard Compass.	Bearing of Distant Object by Standard Compass.	Deviation Required.	Ship's Head by Standard Compass.	Bearing of Distant Object by Standard Compass.	Deviation Required.
North	N. 13° W.		South	N. 23° W.	
N.E.	N. 35 W.		S.W.	N. 2 W.	
East	N. 41 W.		West	N. 5 E.	
S.E.	N. 35 W.		N.W.	North.	

Answer :—Correct magnetic bearing N. 18° 6′ W.

Deviations:—5° W.; 17° E.; 23° E.; 17° E.; 5° E.; 16° W.; 23° W.; 18° W.

II. From the above table construct a NAPIER's curve, and give the courses you would steer by standard compass to make the following courses correct magnetic:—

1. N. ½ W.   2. E. ½ S.   3. N.W. by W. ½ W.   4. W. ¼ S.

Answer:—1. North.   2. N. 73° E.   3. N. 44° W.   4. N. 71° W.

III. You have steered the following courses by standard compass: find the correct magnetic courses from the curve drawn:—

1. E. ½ N.   2. S.E. ¾ E.   3. N. ¾ W.   4. S.W. ¼ S.

Answer:—1. S. 72½° E.   2. S. 35° E.   3. N. 16½° W.   4. S. 27° W.

IV. You have taken the following bearings of distant objects by your standard compass; with the ship's head as given below, find the correct magnetic bearing:—

Ship's Head.	Bearing of Distant Object by Standard Compass.	Ship's Head.	Bearing of Distant Object by Standard Compass.
S.W. ¼ S.	E. ½ S.	N.E. ¼ N.	W. ¼ S.
North	S.E. ¾ S.	E. by S. ½ S. ..	E. ¼ S.

Answer:—Deviations:—14° W.; 5° W.; 16° E.; 22° E.
Correct magnetic bearings, N. 84° E.; S. 42° E.; N. 77° W.; S. 65° E.

EXAMPLE IV.

I. From the following Table find the correct magnetic bearing of the distant object, and thence the deviation:—

(S.S. *Royal Charter*, before compensation.)

Ship's Head by Standard Compass.	Bearing of Distant Object by Standard Compass.	Deviation Required.	Ship's Head by Standard Compass.	Bearing of Distant Object by Standard Compass.	Deviation Required.
North	N. 81° E.		South	N. 16° E.	
N.E.	N. 79 E.		S.W.	N. 2 E.	
East	N. 68 E.		West	N. 8 E.	
S.E.	N. 48 E.		N.W.	N. 51 E.	

Answer:—Correct magnetic bearing N. 44° E.

Answer.—Deviations:—1. 37° W.   2. 35° W.   3. 24° W.   4. 4° W.
5. 28° E.   6. 42° E.   7. 36° E.   8. 7° W.

II. From the above Table construct a NAPIER's curve, and give the courses you would steer by standard compass to make the following courses correct magnetic:—

1. E. ½ N.   2. N. ¾ W.   3. S. by W. ½ W.   4. W. ½ N.

Answer:—1. S. 77° E.   2. N. 29° E.   3. S. 7° E.   4. S. 53° W.

III. Suppose you steer the following courses by standard compass, find the correct magnetic courses from the curve drawn:—

1. N.E. ½ E.   2. S.E. ½ S.   3. South.   4. W. ½ S.

Answer:—1. N. 16½° E.   2. S. 39½° E.   3. S. 28° W.   4. N. 58½° W.

IV. You have taken the following bearings of distant objects by your standard compass as above; with the ship's head N. by E. ½ E., find the correct magnetic bearings.

Standard compass bearings:—S.W. ¼ W. and S. ¾ E.

Answer.—Deviation:—34° W.   Correct magnetic bearings:—S. 6° W. and S. 43° E.

Example V.

I. From the following table find the correct magnetic bearings of the distant object, and thence the deviation. (S.S. *Royal Charter*, after compensation.)

Ship's Head by Standard Compass.	Bearing of Distant Object by Standard Compass.	Deviation Required.	Ship's Head by Standard Compass.	Bearing of Distant Object by Standard Compass.	Deviation Required.
North	S. 88° E.		South	S. 82° E.	
N.E.	South.		S.W.	S. 86 E.	
East	S. 38 E.		West	S. 88 E.	
S.E.	S. 83 E.		N.W.	S. 84 E.	

Answer:—Correct magnetic bearing S. 86° E.

Answer.—Deviations:—1. 2° E. 2. 4° E. 3. 2° E. 4. 3° W. 5. 4° W. 6. 0°. 7. 2° E. 8. 2° E.

II. Given the correct magnetic course, to find the courses to steer by standard compass:—

1. W. ¼ N. 2. S.W. ½ S. 3. E. by S. ½ S. 4. N.N.E. ¼ E.

Answer.—1. N. 89° W. 2. S. 42° W. 3. S. 73° E. 4. N. 22½° E.

III. Given courses by standard compass, to find correct magnetic:—

1. N. by W. ¾ W. 2. W. by N. ¾ N. 3. S. ¼ E. 4. E. by N. ¼ N.

Answer.—1. N. 20° W. 2. N. 70° W. 3. S. 7½° W. 4. N. 78° E.

IV. You have taken the following bearings of distant objects by standard compass with ship's head as given below, find the correct magnetic bearing:—

Ship's head S.S.E. ¼ E., compass bearings S. ½ W. and N.E. by E. ½ E.

Correct magnetic bearings:—S. 1° W. and N. 57½° E.

Example VI.

I. From the following table find the correct magnetic bearing of the distant object, and thence the deviation. (S.S. *Royal Charter*, at Melbourne.)

Ship's Head by Standard Compass.	Bearing of Distant Object by Standard Compass.	Deviation Required.	Ship's Head by Standard Compass.	Bearing of Distant Object by Standard Compass.	Deviation Required.
North	S. 62° W.		South	W. 15° N.	
N.E.	W. 24 S.		S.W.	N. 81 W.	
East	W. 7 S.		West	W. 5 S.	
S.E.	N. 77 W.		N.W.	S. 67 W.	

Answer:—Correct magnetic bearing S. 83¾° W. = S. 84° W.

Deviations:—22° E.; 18° E.; 1° E.; 19° W.; 21° W.; 15° W.; 1° W.; 17° E.

II. From the above table construct a NAPIER's curve, and give the course you would steer by standard compass to make the following courses correct magnetic:—

1. N. ½ E. 2. S. ½ W. 3. E. ¼ S. 4. N.E. ½ E.

Answer.—1. N. 16° W. 2. S. 25° W. 3. S. 86° E. 4. N. 29½° E.

III. You have steered the following courses by standard compass; find the correct magnetic courses from the curve drawn:—

1. N.E. ½ N. 2. S.E. ½ S. 3. W. ¼ N. 4. N.W. ½ N.

Answer:—1. N. 59° E. 2. S. 59° E. 3. N. 87° W. N. 21° W.

IV. You have taken the following bearings of distant objects by your standard compass as above; with the ship's head by standard compass N. ¼ W., find the correct magnetic bearing.

Bearings of distant object by standard compass S.S.E. ½ E. and E. ½ N.

Answer :—Deviation 23° E. Correct magnetic bearings :—S. 2½° E. and S. 73° E.

Example VII.

I. In the following table give the correct magnetic bearing of the distant object, and thence the deviation. (S.S. *Royal Charter*, after return to Liverpool from Melbourne.)

Ship's Head by Standard Compass.	Bearing of Distant Object by Standard Compass.	Deviation Required.	Ship's Head by Standard Compass.	Bearing of Distant Object by Standard Compass.	Deviation Required.
North	S. 17° E.		South	S. 12° W.	
N.E.	S. 14 E.		S.W.	S. 3 W.	
East	South		West	S. 7 E.	
S.E.	S. 11 W.		N.W.	S. 12 E.	

Answer :—Correct magnetic bearing, S. 16° E.

Deviations :—14° E.; 11° E.; 3° W.; 14° W.; 15° W.; 6° W.; 4° E.; 9° E.

II. From the above table construct a NAPIER's curve, and give the course you would steer by standard compass to make the following courses correct magnetic :—

1. N.E. ¼ N. 2. E. ¾ N. 3. S. 8½° W. 4. N.W. ½ W.

Answer.—1. N. 29½° E. 2. N. 81½° E. 3. S. 20½° W. 4. N. 58° W.

III. Given the following courses by standard compass, to find the correct magnetic courses from the curve drawn :—

1. N.E. ½ E. 2. N.N.W. ¼ W. 3. W. by S. ¼ S. 4. S.E. ¾ E.

Answer.—1. N. 60° E. 2. N. 13° W. 3. S. 70½° E. 4. S. 66½° E.

IV. You have taken the following bearings of two distant objects by your standard compass, with the ship's head S.S.E. ¼ E.

St. Catherine's Point, N.E. ½ N. Needles Light, N.N.W. ½ W.

Answer.—Deviation for ship's head, N.N.W. ½ W. is 15½° W.

Correct magnetic bearings :—N. 24° E. and N. 43° W.

Example VIII.

I. In the following table give the correct magnetic bearing of the distant object, and thence the deviation :—

Ship's Head by Standard Compass.	Bearing of Distant Object by Standard Compass.	Deviation Required.	Ship's Head by Standard Compass.	Bearing of Distant Object by Standard Compass.	Deviation Required.
North	N. 60° W.		South	N. 66° W.	
N.E.	N. 80 W.		S.W.	N. 53 W.	
East	N. 84 W.		West	N. 42 W.	
S.E.	N. 78 W.		N.W.	N. 41 W.	

Answer :—Correct magnetic bearing of distant object is N. 63° W.

Deviations :—3° W.; 17° E.; 21° E.; 15° E.; 3° E.; 10° W.; 21° W.; 22° W.

II. From the Table construct a NAPIER's curve (a). and give the courses you would steer by the standard compass to make the following courses correct magnetic:—

1. N. by E. 2. S.W. ½ W. 3. E. ¼ N. 4. N.N.W. 5. E.S.E. 6. S.W. ½ S. 7. N. ¼ E. 8. W. ¾ S. 9. N. ¼ W. 10. S. ¼ W. 11. E. ¾ S. 12. E.N.E.

For the method of constructing the curve see No. 259, page 163; and for the Rule for finding the correct magnetic course see Problem II, Rule LX, page 168.

Answer:—1. N. 8¾° E. 2. S. 65¾° W. 3. E.N.E. 4. N. 12° W. 5. S. 88° E. 6. S. 50° W. 7. N. 3¾° E. 8. N. 75° W. 9. North. 10. S. 1° E. 11. N. 72½° E. 12. N. 49½° E.

III. Suppose you have steered the following courses by standard compass, find the correct magnetic courses from the curve drawn:— (See Problem I, page 167.)

1. N.E. by E. 2. E.S.E. 3. S.W. by S. 4. South. 5. S.E. ½ S. 6. W. by N. ¾ N. 7. S. ¼ W. 8. E. ¾ N. 9. W. ¾ S. 10. N.E. by E. ½ E. 11. N.W. ½ N. 12. S.W. ¾ W.

Answer:—1. N. 75¾° E. 2. S. 49½° E. 3. S. 27¾° W. 4. S. 4° W. 5. S. 26° E. 6. S. 86° W. 7. S. 5¼° W. 8. S. 77½° E. 9. S. 62° W. 10. N. 82° E. 11. N. 60° W. 12. S. 41¾° W.

IV. You have taken the following bearings by your standard compass as above; with the ship's head as given, find the correct magnetic bearings:—

No.	Ship's Head.	Bearing of Distant Object by Standard Compass.	No.	Ship's Head.	Bearing of Distant Object by Standard Compass.
1	W.N.W.	South.	5	N. ¼ E.	S. ¼ E.
2	S.E. by E.	N.W. by N.	6	W. by N. ½ N...	E. ¾ S.
3	N.E. by N.	North.	7	E. ¼ N.	E. ¼ N.
4	E. by S. ½ S. ..	N.N.W.	8	E. ¾ S.	W. by S. ¼ S.

Answer:—1. Corr. mag. bear. S. 24° E. 2. N. 17° W. 3. N. 14° E. 4. N. 3° W. 5. S. 3½° E. 6. N. 72° E. 7. S. 72° E. 8. N. 84° W.

---

266. From the definitions and principles given in pages 84 and 90, are deduced the following formulæ or equations, and these expressed in words constitute the common rules of navigation for finding the *place* of a ship, that is, its latitude and longitude.

FUNDAMENTAL FORMULÆ OF NAVIGATION.

Departure = distance × sine course, [1]

∴ log. dep. = log. dist. + log. sine course − 10.

True diff. lat. = distance × cosine course, [2]

∴ long. true diff. lat. = log. dist. × log. cos. course − 10.

Diff. long. = meridional diff. lat. × tang. course, [3]

∴ log. diff. long. = log. mer. diff. lat. + log. tang. course − 10.

In parallel sailing.

Distance = diff. long. × cos. lat., [4]

∴ log. dist. = log. diff. long. + log. cos. lat. − 10.

In middle latitude sailing.

Dep. (nearly) = diff. long. × cos. mid. lat., [5]

∴ log. dep. (nearly) = log. diff. long. + log. cos. mid. lat. − 10.

NAPIER'S DIAGRAM,

Showing Curve of Deviation.

Ship

West Deviation. East Deviation.

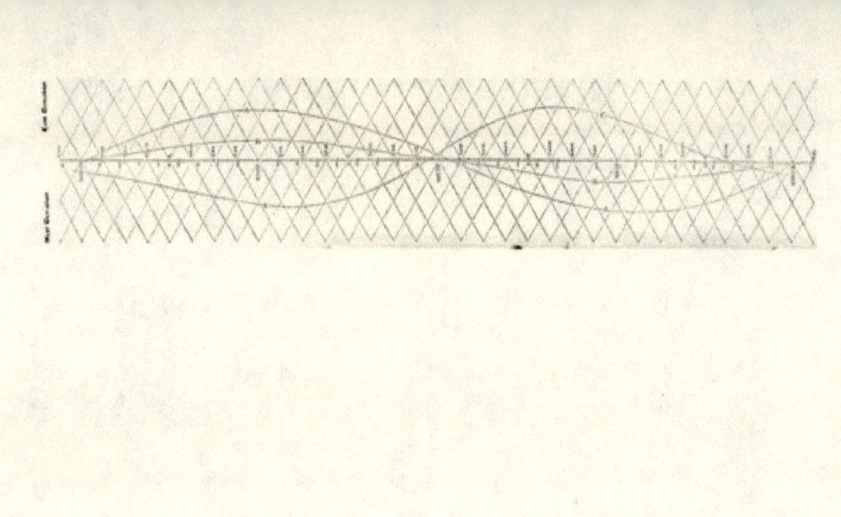

THE TRAVERSE TABLE.

267. In all collections of tables for the use of navigators there is inserted a table containing the *true difference of latitude* and *departure*, corresponding to certain *distances* (at intervals of one mile) up to 300 nautical miles, for every course, at intervals of a quarter point, and also of degrees, from 0° to a right-angle (90°). Tables I and II (RAPER or NORIE).

This table is constructed by solving a right-angled triangle, of which one angle represents the course, and the hypothenuse the distance; by giving these different and successive values, the corresponding values of the other two sides are found, which sides represent the true difference of latitude and departure.

Thus, in the triangle C A B, right-angled at C: if the angle C A B represent a given course, and A B a given distance, the side A C will be the true diff. of lat., and B C the dep. corresponding to that course and distance.

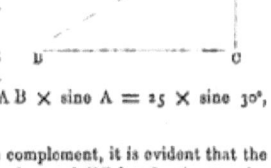

Given course 30°, and distance 25 miles, compute corresponding true diff. lat. and dep.

In the triangle C A B, let A = 30°, and A B = 26°; then true diff. lat. = A C = A B × cos. A = 25 × cos. 30°, ∴ true diff. lat. = 21'·65, and dep. = B C = A B × sine A = 25 × sine 30°, ∴ dep. = 12'·5.

Inasmuch as the sine of an angle is the cosine of its complement, it is evident that the difference of latitude and departure for any course are the dep. and diff. lat. for the complement of that course. Thus, let it be required to find the diff. lat. and dep. for course 60° and dist. 25 miles:—diff. lat. for 60° = 25 × cos. 60° = 25 × sine 30° = dep. for 30°, and dep. for 60° = 25 × sine 60° = 25 × cos. 30° = diff. lat. for 30°. When the diff. lat. and dep. are computed in this manner up to 45°, the diff. lat. and dep. for course *above* 45° may be found by changing the titles to the columns, and hence the table is compactly arranged by interchanging the headings of the columns containing these elements at the top and bottom of the page, and reading the top reading for courses from 0° to 45°, and the bottom reading for courses from 45° to 90°. This table may be used for a great number of problems depending for their solution on the relation of the several parts of a right-angled triangle, and since all the relations between any two quantities may be expressed as functions of some angle in terms of the sine, cosine, or tangent; it may be used, in fact, as a general proportional table.

In these Tables the *course* is found at the top of the Table, when under 4 points or 45°; but at the bottom of the Table, when it exceeds 4 points or 45°. The first column contains the *distance* to 60 miles, the second column contains the *difference of latitude*, expressed in minutes and tenths, and the third column, similarly expressed, contains the *departure;* but if the course exceeds 4 points, or 45°, the second column contains the *departure*, and the third column the *difference of latitude*. The other columns are a continuation of the former, exactly upon the same principle, and extending to 300 miles of distance. (See Tables I and II, NORIE or RAPER.)

THE TRAVERSE TABLE.

267. In all collections of tables for the use of navigators there is inserted a table containing the *true difference of latitude* and *departure*, corresponding to certain *distances* (at intervals of one mile) up to 300 nautical miles, for every course, at intervals of a quarter point, and also of degrees, from 0° to a right-angle (90°). Tables I and II (RAPER or NORIE).

This table is constructed by solving a right-angled triangle, of which one angle represents the course, and the hypothenuse the distance; by giving these different and successive values, the corresponding values of the other two sides are found, which sides represent the true difference of latitude and departure.

Thus, in the triangle C A B, right-angled at C: if the angle C A B represent a given course, and A B a given distance, the side A C will be the true diff. of lat., and B C the dep. corresponding to that course and distance.

Given course 30°, and distance 25 miles, compute corresponding true diff. lat. and dep.

In the triangle C A B, let A = 30°, and A B = 25°; then true diff. lat. = A C = A B × cos. A = 25 × cos. 30°, ∴ true diff. lat. = 21'·65, and dep. = B C = A B × sine A = 25 × sine 30°, ∴ dep. = 12'·5.

Inasmuch as the sine of an angle is the cosine of its complement, it is evident that the difference of latitude and departure for any course are the dep. and diff. lat. for the complement of that course. Thus, let it be required to find the diff. lat. and dep. for course 60° and dist. 25 miles:—diff. lat. for 60° = 25 × cos. 60° = 25 × sine 30° = dep. for 30°, and dep. for 60° = 25 × sine 60° = 25 × cos. 30° = diff. lat. for 30°. When the diff. lat. and dep. are computed in this manner up to 45°, the diff. lat. and dep. for course *above* 45° may be found by changing the titles to the columns, and hence the table is compactly arranged by interchanging the headings of the columns containing these elements at the top and bottom of the page, and reading the top reading for courses from 0° to 45°, and the bottom reading for courses from 45° to 90°. This table may be used for a great number of problems depending for their solution on the relation of the several parts of a right-angled triangle, and since all the relations between any two quantities may be expressed as functions of some angle in terms of the sine, cosine, or tangent; it may be used, in fact, as a general proportional table.

In these Tables the *course* is found at the top of the Table, when under 4 points or 45°; but at the bottom of the Table, when it exceeds 4 points or 45°. The first column contains the *distance* to 60 miles, the second column contains the *difference of latitude*, expressed in minutes and tenths, and the third column, similarly expressed, contains the *departure;* but if the course exceeds 4 points, or 45°, the second column contains the *departure*, and the third column the *difference of latitude*. The other columns are a continuation of the former, exactly upon the same principle, and extending to 300 miles of distance. (See Tables I and II, NORIE or RAPER.)

USE OF THE TABLE.

268. Given the course and distance, to find the difference of latitude and departure.

RULE LXIII.

With the Course *open the Tables, and* under *or* above *the proper number of* points *(or degrees), and* opposite *the* distance, *will be found the* difference of latitude *and* departure.

Obs.—When the course is found at the bottom of the page, care must be taken to see that the diff. of lat. and the dep. are taken from the proper column *above* the words *departure* and *diff. lat.* It must be carefully remembered that when the course is *less* than 4 points or 45°, the *diff. lat. exceeds the dep.;* but when it is *more* than 4 points or 45° the *dep. exceeds the diff. lat.*

EXAMPLES.

Ex. 1. A ship sails N.W. ½ N. a distance of 78 miles: required the difference of latitude and departure by inspection.

The given course is 3½ points; and referring to Table I we find the page assigned to this course to be page 14, NORIE, or page 436, RAPER's Navigation, in which against 78, in column headed *dist.*, stands 60·3 under the head *lat.*, and 49·5 under the head *dep.* We conclude, therefore, that for the given course and distance, the difference of latitude is 60·3 miles, and the departure 49·5 miles.

Ex. 2. Suppose the course to be 5½ points, and the distance 98 miles.

Then, since the course here exceeds four points, we look for it at the foot of the page (page 10, NORIE, or 432, RAPER), and against 98 in the *distance* column we find in the adjacent column (marked at the bottom dep. and diff. lat.) *dep.* 86·4, and *diff. lat.* 46·2, so that the difference of latitude made is 46·2, and the departure 86·4.

Ex. 3. Course N.E. by N., distance 129 miles: find diff. lat. and dep.

Enter Table I, and find 3 points at the top, and in one of the columns marked *dist.* find the distance 129, then in the columns opposite to this, marked *lat.* and *dep.* at the top, stands the difference of latitude 107·3, and departure 71·7.

Ex. 4. Course E. by N. ½ N., distance 264 miles: find diff. of lat. and dep.

Open Table I at 6½ points, found at the bottom, and opposite the distance 264 stands departure 252·6, and difference of latitude 76·6.

Ex. 5. A ship sails N. 40° E., 50 miles: required the diff. of lat. and departure.

The course being less than 45°, is found at the top (Table II), and the distance being under 60 miles, is found in the left hand column; therefore, on the page (56 NORIE) is 40° at the top, and opposite to 50 in the distance column (marked Dist.) is 38·3 *under* Lat., and 32·1 *under* Dep., the difference of latitude and departure required.

Ex. 6. A ship sails N. 64° W., 175 miles: required the diff. of lat. and departure.

The course being more than 45°, is found at the bottom, in page 42, and opposite to the distance 175 miles, is 76·7 *over* Lat., and 157·3 *over* Dep., which was required.

(*a*) To find diff. lat. and dep. when there are tenths in the distance.

Take the distance as an entire number *of miles, i.e., as a whole number, and find the corresponding* diff. lat. *and* dep., *from each of which cut off the right hand figure, or tenths, and remove the decimal point one place to the left hand, which will give the required* diff. lat. *and* dep. *in miles and tenths of a mile.* The tenths, however, must be increased by* 1, *if the figure cut off is* 5, *or upwards.*

---

* The reason of this rule is that the Traverse Table is entered with a distance *ten times* as great as the given distance, and the resulting diff. lat. and dep. is divided by ten. This is done by merely imagining the decimal point to be removed one place to the *right* before entering the table, and then one place to the *left* after taking out the results (see p. 26, (10).

EXAMPLES.

Ex. 1. Course 3½ points, distance 20·3; required the diff. lat. and dep. corresponding thereto.

With course 3½ points, and dist. 20·3, taken as 203, we get the diff. of lat. 156·9, dep. 128·8; now cut off the right hand figure of each (the 9 and 8), and shifting the decimal point one place to the left, we have diff. lat. 15·7, and dep. 12·9. It will be observed that the tenths are increased by 1, in each case, as the figure cut off in both cases exceeds 5.

Ex. 2. Required the diff. lat. and dep. corresponding to course 4½ points, and dist. 24·3 miles.

With course 4½ points, and dist. 24·3 (as 243 miles), we find diff. lat. 154·2, and dep. 187·7: hence we obtain, after dropping the tenths, and removing the decimal point in each one place to the left, 15·4 and 18·8, for the required quantities. The tenths in the dep., it will be observed, are increased by 1, since the figure dropped exceeds 5.

Ex. 3. A ship sails N. 67° E., distance 29'·5: find diff. of lat. and dep. corresponding.

In this case take out for distance 295. Thus, for 67° and distance—

$$295 = 115\cdot 3 \text{ diff. lat., } 271'\cdot 5 \text{ dep.}$$
$$\therefore 29\cdot 5 = 11\cdot 5 \text{ diff. lat., } 27'\cdot 2 \text{ dep.}$$

After dropping the tenths, and removing the decimal point one place to the left, we have diff. lat. 11·5 and dep. 27·2 (the tenths of dep. is increased by 1, as the figure cut off is 5.)

Ex. 4. N. 34° W., and dist. 20·6 miles (as 206), give diff. lat. 170·8, and dep. 115·2 dropping the tenths in each case (the 8 and the 2), and shifting the decimal point one place to the left, we get diff. lat. 17·1 N., and dep. 11·5 W. The tenths of diff. lat. must be increased by 1, as the figure cut off (8) exceeds 5.

Ex. 5. N. 65° E., and dist. 21·5 (as 215), give diff. lat. 90·9, and dep. 194·9, which is diff. lat. 9·1 N., and 19·5 E. It will be observed that the tenths are increased by 1, in each case, as the figure dropped exceeds 5.

(b) *If the distance exceeds the limits of the Traverse Table.*

Take the half, *the* third, &c., *so as to bring it within the limits,* taking care *to* multiply *the* corresponding quantities *by* 2, 3, &c.

Ex. 6. Let the course be 37°, and distance 435: required the corresponding diff. lat. and dep.

435 divided by 3 gives 145.

Course 37°, and dist. 145, give diff. lat. 115·8 and dep. 87·3
$$\times 3 \qquad \times 3$$
Diff. lat. 347·4   Dep. 261·9

If the distance had been 43·5, the diff. lat. would have been 34·7, and the dep. 26·1.

(c) But when the distance is between 300 and 600 we may proceed as follows:—

Take out diff. of lat. and dep. for 300, *and for the* excess *of* 300; *take the* sum *of the quantities thus found, cut off the last* figure, *and* remove *the* decimal point *as before.*

B B

Ex. 7. Course 65°, and distance 526: required the corresponding diff. of lat. and dep.

Course 65°, and dist. 300, give diff. lat. 126·8, and dep. 271·9
  226   95·5   204·8
  ───   ────   ─────
  526   222·3   476·7

If the distance were 52·6 we should proceed as above, and then cutting off the last figure of each, and removing the decimal point one place to the left, the diff. of lat. is 22·2, and dep. 47·8. The tenths are increased by 1 in the last case of dep., as the figure cut off in one exceeds 5.

Ex. 8. Course S. 53° E., dist. 61·1 (take 61·1 as 611).

Course.	Dist.	Diff. Lat.	Dep.
53°	300	180·5	239·6
··	300	180·5	239·6
··	11	6·6	8·8
53	611	367·6	488·0

∴ S. 53° E., dist. 61·1, give diff. lat. 36·8 S., dep. 48·8 E.

Ex. 9. Course S. 44° E., dist. 68·7 (take dist. 68·7 as 687).

Course.	Dist.	Diff. Lat.	Dep.
44°	300	215·8	208·4
··	300	215·1	208·4
··	87	62·6	60·4
44	687	494·2	477·2

∴ Course S. 44° E., dist. 68·7, give diff. lat. 49·4 S., and dep. 47·7 E.

Ex. 10. Find from the table the diff. lat. and dep. for 485·7 on a N. 37° W. course.

Course N. 37° W.

Dist.	Diff. Lat.	Dep.
300	239·6	180·5
185	147·7	111·3
·7	·6	·4
485·7	387·9	292·2

The decimal 7 we take out as 7, which gives diff. lat. 5·6, and dep. 4·2, and shifting the decimal point in each one place to the left we have for diff. lat. 0·6, and for the dep. 0·4.

EXAMPLES FOR PRACTICE.

In each of the following examples find the difference of latitude and departure corresponding to the given course and distance:—

No.	Given Course.	Dist.	No.	Given Course.	Dist.
1.	S.S.E.	30	8.	S. 72° W.	35
2.	E. by S.	48	9.	N. 21 W.	24·5
3.	S.W. ¼ S.	136	10.	S. 65 W.	25·7
4.	W. ¼ N.	84	11.	S. 80 W.	14·7
5.	S.E. by E. ¼ E.	56	12.	N. 27 W.	30·6
6.	S.W. ½ W.	225	13.	W. 10 S.	42·8
7.	E. by N. ¼ N.	183	14.	N. 18 W.	34·9

269. Given the **difference of latitude and departure**, to find the **course and distance**.

RULE LXIV.

Seek in the Traverse Table (Table II) *till the diff. lat. and dep., or the quantities agreeing most nearly with them, are found opposite each other in the proper columns; to the left or abreast of the quantities thus found, and under "Dist," will be the distance made good; and the Course (in degrees) must be taken from the top of the page when diff. lat. is greater than the dep.; but from the bottom of the page when the dep. is greater than the diff. lat.*

The Course *will have the* same name (from N. or S. towards E. or W.) *as the* diff. lat. *and* dep. *made good.*

NOTE.—Always seek for the larger of the two given numbers in the column next the distance, viz., the column marked "Diff. Lat." at the top, and examine page after page, until the smaller number is found by its side in the column marked "Dep." at the top; being careful to *remember that* when diff. lat. *is* greater *than the* dep. the course *will be at the* top; otherwise *it is to be found* at the bottom.

(*a*) When the diff. lat. and dep. on two successive pages of the Traverse Table, appear to be equally near the given diff. lat. and dep., neither page giving values actually corresponding to them, take a course midway between those on the two successive pages as the course actually made good.

(*b*) If the difference of latitude and departure, or any of the sides of the proposed triangle, should exceed the limits of the Traverse Table, they may be divided by any number that will bring them within these limits, and then the results from the table multiplied by the same number will give the required parts of the proposed triangle; observing that the **angle or course must in no case be multiplied or divided,** because the **course will be the same,** whether determined to the whole difference of latitude or departure, or by using an aliquot part of the same.

(*c*) In certain extreme cases, as when diff. lat. and dep. are very small, the course cannot be found correctly as above, since the diff. lat. and dep. may be found to agree in their respective columns for several successive pages of the table. In such cases, the diff. lat. and dep. may be multiplied by 10, and the products, when found to agree in their respective columns, will give the correct course; but the distance from the table, divided by 10, will give the correct distance.

EXAMPLES.

Ex. 1. A ship having sailed between the N. and E., until her difference of latitude is 199 miles, and the departure 144·6: required her course and distance.

In page 52, NORIE, or page 474, RAPER, these quantities will be found to correspond with 246 in the distance column, and with the angle 36° found at the *top* of the table (the diff. lat. being greater than dep.); the course is therefore N. 36° E., and distance 246 miles.

Ex. 2. A ship has made upon one course 36 miles diff. lat. to the northward, and 58 miles dep. to the westward: required the course and distance run.

Look for 58 and 36 in two adjoining columns marked "diff. lat." and "dep." at the top. In the table will be found 57·7 and 36·0 abreast of each other. In the same line at the dist. column will be found 68, the distance sought. As 58, the dep., is greater than diff. lat. 36, the course is taken from the bottom of the page. It is 58°. As the ship has gone to the N. and W., the course she has made is N. 58° W.

Ex. 3. A ship having sailed between S. and W. until her difference of latitude is 40 miles, and her departure 139·4 miles: required the course and distance.

In page 32, NORIE, or page 454, RAPER, the course answering to *diff. latitude* 40 miles, and *departure* 139·4 miles, corresponds with the angle 74°, at the bottom of the table, and opposite the distance 145 miles; the course is therefore S. 74° W., and distance 145 miles, which were required.

Ex. 4. Given the diff. lat. 240'·0 S., and dep. 208'·6 E.; required the corresponding course and distance. (See Rule LXIV (b), page 186.)

As both diff. lat. and dep. exceed the limits of the table, divide them by 2, so as to bring them within these limits, thus:—

Diff. lat.	Dep.
2)240·2	2)208·8
120·1	104·4

Then seeking for 120·1 and 104·4 in two adjoining columns marked "diff. lat." and "dep." at the top. In the Table are found 120'·0 and 104'·3 abreast of each other. In the same line in the distance column will be found 159, which must be multiplied by 2 (the number the diff. lat. and dep. were divided by), the product 318 is the distance sought. As 120·1, the diff. lat., is greater than the dep, the course is taken from the top of the page. It is 41°. As the ship has gone to the S. and E., the course she has made is S. 41° E.

Ex. 5. Given diff. lat. 1'·0 and dep. 0'·1; to find the corresponding course and distance.

Here diff. lat. 1'·0 and dep. 0'·1 are found in Tables to give as course 3°, 4°, 5°, 6°, 7°, and 8°; now which is the correct answer in this case. Multiplying diff. lat. and dep., say by 10, we obtain diff. lat. 10'·0 and dep. 1'·0; these are found in the tables to correspond to course 6° and distance 10 miles. Dividing the distance by 10 gives 1 (dividing by 10 as the tables were entered with *ten* times the diff. lat. and dep.), hence course 6° and distance 1 mile.

EXAMPLES FOR PRACTICE.

In each of the following examples the difference of latitude and departure are given, to find the corresponding course and distance.

No.	Given Diff. Lat.	Dep.	No.	Given Diff. Lat.	Dep.
1.	72·7 S.	25·0 E.	6.	37·9 N.	36·4 E.
2.	72·3 N.	171·7 E.	7.	53·3 S.	76·0 W.
3.	64·0 N.	146·9 W.	8.	160·7 S.	16·5 W.
4.	98·6 S.	37·5 E.	9.	172·6 S.	7·9 W.
5.	415·6 N.	240·0 W.	10.	164·2 N.	262·8 E.

TRAVERSE SAILING.

270. **Traverse Sailing** is the case in plane sailing when the ship makes several courses in succession, the track being zigzag, and the direction of its several parts "traversing," or lying more or less athwart of each other. For all these actual courses and distances run on each, a single equivalent imaginary course and distance may be found, which the ship would have described had she sailed direct for the place of destination. Finding this course is called "Working a Traverse."

In order to do this, the difference of latitude and departure for each distinct course must be found, and the aggregate of the several differences of latitude and departure taken for the single difference of latitude and departure which would be made by sailing from the place left to that reached on a single course. The determination of this course, and the corresponding distance, is then to be effected.*

271. In resolving a traverse, it is usual to take the diff. lat. and the dep. due to each of the component courses from the Traverse Table; hence we proceed by the following

RULE LXV.

1°. *Draw out a form similar to that given in the example following.*

2°. *In the column headed* Courses, *enter each course in succession; and in column* Dist., *enter the distance run on each course.*

In entering the courses in the appropriate column, reckon, in each case, the points, and fractional parts of a point, from the North or South, whichever is nearest, and write them down as in the following example.

3°. *Take out of the* Traverse Tables (Table I or II, RAPER *or* NORIE), *the difference of latitude and departure to each course and distance, and enter the latitude in column* N. *or* S., *and the departure in column* E. *or* W., *according to the name of the course.*

Thus, if the course is S.E. by S., the difference of latitude must be entered in the column S., and the departure in the column E.; if the course is W. ¼ N., the difference of latitude must be entered in the column N., and the departure in the column W.

3*a*. *When the course is* exactly North or South, *the distance and diff. lat. are the same, there is* no departure, *and the whole* distance *is entered as* difference of latitude *in the corresponding column* N. *or* S., *as the case may be; so also when the course is due* East *or* West, *the departure is* identical *with the* distance, *there is* no difference of latitude, *and the* whole distance *run is entered as* departure *in the* E. *or* W. *column.* (See pages 191—192, Exs. 2, 3, and 5.)

---

* The plane sailing formula—
$$\text{Dep.} = \text{Dist.} \times \text{Sine course} \quad (1)$$
$$\text{D. Lat.} = \text{Dist.} \times \text{Cos. course} \quad (2)$$

give for each course and distance the corresponding departure and difference of latitude; and taking the algebraic sum of all the diff. lats. and dep., we get the course from formula

$$\text{Tang course} = \frac{\text{Dep.}}{\text{D. Lat.}}$$

and then the distance from formulæ (1) and (2). The *Traverse Table* is used to obviate the necessity of computations.

4°. Add *the diff. of lats. in each column, and write the sum at the bottom of each, write the less of the two sums under the greater, and take their difference. Do the same with the departure.*

5°. *These differences are the* diff. lat. and dep. *made good on the whole, and each takes the* name *of the* column *it stands in.*

6°. The course and distance are then found by Rule LXIV, page 186.

NOTE.—(*a*). When there is no resulting departure the Traverse Table need not be referred to, as the ship has returned to the same meridian, and the course made good is North (N.) or South (S.), according as the diff. lat. is North or South, and the distance is equal to the diff. lat. (See Example 5, page 192.)

(*b*) Similarly, when there is no resulting difference of latitude, the course made good will be either East (E.) or West (W.) as the departure made good is East or West, and the distance will be of equal value with the difference of departures. See Ex. 4, page 192.

(*c*) Should the difference of latitudes and also the departures balance each other, in which case the ship will have made good neither difference of latitude nor departure, the vessel must be considered to have returned to the place from which she set out.

It may be advisable for a beginner, before he proceeds to take out the quantities from the Traverse Tables, to write a *dash* in all the places *not* to be occupied by a difference of latitude or departure, in order to avoid writing a quantity in the wrong column. Such helps, however, are useless to an expert computer.

EXAMPLES.

Ex. 1. A ship from the Dudgeon Light, in lat. 53° 19′ N., sails S.S.E. ¼ E., 8 miles; E.N.E., 23 miles; N.W. by W. ½ W., 36 miles; E. ¾ N., 48 miles; and N.W. ½ W., 46 miles: required the latitude arrived at, also the course and distance made good.

Courses.	Dist.	Diff. Lat.		Departure.	
		N.	S.	E.	W.
S.S.E. ¼ E..... = S. 1¼° E.*	8	—	7·2	3·4	—
E.N.E......... = N. 6 E.	23	8·8	—	21·3	—
N.W. by W.½ W.= N. 5½ W.	36	17·0	—	—	31·8
E. ¾ N....... = N. 7¼ E.	48	7·0	—	47·5	—
N.W. ½ W..... = N. 4½ W.	46	29·2	—	—	35·6
* The courses are given in this double form merely as an illustration of the method of using them.		62·0 7·2	7·2	72·2 67·4	67·4
		54·8		4·8	

(Table I, RAPER, and Table I, NORIE. See Explanation below.)

Explanation.—The courses and distances are entered in their proper columns, in the same order as they stand in the question; then in Traverse Table I, NORIE or RAPER, the diff. lat. and dep. to each course and distance is found and entered in their proper columns. (See Rule LXV, 3°.)

The sum of the respective columns, N., S., E., and W., is next found, and set down at the bottom of each column, and the difference between the Northing, viz., 62·0, and the Southing, viz., 7·2, is taken, which leaves 54·8 N., which is the Diff. lat. made; the difference between the Easting 72·2, and Westing 67·4, leaves 4·8 E., departure.

We proceed in the next place to find the course and distance made good, thus:—

In Traverse Table II. { Diff. lat. 54·8 N. and Dep. 4·8 E. } give { Course N. 5° E. Dist. 55 miles. } made good.

This is an illustration of the remark, that when the diff. lat. is more than the dep., the course is less than 4 points, or 45° (see No. 161, page 104), and it is named from the N. towards the E., since the diff. lat. is N. and the dep. E.

Lat. left (or sailed from) Dudgeon Light 53° 19′ N.
Diff. lat. 54·8   = 55 N.   Lat. in is found according to
                                                ———   Rule XLIV, 1°, page 103.
Lat. in (or arrived at)   54 14 N.

Ex. 2. A ship from Cape Espicheli, in lat. 38° 25′ N., sails as follows: S.W. by W., 28 miles; W. by N., 55 miles; West, 47 miles; S.E. ¾ S., 25 miles; South, 101 miles; W. ⅞ S., 72 miles: required the latitude in, also the course and distance made good.

Courses.	Dist.	Diff. Lat.		Departure.	
		N.	S.	E.	W.
S. 5° W.	28	—	15·6	—	23·3
N. 7 W.	55	10·7	—	—	53·9
W. •	47	—	—	—	47·0
S. 3¼ E.	25	—	20·1	14·9	—
S. •	101	—	101·0	—	—
S. 7⅞ W.	72	—	10·6	—	71·2
*See Rule LXV, 3°, page 189.		10·7	147·3 10·7	14·9	195·4 14·9
			136·6		180·5

We seek in the Traverse Table till the diff. of lat. 136·6, and dep. 180·5, are found opposite each other, in their respective columns; the nearest to these are 180·5 and 136·0, which give the course (at the bottom of page, dep. being the most) S. 53° W., and distance 226′. This is an illustration of the remark, No. 161, page 104, that when the departure exceeds the difference of latitude, the course is more than 45°.

Lat. left   38° 25′ N.
Diff. lat. 136·6   = 2 17 S.   } The lat. in is found according to Rule XLIV, 2°, page 108.
Lat. in (or arrived at)   36 8 N.

Ex. 3. A ship from lat. 37° 24′ S., sails the following true courses:—S.W. by S., 20 miles; West, 16 miles; N.W. by W., 28 miles; S.S.E., 32 miles; E.N.E., 14 miles; S.W., 36 miles: required the lat. in, also the course and distance made good.

Courses.	Dist.	Diff. Lat.		Departure.	
		N.	S.	E.	W.
S. 3° W.	20	—	16·6	—	11·1
W. •	16	—	—	—	16·0
N. 5 W.	28	15·6	—	—	23·3
S. 2 E.	32	—	29·6	12·3	—
N. 6 E.	14	5·4	—	12·9	—
S. 4 W.	36	—	25·5	—	25·5
*See Rule LXV, 3°, page 189.		21·0	71·7 21·0	25·2	75·9 25·2
			50·7		50·7

We seek in the several pages of the Traverse Table II, for the diff. lat. 50·7; and dep. 50·7; the nearest found to these are diff. lat. 50·9, dep. 50·9, give course S. 45° W., distance 72 miles.

The diff. lat. and dep. being of equal amount, the course is 45°, or 4 points, which illustrates the remark, No. 161, page 104.

Lat. left   37° 24′ S.   } The lat. sailed from being South, and the ship having sailed South, the ship has evidently increased her South lat.; whence the sum of lat. from and diff. lat. is taken to obtain lat. in.—(See Rule XLIV, 1°, page 108.)
Diff. lat. 50·7 =   51 S.
Lat. arrived at   38 15 S.

Ex. 4. A ship from lat. 46° 20' N. sails (all true courses) N. 72° E., 21 miles; N. 38° E., 17 miles; S. 26° W., 13 miles; S. 73° E., 19 miles; S. 1° W., 19 miles; S. 65° E., 48 miles; N. 76° E., 19 miles; N. 48° E., 48 miles: required the lat. in, also the course and distance made good.

Courses.	Dist.	Diff. Lat.		Departure.	
		N.	S.	E.	W.
N. 72° E.	21	6·5	—	20·0	—
N. 38° E.	17	13·4	—	10·5	—
S. 26° W.	13	—	11·7	—	5·7
S. 73° E.	19	—	5·6	18·2	—
S. 1° W.	19	—	19·0	—	0·3
S. 65° E.	48	—	20·3	43·5	—
N. 76° E.	19	4·6	—	18·4	—
N. 48° E.	48	32·1	—	35·7	—
		56·6	56·6	146·3	6·0
			56·6	6·0	
				140·3	

Course due East, and dist. 140·3, the same as the departure. (See No. 161, page 104.)

The Traverse Table being filled up, the sums of the Northings and Southings are both 56·6, and being of contrary directions, show that the ship has returned to the same parallel of latitude which she sailed from. The sum of the Eastings is 146·3, and that of the Westings 6·0; their difference, 140·3, shows that the ship has gained so much to the Eastward, that being the greater. Consequently the *Course* is due *East*, and the *Distance* 140·3, the same as the departure.

Ex. 5. A ship from a place in lat. 1° 5' S., sails the following true courses:—N. 17° E., 13 miles; North, 38 miles; N. 27° E., 18 miles; N. 79° E., 25 miles; S. 83° W., 23 miles; S. 48° E., 25·2 miles; N. 48° W., 27·1 miles; N. 36° W., 21 miles: required the latitude in, also the course and distance made good.

Courses.	Dist.	Diff. Lat.		Departure.	
		N.	S.	E.	W.
N. 17° E.	13	12·4	—	3·8	—
North*	38	38·0	—	—	—
N. 27° E.	18	16·0	—	8·2	—
N. 79° E.	25	4·8	—	24·5	—
S. 83° W.	23	—	2·8	—	22·8
S. 48° E.	25·2	—	16·9	18·7	—
N. 48° W.	27·1	18·1	—	—	20·1
N. 36° W.	21	17·1	—	—	12·3
*See Rule XLV, 3a, page 189.		106·3	19·7	55·2	55·2
		19·7		55·2	
		86·6			

The Traverse Table being completed, the sum of the Northings is 106·3, and the sum of the Southings is 19·7, the difference 86·6, and to that amount the ship has altered her latitude. The miles of departure in the East are 52·2, and those in the West columns are also 52·2; but as the East and West departures destroy one another, there is no resulting departure, and therefore it is not necessary to refer to the Traverse Table. The ship is under the same meridian as she sailed from; consequently, the course is due North, and the distance sailed is equal to the diff. lat., viz., 86·6. This is according to No. 161, page 104.

Latitude left   1° 5' S.
Diff. lat. 6,0)8,6·6

     1 26·6 = 1 27 N.

Latitude in   0 22 N.

The ship being 1° 5', or 65 miles, S. of the equator, must evidently be in N. lat. after making 87 miles of Northing. Thus, in subtracting one of the quantities from the other, the difference takes the *name* of the greater. Rule XLIV, page 108.

The course is North, and dist. 86·6, the same as diff. lat.

Ex. 6. A ship from latitude 46° 10′ N., sails as follows:—S. 48° E., 25 miles; S. 51° E., 18·9 miles; N. 87° E., 12·4 miles; S. 70° E., 14·5 miles; S. 68° E., 21·6 miles; N. 25° W., 16·4 miles; N. 8° E., 7·8 miles; N. 19° E., 13·7 miles; N. 76° E., 39·6 miles: required the lat. in, also the course and distance made good.

Courses.	Dist.	Diff. Lat.		Departure.	
		N.	S.	E.	W.
S. 48° E.	25	—	16·7	18·6	—
S. 51 E.	18·9	—	11·9	14·7	—
N. 87 E.	12·4	0·7	—	12·4	—
S. 70 E.	14·5	—	5·0	13·6	—
S. 68 E.	21·6	—	8·1	20·0	—
N. 25 W.	16·4	14·9	—	—	6·9
N. 8 E.	7·8	7·7	—	1·1	—
N. 19 E.	13·7	13·0	—	4·5	—
N. 76 E.	39·6	9·6	—	38·4	—
		45·9	41·7	123·3	6·9
		41·7		6·9	
		4·2		116·4	

For the method of taking out the course and distance from the Traverse Table when the distance is given in *tenths*, see Rule LXIII, page 184.

EXAMPLES FOR PRACTICE.

1. A ship from the Texel in lat. 52° 58′ N., sails W. by N., 44 miles; S. by E., 45 miles; W. by S., 35 miles; S.S.E., 44 miles; W.S.W. ½ W., 42 miles; find diff. lat. and dep., the course and dist. made good, also the lat. arrived at.

2. A ship from Heligoland, lat. 54° 12′ N., sails W.S.W., 12 miles; N.W., 24 miles; S. by W., 20 miles; N.W. by W., 32 miles; S. by E., 36 miles; W. by N. ½ N., 42 miles; S.S.E. ½ E., 16 miles; W. ¾ N., 45 miles: required diff. lat. and dep., course and dist. made good, also the lat. arrived at

3. A ship sails from lat. 3° 50′ N., sails S.S.W., 112 miles; S. by E., 86 miles; S.S.E., 112 miles; S. by W., 86 miles: find diff. lat. and dep., the course and dist. made good, also the lat. arrived at.

4. Yesterday we were in lat. 19° S., and since then have sailed S.E. ½ S., 13 miles; S. by E., 19 miles; S.E. by E., 22 miles; E. by S. ¼ S., 32 miles; N.N.E., 20 miles; N. by W. ¼ W., 27 miles; N.E. by E. ½ E., 24 miles; S.W. ¼ S., 10 miles.

5. A ship from lat. 1° N., sails East, 8 miles; E. ¼ N., 20 miles; S.E. by E., 33 miles; S. ¾ W., 31 miles; N.E. ½ N., 43 miles; South, 28 miles; S. ¾ E., 21 miles; S. by W. ¼ W., 12 miles: required diff. lat. and dep., course and dist. made good, and also the lat.

6. A ship from lat. 1° 10′ N., sails N. 40° W., 20 miles; S. 56° W., 51 miles; S. 19° W., 19 miles; S. 48° W., 16 miles; N. 85° E., 28 miles; S. 44° E., 15 miles; N. 22° W., 25 miles; S. 9° E., 54 miles: find diff. lat. and dep., course and dist. made good, also the lat. in.

7. A ship from lat. 47° 12′ N., sails S. 31° W., 16 miles; N. 72° E., 13°·1; S. 52° W., 15′; S. 44° E., 15′·1; N. 44° W., 19′·7; N. 77° E., 11′·4; S. 40° W., 16′; S. 14° E., 6′: required the course and dist. made, the lat. arrived at, and the dep. made.

8. Since leaving lat. 34° 11′ N., we have sailed the following courses:—N. 36° W., 27′; N. 24° E., 30′; S. 75° W., 47′; S. 80° W., 29′; N. 72° W., 42′; N. 78° W., 34′; S. 12° E., 28′; required the course and dist. made, the lat. arrived at, and the dep made.

9. Since leaving lat. 36° 35′ S., the ship has sailed N. 84° W., 18′; N. 89° W., 30′·4; N. 67° W., 29′·9; N. 39° W., 33′·9; N. 8° W., 25′·9; N. 73° W., 34′·9; N. 86° W., 44′·7; S. 65° E., 56′; required the lat. arrived at, and the course and dist. made good.

10. A ship sails from lat. 1° 46′ N., on the following compass courses, viz., N. 84° W., 23 miles; S. 6° W., 48 miles; S. 61° W., 37 miles; S. 24° W., 30 miles; N. 75° W., 44 miles; N. 69° W., 37 miles; N. 20° W., 38 miles; and N. 33° W., 36 miles: required the lat. arrived at, and the course and dist. made good, the variation of the compass being 21½° W.

PARALLEL SAILING.

272. When two places lie on the same parallel of latitude, or due East or West of each other, the distance between them estimated along a parallel, or E. and W. (which is all departure) is converted into difference of longitude; or, on the other hand, the difference of longitude is converted into distance by **Parallel Sailing.**

Since the meridians are all parallel at the equator and meet at the poles, the distance between any two meridians, measured East and West, is less as the latter is greater—that is, the *absolute number of miles*, or of feet, in a degree of longitude, is less as the latitude in which they are measured is greater. Hence, also, a given number of miles between two meridians corresponds to a greater difference of longitude, as the latitude in which they are measured is greater. For example, two places in lat. 10°, and distant 60 miles East and West from each other, have 60′·9 diff. long. In lat. 60° N. or S., two places similarly situated have 2° 0′ diff. long., while at 73° the diff. long. is 3° 25′. Questions of this kind are solved by Parallel Sailing.

273. Given the departure made good on a given parallel of latitude, to find the diff. of long. corresponding thereto.

RULE LXVI.

1°. *Take out of the Tables the log. secant of latitude (rejecting* 10 *from index), and the log. of the departure made good.*

2°. *Add these logs. together, and find the nat. number corresponding thereto.* The result is the difference of longitude required.

274. In parallel sailing, the latitude being constant, the difference of longitude bears a constant ratio to the distance, and all problems may be completely solved by the solution of a right-angled plane triangle, and therefore by inspection of the Traverse Table by

RULE LXVII.

With the latitude of the parallel as a course, *and the distance sailed on it as* difference of latitude, *the corresponding* distance, *in the Traverse Table, is the* difference of longitude.

EXAMPLES.

Ex. 1. In lat. 29° 51′ S., the dep. made good 161 miles: required the diff. of long.

Lat. 29° 51′   Secant 0·061815
Dep. 161   Log.   2·206826

            Log. 2·268641
Diff. of long. 185·6.
By Inspection.

In Traverse Table II, lat. 30° as course, dep. 161·1 in lat. column, give diff. long. 186 miles in dist. col.

Ex. 2. A ship sailed 94·6 miles on the parallel 64° 38′ N.: required the diff. long.

Lat. 64° 38′   Secant 0·368141
Dep. 94·6   Log.   1·975891

            Log. 2·344032
Diff. of long. 220·8.
By Inspection.

In Traverse Table II, lat. 64° as course and dep., 91·7 give diff. lat. in dist. column 216 miles; and course 65°, and dep. 94·7, give diff. long. in dist. column 224 miles; therefore, the diff. long. for 64½° will equal 216 + 224 ÷ 2 = 220 miles.

Ex. 3. From long. 0° 59' W., the dep. made was 125 East, on the parallel of 52°: required the long. in.

Lat. 52°	Secant 0·210658
Dep. 125	Log. 2·096910

Log. 2·307568
Diff. long. 6,0)20,3

3 23 = 3° 23' E.
Long. left 0 59 W.

Long. in 2 24 E.

Ex. 4. A ship from long. 179° 20' W. sails 109 miles West, on the parallel of 61° 25': what is the long. in?

Lat. 61° 25'	Secant 0·320176
Dep. 109	Log. 2·037426

Log. 2·357602
Diff. of long. 227·8 W.
6,0)22,8   3° 48' W.
                  179 20 W.
3 48
Long. in 183 8 W.
          360 0
or 176 52 E.

Ex. 5. In lat. 71° 25' N., the dep. made good was 71¼ miles: required the diff. of long.

Lat. 71° 25'	Secant 0·496640
Dep. 71·25	Log. 1·852785

Log. 2·349425
Diff. of long. 223·6 nearly.

Ex. 6. In lat. 80°, the dep. made good was 80 miles: required the diff. of long.

Lat. 80°	Secant 0·760330
Dep. 80	Log. 1·903090

Log. 2·663420
Diff. of long. 460·7.

EXAMPLES FOR PRACTICE.

In each of the following examples the difference of longitude is required:—

No.	Lat. in.	Dep.	No.	Lat. in.	Dep.
1.	6° 7' N.	249' W.	5.	46° 37' S.	352' E.
2.	19 48 S.	324 E.	6.	64 16 N.	265·7 W.
3.	39 57 N.	398 W.	7.	51 28 S.	70·9 E.
4.	60 0 N.	74 W.	8.	60 0 S.	204 E.

275. The method of parallel sailing will apply correctly enough for all practical purposes to cases where the course is nearly East and West (true). In latitudes not higher than 5°, when the distance does not exceed 300 miles, the departure may be used at once for the difference of longitude, the resulting error scarcely exceeding one mile.

276. Given the difference of longitude of two places on the same parallel, to find their distance as measured along the parallel.

RULE LXVIII.

To the log. of the diff. of long. add the cosine of lat.; the sum (neglecting 10) is log. of the distance required.

EXAMPLE.

Ex. 1. Required the distance between St. Abb's Head, in latitude 55° 55' N., longitude 2° 10' W., and Uraniberg in the same latitude, but in longitude 12° 52' E.

Longitude of St. Abb's Head 2° 10' W.
Longitude of Uraniberg   12 52 E.

15 2
60

Difference of longitude   902 miles.

Log. distance = log. diff. long. + log. cosine lat. − 10.

Log. diff. long. 902 = 2·955207
Log. cos. lat. 55° 55' = 9·748497

Log. distance 505·5 = 2·703704

277. Given the meridian distance and difference of longitude, to find the latitude.

RULE LXIX.

From the log. of meridian distance (adding 10 to the index) *subtract the log. of diff. long.; the remainder is the log. cosine of the latitude.*

EXAMPLE.

Ex. 1. From a place in longitude 3° 12' W., a ship sails due East 246 miles, and by observation is found to be in longitude 4° 8' E.: required the latitude of the parallel on which she sailed.

By Calculation.

$$\text{Cos. lat.} = \frac{\text{mer. dist.}}{\text{diff. long.}}$$

Long. left 3° 12' W.
Long. in   4 8 E.

D. long.   7 20 = 440 miles.

Mer. dist. 246 log. (+ 10) 12·390935
Diff. long. 440   log.   2·643453
Lat.   56° 0'   cos.   9·747482

By Inspection.

Since the diff. long. given, 440, exceeds the distance given in the Traverse Table, its half is taken, and also the half of the meridian distance 246, these are 220 and 123 respectively. Entering the Tables with 220 as distance, and 123 as diff. lat., we find, on searching the Table, these quantities, in their respective columns, on the page with 56° at the bottom; hence the latitude sought is 56°.

EXAMPLES FOR PRACTICE.

1. Required the compass course and distance from A to B.

   Given lat. A 52° 15' S.; var. 1½ points W.; long. A 37° 30' W.
         B 52 15 S.; dev. 8° 50' W.;   B 48 18 W.

2. A and B lie on the parallel of 58° 30' N.

   Given long. A 15° 12' E.
          B 13 18 W.

   What is the distance between them in nautical miles?

3. Define a great circle and a small circle of a sphere, giving an example of each. What connection is there between the tropic of Cancer and the Arctic Circle?

4. Required the compass course and distance from A to B.

   Lat. A 28° 40' N.; var. 1¾ points W.; long. A 2° 20' E.
        B 28 40 N.; dev. 8° 50' E.;   B 4 10 E.

5. In what latitudes are the lengths of a degree of longitude 30 and 20 miles respectively?

6. In travelling 35 nautical miles on the parallel of 55° 25' N., how much do I change my longitude?

7. Find the true course and distance from A to B.

   Lat. A 54° 25' S.; long. A 15° 30' E.
        B 54 25 S.;   B 9 15 W.

MIDDLE LATITUDE SAILING.

278. **Middle Latitude Sailing** is a method founded on the principle of parallel sailing, converting Departure into Difference of Longitude, and the Difference of Longitude into Departure, when the ship's course lies obliquely across the meridian, that is, when besides departure she makes difference of latitude.

Suppose a ship, in going on the same course, from latitude 40° to latitude 44°, make 100 miles departure: this departure, if made good altogether in latitude 40°, would give 130·5 difference of longitude by Rule LXVI, page 194; and again, if made good in latitude 44°, it would give 139 difference of longitude. Now, since the ship has sailed between these two parallels, and not on either of them exclusively, her real difference of longitude must be between 130·5 and 139, and therefore we may conclude it to be not far from that which would result from a departure made good altogether in the *middle parallel;* hence the name *Middle Latitude Sailing.* Middle latitude sailing, then, is founded on the consideration that the arc of the parallel of middle latitude of two places intercepted between their meridians is nearly equal to the departure. If we conceive the ship to sail along this middle parallel, we may apply the principle of *parallel sailing* to the cases in point. In parallel sailing, as has been shown, the departure (or distance) and difference of longitude are connected by the relation, dep. = diff. of long. × cos. lat. When the ship's course lies obliquely across the meridian, making good a difference of latitude, a modification of this formula gives the formula for middle latitude sailing, dep. (nearly) = diff. of long. × cos. mid. lat.; or in logarithms, log. dep. = log. diff. of long. + log. cos. mid. lat. − 10. Middle latitude sailing has thus the same two cases as parallel sailing, and accordingly the rules for inspection and computation already given, Rule LXVI, page 194, apply equally to this sailing, observing merely to read *middle latitude* for *latitude.*

279. To find the **latitude** and **longitude in, the course and distance** from a known place being given, by Traverse Table and Middle Latitude.

By working a Traverse the difference of latitude and departure are obtained. Hence, by applying the difference of latitude to the latitude from, we have the latitude in. The middle latitude is then found, and the solution of the problem completed by the aid of the formula above, No. 278, note, viz.:—

Diff. long. = dep. × sec. mid. lat.

the diff. of long. applied to the long. from giving the long. in, hence—

RULE LXX.

1°. *With the given* course *and* distance *enter the Traverse Table, and take out* true difference of latitude and departure (see Rule LXIII, page 184).

2°. *With* difference of latitude *and* latitude from, *find* latitude in (see Rule XLIV, page 108).

3°. *Get the* middle latitude, as directed, Rule XLV, page 108.

4°. *With the middle latitude as* course, *look in the* difference of latitude *column for the* departure, *the corresponding* distance *at the top is the* difference of longitude.

5°. *With* difference of longitude *and* longitude from *get* longitude in, *as in* Rule XLVII, page 110.

NOTE.—When the *departure* to be looked for as *difference of latitude* at the middle latitude, is beyond the limits of the Table, one-half, one-third, &c., must be used, and the resulting *difference of longitude* multiplied by the divisor, in order to get the whole *difference of longitude.*

EXAMPLES.

Ex. 1. A ship from lat. 52° 6' N., long. 35° 6' W., sailed S.W. by W., 256 miles: required her latitude and longitude in.

Course S. 5 pts. W. } give diff. lat. 142·2, and dep. 212·9 (see Rule LXIII, page 184).
Distance 256 miles. }

```
6,0)14,2         Diff. lat.    2° 22' S.                    2)212·9
  ———            Lat. from    52   6  N.                    ———
  2 22                                                      ½ dep. 106·4
                 Lat. in      49  44  N.                         By Inspection.
Diff. long.=dep.×sec.mid.lat. 2)101 50   (See Rules XLIV &   Mid. lat. 51° as course (Table II), and
                                          XLV, page 108).   half dep. 106·4, in diff. lat. column, give
                 Mid. lat.    50  55                         in dist. column 169 miles, the half the
                                                             diff. of long. Then 169 × 2 = diff.
                                                             long. 338.
```

By Calculation.

```
Mid. lat.   50° 55'  sec.  0·200349    6,0)33,8                         5° 38' W.  (See Rule
Dep.           212·9  log. 2·328176     ———          Long. from 35   6  W.   XLVII,
                                        5 38                                         page 110).
Diff. long. 337·7    log. 2·528525                   Long. in   40  44  W.
```

Explanation.—The difference of latitude and departure is found as described in Rule LXIII, page 184. The latitude in is found by Rule XLIV, page 108; and thence the middle latitude, by adding the latitude from and latitude in together, and divided by 2 (see Rule XLV, page 108). The departure exceeding the limits of the Tables, the half is taken. Then with *middle latitude* as a *course*, and *half the departure*, in *difference of latitude* column, half the difference of longitude is found in the *distance* column. This being doubled (as half the departure was taken) and divided by 60, gives the difference of longitude expressed in degrees and minutes. The ship is in *West* longitude, *sailing West*, add difference of longitude to longitude left to obtain longitude in (Rule XLVII, page 110).

This Method by Inspection is the usual case at sea of Working the Day's Work.

Ex. 2. A ship from lat. 48° 27' S., and long. 29° 12' W., sails S.E. by S., 22·5 miles: required the latitude in, also the longitude in.

Course S.E. by S. = 3 pts.; then 3 pts. and dist. 22·5 give diff. lat. 18·7, and dep. 12·5 (see Rule LXIII, page 184).

```
Diff. lat.   0° 19' S.                        By Inspection.
Lat. from   48  27  S.           Mid. lat. 48½° as course, and dep. as diff. lat.,
                                 give in dist. column 19 miles, which is the diff.
Lat. in     48  46  S.           of long.
                                        Diff. long.  0° 19' E.
           2)97  13                     Long. left  29  12  W.
Mid. lat.   48  36
                                        Long. in    28  53  W.
                                 (The long. in is found by Rule XLVII, page 110).
```

By Calculation.

```
Mid. lat.   48° 36'  sec.  0·179594
Dep.           12·5  log.  1·096910

Diff. long.    18·9  log.  1·276504
```

Ex. 3. A ship from the Lizard, in lat. 49° 57' N., sails between the South and West until diff. lat. is 126·7, and dep. 102·6: required the latitude come to, and difference of longitude.

```
6,0)12,6·7                                  By Inspection.
                               Then mid. lat. 48° 54', say 49°, as a course, and
  2 6·7   or   2°  7' S.       dep. 102·3 (nearest), found in the lat. column,
        Lat. left  49 57 N.    opposite to which, in the dist. column, is 156,
                                nearest, the difference of longitude.
        Lat. in   47  50 N.              By Calculation.
                                    Mid. lat.   48° 53'  sec.  0·182042
                2)97  47            Dep.          102·6  log.  2·011147
        Mid. lat. 48  53
                                    Diff. long.   156·0  log.  2·193189
```

Ex. 4. Lat. from 59° 0' N., long. from 3° 33' E., the ship sailed between S. and E., making diff. lat. 81·7 and dep. 172·7.

D. Lat.

6,0)8,1·7

1 21·7   or 1° 22' S.
Lat. from 59 0 N.

Lat. in   57 38 N.

116 38

Mid. lat.   58 19

By Calculation

Mid. lat. 58° 19'   sec. 0·279655
Dep.   172·7   log. 2·237292

log. 2·516947

D. long. 328·8

6,0)32,8·8

5° 28·8

By Inspection.

58°
Dep. 159·0 give D. long. 300
    13·8   ,,   26
    172·8   ,,   326

59°.
Dep. 154·5 give D. long. 300
    18·0   ,,   35
    172·5   ,,   335

D. long. for mid. lat. 58° is 326
       ,,   ,,   59 ,, 335

Diff. for 1° of mid. lat.   9

60' (or 1°) : 19' :: 9
                9

6,0)17,1

2·85

D. long. for mid. lat. 58° is   326
Corr. for 19' (over 58°)   + 2·8

D. long. for mid. lat. 58° 19' is 328·8

Remark.—When the mid. lat. is high and between two whole degrees, and also the dep. great as in this example, the diff. long. is best found by calculation.

Ex. 5. Sailed from A, in lat. 50° 48' N., long. 1° 10' W., S. 41° E., 275 miles.

Entering Traverse Table II with *dist.* 275 miles, and *course* 41°, the *true diff. lat.* is 207'·5, or 3° 27'·5 S.; applying this to *lat. from,* the *lat. in* is 47° 20'·5 N. The corresponding *dep.* is taken out at the same opening, which is 180'·4. The *mid. lat.,* or half sum of *lat. from* and *lat. in,* is 49° to the nearest degree. The *dist.* corresponding to 49° as a course, and 180'·4 in *diff. lat.* column, is found to be 275', in degrees 4° 35' E., which is the *diff. long.* Applying this to the *long. from,* 1° 10' W., we have the *long. in* 3° 25' E.

EXAMPLES FOR PRACTICE.

In each of the examples following, the latitude and longitude arrived at are required to be found, having given the latitude and longitude from, with the course and distance sailed.

	Lat. from.	Long. from.	Course.	Dist.
1.	25° 35' N.	60° 0' W.	E.N.E.	296
2.	32 30 N.	25 24 W.	N.W. by W. ½ W.	212
3.	39 30 S.	74 20 E.	S.W. by W. ¾ W.	210
4.	46 24 S.	178 28 E.	S.E. ¾ E.	278
5.	20 29 N.	179 10 W.	W. by S. ½ S.	333
6.	0 56 N.	29 50 W.	S. 47° E.	168

MERCATOR'S SAILING.

280. Mercator's Sailing, like middle latitude sailing, relates to finding the difference of longitude a ship makes when sailing on any oblique rhumb, and is a perfectly general and rigorously true method, which the other is not.

Mercator's sailing is characterised by the use of the *Table of Meridional Parts*, and the chart constructed by means of it called *Mercator's Chart*. With the assistance of this Table, the rules of plane trigonometry suffice for the solution of all the problems.

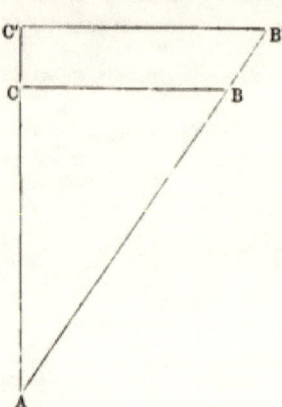

In the triangle ACB let A be the course, AB the distance, AC the true difference of latitude, CB the departure; then corresponding to AC, the Table of Meridional Parts gives AC′, the meridional difference of latitude, and completing the right-angled triangle AC′B′, C′B′ will be the difference of longitude. In addition, then, to the three canons of *plane sailing* which can be deduced from the triangle ACB, the triangle AC′B′ gives the characteristic canon of *Mercator's Sailing* (since C′B′ = AC′ tang. A) diff. long. = mer. diff. lat. × tang. course.

281. Given the latitudes and longitudes of two places, to find the course and distance between them.

RULE LXXI.

1°. *Find the true difference of latitude*, according to Rule XLII, page 106.
2°. *Find the meridional difference of latitude*, Rule XLIII, page 107.
3°. *Next find the difference of longitude*, Rule XLVI, page 109.
4°.* To find the course.—*From the log. of diff. of longitude (increasing its index by 10), subtract the log. of meridional diff. of lat.: the remainder is the tangent of course, which take out of the tables, and place before it the letter of diff. of lat., and after it the letter of diff. of long.*

NOTE.—Be careful to remember that when the *sum* of the longitudes exceeds 180°, it must be taken from 360°, and the *course* must be *named the same* as the *longitude left*.

5°. To find the distance.—*To the secant of course (rejecting 10 from the index), add the log. of diff. of lat.: the sum is the log. of distance, the natural number corresponding to which find in the tables.*

---

* From the formula:—

Tang. course = $\dfrac{\text{Diff. long.}}{\text{Mer. diff. lat.}}$ ∴ log. tang. *course* − 10 = log. diff. long. − log. mer. diff. lat.

Dist. = $\dfrac{\text{True diff. lat.}}{\text{Cos. course.}}$ ∴ log. *dist.* = log. true diff. lat. + log. sec. course − 10.

Mercator's Sailing.

EXAMPLES.

Ex. 1. Required the course and distance from Tynemouth Light to the Naze of Norway.

| Lat. Tynemouth | 55° 1' N. | Mer. parts 3970 | Long. Tynemouth | 1° 25' W. |
| Lat. Naze | 57 58 N. | Mer. parts 4291 | Long. Naze | 7 2 E. |

 2 57
 60
Mer. diff. lat. 321 8 27 / 60

Diff. of lat. 177 N. Diff. of long. 507 E.

To find the Course. | **To find the Distance.**

Diff. long. 507 Log. (+ 10) 12·705008 | Course 57° 40' Secant 0·271773
Mer. diff. lat. 321 Log. 2·506505 | Diff. of lat. 177 Log. 2·247973

 Tang. 10·198503 Log. 2·519746
Course N. 57° 40' E. Distance 331.

Ex. 2. Required the course and distance from A to B.

| Lat. A | 51° 23' N. | Mer. parts 3606 | Long. A | 9° 29' W. |
| Lat. B | 48 23 N. | Mer. parts 3316 | Long. B | 4 29 W. |

 3 0
 60
Mer. diff. lat. 280 5 0 / 60

Diff. of lat. 180 S. Diff. of long. 300 E.

Diff. long. 300 Log. (+ 10) 12·477121 | Course 46° 58½' Secant 0·166014
Mer. diff. lat. 280 Log. 2·447158 | Diff. lat. 180 Log. 2·255273

 Tang. 10·029963 Log. 2·421287
Course S. 46° 58½' E. Distance 163·8.

Ex. 3. Required the course and distance from Cape Bajoli to Cape Sicie.

| Lat. Cape Bajoli | 40° 1' N. | Mer. parts 2624 | Long. Cape Bajoli | 3° 48' E. |
| Lat. Cape Sicie | 43 3 N. | Mer. parts 2867 | Long. Cape Sicie | 5 51 E. |

 3 2
 60
Mer. diff. lat. 243 2 3 / 60

Diff. of lat. 182 N. Diff. of long. 123 E.

Diff. long. 123 Log. (+ 10) 12·089905 | Course 26° 51' Secant 0·049542
Mer. diff. lat. 243 Log. 2·385606 | Diff. lat. 182 Log. 2·160071

 Tang. 9·704299 Log. 2·309613
Course N. 26° 51' E. Distance 204.

Ex. 4. Required the course and distance from Cape Formosa to St. Helena.

| Lat. Cape Formosa | 4° 15' N. | Mer. parts 255 | Long. Cape Formosa | 6° 11' E. |
| Lat. St. Helena | 15 55 S. | Mer. parts 968 | Long. St. Helena | 5 45 W. |

 20 10
 60
Mer. diff. lat. 1223 11 56 / 60

Diff. of lat. 1210 S. Diff. of long. 716 W.

Diff. long. 716 Log. (+ 10) 12·854913 | Course 30° 21' Secant 0·064012
Mer. diff. lat. 1223 Log. 3·087426 | Diff. of lat. 1210 Log. 3·082785

 Tang. 9·767487 Log. 3·146797
Course S. 30° 21' W. Distance 1402.

D D

Ex. 5. Required the course and distance from Bahia to Fernando Po.

Lat. Bahia	13° 1' S.	Mer. parts 788	Long. Bahia	38° 32' W.
Lat. Fernando Po	3 48 N.	Mer. parts 223	Long. Fernando Po	8 43 E.
	16 49	Mer. diff. lat. 1016		47 15
	60			60

Diff. of lat. 1009 N.   Diff. of long. 2835 E.

Diff. long. 2835   Log. (+ 10) 13·452553   Course 70° 17'   Secant 0·471895
Mer. diff. lat 1016   Log.   3·006894   Diff. of lat. 1009   Log.   3·003891

                 Tang. 10·445659   Log.   3·475786
Course N. 70° 17' E.   Distance 2991.

Ex. 6. Required the course and distance from A to B.

Lat. A	44° 44' S.	Mer. parts 3007	Long. A	148° 39' W.
Lat. B	55 55 N.	Mer. parts 4065	Long. B	44 44 E.
	100 39	Mer. diff. lat. 7072		193 23 E.
	60			360 0

Diff. of lat. 6039 N.   166 37
                                                                    60

                                                  Diff. of long. 9997 W.

Diff. long. 9997 Log. (+ 10) 13·999870   Course 54° 43' 0"   Secant 0·238358
Mer. diff. lat. 7072   Log.   3·849542   Parts for 26"   77
                                                           Diff. of lat. 6039   Log.   3·780965
       54° 43'   Tang. 10·150328
                                       210   10457 Log. 4·019400
                              446)11800(26   116
                                         892
                                                                              416)2840(7
                                        2880   Distance 10457 nearly.
                                        2676

Course N. 54° 43' 26" W.

This question worked to the nearest minute of arc gives course N. 54° 43' W., and distance 10455 miles.

Ex. 7. Required the course and distance from Cape East, New Zealand, to Cape Horn.

Lat. Cape East	37° 42' S.	Mer. parts 2445	Long. Cape East	178° 40' E.
Lat. Cape Horn	55 59 S.	Mer. parts 4072	Long. Cape Horn	67 16 W.
	18 17	Mer. diff. lat. 1627		245 56 W.
	60			360 0

Diff. of lat. 1097 S.   114 4
                                                                    60

                                                  Diff. of long. 6844 E.

Diff. long. 6844 Log. (+ 10) 13·835310   Course 76° 37' 0"   Secant 0·635515
Mer. diff. lat. 1627   Log.   3·211388   Parts for 39"   345
                                                           Diff. of lat. 1097   Log.   3·040207
       76° 37'   Tang. 10·623922
                                       558   3·676067
                              935)36400(39   53
                                       2805
                                                                               92)140(2
                                        8350   Distance 4743·2 nearly.
                                        8415

Course S. 76° 37' 39" E.

Examples for Practice.

Required the course and distance from A to B in each of the following examples:—

1.	A 38° 14′ N.	A 2° 7′ E.	9.	A 4° 24′ N.	A 7° 46′ W.		
	B 39 51 N.	B 4 18 E.		B 8 48 S.	B 13 8 E.		
2.	A 49 53 N.	A 6 19 W.	10.	A 57 43 S.	A 10 37 E.		
	B 48 28 N.	B 5 3 W.		B 55 35 S.	B 1 28 W.		
3.	A 64 30 N.	A 4 20 W.	11.	A 6 11 N.	A 80 15 W.		
	B 60 40 N.	B 0 10 E.		B 6 0 S.	B 39 16 W.		
4.	A 54 54 S.	A 60 28 W.	12.	A 55 28 N.	A 1 9 E.		
	B 34 22 S.	B 18 24 E.		B 57 58 N.	B 7 3 E.		
5.	A 45 15 N.	A 35 26 W.	13.	A 35 51 S.	A 138 54 E.		
	B 47 10 N.	B 32 15 W.		B 38 52 N.	B 165 53 W.		
6.	A 34 22 S.	A 18 29 E.	14.	A 15 30 N.	A 176 34 E.		
	B 15 55 S.	B 5 43 W.		B 15 30 S.	B 176 34 W.		
7.	A 49 57 N.	A 5 12 W.	15.	A 22 22 S.	A 122 22 W.		
	B 36 58 N.	B 25 12 W.		B 33 33 N.	B 111 11 E.		
8.	A 35 14 S.	A 75 30 E.	16.	A 17 0 N.	A 180 0 E.		
	B 18 23 S.	B 12 2 E.		B 20 0 N.	B 161 0 E.		

282. To find the **latitude** and **longitude in**, having given the **latitude from**, the **longitude from**, and the **course** and **distance** between the two places by Traverse Table and meridional parts.*

RULE LXXII.

1°. *With given* course *and* distance *enter the Traverse Table and take out the corresponding* true difference of latitude, Rule LXIII, page 184, *from which and* latitude from *find* latitude in, *as in* Rule XLIV, page 108, *and then* meridional difference of latitude, *as in* Rule XLIII, page 107.

2°. *At the given* course *look in the column of the* true difference of latitude *for the* meridional difference of latitude; *the corresponding* departure *will be the* difference of longitude, *from which and the* longitude from *find the* longitude in, *as in* Rule XLVII, page 110.

Examples.

Ex. 1. A ship from lat. 55° 1′ N., long. 1° 25′ W., sails S.S.E. ¼ E., 246 miles: required the lat. in and long. in.

Entering Traverse Table II, with course S. 2½ points E., and distance 246, we obtain diff. lat. 217·0, and dep. 116·0.

$$\begin{array}{c} 6,0)21,7 \\ \hline 3° 37′ \end{array} \quad \begin{array}{l} \text{Lat. left} \\ \text{D. lat.} \\ \text{Lat. in} \end{array} \begin{array}{r} 55° \ 1′ \text{ N.} \\ 3\ 37 \text{ S.} \\ \hline 51\ 24 \text{ N.} \end{array} \left.\begin{array}{c}\\ \\ \\ \end{array}\right\} \begin{array}{c}\text{Rule XLIV,}\\ \text{page 108.}\end{array} \quad \begin{array}{l}\text{Mer. parts}\\ \text{Mer. parts}\\ \hline \text{Mer. diff. lat.}\\ \frac{1}{2} \text{ mer. diff. lat.}\end{array} \begin{array}{r}3970\\ 3607\\ \hline 363\\ 181·5\end{array} \left.\begin{array}{c}\\ \\ \\ \end{array}\right\} \begin{array}{c}\text{Rule XLIII,}\\ \text{page 107.}\end{array}$$

* The general method of solution by "meridional parts," is from the formula:—
True diff. lat. $=$ dist. $\times$ cos. course.
$\therefore$ log. *true diff. lat.* $=$ log. dist. $+$ log. cos. course $-$ 10.
Diff. long. $=$ mer. diff. lat. $\times$ tang. course.
$\therefore$ log. *diff. long.* $=$ log. mer. diff. lat. $+$ log. tang. course $-$ 10.

The course 2½ points, and half mer. diff. lat. 181·5 (in diff. lat. column), the nearest found in the Table is 181·7, the corresponding departure is 97·1, which multiplied by 2 (having divided mer. diff. lat. by 2) gives diff. long. 194·2 miles.

6,0)19,4·2	Long. left	1° 25′ W.	} Rule XLVI, page 115.	The ship being 1° 25′ W., or 85° West of Greenwich, must evidently be in East longitude, after having sailed 194 miles to the Eastward (see Rule XLVII, page 115).
3° 14′	D. long.	3 14 E.		
	Long. in	1 49 E.		

Ex. 2. A ship from lat. 42° 36′ S., long. 178° 43′ E., makes diff. lat. 178·1 S., and dep. 240·2 E.: find lat. in and long. in.

Course 4¾ points, and dist. 299, diff. lat. 178·1, dep. 240·2.

6,0)17,8·1	Lat. left	42° 36′ S.	} See Rule XLIV, page 108.	Mer. parts	1830	} See Rule XLIII, page 107.
2° 58′	D. lat.	2 58		Mer. parts	3078	
	Lat. in	45 34 S.		Mer. diff. lat.	2)248	
					124	

Course 4¾ points, and half mer. diff. lat. 124 (in diff. lat. column), give in dep. column 167·1, which doubled is 334·2, the diff. long.

6,0)33,4·2	Long. left	178° 43′ E.	} See Rule XLVII, page 110.
5° 34′	D. long.	5 34 E.	
		184 17 E.	
		360 0	
	Long. in	175 43 W.	

Ex. 3. From lat. 50° 48′ N., and long. 1° 10′ W., sailed S. 41° E., 275 miles: required the lat. in and long. in.

In the Traverse Table at the distance 275, and course 41°, the corresponding *true diff. lat.* is 207·5, or 3° 27′·5, which being subtracted from 50° 48′ N., the *lat. in* is 47° 20′·5 N.; taking out the *mer. parts* for 50° 48′, and 47° 20′·5, the *mer. diff. lat.* is found to be 317, to *half* which as a *true diff. lat.*, and the course 14°, the dep. is 137·8, twice which is 275·6,—that is, the *diff. long.* is 4° 36′ E.: hence the long. in is 3° 26′ E.

Ex. 4. From lat. 50° 30′ N., and long. 37° 55′ W., sailed S.W. ¾ S., until arrived at lat. 52° 15′ N.

Lat. from 50° 30′ N.	Mer. parts 3521	Course 3¼ points, and mer. diff. lat. in diff. lat. column, give in *dep.* column 125·4, which is the diff. long.
Lat. in 52 15 N.	Mer. parts 3690	
	Mer. diff. lat. 169	Long. left 37° 55′ W.
	6,0)12,5·4	D. long. 2 5 W.
	2° 5′	Long. in 40 0 W.

EXAMPLES FOR PRACTICE.

For examples for practice in this problem take those given in middle latitude sailing at page 199.

REMARKS ON MIDDLE LATITUDE AND MERCATOR'S SAILINGS.

283. "The difference of longitude found by middle latitude is true at the equator, and very nearly true for short distances in all latitudes, especially when the course is E. or W. In high latitudes, when the distance is great and the course oblique, the error becomes considerable: but the result may be made as accurate as we please by sub-dividing the distance run into small portions, and finding the difference of longitude for each portion separately. The difference of longitude deduced by middle latitude sailing is too small:

an estimate of the error for places on the same side of the equator may be formed by the help of a few cases. Suppose the course 4 points or 45°, and the difference of latitude 10° or 600′; then if this difference of latitude is made good in any latitude below 30°, the error of the difference of longitude will not exceed 2′; if made good between the parallels of 40° and 50°, the error will be about 3′; and between 60° and 70° about 19′, or $\frac{1}{3}$ of a degree. For smaller distances the errors will be much less, and for greater distances much greater, as they vary in much more rapid proportion than the distances. It has been observed before that when the course is large, the difference of longitude should be found by middle latitude in preference to Mercator's sailing; because, although the latter is mathematically correct in principle, yet a small error in the course may, when the course is large, produce a considerable error in the difference of longitude. The reason of this is easily shown. In middle latitude sailing we convert the *departure* into difference of longitude. The process increases the departure in a proportion which is less than 2 to 1 in all latitudes below 60°; and exceeds 3 to 1 in all latitudes beyond 70°. The error of the departure, increased in the same proportion, becomes thus the error of difference of longitude. Now when the course is nearly E. or W., the departure is nearly the same as the distance, and an error of some degrees in the course does not affect the departure sensibly; hence in this case the error of the difference of longitude depends on that of the distance alone. But in Mercator's sailing, on the other hand, we convert the *meridional difference of latitude* into difference of longitude, and the process, when the course is large, converts a given meridional difference of latitude into a difference of longitude much greater than itself; and thus increases the error of the meridional difference of latitude in the same proportion. Thus, for example, at the course 80°, the difference of longitude exceeds the meridional difference of latitude in the proportion of 6 to 1; at the course 85° this proportion is 11 to 1. Now, when the course is large, a slight change in it sensibly affects the difference of latitude, and also the meridional difference of latitude, which is deduced directly from it. In high latitudes the meridional parts vary rapidly, and the error of the difference of longitude is increased accordingly; hence the precept more especially demands attention in high latitudes."—*Raper's Practice of Navigation*, pp. 103, 104.

THE DAY'S WORK.

284. This is the process of *finding the ship's place at noon*—that is, its latitude and longitude, having given the latitude and longitude at noon preceding, or a departure taken since, the compass courses and distances run in the interval, the leeway (if any), variation and deviation (if any), direction and rate of current (if any), &c., &c.

RULE LXXIII.*

1°. *Correct each* course *for* leeway, variation, *and* deviation (see Rules XLVIII to LV, pages 120 to 160, *which arrange in the tabular form as in the example following.* Add *together the* hourly distances *sailed on each* course, *and insert the same in the Table, opposite the true course.*

NOTE.—Allow the leeway in points before expressing the course in degrees.

Departure Course.—*When a departure has been taken, consider the* opposite to the bearing *as a* course, *which correct for* variation, *and the* **deviation due to the direction of the ship's head when the bearing was taken,** *and insert in the Table as an actual* course, *with the* distance *of the object as a* distance. *The* departure course *is generally put down in the Tables as the* first course. See No. 251, page 157.

As the ship leaves the land, the bearing (by compass) of some prominent object or known headland is taken, and its distance is generally estimated by the eye; this process is called "taking a departure." The latitude and longitude of the landmark are known; and thence, by supposing the ship to have sailed on a course the *opposite* to the bearing of the object, through the distance that object is off, we thus obtain, on commencing a voyage, a determinate starting point, from whence to reckon the subsequent courses and distances. Thus, supposing, for example, a ship leaving the Tyne observes Tynemouth Light dipping, and setting it, finds its bearing to be W. by N., distant (by estimation) 10 miles. Now in sailing from Tynemouth Light to the present position of the ship, she would have to sail in the *opposite* direction to the bearing of the light, viz., E. by S., 10 miles. At the end of the day, the Day's Work gives us a change of the ship's place as referred to the landmark, *and not the supposed position.* For methods of determining the distance, see *Raper's Practice of Navigation*, on Taking Departures, ch., IV, pp. 114—122.

Current Course.—*The* set *of a current is to be corrected for* variation *only (being correct* magnetic*), and inserted in the Table as a* course; *the* drift *being taken as a* distance. *The* current course *is generally inserted in the Table as the* last course.

2°. *Take out of the Traverse Tables* (Table I or II, RAPER or NORIE) *the* difference of latitude *and* departure *to each* course *and* distance, (see Rule LXIII, page 184), *and proceed to find the* difference of latitude *and* departure made good *as directed in* Rule LXV, page 189, *Traverse Sailing.*

3°. *Find the* course *and* distance made good (see Rule LXIV, page 186.)

4°. *Find the* latitude in *by applying the* difference of latitude *to the* latitude from (see Rule XLIV, page 108).

If a departure has been taken, the difference of latitude is to be applied to latitude of the point of land; if otherwise, to yesterday's latitude.

---

* Nearly the entire process of computing the Day's Work has already been given, and if the learner has thoroughly mastered the rules laid down in the preceding pages, he will find no difficulty in working the Day's Work without reference to them.

NOTE.—When the course is less than 5 points or 56°, the difference of longitude may be found by either or both *Middle Latitude* or *Mercator*'s method, but if the course exceeds 5 points the method of *Middle Latitude* should be used in preference to *Mercator*'s (see *Remarks* in pages 204—205).

5°. To find the difference of longitude.—*By Middle Latitude Sailing.*
(a) *Find the* middle latitude *as directed,* Rule XLV, page 108.
(b) *Next at the page of Traverse Table on which the* degrees *(at top or bottom) correspond to* middle latitude, *find the* departure *in a* difference of latitude column, *then the corresponding* distance *is the* difference of longitude *of the same name as the* departure (see Rule LXX, 4°, and note, page 197.)
(c) Or thus, by calculation:—*To* log. sec. *of middle latitude* add *log. of dep., the sum (rejecting 10 from the index) is the log. of diff. of long.*

When the latitude *left* and latitude *in* are of *contrary* names, that is, in low latitudes, no sensible error can arise from taking the departure itself as the difference of longitude.

6°. *If the ship has made a due* E. *or due* W. *course good, the* difference of longitude *is found by* Parallel Sailing, *thus:*—
With the latitude *as a* course *and the* departure *in a* difference of latitude column, *then the corresponding* distance *is the* difference of longitude (see Rule LXVII, page 194).

7°. To find the difference of longitude.—*By Mercator's Sailing.*
(a) *Find* meridional difference *of latitude* (see Rule XLIII, page 107).
(b) *Then with* course *and* meridional difference of latitude *(in a* latitude column*), find the corresponding* departure, *which is the* difference of longitude (see Rule LXXII, page 203).
(c) **To find the long. in.**—*With the* longitude left *and* difference of longitude *find the* longitude in (see Rule XLVII, page 110).

When a departure has been taken the longitude left is that of the point of land; otherwise that of yesterday.

EXAMPLE I.

H.	COURSES.	K.	$\frac{1}{10}$	WINDS.	LEE-WAY.	DEVIATION.	REMARKS, &c.
					pts.		
1	S.S.E. ½ E.	4	2	East.	2	7° E.	A point of land in lat. 42° 12′ S., long. 42° 58′ W., bearing by compass E. by N. ½ N. dist. 21 miles. Ship's head S.S.E. ½ E.; deviation as per log.
2		4	3				
3		5					
4		5	2				
5		4					
6	N.N.E.	4	1	East.	2¼	9° E.	
7		3	8				
8		3	5				
9		3	2				
10	S.W. ½ W.	3	5	W.N.W.	1¾	7½° W.	
11		3	6				
12		4					
1		4	2				
2	N. ¼ E.	4	3	W.N.W.	2½	1° W.	Variation 20° W.
3		4	4				
4		4	5				
5	S.S.W.	6	2	West.	½	2½° W.	
6		6	4				
7		6	2				
8		6	5				A current set W.S.W correct magnetic 26 miles from the time the departure was taken to the end of the day.
9	N. by W. ½ W.	6	2	West.	¾	10° W.	
10		5	7				
11		5	3				
12		5	4				

The Departure Course.

The opposite point to E. by N. ½ N. is W. by S. ½ S., and the ship's head being S.S.E. ½ E., the deviation is same as given in log. for S.S.E. ½ E., viz., 7° E.

W. by S. ½ S. = 6½ pts. R. of S.
or 73° 8' R. of S.
Deviation 7° R. } 13 L.
Variation 20 L.

True course 60 8 R. of S.
or S. 60° W., distance 21 miles.

This is inserted in the Traverse Table as 1st course.

2nd Course, N.N.E.

The deviation for N.N.E. is 9° E.

	H.	K.
N.N.E. = 2 pts. R. of N.	6	4·1
Leeway 2¼ ,, L.	7	3·8
	8	3·5
0¼ ,, L. of N.	9	3·2
or N. 2° 49' E. of N., L. of N.	14·6

Dev. 9°R. } 11 L.
Var. 20 L.

True course 13 49 L. of N.
or N. 14° W., distance 14·6 miles.

The distance, 14·6, is found by adding up the hourly distances from 6 o'clock until the course is changed at 10 o'clock.

4th Course, N. ¼ E.

The deviation for N. ¼ E. is 1° W.

	H.	K.
N. ¼ E. = ¼ pt. R. of N.		
Leeway 2½ ,, R.	2	4·3
	3	4·4
2¾ ,, R. of N.	4	4·5
or 30° 56' R. of N.	13·2

Dev. 1° L. } 21 0 L.
Var. 20 L.

True Course 9 56 R. of N.
or N. 10° E., distance 13·2 miles.

6th Course, N. by W. ½ W.

The dev. for N. by W. ½ W. is 10° W.

	H.	K.
N. by W. ½ W. = 1½ L. of N.	9	6·2
½ R.	10	5·7
	11	5·3
⅝ L. of N.	12	5·4
or 8° 26' L. of N.	22·6

Dev. 10° L. } 30 0 L.
Var. 20 L.

True course 38 26 L. of N.
or N. 38° W., distance 22'·6.

1st Course, S.S.E. ½ E.

	H.	K.
S.S.E. ½ E. = 2½ pts. L. of S.	1	4·2
Leeway 2 ,, R.	2	4·3
	3	5·0
0½ ,, L. of S.	4	5·2
	5	4·0
or 5° 38' L. of S.		

Dev. 7° R. } 13 L.
Var. 20 L.   22·7

True course 18 38 L. of S.
or S. 19° E., distance 22·7 miles.

The distance 22·7, is found by adding up the hourly distances sailed, until the course is altered at 6 o'clock. Insert this course and distance as 2nd course.

3rd Course, S.W. ½ W.

The deviation for S.W. ½ W. is 7½° W.

	H.	K.
S.W. ½ W. = 4½ pts. R. of S.	10	3·5
Leeway 1¾ ,, L.	11	3·6
	12	4·0
2¾ ,, R. of S.	1	4·2
or 30° 56' R. of S.		

Dev. 7°30'L } 27 30 L.
Var. 20 0 L.   15·3

True course 3 26 R. of S.
or S. 3° W., distance 15·3 miles.

Distance, 15·3, is found by adding up hourly distances from 10 o'clock until 2.

5th Course S.S.W.

The deviation for S.S.W. is 2½° W.

	H.	K.
S.S.W. = 2 pts. R. of S.	5	6·2
Leeway ½ ,, L.	6	6·4
1½ ,, R. of S.	7	6·2
	8	6·5
or 16° 53' R. of S.		

Dev. 2° 30' L. } 22 30 L.
Var. 20 0 L.   25·3

True Course 5° 37' L. of S.
or S. 6° E., distance 25·3 miles.

Current Course, W.S.W.

W.S.W. = 6 pts. R. of S.
or 67° 30' R. of S.
Deviation 20 0 L.

47 30 R. of S.
or S. 48° W., distance 26'.

The corrected courses are written down to the nearest degree, and the work will stand as follows:—

Courses.	Dist.	N.	S.	E.	W.
S. 60° W.	21		10·5		18·2
S. 19 E.	22·7		21·5	7·4	
N. 14 W.	14·6	14·2			3·5
S. 3 W.	15·3		15·3		0·8
N. 10 E.	13·2	13·0		2·3	
S. 6 E.	25·3		25·2	2·6	
N. 38 W.	22·6	17·8			13·9
S. 48 W.	26		17·4		19·3
		45·0	89·9	12·3	55·7
			45·0		12·3
			44·9		43·4

Difference latitude 44·9 } give in Table II { Course S. 44° W.*
Departure 43·4 } { Distance 62½ miles.

Latitude left 42° 12′ S. }
Diff. latitude 45 S. } Rule XLIV. page 108.

Latitude in 42 57 S. }

Meridional parts 2798 } Rule XLIII. page 127.
Meridional parts 2859 }

Mer. diff. lat. 61 }

Sum 2)85 9

Middle lat. 42 34

* The course being less than 45°, the difference of longitude may be found both by middle latitude and Mercator's method.

Course S. 44° W. } give in Table II { Difference of longitude 59·0
Mer. diff. lat. 61·1 } { (in departure column).

Mid. latitude 42½° } give in Table II { Difference of longitude 59′
Dep. 43·4 (as d. lat.) } { (in distance column).

Longitude left 32° 58′ W. }
Diff. longitude 0 59 W. } Rule XLIV. page 108.

Longitude in 43 57 W. }

Previous to opening the Traverse Table to take out the difference of latitude and departure to each course and distance in the above table, fill up the columns not wanted: thus, in the first course, S. 60° W., the S. and W. will be wanted, and the N. and E. will not be wanted; fill up these last two columns by drawing a dash under N. and E. Proceed in the same manner with the other courses.

2. *To find the difference of latitude and departure* to each course and distance by the Traverse Table.

Enter Traverse Table, and take out the difference of latitude and departure corresponding to 60° and distance 21′. Insert them in the columns S. and W.

The second course is S. 19° E., and the distance 22·7. Then, 19 degrees and distance 227 (omitting the decimal point) give difference of latitude 214·6, departure 73·9; now dropping the tenths in each, namely, the 6 and the 9, and shifting the decimal point one place to the left, we have difference of latitude 21·5, departure 7·4, which insert in columns S. and E., the course being marked S. and E.

The third course is N. 14° W., and distance 14·6. Look for 14 degrees and distance 14·6, which give difference of latitude 141·7, departure 35·3; now dropping the tenths, the 7 and the 3, and increasing the preceding figure by 1, in the first case, as the tenths exceed 5, we have, by removing the decimal point one figure to the left, the difference of latitude 142, and departure 3·5.

Proceed in this way with the remaining courses.

Next we find the sum of the four columns, when it appears the ship has sailed 45·0 N., and 89·9 S.; therefore, upon the whole, the difference of latitude is 44·9 S. The sum of the eastings is 12·3, of the westings 55·7, and the departure made good is 43·4 W.

3. *To find the Course and Distance made good.*—The difference of latitude is 44·9 and departure 43·4 found to correspond in their columns, give course S. 44° W., distance $62\frac{1}{2}$ miles (see Rule LXIV, page 186).

4. We next apply the difference of latitude, 45′ S. (44·9), to the latitude left, 42° 12′ S., (the latitude of point of land), taking the *sum*, as they are of the same name, and the latitude 42° 57′ S., takes the name of either (Rule XLIV, page 108).

5. *To find the Difference of Longitude.*—Take out the meridional parts for latitude left, 42° 12′, and also for latitude in, 42° 57′, and take the less from the greater, as the latitudes are of one name. The remainder is meridional difference of latitude (XLIII, page 107).

Or, find middle latitude by adding together latitude left and latitude in, and divide the sum by 2; the quotient is the middle latitude (Rule XLV, page 108).

Then the course 44°, in Table II, and meridional difference of latitude 61′, found in difference of latitude column, gives in departure column 59′, or difference of longitude 59′ (Rule LXXII, 2°, page 203).

Or, the middle latitude $42\frac{1}{2}$° in Table II, and departure 43·4 in difference of latitude column, gives in distance column 59′, the difference of longitude (Rule LXX, 4°, page 197).

Thus:—Mid. lat.	42°	and dep.	43′·4	give in dist. column	$58\frac{1}{2}$
and	" 43	"	43·4	"	$59\frac{1}{2}$
					2)118
				Diff. long.	59

The difference of longitude 59′ W. (that found by Mercator's sailing), added to longitude left 42° 58′ W., gives longitude in 43° 57′ W. (Rule XLVII, page 110).

EXAMPLE II.

H.	Courses.	K.	$\frac{1}{10}$	Winds.	Lee-way.	Deviation.	Remarks, &c.
					pts.		
1	S. by W.	4	1	W. by S.	$2\frac{1}{4}$	0°	A point of land in lat. 62° 18′ N., long. 85° 17′ E., bearing by compass N. by E. $\frac{3}{4}$ E., 16 miles. Ship's head S. by W. Deviation as per log.
2		3	9				
3		4	0				
4		4	0				
5	S.W. $\frac{3}{4}$ W.	3	5	S. by E.	$3\frac{1}{2}$	8° W.	
6		3	4				
7		3	1				
8		3	0				
9	E. $\frac{3}{4}$ S.	5	4	S. by E.	$1\frac{3}{4}$	15° E.	
10		5	6				
11		5	4				
12		5	6				
1	W.N.W.	4	4	North.	3	$18\frac{1}{2}$° W.	
2		4	4				Variation 42° E.
3		4	2				
4		5	0				
5	N.W. $\frac{1}{2}$ N.	9	6	S. by W.	0	$16\frac{1}{2}$° W.	
6		10	2				
7		11	4				
8		11	8				
9	E. $\frac{3}{4}$ N.	3	4	N. by E.	$3\frac{1}{4}$	$17\frac{1}{4}$° E.	
10		3	2				A current set the ship W.S.W. (correct magnetic), 23 miles.
11		3	0				
12		2	4				

W. N. E. S. W. N.

The Departure Course.

The opposite point to N. by E. ¼ E. is S. by W. ¼ W., and the ship's head being S. by W., the deviation is the same as given in the log, for S. by W.

S. by W. ¼ W. = 1¼ pts. R. of S.

or 14° 4′ R. of S.
Var. 42° R. } 42 0 R.
Dev. 0

True Course 56 0 R. of S.
or S. 56° W., distance 16 miles.

This is inserted in the Traverse Table as 1st course.

1st Course, S. by W.

The deviation for S. by W. is 0.

		H.	K.
S. by W. =	1 pt. R. of S.	1	4·1
Leeway	2¼ „ L.	2	3·9
	———	3	4
	1¼ „ L. of S.	4	4
			16

or 14° 4′ L. of S.
Dev. 0 } 42 0 R.
Var. 42° R.

True Course 28 0 R. of S.
or S. 28° W., distance 16′.

The distance 16·6 is found by adding up the hourly distances until the course is changed at 5 o'clock. Insert this course and distance in Traverse Table as the 2nd course.

2nd Course, S.W. ¾ W.

The deviation for S.W. ¾ W. is 8° W.

		H.	K.
S.W. ¾ W. =	4¾ pts. R. of S.	5	3·5
Leeway	3½ „ R.	6	3·4
	———	7	3·1
Sum exc. 8 pts.	8¼ „ R. of S.	8	3
Subtract from	16		
	———		
	7¾ „ L. of N.		13

or 87° 11′ L. of N.
Dev. 8° L. } 34 0 R.
Var. 42 R.

True Course 53 0 L. of N.
or N. 53° W., distance 13′.

The distance is obtained by adding up the hourly distances sailed from 5 o'clock until the course is changed at 9 o'clock.

3rd Course, E. ¾ S.

The deviation for E. ¾ S. is 15° E.

		H.	K.
E. ¾ S. =	7¼ pts. L. of S.	9	5·4
Leeway	1¾ „ L.	10	5·6
	———	11	5·4
Sum exc. 8 pts.	9 „ L. of S.	12	5·6
Subtract from	16		
	———		
	7 „ R. of N.		22·0

or 79° R. of N
Dev. 15° R. } 57 R. of N.
Var. 42 R.

Sum exc. 90° 136 R. of N.
Subtract from 180

True Course 44 L. of S.
or S. 44° E., distance 22′.

Hourly distances sailed from 9 o'clock until course is changed at 1 o'clock being added up, give the distance to enter in the Traverse Table.

4th Course, W.N.W.

The deviation for W.N.W. is 18½° W.

		H.	K.
W.N.W. =	6 pts. L. of N.	1	4·4
Leeway	3 „ L.	2	4·4
	———	3	4·2
Sum exc. 8 pts.	9 „ L. of N.	4	5
Subtract from	16		
	———		
	7 „ R. of S.		18·0

or 78° 45′ R. of S.
Dev. 18°30′ L. } 23 30 R.
Var. 42 R.

Sum. exc. 90° 102 R. of S.
Subtract from 180

True Course 78 L. of N.
or N. 78° W., distance 18′.

The distance is found by adding up hourly distances sailed from 1 o'clock until the course is changed at 5 o'clock.

5th Course, N.W. ½ N.

The deviation for N.W. ½ N. is 16½° W.

		H.	K.
N.W. ½ N. =	3½ pts. L. of N.	5	9·6
Leeway	0	6	10·2
	———	7	11·4
	3½ „ L. of N.	8	11·8

or 39° 22′ L. of N.
Dev. 16°30′ L. } 25 30 R.
Var. 42 R.   43·0

True Course 13 52 L. of N.
or N. 14° W., distance 43′.

Add up hourly distances sailed from 5 o'clock until course is changed at 9 o'clock.

6th Course, E. ¾ N.

The deviation for E. ¾ N. is 17½° E.

```
E. ¾ N. =        7½ pts. R. of N.   II.   K.
Leeway           3½  ,,  R.          9   3·4
                 ———                10   3·2
Sum exc. 8 pts. 10½ ,, R. of N.     11   3·0
Subtract from   16                  12   2·4
                ———                     ———
                 5½ ,, L. of S.         12·0
           or 61° 53′ L. of S.
Dev. 17° 15′ R }
Var. 42        }  59  15  R.
True Course  2 38 L. of S.
   or S. 3° E., distance 12′.
```

Current Course, W.S.W.

```
W.S.W. =         6 pts. R. of S.
              or 67° 30′ R. of S.
Variation       42  R.
Sum exc. 90°   109 30 R. of S.
Subtract from  180
               ————
True Course   70 30 L. of N.
   or N. 71° W., distance 23′.
```

The corrected courses are written down to the nearest degree and the work will stand as follows:—

Courses.	Dist.	N.	S.	E.	W.
S. 56° W.	16		8·9		13·3
S. 28 W.	16		14·1		7·5
N. 53 W.	13	7·8			10·4
S. 44 E.	22		15·8	15·3	
N. 78 W.	18	3·7			17·6
N. 14 W.	43	41·7			10·4
S. 3 E.	12		12·0	0·6	
N. 71 W.	23	7·5			21·7
		60·7	50·8	15·9	80·9
		50·8			15·9
		9·9			65·0

Course N. 81° W., distance 66 miles.

Diff. lat. 9·9 and dep. 65·0 being found to correspond in their columns, give course N. 81° W., distance 66 miles. Course exceeds 56°, diff. long. must be found by middle latitude sailing.

```
Lat. left   62° 18′ N. }  See Rule XLIV., p. 108
Diff. lat.     10  N.  }  & Rule XLIV., p. 108
Lat. in     62  28  N. }
Sum        124  46     }
Mid. lat.   62  23     }
```

The mid. lat. is high and between two whole degrees, therefore we proceed thus:—

Mid. lat. 62° as course (in Table II), and dep. 64·8 (nearest to 65·0) as diff. lat., give in dist. column 138; and mid. lat. 63° and dep. 64·9 (nearest to 65·0) as diff. lat., give in dist. column 143: whence it is evident that for 1° (or 60′) of mid. lat., the diff. long. increases 5′: we next make the proportion

```
Mid. lat. 62° gives D. long. 138
Correction for 23′             2
                             ———
∴ Mid. lat. 62° 23′ gives D. long. 140
```

```
60′ : 23′ :: 5 : x
            5
         ———
       6,0)11,5
```

2 nearly, the correction for 23′.

By Calculation.

```
Mid. lat. 62° 23′   sec.  0·333900
Dep. 65′            log.  1·812913
                         ————————
Diff. long. 140′·2  log.  2·146813
   or 2° 20′·2.
```

```
Long. left  85° 17′ E. }
Diff. long.  2  20  W. }  Rule XLVII., 2°, page 110.
Long. in    82  57  E. }
```

W. N. E. S. W. N.

Departure Course.

The opposite point to bearing N. by W. ½ W. is S. by E. ½ E.
S. by E. ½ E. = 1½ pts. L. of S.
or 16° 53′ L. of S.
Dev. 19° R. }
Var. 14 R. } 33 R.
16 7 R. of S.
or S. 16° W., distance 11′.

1st Course, E. by S., dev. 19° E.
E. by S. = 7 pts. L. of S.
Leeway = ¼ „ L.
7¼ „ L. of S.
or 81° 34′ L. of S.
Dev. 19° R. }
Var. 14 R. } 33 0 R.
True course 48 34 L. of S.
or S. 49° E., distance 56′·6.

2nd Course, E. ½ S., dev. 20° E.
E. ½ S. = 7½ pts. L. of S.
Leeway = ½ „ L.
7¾ „ L. of S.
or 87° 11′ L. of S.
Dev. 20° R. }
Var. 14 R. } 34 0 R.
53 11 L. of S.
or S. 53° E., distance 61′·1.

3rd Course, E.S.E., dev. 18° E.
E.S.E. = 6 pts. L. of S.
Leeway = ¾ „ L.
6¾ „ L. of S.
or 75° 56′ L. of S.
Dev. 18° R. }
Var. 14 R. } 32 0 R.
43 56 L. of S.
or S. 44° E., distance 68′·7.

4th Course, N.E. by E., dev. 19° E.
N.E. by E. = 5 pts. R. of N.
Leeway = ½ „ R.
5½ „ R. of N.
or 61° 53′ R. of N.
Dev. 19° R. }
Var. 14 R. } 33 0 R.
Sum exc. 90° 94 53 R. of N.
Subt. from 180 0
True course 85 7 L. of S.
or S. 85° E., distance 68′·0.

5th Course, S.E. ½ E., dev. 15° E.
S.E. ½ E. = 4½ pts. L. of S.
Leeway = ¾ „ L.
5 „ L. of S.
or 56° 15′ L. of S.
Dev. 15° R. }
Var. 14 R. } 29 0 R.
True course 27 15 L. of S.
or S. 27° E., distance 24′·7.

6th Course, N. by E., dev. 2° E.
N. by E. = 1 pt. R. of N.
Leeway = 5¾ „ L.
4¾ „ L. of N.
or 53° 26′ L. of N.
Dev. 2° R. }
Var. 14 R. } 16 0 R.
37 26 L. of N.
or N. 37° W., distance 4′·7.

Current Course.
N.W. by N. = 3 pts. L. of N.
or 33° 45′ L. of N.
Var. 14 0 R.
19 45 L. of N.
or N. 20° W., distance 13′.

EXAMPLE III.

H	Courses.	K	$\frac{1}{10}$	Winds.	Leeway	Deviation.	Remarks, &c.
					pts		
1	E. by S.	10	6	S. by E.	¼	19° E.	A point of land in lat. 47° 44′ S., long. 170° 7′ E., bearing by compass N. b W. ½ W. dist. 11 miles. Ship's head E. by S. Dev. as per log.
2		11	4				
3		11	6				
4		11	4				
5		12					
6	E. ½ S.	12		S. by E. ½ E.	½	20° E.	
7		12	3				
8		12	4				
9		12					
10		12	3				
11	E.S.E.	13	4	South.	¾	18° E.	
12		13	4				
1		13	6				Variation 14° E.
2		14	3				
3		14	3				
4	N.E. by E.	13	8	N. by W.	½	19° E.	
5		13	8				
6		13	5				
7		13	5				
8		13	4				A current set the ship N.W. by N. correct magnetic 13 mls. from the time the departure was taken to the end of the day.
9	S.E. ½ E.	12	5	SSW ½ W	¾	15° E.	
10		12	5				
11	N. by E.	2	4	E.N.E.	5¾	2° E.	
12		2	6				

Courses.	Dist.	N.	S.	E.	W.
S. 16° W. ..	11		10·6		3·0
S. 49° E. ..	57		37·1	42·7	
S. 53° E. ..	61		36·8	48·8	
S. 44° E. ..	69		49·4	47·7	
S. 85° E. ..	68		5·9	67·7	
S. 27° E. ..	25		21·8	11·1	
N. 37° W. ..	5	3·8			2·8
N. 20° W. ..	13	12·2			4·4
		16·0	161·6	218·0	10·2
			16·0	10·2	
			145·6	207·8	

Diff. lat. 145·6 and dep. 207·8, found to correspond in the columns, give course S. 55° E., and distance 254 miles (see Rule LXIV, page 186).

Lat. left 47° 44′ S.
Diff. lat. 2 26 S.
Lat. in 50 10 S
Sum 2)97 54
Mid. lat. 48 57
} Rule XLIV, page 108, & Rule XLV page 108.

Mid. lat. 49° as course in Table II, and half of the dep. 101·9 (the whole dep. being too large a number to be found in the Table) gives in distance column 158 which multiplied by 2 (as only half the dep. was used in entering the Table) gives diff. long. 316 miles.

Mid lat. 48° 57′ sec. 0·182621
Dep. 207·8 log. 2·317645
D. long. 316′·4 log. 2·500266
or 5° 16′·4

Long. left 179° 7′ E.
D. long. 316′ 5 16 E.
Sum exc. 180° 184 23 E.
Subt. from 360 0
Long. in 175 37 W.
} Rule XLVII, page 110.

W. N. E. S. W. N.

Departure Course.

Opposite to bearing E. by S. ½ S.
6½ pts. L. of S.
or 73° 8′ L. of S.
Dev. 17° 45′ R. } 3 30 L.
Var. 21 15 L. }
75 38 L. of S.
or S. 77° E., distance 15′.

1st Course, E. by N.
7 pts. R. of N.
Leeway ¼ „ L.
6¾ „ R. of N.
or 75° 56′ R. of N.
Dev. 17° 45′ R. } 3 30 L.
Var. 21 15 L. }
72 26 R. of N.
or N. 72° E., distance 49′.

2nd Course, E.S.E.
6 pts. L. of S.
Leeway ½ „ L.
6½ „ L. of S.
or 73° 8′ L. of S.
Dev. 13° 30′ R. } 7 45 L.
Var. 21 15 L. }
80 53 L. of S.
or S. 81° E., distance 42′.

3rd Course, N.E. by E.
5 pts. R. of N.
Leeway 1 „ L.
4 „ R. of N.
or 45° R. of N.
Dev. 17° 15′ R. } 4 L.
Var. 21 15 L. }
41 R. of N.
or N. 41° E., distance 33′.

4th Course, S.S.E.
2 pts. L. of S.
Leeway 1¼ „ R.
¾ „ L. of S.
or 8° 26′ L. of S.
Dev. 5° 30′ R. } 15 45 L.
Var. 21 15 L. }
24 11 L. of S.
or S. 24° E., distance 29′.

5th Course, S.E. by S.
3 pts. L. of S.
Leeway 2 „ R.
1 „ L. of S.
or 11° 15′ L. of S.
Dev. 8° 30′ R. } 12 45 L.
Var. 21 15 L. }
24 0 L. of S.
or S. 24° E., distance 22′.

6th Course, E.S.E.
6 pts. L. of S.
Leeway 2¼ „ R.
3¾ „ L. of S.
or 42° 11′ L. of S.
Dev. 13° 30′ R. } 7 45 L.
Var. 21 15 L. }
49 56 L. of S.
or S. 50° E., distance 19′.

Current Course, S.S.W. ½ W.
2½ pts. R. of S.
or 28° 7′ R. of S.
Var. 21 15 L.
6 52 R. of S.
or S. 7° W., distance 18′.

EXAMPLE IV.

H	Courses.	K	$\frac{1}{10}$	Winds.	Leeway. pts	Deviation.	Remarks, &c.
1	E. by N.	12	4	S.E. by S.	¼	17¾° E.	A point, Tynemouth in lat. 55° 1′ N., long. 1° 25′ W., bearing by compass W. b N. ½ N. dist. 15 miles. Ship's head E. by N. Dev. as per log.
2		12	2				
3		12	2				
4		12	2				
5	E.S.E.	10	6	South.	½	13½° E.	
6		10	5				
7		10	4				
8		10	5				
9	N.E. by E.	8	2	S.E. by E.	1	17¼° E.	
10		8	3				
11		8	3				
12		8	2				
1	S.S.E.	7	4	East.	1¼	5½° E.	Variation 21¼° W.
2		7	2				
3		7	2				
4		7	2				
5	S.E. by S.	5	8	E. by N.	2	8½° E.	
6		5	6				
7		5	4				
8		5	2				
9	E.S.E.	5	4	N.E.	2¼	13½° E.	A current set the ship S.S.W. ½ W. correct magnetic 18 mls. from the time the departure was taken to the end of the day.
10		4	6				
11		4	5				
12		4	5				

Courses.	Dist.	N.	S.	E.	W.
S. 77° E. ..	15		3·4	14·6	
N. 72 E. ..	49	15·1		46·6	
S. 81 E. ..	42		6·6	41·5	
N. 41 E. ..	33	24·9		21·6	
S. 24 E. ..	29		26·5	11·8	
S. 24 E. ..	22		20·1	8·9	
S. 50 E. ..	19		12·2	14·6	
S. 7 W. ..	18		17·9		2·2
		40·0	86·7	159·6	2·2
			40·0	2·2	
			46·7	157·4	

Diff. lat. 46′·7 } give in Table II { Course S. 73½° E.
Departure 157·4 } { Dist. 164½′

Lat. left. 55° 1′ N.
Diff. lat. 47 S.
Lat. in 54 14 N.
2)109 15
Mid. lat. 54 37
} Rules XLIV and XLV, page 108.

To find D. long. (1) *By Calculation*

Lat. 54° 37′	sec.	0·237288
Dep. 157·4	log.	2·197005
D. long. 271·8	log.	2·434293

Mid. lat. 54° and dep. 157′·5 in diff. lat. column (the nearest in Table to 157·4), gives in dist. column 268 for diff. long.; and mid. lat. 55° and dep. 157′·7 in diff. lat. column give in dist. column diff. long. 275; whence it is evident that for 1° change of mid. lat. we have (275 − 268) = 7′ change in diff. long., thus:—60 : 37 :: 7 : 4·3.

Mid. lat. 54°, dep. 157 give D. long.	268	Long. left	1° 25′ W.
Correction for 37′	+ 4·3	D. long. 272′ or	4 32 E.
∴ Mid. lat. 54° 37′, dep. 157·4 give	272·3	Long. in	3 7 E.

} Rule 47, page 110

The Day's Work.

EXAMPLE V.

```
W. N. E, S. W. N.
Departure Course.
The opposite point to N.E. by N.
is S.W. by S.
S.W. by S.    = 3 pts. R. of S.
         or 33° 45' R. of S.
Dev. 11°L. ⎫
Var. 25 L. ⎬ 36 0 L.
True Course   2 15 L. of S.
or S. 2° E., distance 17 miles.
     1st Course.
W. by N.   = 78° 45' L. of N.
Leeway 0
Dev. 11° L. ⎫
Var. 25 L.  ⎬ 36 0 L.

Sum exc. 90°  115  0
Subtract from 180  0

True Course  65  0 R. of S.
or S. 65° W., distance 25 miles.
     2nd Course, W.S.W.
W.S.W.     = 6 pts. R. of S.
Leeway(port tack) ½ " R.

            6½ " R. of S.
         or 73° R. of S.
Dev. 9° L. ⎫
Var. 25 L. ⎬ 34 L.

True Course  39 R. of S.
or S. 39° W., distance 22 miles.
     3rd Course, W.N.W.
W.N.W.     = 6 pts. L. of N.
Leeway(port tack) 1 " R.

            5 " L. of N.
         or 56° L. of N.
Dev. 9° L. ⎫
Var. 25 L. ⎬ 34 L.

True Course  90 L. of N.
or West, distance 19 miles.
     4th Course, S.W.
S.W.       = 4 pts. R. of S.
Leeway (star. tack)2¼ " L.

            2¼ " R. of S.
         or 31° R. of S.
Dev. 6° L. ⎫
Var. 25 L. ⎬ 31 L.

            0
or South, distance 16 miles.
     5th Course, S.W. by W.
S.W. by W. = 5 pts. R. of S.
Leeway(star. tack)1½ " R.

            3½ " R. of S.
         or 37° R. of S.
Dev. 8° L. ⎫
Var. 25 L. ⎬ 33 L.

True Course  4 R. of S.
or S. 4° W., distance 13 miles.
     6th Course, South.
South     = 0 pts.
Leeway (star. tack) 2¼ " L.

            2¼ " L. of S
         or 25° L. of S.
Dev. 0° ⎫
Var. 25 L. ⎬ 25 L.

True Course  50 L. of S.
or S. 50° E., distance 13 miles.
Current Course, N.W. ¾ W.
N.W. ¾ W.  = 53° L. of N.
Variation      25 L.

True Course  78 L. of N.
or N. 78° W., distance 6 miles.
```

H	Courses.	K	½	Winds.	Leeway	Deviation.	Remarks, &c.
1	W. by N.	6	3	E.S.E.	0	11° W.	A point, Lizard, in lat. 49° 58' N., long. 5° 12' W., bearing by compass N.E. by N., distance 17 miles. (Ship's head W. b N) Dev. as per log.
2		6	3				
3		6	4				
4		6					
5	W.S.W.	5	6	S.	½	9° W.	
6		5	5				
7		5	5				
8		5	4				
9	W.N.W.	5		S.W.	1	9° W.	
10		4	8				
11		4	6				
12		4	6				
1	S.W.	4		W.N.W.	1¼	6° W.	Variation 25° W.
2		4	4				
3		4	2				
4		3	8				
5	S.W. b W.	3	6	N.W. b W.	1¾	8° W.	
6		3	4				
7		3					
8		3					
9	S.	3	3	W.S.W.	2¼	0°	A current set (correct magnetic) N.W. ¾ W., 6 miles, from the time the departure was taken to the end of the day.
10		3	3				
11		3	2				
12		3	2				

Courses.	Dist.	N.	S.	E.	W.
S. 2° E.	17		17·0	0·6	
S. 65° W.	25		10·6		22·7
S. 39° W.	22		17·1		13·8
West	19				19·0
South	16		16·0		
S. 4° W.	13		13·0		0·9
S. 50° E.	13		8·4	10·0	
N. 78° W.	6	1·2			5·9
		1·2	82·1	10·6	62·3
			1·2		10·6
			80·9		51·7

Diff. lat. 80'·9 S. } give in Table II { Course S. 32½° W.
Departure 51'·7 W. }   { Distance 95 miles.

Lat. left   49° 58' N. ⎫
Diff. lat.   1 21 S.   ⎬ Rules XLIV & XLV, page 108.
Lat. in.   48 37 N.   ⎭
Sum   2)98 35
Middle lat. 49 17   Mer. parts 3471 ⎫
                                       Mer. parts 3347 ⎬ Rule XLIII, page 107.
Course S. 32½° W.   Mer. diff. lat. 124 ⎭
Mer. diff. lat. 124 } give in Table II { Diff. long. 79'
(D. lat. col.)   }   { (In departure column).

Mid. lat. 49°   } give in Table II { Diff. long. 79'
Dep 51'·8(as diff.lat.)   { (In distance column)

Longitude left 5° 12' W. ⎫
Diff. longitude 1 19 W. ⎬ Rule XLVII, page 110.
Longitude in   6 31 W. ⎭

EXAMPLE VI.

H.	Courses.	K.	1/10	Winds.	Lee-way.	Deviation.	Remarks, &c.
1	N.E. ½ E.	6		N.N.W.	2¼	16¼° E.	A point of land in lat. 47° 35′ S., long. 179° 26′ E., bearing by compass S.E. ¾ E., dist. 14 miles. Ship's head N.E. ½ E. Dev. as per log.
2		6					
3		6	6				
4		6	4				
5	N.E. by E. ½ E.	5	7	S.E. ½ E.	2¼	17¾° E.	
6		5	8				
7		6	3				
8		6	2				
9	S.E. ¾ E.	12		N.E. by N.	0	11° E.	
10		12	4				
11		12					
12		11	6				Variation 25° E.
1	South.	4	6	E.S.E.	2	2¼° W.	
2		4	6				
3		4	6				
4		5	2				
5	N.W.bW.½W.	4	4	S.W. ½ W.	1¾	18½° W.	
6		4	5				A current set the ship N.E. by E. ¾ E., correct magnetic, 36 miles, from the time the departure was taken to the end of the day.
7		4	5				
8		4	6				
9	N.W. ½ W.	12	6	S.W. b W. ½ W.	¼	18° W.	
10		12	5				
11		12	4				
12		12	5				

Courses.	Dist.	N.	S.	E.	W.
N. 12° W.	14	13·7			2·9
S. 63 E.	25		11·3	22·3	
N. 79 E.	24	4·6		23·6	
S. 17 E.	48		45·9	14·0	
S. 45 W.	19		13·4		13·4
N. 36 W.	18	14·6			10·6
N. 41 W.	50	37·7			32·8
East	36			36·0	
		70·6	70·6	95·9	59·7
		70·6		59·7	
				36·2	

Having filled up the Traverse Table, the sum of the Northings and Southings are equal, consequently the latitude remains unaltered, or, the ship, after sailing the foregoing courses and distances, has returned to the same parallel. Altogether, the vessel has sailed 95′·9 towards the east on four courses, while she has made 59′·7 Westing on the other four, leaving 36′·2 of progress towards the East; hence

The Course is East, distance 36′·2 (see No. 161, page 104).

To find the Latitude and Longitude in.

The ship not having altered her latitude, the latitude arrived at is the same as the latitude left, viz., 47° 35′ S., and consequently the diff. of long. made good is to be found by Parallel Sailing, Rule LXVII, page 194, thus:—

Lat. 47½° (as course) and Dep. (in diff. lat. column) } give in Table II { Diff. long. 53½′. (in distance column).

Long. left	179° 26′ E.	Rule XLVII, page 110.
Diff. long.	+ 53·5 E.	
Long. in	180 19·5 E.	
Subtract from	360 0	
Long. in	179 40·5 W.	

By Calculation.

Lat. 47° 35′	sec.	0·171007
Dep. 36′·2	log.	1·558709
Diff. long. 53·7	log.	1·729716

EXAMPLES FOR PRACTICE.

EXAMPLE I.

H.	Courses.	K.	$\frac{1}{10}$	Winds.	Lee-way.	Deviation.	Remarks, &c.
					pts.		
1	S.E. by S.	12	5	E. by N.	¼	3° E.	A point of land in lat. 35° 15' N., long. 75° 30' W., bearing by compass W. by N., dist. 19 miles. Ship's head S.E. by S. Deviation as per log.
2		12	5				
3		12	6				
4		12	4				
5	East.	9	8	N.N.E.	1	23° E.	
6		9	4				
7		9	4				
8		9	4				
9	N.E.	10	4	N.N.W.	¾	17° E.	
10		10	6				
11		10	4				
12		10	6				
1	North.	11		W.N.W.	¾	4° E.	Variation 15° W.
2		10	4				
3		10	2				
4		10	4				
5	N.N.E.	11		East.	½	13° E.	
6		10	4				
7		10	4				
8		10	2				
9	E.N.E.	9	5	North.	1	18½° E.	A current set N.E. by E. correct magnetic 52 miles from the time the departure was taken to the end of the day.
10		8	8				
11		9	4				
12		9	3				

EXAMPLE II.

H.	Courses.	K.	$\frac{1}{10}$	Winds.	Lee-way.	Deviation.	Remarks, &c.
					pts.		
1	East.	9	4	S.S.E.	¾	16° E.	A point, Flambro' Head, lat. 54° 7' N., long. 0° 5' W., bearing by compass N.W. b W. dist. 17 miles. Ship's head E.S.E. Deviation 13° E.
2		9	6				
3		9					
4	S.E. by E.	10	4	S. by W.	½	12° E.	
5		10	2				
6		10	4				
7	E. ½ S.	6	7	S. by E. ½ E.	1¼	15° E.	
8		6	6				
9		6	7				
10	S. by W.	5		S.E. by E.	2	0°	
11		4	8				
12		4	6				
1		4	6				Variation 25° W.
2	South.	4	4	E.S.E.	2½	2° E.	
3		4	4				
4		4	2				
5	S.E. by S.	3	5	E. by N.	2¾	8° E.	
6		3	5				
7		3					
8	E.N.E.	3		S.E.	3¼	18° E.	A current set (correct magnetic) N.N.E., 6 miles, from the time the departure was taken to the end of the day.
9		3					
10		3					
11		3					
12		3					

F F

EXAMPLE III.

H.	Courses.	K.	$\frac{1}{10}$	Winds.	Lee-way.	Deviation.	Remarks, &c.
					pts.		
1	N.N.E.	9	5	East.	½	8° E.	A point of land in lat. 43° 47′ N., long. 7° 51′ W., bearing by compass S.W. by S., dist. 13 miles. Ship's head N.N.E. Deviation as per log.
2		9	5				
3		9	6				
4		9	4				
5	E.N.E.	3	6	S.E.	2¾	15¼° E.	
6		4	4				
7		5					
8		5					
9	E.S.E.	6		South.	2	13½° E.	
10		7					
11		6	4				
12		5	6				
1	W. by N. ½ N.	6		S.W. ½ S.	1½	15¾° W.	Variation 25° W.
2		5	6				
3		5	4				
4		6					
5	S.S.E.	5	6	S.W.	½	5½° E.	
6		6	2				
7		6	4				A current set the ship N. by W. correct magnetic, 21 miles, from the time the departure was taken to the end of the day.
8		7					
9	N.N.W.	6	5	West.	1	12° W.	
10		6	4				
11		7					
12		7	2				

EXAMPLE IV.

H.	Courses.	K.	$\frac{1}{10}$	Winds.	Lee-way.	Deviation.	Remarks, &c.
					pts.		
1	S.W. ½ W.	6	5	S. by E. ½ E.	2	14½° W.	A point of land in lat. 46° 12′ S., long. 2° 10′ W. bearing by compass E. by S. ½ S., 20 mls. (Ship's head W. by compass). Deviation 9° E.
2		6	2				
3		6	6				
4		6	7				
5	N. ¾ E.	8	2	E.N.E.	¼	20° E.	
6		8	2				
7		8	6				
8		9					
9	S. by E. ½ E.	6	3	S.W. ½ W.	2¾	20° W.	
10		5	5				
11		5	6				
12		4	6				
1	W. by S.	6	4	S. by W.	2¼	5° W.	
2		6	5				Variation 14° E.
3		5	6				
4		6	5				
5	E.N.E.	6		S.E.	2½	13½° E.	
6		6	4				
7		7	6				
8		6					
9	S.S.W. ½ W.	6	7	S.E.	1¾	18° W.	A current set the ship S.W. ¼ W. by compass 22½ miles these last 5 hours.
10		6					
11		5	6				
12		5	7				

EXAMPLE V.

H.	Courses.	K.	$\tfrac{1}{10}$s	Winds.	Lee-way.	Deviation.	Remarks, &c.
					pts.		
1	W.S.W.	9	4	N.W.	½	10° W.	A point, lat. 35° 10′ N. long. 5° 36′ W., bearing by compass E. by S. Ship's head N.N.E., dist. 9 miles. Deviation 9° E.
2		9	6				
3		9	4				
4		9	6				
5	North.	11	4	W.N.W.	¼	3° W.	
6		11					
7		11	2				
8		11	4				
9	N.W.	8	4	W.S.W.	¾	17° W.	
10		8	6				
11		8	4				
12		8	6				
1	S.W. by S.	11	7	W. by N.	½	5° W.	
2		11	5				Variation 23° W.
3		11	4				
4		11	4				
5	W.S.W.	9	6	N.W.	½	10° W.	
6		9	5				
7		9	4				
8		9					
9	East.	6	4	S.S.E.	1	15° E.	
10		6	4				A current set the ship (correct magnetic) S.E. by E., 15 miles.
11		6					
12		6					

EXAMPLE VI.

H.	Courses.	K.	$\tfrac{1}{10}$s	Winds.	Lee-way.	Deviation.	Remarks, &c.
					pts.		
1	N.N.W. ½ W.	3	5	N.E.	1¾	2° W.	A point, lat. 29° 59′ N. long. 32° 54′ E., bearing by compass N.N.E. ½ E. dist. 15 miles. Ship's head N.W. by W. Deviation 6° W.
2		4	2				
3		4	3				
4	E.S.E.	2	7	N.E.	2	7° E.	
5		3					
6		3	3				
7		4					
8	S. ¼ E.	5	4	E.S.E.	2¼	2° W.	
9		5					
10		5	5				
11		4	5				
12		4	6				
1	N.E. ¼ N.	4	7	E.S.E.	1½	8° E.	Variation 25° W.
2		4	2				
3		4	4				
4		3	7				
5		3					
6	W. ½ N.	3	5	S.S.W. ½ W.	1¼	9° W.	
7		4	3				A current set the ship (correct magnetic) N.E., 30 miles, from the time the departure was taken to the end of the day.
8		3	6				
9		3	6				
10	N. by E.	8	5	E. by N.	½	6° E.	
11		9	3				
12		9	2				

EXAMPLE VII.

H.	Courses.	K.	$\frac{1}{10}$	Winds.	Lee-way.	Deviation.	Remarks, &c.
1	N.N.W.	10	2	West.	pts. $\frac{3}{4}$	$19\frac{1}{4}°$ E.	A point, lat. 44° 20′ S., long. 176° 49′ W., bearing by compass E. by N. $\frac{1}{4}$ N., distance 18 miles. Ship's head N.N.W. Deviation as per log.
2		9	4				
3		9	4				
4		9					
5	West.	8	6	N.N.W.	1	$1\frac{1}{4}°$ W.	
6		8	4				
7		8	4				
8		8	6				
9	W. by S.	11	6	N.W. by N.	$\frac{1}{2}$	$4\frac{3}{4}°$ W.	
10		12	2				
11		11	8				
12		12	4				
1	S.S.W. $\frac{1}{2}$ W.	6	3	West.	$1\frac{1}{2}$	$17\frac{1}{4}°$ W.	Variation 15° E.
2		6	3				
3		6	4				
4		6					
5	South.	9	6	E.S.E.	$\frac{3}{4}$	$21°$ W.	
6		9	4				
7		9	5				
8		9	5				A current set the ship (correct magnetic) N. by E., 18 miles, from the time the departure was taken to the end of the day.
9	S. by E. $\frac{1}{2}$ E.	12	5	E. by S.	$\frac{1}{4}$	$20°$ W.	
10		12	6				
11		12	5				
12		12	4				

EXAMPLE VIII.

H.	Courses.	K.	$\frac{1}{10}$	Winds.	Lee-way.	Deviation.	Remarks, &c.
1	E.S.E.	12	4	N.E.	pts. $\frac{1}{4}$	$13°$ E.	A point, lat. 62° 18′ N. long. 63° 17′ W., bearing by compass W.N.W., distance 21 miles. Ship's head E.S.E. Deviation as per log.
2		12	5				
3		12	5				
4		12	6				
5	E. $\frac{1}{4}$ N.	4		N.N.E.	$3\frac{3}{4}$	$17°$ E.	
6		4	4				
7		4	3				
8		4	3				
9	E. $\frac{1}{4}$ S.	8	4	S.S.E.	$1\frac{1}{2}$	$13°$ E.	
10		8	5				
11		8	6				
12		8	5				
1	S.W. $\frac{1}{4}$ W.	3	5	S. by E.	$3\frac{3}{4}$	$8°$ W.	Variation 60° W.
2		3	5				
3		3	4				
4		3	6				
5	S. by W.	5	3	W. by S.	$2\frac{1}{4}$	$0°$	
6		5	3				
7		5	4				
8		5	5				A current set the ship (correct magnetic) E. by S. $\frac{1}{2}$ S., 49 miles, from the time the departure was taken to the end of the day.
9	W.N.W.	4	2	North.	3	$17°$ W.	
10		4	2				
11		4	2				
12		4	4				

EXAMPLE IX.

H.	Courses.	K.	$\frac{1}{10}$	Winds.	Lee-way.	Deviation.	Remarks, &c.
					pts.		
1	South.	5	3	E.S.E.	1¼	2° E.	A point, lat. 59° 49′ N., long. 43° 54′ W., bearing by compass N.E. ½ N., distance 14 miles. Ship's head South. Deviation as per log.
2		4	8				
3		4	5				
4		4	4				
5	N.E. ½ N.	6	6	E. by S. ½ S.	1	14° E.	
6		6	4				
7		6					
8		6					
9	S.S.W. ½ W.	5	5	S.E. ½ S.	1½	5° W.	
10		5	6				
11		5	4				
12		5	5				
1	E. ½ S.	8		S. by E. ½ E.	¾	17° E.	Variation 53° W.
2		8	4				
3		8	4				
4		8	2				
5	S.W. ½ S.	4	6	S.S.E. ½ E.	2	5° W.	
6		5	4				
7		4	4				A current set the ship (correct magnetic) S.E. ¼ E., 41 miles, from the time the departure was taken to end of the day.
8		4	6				
9	S.E. ½ S.	6	4	E. by N. ½ N.	1	10° E.	
10		6	3				
11		6					
12		6	3				

EXAMPLE X.

H.	Courses.	K.	$\frac{1}{10}$	Winds.	Lee-way.	Deviation.	Remarks, &c.
					pts.		
1	S.W. ½ W.	4	8	S. by E.	2¼	31½° E.	A point of land, lat. 36° 10′ S., long. 110° 10′ W., bearing by compass E. by N., dist. 14 miles. Ship's head S.W. ½ W. Deviation as per log.
2		5	2				
3		5	2				
4		5	3				
5	W. by S. ½ S.	4	3	S. by W.	2½	25½° E.	
6		4	3				
7		4	3				
8		4	3				
9	W. by N. ¾ N.	6		S.W.	2	17½° E.	
10		5	4				
11		6	2				
12		6	4				
1	N.W. ½ W.	5	6	W. by S. ¾ S.	2½	7½° W.	
2		5	4				Variation 20° E.
3		5	5				
4		5	8				
5	W. by S.	7	4	S. ½ W.	1½	24° E.	
6		7	4				
7		8	2				
8		8	2				
9	S.W.	5	2	S. by E.	2½	33° E.	A current set the ship the last 8 hours E. ½ S. (correct magnetic) 2 miles an hour.
10		4					
11		5	7				
12		5	4				

EXAMPLE XI.

H.	Courses.	K.	$\frac{1}{10}$	Winds.	Leeway.	Deviation.	Remarks, &c.
1	N.W. by W.	8	4	N. by E.	pts. $\frac{3}{4}$	15° E.	A point of land in lat. 55° 59' S., long. 67° 16' W., bearing by compass E.S.E., dist. 17 miles. Ship's head N.W. by W. Deviation 15° E.
2		8	4				
3		8	4				
4		8	4				
5	North.	6		E.N.E.	1	22° E.	
6		5	6				
7		5	6				
8		5	6				
9	N.W. by N.	11	4	N.E. by N.	$\frac{1}{2}$	17° E.	
10		11					
11		11	2				
12		11					
1	West.	11	4	S.S.W.	$\frac{1}{2}$	1$\frac{1}{2}$° W.	Variation 23° W.
2		11	6				
3		12	4				
4		12	4				
5	N.N.E.	7	3	East.	1	19° E.	
6		7	4				
7		7	4				
8		7	4				A current set the ship N. by W. correct magnetic 27 miles, from the time the departure was taken to the end of the day.
9	S.S.E.	9	5	East.	$\frac{1}{2}$	20° W.	
10		9	5				
11		9	4				
12		9	4				

EXAMPLE XII.

H.	Courses.	K.	$\frac{1}{10}$	Winds.	Leeway.	Deviation.	Remarks, &c.
1	W. by N.	9		W. by S.	pts. $\frac{1}{2}$	11$\frac{1}{2}$° W.	A point of land in lat. 44° 20' S., long. 176° 49' W., bearing by compass E. by N., dist. 16 miles. Ship's head W. by N. Deviation as per log.
2		10	3				
3		10	4				
4		10	3				
5	W. $\frac{1}{2}$ S.	10	5	S.S.W.	1	1° W.	
6		9	5				
7		8	6				
8		8	4				
9	W.S.W.	8	5	South.	$\frac{1}{2}$	9$\frac{1}{4}$° W.	
10		9	5				
11		9	5				
12		10	5				
1	S.W. $\frac{1}{4}$ W.	8	4	S. by E.	1$\frac{1}{2}$	7° W.	Variation 22$\frac{1}{2}$° E.
2		7	6				
3		7	5				
4		7	5				
5	S. $\frac{1}{2}$ W.	7	4	S.E. by E.	2	1° W.	
6		7	6				
7		8					
8		8					A current set the ship W.S.W. (correct magnetic), 28 miles, from the time the departure was taken to the end of the day.
9	West.	9	6	East.	0	12° W.	
10		10	6				
11		11	2				
12		10	6				

PRELIMINARY RULES IN NAUTICAL ASTRONOMY.

CIVIL AND ASTRONOMICAL DAY.

285. The **Civil Day**, or common method of reckoning time, begins at midnight, and ends the following midnight, the interval being divided into two periods of 12 hours each; the first twelve hours, from midnight to noon, are denoted by A.M. *(ante meridian)*; the latter, from noon to midnight, are styled P.M. *(post meridian)*; thus we say 10 A.M. when an event occurred at 10 o'clock in the morning, and 10 P.M. when it occurred at 10 o'clock in the evening.

286. The **Astronomical Day** begins at noon and ends at the following noon, and is later than the civil day by twelve hours. The hours are reckoned throughout, or continuously from 0^h to 24^h. The distinction of A.M. and P.M. is not recognised in astronomical time. Thus, 11 o'clock in the forenoon of the second of January in the civil reckoning of time corresponds to January 1 day 23 hours in the astronomical reckoning; and 1 o'clock in the afternoon of the former to January 2 days 1 hour of the latter reckoning.

287. Since the civil day commences at the midnight *preceding* the noon which commences the astronomical day, it is evident that the civil mode of reckoning is always twelve hours in advance of the astronomical reckoning, and hence we have the following Rule for converting civil into astronomical time.

Given civil time at ship, to reduce it to astronomical time.

RULE LXXIV.

$1°$. *If the civil time at ship be* P.M., *it will also be astronomical time*, P.M. *being omitted.*

$2°$. *If the civil time be* A.M., *add twelve to the hours and subtract one from the days of the month; also omit* A.M. The result in each case is the Astronomical Date.

EXAMPLES.

Ex. 1. May 10th, at 5^h 30^m P.M., civil time is 5^h 30^m astronomical time of the same date; because the 10th astronomical day begins at noon of the 10th civil day, and 5^h 30^m have elapsed since that noon. But 5^h 30^m A.M. civil time on May 10th is 17^h 30^m *astronomical* time on the 9th of May, for the 9th day of the month, according to the astronomical reckoning, commences at noon of the 9th civil time, and ends at noon of the 10th civil day (the hours being reckoned up to 24), and 5^h 30^m A.M. of the 10th is 17^h 30^m from noon of the 9th.

Ex. 2. October 7th, at 3^h 10^m P.M., civil time, is October 7th, at 3^h 10^m astronomical time. (See $1°$ of Rule LXXIV.)

Ex. 3. October 7th, at 3^h 10^m A.M., civil date, is October 6^d 15^h 10^m astronomical date; since 7^d less 1^d is 6^d, and 12^h added to 3^h 10^m is 15^h 10^m. (See $2°$ of Rule.)

Ex. 4. January 31st, at 7^h 20^m P.M., civil time, is January 31st, at 7^h 20^m astronomical time. (Rule LXXIV, $1°$.)

Ex. 5. February 1st, at $6^h\ 18^m$ A.M., civil date, is January $31^d\ 18^h\ 18^m$ astronomical date; since February 1^d, diminished by 1^d, gives January 31^d, and 12^h added to $6^h\ 18^m$ is $18^h\ 18^m$. (Rule LXXIV, $2°$).

Ex. 6. What is the astronomical date corresponding to 1873, January 1st, 8^h A.M. The corresponding astronomical date is 1872, December $31^d\ 20^h$. In this case the year is diminished by 1, since in diminishing the day of the month by 1, the reckoning throws us back into the last month of the previous year, i.e., the day before January 1st, 1873, also 12^h added to 8^h is 20^h.

EXAMPLES FOR PRACTICE.

Express the following dates in astronomical time.

1.	Jan. 2nd,	4^h	38^m	9^s A.M.	7.	Dec. 31st,	$6^h\ 18^m$	34^s	P.M.
2.	Feb. 27th,	8	12	0 P.M.	8.	July 1st,	8 3	24	P.M.
3.	Aug. 14th,	6	28	40 P.M.	9.	July 1st,	11 30	10	A.M.
4.	April 1st,	7	54	19 A.M.	10.	Oct. 1st,	0 10	12	P.M.
5.	June 4th,	4	18	3 A.M.	11.	1872, Jan. 1st,	8 9	50	A.M.
6.	Sept. 1st,	8	10	52 A.M.	12.	1873, Jan. 1st,	0 44	12	A.M.

288. **Given the astronomical date, to find the corresponding civil date.**

RULE LXXV.

If the hours of astronomical time be less than 12^h *write* P.M. *after it, and it will be the required civil time; but if the* astronomical time *be greater than* 12^h, add 1 to the days, diminish the hours by 12 and write A.M. after it: the result will be the required civil time.

Express the following astronomical dates in civil time:—

1.	Jan. 10th,	$16^h\ 31^m\ 15^s$	2.	Oct. 14th,	$15^h\ 17^m\ 13^s$			
	Feb. 3rd,	11 28 56		Dec. 3rd,	5 16 12			
3.	May 17th,	7 15 11	4.	Mar. 31st,	23 10 16			
	Mar. 13th,	23 15 7		Mar. 21st,	7 24 12			
5.	Sept. 1st,	8 10 54	6.	1872, Jan. 1st,	9 50 41			
	Aug. 31st,	20 10 54		1872, Dec. 31st,	22 41 56			

LONGITUDE IN ARC AND LONGITUDE IN TIME.

289. The earth rotates uniformly on her axis once in twenty-four hours, and thus every spot on her surface describes a complete circle, or $360°$, in that space of time; hence the longitude of any place is proportional to the time the earth takes to revolve through the angle between the first meridian and the meridian of the place, and thus the longitude of a place may be expressed either in *arc* or in *time*.* Longitude in arc and longitude in time are easily convertible, for since $360°$ is equivalent to 24^h ($360 \div 24 = 15°$), $15°$ is equivalent to 1^h, $15'$ to 1^m, and $15''$ to 1^s; whence

$1°$ is equivalent to 4^m (i.e., the 15th part of 1 hour or $60°$)
$1'$,, 4^s (i.e., the 15th part of 1 minute or $60'$)
$1''$,, 4^t (i.e., the 15th part of 1 second or 60^t)†

and the following rules are sufficiently clear.

---

* In reckoning by arc, each degree is divided into sixty minutes, and each minute into sixty seconds. In reckoning by time, each hour is also divided into sixty minutes, and the minutes into sixty seconds. But a distinct notation for each of these has been adopted, degrees, minutes, and seconds being represented by $°\ '\ ''$, and hours, minutes, and seconds by $^h\ ^m\ ^s$; and care should be observed not to use the same marks for both, great confusion arising from so doing.

† A third is the name given to the sixtieth part of a second.

290. **To convert arc (or longitude) into time.**

RULE LXXVI.

Multiply the degrees, minutes, &c., by 4; this turns the degrees (°) into minutes (^{m}) of time, minutes (') into seconds (^{s}) of time, and the seconds (") into thirds (t) of time; or in other words, mark the resulting figures thus:—*Those under seconds (") thirds (t), those under minutes (') seconds (^{s}), those under degrees (°) minutes (^{m}), and those to the left of the latter, hours (^{h}).*

NOTE.—Instead of thirds it is customary to use tenths of seconds, in which case the thirds must be reduced to tenths by dividing by 60 (see Rule XVI, page 54).

EXAMPLES.

Ex. 1. Convert 12° 18' 15" into time.

$$12° \ 18' \ 15''$$
$$4$$
$$49^m 13^s \ 0^t$$

Four times 15" are 60", which contains 60 once and 0 over; write this remainder down under the seconds (") and mark it thirds (t) as directed in the Rule, carrying the 1. Again, 4 times 18' are 72, and the 1' carried makes 73; 60 goes in 73 once and 13 over; write this remainder (13) under the minutes (') and call them seconds (^{s}) and carry the 1. Again, 4 times 12 are 48, and 1 carried makes 49; write this under degrees (°) and mark it minutes (^{m}): whence the time corresponding to arc 12° 18' 15" is 49^m 13^s 0t.

Ex. 2. Convert 25° 15' 16" into time.

$$25° \ 15' \ 16''$$
$$4$$
$$1^h 41^m 1^s 4^t$$

Four times 16" are 64", which contains 60 once and 4 over, and according to Rule this remainder placed under seconds (") becomes thirds (t), and the 1 is to be carried. Again, four times 15' are 60 and 1 carried makes 61, which contains 60 once and 1 over; write the remainder 1 under minutes ('), and carry 1: four times 25 are 100 and 1 carried gives 101, and 60 into 101 goes once and 41 remainder, which remainder being placed under degrees (°) gives minutes (^{m}) and the 1 carried on being placed to the left of the latter is marked hours (^{h}); whence 1^h 41^m 1^s 4t is the time corresponding to the arc 25° 15' 16".

Ex. 3. Turn 77° 2' 10" into time.

$$77° \ 2' \ 10''$$
$$4$$
$$5^h \ 8^m \ 8^s \ 40^t \qquad 60)40^t \cdot 0$$
$$\qquad \qquad \qquad \qquad \cdot 66$$

or, $5^h \ 8^m \ 8^s \cdot 66$

Ex. 4. What time corresponds to 127° 32' 40"?

$$127° \ 32' \ 40''$$
$$4$$
$$8^h \ 30^m \ 10^s \ 40^t \qquad 60)40^t \cdot 0$$
$$\qquad \qquad \qquad \qquad \cdot 66$$

or, $8^h \ 30^m \ 10^s \cdot 66$

Ex. 5. What time is equivalent to 15° 47' 58"?

$$15° \ 47' \ 58''$$
$$4$$
$$1^h \ 3^m \ 11^s \ 52^t$$

or, $1^h \ 3^m \ 11^s \cdot 86$

Ex. 6. Convert 178° 45' 53" into time.

$$178° \ 45' \ 53''$$
$$4$$
$$11^h \ 55^m \ 3^s \ 32^t \qquad 60)32^t \cdot 00$$
$$\qquad \qquad \qquad \qquad \cdot 53$$

or, $11^h \ 55^m \ 3^s \cdot 53$

EXAMPLES FOR PRACTICE.

Reduce the following arcs into time:—

1. 18° 54'; 12° 40' 45"; 137° 27'; 96° 10' 45"; and 89° 16'.
2. 67° 42'; 76° 20' 30"; 1° 25'; 140° 32' 10"; and 69° 29'.
3. 0° 58'·6; 49° 4' 20"; 0° 26'·8; 14° 2' 30"; and 130° 19'.
4. 9° 14'; 163° 2' 48"; 0° 37' 4"; 2° 18' 12"; and 170° 15'.
5. 108° 37'; 10° 27' 14"; 2° 29'; 84° 42' 30"; and 0° 34½'.
6. 0° 13'·5; 51° 10' 12"; 156° 52'; 178° 49' 45"; and 0° 41"·7.

TO CONVERT TIME INTO LONGITUDE.

291. It has been shown (No. 289, page 224) that 4^m of time are equivalent to 1° of arc; hence it is evident that if we bring any given time into minutes, and divide by 4, we shall have the corresponding arc in degrees, minutes, and seconds. This is the reverse of the last process.

RULE LXXVII.

Reduce the hours and minutes into minutes, after which place the seconds, &c., then divide all by 4, and the quotient will be the degrees, minutes, &c., of the corresponding arc; or, in other words, after dividing by 4, mark the resulting figures thus:—Those under minutes (ᵐ) degrees (°), those under seconds (ˢ) minutes ('), those under thirds (ᵗ) seconds (").

EXAMPLES.

Ex. 1. Turn $1^h\ 5^m\ 12^s$ into arc.

$$\begin{array}{r}1^h\ 5^m\ 12^s \\ 60 \\ \hline 4)65^m\ 12^s\ 0^t \\ \hline 16°\ 18'\ 0'' \end{array}$$

Multiply 1^h by 60, add the minutes (5) and divide by 4, the quotient is 16° with remainder 1. Multiply this remainder by 60, and to the product add the 12 seconds; the sum is 72. Again, the quotient of 72, divided by 4, is 18, which is minutes ('); whence the arc corresponding to the time $1^h\ 5^m\ 12^s$ is 16° 18'.

Ex. 2. Reduce $6^h\ 24^m\ 43^s$ into arc.

$$\begin{array}{r}6^h\ 24^m\ 43^s \\ 60 \\ \hline 4)384^m\ 43^s\ 0^t \\ \hline 96°\ 10'\ 45'' \end{array}$$

Multiplying 6^h by 60, and adding the 24^m to the product, gives 384 as the sum; the quotient of this, divided by 4, is 96°, with no remainder. 43^s divided by 4 gives quotient 10' with remainder 3; remainder 3 multiplied by 60 gives 180, which divided by 4 gives quotient 45''; therefore, 96° 10' 45'' is the arc which corresponds to $6^h\ 24^m\ 43^s$.

Ex. 3. What arc corresponding to $0^h\ 47^m\ 36^s$?

$$\begin{array}{r}4)0^h\ 47^m\ 36^s \\ \hline 11°\ 54' \end{array}$$

In this instance it is not necessary to multiply by 60, as there are no hours to reduce into minutes; we divide 47ᵐ at once by 4.

Ex. 4. What is the equivalent arc to $9^h\ 25^m\ 37^s$?

$$\begin{array}{r}9^h\ 25^m\ 37^s \\ 60 \\ \hline 4)565^m\ 37^s\ 0^t \\ \hline 141°\ 24'\ 15'' \end{array}$$

Ex. 5. Convert $8^h\ 17^m\ 35^{s\cdot}5$ into arc.

$$\begin{array}{r}8^h\ 17^m\ 35^{s\cdot}5 \\ 60 \\ \hline 4)497^m\ 35^s\ 30^t \\ \hline 124°\ 23'\ 52''\cdot5 \end{array} \qquad \begin{array}{r}\cdot5 \\ 6 \\ \hline 30 \end{array}$$

Multiply the hours (8^h) by 60, and adding the minutes (17^m) to the product gives 497^m; divide the result by 4, the quotient is 124°, with remainder 1. Again, multiply the remainder just obtained ($1°$) by 60, and to the product add the seconds of time, viz., 35ˢ, the sum is 95; then divided by 4, the quotient is 23' (minutes of arc) with remainder 3. Next multiply this last remainder by 60, the product is 180, to which add the 30ᵗ; and the sum 210, divided by 4, gives 52'' of arc, and remainder 2, to which annex a cypher and divide by 4, the quotient is ·5 of seconds of arc.

Ex. 6. Convert $11^h\ 39^m\ 50^{s\cdot}7$ into arc.

$$\begin{array}{r}11^h\ 39^m\ 50^{s\cdot}7 \\ 60 \\ \hline 4)699^m\ 50^s\ 42^t \\ \hline 174°\ 57'\ 40''\cdot5 \end{array} \qquad \begin{array}{r}\cdot7 \\ 6 \\ \hline 42 \end{array}$$

Multiply 11^h by 60, and to product 660 add 39ᵐ, dividing the sum, viz., 699 by 4 gives 174°, with remainder 3; this remainder (3) multiplied by 60, and 50ˢ added to product gives 230; this sum divided by 4 gives 57', with remainder 2; remainder 2 multiplied by 60, and 42ᵗ added, gives sum 162, which divided by 4 gives 40'', and remainder 2; remainder 2, with a cypher annexed and divided by 4, gives quotient ·5; whence the arc corresponding to $11^h\ 39^m\ 50^{s\cdot}7$ is 174° 57' 40''·5.

EXAMPLES FOR PRACTICE.

Convert the following times into arc:—

1.	$1^h\ 13^m\ 52^s$	6.	$9^h\ 49^m\ 38^s$	11.	$0^h\ 21^m\ 30^s\cdot9$	16.	$0^h\ 20^m\ 41^s$
2.	3 52 4	7.	0 34 58·2	12.	11 41 6·66	17.	8 36 56
3.	0 42 12	8.	1 41 1·6	13.	0 3 52	18.	5 0 51
4.	11 15 21	9.	5 59 4	14.	0 9 56	19.	11 59 57
5.	4 29 5	10.	8 17 6	15.	0 0 52	20.	0 1 52

GREENWICH DATE.

REDUCTION OF GREENWICH DATE.

292. **Def.**—The Greenwich Date is the day and time (reckoned astronomically) at Greenwich corresponding to a given day and time elsewhere. It is necessary to find the Greenwich date before the information contained in the Nautical Almanac can be made available, because all the elements there tabulated are given for time at the meridian of Greenwich.

As in almost every computation of nautical astronomy we are dependent for some data upon the Nautical Almanac—and these are commonly given for Greenwich—it is generally the first step in such a computation to deduce an exact or, at least, an approximate value of the Greenwich astronomical time. It need hardly be added that the Greenwich time should never be otherwise expressed than astronomically.

The Greenwich Date is found at once from a chronometer, the error and the rate of which is known; but it can also be found by means of the approximate time at place and the approximate longitude.

293. To find the Greenwich Date, the time at any other place and the longitude being given.

RULE LXXVIII.

1°. *Express the ship time astronomically* (Rule LXXIV, page 223).

2°. *Convert the longitude into time* (Rule LXXVI, page 225).

3°. **In West longitude.**—ADD *longitude in time to ship time; the sum, if less than 24 hours, is the corresponding Greenwich date on the same day with the ship date; if greater than 24 hours, reject the 24 hours, and put the day one forward.*

4°. **In East longitude.**—*From ship astronomical time* SUBTRACT *longitude in time, if less than the hours, minutes, &c., of the ship date; the remainder is the corresponding Greenwich date of the same day as the ship date; if the longitude in time be greater than the hours, minutes, &c., of ship astronomical,* ADD 24 *hours to the latter, and put the day one back before the subtraction is made.*

5°. **When it is noon at the place.**—*The longitude in time, if West, is the Greenwich date (apparent time); but if* East, SUBTRACT *the longitude in time from* 24 *hours; the remainder is the Greenwich date (apparent time) after noon of the preceding day.*

(*a*) From this last it is evident that when the sun is on a meridian in *West* longitude, the Greenwich *apparent* time is precisely equal to the longitude, that is, the Greenwich apparent time is *after* the noon of the same date with the ship date, by a number of hours, &c., equal to longitude. When the sun is on a meridian in *East* longitude, the Greenwich apparent time is *before* the noon of the same date as the ship date, by a number of hours equal to the longitude in time.

NOTE.—A bad habit prevails in writing dates, of separating the month and day from the hours, minutes, and seconds. *The day of the month should always precede the minor divisions of time* which give the precise instant of the day intended.

EXAMPLES.

Ex. 1. November 9th, at $4^h\ 10^m$ P.M., apparent time at ship, longitude $32°\ 45'$ W.: required the corresponding time at Greenwich, or the Greenwich date.

```
Ship date (A.T.) Nov. 9ᵈ 4ʰ 10ᵐ          Longitude  32° 45'
Long. in time        +   2 11                            4
                  ─────────────                 ──────────────
Greenwich date, Nov. 9  6 21             6,0)13,1  0
                                         ──────────────
                                         2ʰ 11ᵐ 0ˢ
```

Ex. 2. June 5th, at $7^h\ 15^m$ A.M., app. time at ship, longitude $140°\ 30'$ E.: find corresponding Greenwich date.

```
Ship date (A.T.)     June 4ᵈ 19ʰ 15ᵐ
Longitude in time         −    9 22
                     ──────────────
Green. date (A.T.)   June 4    9 53
```

Ex. 3. January 3rd, at $8^h\ 12^m$ P.M., mean time at ship, long. $50°\ 45'$ E.: find Greenwich date.

```
Ship date (M.T.)     Jan. 3ᵈ 8ʰ 12ᵐ
Longitude in time         −  3 23
                     ──────────────
Green. date (M.T.)   Jan. 3  4 49
```

Ex. 4. April 27th, at $5^h\ 35^m\ 45^s$ A.M., app. time at ship, long. $122°\ 13'$ W.: what is corresponding Greenwich date?

```
Ship date (A.T.)   April 26ᵈ 17ʰ 35ᵐ 45ˢ
Longitude 122° 13' W.     +  8  8 52
                         ──────────────
                            26 25 44 37
                          −        24
                         ──────────────
Green. date (A.T.)  April 27  1 44 37
```

Ex. 5. July 20th, at $3^h\ 35^m\ 7^s$ P.M., mean time at ship, long. $85°\ 24'$ E.: find corresponding Greenwich date.

```
Ship date (M.T.)   July 20ᵈ 3ʰ 35ᵐ 7ˢ
                         + 24
                        ──────────────
                      or 19 27 35  7
Longitude 85° 24' E.   −    5 41 36
                        ──────────────
Green. date (M.T.)       19 21 53 31
```

In Ex. 4, the added longitude advances the day of the month. (This illustrates latter part of 3° of the Rule). In Ex. 5, a day (or 24 hours) is borrowed before the subtraction is made, since the longitude in time exceeds the astronomical ship date, thus making the day at Greenwich one less than at the place. (This illustrates the latter part of 4° of the Rule).

Ex. 6. 1873, January 1st, $3^h\ 40^m\ 20^s$ P.M., mean time at ship, long. $95°\ 7'$ E.: find the Greenwich date.

```
Ship date (M.T.) 1873  Jan. 1ᵈ 3ʰ40ᵐ20ˢ
Longitude 95° 7' E.        −   6 20 28
                       ──────────────
Green. date(M.T.)1872, Dec. 31 21 19 52
```

Ex. 7. 1872, January 1st, $2^h\ 1^m$ A.M., mean time at ship, long. $107°\ 4'$ W.: find the Greenwich date.

```
Ship date (M.T.) 1871  Dec. 31ᵈ21ʰ1ᵐ 0ˢ
Longitude 107° 4' W.       +   7 8 16
                       ──────────────
Green. date (M.T.) 1872, Jan. 1  4 9 16
```

Ex. 8. 1872, June 12th, $6^h\ 40^m$ A.M., app. time at ship, long. $42°\ 16'$ W.: find the Greenwich date.

```
Ship date (A.T.)    June 11ᵈ18ʰ40ᵐ0ˢ
Longitude 42° 16' W.     + 2 49 4
                    ──────────────
Green. date (A.T.)  June 11 21 29 4
```

Ex. 9. 1872, October 1st, long. $2°$ W., the sun on meridian: required Greenwich date (app. time).

```
Ship date (A.T.)    October 1ᵈ 0ʰ 0ᵐ
Longitude 2° W.          +     8
                    ──────────────
Green. date (A.T.)  October 1  0 8
```

Ex. 10. Required the Greenwich date when the sun is on the meridian of a place in long. $80°\ 44'$ E., on January 12th.

The sun being on the meridian, it is app. noon: hence

```
Ship date (A.T.)     Jan. 12ᵈ 0ʰ 0ᵐ 0ˢ
Longitude 80° 44' E.       −  5 22 56
                     ──────────────
Green. date (A.T.)   Jan. 11 18 37 4
```

Ex. 11. What is the Greenwich date when the sun is on the meridian of a place in long. $155°\ 19'$ W., on March 31st?

```
Ship date (A.T.)    March 31ᵈ 0ʰ 0ᵐ 0ˢ
Longitude 155° 19' W.      + 10 21 16
                    ──────────────
Green. date (A.T.)  March 31 10 21 16
```

In Ex. 10, the hours, &c., of longitude to be subtracted are to be taken from a borrowed day, or 24 hours, thus making the day of the month at Greenwich one less than at the place. (See 5° of Rule).

Ex. 12. 1882, February 1st, long. 135° E.: find the Greenwich date when the sun is on the meridian.

Ship date (A.T.)   February   1ᵈ   1ʰ   0ᵐ
Longitude 135° E.   − 9   0

Green. date (A.T.)   January 31   15   0

Ex. 13. 1883, January 1st, the ship in long. 160° 30′ E.: required the Greenwich date when the sun is on the meridian.

Ship date (A.T.) 1883   Jan.   1ᵈ   0ʰ   0ᵐ
Longitude 160° 30′ E.   − 10   42

Green. date (A.T.) 1882 Dec. 31   13   18

EXAMPLES FOR PRACTICE.

Required the Greenwich date in each of the following examples:—

1.	1882,	January 6th	at 3ʰ 40ᵐ 16ˢ P.M.	apparent time,	long. 66° 56′ 0″ W.
2.	,,	February 13th	at 8 40 3 A.M.	apparent time,	long. 21 4 0 W.
3.	,,	February 1st	at 5 10 50 A.M.	mean time,	long. 145 20 30 E.
4.	,,	March 15th	at 9 16 22 P.M.	apparent time,	long. 17 4 0 E.
5.	,,	May 15th	at 8 38 35 A.M.	apparent time,	long. 141 51 15 W.
6.	,,	November 1st	at 5 0 10 P.M.	mean time,	long. 114 30 0 E.
7.	,,	December 1st	at 8 0 5 A.M.	mean time,	long. 158 10 0 W.
8.	,,	July 1st	at 4 0 33 P.M.	apparent time,	long. 170 55 15 E.
9.	,,	August 4th	at 6 31 32 P.M.	apparent time,	long. 100 17 30 E.
10.	,,	September 1st	at 8 29 1 A.M.	mean time,	long. 148 47 30 W.
11.	,,	December 28th	at 2 42 10 P.M.	mean time,	long. 50 40 0 E.
12.	,,	July 8th	at 0 4 36 A.M.	apparent time,	long. 178 51 0 W.
13.	,,	February 1st	at noon	apparent time,	long. 153 40 0 E.
14.	,,	June 1st	at noon	apparent time,	long. 83 50 0 E.
15.	,,	March 2nd	at noon	apparent time,	long. 1 25 0 W.
16.	1883,	January 1st	at noon	apparent time,	long. 149 10 0 E.

REDUCTION OF ELEMENTS FROM NAUTICAL ALMANAC.

294. The *Nautical Almanac*[*] or *Astronomical Ephemeris* contains the right ascension, declination, &c., of the principal heavenly bodies for given equidistant instants of Greenwich time; the right ascension and declination of the sun and planets, for example, being given for every day at noon (0ʰ 0ᵐ 0ˢ) at Greenwich, while for the moon these elements are given for every hour. Before we can find from the *Almanac* the value of any of these quantities for a given local or ship time, we must find the corresponding Greenwich date (Rule LXXVIII, page 227). Where this time is exactly one of the instants for which the required quantity is put down in the Ephemeris, nothing more is necessary than to transcribe the quantity as there put down. But when, as is mostly the case, the time falls between two of the times in the Ephemeris, we must obtain the required quantity by interpolation, it being requisite to apply a correction to that taken from the *Almanac*, in order to reduce it to its value at the given instant. To facilitate this interpolation the Almanac

---

[*] The French Ephemeris, *La Connaissance des Temps*, is computed for the meridian of Paris, the German *Berliner Astronomisches Jahrbuch* for the meridian of Berlin. All these works are published annually several years in advance.

contains the rate of change, or difference of each of the quantities in some unit of time, or, which is in general the simplest method, we may make use of certain tables computed for the purpose, called Tables of *Proportional Logarithms*.

To use the difference columns with advantage, the Greenwich time should be expressed in that unit of time for which the difference is given: thus, when the difference is for one hour, the time must be expressed in hours and decimals of an hour; when the difference is for one minute of time, the time should be expressed in minutes and decimals of a minute.

295. **Simple Interpolation.**—In the greater number of cases in practice, it is sufficiently exact to obtain the requisite quantities by *simple* interpolation; that is, by assuming that the difference of the quantities are proportional to the differences of the times, which is equivalent to assuming that the differences in the Ephemeris are constant. This, however, is never the case; for example, referring to the *Nautical Almanac* for 1882, the variation in declination for 1 hour for

$$
\begin{array}{lll}
\text{1st January is} & 12''\cdot66 \\
\text{2nd} \quad ,, & 13''\cdot80 \\
\text{3rd} \quad ,, & 14''\cdot94 \\
& \&\text{c}.
\end{array}
$$

But the error arising from the assumption will be smaller the less the interval between the times in the Ephemeris; hence, those quantities which vary most irregularly, as the Moon's Right Ascension and Declination, are given for every hour of Greenwich time; others, as the Moon's Parallax and Semidiameters, for every twelfth hour, or for noon and midnight; others, as the Sun's Right Ascension, &c., for each noon; others, as the right ascensions and declinations of the fixed stars, for every tenth day of the year.

TO REDUCE SUN'S DECLINATION.

296. The declination of the sun is given in the "Nautical Almanac," pages I and II of each month, for every day both for apparent and mean noon at Greenwich. The *difference of declination* for one hour *("Var. in 1 hour")* is always annexed, and is intended to facilitate the reduction of the quantities from noon to any other time. In general it is necessary to take out the required quantities for the nearest Greenwich time to the given time, and interpolate in either direction to the given instant of Greenwich time.

Method I.—By hourly difference.*

RULE LXXIX.

1°. *Get a Greenwich date by means of ship time, expressed astronomically, and longitude* (see Rule LXXVIII, page 227), *or by means of chronometer.*

To express the Greenwich time in hours and decimals of an hour. Annex a cypher to the minutes and divide by 60, or divide the minutes by 6, *and consider the quotient as tenths* of an hour, and to this prefix the hours. For example, let it be required to express $7^h 18^m$ in hours and decimals of an hour. Then 6 is contained in 18 three times; to this prefix the hours (7) and we have 7·3 hours.—(See Ex. 3, page 54).

---

* This method of reducing the sun's declination is the one required to be used at the Local Marine Board Examinations.

2°. *Take out of* Nautical Almanac *the declination for the* nearest noon *to the given Greenwich date, noting whether the declination is* increasing—*in which case affix the sign* + *to it—or* decreasing—*when the sign* — *must be affixed—and a little to the right place the* "*Var. in* 1 *hour*" *found in page* I of the month in the Nautical Almanac.

(*a*) When Greenwich date is given in *apparent* time, use page I of the month, but for *mean* time use page II of the month.

(*b*) The tenths of seconds (′) of declination as given in the *Nautical Almanac* may be rejected when less than five, but call them 1″ when they amount to five or above—thus, 6″·4 is put down 6″, but 42″·7 will be put down 43″; and it may be here observed, that whenever a decimal is rejected in a final result if the first decimal figure be 5, or above it, add 1 to the last figure of the result.

(*c*) When the seconds of time (in Greenwich date) are less than 30ˢ, they may be rejected; but if above 30ˢ, increase the minutes of time by 1ᵐ; thus, Greenwich time 2ʰ 35ᵐ 40ˢ would be called 2ʰ 36ᵐ.

3°. *Multiply the* "*Var. in* 1 *hour*" *by the* hours, *and fractional parts of an hour,* that have elapsed since, *or must elapse before* that noon, *as the case may be; the product reduced to minutes and seconds is the change of declination in the time from* noon.

4°. *Apply this correction to the declination for the* nearest noon *to the given time, i.e., to the declination of the* same *noon as that for which the* "*Var. in* 1 *hour*" *has been taken as follows:—*

(a) *When the Decl. is* **increasing** *(or has* + *affixed), the correction for the* time elapsed since noon *is* **additive**, *but the correction for the* time that must elapse *is* **subtractive**.

(b) *When the Decl. is* **decreasing** *(or has the sign* — *affixed), the correction for the* time elapsed since noon *is* **subtractive**, *but the correction for the* time that must elapse before noon *is* **additive**.

The result is the declination sought.

NOTE.—It must be remembered that when the declination is taken out of the *Nautical Almanac* for the noon of the *day following* that of given Green. date, the correction is applied the *contrary* way to the sign affixed.

EXAMPLES.

Ex. 1. Greenwich date, Jan. 10ᵈ 6ʰ; in this case take 22″·08 the Diff. for hour on the 10th, which multiplied by 6 gives the correction of the Decl. for the 10th day—to be *subtracted* because the Declination is decreasing, and we have multiplied by the number of hours that have elapsed *since* noon.

Ex. 2. Greenwich date, Jan. 10ᵈ 19ʰ; in this case take 23″·14, the "Var. in 1 hour" on the 11th, which multiplied by the difference between 24ʰ and 19ʰ gives the correction of the Decl. for the 11th day—to be *added* because the Decl. is *decreasing*, and we have multiplied by the number of hours that must elapse before noon of the 11th day.

Ex. 3. Greenwich date, April 2ᵈ 6½ʰ; in this case take 57″·49, the "Var. in 1 hour" on the 2nd, which multiplied by 6½ʰ gives the correction of the declination for the 2nd April —to be *added* because the Decl. is *increasing*, and we have multiplied by 6½ hours the time that has elapsed since noon, April 2nd.

Ex. 4. Greenwich date, April 2^d $17\frac{1}{2}^h$; in this case take $57'\cdot 25$, the "Var. in 1 hour" on the 3rd, which multiplied by $6\frac{1}{2}$ (the difference between 24^h and $17\frac{1}{2}^h$) gives the correction of the Decl. for the 3rd day—to be *subtracted* because the Decl. is *increasing*, and we have multiplied by the number of hours that must elapse before noon, 3rd.

5°. *If the correction when* subtractive *exceeds the* declination *itself*, subtract *the* declination *from the* proportional part; *the remainder is the declination of the contrary name.*

In March, when the declination changes from South to North, and in September, when it changes from North to South, if the correction, by being subtractive, exceed the declination, subtract the declination from the correction, and call the remainder *N.* in March, but *S.* in September. (See Ex. 3.)

Method II.—By proportional logarithms.

RULE LXXX.

1°. *Find a Greenwich date.*

2°. *Take out of the* Nautical Almanac *the declination for the noon at Greenwich, and that following it.*

3°. *When the declinations are of like names, take their difference; but when of different names, take the sum: this is the daily variation of declination.*

(a) *When the declination is* increasing, *place the sign of* addition (+) *before the daily variation; but when the declination is* decreasing, *place the sign of* subtraction (−) *before it.*

4°. *Under the daily variation place the hours and minutes of Greenwich time, and take from the table* (Table **XXI A**, RAPER, or **XXXIII**, NORIE) *log. of change of declination in* 24 *hours and log. of hours and minutes of Greenwich time; the sum of these logs. found in the table will give the* proportional part *of daily change of declination.*

In using Table XXI A, RAPER, or NORIE XXXIII, minutes (′) of declination, and hours of time (ʰ), are found at the top of the columns; seconds (″) of declination, and minutes (ᵐ) of time at the side columns.

5°. *Apply the proportional part to the declination at the first noon*, adding *when the declination is* increasing, *but* subtracting *when the declination is* decreasing.

The result is the declination at the time required.

EXAMPLES.

Ex. 1. 1882, January 13th, at 3^h 54^m 16^s P.M., app. time at ship, long. 30° 4′ E.: find the sun's declination.

```
Ship date (A.T.) January       13ᵈ 3ʰ54ᵐ16ˢ              Longitude  30°  4′
Longitude (30° 4′ E.) in time  --  2  0 16                            4
                               ────────────     54ᵐ = 54    ─────────────
Green. date (A.T.) January     13ᵈ 1ʰ54ᵐ 0ˢ     60)54·0     6,0)12,0 16
                                 or, 1ʰ·9          ·9          2ʰ 0ᵐ 16ˢ
```

Method I.

Decl., page I, N.A., for January 13th, app. noon, is 21° 26′ 44″ S., *decreasing* (—), and var. for 1 hour is 25″·82.

Var. for 1^h, 13th, noon	25″·82
Green. time $1^h\ 54^m = 1^h·9$	× 1·9
	23238
	2582
Correction	— 49·058

Decl., noon, Jan. 13th	21° 26′ 44″ S. (—)
Correction	— 49
Red. decl.	21 25 55 S.

Method II.

Decl. app. noon, page I, N.A.

Jan. 13th,	21° 26′ 44″ S.		
14th,	21 16 12 S.		
Daily var.	— 10 32	log.	3576
Green. time	$1^h 54^m$	log.	1·1015
Correction	— 0 50	log.	1·4591
13th, at noon	21 26 44 S.		
Red. decl.	21 25 54 S.		

The correction 49″, which is for a time elapsed since the noon for which the declination is taken out, is subtracted from declination at noon, because the declination is decreasing.

Having found the Greenwich date, the sun's declination is taken from the Nautical Almanac, where it is found in page I of the month (the Greenwich date being in app. time), and on the same page and in column headed "Var. in 1 hour" is found the change of declination for 1 hour past noon; next observe that the declination is *decreasing*, and make a note of it. Now, since the declination changes 25″·82 in 1 hour past noon, how much does it change in the Greenwich time past noon, viz., $1^h\ 54^m$? First annex a cypher to the minutes (54^m) and divide by 60; thus 60 is contained in 540 nine times and nothing over. To this we prefix the hour, and we then have the Greenwich time $1^h\ 54^m = 1^h·9$ expressed in hours and decimals of an hour. (See Rule XVI, page 54.) Set this under the hourly diff. and then proceed as in multiplication of decimals, the resulting figures are 49058, but as we have two decimals in the multiplicand and one in the multiplier, in all three places, *three* figures are to be marked off from the right hand, leaving 49″ (see Rule XIII, page 47).

Ex. 2. 1882, May 21st, at $7^h\ 50^m$ A.M., mean time at ship, long. 149° 30′ E.: required the sun's declination.

Ship date (M.T.) May	$20^d\ 19^h\ 50^m$
Longitude 149° 30′ E.	— 9 58
Green. date (M.T.) May	20 9 52

Decl., page II, N.A., May 20th, at noon, is 20° 1′ 1″ N. *(increasing)*, var. for 1^h at noon, May 20th, 31″·18.

Var. for 1^h	+ 31″·18
$9^h\ 52^m = 9^h·87$ nearly	× 9·87
	21826
	24944
	28062
	6,0)30,7·7466

Correction + 5′ 8 nearly.

Decl., May 20th, noon	20° 1′ 1″ N. +
Corr. for $9^h\ 52^m$ =	+ 5 8
Red. decl.	20 6 9 N.

By Hourly Diff.

$52^m = \frac{52}{60}$)520

866 or ·87 nearly.

Decl. mean noon, page II, N.A.

May 20th,	20° 1′ 1″ N.		
21st,	20 13 19 N.		
Daily variation	+ 12 18	log.	2903
Green. time	$9^h 52^m$	log.	3860
Correction	+ 5′ 3″	log.	6763
Decl. 20th, noon	20 1 1 N.		
Red. decl.	20 6 4 N.		

The correction 5′ 3″, which is for the time elapsed since noon, is *added* to the declination at noon, because the declination is *increasing*.

Explanation.—Having found the Greenwich date; with this date the sun's declination is taken out of the Nautical Almanac, page II for May (the Greenwich date being *mean time*), and in page I of the month, in the column headed "diff. for 1 hour" is found the change for 1 hour past noon; next observe whether the declination is increasing or decreasing. In this instance it is increasing, and we note this. Now, since the declination *increases* 31″·18 in 1^h, what will be the change in Green. time past noon, viz., $9^h\ 52^m$? We have now to express this in hours and decimal parts of an hour. Then $52^m = \frac{52}{60}$ of an hour, and annexing cyphers to 52 and dividing 60, we have 60 in 520, or 6 in 52 goes 8 times and 4 over; 6 is contained in 40 (the remainder and a cypher annexed) six times and 4 over, or seven times nearly—two places of decimals only being used. Set the ·87 under the hourly difference to which prefix the 9^h, and then proceed as in common multiplication. The resulting figures are 3077466, but as we have two decimal places in the multiplicand, and two decimal places in the multiplier, in all four, *four* figures are to be marked off from the right hand, leaving 307″; but since the first decimal figure (7) exceeds 5, we increase the seconds by 1″ in consequence, and the correction is 308″, which divided by 60 gives + 5′ 8″, the correction of declination.

Ex. 3. 1882, Greenwich date, March $13^d\ 14^h\ 24^m\ 18^s$, mean time: required the sun's declination.

Green. date (M.T.) March	$13^d\ 1^h;{}^h 24^m\ 18^s$	$35^m\ 42^s$ or $36^m = \frac{3.6}{6.0})36^\cdot 0$
Subtract from	25	
Time *before* noon March 14th	9 35 42	·6
	or $9^{h\cdot}6$	

Decl. page II, N.A., March 14th, at noon is $2°\ 27'\ 14''$ S. (—), and var. in $1^h = 59''\cdot 16$.

By Aliquot parts.

Var. in $1^h =$	$59''\cdot 16$
Time before noon	$9\cdot 6$
	35496
	53244
	6,0)56,7·936
Correction	$+\ 9'\ 28''$

Var. in $1^h =$		$59''\cdot 16$
		9
30^m	$\frac{1}{2}^h$	53244
6	$\frac{1}{10}$	2958
	$\frac{1}{100}$	591
		6,0)56,·793
Correction		9·28

Decl., March 14th, noon $2°\ 27'\ 14''$ S. (—)

Red. decl. $2\ 36\ 42$ S.

In this example the Greenwich noon of March 14th is nearer the given time (which exceeds 12^h) than the Greenwich noon, March 13th; therefore take out the declination from page II of the month, *Nautical Almanac* (because *mean* time), for noon March 14, and from page I of the month, also the "Var. in 1 hour" corresponding to this declination, viz., $59''\cdot 16$. Next subtract the hours and minutes of Greenwich time from 24^h, the remainder $9^h\ 35^m\ 42^s$ or $9^h\ 36^m$ is the time that must elapse before noon 14th. Divide the minutes of this last by 60 to get decimals of an hour; thus 6 is contained in 36 six times, hence we have ·6 (see Rule XVI, page 54), to this we prefix the hours (9) and we then have $9^h\cdot 6$. Next multiply the hourly difference by this, and the resulting figures are 567936, then three figures marked off from the right hand, leaves 567", which being increased by 1 in consequence of the first figure on the right of the decimal point exceeding 5, gives for the correction 568", which divided by 60 gives $9'\ 28''$. And since the declination at noon 14th is *decreasing*, it is evident that the declination at $9^h\ 36^m$ *before* that noon will be more than at noon 14th, and the correction $9'\ 28''$ is, therefore, to be *added*; whence the reduced declination is $2°\ 36'\ 42''$ S.

Ex. 4. 1882, March 20th, $6^h\ 39^m$, app. time at Greenwich: required the sun's declination.

Green. date (A.T.), March	$20^d\ 6^h 39^m$	$39^m = \frac{3.9}{6.0})39\cdot 0$
	$6^h\cdot 65$	·65

Decl., page I, N.A., March 20th, at noon, is $0°\ 4'\ 52''\cdot 6$ S. (—), and var. in $1^h = 59''\cdot 25$.

By Aliquot parts.

Var. in $1^h =$	$59''\cdot 25$
	$6\cdot 65$
	29625
	35550
	35550
	6,0)39,4·0125
Correction —	6·34

Var. in $1^h =$		$59''\cdot 25$
		6
30^m	$\frac{1}{2}^h$	35550
6	$\frac{1}{10}$	2962
3	$\frac{1}{20}$	592
		296
		6,0)39,4·00
		— 6·34

March 20th, at noon	$0°\ 4'\ 53''$ S. (—)
Correction for $6^h\cdot 65$	— 6 34
Red. decl.	0 1 41 N.

In this example the correction $6'\ 34''$, which is that due to the time elapsed since the noon for which the declination is taken out, is *subtractive* and the declination *decreasing*; but since the correction *exceeds* the declination itself, *subtract* the declination from the correction; the remainder is the reduced declination, and of a *contrary name* to that at noon. We may consider the sun moving northward, and therefore the correction may be marked N.; and since it is greater than the decl. it shows that the sun has crossed the equator, and has now North decl. Hence the difference between the decl. and the correction must be taken, and marked with the name of the greater.

Ex. 5. 1882, February 11th, at 8ʰ 54ᵐ 47ˢ P.M., apparent time, long. 11° 4′ W.: find the declination.

```
Ship date (A.T.) February 11ᵈ   8ʰ 54ᵐ 47ˢ           Longitude   11° 4′
Longitude in time               + 44  16                             4
                                                                ─────────
Green. date (A.T.) February 11ᵈ  9  39  3                       44ᵐ 16ˢ
                                ─────────
                                 or 9·65
```

Hourly Diff., page I, N.A.   By Aliquot Parts.

```
Feb. 11th at noon     49″·48                                      49″·48
                       9·65                                            9
                     ────────                                     ──────
                  6,0)47,7·4820                        30 | ½ |  445·32
                     ────────                           6 | ⅕ |   24·74
Correction            7·57                              3 | ½ |    4·94
                                                                   2·47
       Decl., page I, N.A.                                      ────────
Feb. 11th, at noon  13° 56′ 59″ S. (−)              6,0)47,7·47
Corr. for 9ʰ 39ᵐ       −   7  57                        ────────
                    ─────────────                         7·57
Reduced decl.       13  49   2  S.
```

297. **To find the declination of the sun at the time of its transit over a given meridian.**

When the sun is on a meridian in WEST longitude, the Greenwich *apparent time* is precisely equal to the longitude; that is, the Greenwich apparent time is *after* the noon of the same date with the ship date by a number of hours, equal to the longitude in time. When the sun is on a meridian in EAST longitude, the Greenwich apparent time is *before* the noon of the same date as the ship date by a number of hours, equal to the longitude in time. Hence, to obtain the sun's declination for apparent noon at any meridian we have

RULE LXXXI.

Take the declination from the Nautical Almanac (page I of the month) *for Greenwich apparent noon of the same date as the ship date, and apply a correction equal to the hourly difference multiplied by the longitude,* observing *to* add *or* subtract *this correction according as the numbers in the* Nautical Almanac *may indicate for a time* **before** *or* **after** *noon.*

EXAMPLES.

Ex. 1. 1882, September 10th, the sun on the meridian, long. 100° 35′ E.: required the sun's declination.

```
Longitude  100° 35′ E.
              4
           ──────────
         6,0)40,2  20
           ──────────
           6ʰ 42ᵐ 20ˢ.
```

The longitude being 6ʰ 42ᵐ 20ˢ East, the Green. A.T. is 6ʰ 42ᵐ *before* the noon of September 10th—the same date as the ship date. The decl. is taken out of the *Nautical Almanac*, page I of the month; also take out at the same time the hourly diff.; the work will stand thus:—

```
Sun's decl., page I, N.A.                       Var. in 1ʰ, page I, N.A.
Sept. 10th, noon   4° 53′  8″ N. (−)        Sept. 10th, noon      56″·95
Corr. for 6ʰ 42ᵐ    +  6  22          Time from noon 10th, 6ʰ 42ᵐ ≡ 6·7
                   ─────────────                                 ────────
Reduced decl.      6  59  30  N.                             6,0)38,1·565
                                                                 ────────
As the declination is decreasing, the declination at    Correction  6·22
6ʰ 42ᵐ before noon will be greater than that for noon.
```

Ex. 2. 1882, June 1st, the sun on the meridian, long. 75° W.: required the sun's decl.

Long. 75° W.
$$4$$
6,0)30,0
$$5^h$$

The longitude being 5^h West, the Greenwich A.T. is 5^h *after* the noon of June 1st—the same date as the ship date. The decl. is taken out of the *Nautical Almanac*, page I of the month; also take out at the same time the var. in 1^h; the work will stand thus:—

Sun's decl., page I, N.A.		Var. in 1^h, page I, N.A.	
June 1st, at noon	22° 4′ 57″ N. +		20″·25
Correction	+ 1 41		5
Red. decl.	22 6 38 N.		6,0)10,1·25
		Correction	1·41

As the decl. is *increasing*, the decl. at 5^h *after* noon will be greater than that for noon.

Ex. 3. In the last question suppose the longitude to be 75° E.

The longitude being 5^h E, the Green. A.T. is 5^h *before* the noon of June 1st—the same date as the ship date. The decl. is taken out of the *Nautical Almanac*, page I of the month; also the var. in 1^h, and the work is as follows:—

Sun's decl., page I, N.A.		Var. in 1^h =	20″·25
June 1st, at noon	22° 4′ 57″ N. +		5
Correction	− 1 41		
Red. decl.	22 3 16 N.		6,0)10,1·25
			1·41

As the decl. is *increasing*, the decl. at 5^h *before* noon will be less than that for noon.

298. Interpolation by Second Differences.—The differences between the successive values—given in the *Nautical Almanac* as functions of time—are called the *first differences*; the differences between these successive differences are called the *second differences*; the differences of the second differences are called the *third differences*, &c. In simple interpolation we assume the function to vary uniformly; that is, we regard the first difference as constant, neglecting the second difference, which is, consequently, assumed to be zero. In interpolation by second differences we take into account the variation in the first difference, but we assume *its* variations to be constant; that is, we assume the second difference to be constant, and the third difference to be constant.

When the *Nautical Almanac* is employed we can take the second differences into account in a very simple manner. In this work, since the year 1863, the difference given for a unit of time is always the difference belonging to the instant of Greenwich time against which it stands, and it expresses, therefore, the rate at which the function is changing *at that instant*. This difference, which we may here call the first difference, varies with the Greenwich time, and (the second difference being constant) it varies uniformly, so that its value for any intermediate time may be found by simple interpolation, using the second differences as first differences. Now, in computing a correction for a given interval of Greenwich time, we should employ the *mean*, or average value, of the first difference for the interval, and this mean value, when we regard the second differences as constant, is that which belongs to the middle of the interval. Hence, to take into account the second differences, we have only to observe the very simple rule—*employ that (interpolated) value of the first difference which corresponds to the middle of the interval for which the correction is to be computed.*

299. **Degree of Dependence.**—The sun's declination changes nearly 1' an hour, or 1" in 1^m, in March and September; hence to insure it to 1" in the extreme case, the Greenwich date must be true to 1^m.

EXAMPLES FOR PRACTICE.

Required the sun's declination in each of the following examples:—

[These are preparatory to working Amplitudes, Azimuths, &c.]

1.	1882, January 5th,	6^{h}23^{m}32^s A.M.	app. time at ship	long. 103° 7' W.
2.	,, February 2nd,	3 9 0 P.M.	app. time at ship	long. 52 45 W.
3.	,, March 31st,	6 2 12 P.M.	app. time at ship	long. 156 3 E.
4.	,, March 26th,	7 8 22 A.M.	mean time at ship	long. 72 47 E.
5.	,, May 16th,	9 17 20 A.M.	mean time at ship	long. 45 40 W.
6.	,, April 29th,	2 26 52 P.M.	mean time at ship	long. 110 57 W.
7.	,, June 10th,	8 45 0 P.M.	app. time at ship	long. 129 30 E.
8.	,, November 1st,	10 20 16 A.M.	mean time at ship	long. 11 17 E.
9.	,, September 1st,	8 20 40 A.M.	app. time at ship	long. 172 9 E.
10.	,, October 1st,	6 11 50 A.M.	mean time at ship	long. 68 15 W.
11.	,, December 16th,	4 35 32 A.M.	app. time at ship	long. 4 8 E.
12.	,, November 14th,	6 45 8 P.M.	mean time at ship	long. 100 2 E.

In each of the following examples it is required to find the sun's declination when the sun is on the meridian (at apparent noon):—

13.	1882, Jan. 19th, long. 86° 57' W.	19.	1882, July 28th, long. 2° 0' W.
14.	,, Feb. 16th, long. 72 59 E.	20.	,, Sept. 23rd, long. 156 0 E.
15.	,, Mar. 20th, long. 168 3 W.	21.	,, Oct. 1st, long. 170 58 E.
16.	,, May 8th, long. 10 35 W.	22.	,, Dec. 22nd, long. 179 52 E.
17.	,, June 21st, long. 167 15 E.	23.	1883. Jan. 1st, long. 156 48 E.
18.	,, Mar. 20th, long. 129 0 W.	24.	1882, Sept. 23rd, long. 174 15 E.

300. **The Polar Distance** of a heavenly body is its angular distance from the elevated pole of the heavens; it is measured by the intercepted arc of the hour circle passing through it, or by the corresponding angle at the centre of the sphere. According as the North or South pole is elevated, we have the *North Polar Distance*, or the *South Polar Distance*.

301. To find the Polar distance of a celestial object, proceed according to the following rule:—

RULE LXXXII.

When the latitude *of the place, and* declination *of the object, are of the same name* subtract the declination *from* 90°; *but when the* latitude *and* declination *are of* contrary names, add the declination *to* 90°; the result in either case is the polar distance.

When the latitude is 0, the declination, either *added* to or *taken* from 90°, is the polar distance.

EXAMPLES.

Lat.	Declination.	Polar distance.
N.	8° 12' 18" S.	98° 12' 18"
N.	22 30 0 N.	67 30 0
S.	2 31 15 S.	87 28 45
N.	30 23 15 S.	120 23 15
S.	7 22 32 N.	97 22 32
S.	26 42 12 S.	63 17 48
0	12 48 2 N.	{ 102 48 2 or 77 11 58

TO FIND THE EQUATION OF TIME.

302. Apparent Solar Day is the interval between two successive transits of the actual sun's centre over the same meridian; it begins when that point is on the meridian. The apparent solar day is variable in length from two causes; first, the sun does not move uniformly in the ecliptic—its apparent path sometimes describing an arc of $57'$, and at other times an arc of $61'$ in a day; second, the ecliptic twice crosses the equinoctial—the great circle whose plane is perpendicular to the axis of rotation—and hence is inclined to it in its different parts; at the point of intersection the inclination is about $23°\ 27'$, at two other limiting points they are parallel. A uniform measure of time is obtained by the invention of the *Mean Solar Day*.

303. Mean Solar Day is the interval between two successive transits of the *mean sun* over the same meridian; it begins when the mean sun is on the meridian. This fictitious body is conceived to move in the equinoctial with the mean motion of the actual sun in the ecliptic. The length of the mean solar day is the average length of the *apparent solar days* for the space of a solar year.

304. Equation of Time is the difference between apparent and mean time. It is measured by the angle at the pole of the heavens between two circles passing, the one through the apparent sun's centre, the other through the mean sun. The Equation of Time is so called because it enables us to reduce apparent to mean, or mean to apparent time. In consequence of the motion of the sun in the ecliptic being variable, and the ecliptic not being perpendicular to the axis of the earth's rotation, apparent time is variable, and this fluctuation is considerable, amounting to upwards of half an hour—apparent noon sometimes taking place as much as 16^m before mean noon, and at others as much as $14\frac{1}{2}^m$ after. These are the greatest values of the equation of time; it vanishes altogether four times a year—this occurring about April 15th, June 15th, September 1st, and December 24th. It is calculated and inserted in the *Nautical Almanac* for every day in the year. On page I of each month the equation of time given is that to be used in deducing mean from apparent time; that on page II is to be used in deducing apparent from mean time. The difference in the value of the two arises from the one being that at apparent noon, and the other that at mean noon. As these may be separated by an interval of more than a quarter of an hour, the equation of time given in pages I and II may differ by a quarter of the "Var. in 1 hour" given in the adjoining column. The equation of time is itself a portion of mean time.

305. To Reduce Equation of Time to Greenwich date.—The method of correcting the equation of time for the Greenwich date is similar to that for correcting the sun's declination, and the "Variation in 1 hour" may be used for the purpose.

RULE LXXXIII.

1°. *Get a Greenwich date, as before.*

NOTE.—The time by chronometer when error and rate are applied to it, gives Mean Time at Greenwich.

2°. *Take out of* Nautical Almanac *page II of the month*) *the Equation of Time for the noon of* Greenwich *date, and mark it* additive *or* subtractive, *according to the heading of Equation of Time at the top of the column in page I of the month*; *also note whether it is* increasing—*when affix the sign* +; *or* decreasing—*affixing the sign* —; *at the same time take from the column in* page I *the* "*Var. in* 1 *hour.*"*

NOTE.—It sometimes happens that the precept for applying the Eq. of Time changes in the course of the month. Thus in April, 1882, a black line is placed between the Eq. T. for the 14th and that for the 15th, indicating that a change of precept occurs between those days. The Equations above the line, page I, have to be added, those below have to be subtracted.

3°. *Multiply the* "*Var. in* 1 *hour*" *by the hours of* Greenwich *time, and when great precision is necessary, by the fractional parts of an hour also. The result is the correction to be applied to the equation of time taken from the* Nautical Almanac, *and is to be* added *when equation of time is* increasing, *but* subtracted *when equation of time is* decreasing; *the result is the Equation of Time sought*.

NOTE.—We may, as in reducing the declination (see preceding Rule LXXXIII), take the Eq. T. and "Var. in 1ʰ" from the *Nautical Almanac* for the *nearest* noon to the Greenwich time, and multiplying the "Var. in 1ʰ" by the time that must elapse before noon; the correction thus obtained must be applied to the Eq. of T. taken out of *Nautical Almanac* in a *contrary* way to that directed above, that is to say, when correcting *backwards* the rule is Eq. T. *increasing,* subtract—Eq. T. *decreasing,* add. (See Exs. 3. 4. and 5).

(*a*) *When the correction, being* subtractive, *exceeds the equation of time itself*, *subtract the equation of time from the correction; the remainder is the reduced equation of time sought—and it is to be* subtracted *from apparent time when equation of time at noon is directed to be* added, *but* added *to apparent time when equation of time at noon is directed to be* subtracted; *i.e., the Equation has to be applied to A.T. according to the precept for the day following the given day.*

EXAMPLES.

Ex. 1. 1882, January 29th, 6ʰ 53ᵐ 49ˢ mean time at Greenwich; find Equation of time to be applied to apparent time in working the chronometer.

Eq. of Time, page II, N.A.		Hourly Diff., page I, N.A.	
Jan. 29th, *add*	13ᵐ24ˢ·6 (+)	Jan. 29th, at noon	0ˢ·423
Corr. for 6ʰ·9	+ 2·9	6ʰ 54ᵐ is 6ʰ·9	6·9
Red. Eq. Time	13 27·5		3807
(To be *added* to app. time.)			2538
		Correction	2·9187
		or,	2ˢ·9

In working this example the "Var. for 1 hour" is taken from the *Nautical Almanac* from the column in page I of the month, and against the given day. The Greenwich date being mean time, take the equation o time from page II of the month, and mark it *additive* to app. time as directed at the top of the column in page I; also note that the equation is increasing. The Green. time being 6ʰ 54ᵐ or 6ʰ·9; hourly difference is multiplied by 6·9 giving the product 29187; and since there are three decimal figures in H.D. (·423) and one in Green. time (·9) in all four, *four* decimal places are marked off from the right hand of the product, the result 2ˢ·9187 or 2ˢ·9 is the correction to be applied to the Eq. of time at noon, and is to be *added* to it because it is that due to time *elapsed since* noon while the Eq. T. is *increasing.*

* As the equation of time is not a uniformly varying quantity, it is not quite accurate to compute its correction as above, by multiplying the given hourly difference by the number of hours in the Greenwich time; for that process assumes that this hourly difference in the same for each hour. The variations in the hourly difference are, however, so small that it is only when extreme precision is required that recourse must be had to the more exact method of interpolation for second differences.

Ex. 2. 1882, September 30th, $10^h\ 15^m$ mean time at Greenwich: find the Equation of time to be applied to app. time in working the chronometer.

Eq. of Time, page II, N.A.
Sept. 30th, noon, *subt.* $10^m\ 2^s\cdot0\ (+)$
Corr. for $10\frac{1}{4}^h$ $+\ 8\cdot3$

Red. Eq. T. $10\ \ 10\cdot3$

(To be *subtracted* from A.T.)

Diff. for 1^h, page I, N.A.
Sept. 30th, at noon $0^s\cdot810$
                            $10\frac{1}{4}$

Diff. for 10 hours 8100
Diff. for $\frac{1}{4}$ hour 202

Correction $8\cdot302$

          or, $8^s\cdot3$

Ex. 3. 1881, December 24th, $10^h\ 54^m$ mean time at Greenwich: find the Equation of time to be applied to apparent time in working the chronometer.

Greenwich date (M.T.) Dec. 24th, $10^h\ 54^m = $ Dec. $24^d\ 10^h\cdot9$.

Eq. of Time, page II, N.A.
Dec. 24th, noon, *subt.* $0^m\ 9^s\cdot1\ (-)$
Corr. for $10^h\cdot9$ $-\ 13\cdot5$

Red. Eq. T. $0\ \ \ 4\cdot4$

(To be *added* to A.T.)

Var. 1 hour $=\ 1^s\cdot243$
                          $10\cdot9$

                          11187
                          1243

                          $13\cdot5487$

The Eq. T. is taken out for noon of Dec. 24th, page II, N.A., the sign — being affixed because Eq. T. is *decreasing*, and according to the precept at the top of the column in page I the Eq. T. at the Green. noon is *subt.* from A.T. The "Var. in 1 hour" multiplied by the time from noon—expressed in hours and decimals of an hour—gives the correction, which is *subtractive* from the Eq. T. at noon, because the latter is *decreasing*; but since the correction *exceeds* the Eq. T. itself, take this latter from the correction, the result is the reduced Eq. T., and is to be applied to A.T. the *contrary* way to the Eq. T. at noon, i.e., according to the precept for the day following the given day, and is therefore *additive* to apparent time.

Ex. 4. 1882, August 31st, $15^h\ 42^m\ 15^s$, mean time at Greenwich: find Equation of time to be applied to apparent time.

Eq. of Time, page II, N.A.
Aug. 31st, at noon, *add* $0^m 10^s\cdot7\ (-)$
Corr. for $15^h\ 42^m$ $-\ 12\cdot2$

Red. Eq. T., *subt.* $0\ \ 1\cdot5$

(To be *subtracted* from A.T.)

Hourly Diff., page I, N.A.
Aug. 31st, noon $0^s\cdot775$
$15^h\ 42^m$ $=\ 15\cdot7$

                        5425
                        3875
                        775

Correction $12\cdot1675$

          or, $12^s\cdot2$

In this case the correction is *subtractive*, and exceeds in amount the equation of time at noon, therefore the equation of time is taken from the correction, and the remainder is the reduced equation of time to be *subtracted* from A.T., according to the precept for the day following the given day—a change of precept occurring between Aug. 31st and Aug. 32nd (Sept. 1st)—which change is shown by means of a black line drawn between the equations for the two named days.

Ex. 5. 1882, June 14th, $22^h\ 25^m\ 21^s$, mean time at Greenwich: find Equation of time to be applied to apparent time in working the chronometer.

Greenwich date, June 14th, $22^h\ 25^m$

           or, $22^h\cdot4$

Eq. T., page II, N.A.
June 14th, noon, *subt.* $0^m\ 3^s\cdot55\ (-)$
Correction for $22^h\cdot4$ $-\ 11\cdot87$

Red. Eq. T., *add* $0\ \ 8\cdot32$

(To be *added* to A.T.)

Hourly Diff., page I, N.A.
June 13th, noon $0^s\cdot530$
                  $22\cdot4$

                  2120
                  1060
                  1060

Correction $11\cdot8720$

          or, $11^s\cdot87$

In this case also, the correction is *subtractive*, and *exceeds* the equation itself, therefore, the equation is *subtracted* from the *correction* and the difference is the deduced Eq. T., which is to be applied to apparent time according to the precept for the day following the given day.

By using the Eq. T. corresponding to the nearest Greenwich noon, viz., that for June 15th, the work will stand thus:—

Green. date, June 14th,	22ʰ25ᵐ		Hourly Diff., page I, N.A.		
Subtract from	24		June 15th, at noon,		0ˢ·535
	——		Time from noon 15th		1·6
Time from noon, June 15th	1·35				——
or,	1ʰ·6 nly.				3210
Eq. T. page II, N.A.					535
June 15th, at noon, *add*	0ᵐ9ˢ·25 +		Correction		·8560
Correction for 1ʰ·6	− 0·86				——
	——			or,	0ˢ·86
Red. Eq. T. add	0 8·39				
(To be *added* to A.T.)					

The Eq. T. would be *less* at 1ʰ·6 *before* noon than what it is at noon, the correction is therefore subtracted from the noon Eq. of Time.

EXAMPLES FOR PRACTICE.

In each of the following examples it is required to find the equation of time corresponding to the given Greenwich date:—

1. 1882, Jan. 5th, at 4ʰ33ᵐ 0ˢ M.T.
2. „ Feb. 18th, at 8 20 0 M.T.
3. „ Mar. 24th, at 3 4 8 M.T.
4. „ April 14th, at 22 30 10 M.T.
5. „ May 19th, at 6 56 0 M.T.
6. „ June 14th, at 11 49 50 M.T.
7. „ July 16th, at 1 14 0 A.T.
8. „ Aug. 31st, at 21 14 40 A.T.
9. „ Sept. 18th, at 0 53 10 M.T.
10. „ Oct. 5th, at 19 19 2 A.T.
11. 1882, June 14th, at 12ʰ52ᵐ 0ˢ M.T.
12. „ Aug. 31st, at 15 54 0 A.T.
13. „ May 14th, at 9 36 0 A.T.
14. „ April 14th, at 22 36 53 M.T.
15. „ Nov. 14th, at 21 35 0 A.T.
16. „ July 20th, at 20 57 16 M.T.
17. „ Dec. 24th, at 18 2 54 M.T.
18. „ Oct. 26th, at 7 56 21 M.T.
19. „ Dec. 24th, at 7 18 0 A.T.
20. „ June 14th, at 6 41 20 M.T.

CORRECTION OF THE OBSERVED ALTITUDE.

306. **Def.**—The **Altitude** of a celestial body is the angular distance of the body from the horizon. It is measured by the arc of a circle of Azimuth (which is hence generally called a circle of altitude) passing through the plane of the body, or by the corresponding angle at the centre of the sphere.

307. The corrections necessary to reduce an altitude observed from the sea-horizon with a quadrant or sextant, &c., to the *true* altitude, consist of the index correction, the dip, the correction of altitude, or the joint effect of refraction and parallax, and, in certain cases, of the semi-diameter.

The altitudes of heavenly bodies are observed from the deck of a ship at sea, with the sextant, for the purpose of finding latitude, longitude, &c. Such an altitude is called the "*observed altitude*." There are certain instrumental and circumstantial sources of error by which this is affected:—(*a*) The sextant (supposed otherwise to be in adjustment) may have an index error; (*b*) The eye of the observer being elevated above the surface of the sea, the horizon will appear to be depressed, and the consequent altitude in reality too great; and (*c*) One of the limbs of the body may be observed instead of its centre. When the correction for these errors and method of observing are applied—"the index correction," "correction" for dip, and "semi-diameter,"—the observed is reduced to the *apparent altitude*. But, again, for the sake of comparison and computation, all observations must be transformed into what they would have been had the bodies been viewed through a uniform medium, and from one common centre—the centre of the earth. The altitude supposed to be so taken is called the "*true altitude;*" it may be deduced from the apparent altitude by applying the corrections called "corrections for refraction" (Table V, NORIE, or XXXI, RAPER), and "correction for parallax" (Table VI, NORIE, or XXXIV, RAPER), which, however, are sometimes given in tables combined under the names "correction of altitude" (Table XVIII, NORIE). (*a'*) "Correction for refraction;" when a body is viewed through the atmosphere, refraction will cause the apparent to be greater than the true altitude; hence the correction for refraction is subtractive in finding the true from the apparent altitude. (*b'*) "Correction for parallax;" the position of the observer on the surface, especially for near bodies, will cause the apparent to be less than the true altitude; hence the correction for parallax is additive in finding the true from the apparent altitude.

TO CORRECT THE SUN'S ALTITUDE.
RULE LXXXIV.

1°. *Correct the observed altitude of the sun for index error, if any.*

2°. *Subtract the dip answering to height of eye* (Table V, NORIE, and Table XXX, RAPER); *the remainder is the apparent altitude of the limb observed.*

3°. *Subtract the refraction* (Table IV, NORIE, and XXXI, RAPER), *add the parallax* (Table VI, NORIE, and XXXIV, RAPER); *or take out the "correction in altitude of sun"* (Table XVIII, NORIE), *and subtract it; the remainder is the true altitude of the observed limb.*

4°. *Take from page II of the month in the* Nautical Almanac *the sun's semi-diameter, adding it when the sun's* lower limb (L.L.) *is observed, but subtracting it when the sun's* upper limb (U.L.) *is observed;* the result thus obtained is the true altitude of the sun's centre.

Table 9, NORIE, and Table 38, RAPER, contain the gross correction of altitude, or the corrections for dip, refraction, sun's semi-diameter, and parallax—exclusive of index error—which are sometimes used in solving questions in nautical astronomy when great precision is not necessary.

EXAMPLES.

Ex. 1. 1882, January 6th, the observed altitude sun's L.L. 39° 8′ 30″, index correction + 33″, height of eye 19 feet: required the true altitude.

Raper.				Norie.			
Obs. alt. sun's L.L.	39°	8′	30″	Obs. alt. sun's L.L.	39°	8′	30″
Index correction	+		33	Index correction	+		33
	39	9	3		39	9	3
Dip. (Table 30)	−	4	15	Dip (Table 5)	−	4	11
App. alt. sun's L.L.	39	4	48	App. alt. sun's L.L.	39	4	52
Ref. (Table 31) } —Par. (Table 34) }	−	1	5	Corr. alt. (Table 18)	−	1	3
True alt. sun's L.L.	39	3	43	True alt. sun's L.L.	39	3	49
Semi-diameter	+	16	18	Semi-diameter	+	16	18
True altitude	39	20	1	True altitude	39	20	7

Correction of the Observed Altitude.   243

Ex. 2. 1882, June 18th, the observed altitude sun's L.L. 71° 19′ 20″, index correction + 3′ 46″, height of eye 18 feet; required the true altitude.

Obs. alt. sun's L.L.	71° 19′ 20″	Obs. alt. sun's L.L.	71° 19′ 20″
Index correction	+ 3 46	Index correction	+ 3 46
	71 23 6		71 23 6
Dip. (Table 30)	− 4 10	Dip (Table 5)	− 4 4
	71 18 56		71 19 2
Ref. − 0′ 20″ } Par. + 3 }	− 17	Corr. of alt. (Table 18)	− 17
	71 18 39	Semi-diameter	71 18 45
Semid., p. II, N.A.	+ 15 46		+ 15 46
True altitude	71 34 25	True altitude	71 34 31

Ex. 3. 1882, October 8th, the observed altitude sun's L.L. 19° 50′ 10″, index correction + 50″, height of eye 16 feet.

Obs. alt. sun's L.L.	19° 50′ 10″	Obs. alt. sun's L.L.	19° 50′ 10″
Index correction	+ 50	Index correction	+ 50
	19 51 0		19 51 0
Dip	− 4 0	Dip	− 3 50
	19 47 0		19 47 10
Ref. − 2′ 41″ } Par. + 8 }	− 2 33	Correction of altitude	− 2 29
	19 44 27	Semi-diameter	19 44 41
Semi-diameter	+ 16 3		+ 16 3
True altitude	20 0 30	True altitude	20 0 44

Ex. 4. 1882, August 8th, observed altitude sun's U.L. 12° 52′ 30″, index correction + 3′ 10″, height of eye 17 feet.

Obs. alt. sun's U.L.	12° 52′ 30″	Obs. alt sun's U.L.	12° 52′ 30″
Index correction	+ 3 10	Index correction	+ 3 10
	12 55 40		12 55 40
Dip 17 feet	− 4 5	Dip	− 3 57
	12 51 35		12 51 43
Ref. − 4′ 11″ } Par. + 8 }	− 4 3	Correction altitude	− 3 56
	12 47 32	Semi-diameter	12 47 47
Semi-diameter	− 15 49		− 15 49
True altitude	12 31 43	True altitude	12 31 58

EXAMPLES FOR PRACTICE.

1.	1882,	Jan. 29th,	Obs. alt. sun's L.L.	17° 44′ 30″	Index corr.	− 1′ 25″	Eye 16 feet.
2.	,,	Feb. 18th,	,,	48 4 10	,,	+ 0 55	,,	12	,,
3.	,,	Mar. 24th,	,,	29 50 30	,,	+ 1 3	,,	17	,,
4.	,,	April 20th,	,,	76 3 0	,,	− 1 27	,,	10	,,
5.	,,	May 8th,	,,	58 38 20	,,	− 1 10	,,	18	,,
6.	,,	June 19th,	,,	24 48 30	,,	− 1 14	,,	20	,,
7.	,,	July 16th,	,,	65 1 0	,,	+ 0 17	,,	14	,,
8.	,,	Aug. 7th,	,,	85 13 20	,,	− 2 10	,,	18	,,
9.	,,	Sept. 2nd,	,,	U.L. 28 16 20	,,	− 4 8	,,	10	,,
10.	,,	Oct. 11th,	,,	U.L. 67 44 0	,,	− 1 38	,,	15	,,
11.	,,	Nov. 15th,	,,	U.L. 14 3 40	,,	+ 4 1	,,	12	,,
12.	,,	Dec. 14th,	,,	U.L. 12 10 5	,,	− 0 49	,,	12	,,

TO FIND THE LATITUDE BY A MERIDIAN ALTITUDE OF THE SUN.

RULE LXXXV.

1°. *With the ship's date and longitude in time, find the Greenwich date* in apparent time (Rule LXXVIII, 5°, page 227).

2°. *Take the sun's declination from* Nautical Almanac, (page I of the month), *and correct it for the Greenwich date* (Rule LXXX, page 232).

Instead of proceeding according to 1° and 2° the declination may be found thus:—1. Take the sun's declination from the *Nautical Almanac*, for apparent noon, page I; and also the corresponding hourly difference. 2. Multiply the hourly diff. by long. in time, expressed in hours and decimals of an hour. 3. When the declination is *increasing* the correction is to be *added* in *West*, but *subtracted* in *East* longitude; but when the declination is *decreasing*, *subtract* in *West* but *add* in *East* longitude. See Rule LXXXI, page 235.

3°. *Correct the observed altitude for index error, dip, semi-diameter, and refraction and parallax, and thus get the true altitude* (Rule LXXXIV, page 242); *subtract true altitude from* 90°: *the result will be the true zenith distance.**

4°. *Call the zenith distance* N. *when the observer is North of sun, or when the sun bears South; call zenith distance* S. *when the observer is South of sun, or when it bears North.*

5°. Add *together the declination and zenith distance, when they have the* same *name* (see Exs. 1 and 3); *but take the* difference *if their names be* unlike (see Exs. 2, 5, and 6); *the latitude is* N. *or* S., *as the greater is*.

6°. *When the declination is* 0°, *the zenith distance is the latitude, and of the same name as the* zenith distance (see Ex. 7); *and when the zenith distance is* 0°, *the declination is the latitude, which is of the same name as the* declination (see Ex. 4).

EXAMPLES.

Ex. 1. 1882, January 15th, in longitude 72° 42′ W., the observed meridian altitude of the sun's L.L. (lower limb) was 59° 42′ 10″, bearing North; index error + 2′ 10″, height of eye 14 feet: required the latitude.

The observation was made when the sun was on the meridian, that is, at apparent noon; the date, therefore, at the place of observation is January 15th, 0ʰ 0ᵐ 0ˢ. But the meridian of the place of observation is 72° 42′ W. of meridian of Greenwich, and therefore the sun is 72° 42′ W. of meridian of Greenwich; or, in time 4ʰ 50ᵐ 48ˢ, since 72° 42′ is equivalent to 4ʰ 50ᵐ 48ˢ (see below). It is, therefore, 4ʰ 50ᵐ 48ˢ *past* apparent noon at Greenwich, and the Greenwich date is found by *adding* 4ʰ 50ᵐ 48ˢ to the time of apparent noon at ship, January 15th, thus:—

Ship date, January	15ᵈ 0ʰ 0ᵐ 0ˢ	72° 42′
Longitude 72° 42′ W. +	4 50 48	4
Greenwich date Jan.	15 4 50 48	4ʰ 50ᵐ 48ˢ

With this date the sun's declination must be taken out of *Nautical Almanac*, where it will be found in page I for January. It may be reduced to Greenwich date by means of the Tables, or by "hourly diff.," thus:—

---

* When *true* altitude exceeds 90°, subtract 90° from it.

Latitude by Sun's Meridian Altitude.

Decl., page I, N.A.		Hourly diff., page I, N.A.	
Jan. 15th, at noon	21° 5′ 15″ S. (−)	Jan. 15th	− 27″·87
Corr. for 4ʰ 51ᵐ	2 15	4ʰ 51ᵐ = 4ʰ·85	× 4·85
Red. decl.	21 3 0 S.		13935
			22296
			11148

In working this example the H. diff. for the noon of the day is taken. We divide the minutes of Greenwich time by 6; thus, 6 is contained in 51 eight times and three over, 6 is contained in 30 (the remainder) with a c added) five times; hence we have the decimal ·85, to this we prefix the hours (4), and we then have 4ʰ·85 to multiply by. As the Greenwich date wants 10ᵐ of 5 hours, we might have multiplied the hourly diff. by 5, and deducted one-sixth of hourly diff. from the product.

			(6,0)13,5·1695
		Correction	− 2 15

Raper.

		Norie.	
Obs. alt. sun's L.L.	59° 42′ 10″ N.	Obs. alt. sun's L.L.	59° 42′ 10″ N.
Index error	+ 2 10	Index error	+ 2 10
	59 44 20		59 44 20
Dip (Table 30)	− 3 40	Dip (Table 5)	− 3 36
App. alt. sun's L.L.	59 40 40		59 40 44
Refraction (Table 31)	− 34	Corr. alt. (Table 18)	− 29
	59 40 6		59 40 15
Parallax (Table 34)	+ 4	Semi-diameter	+ 16 18
True alt. sun's L.L.	59 40 10	True altitude	59 56 33
Semi-diameter	+ 16 18		90 0 0
True altitude	59 56 28	Zenith distance	30 3 27 S.
	90 0 0	Declination	21 3 0 S.
Zenith distance	30 3 32 S.	Latitude	51 6 27 S.
Declination	21 3 0 S.		
Latitude	51 6 32 S.		

The meridian zenith distance and declination are added, because they are of the *same* name. (This is according to No. 5° of the Rule.)

Ex. 2. 1882, February 3rd, in longitude 139° 42′ W., the observed meridian altitude of the sun's L.L. 56° 56′ 56″, bearing South; index correction − 3′ 4″; height of eye 14 feet.

Ship date, February	3ᵈ 0ʰ 0ᵐ 0ˢ	Norie.	
Long. 139° 42′ W.	+ 9 18 48	Obs. alt. sun's L.L.	56° 56′ 56″ S.
Green. date, February	3 9 18 48	Index correction	− 3 4
Decl., page I, N.A., Feb. 3rd = 16° 27′ 20″ S., *decr.* Hourly diff. 44″·31.		Dip (Table 5, *Norie*)	56 53 52 − 3 36
Hourly diff. Feb. 3rd, noon	44″·31	App. alt. sun's L.L.	56 50 16
Time from noon 9ʰ 18ᵐ	× 9·3	Corr. alt. (Table 18)	− 0 32
	13293	True alt. sun's L.L.	56 49 44
	39879	Semi-diameter	+ 16 16
	(6,0)41,2·083	True altitude	57 6 0
			90 0 0
Correction	− 6 52		
Decl., noon, Feb. 3rd	16 27 20 S −	Zenith distance	32 54 0 N.
		Declination	16 20 28 S.
Red. decl.	16 20 28 S.	Latitude	16 33 32 N.

By *Raper*: index. corr. − 3′ 4″; dip − 3′ 40″; refr. − 0′ 38″; par. + 4″; semid. + 16′ 16″; true alt. 57° 5′ 54″; latitude 16° 26′ 46″ N.

The difference of zenith distance and declination is taken because they are of *contrary* names. (See No. 5° of Rule.)

246 *Latitude by Sun's Meridian Altitude.*

Ex. 3. 1882, March 20th, longitude 138° 5' W., observed meridian altitude of the sun's L.L. 52° 52' 50", bearing South; index correction + 1' 5"; height of eye 12 feet.

Green. date, March 20^d 9^h 12^m		Obs. alt. sun's L.L.	52° 52' 50" S.	
or, 9^{h}·2		Index correction	+ 1 5	
			52 53 55	
Hourly diff., noon, 20th 59"·25		Dip *(Norie)*	− 3 19	
T. from noon, 20th, at 9^h 12^m × 9·2		App. alt. sun's L.L.	52 50 36	
	11850	Corr. of alt.	− 38	
	53325	True alt. sun's L.L.	52 49 58	
	6,0)54,5·100	Semi-diameter	+ 16 5	
Correction	− 9 5	True altitude	53 6 3	
			90 0 0	
Decl., March 20th, noon 0° 4' 53" S., *decr.*		Zenith distance	36 53 57 N.	
Correction − 9 5		Declination	0 4 12 N.	
Red. decl. 0 4 12 N.		Latitude	36 58 9 N.	

By *Raper*: dip − 3' 20"; refr. − 0' 44"; par. + 5"; semid. + 16' 5". True alt. 53° 6' 1", and latitude 36° 58' 11" N.

Ex. 4. 1882, April 16th, longitude 139° 50' E., observed meridian altitude sun's L.L. 89° 46' 10", bearing North; index correction + 1' 56"; height of eye 18 feet.

Ship date (A.T.), April 16^d 0^h 0^m 0^s		Obs. alt. sun's L.L.	89° 46' 10" N.	
Long. in time − 9 19 20		Index correction	+ 1 56	
Green. date, April 15^d 14 40 40			89 48 6	
		Dip *(Norie)*	− 4 4	
Time from noon, April 16^d 9 19 20			89 44 2	
Decl., page I, N.A.		Corr. alt.	0	
H. diff., noon, April 16th			89 44 2	
Green. time 9^h 19^m = 53"·11		Semi-diameter	+ 15 58	
× 9·3			90 0 0	
15933			90 0 0	
47799				
6,0)49,3·123		Zenith distance	0 0 0	
Correction − 8 14		Declination	10 2 53 N.	
Decl., noon, 16th 10° 11' 7" N. *incr.*		Latitude	10 2 53 N.	
Correction − 8 14				
Red. decl. 10 2 53 N.				

By *Raper*: index corr. + 1' 56"; dip − 4' 10"; ref., &c., 0'; semid. + 15' 58". True alt. 89° 59' 54", lat. 10° 2' 47" N.

Instead of finding Green. date as above, we may proceed as follows:—Since the longitude is 9^h 19^m 20^s *East*, the Greenwich date is therefore that amount *before* the noon of April 16th (the noon of the ship date), then the decl. and hourly diff. is taken out of the *Nautical Almanac*, page I, for the *nearest* noon to Greenwich date, viz., noon of April 16th, and hourly diff. 53"·11 is multiplied by 9^h·3; the resulting figures are 493"·123, or 8' 14", the correction. The declination at noon *increasing* will evidently be less 9^h 19^m *before* noon; therefore the correction is 8' 14" to *subtract* (see Rule LXXXV, 2°, note, page 244).

Latitude by Sun's Meridian Altitude.

Ex. 5. 1882, July 13th, longitude 100° W., observed meridian altitude sun's L.L. 68° 2′ 0″, bearing North; index correction − 25″, height of eye 17 feet.

Norie.

Ship date, July	13ᵈ 0ʰ 0ᵐ 0ˢ	
Long. in time	6 40 0	
Green. date, July	13ᵈ 6 40 0	

Obs. alt. sun's L.L.	68° 2′ 0″N.
Index correction	− 25
	68 1 35
Dip (Table 5)	− 3 57
	67 57 38
Corr. alt. (Table 18)	− 20
	67 57 18
Semi-diameter (N.A.)	+ 15 46
True alt.	68 13 4
	90 0 0
Zenith distance	21 46 56 S.
Declination	21 46 56 N.
Latitude	0 0 0

Decl., p. I., N.A., July 13th, 21° 49′ 22″ N. —
Hourly diff., July 13th, noon 21″·93
Time from noon 6ʰ 40ᵐ

```
              30ᵐ  | ½ʰ | 131·58
               10  | ⅓  |  10·96
                               3·65
                          6,0)14,6·19
Correction              −  2 26
Decl. noon, 13th        21 49 22 N.
Red. decl.              21 46 56 N.
```

By *Raper*: Index corr., − 25″; dip, − 4′ 5″; corr. alt., − 21″; semid., + 15′ 46″; true alt., 68° 12′ 55″; latitude 0° 0′ 9″ S.

The ship is on the equator.

When the zenith distance and declination are numerically equal, and of contrary names, the ship is on the equator.

Ex. 6. 1882, December 17th, longitude 175° 45′ E., observed meridian altitude sun's L.L. 89° 54′ 20″ bearing North; index correction + 4′ 4″, height of eye 24 feet.

Ship date Dec.	17ᵈ 0ʰ 0ᵐ 0ˢ	
Long. in time	11 43 0	
Green. date (A.T.) Dec.	16ᵈ 12 17 0	
T. from noon Dec.	17ᵈ 11 43 0	
	or 11ʰ·7	

Obs. alt. sun's L.L.	89° 54′ 20″N.
Index correction	+ 4 4
	89 58 24
Dip (Table 5) 24 feet	− 4 42
	89 53 42
Corr of alt. (Table 18)	0 0
	89 53 42
Semi-diameter	+ 16 18
True altitude	90 10 0
Zenith distance	0 10 0 N.
Declination	23 21 33 S.
Latitude	23 11 33 S.

Decl., page I, N.A., Dec. 17th, at noon, is 23° 22′ 34″ S. (+).

```
       Var. in 1ʰ        5″·19
       T. from noon, 17th  11·7
                         3633
                         5709
                      6,0)6,0·723

             Correction        1 1
Decl., noon, Dec. 17th   23° 22′ 34″ S. +
Corr. for 11ʰ 43ᵐ        −  1  1
Red. decl.               23 21 33 S.
```

The true altitude by *Raper's* Tables is 90° 9′ 52″, zenith dist. 0° 9′ 52″, latitude 23° 11′ 41″ S.

90° is subtracted from the true altitude; the remainder is the zenith distance, North.

The declination is taken out for the *nearest* noon to Green. date, viz., Dec. 17th, and corrected for the interval between it and the Green. time, which is equal to the longitude in time, viz., 11ʰ 43ᵐ (= 11ʰ·7).

Latitude by Sun's Meridian Altitude.

Ex. 7. 1882, September 23rd, long. 123° 45' E., observed meridian altitude sun's L.L. 40° 9', bearing North; index correction + 20", height of eye 18 feet.

Green. date (A.T.) Sept. 22nd, 15ʰ 45ᵐ		Obs. alt. sun's L.L.	40° 9' 0" N.
Time from noon 23rd, 8 15		Index correction	+ 20
			40 9 20
Decl., page I, N.A., Sept. 23rd, at noon is 0° 8' 2" S., *increasing*, hourly diff. 58"·47.		Dip 18 feet (Table 5)	− 4 4
			40 5 16
H. diff. Sept. 23rd, noon 58"·47		Corr. alt. (Table 18)	− 1 1
Time from noon, Sept. 23rd 8ʰ·25			
			40 4 15
29235		Semi-diameter, N.A.	+ 15 59
11694			
46776		True altitude	40 20 14
			90 0 0
6,0)48,2·3775			
− 8 2		Zenith distance	49 39 46 S.
Decl., Sept. 23rd 0 8 2 S.		Declination	0 0 0
Red. decl. 0 0 0		Latitude	49 39 46 S.

By *Raper:* index correction + 20"; dip − 4' 10"; refr. − 1' 9"; par. + 7"; semid. + 15' 59"; true alt. 49° 39' 54"; latitude 49° 39' 54" S.

Ex. 8. 1882, June 25th, longitude 59° 15' E., observed meridian altitude sun's U.L. 60° 24' 10" (zenith South of observer); index correction − 3' 17"; height of eye 30 feet.

Ship date (A.T.) June 25ᵈ 0ʰ 0ᵐ		Obs. alt. sun's U.L.	60° 24' 10" N.
Long. 59° 15' E. − 3 57		Index correction	− 3 17
			60 20 53
Green. date (A.T.) June 24th 20 3		Dip 30 feet (Table 5)	− 5 15
Time from noon, June 25th 3 57			
			60 15 38
or, 3ʰ·95		Corr. of alt. (Table 18)	− 29
Decl., p. I, N.A., 25th, 23° 23' 29" N. *decr.*			
H. diff., June 25th, noon 4"·07			60 15 9
T. from noon, 25th, is 3ʰ 57ᵐ = 3·95		Semi-diameter	− 15 46
2035		True altitude	59 59 23
3663			90 0 0
1221			
		Zenith distance	30 0 37 S.
Correction 16·0765		Declination	23 24 13 N.
Decl., June 25th, at noon 23° 23' 57" N. *decr.*			
Correction + 16		Latitude	6 36 24 S.
Reduced declination 23 24 13 N.			

The decl. is taken out for the *nearest* noon to Green. date, viz., June 25th, at noon, and corrected for the interval between it and the Greenwich time, which is equal to the long. in time, viz., 3ʰ 57ᵐ (= 3·95 hrs.) We might have found the correction for 4ʰ, and taken from this result the change for 3ᵐ, or one-twentieth of the hourly difference.

By *Raper:* Index corr. − 3' 17"; dip − 5' 20"; refr. − 33"·4; par. + 4"·2; Semid. − 15' 46"; true alt. 59° 59' 18"; latitude 6° 36' 29" S.

Ex. 9. 1882, August 23rd, longitude 168° 25' W., observed meridian altitude sun's L.L. 40° 5' 30"; observer N. of sun; index corr. — 54"; height of eye 12 feet.

Green. date, Aug. 23rd, 11ʰ 13ᵐ 40ˢ.

Decl., page I, N.A., August 23rd, at noon, 11° 24' 12" N., *decreasing*, hourly diff. 50"·96 × 11ʰ·23 nearly = 572"·2808 or 9' 32", the *corr.* to be *subtracted*; whence red. decl. = 11° 14' 40" N.

By *Norie*: index corr. — 54"; dip — 3' 19"; corr. of alt. — 1' 1"; semid. + 15' 52"; true alt. 40° 16' 8".

True altitude	40° 16' 8"
	90 0 0
Zenith distance	49 43 52 N.
Declination	11 14 40 N.
Latitude	60 58 32 N.

By *Raper*: index corr. — 54"; dip — 3' 20"; refr. — 1' 9"·5; par. + 7"; semid. + 15' 52"; true alt. 40° 16' 5"; latitude 60° 58' 35" N.

Ex. 10. 1883, January 1st, longitude 150° E., observed meridian altitude sun's L.L. 70° 20' (zenith N. of sun); index corr. — 30"; height of eye 19 feet.

Green. date, 1882, Dec. 31st, 14ʰ 0ᵐ.

Time from noon, Jan. 1st, 1883, or Dec. 32nd = longitude in time 10ʰ 0ᵐ.

Decl., 1882, December 32nd, 23° 0' 50" S., *decreasing*, hourly diff. 12"·39 × 10ʰ (long. in time E.) = 123"·90 or 2' 4"; whence red. decl. 23° 2' 54" S.

By *Norie*: index corr. — 30"; dip — 4' 11"; corr. alt. — 18"; semid. + 16' 18"; true altitude 70° 31' 19".

True altitude	70° 31' 19"
	90 0 0
Zenith distance	19 28 41 N.
Declination	23 2 54 S.
Latitude	3 34 13 S.

EXAMPLES FOR PRACTICE.

In each of the following examples the latitude is required:—

No.	Civil date.	Longitude.	Obs. alt. sun's L.L.	Index corr.	Eye.
1.	1882, Jan. 10th,	49° 51' W.	68° 39' 40" N.	+ 5' 10"	13 ft.
2.	„ Feb. 1st,	39 51 E.	72 43 50 S.	+ 1 42	13
3.	„ March 8th,	89 48 E.	51 49 30 S.	— 3 17	15
4.	„ April 28th,	165 23 W.	U.L. 82 51 10 N.	+ 4 10	18
5.	„ May 2nd,	32 3 E.	U.L. 46 18 0 S.	0	20
6.	„ June 11th,	62 57 E.	L.L. 42 24 45 N.	+ 2 15	21
7.	„ July 20th,	156 38 W.	51 58 30 N.	— 2 39	16
8.	„ Aug. 19th,	82 30 W.	57 41 0 S.	— 1 3	22
9.	„ Sept. 23rd,	166 30 E.	41 36 10 S.	— 4 41	17
10.	„ Oct. 23rd,	90 12 W.	54 40 40 S.	— 0 49	18
11.	„ Nov. 15th,	80 11 E.	67 43 0 S.	+ 1 38	15
12.	„ Dec. 10th,	55 20 E.	25 52 15 S.	+ 2 0	17
13.	„ Sept. 21st,	60 1 E.	56 26 0 N.	0	20
14.	„ March 20th,	103 30 W.	61 49 30 S.	— 3 17	15
15.	„ April 7th,	139 45 W.	89 55 50 S.	+ 5 10	12
16.	„ Sept. 23rd,	123 45 E.	83 40 30 S.	0	18
17.	„ Nov. 3rd,	106 0 E.	70 29 45 N.	+ 1 22	19
18.	„ Sept. 23rd,	173 58 E.	71 19 20 S.	+ 3 40	18
19.	„ Feb. 12th,	8 12 W.	29 55 20 S.	— 1 10	19
20.	„ March 20th,	77 45 E.	76 58 15 N.	— 2 20	21
21.	1883, Jan. 1st,	125 32 E.	U.L. 54 57 20 S.	+ 2 10	22
22.	1882, Oct. 1st,	71 20 E.	U.L. 82 0 15 N.	— 3 15	14

ON AMPLITUDES.

308. **The Correction or Error of Compass** is found by comparing the bearing of the sun or other celestial body, as shown by the compass, with the true bearing, as found by calculation.

309. **The True Amplitude** is the bearing of a celestial body at rising or setting (*i.e.*, when its centre is on the *rational* horizon), from the *true* East or West point, found by calculation, from the latitude of the place and declination of the body, or taken by inspection from a table, of which these quantities are the arguments (Table XLII, Norie, or LIX, Raper).

310. **The Magnetic Amplitude** is the bearing of a celestial body at rising or setting from the compass East or West points, found by direct observation with an instrument fitted with a magnetic needle, as the Azimuth Compass.

The magnetic amplitude is distinguished as *observed*, or *apparent*, and *corrected*. The *observed* or apparent magnetic amplitude of a celestial body is its bearing from the compass East or West point, when it appears in the sea-horizon of an observer standing on the deck of a ship. The *corrected* magnetic amplitude is the bearing of the body from the compass East or West point, when on the rational horizon, as it would appear to a spectator at the centre of the sphere through an uniform medium. The diurnal circles of the celestial bodies being, except at the equator, inclined to the horizon, and more and more the higher the latitude, any cause which affects the time of rising will affect the apparent amplitude, and in a greater degree as the latitude increases. The following are the causes:—1. The elevation of the observer depresses the sea-horizon, while it does not affect the place of the celestial body—hence by reason of the *dip* the body appears to rise before it is truly on the sensible horizon. 2. The great *refraction* at the horizon causes the body to appear to rise considerably before it comes to the sensible horizon. 3. When a body is in the sensible horizon, to an eye at the centre of the sphere it has already passed the rational horizon. This being the effect of *parallax*, is only of importance in the case of the moon. These corrections will be found in Table 59 A, Raper.

RULE LXXXVI.

1°. *With the ship date and longitude in time, find the Greenwich date* (see Rule LXXVIII, page 227).

The time of sunrise and sunset is generally given in apparent time.

2°. *Take out of* Nautical Almanac, page I, *the sun's declination and correct it for this date* (see Rule LXXIX, page 230).

3°. *Take from the Table the log. sine of declination, and log. secant of latitude (rejecting 10 from the index); the sum of these is log. sine of true amplitude, which take out of Tables.* (Table XXV, Norie, or LXVIII, Raper).

4°. **To name the True Amplitude.**—*If the body is rising, or* A.M., *mark true amplitude* East; *if it is setting, or* P.M., *mark it* West; *mark it also* North *when declination is* North; *or* South *when declination is* South.

The time of sun rising is always A.M., and of sun setting P.M.

(a) *When the declination is* 0, *the true amplitude is* 0; *that is, it is* East *if the object is rising,* West *if it is setting.*

(b) *When the latitude is* 0, *the true amplitude is of the same amount as the declination.*

5°. **Correction or Error of the Compass for the Position of Ship's Head.**— *Under the true amplitude write the observed amplitude; then*—

(a) *If both amplitudes are* North *or both* South, *take their difference.*

(b) *When one is* North *and the other* South, *take their sum.*

(c) *If one is reckoned from* East *and the other from* West, *take the* True Amplitude *from* 180°, *and change the name from* East *to* West, *or from* West *to* East; *the name as to* North *or* South *remains unaltered; then take their difference.*

The sum *or* difference *(as the case may be) is the* entire correction, *or* error of the compass.*

The observed amplitude must be reckoned from East or West towards the North or South, and then expressed in degrees and minutes before it is placed underneath the true. Thus, the magnetic amplitude S.E. by E. ½ E. is E. 2½ points S., or E. 28° 7′ 30″ S.

6°. **To name the Error of Compass.**—*The correction is named* East *when the true amplitude is to the* right *of observed amplitude;* West *when true is to the* left *of magnetic:* the observer being supposed looking from the centre of the compass in the direction of the observed amplitude.

NOTE.—The learner will find it very useful to draw a figure, thus:—

Make a rough sketch of the compass by drawing two lines crossing at right-angles, the ends of which will represent the four cardinal points, which mark N., S., E., W. (see Fig., Ex. 1); then to name the error of the compass proceed as follows:—Consider the cardinal point from which the amplitude is reckoned as the origin, and draw two straight lines from the centre to represent the true and magnetic amplitudes, and mark their extremities T and M respectively—taking care to place the line T further from the origin if the true be greater than the observed (or magnetic) amplitude, but nearer the origin if the true is less. The arc between M and T is the error which will be East when T is to the right of M, but West if to the left. It is easily seen whether the error of the compass is the sum or difference of T and M.

7°. **To find the Deviation.**—*Under the error of the compass place the variation: then*

(a) *If they are of* like *names, i.e., are both* East *or both* West, *take their difference.*

(b) *But if they have* unlike *names, i.e., if one is* East *and the other is* West, *take the* sum.

The sum *or* difference *(as the case may be) is the* deviation.

(c) *If the variation is* 0, *the error of the compass is also the deviation.*

(d) *If the error of the compass is* 0, *the deviation is of the same amount as the variation.*

8°. **To name the Deviation.**—*The deviation is of the same name as the error*, UNLESS *the error has been subtracted from the variation, in which case the deviation is of a contrary name to the error*, i.e., *the deviation is* E. *when the error is* W., *but* W. *when error is* E.

---

* The result as deduced above is generally called the variation, but the effects of the iron in the ship modify the bearing by compass. Every error determined on board ship is compounded of variation proper and deviation, and is the entire correction necessary to be applied to every bearing taken, and course steered, but will vary with the position of the ship's head and heel of the ship. If the iron of the vessel exercise no influence on the compass, the result obtained is only variation, and ought to agree with that registered on the chart.

Also, when the error *is* o, *the* deviation *is of* opposite name *to* variation; *when* variation *is* o, *the* deviation *is of* same name *as* error: thus—

Error 14° 10′W.	Error 14° 10′W.	Error 14° 10′W.	Error 0° 0′W.	Error 14° 10′W.
Var. 2 25 E.	Var. 2 25 W.	Var. 22 25 W.	Var. 22 25 W.	Var. 0 0
Dev. 16 35 W.	Dev. 11 45 W.	Dev. 8 15 E.	Dev. 22 25 E.	Dev. 14 10 W.

9°. Otherwise the observer must suppose himself in the *centre* of the compass, looking in the direction of the variation,—then *the deviation is* East *when the* error of compass *is to the right of the* variation; West *when the* error of compass *is to the left of the* variation—*both the error of the compass and the variation being reckoned from the* North *point of the compass*.

NOTE.—It will be convenient for beginners to draw a figure for the deviation, thus:— (See Fig. 2, Ex. 1.)

Make a rough sketch of the compass; the upper part of the vertical line being taken to represent the origin, which mark N., and mark the extremities of the horizontal line W. and E. respectively. Then from the centre of the compass draw two lines to represent the error of compass and the variation, calling them E and V respectively. The line E must be drawn to the right of N. if the error of compass is E., but to the left of N. if the error be W.; similarly, the line V must be to the right of N. when the variation is E., but to the left of N. if the variation is W. Take care to draw E further from N. than V if the error of compass is greater than the variation, but nearer to N. if the error is the less. The deviation is the distance from V to E, and is East when E is to the right of V, but West when E is to the left of V. It is easily seen whether the deviation is the sum or difference of E and V.

NOTE.—In the following examples the seconds of declination are rejected. When the seconds are 30, or above, 1 is added to the minutes; but when they are below 30 nothing is added to the minutes.

EXAMPLES.

EX. 1. 1882, January 6th, at 4ʰ 44ᵐ 27ˢ A.M., apparent time at ship, lat. 37° 59′ S., long. 36° 24′ W., the sun's magnetic amplitude was S.E. by E. ¼ E.: required the true amplitude and error of compass; and supposing the variation to be 3° 40′ E., required the deviation for the position of the ship's head at the time of observation.

```
Ship date (A.T.) Jan.   5ᵈ 16ʰ 44ᵐ 27ˢ              Or thus, H.D.
Long. in time            +   2  25  36                   18″·30
                        ─────────────                       5
Green. date (A.T.) Jan.  5   19  10   3          ─────────────
                                                 10ᵐ | ¾ʰ  91·50
Time from noon Jan. 6th      4  50 = 4ʰ·83                 −3·05
                                                 6,0 | 8,8·45
H. diff., noon, Jan. 6th      − 18″·30                      1 28
Time from noon, 4ʰ 50ᵐ        ×   4·83
                             ──────────
                                5490
                               14640
                                7320
                             ──────────
                             6,0)8,8·3890
Correction                   +   1 28
Decl., 6th, noon             22 28 38 S. (−)
                             ──────────
Red. decl.                   22 30  6 S.
```

The declination is here taken for the nearest noon, viz., the 6th, and since the Green. time wants only 4ʰ 50ᵐ of being noon of 6th, (2¼ʰ 0ᵐ − 19ʰ 10ᵐ = 4ʰ 50ᵐ), multiply the hourly diff. by this quantity, and apply the resulting correction the opposite way, since the declination is *decreasing* the declination at 4ʰ 50ᵐ before noon will be more than it is at noon, hence we add the correction. Or, multiply by 5ʰ, then since 5ʰ is 10ᵐ in excess of 4ʰ 50ᵐ, deducting 1-6th of H.D. from the product above, the result is correction.

Decl.   22° 30   sine   9·582840   Fig. 1.   N.
Lat.   37 59   secant   0·103369

                   sine   9·686209

(A.M. and S. decl.) True amp. E. 29° 3′ S.
(S.E. by E. ½ E.) Mag. amp. E. 28 7½ S. = E. 2½ pts.

    Error of compass 0 55½ E., because *true*
    amplitude is to the *right* of *magnetic* amplitude.

To find the Deviation.

Error of compass   0° 55½′ E.   Error and var. *same*   Fig. 2.   N V
Variation by chart   3 40   E.   name, take the *difference*.

    Deviation   2 44½ W., because
the *error* of compass is to the *left* of the variation.

Make a rough sketch of the compass as in Fig. 1 in the above example. In this example the magnetic amplitude is reckoned from E. towards S. (S.E. by E. ½ E. = E. 2½ pts. S. = E. 28° 7½′ S.) To represent this, draw a line from the centre of the compass to a point M, somewhere between E. or S. Again, the true amplitude is reckoned from E. towards S. To represent this, draw a line from the centre of the compass to point T, further from E than M is from E, because the true amplitude is greater than the magnetic amplitude. Then it is evident that the line T, or the true amplitude, is to the *right* of the line M, or the magnetic amplitude. Hence, by Rule, 6°, the error of the compass is East.

Again, to name the deviation:—Draw a figure (see Fig. 2 above) and mark the end of the vertical line N, to represent the true meridian (or true North point), and the extremities of the horizontal line W and E respectively, to represent West and East. Next, from the centre of the compass draw a line E (see Fig. 2) to the *right* of North, to represent the Error of the Compass, which is E.; and since the variation is also East, draw another line V to the *right* of North, but further from N than E is, because the variation is *greater* than the error. (See Fig. 2.) It is evident that the deviation is the angle included between E and V, and is East because E, the error, is to the *right* of V, the variation (Rule, 7° and 9°). It is evident too that in this instance the deviation is the difference of E and V.

Otherwise:—By 8° of Rule, the error being *subtracted from* the variation, the deviation is of the *opposite* name to the error, *i.e.*, the error being E., the deviation is named W.

Ex. 2. 1882, February 16th, at 4ʰ 58ᵐ P.M., apparent time at ship, latitude 51° 9′ N., longitude 15° W., sun's observed amplitude W. ½ N.: required the true amplitude and error of the compass; and supposing the variation to be 28° 30′ W.: required the deviation for the position of the ship's head at the time of observation.

Ship date (A.T.), Feb.   16ᵈ 4ʰ 58ᵐ   H. diff., noon, Feb. 16th — 52″·15
Long. 15° 0′ W.   +   1   0   5·97

Green. date (A.T.), Feb. 16   5 58   36505
                                                                     46935
                          or,   5ʰ·97   26075

H.D. = 52ᵐ·15   (6,0)31,1′3355
             6ʰ
                                                                    5 11·3
2ᵐ {ᵃ·ʰ   31290
            −1·74   Correction − 5 11
       6,0 | 31,1·16   Decl., noon, Feb. 16th = 15 17 S. decr.

Corr.   5 11   Red. decl.   12 10 6 S.

```
Declination   12° 10'      sine     9·323780
Latitude      51   9       secant   0·202536
                           ─────────────────
                           sine     9·526316
```

(P.M. and S. decl.) True amp. W. 19° 38' S.   Amplitudes *same*
(W. ¼ pt. N.)   Mag. amp. W. 2 49 N.   name, take the *dif-*
                                  ─────────   *ference.*
               Error of compass 22 27 W., the *true*
               amplitude being to the *left* of *magnetic*.

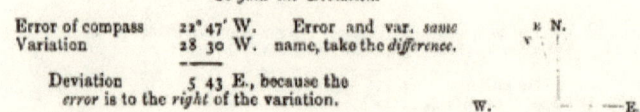

To find the Deviation.

```
Error of compass   22° 47' W.   Error and var. same
Variation          28  30  W.   name, take the difference.
                   ──────────
Deviation           5  43  E., because the
  error is to the right of the variation.
```

Otherwise:—Error being *subtracted* from variation, the deviation is of *contrary* name to the error, *i.e.*, deviation is E.

Make a rough sketch of the compass by drawing two lines crossing at right-angles; and since the magnetic amplitude is reckoned from W. and N., draw a line M somewhere between W. and N. to represent it. Again, the true amplitude is reckoned from W. towards S.; draw another line T somewhere between W. and S. to represent the true amplitude. It is easily seen that the error of the compass is the angle included between M and T, *i.e.*, the sum of the true and magnetic amplitudes; and it is evident T, the true amplitude, is to the *left* of the magnetic amplitude, the observer being supposed looking from the centre of the compass in the direction of magnetic amplitude; whence, according to Rule, 6°, the error of the compass is marked West.

To name the Deviation.—Draw another compass, and taking N. as the origin, and to represent the error of compass, draw a line E from the centre of the compass, but to the left of N., because the error is West. Again, the variation also being West, draw another line V to the left of N., but further than E is from N., because the variation is greater than error. It is easily seen that, in this instance, the deviation is the difference of V and E, and the deviation is named East, because the error is to the right of the variation; the observer being supposed looking from the centre of the compass, in the direction of the variation.

Ex. 3. 1882, April 13th, at 5ʰ 47ᵐ 20ˢ A.M., apparent time at ship, latitude 20° 2' N., longitude 107° 56' E., sun's magnetic amplitude E. ¼ N., variation 1° 40' E.

```
Ship date (A.T.)  April   12ᵈ 17ʰ 47ᵐ 20ˢ     H. diff., April 12th, noon + 54"·67
Long. 107° 56' E.            −   7  11  44                                  10·6
                          ─────────────────                               ───────
Green. date (A.T.), April 12ᵈ 10  35  36                                   32802
                                                                            5467
                            or, 10ʰ·6                                     ───────
                                                                       6,0)579·502
Declination 8° 54½'       sine    9·189922                                ───────
Latitude    20  2         secant  0·027106                                 9 39·5
                          ─────────────────
                          sine    9·217028    Correction         + 9' 39"·5
                                              Decl., 12th, noon   8° 44  51'·4 N.
(A.M. and N. decl.) True amp. E.  9° 29' N.                       ────────────
(E. ¼ pt. N.)       Mag. amp. E.  2  49  N.   Red. decl.          8  54  31   N.
                                 ──────────
               Error of compass   6  40  W., the true
               amplitude being to the left of magnetic.
```

To find the Deviation.

Error of compass	6° 40' W. }	Different names,	x N. v
Variation	1 40 E. }	take the *sum*.	
Deviation	8 20 W., because the		
	error is to the *left* of the *variation*.	W. — — E.	

Ex. 4. 1882, June 10th, at 4ʰ 45ᵐ P.M., apparent time at ship, latitude 36° 42' S., longitude 109° 30' E., magnetic amplitude W. 29° 12' N., variation 7° 20' W.: required the deviation for the position of the ship's head at the time of observation.

Ship date (A.T.), June	10ᵈ 4ʰ 45ᵐ	H. diff., noon, June 10th,	+ 11″·35
		T. from noon, 2ʰ 33ᵐ = 2ʰ·55	2·55
or June	9 28 45		
Long. 109° 30' E.	− 7 18		5675
			5675
Green. date (A.T.), June	9 21 27		2270
Time from noon, June	10 2 33 = 2ʰ·55		28·9425
Declination 23° 1½'	sine 9·592324	Correction	− 0' 29″
Latitude 36 42	secant 0·095947	Decl., 10th, noon	23° 1 59 N.
	sine 9·688271	Red. decl.	23 1 30 N.

(P.M. and N. decl.) True amp. W. 29° 12' N. } Same name,
Mag. amp. W. 29 12 N. } take their difference.

Error of compass	0 0	*Obs.*—In this instance the error of compass is 0, and the dev. is equal in amount to the variation but of an *opposite name*.
Variation	7 20	
	——— W.	
Deviation	7 20 E., because the *error* is to the *right* of the *variation*.	

Ex. 5. 1882, July 31st, at 4ʰ 26ᵐ A.M., apparent time at ship, latitude 46° 3' N., longitude 165° 58' W., sun's magnetic amplitude N.E. by E., variation 13° 0' W., ship's head E. by N.

Ship date (A.T.)	July 30ᵈ 16ʰ 26ᵐ 0ˢ	H. diff., July 31st, noon	37″·11
Long. in time	+ 11 3 52	3ʰ 30ᵐ =	3·5
Green. date	July 31 3 29 52		18555
			11133
	or 3ʰ·5		
Declination 18° 12½'	sine 9·494813		6,0)129,885
Latitude 46 3	secant 0·158622		
		Correction	− 2' 10″
	sine 9·653435	Decl., 31st, noon	18° 14 47 N decr

(A.M. and N. decl.) True amp. E. 26° 45½' N.
(E. 2¾ pts. N.) Mag. amp. E. 30 56 N.   Red. decl. 18 12 37 N.

Error of compass	4 10½ E.,	the true amplitude being to the *right* of the *magnetic*.
Variation	13 0 W.	
Deviation	17 10½ E., because the *error* is to the *right* of variation.	

Ex. 6. 1882, Sept. 22nd, at 6ʰ 0ᵐ P.M., apparent time at ship, latitude 24° 40' S., longitude 146° 15' W., sun's magnetic amplitude W. 2° 50' N., variation 7° 40' E.

Ship date, Sept.	22^d 6^h 0^m 0^s	H. diff., noon, Sept. 22nd —	58″·47
Long. 146° 15′ W., in time	+ 9 45 0		8¼^h

Green. date, Sept.   22 15 45 0

Time from Sept. 23rd   8 15

$$46776$$
$$1462$$
$$6,0)48,2\cdot 38$$

The decl. being 0°, the true amplitude is 0°, or W. 0° 0′, whence the error of compass is 2° 50′ W., because the *true* amplitude is to *left* of magnetic.

Correction   — 8 2
Decl., 22nd, noon   0 8 2 S. *incr*
Red. decl.   0 0 0

To Find the Deviation.

Error of compass   2° 50′ W.
Variation   7 40 E.

Deviation   10 30 W., because the error of compass is to the *left* of variation.

Ex. 7. 1882, December 9th, at 8^h 27^m A.M., apparent time at ship, latitude 54° 35′ N., longitude 53° 15′ W., sun's magnetic amplitude S.E. ½ E., variation 36° 20′ W., ship's head S.W. by W.

Ship date (A.T.), Dec.   8^d 20^h 27^m 0^s Decl. at noon, Dec. 9th, 22° 51′ 6″ S.
Long. in time   + 3 30 0

Green. date, Dec.   8 24 0 0   The Green. date being noon, Dec. 10th, one of the instants for which the decl. is put down in the Almanac, nothing more is necessary than to transcribe the quantity as there put down.

or Dec. 9 0 0 0

Declination   22° 51′   sine   9·589190
Latitude   54 35   secant   0·236933

sine   9·826123

(A.M. and S. decl.) True amp. E. 42° 4′ S.
(E. 3½ pts. S.)   Mag. amp. E. 39 22 S.

Error of compass   2 42 E., the true amplitude being to
Variation   36 20 W. the *right* of magnetic.

Deviation   39 2 E., because the error is to the *right* of variation.

Ex. 8. 1882, December 21st, at 4^h 31^m P.M., apparent time at ship, latitude 41° 12′ N., longitude 110° 45′ E., sun's setting amplitude S.W. ¼ W., variation 0.

Green. date, Dec. 20^d 21^h 8^m.   Decl., noon, Dec. 21st, 23° 27′ 7″ S., *incr*.
Declination 23° 27′   sine   9·599827   H. diff., noon, 21st + 0″·49 × time
Latitude   41 12   secant   0·123543   from noon, 2^h 52^m (= 2^h·87) = 1′·41 =

sine   9·723370   corr. — 1″; red. decl. 23° 27′ 6″.

(P.M. and S. decl.) True amp. W. 31° 56′ S.
(W. 3¾ pts. S.)   Mag. amp. W. 42 11 S.

Error of compass   10 15 E., the *true* amplitude being to the
Variation   0 0   *right* of magnetic.

Deviation   10 15 E.

The Variation being 0, the Error of the compass is also the Deviation.

Ex. 9. 1882, June 19th, at $9^h\ 40^m$ P.M., apparent time at ship, lat. 62° 31′ N., long. 60° 24′ W., sun's magnetic amplitude, N.N.E.; and supposing the variation of the compass is 57° 50′ W., required the deviation for the position of the ship's head at the time the observation was taken.

```
Ship date (A.T.),      June 19d 9h 40m 0s     H. diff., 19th           2"·12
Long. (60° 24′ W.) in time   +  4  1 36       T. from noon             × 13·7
Green. date (A.T.) June 19th    13 41 36      Corr. of decl.           29·044
                                 13h·7        Decl., June 19th, noon  23° 16′ 18″ N. incr.
                                                        Corr.             29
   Declination   23° 27′    sine   9·599827
   Latitude      62  31     secant 0·335837   Red. decl.              23 16 47 N.
                            sine   9·935664
```

(P.M. and N. decl.) True amplitude W. 59° 35′ S.
                                      180 0
                                      E. 120 25 N.
(E. 6 points N.) Mag. amplitude E. 67 30 N. = N.N.E.

Error of compass   52 55 W., because *true* amplitude is to the
Variation   57 50 W.   *left* of mag. amplitude.

Deviation   4 55 E., because error is to the *right* of
                                  variation.

The true and observed amplitudes must both be reckoned *from the same point* of the compass, E. or W., but in this instance one is reckoned from W. and the other from E; therefore, by taking either of them from 180°, they would both be reckoned from the same point —the true amplitude, in this example, is taken from 180°, and it is then reckoned from E. instead of W. Next take the difference of the amplitudes, as they are both marked N.; and since the true amplitude is to the *left* of the magnetic—looking from the centre of the compass in the direction of the magnetic—the error of compass is W. The error of compass and variation being of the same name, take their difference for the deviation, which mark E., because the error of the compass is to the right of variation, looking from the centre of the compass in the direction of the variation.

Ex. 10. 1882, July 1st, at $8^h\ 36^m$ P.M., lat. 56° 4′ N., long. 64° 50′ W., sun's magnetic amplitude North, variation 36° 0′ W.

```
Green. date, July 1d 12h 55m 20s     Decl., page I, N.A., July 1st, at noon,
           or        12h·92          23° 6′ 49″ N. decr., H.D. 10″·19 × 12·92
                                     gives correction — 1′ 12″, whence Red.
                                     Decl. is 23° 4′ 37″ N.
   Declination  23° 4½′   sine   9·593215
   Latitude     56  4     secant 0·153188
                          sine   9·846403
```

(P.M. and N. decl.) True amplitude W. 44° 36′ N.
(N., or W. 8 pts. N.) Mag. amplitude W. 90 0 N. (8 pts. = 90°)

                     Error of compass   45 24 W.
                     Variation   36 0 W.

                     Deviation   9 24 W.

L T.

Examples for Practice.

In each of the following examples the Error of Compass and Deviation are required for the position of the ship's head at the time of observation.

No.	Civil Date.	App. Time.	Latitude.	Longitude.	Sun's Mag. Amp.	Variation.
	1882.	h m s	° ′	° ′		° ′
1	Jan. 27th....	6 55 40 A.M.	35 42 N.	12 52 W.	S.E. by S.	21 50 W.
2	Feb. 17th ..	6 48 0 P.M.	34 57 N.	40 8 E.	S.W. by W.....	7 40 E.
3	March 29th ..	5 50 0 A.M.	25 50 S.	127 35 W.	E.S.E.	23 40 W.
4	April 5th	6 15 0 P.M.	20 10 S.	155 30 E.	W. 6° 40′ N. ..	6 40 E.
5	Nov. 7th	5 25 0 A.M.	27 41 S.	70 2 W.	E. ¼ N.	13 50 E.
6	May 26th....	7 56 0 A.M.	51 22 S.	48 0 E.	E. ¾ S.	35 20 W.
7	June 2nd	8 8 2 P.M.	52 30 N.	27 6 W.	N.N.W. ½ W. ..	37 20 W.
8	July 14th ..	6 50 58 A.M.	28 59 S.	111 11 W.	N.E. ¼ N.	11 40 E.
9	Aug. 27th ..	5 44 0 P.M.	21 4 S.	36 19 E.	N.W. ¾ W.	23 10 W.
10	Sept. 8th	5 47 0 A.M.	24 22 N.	57 30 W.	East	0 0
11	Oct. 1st	5 48 50 A.M.	42 44 S.	175 15 W.	E. ¾ N.	18 50 E.
12	Sept. 23rd ..	6 0 0 A.M.	56 41 S.	179 42 E.	E. ¼ S.	11 0 E.
13	Nov. 3rd	6 34 0 P.M.	29 20 S.	136 35 E.	W.S.W.	2 50 W.
14	Dec. 4th	7 56 48 P.M.	49 59 S.	160 45 E.	S.W. by W.	16 0 E.
15	March 20th ..	6 0 0 P.M.	55 10 N.	15 54 E.	West	15 0 E.
16	Sept. 23rd ..	6 0 0 A.M.	60 1 S.	33 45 W.	East	21 50 W.
17	June 9th	6 0 0 A.M.	0 0	10 21 W.	E. ½ N.	20 15 W.
18	Feb. 26th....	7 49 0 A.M.	62 5 N.	12 52 W.	S.S.E.	35 45 W.
19	April 30th ..	6 28 12 P.M.	24 58 N.	138 52 W.	W. by N. ½ N...	10 0 E.
20	May 27th....	7 40 0 P.M.	47 40 N.	148 3 W.	W. by N.	20 15 E.
21	June 18th ..	1 47 0 A.M.	63 54 N.	174 20 W.	N. by W. ¼ W. .	25 0 E.
22	Mar. 6th	6 14 0 P.M.	31 24 S.	2 10 E.	W. 16° 52′ N. ..	17 50 W.
23	April 10th .	6 45 0 P.M.	53 58 N.	178 33 E.	W. ¾ S.	16 10 E.
24	Dec. 14th....	4 35 0 A.M.	42 0 S.	74 56 E.	South	19 20 W.

ON FINDING THE TIME OF HIGH WATER.

"BY THE ADMIRALTY TIDE TABLES."

311. These Tide Tables, published annually, give the *time* (A.M. and P.M.) *of high water, and the height for every day in the year*, at the following places, viz.:—Brest, Devonport, Portsmouth, Dover, Sheerness, London, Harwich, Hull, Sunderland, North Shields, Leith, Thurso, Greenock, Liverpool, Pembroke, Weston-super-mare, Holyhead, Kingston, Belfast, Londonderry, Sligo Bay, Galway, Queenstown, and Waterford.

312. **To find the times of high water from the Tide Tables if the place is one of the Standard Ports, proceed by**

RULE LXXXVII.

Turn to the month in the Tide Tables and find the given place; then opposite *the* given date *will stand the* morning (A.M.) *and* afternoon (P.M.) *times of high water required.*

Note.—When the mark — occurs it shows that there is but one tide during that day; no high water, therefore, takes place in the morning or afternoon in which the mark appears.

Thus, wishing to know the time of high water at North Shields on the 9th of February, 1880—on turning to February under the head of North Shields (see page 13), it is seen at a glance that high water takes place at $2^h\ 26^m$ A.M., and that the height of tide is 12 ft. 1 in. *above the mean low water level of spring tides*, and that the time of high water on the afternoon of same day is $2^h\ 51^m$, while the height of tide above the low water level of spring tides is 12 ft. 6 in. Similarly, desiring to know the particulars of the tide at Brest on the morning of April 20th, 1880 (see page 26), the mark — shows that no tide occurs on the morning of that day; there will be a high water at $11^h\ 32^m$ P.M. on the 19th, and again at $0^h\ 7^m$ P.M. (*i.e.*, 7^m past noon) of the 20th April, but none in the interval.

Again, if it be required to know the times of high water on May 1st, 1880, at Weston-super-mare—on turning to May, and under Weston-super-mare (see page 39), and opposite the 1st we find that the times of high water are $11^h\ 24^m$ A.M., and $11^h\ 57^m$ P.M. respectively.

313. If the place at which the time of high water is required be not a **Standard Port**, it is to be referred (if in the west of Europe) to a **Standard Port**, by adding or subtracting a certain constant to the time of that Standard Port, as directed in the Tables.

In pages 103 to 108 of the Admiralty Tables, 1880, will be found upwards of two hundred ports on the coasts of the United Kingdom, and in Europe, for which Standard Ports of Reference are given, and the time which is to be added to or subtracted from the time of high water at such Standard Port.

314. **To find the times of high water by the Tidal Constants.**

RULE LXXXVIII.

1°. *Seek in the* "*Tide Tables,*" *pages* 104—108, *in the* left-hand column *for the* given place, *and in the column headed* "Standard Port for Reference," *will be found the* Standard Port *for the given place; also, from the column headed* "Time," *and* opposite *the* given place, *take out the* "Constant," *being careful to note whether it is* additive (marked +), *or* subtractive (marked —).

2°. *Take out of* "*Admiralty Tide Tables,*" *pages* 1—97, *the* morning (A.M.) *and* afternoon (P.M.) *times of high water at the* "Standard Port for Reference," *being careful to annex the letters* A.M. *or* P.M. *to the tides so taken out.*

(a) *If a blank* (——) *occurs in either* morning (A.M.) *or* afternoon (P.M.) *column, use the* preceding *time of high water instead when the Constant is marked* additive (+), *but use the* time of high water following *the blank* (——) *when the difference is marked* subtractive (—).

3°. *To the times of high water at the* Standard Port *just* taken out, apply *the* Constant (No. 2°), adding *or* subtracting *said* Constant according as it is marked + or —; *the result in each case, if less than* 12^h, *is respectively the* morning (A.M.) *and* afternoon (P.M.) *times of high water required.* (See Exs. 1 and 10).

(a) *When the* sum *of the* Constant *and the* morning (A.M.) *time of high water at the* Standard Port *exceeds* 12^h, *deduct* 12^h, *the remainder is the* afternoon (P.M.) *time of high water at the given place.* To obtain the morning (A.M.) time of high water at the given place, *if any*, add *the* Constant *to the* preceding **afternoon** (P.M.) *time of high water at the* Standard Port, *and if the* sum exceeds 12^h, deduct 12^h, *the remainder is the* morning (A.M.) tide sought (Ex. 3), *but if the* sum be less than 12^h, *it is the* afternoon (P.M.) tide of the day before, *and* there is no morning (A.M.) tide that day at the given place. (Ex. 4).

(b) *When the Constant* added *to the* morning (A.M.) *time of high water at the* Standard Port *is* **less than** 12^h (i.e., *gives* morning (A.M.) *tide at given place*), *but when* added *to the* afternoon (P.M.) *tide at the* Standard Port *is* greater *than* 12^h*, there is* only a morning (A.M.) *tide at the given place on that day.* (Ex. 5).

NOTE.—When the sum of the Constant and the tide taken from the Tables is less than 12^h, it remains a tide of the *same name as that used*, but when the sum exceeds 12^h, the time over 12^h will be a tide of *the name following* that taken out; consequently, in such a case take from the Tables the tide immediately preceding the one you require.

(c) *When the Constant is* subtractive, *and* **exceeds** *the* morning (A.M.) *time of high water at the* Standard Port, reject *this last* and use *the following* afternoon (P.M.) *tide at the* Standard Port. *If the* subtractive Constant **exceeds** the afternoon (P.M.) *tide at* Standard Port, 12^h *must be* added *to this last before* subtraction *is made, the remainder will be the* morning (A.M.) *tide at the given place. For the* afternoon (P.M.) *tide use the following tide at* Standard Port, *that is, the* morning (A.M.) *time of high water next day, borrowing* 12^h *if Constant exceeds it, the remainder is* afternoon (P.M.) *tide at the given place.*

(d) *If* Constant *being* subtractive, exceeds *the* Standard morning (A.M.) high water, *but is less than the* Standard afternoon (P.M.) *tide, there is only an* afternoon (P.M.) *tide at the given place on that day.*

(e) *If when the* Constant *is* subtractive, *the* Standard afternoon (P.M.) tide *has to be* increased 12^h, *but* Constant *is* less *than the* Standard morning (A.M.) tide following; *there is* only a morning (A.M.) *tide at the given place that day.*

EXAMPLES.

EX. 1. 1880, January 12th: find the times of high water at Scarborough.

Turning to Admiralty Tide Tables for 1880, at page 106, in the left-hand column, and under the head of "Ports of Great Britain," we find Scarborough, and in the right-hand column, immediately abreast, we find that the Standard Port for Reference in this instance is Sunderland, and in the column under Time we have the Constant $+ 0^h 49^m$, that is, we have to *add* $0^h 49^m$ to the time of high water at Sunderland on any day in order to obtain the corresponding time of high water at Scarborough. The work will stand as follows:—

Port for Reference—Sunderland, Jan. 12th $3^h 18^m$ A.M.   Jan. 12th $3^h 42^m$ P.M.
                 Constant for Scarborough $+$ 0 49 $+$ 0 49

Time of H.W. Scarborough, Jan. 12th   4 7 A.M.   Jan. 12th 4 31 P.M.

EX. 2. 1880, Feb. 17th: find times of high water A.M. and P.M. at Bordeaux.

Turning to page 107, Admiralty Tide Tables, it is seen that the Port for Reference for Bordeaux is Brest, and the Constant is $+ 3^h 3^m$, that is, the Bordeaux tides are $3^h 3^m$ later than the Brest tides, and consequently $3^h 3^m$ must be added to the time of high water at Brest on any day, to obtain the corresponding time of high water at Bordeaux.

Port for Reference—Brest, H.W., Feb. 17th $7^h 45^m$ A.M.   Feb. 17th $8^h 6^m$ P.M.
                Constant for Bordeaux $+$ 3 3 $+$ 3 3

Times H.W. Bordeaux, Feb. 17th   10 48 A.M.   Feb. 17th 11 9 P.M.

It may be here remarked, that on adding a *Constant* to the time at the Standard Port for Reference, a morning tide frequently becomes an afternoon tide, and an afternoon tide may become a morning tide *for the next day*, in which case the afternoon tide of the previous day must be employed to find the morning tide at the given port. (See 3° (a) of Rule).

Ex. 3. 1880, May 19th: find times of high water, A.M. and P.M. at Cherbourg.

The Standard Port for Reference for Cherbourg (see page 108, Admiralty Tide Tables) is Brest, and the Constant is $+ 4^h 2^m$, that is, for the times of high water at Cherbourg we must always add $4^h 2^m$ to the times of high water at Brest.

In this instance, high water at Brest, May 19th, occurs at $11^h 8^m$ A.M. (i.e., 52^m before noon); consequently, $4^h 2^m$ (the Cherbourg Constant) added to that time must evidently give a P.M. tide at Cherbourg; the A.M. high water at Cherbourg must, therefore, be sought from the previous (P.M.) tide at Brest.

In this example it will be seen that when the *additive* constant is applied to the preceding afternoon (P.M.) tide the *sum* exceeds 12^h, consequently the tide flows past midnight—the excess of 12^h being evidently the morning (A.M.) tide at Cherbourg. The work stands as follows:—

Port for Reference—Brest, H.W., May 19th	$11^h\ 8^m$ A.M.	May 18th	$10^h\ 35^m$ P.M.
Constant for Cherbourg	$+\ 4\ \ 2$		$+\ 4\ \ 2$
	$15\ 10$		$14\ 37$
	$-\ 12$		$-\ 12$
Times H.W. Cherbourg, May 19th	$3\ 10$ P.M.	May 19th	$2\ 37$ A.M.

If the morning tide, by **adding a Constant**, is more than 12^h, and thus becomes an afternoon tide, but the afternoon tide of the day before *remains less than* 12^h when the Constant is added, then there is **no morning high water** at the required port, thus:—

Ex. 4. 1880, June 22nd: find A.M. and P.M. tides at Flushing.

The Standard Port for Reference in this case is Dover, and the Constant $+\ 1^h\ 42^m$.

In this case it is high water at Dover, June 22nd, at $10^h 24^m$ A.M. (i.e., $1^h 36^m$ before noon), and the Constant $1^h 42^m$ added to that will evidently give a P.M. tide at Flushing. The preceding time of high water at Dover i.e., the time of high water in the afternoon (P.M.) of the previous day must be employed to obtain the morning (or A.M.) tide at Flushing—if any. In this example it will be seen that when the *additive* constant is applied to the preceding afternoon tide at Dover, the sum is less than 12^h, consequently, the tide does not flow *past midnight*—the result being P.M. tide of June 21st. There is, therefore, no A.M. tide on the 22nd of June at Flushing.

Time H.W. Dover, June 22nd	$10^h 24^m$ A.M.	June 21st	$9^h 54^m$ P.M.
Constant	$+\ 1\ 42$		$+\ 1\ 42$
	$12\ \ 6$	June 21st	$11\ 36$ P.M.
	$-\ 12$		
Time H.W. Flushing, June 22nd	$0\ \ 6$ P.M.	(No A.M. tide.)	

So, also, if, when the Constant be added to the morning (A.M.) time of high water, the time is less than 12^h, but when added to the afternoon time is greater than 12^h, there is **only a morning time of high water at the given port**.

Ex. 5. 1880, April 18th: find the times of high water A.M. and P.M. at Lyme Regis.

Turning to page 107 of the Admiralty Tide Tables for 1880, it will be seen that the Standard Port for Reference in this instance is Devonport, and the Constant is $+ 0^h 38^m$, i.e., in order to obtain the time of high water on any given day we must *add* $0^h 38^m$ to the times of high water at Devonport on that day.

Port for Reference—Devonport, April 18th	$10^h 48^m$ A.M.	April 18th	$11^h 27^m$ P.M.
Constant	$+\ 0\ 38$	Constant	$+\ 0\ 38$
H.W. Lyme Regis, April 18th	$11\ 26$ A.M.	April 18th	$12\ \ 5$ P.M.
		or April 19th	$0\ \ 5$ A.M.

H.W. Lyme Regis, April 18th, $11^h 26^m$ A.M.; no P.M.

It also frequently occurs that *when there is but one high water at the Standard Port*—a blank (—) occurring in one of the columns—there may *be two at the given port*.

(a) When there is only a **morning time** of high water at the Standard Port, *i.e.*, a blank occurs in P.M. column, and the *Constant is additive*, the morning tide may become an afternoon tide, and the afternoon tide of the day before may become the morning tide required, thus:—

Ex. 6. 1880, August 1st: find times of high water A.M. and P.M. at Filey Bay.

Referring to page 106, Admiralty Tide Tables for 1880, it will be seen that the Standard Port for Reference in this case is Sunderland, and the Constant is $+0^h 58^m$. There is only one high water at the Standard Port, which occurs at $11^h 43^m$ A.M., *i.e.*, $0^h 17^m$ *before noon*. The constant $+ 0^h 58^m$ being *added* to this morning time evidently gives the afternoon (P.M.) tide (see Rule 3° (a)); and the morning (A.M.) tide is sought by applying the constant to the *previous* afternoon (P.M.) tide, which occurs on July 31st, (the last day of the previous month) at $11^h 5^m$ P.M., *i.e.*, $0^h 55^m$ before midnight; the result is July 31st, $12^h 3^m$ P.M., which is equivalent to August 1st, $0^h 3^m$ A.M.

Sunderland, H.W., August 1st	$11^h 43^m$ A.M.		Sunderland, July 31st	$11^h 5^m$ P.M.
Constant for Filey	$+ 0\ 58$		Constant	$+ 0\ 58$
	12 41			12 3 P.M.
	− 12			− 12
Filey Bay, H.W., August 1st	0 41 P.M.		H.W., August 1st	0 3 A.M.

This example shows the method of proceeding when the first tide has to be taken from the first day of one month and the second tide from the last day of the preceding month.

(b) When there is only an **afternoon** (P.M.) time of high water at the Standard Port, *i.e.*, when a blank (—) occurs in the morning (A.M.) column at Standard Port, and the *Constant is additive*, the *previous* afternoon (P.M.) time of high water at the Standard port may become the morning (A.M.) time of high water at the given port, while the afternoon (P.M.) time of high water at the Standard Port may give the afternoon (P.M.) time of high water required, thus:—

Ex. 7. 1880, August 12th: find A.M. and P.M. times of high water at Stromness.

The Standard Port for Reference is Thurso, and the constant is $+ 0^h 32^m$ (page 106, Admiralty Tide Tables, 1880). A blank occurring in the morning (A.M.) column of the Standard Port for Reference, shows that there is *no morning* tide at that place, only an *afternoon* time of high water. The Constant being *additive*, we must take out the high water for the *previous* afternoon (P.M.), which is $11^h 46^m$ P.M. on the 11th, *i.e.*, $0^h 14^m$ *before* midnight on the 11th; the Stromness Constant, $+ 0^h 32^m$, *added*, evidently gives the *next morning* (A.M.) tide at Stromness, since $12^h 18^m$ past noon August 11th is evidently $0^h 18^m$ A.M. on August 12th. The afternoon (P.M.) time of high water at Stromness is found by *adding* the Stromness Constant to the afternoon (P.M.) time of high water at Thurso. The work stands thus:—

Standard Port for Reference—Thurso; Constant $+ 0^h 32^m$.

Thurso, H.W., August 11th	$11^h 46^m$ P.M.		H.W., August 12th	$0^h 9^m$ P.M.
Constant	$+\ 0\ 32$		Constant	$+\ 0\ 32$
August 11th	12 18 P.M.		H.W., August 12th	0 41 P.M.
	12			
Stromness, August 12th	0 18 A.M.			

It sometimes happens that when there is but one high water at the Standard Port, there may be but one at the given port.

(a) When a blank (———) occurs in the morning column for the given day at the Standard Port, *i.e.*, when there is *no morning* tide at that place, and the Constant being *additive*, the *previous* afternoon (P.M.) time of high water at the Standard Port may give the afternoon (P.M.) tide at the given port for the previous day, and the afternoon (P.M.) time of high water at the Standard Port for the given day gives the afternoon (P.M.) time of high water for the same day at the given port.

Finding the Time of High Water. 263

Ex. 8. 1880, July 23rd: find times of high water A.M. and P.M. at Boulogne.

The Standard Port for Reference for Boulogne is Dover, and the Constant is $+ 0^h 13^m$ (see Admiralty Tide Tables, 1880, page 108). A blank occurring in the morning (A.M.) column of Port for Reference (Dover), and the Constant being *additive* ($+$), the preceding time of high water, *i.e.*, that for the afternoon (P.M.) of the *previous* day is used to determine the morning tide at Boulogne—if any. In this example it will be seen that on applying the Constant to the previous P.M. tide at Dover, the sum ($11^h 52^m$) is *less* than 12^h, consequently it remains a P.M. tide of the preceding day—the 22nd.

There is no A.M. tide. The Constant added to the afternoon tide of given day at the Standard Port gives the required afternoon tide at Boulogne.

Time H.W. Dover, July 22nd	$11^h 39^m$ P.M.		Dover, July 23rd	$0^h 5^m$ P.M.
Constant for Boulogne	$+\ 0\ 13$		Constant	$+\ 0\ 13$
Time H.W. Boulogne, July 22nd	$11\ 52$ P.M.		Boulogne, July 23rd	$0\ 18$ P.M.

No A.M. tide; $0^h 18^m$ P.M.

Ex. 9. 1880, January 8th: find A.M. and P.M. times of high water at Whitby.

The Standard Port for Reference for Whitby is Sunderland, and the Constant $+ 0^h 23^m$. There is no afternoon tide on Jan. 8th, at Standard Port. The morning time of high water at Sunderland is $11^h 37^m$ A.M., or $0^h 23^m$ *before* noon, and the Constant $0^h 23^m$ *added* gives noon as the time of high water at the given port.

Time of H.W. Sunderland, January 8th	$11^h 37^m$ A.M.
Constant for Whitby	$+\ 0\ 23$
	$12\ 0$ A.M.
	or Noon

Ex. 10. 1880, January 23rd: find the times of high water A.M. and P.M. at the Needles Point.

Turning to the "Admiralty Tide Table" for 1880, at page 107, in the left-hand column, we find Needles Point, and in the right-hand column, immediately abreast, we find that the Standard Port for Reference which in this instance is Portsmouth, and in the column under Time we have the Constant — $1^h 55^m$, that is, we have to *subtract* $1^h 55^m$ from the time of high water at Portsmouth on any day in order to obtain the corresponding time of high water at Needles Point. The work will stand as follows:—

Port for Reference—Portsmouth, Jan. 23rd,	$8^h 26^m$ A.M.		Jan. 23rd,	$9^h 4^m$ P.M.
Constant for Needles	$-\ 1\ 55$			$-\ 1\ 55$
Times H.W. Needles, Jan. 23rd,	$6\ 31$ A.M.		Jan. 23rd,	$7\ 9$ P.M.

When the Constant is subtractive, the morning tide at the Standard Port frequently becomes an afternoon tide of the day before, and the afternoon tide of the given day becomes a morning tide, in which case the morning tide of the succeeding day must be employed to find the afternoon tide at the given port, as in the example following:—

Ex. 11. 1880, May 16th: find A.M. and P.M. tides at Portland Breakwater.

In this case the Standard Port for Reference is Portsmouth, and the first tide at Portsmouth occurs at $3^h 34^m$ A.M. (*i.e.*, $3^h 34^m$ after midnight), consequently, since Portland Constant shows that high water occurs there $4^h 40^m$ earlier than at Portsmouth, and since that quantity, subtracted from May 16th, $3^h 34^m$ A.M., would give a P.M. tide on the 15th at Portland; we therefore use the Portsmouth tide of the 16th P.M., and of the 17th A.M. thus:— (See Rule).

Time H.W. Portsmouth, May 16th	$3^h 58^m$ P.M.		May 17th	$4^h 25^m$ A.M.
	$+\ 12$			$+\ 12$
	$15\ 58$			$16\ 25$
Constant for Portland	$-\ 4\ 40$			$-\ 4\ 40$
Times H.W. Portland B'kwater, May 16th	$11\ 18$ A.M.			$11\ 45$ P.M.

When at the Standard Port there is only an afternoon high water, and the Constant is *subtractive*, and *greater* than the given time, the afternoon tide for the *given day* will give the morning tide required, and the morning time of high water for the succeeding day must be employed to determine the afternoon time, thus:—

Ex. 12. 1880, July 17th: find A.M. and P.M. tides at Falmouth.

The Standard Port for Reference is Devonport, and the Constant is — $0^h\ 46^m$. A blank (—) occurs in the morning column of the 17th, we therefore use the next tide (as the Constant is *subtractive*), viz., the P.M. tide, 12^h being added to make the subtraction, thus :—

```
Time H.W. Devonport, July 17th   0h 16m P.M.      (next tide) July 18th    0h 57m A.M.
                               + 12                     Constant —  0  46
                                 ─────                                ─────
                                 12 16                  July 18th    0  11  A.M.
        Constant for Falmouth  — 0 46
                                 ─────
Times H.W. Falmouth, July 17th 11 30 A.M.
```

Here there is no P.M. tide on July 17th at Falmouth.

The following example shows the mode of procedure when the first tide is taken out for the last day of one month, and the second tide on the first day of the succeeding month.

Ex. 13. 1880, February 29th: find A.M. and P.M. times of high water at Aberystwyth.

The Standard Port for Reference is Holyhead, and the Constant is — $2^h\ 40^m$. A blank (—) occurs in the morning column of February 29th at the Standard Port; there is, therefore, only an afternoon tide at that port on the 29th, and the Constant being *subtractive*, we use the *afternoon* time of high water at Holyhead (increased by 12^h) to obtain the morning (A.M.) time of high water at Aberystwyth; and the next time of high water at Holyhead, i.e., the time of high water at Holyhead on the morning (A.M.) of March 1st must be employed to determine the P.M. tide at Aberystwyth. The work will stand as below :—

```
H.W. Holyhead, Feb. 29th,   0h 5m P.M.        (next tide) H.W., March 1st,   0h 25m A.M.
                          + 12                                             + 12
                            ─────                                            ─────
                            12  5                                            12 25
         Constant        —  2 40                    Constant             —  2 40
                            ─────                                            ─────
H.W. Aberystwyth, Feb. 29th 9 25 A.M.           H.W. February 29th,        9 45  P.M.
```

Ex. 14. 1880, July 1st: find A.M. and P.M. times of high water at Milford Haven (entrance).

The Standard Port for Reference is Pembroke, and the Constant — $0^h\ 20^m$ (see page 105, Admiralty Tide Tables). Constant *exceeds* Standard A.M. tide, therefore reject it; but it is less than P.M.; there is only a P.M. tide at Milford Haven.

```
Time H.W. Pembroke, July 1st      0h  6m A.M.              0h 35m P.M.
              Constant           — 20                     — 20
                                                            ─────
                                                            0 15 P.M.
```

No A.M. tide on July 1st at Milford Haven.

Ex. 15. 1880, July 17th: find A.M. and P.M. times of high water at Skull.

The Port for Reference for Skull is Queenstown, and the Constant is — $0^h\ 59^m$ (see page 104, Admiralty Tide Tables, 1880). Here the first tide at Queenstown occurs at $11^h\ 50^m$ A.M., and there is no afternoon (P.M.) tide that day, as a blank occurs in that column; but the next tide occurs July 18th, $0^h\ 32^m$ A.M., and since the Constant for Skull shows that the time of high water there takes place $0^h\ 59^m$ *earlier* than at Queenstown, that quantity *subtracted* from July $18^d\ 0^h\ 32^m$ A.M., will give a P.M. tide at Skull; therefore use times of high water at Queenstown on the morning (A.M.) of July 17th, and the morning (A.M.) of July 18th; thus :—

```
Queenstown, H.W., July 17th  11h 50m A.M.       H.W., July 18th      0h 32m A.N.
              Constant     —  0  59                                 + 12
                              ─────                                   ─────
Skull, H.W., July 17th       10 51  A.M.                              12 32
                                                     Constant    —    0 59
                                                                      ─────
                                            Skull, H.W., July 17th   11 33 P.M.
```

Ex. 16. 1880, October 28th: find A.M. and P.M. times of high water at Ballycottin.

The Standard Port for Reference is Waterford, and the Constant = 0^h 20^m. In this example there is a blank in the morning column at Standard Port, we therefore take the afternoon tide, and the Constant being of the same value precisely as the time of high water, the result is a noon tide at Ballycottin.

(Only tide).   Time H.W. Waterford, Oct. 28th   0^h 26^m P.M.
                Constant   — 26

                Time H.W. Ballycottin, Oct. 28th   0   0

There is only one high water on the 28th October, and this occurs at Noon.

EXAMPLES FOR PRACTICE.

In each of the following examples it is required to find the time of high water A.M. and P.M.

1.	1880, Jan. 8th,	Maryport.	16.	1880, Jan. 24th,	Southampton.
2.	,, Feb. 29th,	Cardigan.	17.	,, June 2nd,	Abervrach.
3.	,, March 18th,	Aberystwyth.	18.	,, June 17th,	Ballycottin.
4.	,, April 14th,	Lerwick.	19.	,, June 19th,	Valentia Harbour
5.	,, May 30th,	Portland Bk'water.	20.	,, Jan. 22nd,	Bayonne.
6.	,, June 14th,	Ballycastle.	21.	,, May 23rd,	Aberdeen.
7.	,, July 8th,	Boulogne	22.	,, Dec. 29th,	Tay Bar.
8.	,, August 13th,	Wexford.	23.	,, June 16th,	Bordeaux.
9.	,, Sept. 1st,	Cadiz.	24.	,, June 10th,	Douglas.
10.	,, Oct. 11th,	Torbay.	25.	,, June 10th,	Wicklow.
11.	,, Nov. 10th,	Stromness.	26.	,, March 1st,	Poole.
12.	,, Dec. 12th,	Aberdeen.	27.	,, Nov. 11th	Ilfracombe.
13.	,, Jan. 22nd,	Gibraltar.	28.	,, Dec. 7th,	Port Rush.
14.	,, Feb. 7th,	Blyth.	29.	,, Feb. 18th,	Coleraine.
15.	,, March 25th,	Peterhead.	30.	,, Feb. 5th,	Bantry Harbour.

315. In pages 151 to 232 of Admiralty Tide Tables for 1880, are given the times of high water at full and change of a great number of ports, by which we are enabled to calculate approximately the time of high water on each day. The Constant is found by taking Brest as the Standard Port, at which place the time of high water, full and change, is 3^h 47^m. The difference between the full and change at the given port and Brest will be the Constant to be employed (as in the preceding Rules), except there be a great difference of longitude, in which case the correction for the moon's meridian passage must be employed, since for the greatest longitude this correction may amount to half an hour. Should the longitude, however, not exceed 5°, it may be neglected, as doing so will scarcely make more than a difference of one minute. It must also be observed that the longitude of Brest is 4½° W. of Greenwich, and in strictness, therefore, in determining this correction 4° should be subtracted, if the longitude of the place be East, or added if it be West. The correction is found in Table XVI, NORIE, or TABLE XXVIII, RAPER. Hence:

316. **To find the time of high water at Foreign Ports whose Constants are not given in the Tide Tables.**

RULE LXXXIX.

1°. **To find the Constant.**—*In the* Alphabetical List of Ports *at the end of the* Admiralty Tide Tables *for* 1880, page 189—232, *find the time of high water* **Full and Change**, *at* Brest, *and also that corresponding* to the given port; subtract *the less from the greater of these two times, and the remainder will be the* CONSTANT, **additive** if the full and change (F. & C.) *at the given port is greater than that of* Brest, *but* **subtractive** *if less.*

2°. *Take out the times of high water at* Brest *for the given day, and apply the Constant as directed in the preceding* Rule LXXXVIII, *pages* 259—260; *the result is the time of high water* (nearly) *at the given place.*

3°. *Take out the longitude of* Brest *and also of the given place; take the sum if the names are alike, but take the difference if the names are unlike.*

4°. *Take out* (from the column to the left of those containing the times of high water at Brest) *the moon's transit for the* proposed *day and the following one, if the long. is West; but for the given day and the preceding one if the long. is East.* Their difference, in either case, is the Daily Variation, or Retardation.

5°. *Take from* Table 28, RAPER, *or* Table 16, NORIE, *the correction corresponding to the daily variation and longitude.*

6°. *Apply this correction by* addition *in* West *longitude, but by* subtraction *in* East *longitude, to the approximate times of high water already found,* the result is the times of high water on the proposed day at the given place.

EXAMPLES.

Ex. 1. 1880, March 19th: required the time of high water at Victoria River, Turtle Point (N.W. coast of Australia), longitude 130° E.

```
Time of H.W. full and change, Victoria River   7h 15m (p. 231)
         „              „        Brest         3  47  (p. 196)
                                         Constant + 3 28

☽'s transit, March 19th, 6h 43m P.M.    Long. Victoria River  130° E.
              18th,     5  53                  „  Brest         4  W.
                        ─────                                   ────
                        50                                      126
```

Under 50m and against 126° longitude, in RAPER, Table 28, or NORIE, 16, we find 17m to be subtracted, because the longitude is E.

```
Time H.W. Brest, March 19th   8h 52m A.M.    Time H.W. Brest, March 18th   8h 22m P.M.
          Constant           + 3 28                    Constant           + 3 28
                             ───────                                      ───────
                              12 20                                        11 50 P.M.
Correction for longitude     − 0 17          Correction for longitude     − 0 17
                             ───────                                      ───────
Time H.W. at Victoria }      12  3 A.M.      Time H.W. at Victoria  }     11 33 P.M.
River, March 19th    }                       River, March 18th      }
                              0  3 P.M.                                   No A.M. tide.
```

Ex. 2. 1882, October 9th: find the times of high water at Sandy Hook, long. 74° W.

```
Time of H.W. full and change, Sandy Hook   7h 29m (p. 225)
         „              „        Brest     3  47  (p. 196)
                                     Constant + 3 42

☽'s transit, Oct.  9th   4h 42m       Long. Sandy Hook,  74° W.
                  10th   5  42             „  Brest      + 4  W.
                         ─────                           ─────
                         1  0                            78
                         60m
```

Under 60m and against 78° in longitude, in RAPER, Table 28, or NORIE 16, we find 13m to be added, because the longitude is West.

```
Time H.W. Brest, Oct. 9th   6h 46m A.M.    Time H.W. Brest, Oct. 9th    7h 13m P.M.
          Constant         + 3 42                    Constant          + 3 42
                           ───────                                     ───────
                            10 28 A.M.                                  10 55 P.M.
Correction for longitude   + 13            Correction for longitude    + 13
                           ───────                                     ───────
T. H.W. Sandy Hook, Oct. 9th  10 41 A.M.   T. H.W. Sandy Hook, Oct. 20th  11  8 P.M.
```

Ex. 3. 1882, May 25th: required the times of high water at Nelson, New Zealand, longitude 173° E.

```
    Time of H.W. full and change, at Nelson      9h 50m  (p. 217)
         "             "             Brest       3  47   (p. 196)
                                 Constant      + 6   3
                  ☽'s transit, May 25th,   0h 41m A.M.
                              23rd,       11  36  P.M.
                                          ─────────
                                           1    5
                                          ─────────
                                           65m
```

Under 65m and opposite 169° (173° − 4°) in Table 16, NOKIE, or 28, RAPKI, stands the correction 30m to be subtracted.

Time H.W. Brest, May 25th 4h 3m A.M.	Time H.W. Brest, May 25th 4h 28m P.M.
Constant + 6 3	Constant + 6 3
10 6 A.M.	10 31 P.M.
Correction for longitude − 30	Correction for longitude − 30
Time H.W. Nelson, May 25th 9 36 A.M.	Time H.W. Nelson, May 25th 10 1 P.M.

Ex. 4. 1882, August 8th: find the times of high water at Cape Virgin, Straits of Magellan, longitude 68° W.

```
    Time of H.W. full and change, Cape Virgin    8h 30m  (p. 232)
         "             "             Brest       3  47   (p. 196)
                                 Constant      + 4  43
            ☽'s transit, August 8th,   1h 46m   54m and long. 72° W. (68° + 4°) give corr. + 8m.
                          9th         2  30
                                     ────────
                                        44
```

Time H.W. Brest, Aug. 8th, 4h 56m A.M.	Time H.W. Brest, Aug. 8th 5h 12m P.M.
Constant + 4 43	Constant + 4 43
9 39	9 55
Correction for longitude + 8	Correction for longitude + 8
Time H.W. C. Virgin, Aug. 8th 9 47 A.M.	Time H.W.C.Virgin, Aug. 8th 10 3 P.M.

EXAMPLES FOR PRACTICE.

On the dates given, find the times of high water at the undermentioned places.

1. 1882, January 12th: Rio Janeiro, longitude 43° 9′ W.
2.   "   August 1st: Caraccas River, Ecuador, longitude 67° W.
3.   "   September 26th: Auckland, New Zealand, longitude 175° E.
4.   "   May 20th: Point de Galle, Ceylon, longitude 80° E.
5.   "   February 28th: San Francisco Bay, longitude 122° W.
6.   "   September 27th: Malacca Fort, longitude 102° 15′ E.
7.   "   July 12th: Port Jackson, North Head, longitude 151° 16′ E.
8.   "   July 31st: St. Julian, longitude 67° 38′ W.
9.   "   February 4th: Awatska Bay, longitude 158° 47′ E.
10. "   February 10th: Cape Cod, longitude 70° 6′ W.
11. "   January 2nd: Point de Galle, longitude 80° E.
12. "   August 24th: Macao, longitude 113° 34′ E.

GREENWICH DATE BY CHRONOMETER.

317. **The Error of Chronometer on Mean Time** at any place is the difference between the time indicated by the chronometer and the mean time at that place. *The error of chronometer on Mean Time at Greenwich* is the difference between the time indicated by the chronometer and the mean time at Greenwich. The error is said to be *fast* or *slow* as the chronometer is in advance of or behind the mean time at Greenwich.

318. **Rate of Chronometer** is the daily change in its error, or the interval it shows more or less than twenty-four hours in a mean solar day. If the instrument is going too fast, the rate is called *gaining*; if too slow, *losing*.

319. **To find the Rate.**—The rate of a chronometer is determined by comparing its errors for mean time, as found by observation at a given place, on different days. Thus, if by observation a chronometer is found 20^s *slow*, and at the end of ten days is found to be 50^s slow for mean time at the same place, it has evidently lost 30^s in ten days, whence its mean daily rate is 3^s *losing*. If on a given day, chronometer be 12^s *fast*, and at the end of thirteen days 57^s *fast* for mean time at any place, it must have gained 45^s in thirteen days, or its rate is about 3^s·5 a day *gaining*. Hence the amount of the daily rate (supposed uniform) is found by the following

RULE XC.

Write one error under the other, then
   Both errors fast, or both slow, take their difference.
   One error fast and the other slow, take their sum.
Bring the sum *or remainder into seconds, and divide by the* number *of days between the dates of the two errors; the result will be the daily rate in seconds and tenths, or (perhaps) tenths only.*

320. **To name the Rate.**—When the chronometer is *fast* either on Greenwich mean time, or on the time at place, if the error is *increasing*, the rate is *gaining*; if *decreasing*, the rate is *losing*. When the chronometer is *slow*, if the error is *increasing*, it is *losing*; if *decreasing*, it is *gaining*. When the chronometer is *fast*, and the error changes to *slow*, the rate is *losing*; if the error changes from *slow* to *fast*, the rate is *gaining*.

EXAMPLES.

Ex 1. A chronometer was 25^m 20^s *slow* for mean time at Greenwich on Nov. 20th, and on November 30th was 24^m 45^s *slow* on Greenwich mean time: required the daily rate.

November 20th, chronometer *slow*	25^m 20^s
November 30th, ,, *slow*	24 45
Change of error in 10 days	35
Rate for 1 day	3·5 *gaining*.

In this example the chronometer is *slow* on November 20th, and the error is *decreasing*, therefore the chronometer is *gaining*.

Ex. 2. A chronometer was *slow* $28^m\ 5^s$ on mean time at Greenwich, Feb. 27th, 1880, and on March 11th was *slow* $29^m\ 36^s$ on mean time at Greenwich: find daily rate.

1880, February 27th, chronometer *slow*	$28^m\ 5^s$	Feb.	29 (leap year)
1880, March 11th, ,, *slow*	29 36	Feb.	27
Change of error in 13 days	1 31	March	2
	13)91(7^s·0		11
	91	Int.	13^d

The error of chronometer, which is *slow*, is *increasing*, it is therefore *losing* 7^s·0.

Ex. 3. A chronometer was *fast* $1^m\ 23^s$ on mean time at Greenwich, June 2nd, and on July 1st was *fast* $1^m\ 37^s$·5 on mean time at Greenwich: find daily rate.

June 2nd, chronometer *fast*	$1^m 23^s$	June	30
July 1st, ,, *fast*	1 37·5	June	2
Change in 29 days	14·5	July	28
	29)14·5(0·5		1
	14·5	Int.	29^d

The error of chronometer is *fast* and *increasing*, hence the daily rate is 0^m·5 *gaining*.

Ex. 4. A chronometer was *fast* $1^m\ 51^s$ on mean time at Greenwich, May 1st, and on May 15th was 41^s *fast* on mean time at Greenwich: find daily rate.

May 1st, chronometer *fast*	$1^m 51^s$	May	1
May 15th, ,, *fast*	0 41	May	15
Change in 14 days	1 10	Int.	14

$$14 \begin{cases} 2 \\ 7 \end{cases} \begin{array}{|c} 70 \\ \hline 35 \\ \hline 5 \end{array}$$

In this example the chronometer is *fast* and the error *decreasing*, the rate therefore is *losing*.

Ex. 5. July 28th, at 3^h P.M., the chronometer was $0^m\ 6^s$·0 *fast*, and on August 4th at same time, it was $0^m\ 17^s$·1 *slow*: required the daily rate.

July 28th, at 3^h P.M., chronometer *fast*	$0^m\ 6^s$·0
August 4th, ,, ,, *slow*	0 17 ·1
Change of error in 7 days	23 ·1
	3 ·3 *losing*.

In this example the error has changed from *fast* to *slow*, the chronometer therefore is *losing*.

Ex. 6. A chronometer was *slow* $1^m\ 4^s$ on mean time at Greenwich, March 1st, and on March 23rd, was $0^m\ 19^s$·6 *fast* on mean time at Greenwich: required the rate of chronometer.

March 1st, chronometer *slow*	$1^m\ 4^s$	March	23
March 23rd, ,, *fast*	0 19·6	March	1
Change of error in 22 days	1 23·6	Int.	22^d

$$22 \begin{cases} 2 \\ 11 \end{cases} \begin{array}{|c} 83\cdot6 \\ \hline 41\cdot8 \end{array}$$

Rate 3·8 *gains*.

In this example the error of chronometer has changed from *slow* to *fast*, it is evident, therefore, that the chronometer is *gaining*.

EXAMPLES FOR PRACTICE.

1. A chronometer was *slow* $2^m\ 14^s$ on mean time at Greenwich, March 3rd, and on March 25th was *slow* $50^s\cdot4$ on mean time at Greenwich: find the daily rate.

2. A chronometer was *slow* $5^m\ 19^s$ on mean time at Greenwich, January 30th, and on February 17th was *slow* $6^m\ 13^s$ on mean time at Greenwich: find the daily rate.

3. A chronometer was $2^m\ 2^s$ *fast* on mean time at Greenwich, January 24th, and on February 10th was *fast* $3^m\ 18^s\cdot5$ for mean time at Greenwich: find the daily rate.

4. A chronometer was *slow* $49^s\cdot3$ on mean time at Greenwich, March 17th, and on April 1st was $1^m\ 58^s\cdot7$ *fast* for mean time at Greenwich: find daily rate of chronometer.

5. A chronometer was *fast* $1^m\ 4^s$ on mean time at Greenwich, January 10th, and on February 10th was $1^m\ 6^s\cdot2$ *slow* for mean time at Greenwich: required the daily rate.

6. A chronometer was *fast* $1^m\ 29^s$ on mean time at Greenwich, July 1st, and on July 23rd was *fast* $1^m\ 5^s\cdot9$ on mean time at Greenwich: find daily rate.

7. A chronometer was *fast* 48^s on mean time at Greenwich, February 28th, and on March 15th was *slow* 48^s on mean time at Greenwich: find daily rate.

8. A chronometer was *slow* 20^s on mean time at Greenwich, September 1st, and on September 15th was *fast* $1^m\ 18^s$ on mean time at Greenwich: find daily rate.

321. To find the accumulated rate proceed thus:—

EXAMPLES.

Ex. 1. If a chronometer gains $2^s\cdot6$ in a day, what will it gain in $32^d\ 16^h$?

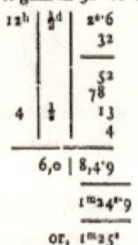

or, $1^m 25^s$.

Explanation.—Multiply the decimal $2\cdot6$ by the number of whole days, namely, 32. Next consider that 12 hours is the $\frac{1}{2}$ of 1 day, and 4 hours is the $\frac{1}{3}$ of 12 hours. 12 and 4 make up the whole number of hours, namely, 16. Divide $2\cdot6$ by 2 and the quotient $1\frac{1}{3}$ by 3 (see example). Add the products and quotients together; its sum is $84\cdot9 = 1^m\ 24^s\cdot9$; and observe that the decimal is rejected, and since it is above $\cdot5$, therefore 1 is added to the seconds.

Ex. 2. If a chronometer loses $9^s\cdot4$ in a day, what will be the accumulated loss in $12^d\ 9^h\ 34^m$?

$12^d\ 9^h\ 34^m$ may be reckoned as $12^d\ 9^h\ 30^m$

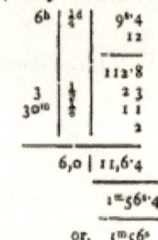

or, $1^m 56^s$.

Explanation.—Multiply by 12; then 6 hours is $\frac{1}{4}$ of 1 day, and 3 hours $\frac{1}{2}$ of 6 hours, and 30 minutes is 1-6th of 3 hours. Divide the daily rate, $9^s\cdot4$, by 4, which will give $2^s\cdot3$, the proportional part of rate in 6 hours ($\frac{1}{4}$ of a day); next, divide $2^s\cdot3$ by 2, which gives the rate for 3 hours ($\frac{1}{2}$ of 6 hours); again, divide $1^s\cdot1$ by 6, which gives the change for 30 minutes (1-6th of 3 hours); then, add the product and several quotients together, the result is the accumulated rate for the interval.

322. The accumulated rate may also be found by decimals, thus:—

RULE XCI.

1°. *Affix two cyphers to the* hours, *and divide the result by* 6, *and the quotient by* 4 (*i.e.*, divide by 24); *the last quotient is the hours expressed as decimals of a day*. (See Rule XVI, page 54).

2°. Multiply *the* days *and* decimals *of a day by the* seconds *and* decimals *of a second* (if any) *for the daily rate;* the product is the accumulated rate.

EXAMPLES.

Ex. 1. A chronometer gains 2^s·6 in a day; what does it gain in 32^d 16^h?

$$24\begin{cases} 6)16·00 \\ \overline{} \\ 4)2·66 \end{cases}$$

Prefixing the days to the decimals of a day.

32·66
2·6

19596
6532

6,0)8,4·916

1^{m}24^s·9

Ex. 2. If a chronometer loses 9^s·4 in a day, what is its loss in 12^d 9^h 34^m = 12^d 10^h (nearly)?

$$24\begin{cases} 6)10 \\ \overline{} \\ 4)1·66 \end{cases}$$

Prefixing the days to the decimals of a day.

12·41
9·4

4964
11169

6,0)11,6·654

1^{m}56^s·7
or 1^{m}57^s.

The result obtained by this rule in these examples is a little more than by the previous one of aliquot parts, as we have taken 9^h 34^m as 10^h in this, while in the other it was reckoned 9^h 30^m.

323. Before going to sea, the error of the chronometer on Greenwich mean time, and its daily rate, are supposed to have been accurately determined, either at an observatory by means of daily comparison with an astronomical clock, or by observations taken by a sextant at a place whose longitude is known.

324. When the error of a chronometer on Greenwich mean time, and also its daily rate, are known, we may determine Greenwich mean time at some other instant, as when an observation is taken, by the following

RULE XCII.

1°. *To the time by chronometer apply the original error,* adding *it if the chronometer was slow, rejecting* 24^h *if greater than* 24^h, *and putting the day one forward; but if the chronometer is* fast, subtract *original error,* increasing *time shown by chronometer by* 24^h, *if necessary, and putting the day one back.*

2°. *Find the number of* days *and parts of a day, to the nearest hour, elapsed since the original error was ascertained.*

3°. *Multiply the daily rate of chronometer by the elapsed time, and add thereto the proportionate part for the fraction of a day, found by proportion or otherwise;* the result is the accumulated rate in the interval.

4°. *To the result found by* 1° add *the accumulated rate, if chronometer is* losing; *but* subtract *if* gaining; *the result will be mean time at Greenwich, at the instant of observation.*

Examples.

Ex. 1. 1882, Jan. 30th, P.M. at ship, time by a chronometer, Jan. $29^d\,15^h\,47^m\,48^s\cdot3$, which was $9^m\,19^s\cdot6$ *slow* for Greenwich mean time Dec. 1st, 1881, and on Jan. 1st, 1882, was $10^m\,24^s\cdot7$ *slow* on Greenwich mean time.

```
    1881, Dec. 1st, slow      9m 19s·6            Dec. 31d
    1882, Jan. 1st, slow     10   24·7            Dec.  1
                             ─────────
    Change of error in 31 days  1    5·1                30
                               60                Jan.   1
                             ─────────           Int.  31d
                          31) 65·1 (2s·1 losing.
                              62
                             ─────
                              31
                              31
```

The chronometer being *slow* and the error *increasing*, the rate must be marked *losing*.

```
Time by chron., Jan.    29d 15h 47m 48s·3    ⎧ Daily rate    2s·1
Original error            + 10   24·7        ⎪                  28
                       ─────────────────     ⎪              ─────
                 Jan.   29 15  58  13·0      ⎪                 468
Accumulated rate           +    1   0·1      ⎪                  42
                       ─────────────────     ⎪              ─────
Greenwich date, Jan.    29 15  59  13·1      ⎨        h  d
                                             ⎪       12  ½    58·8
                                             ⎪        4  ⅛     1·0
                                             ⎪                  ·3
                                             ⎪              ─────
                                             ⎪          6,0) 6,0·1
                                             ⎩ Acc. rate   1m  0s·1
```
Interval from January 1st to January 19th 15h 58m is 28d 16h nly.

Ex. 2. 1880, March 20th, P.M. at ship, an observation was made when the time by chronometer was March $20^d\,0^h\,7^m\,55^s$, which was $50^m\,51^s$ *fast* on Greenwich mean time, November 22nd, 1879, and on December 21st, 1879, was *fast* $47^m\,33^s\cdot8$ for mean noon at Greenwich: required the Greenwich date by chronometer.

```
November 22nd, chron. fast    50m 51s·0      Nov. 30d
December 21st,        fast    47    33·8     Nov. 22
                             ─────────       ───────
Change of rate in 29d          3    17·2            8
                              60             Dec.  21
                             ─────────       ───────
                         29) 197·2 (6s·8 losing.  Int. 29d
                             174
                             ─────
                              232
                              232
```

The chronometer is *fast* and the *error decreasing*, the rate is therefore *losing*.

```
Time by chron., March  20d  0h  7m 55s    Dec.  31   ⎧ Rate        6s·8
                       ─────────────         21      ⎪               89
                      or 19  24   7  55              ⎪             ─────
Original error              −  47  33·8   Jan.  31   ⎪              612
                          ─────────────   Feb.  29   ⎪              544
                        23 20   21·2      March 19 23⎨   h  d     ─────
Accumulated rate         +   10  11·7                ⎪  12  ½     605·2
                       ─────────────────  Intr. 89 23⎪   8          3·4
Greenwich date, March   19 23   30  32·9             ⎪   3  ⅛       2·3
                                                     ⎪                ·8
                                                     ⎪              ─────
                                                     ⎪          6,0) 61,1·7
                                                     ⎩ Acc. rate  10m 11s·7
```

In finding accumulated rate, as the interval is within half an hour of 90 days (23½), we might have multiplied by 90, and deducted 1-48th (⅛ is 1-48th of a day) from the daily rate.

Ex. 3. Time by a chronometer, Sept. $7^d\ 23^h\ 16^m\ 28^s$, which was $57^m\ 47^s$ *slow* on Greenwich mean time, June 30th, and on July 12th, was $56^m\ 53^s$ *slow* on mean time at Greenwich.

```
    June 30th, chron. slow    57 47              June   30
    July 12th,    ,,   slow   56 53                     30
                             ——                         ——
    Change in rate in 12 days    54                      0
                                                July   12
                             Daily rate   4·5 gaining.  ——
                                                 Int.  12
```

In this instance the chronometer is *slow* and the error *decreasing*, the rate 4"·5 is therefore to be marked *gaining*.

```
    Time by chron. Sept.   7 23 16 28    July  31   ⎧ Rate        4·5
    Original error            + 56 53          12   ⎪             58
                              ————           ——    ⎪            ——
                    Sept.  8  0 13 21            19 ⎪            360
    Accumulated rate          −  4 21    Aug.  31   ⎨            225
                              ————          Sept. 8 ⎪            ——
    Greenwich date, Sept.  8  0  9  0                ⎪ 6,0)26,1·0
                                             Int. 58 ⎪            ——
                                                     ⎩ Acc. rate  4 21·0
```

EXAMPLES FOR PRACTICE.

1. 1882, February 16th, A.M. at ship, an observation was taken, when the corresponding time by a chronometer was Feb. $16^d\ 8^h\ 59^m\ 25^s$, which was $1^h\ 20^m\ 23^s\cdot4$ *fast* on Greenwich mean time, December 1st, 1881, and on January 23rd, 1882, was $1^h\ 14^m\ 23^s$ *fast* on Greenwich mean time: required the Greenwich date by chronometer.

2. A chronometer showed April $29^d\ 5^h\ 0^m\ 0^s$, which was *fast* $33^m\ 30^s\cdot3$ on Greenwich mean time, March 19th, and on March 26th was $34^m\ 20^s$ *fast* for mean time at Greenwich: required the Greenwich date by chronometer.

3. A chronometer showed May $7^d\ 6^h\ 9^m\ 48^s$, which was *slow* $11^m\ 9^s\cdot4$ on Greenwich mean time, February 16th, and on February 26th was $11^m\ 41^s\cdot6$ *slow* for Greenwich mean time: required the Greenwich date by chronometer.

4. The chronometer showed June $25^d\ 21^h\ 29^m\ 53^s$, which was $30^m\ 12^s$ *fast* on Greenwich mean time, March 31st, and on April 15th was $30^m\ 45^s$ *fast* for mean time at Greenwich: required the Greenwich date by chronometer.

5. 1882, October 25th, P.M. at ship, time by chronometer Oct. $25^d\ 8^h\ 31^m\ 10^s$, which was $12^m\ 9^s\cdot2$ *slow* on Greenwich mean time, July 20th, and on August 13th was $10^m\ 2^s$ *slow* for Greenwich mean time: required the Greenwich date by chronometer.

6. Time by chronometer January $19^d\ 13^h\ 21^m\ 25^s$, which was $53^m\ 47^s$ *fast* on mean time at Greenwich, October 24th, and on October 31st was $53^m\ 19^s$ *fast* for mean time at Greenwich: required the Greenwich date by chronometer.

7. Time by chronometer November $8^d\ 16^h\ 2^m\ 3^s$, which was $33^m\ 0^s$ *slow* on mean time at Greenwich, July 31st, and on August 12th was $32^m\ 2^s\cdot4$ *slow* on mean time at Greenwich.

8. Time by chronometer August $1^d\ 0^h\ 3^m\ 0^s$, which was $6^m\ 4^s$ *fast* on mean noon at Greenwich, May 31st, and on June 14th was $4^m\ 2^s\cdot2$ *fast* for Greenwich mean time.

9. Time by chronometer May $1^d\ 13^h\ 23^m\ 10^s$, chronometer *slow* $3^m\ 23^s$ on mean time at Greenwich, February 2nd, and on February 28th was $3^m\ 49^s\cdot0$ *slow* on Greenwich mean time.

10. Time by chronometer January $20^d\ 0^h\ 4^m\ 21^s$, which was 20^s *fast* on mean time at Greenwich, November 20th, 1881, and on December 10th, 1881, was 4^s *fast* on mean time at Greenwich: required the Greenwich date by chronometer.

11. Time by chronometer September $27^d\ 16^h\ 34^m\ 31^s$, which was $0^m\ 20^s$ *fast* on mean time at Greenwich, April 19th, and on May 9th was $0^m\ 18^s$ *slow* for Greenwich mean time: required the Greenwich date by chronometer.

12. Time by chronometer April $16^d\ 5^h\ 36^m\ 12^s$, which was $1^m\ 2^s$ *slow* for mean time at Greenwich, January 24th, and on February 28th was 29^s *fast* for Greenwich mean time.

325. When the "chronometer question" is given in a form similar to that below, we have to determine for ourselves the day of the month at Greenwich, that is, if the time shown by chronometer was 1^h, 2^h, 3^h, &c., on the civil or on the astronomical day; for, a frequent source of embarrassment in interpreting the indications of a chronometer arises from the division of its face into twelve instead of twenty-four parts, so that the same position of the pointer represents two periods of the day twelve hours distant. Thus, at 2^h past noon, and again at 14^h past noon the hands are in the same place, and it is necessary to determine whether it should be read as 2^h or 14^h, 5^h or 17^h, 6^h or 18^h past noon, and so on. To determine this point proceed according to this rule:—

RULE XCIII.

1°. *Get an approximate Greenwich date by means of ship mean time nearly and the longitude by account* (Rule LXXVIII, page 227).

2°. *Proceed, as directed in* Rule XCII, page 271, *to apply the original error and accumulated rate to the time by chronometer.*

If the difference between Greenwich dates thus found by the two methods is nearly 12^h, then the Greenwich date by *chronometer*, found as above, must be increased by 12^h, and the day put one back, so as to make the two dates agree both in the day and hour nearly.

EXAMPLES.

Ex. 1. August 3rd, at about 3^h P.M., longitude by acct. 75° W., the chronometer marks $8^h 11^m 7^s$, and is $6^m 10^s$ *fast* on Greenwich mean time: what is the Greenwich astronomical date?

Approx. T. at ship, Aug.	3^d 3^h 0^m		Time by chron.	$8^h 11^m 7^s$
Longitude 75° W.	$+$ 5 0		Error of chron. *fast*	$-$ 6 10
Approx. Green. date, Aug.	3 8 0		Green. date, Aug. 3rd	8 4 57

In this example the approximate Greenwich time is 8^h, it is evident that the chronometer must have shown 8^h from noon also.

Ex. 2. June 18th, at $10^h 52^m$ P.M., mean time at ship nearly, long. 60° W., an observation was taken, when a chronometer showed $2^h 48^m 40^s$, on June 6th its error was known to be $3^m 10^{s\cdot}2$ *fast* on Greenwich mean time, and its mean daily rate $3^{s\cdot}5$ *gaining*: required the mean time at Greenwich when the observation was taken.

Approx. Ship date, June	$18^d 10^h 52^m$	Interval from	Daily rate			$3^{s\cdot}5$
Longitude 60° W.	$+$ 4 0	June 6th to				12
Approx. Green. date	18 14 52	June 18th				—
		$14^h 52^m$ is	h	d		$42{\cdot}0$
		$12^d 14^h 52^m$	12	$\tfrac{1}{2}$		$1{\cdot}7$
		$= 12^d 15^h$	3	$\tfrac{1}{2}$		$\cdot 4$
		nearly.				
			Acc. rate			$44{\cdot}1$

Chronometer showed	$2^h 48^m 40^s$	
Original error	$-$ 3 $10{\cdot}2$ *fast*	
	2 45 $29{\cdot}8$	
Accumulated rate	$-$ $44{\cdot}1$ *gaining*	
June	19^d 2 44 $45{\cdot}7$ A.M.	
	$-$ 12 0 0	

$\therefore$ Green. date, June 18 14 44 $45{\cdot}7$

In this instance we see that 2^h by the chronometer must be reckoned as 14^h, that is, 12^h must be added to the indication of chronometer, and the day put one back.

TO FIND THE HOUR-ANGLE.

326. Given the true altitude of an object, its declination, and the latitude of the observer, to find the meridian distance or hour-angle.

RULE XCIV.

1°. *Find the polar distance by* Rule LXXXII, *page* 237.

2°. *Add together the true altitude, latitude, and polar distance; take half their sum, and from the half sum subtract the true altitude, which call the remainder.*

3°. *Add together the secant of latitude, cosecant of polar distance, cosine of half sum, and sine of remainder; the sum of these logs. (rejecting* 10 *from the index), will be the log. of sun's hour-angle* ('Table 31, NORIE); *or sine square of sun's hour-angle* (Table 69, RAPER).

When the polar distance exceeds 90°, take out the secant of reduced declination; or subtract the polar distance from 180°, and take the cosecant of the remainder. (See Rules XXXVIII and XXXIX, pages 89—90).

(a) *When both the latitude and declination are* 0, *take the true altitude from* 90°, *and so get the zenith distance, which convert into time by* Rule LXXVI, *page* 225, *or by* Table 19, NORIE, *or* Table 17, RAPER: *the result is the hour-angle.*

The hour-angle can also be found without a special table, as follows:—Find the sum of the four logs. as above, and divide by 2: the result is the log. sine of *half* the hour-angle in *arc*. From the Table of log. sines find the arc corresponding thereto, which multiplied by 2, and converted into time (Rule LXXVI, page 225), is the hour-angle sought. It is thus evident that the complete solution may be obtained by means of the Table of log. sines, &c., alone.

EXAMPLES.

Ex. 1. Given the true alt. 25° 23′ 41″, lat. 31° 17′ N., decl. 17° 9′ 8″ S., whence pol. dist. 107° 9′ 8″: find the hour-angle.

Altitude	25° 23′ 41″		
Latitude	31 17 0	sec.	0·068232
Polar dist.	107 9 8	cosec.	0·019758
	2)163 49 49		
	81 54 54	cos.	9·148115
	56 31 13	sine	9·921208
Hour-angle =	2ʰ 58ᵐ 11ˢ	log.	9·157311·3
			15724
			7 = 1

NOTE.—In NORIE, Table XXXI, the next less log. to 15731 is 15724, which gives 2ʰ 58ᵐ 10ˢ, and diff. 7 gives 1ˢ to add, whence the term corresponding to 9·15731 is 2ʰ 58ᵐ 11ˢ.

The pol. dist. being greater than 90°, take the secant of decl. for the cosec. of pol. dist., and add the prop. part for 8″.

107° 9′ 0″	cosec. 0·019753	Diff. 65
Parts for 8	+ 5	8
107° 9′ 8″	cosec. 0·019758	5,20

Observation.—Always cut off two figures when working for seconds of arc.

81° 54′ 0″	cos. 9·148915	Diff. 1481
Parts for 54	— 800	54
81° 54′ 54″	cos. 9·148115	5924
		7405
		799·74
56° 31′ 0″	sine 9·921190	Diff. 139
Parts for 13	+ 18	13
56° 31′ 13″	sine 9·921208	417
		139
		18,07

Ex. 2. Given the true altitude 17° 16′ 12″, latitude 50° 42′ S., reduced declination 20° 6′ 17″ S. (when polar dist. is 69° 53′ 43″): find the hour-angle.

Altitude	17° 16′ 12″		
Latitude	50 42 0	sec.	0·198335
Polar dist.	69 53 43	cosec.	0·027304

Sum	137 51 55		
Half sum	68 55 57	cos.	9·555660
½ sum − alt.	51 39 45	sine	9·894521
Hour-angle	5ʰ 48ᵐ 6ˢ	log.	9·67582,0
			79
			1ˢ = 3

In Norie, Table 31, we seek for the nearest log. to 0·67582, the nearest to which is 0·67570, which corresponds to 5ʰ 48ᵐ 5ˢ; then in column prop. part we seek for 3, which gives 1ˢ to add, whence hour-angle is 5ʰ 48ᵐ 6ˢ.

Ex. 4. Latitude 0°, declination 0°, true altitude 30°: required the hour-angle.

True altitude	30° 0′
	90 0
Zenith distance	60 0
	4
	6,0)24,0 0
Hour-angle	4ʰ 0ᵐ 0ˢ

Ex. 3. Given the true altitude 13° 28′ 42″, latitude 10° 35′ 0″ S., reduced declination 23° 23′ 54″ N. (or polar distance 113° 23′ 54″): find the hour-angle.

Altitude	13° 28′ 42″		
Latitude	10 35 0	sec.	0·007451
Polar dist.	113 23 54	cosec.	0·037268

Sum	137 27 36		
Half sum	68 43 48	cos.	9·559623
½ sum − alt.	55 15 6	sine	9·914694
Hour-angle	4ʰ 40ᵐ 41ˢ	log.	9·51903,6
			898
			1ˢ = 6

The nearest log. to 9·51904 is 9·51808, which gives 4ʰ 42ᵐ 40ˢ, the diff. 6 found at right hand in prop. parts gives 1ˢ, whence hour-angle is 4ʰ 40ᵐ 41ˢ.

Ex. 5. Given true altitude 75°, latitude 0°, declination 0°: find the hour-angle.

True altitude	75° 0′
	90 0
Zenith distance	15 0
	4
	6,0)6,0 0
Hour-angle	1ʰ 0ᵐ 0ˢ

EXAMPLES FOR PRACTICE.

Required the hour-angle or meridian distance in each of the following examples :—

	True altitude		Latitude		Declination
1.	11° 21′ 29″	,,	30° 15′ S.	,,	15° 21′ 4″ N.
2.	,, 30 2 4	,,	39 27 S.	,,	5 48 23 N.
3.	,, 27 48 22	,,	40 10 N.	,,	23 26 44 N.
4.	,, 34 49 46	,,	39 20 S.	,,	21 15 7 S.
5.	,, 25 38 11	,,	0 29 N.	,,	23 1 55 N.
6.	,, 15 59 13	,,	60 5 N.	,,	7 25 38 S.
7.	,, 29 2 27	,,	0 0	,,	0 0 0
8.	,, 20 34 4	,,	0 0	,,	23 27 21 N.

LONGITUDE BY CHRONOMETER,

FROM AN OBSERVED ALTITUDE OF THE SUN.

In the following problem two chronometer *errors* are given, which, by means of the elapsed time, give a daily rate : hence proceed as follows :—

RULE XCV.

1°. *Write down the time by chronometer, apply its second error*, adding *if it is* slow, subtracting *if it is* fast ; *then apply the accumulated rate*, adding *if* losing, subtracting *if* gaining (see Rule XCII) : the result is the Greenwich date at the instant of observation.

2°. *Take out of* Nautical Almanac, page II, *the sun's declination and the equation of time for the noon of Greenwich date, and the corresponding hourly difference for each : also take out the sun's semi-diameter.*

3°. *Reduce the sun's declination and equation of time to the Greenwich time* (Rules LXXIX and LXXXIII).

4°. **For the Polar Distance.**—Subtract *the reduced declination from* 90°, *if* latitude *and* declination *of* same name ; *but if of* different names add 90° *to* declination.

5°. **For the True Altitude.**—*Correct observed altitude for index error, dip, refraction and parallax, or correction in altitude, and semi-diameter, and thus get the true altitude* (Rule LXXXIV).

6°. *Find the hour-angle or meridian distance by* Rule XCII.*

7°. *When the observation is made in the afternoon, the hour-angle is* apparent time *past noon of the given day* at ship—*before which write the date at ship, but if the observation is made in the morning, take the hour-angle from* 24^h, *the remainder is* apparent time at ship *reckoned from* noon *of the preceding day, the time at place in both instances being expressed in astronomical time.*

EXAMPLES.

Ex. 1. January 6th, P.M. at ship; suppose the sun's hour-angle to be 3^h 40^m 18^s : what is the apparent time at ship?

Here the time being P.M., we have the ship date app. time, January 6^d 3^h 40^m 18^s.

Ex. 2. January 6th, A.M. at ship; suppose the sun's hour-angle to be 3^h 40^m 18^s : what is the apparent time at ship?

Here the hour-angle is   3^h 40^m 18^s
                              24 0 0

Ship date app. T., Jan. 5th   20 19 42

---

* In finding longitude by chronometer the logs. used in finding the hour-angle are required to be taken out for seconds of arc.

Ex. 3. June 1st, P.M. at ship; suppose the hour-angle to be $3^h 54^m 39^s$: required the apparent time at ship.

Here the time being P.M., we have the app. time at ship, June $1^d 3^h 54^m 39^s$.

Ex. 4. June 1st, A.M. at ship; suppose the hour-angle to be $3^h 54^m 39^s$: what is the apparent time at ship?

Hour-angle $3^h 54^m 39^s$
                 24   0   0

App. T. at ship, May 31st   20   5   21

On comparing these examples with paragraph 7°, which they are intended to illustrate, the seaman will have no difficulty in understanding that, since the sun's Hour-angle is the Distance (in time) of the object from the meridian, if the observation is made in the afternoon (P.M.), as in Ex. 1, the time will be $3^h 40^m 18^s$ *past noon* of the 6th day; that is, the ship date (astronomical time) is January $6^d 3^h 40^m 18^s$—the astronomical day commencing always at noon; but if the observation be made in the morning (A.M.), the hour-angle will be the *time before noon* of the 6th day; or, as shown in Ex. 2, $20^h 19^m 42^s$ past noon of the day before, —that is, January $5^d 20^h 19^m 42^s$. In Ex. 3, similarly, the observation being P.M., the time will be $3^h 54^m 39^s$ past noon of June 1st, while in Ex. 4, the observation being A.M., the time will be $3^h 54^m 39^s$ *before noon* of June 1st, i.e., $20^h 5^m 21^s$ past noon, May 31st.

Obs.—In the new edition of NORIE's Tables the hour-angle is so arranged that when the observation is made P.M. at ship it is read from the top of the page; when A.M., from the bottom, using the next greatest log. to the given one, and it will be the apparent time at ship, reckoning from the day before the ship date; in which case the necessity of deducting from 24^h (as explained in paragraph 7°) is obviated.

7°. *To apparent time apply the reduced equation of time*, adding or subtracting as directed in page I, Nautical Almanac, *and so get mean time*.

8°. *Under ship mean time put Greenwich mean time*—not forgetting the day in each case:—*subtract the less from the greater; the remainder is longitude in time, which convert into arc* $°'''$; see Rule LXXVII, page 226, or Table 17, RAPER, or Table 19, NORIE.

In taking the difference of Greenwich mean time and ship mean time, if the days of the month be different, it will be necessary to add 24 to the hours of the more advanced (that is, the one whose days are most), in order to enable the subtraction to be made.

9°. *Call the longitude* West *when Greenwich time is* greater *than ship mean time; but* East *when Greenwich mean time is least*.

NOTE.—When the latitude at noon is given, the latitude in at the time of observation must be found by means of the course steered and distance sailed. The diff. of lat. *from noon* is to be named North or South, according as the ship at the time of observation is North or South of her latitude at noon. When the longitude is found, as in Exs. 1 to 10 (or according to paragraphs 8° and 9°, above), the diff. of long. between the ship at the time of observation and noon must be applied to find the longitude at noon. The diff. of long. is to be named East or West, according as the ship is East or West of its position at noon.

EXAMPLES.

Ex. 1. 1882, January 11th, P.M. at ship, latitude 49° 30' N., the observed altitude sun's L.L. was 12° 20' 30", height of eye 18 feet, time by a chronometer January $11^d 6^h 44^m 36^s$ (being P.M. at Greenwich), which was $6^m 8^s \cdot 3$ *fast* for mean noon at Greenwich, September 1st, 1881, and on September 30th, 1881, was $8^m 42^s$ *fast* on Greenwich mean time: required the longitude by chronometer.

     Sept. 1st, chron. *fast* $6^m 8^s \cdot 3$ Sept. 30
        30th,   ,,   *fast* $8\ 42 \cdot 0$ Sept.   1
     Change in 29 days $2\ 33 \cdot 7$ Int.   29

$29) 153 \cdot 7 (5^s \cdot 3$

```
T. by chron., Jan.  11ᵈ 6ʰ44ᵐ36ˢ      (a) Interval from       Obs. alt. ☉'s L.L.  12° 20' 30"
Original error     = —   8 42         Oct. 1st to Jan. 11th    Dip                —    4  4
                                      6½ʰ is 103ᵈ 6ʰ.
                      11  6 35 54                                                   12 16 26
Accumulated rate      —    9  7       Daily rate      5ˢ·3     Corr. alt.           4  9
                                                      103
Green. date, Jan.  11  6 26 47                                                      12 12 17
          Sept.  30                                   159      Semi-diameter     + 16 18
                 30                                   530
                                                               True alt.           12 28 35
                  0                                6  ⅔  545·9
          Oct.   31                                ⅓  1/12  1·3   By Raper's Tables: dip —
          Nov.   30                                        ·1    4' 10", rofr. — 4' 23," par. +
          Dec.   31                                              9", semid. + 16' 18", and true
          Jan.   11                                6,0)54,7·3    alt. 12° 28' 24".

                 103ᵈ                               Acc. rate   9 7·3
   (a) The chronometer having been rated, September 30th, there are no days left in September.

                                                   H.D.                                H.D.
Decl., page II, N.A.                                          Eq. T. page II, N.A.
Jan. 11th, noon    21° 46' 37" S. decr.            23·72      Jan. 11th, noon  8ᵐ14ˢ·1 incr.   ·980
Correction         —   1 33                         6·45      Correction       +  6·3         6·45

Red. decl.         21 44  4 S.                     11860      Red. Eq. T.        8 20·4        490
                                                    9488      To be added to A.T.              392
Polar dist.       111 44  4                        14232                                       588

                                                  6,0)15,2·9940                              6·32,0

                                  Corr. — 2 33
Altitude    12° 28' 35"                             App. T. at ship, Jan.    11ᵈ 2ʰ17ᵐ42ˢ
Latitude    49 30  0     sec.     0·187456          Eq. time                 +      8 20
Polar dist. 111 44  4    cosec.   0·031026
                                                    M.T. ship, Jan.          11   2 26  2
            173 42 39                               M.T. Green., Jan.        11   6 26 47

             86 51 19    cos.     8·739240          Long. in time                 4  0 45
                                                     Longitude   60° 11' 15" W.
             74 22 44    sine     9·983655          By Raper:  log. sine sq. (sum of logs.)
                                                    8·942574 gives hour-angle 2ʰ 17ᵐ 44ˢ, long.
Hour-angle  2ʰ17ᵐ42ˢ     log.     8·942377          60° 10' 45" W.
```

The observation having been made in the afternoon (P.M.) at ship, the hour-angle 2ʰ 17ᵐ 42ˢ will be 2ʰ 17ᵐ 42ˢ *past noon* of the 11th day, that is, the apparent time at ship (astronomical time) is January 11ᵈ 2ʰ 17ᵐ 42ˢ (see Rule XCV, 7°, page 277).

Ex. 2. 1882, May 20th, P.M. at ship, latitude 50° 43' N., obs. alt. sun's L.L. 17° 10', index corr. — 1' 39", height of eye 18 feet, time by a chronometer May 20ᵈ 0ʰ 19ᵐ 53ˢ (or 0ʰ 19ᵐ 53ˢ P.M.), which was 33ˢ *fast* on mean noon at Greenwich, March 20th, and on April 1st was 23ˢ·4 *fast* on Greenwich mean time.

```
             March     31             March 20th, chron. fast       0ᵐ 33ˢ
                       20             April 1st     ,,    fast       0  23·4

                       11             Change in 12 days            12)9·6
             April     1
                       —              Daily rate                     0·8 losing.
             Int.     12ᵈ

T. by chron., May 20ᵈ  0ʰ 19ᵐ 53ˢ      Interval from            Obs. alt. ☉'s L.L.  17° 10'  0"
Original error fast    —     23·4      April 1st to May         Index corr.         —   1 39
                                       20th is 49ᵈ.
                       0 19 29·6                                                      17  8 21
Accumulated rate       +    39·2       Rate           0ˢ·8      Dip                 —    5  5
                                       Interval    × 49
Green. date, May 20  0 20  9                                                          17  3 16
                                       Acc. Rate     39·2       Corr. alt.         —    2 55

                                                                                      17  0 21
   By Raper: dip — 5' 10", rofr. — 5' 9", par.                  Semi-diameter      + 15 50
+ 8", semid. + 15' 50", true alt. 17° 16' 0".
                                                                True alt.             17 16 11
```

280   *Longitude by Chronometer.*

H. diff., May 20th	$+31''\cdot18$	H. diff., May 20th	$-0''\cdot139$
$20^m \mid \frac{1}{3}^h \mid$	$31\cdot18$	$20^m \mid \frac{1}{3}^h \mid$	$0\cdot139$
	$+10\cdot39$		$-\cdot046$
Correction		Correction	
Decl., May 20th, noon	20 1 1 N. inc.	Eq. Time, May 20th, noon	$3^m 42^s\cdot79$
Red. decl.	20 1 11 N.	Red. Eq. time	3 42·75
	90	(To be *subtracted* from A.T.)	
Polar dist.	69 58 49		

Altitude	17° 16′ 11″			App. T. Ship, May 20th $5^h 47^m 46^s$
Latitude	50 43 0	sec.	0·198489	Eq. Time $-$ 3 43
Polar dist.	69 58 49	cosec.	0·027168	
	137 58 0			Mean T. Ship, May 20th 5 44 3
				Mean T. Green., May 20th 0 20 9
Half sum alt.	68 59 0	cos.	9·554658	Long. in time 5 23 54
	51 42 49	sin.	9·894827	Longitude 80° 58′ 30″ E.
Hour-angle	$5^h 47^m 46^s$	log.	9·675142	By *Raper*: Log. sine sq. (sum of logs.) 9·675184 gives hour-angle $5^h 47^m 47^s$, long. 80° 58′ 45″ E.

Since the sun's hour-angle is the distance (time) of the object from the meridian, and the observation is
P.M. in this example, the time (or hour-angle) will be $5^h 47^m 46^s$ past noon of the 20th day; hence the app.
time at ship is May $20^d 5^h 47^m 46^s$ (see Rule XCV, 7°, page 277).

Ex. 3. 1882, July 3rd, A.M. at ship, latitude 32° 10′ S., obs. alt. sun's L.L. 14° 10′ 15″,
index corr. $+$ 1′ 22″, height of eye 19 feet, time by a chronometer July $2^d 16^h 33^m 22^s$ (being
$3^d 4^h 33^m 22^s$ A.M. at Greenwich), which was $17^m 16^s \cdot 4$ *fast* for Greenwich mean noon, May
16th, and on June 1st was $16^m 22^s$ *fast* for mean time at Greenwich: required the longitude.

May 16th, Chronometer *fast* $17^m 16^s \cdot 4$
June 1st,   ,,   *fast* 16 22·0

Change in 16 days   54·4
Daily rate   3·4 *losing.*

T. by chron., July	$2^d 16^h 33^m 22^s$	Interval from	Obs. alt. ☉′ L.L.	14° 10′ 15″
Original error	$-$ 16 22	June 1st to July	Index corr.	$+$ 1 22
		2nd, 16^h, is 31^d 16^h.		
	2 16 17 0			14 11 37
Accumulated rate	$+$ 1 48	Daily rate 3·4	Dip	$-$ 4 11
		Interval 31		
Green. date, July	2 16 18 48			14 7 26
		34	Corr. alt.	$-$ 3 35
		102		
				14 3 51
		12 $\mid \frac{1}{2} \mid$ 1054	Semi-diameter	$+$ 15 46
		4 $\mid \frac{1}{8} \mid$ 17		
		6	True alt.	14 19 37
			By *Raper*: index corr. $+$	
		6,0)10,7·7	1′ 22″, dip $-$ 4′ 15″, refr. $-$	
			3′ 47″, par. $+$ 8″, semid. $+$	
		Acc. rate 1 47·7	15′ 46″, true alt. 14° 19′ 29″.	

Decl., page II, N.A.		H.D.	Eq. T., page II, N.A.	H.D.
July 2nd, noon 23° 2′ 33″ N. *decr.*		$-$ 11″·20	2nd, noon, *add* $3^m 43^s \cdot 3$	$+$ 0^s·468
Correction $-$ 3 3		16·3	Correction $+$ 7·6	$\times$ 16·3
Red. decl. 22 59 30 N.		336	Red. Eq. Time 3 50·9	1404
90		672	(To be *added* to A.T.)	2808
		112		468
Polar dist. 112 59 30		6,0)18,2·56		Corr. $+$ 7·6284
		Corr. $-$ 3 3		

Altitude	14° 19' 37"			App. Time ship, July 2nd	20ʰ 23ᵐ 2ˢ
Latitude	32 10 0	sec.	0·072371	Eq. Time	+ 3 51
Polar dist.	112 59 30	cosec.	0·035947		
				Mean time ship, July 2nd	20 26 53
	159 29 7			Mean time Green., July 2nd	16 18 48
	79 44 33	cos.	9·250597	Longitude in time	4 8 5
	65 24 56	sine	9·958731	Longitude	62° 1' 15" E.
Hour-angle	3ʰ 36ᵐ 58ˢ	log.	9·317646	By *Raper*: Log. sine sq. 9·317694 gives hour-angle 3ʰ 36ᵐ 58ˢ, longitude 62° 1' 15" E.	

A.T. ship, July 2ᵈ 20 23 2

In this example the time of observation is A.M. at ship, the hour-angle is therefore the time *before noon* of July 3rd, and as all calculations are made in the astronomical day, we take the hour-angle from 24ʰ, the remainder is A.T. at ship, reckoned from *noon of the preceding day*, viz., that of July 2nd. Since the observation is made in the forenoon at ship we may take the hour-angle from the bottom of the table, and the time so found is A.T. at ship, reckoned from noon of the preceding day. (See Rule XCV, page 277).

Ex. 4. 1882, April 14th, A.M. at ship, latitude 52° 10' N., observed altitude sun's L.L. 18° 20' 25", index corr. + 55", height of eye 12 feet, time by chron. 14ᵈ 5ʰ 5ᵐ 5ˢ (being P.M. at Greenwich), which was *fast* 5ᵐ 52ˢ·4 for mean noon at Greenwich, February 14th, and on February 26th was *fast* 6ᵐ 38ˢ for mean noon at Greenwich.

February 14th, chronometer *fast* 5ᵐ 52ˢ
February 26th, ,, *fast* 6 38

Int. 12 12) 46

3·8 *gaining*.

T. by chron., Ap.	14ᵈ 5ʰ 5ᵐ 5ˢ	Interval from Feb. 26th to April 14th 5ʰ, is 48ᵈ 5ʰ.		Obs. alt. ☉'s L.L.	18° 20' 25"
Original error	− 6 38			Index corr.	+ 55
	14 4 58 27	Rate	3·8		18 21 20
Accumulated rate	− 3 3	Interval	48	Dip	− 3 19
Green. date, April	14 4 55 24				18 18 1
		4 ¾ 182·4		Corr. alt.	− 2 42
		1 ¾ ·6			
		·1			18 15 19
By *Raper*: Index corr. + 55", Dip. − 3' 20", refr. − 2' 55", par. + 8", semid. + 15' 58", true alt. 18° 31' 11".					+ 15 58
		3 3·1		True alt.	18 31 17
Decl., page II, N.A.		H.D.		Eq. T., page II, N.A.	H.D.
Ap. 14th, noon 9° 29' 18" N. *incr*.		+ 53·92		14th, noon, *add* 0ᵐ 13ˢ·98 *decr*.	0ˢ·629
Correction + 4 24		× 4·9		Correction − 3·08	× 4·9
Red. decl.	9 32 42 N.	6,0)26,4·208		Red. Eq. T. 0 10·90 (To be *added* to A.T.)	3·0821
Polar dist.	80 27 18	+ 4 24 corr.			
Altitude	18° 31' 17"			App. T. ship, April	13ᵈ 19ʰ 11ᵐ 55ˢ
Latitude	52 10 0	sec.	0·212280	Eq. Time	+ 11
Polar dist.	80 27 18	cosec.	0·006055		
				Mean T. ship, April	13 19 12 6
	151 8 35			Mean T. Green., April	14 4 55 24
	75 34 17	cos.	9·396501	Longitude in time	9 43 18
	57 3 0	sine	9·923827	Longitude	145° 49' 30" W.
Hour-angle	4ʰ 48ᵐ 5ˢ	log.	9·538673	By *Raper*: Log. sine sq. 9·538703 gives hour-angle 4ʰ 48ᵐ 6ˢ, long. 145° 49' 45" W.	
	24				

A.T.S.Ap.13ᵈ 19 11 55

The observation having been made A.M. at ship, the hour-angle is the time *before noon* at ship, viz., that of April 14th; therefore take the hour-angle from 24ʰ, and the remainder is apparent time at ship, reckoned from the preceding noon, April 13th; or if the hour-angle is taken from the bottom of the table (it being A.M. at ship) it is the apparent time—April 13ᵈ 19ʰ 11ᵐ 54ˢ—reckoning from the preceding noon at ship. The Greenwich and ship dates being different (the Greenwich date being the greater, or in advance of ship date), therefore subtract the latter from the former (borrowing 24ʰ to enable the subtraction to be made).

Ex. 5. 1882, March 6th, P.M. at ship, latitude 40° 20′ S., obs. alt. sun's L.L. 16° 20′ index corr. + 30″, height of eye 18 feet, time by a chronometer 5ᵈ 20ʰ 10ᵐ (being 6ᵈ 8ʰ 10ᵐ A.M. at Greenwich), which was 6ᵐ 14ˢ *fast* for mean noon at Greenwich, January 30th, and on February 13th was *fast* 4ᵐ 29ˢ for mean time at Greenwich.

```
        January 30th, chronometer fast  6ᵐ 14ˢ              ( 2)105ˢ
        Feb.    13th,       ,,    fast  4  29          14 {
        Change in 14 days.              1  45              ( 7) 52ˢ5
                                          60
                                        ─────                7ˢ5 losing.
                                        or 105ˢ         because fast and less fast.
```

```
T. by chron., March   5ᵈ 20ʰ 10ᵐ 0ˢ     Interval  from     Obs. alt. ☉'s L.L. 16° 20′  0″
Original error           −    4 29      Feb. 13th to March  Index corr.         +    30
                         ─────────      5th 20ʰ, is 21ᵈ 20ʰ.                    ─────────
                      5 20   5 31           Rate     7ˢ5                        16 20 30
Accumulated rate      +      2 44                     21    Dip                 −    4  4
                         ─────────                                              ─────────
Green. date, March    5 20   8 15    12 | ¾ | 157·5                             16 16 26
                                      8 | ⅓ |   3·7    Corr. alt.               −    3  5
T. from noon, 6th            3ʰ 52ᵐ          2·5                                ─────────
                                                                                16 13 21
By Raper: Index corr. + 30″,          6,0)16,3·7        Semid.                  +   16  9
   dip −4′ 10″, refr. 3′ 18″, par. + 8″,                                        ─────────
   semid. + 16′ 9″, true alt. 16°29′19″.    2 43·7      True alt.               16 29 30
```

```
Decl., page II, N.A.                      H.D.      Eq. time, page II, N.A.                  H.D.
6th noon      5° 35′ 13″S. decr.          58·14     6th noon     + 11ᵐ 24ˢ·7  decr.          0·598
Correction    +    3 45                    3·87     Correction   +       2·4                 ×   4
              ─────────                             ─────────────────────
Red. decl.    5 38 58 S.  6,0)22,5·0018             Red. Eq. time     11 27·1                2·392
                                                     To be added to A.T.)
Polar dist.   84 21  2   Corr. + 3 45
```

Declination and equation of time are both taken out for the nearest noon at Greenwich, viz., 6th, and corrected for the time wanting to noon, that is, for 3ʰ 52ᵐ or 3ʰ·87.

```
Altitude    16° 29′ 30″                          App. T. at ship, March    6ᵈ  4ʰ52ᵐ31ˢ
Latitude    40 20   0    sec.  0·117879          Eq. time                  +     11 27
Polar dist. 84 21   2    cosec. 0·002115                                   ─────────────
           ───────────                           M.T. ship, March           6  5  3 58
            141 10  32                           M.T. Green., Jan.          5 20  8 15
           ───────────                                                     ─────────────
             70 35  16   cos.  9·521611          Long. in time                 8 55 43
             54  5  46   sine  9·908486          Longitude   133° 55′ 45″ E.
           ───────────                ─────
A.T.S. Mar. 6ᵈ4ʰ52ᵐ31ˢ   log.  9·550091          By Raper: log. sine sq. 9·550134 gives
                                                 hour-angle 4ʰ 52ᵐ 32ˢ, long. 133° 56′ 0″ E.
```

The observation having been made P.M. at ship, the hour-angle is the app. time at ship, before which write the date at ship, viz., March 6th; then mean time at ship being one day in advance of mean time at Greenwich, we subtract the latter from mean time at ship (borrowing 24ʰ) to enable us to complete the subtraction.

Ex. 6. 1882, June 15th, P.M. at ship, latitude 13° 54′ S., obs. alt. sun's L.L. 16° 16′ 16″, index corr. + 0′ 16″, height of eye 16 feet, time by a chronometer 15ᵈ 0ʰ 16ᵐ 16ˢ (or 0ʰ 16ᵐ 16ˢ P.M.) which was 2ʰ 13ᵐ 37ˢ *fast* for mean noon at Greenwich, April 1st, and on April 16th was 2ʰ 16ᵐ 16ˢ *fast* on mean time at Greenwich.

```
April 1st, chron. fast   2ʰ 13ᵐ 37ˢ            2 39
      16th,    ,,  fast  2  16  16             60
                         ─────────            ─────
     Change in 15 days   2      39          15)159(10ˢ·6
                                              15
            Daily rate   10ˢ·6 gaining.       ───
                                               90
                                               90
```

T. by chron., June	15ᵈ 0ʰ16ᵐ16ˢ	Interval from April 16th to June 14th 22ʰ, is 59ᵈ 22ʰ.		Obs. alt. ☉'s L.L.	16° 16′ 16″	
Original error	− 2 16 16			Index corr.	+ 0 16	
	14 22 0 0	Daily rate	10ˢ·6		16 16 32	
Accumulated rate	− 10 35	Interval	59	16 feet	− 3 50	
Green. date, June	14 21 49 25		954		16 12 42	
			530	Corr. alt.	− 3 5	
T. from noon June	15 2 11				16 9 37	
By *Raper* : Index corr. + 0′ 16″, dip − 4′ 0″, refr. − 3′ 18″, par. + 8″, semid. + 15′ 47″, true alt. 16° 25′ 9″.		12 ½ 8 ¼ 2 ¼	625·4 5·3 3·5 ·9	Semi-diameter True alt.	+ 15 47 16 25 24	
		6,0)63,5·1				
		10 35·1				

Decl., page II, N.A.		H.D.		Eq. T., page II, N.A.		H.D.
June 15th, noon, 23° 19′ 36″ N. +		6·24		15th, noon, add 0ᵐ 9ˢ·2 +		+ 0ˢ·535
Correction	− 14	2·2		Correction	− 1·1	× 2ʰ
Red. decl.	23 19 22 N.	1248 1248		Red. Eq. T.	0 8·1	1·070
Polar dist.	113 19 22	13·728		(To be *added* to A.T.)		

Altitude	16° 25′ 24″			App. T. ship, June	15ᵈ 4ʰ19ᵐ41ˢ
Latitude	13 54 0	sec.	0·012908	Eq. Time	+ 0 8
Polar dist.	113 19 22	cosec.	0·037021	Mean T. ship, June	15 4 19 49
	143 38 46			Mean T. Green., June	14 21 49 25
Half sum alt. 71 49 23		cos.	9·494089	Long. in time	6 30 24
	55 23 59	sin.	9·915471	Longitude	97° 36′ 0″ E.
Hour-angle 4ʰ 19ᵐ 41ˢ		log.	9·459489	By *Raper*: Log. sine sq. 9·459550, hour-angle 4ʰ 19ᵐ 42ˢ, long. 97° 36′ 15″ E.	

In this example the observation is made P.M.: hence the hour angle is apparent time at ship, before which we write the date at ship, viz., June 15th (see head of question); and since the mean time at ship is June 15th 4ʰ 19ᵐ 52ˢ, which is in advance of mean time at Greenwich, the latter being June 14ᵈ 21ʰ 49ᵐ 25ˢ, we subtract mean time at ship from mean time at Greenwich, and 24ʰ is borrowed in subtracting.

Ex. 7. 1882, June 23rd, P.M. at ship, latitude 0°, observed altitude sun's L.L. 20° 25′, height of eye 10 feet, time by chronometer June 23ᵈ 6ʰ 4ᵐ 40ˢ, which was *fast* 13ᵐ 11ˢ on Greenwich mean time, April 6th, and on May 1st was 12ᵐ 1ˢ *fast* for mean noon at Green.

Interval from April 6th to May 1st is 25 days; the change in rate in that time is 70ˢ: then 70ˢ ÷ 25 = 2ˢ·8 *losing*.

T. by chron., Jan.	23ᵈ 6ʰ 4ᵐ40ˢ	Interval from May 1st to June 23rd 6ʰ, is 53ᵈ 6ʰ.		Obs. alt. ☉'s L.L.	20° 25′ 0″	
Original error	− 12 1			Dip	− 4 17	
	23 5 52 39	Daily rate	2ˢ·8		20 20 43	
Accumulated rate	+ 2 29	Interval	53	Corr. alt.	− 2 25	
Green. date, June	23 5 55 8		84 140		20 18 18	
By *Raper* : dip − 4′ 20″, refr. − 2′ 36″, par. + 8″, semid. + 15′ 46″, and true alt. 20° 33′ 58″.		6 ¼ ½	148·4 ·7	Semi-diameter True altitude	+ 15 46 20 34 4	
		6,0)14,9·1				
		2 29·1				

	Decl., page II, N.A.		H.D.		Eq. T., page II, N.A.		H.D.
June 23rd, noon	23° 26′ 23″ N. —		2·01	23rd, noon	1ᵐ 53ˢ·4 +		·538
Correction	— 12		5·9	Correction	+ 3·2		5·9
Red. decl.	23 26 11 N.		1809	Red. Eq. T.	1 56·6		4842
			1005				2690
Polar dist.	66 33 49		11·859	(*Added* to A.T.)			3·1742
or	113 26 11						

Altitude	20° 34′ 4″				Altitude	20° 34′ 4″			
Latitude	0 0 0		sec.	0·000000	Latitude	0 0 0		sec.	0·000000
Polar dist.	66 33 49		cosec.	0·037393	Polar dist.	113 26 11		cosec.	0·037393
	87 7 53					134 0 15			
	43 33 56·5		cos.	9·860089		67 0 7·5		cos.	9·591840
	22 59 52·5		sine	9·591840		46 26 3·5		sine	9·860089
Hour-angle	4ʰ 29ᵐ 57ˢ		log.	9·48932,2	Hour-angle	4ʰ 29ᵐ 57ˢ		log.	9·48932,2
Eq. Time	+ 57				Eq. Time	+ 1 57			
M.T. ship 23ᵈ	4 31 54				M.T. ship 23ᵈ	4 31 54			
M.T. G. 23ᵈ	5 55 8				M.T. G. 23ᵈ	5 55 8			
Long. in T.	1 23 14				Long. in Time	1 23 14			
Longitude	20° 48′ 30″ W.				Longitude	20° 48′ 30″ W.			

By *Raper* the answer comes out the same.

Ex. 8. 1882, October 10th, P.M. at ship, latitude at noon 20° 41′ S., ship had sailed N.E. (true) 54 miles since noon, obs. alt. sun's L.L. 18° 45′, height of eye 15 feet, time by chronometer October 9ᵈ 16ʰ 18ᵐ 42ˢ, which was *slow* 11ᵐ 44ˢ on mean time at Greenwich, August 26th, and on September 10th was *slow* 10ᵐ 26ˢ: required the longitude at time of observation and also at noon.

To find the difference of latitude and difference of longitude.—The course 4 points and distance 54 miles give diff. lat. 38′·2 and dep. 38′·2. The diff. of lat. is marked North, because the ship at the time of sights was to the North of the position at noon; and is subtracted from the lat. at noon, viz., 20° 41′ S. to get the lat. at sights, the result is 20° 2′ 48″ S., and the dep. is named East, because the ship at the time of sights was to the East of the position at noon. The mid. lat. 20° 22′ as a course, and dep. 38′·2 as a diff. lat., give the distance 41′ as diff. of long. and is named East, because the ship at time of sights is east of her position at noon.

The daily rate is 5ˢ·2 *gaining*; the interval 29ᵈ 16ʰ × 5ˢ·2 gives accumulated rate 2ᵐ 34ˢ·3; Greenwich date October 9ᵈ 16ʰ 36ᵐ 34ˢ; polar distance 83° 24′ 13″, red. eq. T. 12ᵐ 53ˢ·9 *subt.* from A.T.; true alt. (Nouie) 18° 54′ 43″; latitude in at sights 20° 2′ 48″ S.; hour-angle 4ʰ 48ᵐ 56ˢ; mean time at ship Oct. 10ᵈ 4ʰ 36ᵐ 2ˢ; long. at time of observation 179° 52′ 0″ E.; diff. long. 41′; also longitude at noon 179° 27′ 0″ W.

EXAMPLES FOR PRACTICE.

1. 1882, January 2nd, A.M. at ship, latitude 36° 59′ S., observed altitude sun's L.L. 49° 10′, index correction — 2′ 40″, height of eye 14 feet, time by a chronometer 1ᵈ 19ʰ 8ᵐ 50ˢ (being 7ʰ 8ᵐ 50ˢ A.M. at Greenwich), which was *slow* 18ᵐ 2ˢ for mean noon at Greenwich, November 30th, 1881, and on December 7th, 1881, was 19ᵐ 10ˢ·6 *slow* for mean time at Greenwich: required the longitude.

2. 1882, February 19th, A.M. at ship, latitude 38° 18′ S., observed altitude sun's L.L. 21° 30′ 40″, index correction − 6′ 45″, height of eye 14 feet, time by a chronometer 18ᵈ 19ʰ 53ᵐ 37ˢ·6 (being 7ʰ 53ᵐ 37ˢ·6 A.M. at Greenwich), which was 4ᵐ 16ˢ·6 *fast* for mean noon at Greenwich, January 23rd, and on January 30th was 5ᵐ 9ˢ·8 *fast* for mean time at Greenwich.

3. 1882, March 28th, P.M. at ship, latitude 20° 19′ S., observed altitude sun's L.L. 30° 14′, index correction − 2′ 10″, height of eye 20 feet, time by chronometer 28ᵈ 0ʰ 10ᵐ (being 0ʰ 10ᵐ P.M. at Greenwich), which was 54ᵐ 48ˢ *fast* for mean noon at Greenwich, October 20th, 1881, and on December 2nd, 1881, was 51ᵐ 56ˢ *fast* for mean noon at Greenwich.

4. 1882, April 6th, A.M. at ship, latitude 53° 5′ N., observed altitude sun's L.L. 16° 8′ 40″, index correction − 40″, height of eye 15 feet, time by a chronometer 5ᵈ 19ʰ 18ᵐ 49ˢ (being 7ʰ 18ᵐ 49ˢ A.M. at Greenwich), which was 0ᵐ 4ˢ·4 *slow* for mean noon at Greenwich, February 11th, and on March 11th was 2ᵐ 38ˢ *fast* for mean noon at Greenwich.

5. 1882, May 19th, P.M. at ship, latitude 2° 58′ S., observed altitude sun's L.L. 30° 30′, index correction + 52″, height of eye 19 feet, time by chronometer 19ᵈ 0ʰ 23ᵐ 58ˢ, which was 28ˢ *fast* for mean noon at Greenwich, January 3rd, and on January 31st was 42ˢ *slow* on mean time at Greenwich.

6. 1882, June 15th, A.M. at ship, latitude 12° 11′ N., observed altitude sun's L.L. 39° 39′ 40″, index correction + 20″, height of eye 17 feet, time by a chronometer 14ᵈ 17ʰ 59ᵐ 30ˢ (being 5ʰ 59ᵐ 30ˢ A.M. at Greenwich) which was *slow* 5ᵐ 56ˢ·3 for mean time at Greenwich, April 20th, and on May 12th was 2ᵐ 29ˢ·5 *slow* for mean noon at Greenwich.

7. 1882, July 5th, A.M. at ship, latitude 23° 48′ N., observed altitude sun's L.L. 48° 36′ 50″, index correction − 50″, height of eye 17 feet, time by chronometer 5ᵈ 0ʰ 42ᵐ 38ˢ (being 0ʰ 42ᵐ 38ˢ P.M. at Greenwich), which was *fast* 4ᵐ 47ˢ·8 for mean noon at Greenwich, May 6th, and on June 1st was *fast* 6ᵐ 50ˢ for mean noon at Greenwich.

8. 1882, August 13th, A.M. at ship, latitude 30° 46′ S., observed altitude sun's L.L. 27° 15′, index correction − 1′ 15″, height of eye 21 feet, time by a chronometer 13ᵈ 2ʰ 0ᵐ (being 2ʰ 0ᵐ P.M. at Greenwich), which was *slow* 16ᵐ 7ˢ·6 for mean noon at Greenwich, April 10th, and on May 1st was *slow* 25ᵐ 13ˢ for mean noon at Greenwich.

9. 1882, September 1st, P.M. at ship, latitude 35° 49′ N., observed altitude sun's L.L. 44° 32′ 10″, index correction + 1′ 46″, height of eye 20 feet, time by chronometer August 31st 19ʰ 24ᵐ 57ˢ (being 7ʰ 24ᵐ 57ˢ A.M. at Greenwich), which was *fast* 11ᵐ 57ˢ·4 for mean noon at Greenwich, July 3rd, and on July 31st was *fast* 12ᵐ 17ˢ for mean noon at Greenwich.

10. 1882, October 25th, P.M. at ship, latitude 51° 30′ S., observed altitude sun's L.L. 40° 22′, index correction − 1′ 50″, eye 20 feet, time by chronometer 25ᵈ 8ʰ 22ᵐ 1ˢ (or 8ʰ 22ᵐ 1ˢ P.M.), which was *slow* 24ᵐ 8ˢ·2 for mean noon at Greenwich, June 14th, and on July 20th was *slow* 21ᵐ 19ˢ for mean noon at Greenwich.

11. 1882, November 27th, A.M. at ship, latitude 39° 20′ S., observed altitude sun's L.L. 34° 37′ 55″, index correction + 1′ 15″, eye 18 feet, time by a chronometer 27ᵈ 7ʰ 41ᵐ 30ˢ (being P.M. at Greenwich), which was *fast* 31ᵐ 54ˢ for mean noon at Greenwich, October 20th, and on November 9th was 29ᵐ 40ˢ *fast* on mean noon at Greenwich.

12. 1882, December 25th, A.M. at ship, latitude 9° 59′ S., observed altitude sun's L.L. 10° 38′ 45″, index correction − 3′ 12″, eye 18 feet, time by a chronometer 24ᵈ 17ʰ 36ᵐ 0ˢ (being A.M. at Greenwich), which was *slow* 34ᵐ 19ˢ·1 for mean noon at Greenwich, July 1st, and on July 29th was *slow* 38ᵐ 39ˢ·5 for mean noon at Greenwich.

13. 1882, January 1st, P.M. at ship, latitude 38° 28′ S., observed altitude sun's L.L. 39° 0′, index correction − 2′ 15″, eye 12 feet, time by chronometer 1ᵈ 11ʰ 58ᵐ 29ˢ (being P.M. at Greenwich), which was *slow* 1ʰ 49ᵐ 19ˢ for mean noon at Greenwich, September 12th, 1881, and on October 13th was 1ʰ 52ᵐ 53ˢ *slow* for mean noon at Greenwich.

14. 1882, February 11th, A.M. at ship, latitude 53° 12' N., observed altitude sun's L.L. 12° 10', index correction − 49", eye 12 feet, time by chronometer 10^d 22^h 22^m 22^s (being A.M. at Greenwich), which was *fast* 34^m 41^s·7 for mean noon at Greenwich, October 31st, and on December 1st, 1881, was *fast* 38^m 59^s for mean noon at Greenwich.

15. 1882, October 26th, A.M. at ship, latitude 28° 10' N., observed altitude sun's U.L. 25° 32' 20", index correction 0", eye 17 feet, time by chronometer 0^h 54^m 6^s (being P.M. at Greenwich), which was *fast* 31^m 31^s on mean time at Greenwich, August 1st, and on Sept. 4th was *fast* 30^m 6^s for mean noon at Greenwich.

16. 1882, February 6th, P.M. at ship, latitude 6° 58' N., observed altitude sun's U.L. 21° 43' 40", index correction 0", eye 18 feet, time by a chronometer 11^h 40^m 26^s (being A.M. at Greenwich), which was *slow* 16^m 4^s·8 on mean noon at Greenwich, January 2nd, and on January 20th was *slow* 17^m 42^s on mean noon at Greenwich.

17. 1882, May 1st, P.M. at ship, latitude 21° 8' N., observed altitude sun's L.L. 28° 5' 30", index correction + 2' 50", height of eye 16 feet, time by a chronometer April 30^d 13^h 50^m 29^s·4 (being 6^h 50^m 29^s·4 A.M. at Greenwich), which was 10^m 12^s *slow* for mean noon at Greenwich, December 31st, 1881, and on February 17th, 1882, was 7^m 33^s·6 *slow* for mean noon at Greenwich.

18. 1882, April 21st, P.M. at ship, latitude at noon 0° 20' N., observed altitude sun's U.L. 32° 21' 10", index correction − 1' 10", eye 12 feet, time by a chronometer 3^h 44^m 1^s (being A.M. at Greenwich) which was *slow* 9^m 7^s for mean noon at Greenwich, November 14th, 1881, and on January 11th, 1882, was *slow* 7^m 34^s·2 for mean noon at Greenwich, course since noon S.W. by W. (true), distance 36 miles: required the longitude at the time of observation, and also at noon.

19. 1882, August 21st, A.M. at ship, latitude at noon 0° 20' S., observed altitude sun's L.L. 33° 49', index correction + 2' 10", eye 15 feet, time by chronometer 8^h 14^m 0^s (being P.M. at Greenwich), which was *slow* 4^m 40^s for mean noon at Greenwich, March 13th, and on April 30th was *slow* 5^m 40^s for mean noon at Greenwich, course till noon S.W. by W., distance 36 miles: required the longitude at time of sights, and also at noon.

20. 1882, March 20th, A.M. at ship, latitude 0°, observed altitude sun's L.L. 28° 50' 10", index correction + 1', eye 23 feet, time by chronometer 20^d 1^h 35^m (being P.M. at Greenwich), which was 1^m 59^s·9 *fast* for mean noon at Greenwich, February 1st, and on February 28th was *fast* 2^m 8^s for mean noon at Greenwich.

21. 1882, September 23rd, A.M. at ship, latitude 0°, observed altitude sun's L.L. 28° 52', height of eye 17 feet, time by chronometer September 22^d 15^h 39^m 41^s, which was *slow* 15^s·6 on Greenwich mean time, April 30th, and on June 1st was *fast* 10^s on Greenwich mean time.

SUMNER'S METHOD.

ON FINDING SHIP'S LINE OF POSITION OR PARALLEL OF EQUAL ALTITUDE.

Problem I.

TO FIND THE SHIP'S CURVE OF POSITION OR PARALLEL OF EQUAL ALTITUDE.

327. In the preceding chapters of this work we have treated of methods of finding the position of a point on the earth's surface by the two co-ordinates *latitude* and *longitude;* and, therefore, in all these methods the required position is determined by the intersection of two circles, one a parallel of latitude, and the other a meridian. It is evident that if we could find any two other lines passing through the ship, their intersection would also determine the position of the ship, and thus answer the same purpose as parallels of latitude and meridians. In the following method the position of the ship is determined by circles *oblique* to the parallels of latitude and the meridian. The principle which underlies the method has often been applied, but its value as a practical nautical method was first clearly shown by Capt. T. H. SUMNER.

328. Let an altitude of the sun, or any other object, be observed *at any time*, the time being noted by a chronometer regulated to Greenwich time.

Suppose that at this Greenwich time the sun is vertical to an observer at the point M of the globe (Fig. 1), having there an altitude of 90°, and if a small circle A A' A'' be supposed to be described about the point M as a pole, with a distance MA, equal to the zenith distance or complement of the altitude of the sun. It is evident that at all places *within* this circle an observer would, at the given time, observe a smaller zenith distance, and at all places without this circle a greater zenith distance; and, therefore, the observation fully determines the observer to be *on* the circumference of the small circle A A' A''. If, then, the seaman can project this small circle upon an artificial globe or a chart, *the knowledge that he is upon this circle will be just as valuable to him to avoid dangers as the knowledge of either his latitude alone, or his longitude alone*, since one of the latter elements only determines a point to be in a certain circle without fixing upon any particular point of that circle.

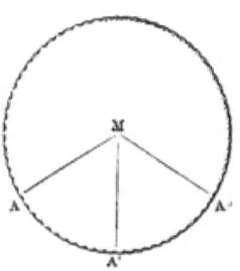

The small circle of the globe, described from the projection of the celestial object as a pole, we shall call a *circle of position*.

329. A small portion of this circle passing through two positions near together may be considered as a straight line, and this line is evidently perpendicular to the direction of the observed object, and it is this line which has been called the *line of position*.

330. In finding the longitude by chronometer, an error in the latitude from which the time is computed, affects the resulting longitude in a corresponding degree. If several latitudes, not differing very greatly from the truth, be successively employed in calculating the longitude, each latitude will give a different longitude; and if with each latitude and the longitude deduced from it a point is found on the chart, it will be seen that these points lie very nearly in a straight line,* whose direction is the complement to the bearing of the observed object at the time of observation; in other words, the *line of position* is perpendicular to the direction of the object at the time of observation.

331. Two assumed latitudes, with the resulting longitudes, will be sufficient to determine the *line of position*, provided the latitudes do not greatly differ from the truth. Lay off on the chart these two positions, and join them by a straight line, somewhere on this line, or very near it, the observer's place is to be found; and thus, if this line (being produced) passes through a point of land or other object, the *bearing* of such object is known, though the ship's *place* on the line of its direction is not known.

332. After the observation the ship changes her place, which is usually the case at sea; but, if from any point in the line of position the ship's run be laid off in proper direction according to the ship's course, and through the point thus found a line be drawn parallel to the line of position, the ship's actual place is still somewhere on *this line*.

333. If a second observation of the same or any other celestial body be taken, separated from the first by a suitable difference of bearing, and treated in the same manner, another *line of position* may be drawn, which will also pass through the place of observation; and, therefore, the intersection of the two lines will give the ship's place, which may thus be obtained by projection on the chart, without the computation of a double altitude. The intersection is most distinctly marked when the lines cross at right-angles, and they must not in any case cut too acutely, or the unavoidable errors of observation will be considerably magnified. The difference of bearings *should not be less than* 45°, *nor greater than* 135°.

---

* The direction of this line at any point is the direction of the tangent to the curve at this point. The direction of this tangent, as just stated, will be at right-angles to the bearing of the sun at that point. Hence, *by a line of position we may determine the azimuth of the sun*. By reference to the principles of construction of the Mercator's chart, it will be seen that only the loxodromic curve plots as a straight line. The circle of position would plot as an irregular figure, its greatest diameter coinciding with the arc of the meridian. The whole figure could not indeed be plotted on ordinary charts unless the zenith distances MA, MA′, &c., were very small.

It is customary to plot only the small portion of the curve lying between the assumed latitudes, as that is all that is required. For small differences of latitude, as before observed, this would be practically a straight line.

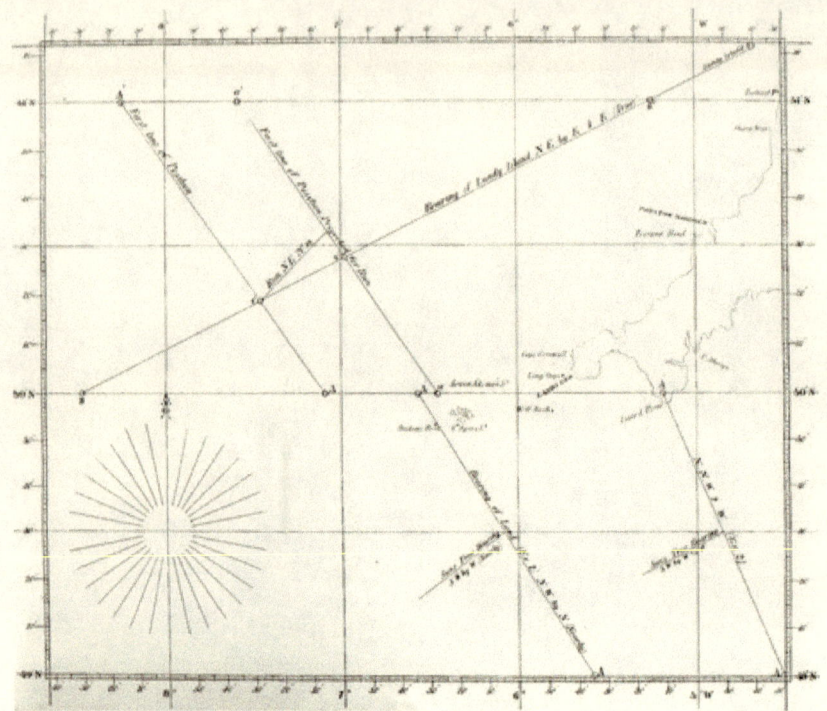

The foregoing remarks contain the elementary principles of Sumner's Method of finding the ship's position at sea.

334. The latitude, longitude, and time at ship being uncertain, it is required, from an altitude of the sun and the Greenwich time, as found from the chronometer, to project on Mercator's chart a *line of position* or a *parallel of equal altitude*, which shall pass through the position of the ship, and show—

    1st—The bearing of the land.
    2nd—The sun's true bearing or azimuth.
    3rd—The error in the longitude corresponding to any error in the latitude.

335. The method of finding the approximate projection of a circle of position on the Mercator's chart is as follows:—

SOLUTION BY CONSTRUCTION ON MERCATOR'S CHART.

The Observation.—Take an altitude of the sun and note the corresponding time by chronometer.

RULE XCVI.

1°. **The Calculation.**—*To the time by chronometer apply its original error and accumulated rate, as directed in* Rule XCII, page 271; the result is the Greenwich date at the instant of observation.

2°. *Take out of the* Nautical Almanac, page II, *the sun's declination and the equation of time for the noon of Greenwich date, and the corresponding hourly difference for each; also, take out the sun's semi-diameter.*

3°. *Reduce the sun's declination and equation of time to the Greenwich time, either by multiplying each hourly difference by the* hours and decimals *of an hour in the Greenwich time; or, by multiplying each hourly difference by the hours of Greenwich time, and taking aliquot parts of an hour for the minutes of the same:* the result is the correction to be applied to the sun's declination and equation of time respectively (Rules LXXIX and LXXXIII, pages 230—238).

4°. *Next find the polar distance* (Rule LXXXII, page 237).

5°. *Correct the observed altitude for index error* (if any), *dip, correction in altitude and semi-diameter, and thus get the true altitude* (Rule LXXXIV p. 242).

6°. *Assume two latitudes* differing *about a* degree or less, *and embracing the supposed latitude of the ship.*

NOTE.—It may be convenient to assume two latitudes from 10′ to 30′ on each side of the latitude by account. In the practical application of this problem at sea, it is customary to assume lats. 10′ to 20′ on each side of the supposed one, and determine the corresponding longitudes. In general, assume the latitudes to cover any supposed error in the latitude.

7°. *Compute the longitude* (by the usual method, see Rule XCV, pages 277—278) *with each assumed latitude, first from one assumed latitude and then from the other,* as follows:—

  (a) *With the altitude, the* less *assumed latitude and the sun's polar distance.*

  (b) *With the altitude, the* greater *assumed latitude, and the sun's polar distance.*

P P

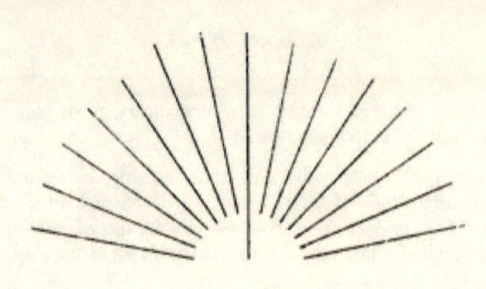

The foregoing remarks contain the elementary principles of Sumner's Method of finding the ship's position at sea.

334. The latitude, longitude, and time at ship being uncertain, it is required, from an altitude of the sun and the Greenwich time, as found from the chronometer, to project on Mercator's chart a *line of position* or a *parallel of equal altitude*, which shall pass through the position of the ship, and show—

    1st—The bearing of the land.
    2nd—The sun's true bearing or azimuth.
    3rd—The error in the longitude corresponding to any error in the latitude.

335. The method of finding the approximate projection of a circle of position on the Mercator's chart is as follows:—

SOLUTION BY CONSTRUCTION ON MERCATOR'S CHART.

The Observation.—Take an altitude of the sun and note the corresponding time by chronometer.

RULE XCVI.

1°. **The Calculation.**—*To the time by chronometer apply its original error and accumulated rate, as directed in* Rule XCII, page 271; the result is the Greenwich date at the instant of observation.

2°. *Take out of the* Nautical Almanac, page II, *the sun's declination and the equation of time for the noon of Greenwich date, and the corresponding hourly difference for each; also, take out the sun's semi-diameter.*

3°. *Reduce the sun's declination and equation of time to the Greenwich time, either by multiplying each hourly difference by the* hours and decimals *of an hour in the Greenwich time; or, by multiplying each hourly difference by the hours of Greenwich time, and taking aliquot parts of an hour for the minutes of the same:* the result is the correction to be applied to the sun's declination and equation of time respectively (Rules LXXIX and LXXXIII, pages 230—238).

4°. *Next find the polar distance* (Rule LXXXII, page 237).

5°. *Correct the observed altitude for index error* (if any), *dip, correction in altitude and semi-diameter, and thus get the true altitude* (Rule LXXXIV p. 242).

6°. *Assume two latitudes* differing *about a* degree or less, *and embracing the supposed latitude of the ship.*

NOTE.—It may be convenient to assume two latitudes from 10' to 30' on each side of the latitude by account. In the practical application of this problem at sea, it is customary to assume lats. 10' to 20' on each side of the supposed one, and determine the corresponding longitudes. In general, assume the latitudes to cover any supposed error in the latitude.

7°. *Compute the longitude* (by the usual method, see Rule XCV, pages 277—278) *with each assumed latitude, first from one assumed latitude and then from the other,* as follows:—

  (a) *With the altitude, the less assumed latitude and the sun's polar distance.*

  (b) *With the altitude, the* greater *assumed latitude, and the sun's polar distance.*

Thus, two longitudes will be found which may be designated A and A' respectively, and the ship's position may now be conveniently projected on a chart of large scale. Thus:—

8°. *On the less assumed parallel mark off the longitude* A, *and on the greater assumed parallel mark off the longitude* A'.

9°. *Join the points* A *and* A' *by a straight line cutting both parallels, which will be the projection of the line of position or parallel of equal altitudes passing through the true place of the ship.*

10°. *The parallel of equal altitude or line of position being produced, if necessary, to meet the coast; the land such a line passes through bears from the ship in the same direction as the line lies on the chart.*

To find the Sun's True Azimuth by a Position Line projected on a chart.

11°. *On any point on the line* A A', *and on that side of the meridian upon which the sun is known to be at the time of observation, erect a perpendicular, and it will be in the direction of the sun's true bearing, and the angle it makes with the meridian is the sun's true azimuth, which is therefore known on being referred to the compass on the chart.*

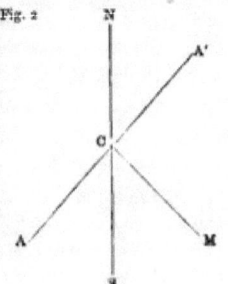

Fig. 2

NOTE 1.—Let A A' (Fig. 2) be a line of position on the chart derived from an observed altitude. At any point C of this line erect C M perpendicular to A A', and let N C S be the meridian passing through C; then SCM is evidently the sun's azimuth. The line C M is, of course, drawn on that side of the meridian M S upon which the sun is known to be at the time of observation. The solution is but approximate, since A A' should be a curve line, and the azimuth of the normal to it would be different for different points of A A'. It is, however, quite accurate enough for the purpose of determining the error of the compass at sea, which is the only practical application of the problem.

NOTE 2.—The difference of longitude between A and A' will show the error of longitude arising from an error equal to the difference between the assumed latitudes; and hence, by proportion, the error of longitude corresponding to any other error of latitude may be found.

EXAMPLES.

Ex. 1. 1880. If at sea on August 28th, P.M. at ship, and uncertain of my position, when the chronometer showed August 28^d 3^h 20^m 48^s M.T.G.; suppose the observed altitude of the sun's L.L. to be 35° 42' 40", height of eye 16 feet: required the line of bearing when the altitude was taken, assuming the ship to be between latitudes 49° 0' N. and 50° 0' N.

It will be observed that in this question the Greenwich date, mean time, is given—in this case August 28^d 3^h 20^m 48^s; therefore there is no chronometer time to correct for error and rate, but from the Nautical Almanac, page II, the Decl. and Eq. T. is taken and corrected in the usual way.

Greenwich date, August	28^d 3^{h}20^{m}48^s	Obs. alt. ☉'s L.L.	35° 42' 40"
	or 3^h 35	Dip	− 3 50
	6,0)21^m·0	Corr. of alt.	35 38 50
			− 1 12
	3^h 35		35 37 38
		Semi-diameter	+ 15 53
		True alt.	35 53 31

Decl., page II, N.A.		H.D.	Eq. T., page II, N.A.		H.D.
Aug. 28th, noon	9° 29' 58" N. *decr.*	53"·16	0ᵐ 55ˢ·78 *decr.*		0ˢ·736
Corr. for 3ʰ 21ᵐ	− 2 58	3ʰ·35	2ˢ·47		3ʰ·35
Red. decl.	9 27 0 N.	26630	Red.Eq.T. 0 53·31		3680
		15978	('To be *added* to A.T.)		2208
Polar dist.	80 33 0	15978			2208
		6,0)17,8·4210		Corr.	2·46560
		2 58			

It will be further noticed that at the end of the question two assumed latitudes are given—in this case 49° 0' N. and 50° N. *For each lat.*, with the true alt. and pol. dist. (as above) next proceed in the usual way to find the longitudes.

True alt.	35° 53' 31"			Altitude	35° 53' 31"		
Latitude	49 0 0	sec.	0·183057	Latitude	50 0 0	sec.	0·191933
Polar dist.	80 33 0	cosec.	0·005934	Polar dist.	80 33 0	cosec.	0·005934
	165 26 31				166 26 31		
	82 43 15	cos.	9·102790		83 13 15	cos.	9·072040
	46 49 44	sine	9·862914		47 19 44	sine	9·866438
A.T.S. Aug. 28ᵈ 2ʰ 57ᵐ 37ˢ		log.	9·154695	A.T.S. Aug. 28ᵈ 2ʰ 53ᵐ 43ˢ		log.	9·136345
Eq. time + 0 53				Eq. time + 0 53			
M.T.S. Aug. 28ᵈ 2 58 30				M.T.S. Aug. 28ᵈ 2 54 36			
M.T.G. Aug. 28ᵈ 3 20 48				M.T.G. Aug. 28ᵈ 3 20 48			
Long. in time 0 22 18				Long. in time 0 26 12			
Longitude (A) 5° 34' 30" W.				Longitude (A') 6° 33' 0" W.			

Thus, then, has been obtained two positions, viz.:—

A. In lat. 49° 0' N., long. 5° 34' 30" W.
A'. In lat. 50° 0' N., long. 6° 33' 0" W.

On a Mercator's chart mark the point A in latitude 49° N., and longitude 5° 34½' W. (exactly as a ship's position is pricked off on the chart by the aid of dividers and parallel rules); also mark the point A' in latitude 50° N. and longitude 6° 33' W. Join A and A' by a straight line, and the ship will be on this line, it being the projection of *the line of position of the ship*, or, as sometimes called, the *parallel of equal altitude of the ship*. The line will be found to pass through the Bishop Rock, and when referred to the compass on the chart, the light will be found to bear N.W. by N. (nearly). Consequently, although the *exact* position of the ship is not known, it is certain that by steering a true N.W. by N. course the ship will make the Bishop Rock. (See Ex. 1, plate).

To find the Sun's True Azimuth.—From any point on AA' draw a perpendicular to AA', and this will give the true bearing of the sun, which will be found to be S.W. by W. (nearly). If this line be produced it will give the perpendicular on the other side of AA', bearing N.E. by E. (nearly); the former, however, lying in the direction of the sun, and as the observation is made in the afternoon, gives the sun's true azimuth.

NOTE 1.—The above calculations should be carried on together, that is, the formulæ for each should be prepared, and the logarithms, &c., filled in, taking care to take out of the tables, at the same time, those logarithms which are found at the same opening of the tables. Thus, after taking out log. secant of lat. 49°, take log. secant of lat. 50°; log. cosecant of polar distance is the same in both instances; cosine of 82° 43' 15" is then taken out, and at the same time cosine of 83° 13' 15"; the sine of 46° 49' 44" and sine of 47° 19' 44" may be taken out at the same time, and lastly, the hour-angle in both cases is taken out at the same time. Consequently, the two calculations required the tables to be opened only five times, the same as if there were only one calculation to be performed. The observation here made is, of course, applicable to all the following examples.

NOTE 2.—The longitude of A and A' being 5° 34½' and 6° 33' W., the difference is 59', and the difference of their latitudes is 1° = 60'. Hence, an error of 60' in latitude gives an error of 59' in longitude; and, if we state 60' : 10' :: 59' : 10' (nearly), we find that an error of 10' in latitude gives an error of 10' in longitude. That is, if we compute the longitude by chronometer from the altitude of the sun, in this example, and assume a latitude which is 10' in error, the longitude thus found will be 10' (nearly) in error.

Ex. 2. 1880. If at sea on August 10th, P.M. at ship, and uncertain of my position when the chronometer showed August 10^d 3^h 35^m 36^s mean time at Greenwich, suppose the observed altitude of sun's L.L. 38° 17', height of eye 18 feet: required the line of bearing when the altitude was taken, assuming the ship to be between latitudes 49° 0' N. and 50° 0' N.

Greenwich date, August	10^d 3^h 35^m 36^s		Obs. alt. ⊙'s L.L.	38° 17' 0"
	or 3^h·6		Dip	− 4 4
	6,0)36^m·0			38 12 56
	3^h·6		Corr. alt.	− 1 5
			Semi-diameter	38 11 51
				+ 15 49
			True alt.	38 27 40

	Decl., page II, N.A.		H.D.	Eq. T., page II, N.A.		H.D.
Aug. 10th, noon	15° 23' 39" N. *decr.*		44^m·29	Aug. 10th, noon	5^m 4^s·38 *decr.*	0·381
Corr. for 3^h 36^m	− 2 39		3·6	Corr. for 3^h 36^m	− 1·37	3·6
	15 21 0 N.		26574	Rted. Eq. T.	5 3	2286
	90		13287	(To be *added* to A.T.)		1143
Polar dist.	74 39 0		6,0)15,9'·444			1·3716
			2 39·4			

To find point A in lat. 49° N.				To find point A' in lat. 50° N.			
Altitude	38° 27' 40"			Altitude	38° 27' 40"		
Latitude	49 0 0	sec.	0·183057	Latitude	50 0 0	sec.	0·191933
Polar dist.	74 39 0	cosec.	0·015776	Polar dist.	74 39 0	cosec.	0·015776
	162 6 40				163 6 40		
	81 3 20	cos.	9·191665		81 33 20	cos.	9·166875
	42 35 40	sine	9·830464		43 5 40	sine	9·834550
A.T.S. Aug. 10^d 3^h 12^m 33^s		log.	9·220962	A.T.S. Aug. 10^d 3^h 9^m 47^s		log.	9·209134
Eq. time	+ 5 3			Eq. time	+ 5 3		
M.T.S. Aug. 10^d 3 17 36				M.T.S. Aug. 10^d 3 14 50			
M.T.G. Aug. 10^d 3 35 36				M.T.G. Aug. 10^d 3 35 36			
Long. in time	0 18 0			Long. in time	0 20 46		
Longitude (A)	4° 30' W.			Longitude (A')	5° 11' 30" W.		

Project on a Mercator's chart (see Ex. 2, plate) the point A corresponding to longitude 4° 30' W., and the *less* assumed latitude 49° N.; also, project the point A' corresponding to longitude 5° 11½' W., and the *greater* assumed latitude 50° N. Join the two positions by a straight line, which, being prolonged to meet the land, will be found to pass through the Lizard lights, and when referred to the compass on the chart the lights will be found to bear N.N.W. ¼ W. Consequently, although the exact position of the ship is not known, it is certain that by steering on a true N.N.W. ¼ W. course the ship will make the Lizard lights.

For the Sun's True Azimuth.—From any point on A A' draw a perpendicular to A A', and this will give the true bearing of the sun S.W. by W. ¾ W. If this line be produced it will give the perpendicular on the other side of A A', bearing N.E. by E. ¾ E.; the former, however, being set off in the direction of the sun, and as the observation was made in the afternoon, gives the sun's true azimuth.

Sumner's Method.

Ex. 3. 1880, April 24th, A.M. at ship, lat. D.R. 49° 40′ N., long. D.R. 5° 20′ W. The obs. alt. ☉'s L.L. was 30° 24′ 40″, height of eye 18 feet, time by chronometer, April 23rd, 20ʰ 33ᵐ 44ˢ, which was 3ᵐ 12ˢ *fast* for mean time at Greenwich on March 11th, and on March 31st was 4ᵐ 12ˢ *fast* for mean time at Greenwich.

Required to project a line of position, and also a line sun's bearing.

Space for minor corrections.					
		Green. date,	April 23ᵈ 20ʰ28ᵐ32ˢ		
			24 0		
		Time from noon, April 24ᵈ	3 31		
			3 5		
		Decl., page II, N.A.	H.D.	Eq. T., page II, N.A.	H.D.
76°57′ 0″= 0·011364 cosec. Pts. for 32 = — 26		April 24ᵈ, noon 13′ 5′ 0″ N. *incr.* Corr. for 3ʰ·5 — 2 52	49″·02 3·5	April 24ᵈ, noon 2ᵐ 0″·9 Corr. for 3ʰ·5 — 1·6	0·454 3·5
76 57 52 0·011338		Red. decl. 13 2 8 N. 90 0 0	24510 14706	Red. Eq. time 1 59·3 (To be *subt.* from A.T.)	2270 1362
Tab. diff. = 52 52					
Corr. = 2ˢ·00		Polar dist. 76 57 52	(6,0)17,1·570		1·5890
			2 51·6		
cosine. 78° 51′ 0″= 9·299034 Pts. for 27 = — 280		To find point A, latitude 49° 30′ N.			
		Obs. alt. ☉'s L.L. 30° 24′ 40″	Altitude 30° 35′ 2″		
78 31 27 9·298754		Dip — 4 4	Latitude 49 30 0 Polar dist. 76 57 52	sec. 0·187456 cosec. 0·011338	
Tab. diff. 1036 27		30 20 36			
		Corr. — 1 30	157 2 54		
7252 2072		30 19 6	78 31 27	cos. 9·298754	
Corr. = 279·72		Semi-diameter + 15 56	47 56 25	sine 9·870666	
sine. 47° 56′ 0″= 9·870618 Pts. for 25 = + 48		True alt. 30 35 2			
			Hour-angle 3ʰ51ᵐ 9ˢ 24 0 0	log. 9·368214 800	
47 56 25 9·870666					
Tab. diff. 190 25			App. T. ship, April 23ᵈ 20 8 51 Eq. time — 1 59	gives 4ˢ 21	
950 380			M.T. ship, April 23ᵈ 20 6 52 M.T. Green., April 23ᵈ 20 28 32		
Corr. = 47·56			Longitude in time 21 40 = 5° 25′ W. (A)		
cosine. 78° 41′ 0″= 9·292768 Pts. for 27 = — 284		Altitude 30° 35′ 2″	To find point A′ in latitude 49° 50′ N.		
		Latitude 49 50 0 Polar dist. 76 57 52		sec. 0·190431 cosec. 0·011338	
78 41 27 9·292484					
Tab. diff. 1053 27			157 22 54		
7371 2106			78 41 27	cos. 9·292484	
Corr. = 284·31			48 6 25	sine 9·871802	
sine. 48° 6′ 0″= 9·871755 Pts. for 25 = + 47		Hour-angle	3ʰ50ᵐ31ˢ 24	log. 9·366055	
48 6 25 9·871802		App. T. ship, April 23ᵈ 20 9 29 Eq. time — 1 59			
Tab. diff. 189 25		M.T. ship, April 23ᵈ 20 7 30 M.T. Green., April 23ᵈ 20 28 32			
945 378					
Corr. = 47·25		Longitude in time	21 2 = 5° 15½′ W. (A′)		

On Mercator's chart mark the point A in latitude 49° 30′ N., long. 5° 25′ W.; also, mark the point A′ in latitude 49° 50′ N., long. 5° 15½′ W. Join A and A′ by a straight line and the ship will be on this line *somewhere*, it being the projection of the *line of position of the ship*, or as it is also sometimes called, the *parallel of equal altitude of the ship*. The line AA′ being produced until it meets the land it will be found to intersect the land near the Lizard Point; and when referred to the compass on the chart it will be found to bear N. by E. ½ E. (true). Consequently, although the *exact* position of the ship is not known, it is certain that by steering N. by E. ½ E. the ship will make the Lizard.

For the Sun's True Azimuth.—From any point on AA′ draw a perpendicular to AA′ on the side of it the sun is known to be at the time of observation, and the sun's bearing will be found to be S.E. ½ E. (nearly).

Problem II.

336. **To find the latitude and longitude of a ship by circles of position projected on Mercator's chart.**

Having given two altitudes of the sun, the times of observation by chronometer (whose error for Greenwich mean time is known), also the sun's declination at both times, and the latitude by dead reckoning, to find the true latitude and longitude of the ship.

RULE XCVII.

1°. *Find the Greenwich date of each observation by correcting the times shown by chronometer for error and accumulated rate.*

2°. *Find the declination and polar distance for each of the Greenwich dates.*

3°. *Find the equation of time corresponding to each Greenwich date.*

4°. *Correct each observed altitude for index error* (if any), *dip, refraction, parallax, and semi-diameter.*

5°. *Assume two latitudes, differing about a degree or less, and embracing the supposed latitude of the ship; thus it may be convenient to assume the latitudes from 10′ to 30′ on each side of the supposed latitude of the ship.**

Compute the longitude (by the usual method) *with each assumed latitude, first from one altitude and then with the other*, as follows:—

(a) *With the 1st altitude, the less assumed latitude, and the sun's polar distance, corresponding to the time of the first observation:* call this A.

(b) *With the 1st altitude, the greater assumed latitude, and the sun's polar distance, corresponding to the first observation;* call this A′.

(c) *With the 2nd altitude, the less assumed latitude, and the sun's polar distance, at the time of taking the second observation:* call this B.

(d) *With the 2nd altitude, the greater assumed latitude, and the sun's polar distance, corresponding to the time of the second observation:* call this B′.

Thus, four longitudes will be found which may be designated as A, A′, B, and B′ respectively, to facilitate reference to them.

---

* It is not, however, essential that the same assumed latitudes should be used in computing both lines of position; it is only more convenient to do so, as it saves some logarithms. In the case of two altitudes of the same object, as of the sun in this case, where a course and distance have been made in the interval, if the course has been nearly North or South, it would be better to assume two latitudes differing from those used for the first observation, and such that they may be more in accordance with the altered position of the ship.

7°. On *Mercator's chart project* A A', *the* ship's line of position, *or parallel of equal altitude, corresponding to the first altitude.* Thus,

(a) *On the* less *assumed parallel of latitude mark the longitude* A, *and in the* greater *assumed parallel of latitude mark the longitude* A'.

(b) *Join the points* A *and* A' *by a straight line, which will be the projection of the* ship's line of position *at the first observation,* and on or near this or this produced the ship is situated.

8°. *Project, in exactly the same manner,* B B', *the* ship's line of position, *corresponding to the second altitude.* That is,

(a) *On the* less *assumed latitude mark the spot corresponding to longitude* B, *and on the* greater *assumed latitude mark the corresponding longitude* B'.

(b) *And the line joining the points* B B', *will be a second* ship's line of position, *and on or near this or this produced the ship is also situated.*

The position of the ship *is at the* intersection *of the* two lines *thus produced,*[*] *from which its* latitude *and* longitude *may be easily found.*

Or proceed thus :—

9°. *On the* less *assumed latitude mark the* longitudes A *and* B ; *on the* greater *assumed latitude mark the* longitudes A' *and* B' ; *draw a line through the* points A *and* A', *cutting both parallels ; also draw another line through the* points B *and* B': *the position of the ship is at the* intersection *of these lines.*

The following example will illustrate the last paragraph of the rule :—

Ex. On the 1st day of August, 1880, at about 10^h apparent time at ship, an altitude of the sun was observed, the latitude by dead reckoning being 50° 16' North. The longitudes as ascertained by working with the assumed latitudes 50° N. and 51° N. are as follows :—

Longitude corresponding to 50° N. is 7° 5' W. (A).
,,   ,,   51° N. is 8° 15' W. (A').

Proceed as in the preceding rule to project the ship's line of position.

Mark on the chart the spot (A) corresponding to latitude 50° N. and longitude 7° 5' W. (see plate). Again, mark the spot corresponding to latitude 51° N. and longitude 8° 15' W. on the chart (A'). Draw a line connecting these two spots, A A'.

At 3^h P.M. of the same day an altitude of the sun was observed, the ship having remained stationary between the observations, working with the same assumed latitudes.

Longitude corresponding to 50° N. is 8° 30' W. (B).
,,   ,,   51° N. is 5° 15' W. (B').

Project, in exactly the same manner as above, the ship's line of position corresponding to the second altitude by marking off on the parallel of 50° N., the longitude of B = 8° 30' W., and the longitude B B' = 5° 15' W. on the parallel of 51° N. Connect the two points B and B' by a straight line.

Then A A' in the diagram (see plate) shows the two longitudes corresponding to the *first observation,* and B B' the longitudes corresponding to the *second observation.*

The intersection of these lines at C shows the position of the ship.

The latitude is 50° 19' N. instead of 50° 14' N., as by dead reckoning ; the longitude is 7° 27½' W.

The line B B' prolonged until it meets the land, gives the bearing of Lundy Island N.E. by E. ¾ E. (true).

---

* If the lines of position when computed and projected intersect considerably beyond one or other of the assumptions, then take another latitude, a little beyond that of the intersecting point, compute anew for this, and so project again. The position will be more accurately determined in this manner.

10°. **If the ship has changed her station between the observations.**—*Set off from any part of the line* AA' *the* distance *sailed between the observations in the direction of the true course made good. Through the point thus found draw a parallel to* AA', *and produce it till it meets* BB'; *and the* intersection *of its* new position *will give the* place of the ship *at the* second observation.

NOTE.—When the ship is steering a straight course, the line indicating the track can, from any point in AA', be drawn, and the distance laid off, on the completion of the second observation. This evidently accomplishes the same result as correcting the altitudes. It possesses the advantage of being simple, and when the chart has the magnetic compass plotted upon it, the compass course can be laid off between the observations. The convenience of allowing in this manner for the run of the ship in the interval is worthy of special attention, as one-half the computation is performed without waiting until the second observation is taken.

The following examples will illustrate this part of the rule:—

EXAMPLES.

Ex. 1. Suppose in the forenoon (see preceding example) an altitude of the sun was observed, the latitude by dead reckoning being 50° 14' N. The longitudes, as ascertained by working with the assumed latitudes 50° N. and 51° N., are as follows:—

Longitude corresponding to 50° N. is 7° 5' W. (A).
,, ,, 51° N. is 5° 15' W. (A').

In the afternoon of the same day a second altitude of the sun was observed, the ship in the interval having sailed N.E. (true) 20 miles. Working with the same assumed latitudes as before, the

Longitude corresponding to 50° N. is 8° 30' W. (B).
,, ,, 51° N. is 5° 15' W. (B').

Project A A', the ship's line of position, or line of equal altitudes, corresponding to the *first* altitude. The ship having moved in the interval, the projection of the first line of equal altitudes, must now be moved parallel to itself in the direction of and through the distance of the run. Set off from any part of the line A A', 20 miles due N.E. Draw through this point a line *a a'* parallel to A A', this will be the line of equal altitude at the time of the first observation, corrected for change of station. (See Plate.) Next project B B', the line of equal altitude corresponding to the second observation. The intersection at C' of the line B B' with *a a'* or that corresponding to the first observation as corrected for change of station, shows the position of the ship. The latitude is 50° 28' N., instead of 50° 14' N., as by dead reckoning; the longitude is 6° 58' W.

It will be evident that the second observation will not give the same altitude as if the ship had remained stationary, but will give a line of position or line of equal altitude 20 geographical miles further to N.E. Be careful not to confound these with longitudinal miles, as in this lat. 20 geographical miles equal 31 miles of longitude.

Ex. 2. It is required to project two observations, showing by the intersection of lines of position the place of the ship, supposing her to have sailed in the interval between the first and second observations 13 miles N.W. ¼ W. (true).

The longitudes at the first observation, as ascertained by working with the assumed latitudes 49° N. and 50° N., were as follows:—

Longitude corresponding to latitude 49° N. is 4° 33½' W. (A).
,, ,, ,, 50° N. is 3° 49' W. (A'.)

Mark on the chart the spot A in latitude 49° N. and long. 4° 33½' W., and also the point A' in lat. 50° N. and long 3° 49' W. Connect these two points A and A' by a straight line, this line lies N.E. by N. and S.W. by S. and produced to meet the land it will be found to cut the Eddystone, bearing N.E. by N. Next project this first line of position for the run of the ship, by setting off from any point in A A' the distance, 13 miles N.W. ½ W. (true), and draw a line aa' through this point parallel to A A'; this will be the line of position at the time of the first observation, corrected for change of station.

After a suitable change of bearing of the sun an altitude of the sun is observed, and the longitudes as ascertained by working with the same assumed latitudes, as before, were as follows:—

Longitude corresponding to 49° N. is 3° 42½' W. (B).
„   „   „   50° N. is 5° 35' W. (B').

Mark the point B on the chart in lat. 49° N. and long. 3° 42½' W., also the point B' in lat. 50° N. and long. 5° 35' W. Connect these two points B and B' by a straight line; this line is the ship's line of position at the second observation, and its intersection at C of the line aa' the position of the ship at the second observation, in lat. 49° 27¼' N., long. 4° 33½' W.

The line of position at the second observation, prolonged to meet the land, cuts the Land's End, bearing N.W. ¼ W.

Ex. 3. It is required to project two observations of the sun, showing by the intersection of their lines of position the place of the ship, supposing the ship to have sailed N.E. by E. ½ E. (true), 19 miles, in the interval between the observations.

The longitudes at the first observation, by working with the assumed latitudes 49° 40' N. and 50° 0' N., were as follows:—

Longitude corresponding to latitude 49° 40' N. is 177° 56¾' E. (A).
„   „   „   50° 0' N. is 179° 27¼' W. (A').

Working with the same assumed latitudes, the corresponding longitudes at the second observation were as follows:—

Longitude corresponding to lat. 49° 40' N. is 179° 48½' W. (B).
„   „   „   50° 0' N. is 179° 48' W. (B').

Mark on the chart the spot corresponding to A, lat. 49° 40' N., long. 177° 56¾' E.; also the point A', lat. 50° 0' N., long. 179° 27¼' W. Connect these two points by a straight line; this is the ship's line of position at the first observation. Set off from the line A A' 19 miles in the direction of the course, N.E. by E. ½ E. (true), and draw a line aa' through this point parallel to AA', which will be a projection of the line of position at the first observation for the run of the ship.

Again, mark the spot B in lat. 49° 40' N., long. 179° 48½' W., and the spot B' in lat. 50° 0' N., long. 179° 48' W.; these two points being joined by a straight line are a projection of the ship's line of position at the second observation, and the intersection of this line B B' with the line aa' (or the line of position at the first observation corrected for change of station) gives the place of the ship at the second observation, lat. C 50° 4' N., long. 179° 48' W.

The first line of position trends E. by N. and W. by S., and the sun's true bearing is S. by E.

We shall now proceed to work out in full detail an example showing how to determine the ship's position by the intersection of lines of position corresponding to two altitudes of the sun, having given the ship's course and distance in the interval.

Ex. 4. 1880, February 29th, A.M. at ship, and uncertain of my position, when the chron. showed February 28^d 23^h 16^m 55^s M.T.G.; suppose the obs. alt. sun's L.L. 29° 46' 50", and again P.M. same day, when the chronometer showed February 29^d 4^h 13^m 50^s, the obs. alt.

sun's L.L. 15° 47' 10", the ship having made 36 miles on a true E. by S. course, height of eye 20 feet: required the line of bearing when the first altitude was taken, and the position of the ship by Sumner's Method when the second altitude was taken.

Suppose the ship's position at the time of taking the second altitude to be as follows:—

B = lat. 49° 0' N., long. 5° 30' W.; B' = lat. 50° 0' N., long. 6° 30½' W.

Space for minor corrections.

First Observation.

Green. date, Feb. 28^d 23^h 16^m 55^s

Time from noon, 29^d 0 44

0^h·73

	Decl., page II, N.A.		H.D.	Eq. T., page II, N.A.		H.D.
cosec.	Feb. 29^d, noon	7° 42' 9" S. *decr.*	56"·83	Feb. 29^d,	12^m 37^s·7 *decr.*	0"·490
97° 42' 0"= 0·003[1034]	Corr. for 0^h·73 + 0 41		0"·73	Corr. for 0^h·7 + 0"·3		7
Pts. for 50 = + 14						
97 42 50 0·00·[648]	Red. decl.	7 42 50 S.	17064	Red. Eq. time 12 38·0 0^s·3430		
Tab. diff. = 28			39816	(To be *added* to A.T.)		
50	Polar dist.	97 42 50				
Corr. = 11·00			41·5224			

Correction of observed altitude.   Computation of time and longitude, with assumed latitude 49° N.

	Obs. alt. ☉'s L.L.	29° 46' 50"	Altitude	29° 57' 11"		
	Dip	− 4 17	Latitude	49 0 0	sec.	0·183057
sine.			Polar dist.	97 42 50	cosec.	0·003948
58° 22' 0" = 9·930145		29 42 33				
Pts. for 49 = + 64	Corr.	− 1 32		176 40 1		
58 22 49 9·930209						
Tab. diff. 130		29 41 1		88 20 0	cos.	8·463665
49	Semi-diameter	+ 16 10				
1170				58 22 49	sine	9·930209
520	True alt.	29 57 11				
Corr. = 63·70			Hour-angle	1^h 30^m 3^s	log.	8·580879
						47
			App. T. ship, Feb. 28^d	22 29 57		
			Eq. time	+ 12 38		41
					gives 3^s	
			M.T. ship, Feb. 28^d	22 42 35		
			M.T. Green., Feb. 28^d	23 16 55		
			Longitude in time	34 20		
			Long. A	8° 35' W.		

With assumed latitude 50° N.

	Altitude	29° 57' 11"			
	Latitude	50 0 0	sec.	0·191933	
	Polar dist.	97 42 50	cosec.	0·003948	
sine.		177 40 1			
58° 52' 0" = 9·932457					
Pts. for 49 = + 62		88 50 0	cos.	8·308794	
58 52 49 = 9·932519					
Tab. diff. 127		58 52 49	sine	9·932519	
49	Hour-angle	1^h 16^m 11^s	log.	8·437194	
1143				10	
508	App. T. ship, Feb. 28^d	22 43 49		—	
Corr. = 62·23	Eq. time	+ 12 38		9 = 1^s	
	M.T. ship, Feb. 28^d	22 56 27			
	M.T. Green., Feb. 28^d	23 16 55			
	Longitude in time	20 28			
	Long. A'	5° 7' W.			

Project A A', the ship's line of position, or line of equal altitudes, corresponding to the *first* altitude. Next, project B B', the line of equal altitudes corresponding to the second observation. The ship having moved in the interval between the observations, the projection of the first line of equal altitudes must now be moved parallel to itself in the direction of and through the distance of the run. Set off from any part of A A', 20 miles due N.E., as follows:—

From the graduated meridian opposite the line extending from A to A' take off the distance 20 miles with the dividers; lay the dividers down, and taking the parallel rules, place them on the compass in the corner of the chart, over N.E. and S.W.; then work the parallels (strictly preserving the direction) towards the line of the first position A to A'; having reached that line, draw another extending from any part of it in the direction of the course N.E., and on this last line lay off the 20 miles of distance already taken with your dividers. Lay the edge of the parallels on the line A A', and working them (preserving the direction) to the point just laid off; draw through this point another line $a a'$ parallel to A A'. The intersection at C' of the line B B' with $a a'$, or that of the first observation as corrected for change of station, shows the position of the ship. The latitude is 50° 28' N., the longitude is 6° 58' W.

Examples for Practice.

1. 1882. If at sea, March 1st, A.M. at ship, and uncertain of my position, when the chronometer showed February 28^d 10^h 50^m 54^s M.T.G.; suppose the obs. alt. sun's L.L. 29° 46' 50"; and again, P.M. on the same day, when the chronometer showed February 28^d 15^h 50^m 49^s M.T.G., obs. alt. sun's L.L. 15° 47' 10", the ship having made 32 miles on a true E. by N. course, height of eye 19 feet.

Required the line of bearing when the first altitude was taken, and the position of the ship by Sumner's Method when the second altitude was taken, assuming the ship to be between latitude 49° 10' N. and 49° 40' N.

On looking at the question it will be noticed that there is a second Greenwich date mean time; also a second observed altitude; in this case February 28^d 15^h 50^m 49^s M.T.G., and obs. alt. 15° 47' 10"; these are the basis of the two positions which will be furnished by the Examiners, and respecting which the circular says—"Candidates will not be for the present obliged to perform the calculations."*

Suppose the ship's calculated position at the time of taking the second altitude to be as follows:—

 B = Lat. 49° 10' N.   Long. 179° 50' W.
 B' = Lat. 49° 40' N.   Long. 179° 35' E.

2. 1882, May 19th, A.M. at ship and uncertain of my position, when the chronometer showed May 18^d 22^h 1^m 39^s M.T.G., obs. alt. sun's L.L. 48° 57', and again, P.M. the same day, when the chronometer showed May 19^d 3^h 37^m 18^s M.T.G., obs. alt. sun's L.L. 40° 45' 15", the ship having made between the observations 25 miles on a true N. 45° E. course, height of eye 18 feet: required the line of bearing when the first altitude was taken, and the position of the ship by Sumner's Method when the second altitude was observed, assuming latitudes 48° 30' N. and 49° 0' N.

Suppose the ship's calculated position at the time of taking the second altitude to be as follows:—

 B = Lat. 48° 30' N.   Long. 5° 18¼' W.
 B' = Lat. 49° 0' N.   Long. 5° 33' W.

3. 1882. If at sea, November 8th, A.M. at ship, and uncertain of my position, when the chronometer showed November 7^d 22^h 29^m 2^s; suppose obs. alt. sun's L.L. 19° 49'; and again, P.M. on the same day, when the chronometer showed November 8^d 3^h 48^m 57^s M.T.G., obs. alt. sun's L.L. 10° 52' 30", the ship having made 41 miles on a true E. ½ N. course in the

---

* For the sake of practice it may be advisable to verify the positions of B and B'.

interval, height of eye 19 feet: required the line of bearing when the first altitude was taken, assuming the ship to be between lats. 48° 10′ N. and 48° 40′ N.

Suppose the ship's calculated position at the time of taking the second altitude to be as follows:—

$$B = \text{Lat. } 48° \text{ 10}' \text{ N.} \qquad \text{Long. } 10° \text{ 33}\tfrac{3}{4}' \text{ W.}$$
$$B' = \text{Lat. } 48° \text{ 40}' \text{ N.} \qquad \text{Long. } 11° \text{ 13}\tfrac{1}{2}' \text{ W.}$$

4. 1882, February 3rd, A.M. at ship, and uncertain of my position, when the chronometer showed February 2ᵈ 22ʰ 17ᵐ 16ˢ M.T.G., obs. alt. sun's L.L. 19° 49′ 50″; and again, P.M. same day, when the chronometer showed February 3ᵈ 3ʰ 37ᵐ 11ˢ M.T.G., obs. alt. sun's L.L. 10° 53′ 30″, the ship having made 39 miles on a true E. by N. ½ N. course in the interval, height of eye 19 feet: required the line of bearing when the first altitude was taken, and the position of the ship by Sumner's Method when the second altitude was observed, assuming the ship to be between lats. 48° 10′ N. and 48° 40′ N.

Suppose the ship's calculated position at the time of taking the second altitude to be as follows:—

$$B = \text{Lat. } 48° \text{ 10}' \text{ N.} \qquad \text{Long. } 0° \text{ 21}\tfrac{1}{4}' \text{ E.}$$
$$B' = \text{Lat. } 48° \text{ 40}' \text{ N.} \qquad \text{Long. } 0° \text{ 17}\tfrac{3}{4}' \text{ W.}$$

5. 1882, January 24th, A.M. at ship, and uncertain of my position, when the chronometer showed January 23ᵈ 22ʰ 19ᵐ 1ˢ M.T.G., obs. alt. sun's L.L. 9° 40′ 15″; and again, P.M. same day, when the chronometer showed January 24ᵈ 2ʰ 19ᵐ 5ˢ M.T.G., obs. alt. sun's L.L. 18° 20′ 10″, the ship having sailed 18 miles on a (true) S.E. by E. ½ E. course in the interval between the observations, height of eye 21 feet, required the time of bearing when the first altitude was taken, and the position of the ship by Sumner's Method when the second altitude was taken, assuming the ship to be between latitudes 51° 15′ N. and 50° 45′ N.

Suppose the ship's position (by calculation) at the time of second observation to be as follows:—

$$B = \text{Lat. } 51° \text{ 15}' \text{ N.} \qquad \text{Long. } 17° \text{ 16}\tfrac{1}{4}' \text{ W.}$$
$$B' = \text{Lat. } 50° \text{ 45}' \text{ N.} \qquad \text{Long. } 14° \text{ 27}\tfrac{1}{4}' \text{ W.}$$

6. 1882, March 27th, A.M. at ship, and uncertain of my position, when chronometer showed March 26ᵈ 10ʰ 26ᵐ 24ˢ G.M.T., obs. alt. sun's L.L. 36° 38′ 30″; again, P.M. same day, when chronometer showed March 26ᵈ 16ʰ 37ᵐ 10ˢ, obs. alt. sun's L.L. 15° 15′ 20″, height of eye 22 feet, course and distance sailed in the interval between the observations S. by E. ¾ E. (true), distance 37 miles: required the line of bearing when the first altitude was taken, and the position of the ship by Sumner's Method when the second altitude was observed, assuming the ship to be between latitudes 51° 40′ N. and 51° 10′ N.

Suppose the ship's position (by calculation) at the time of second observation to be as follows:—

$$B = \text{Lat. } 51° \text{ 40}' \text{ N.} \qquad \text{Long. } 179° \text{ 45}' \text{ W.}$$
$$B' = \text{Lat. } 51° \text{ 10}' \text{ N.} \qquad \text{Long. } 178° \text{ 31}\tfrac{1}{4}' \text{ W.}$$

7. 1882, March 1st, A.M. at ship, and uncertain of my position, when the chronometer showed February 28ᵈ 22ʰ 48ᵐ 15ˢ M.T.G., the obs. alt. sun's L.L. 29° 44′ 30″; and again, P.M. same day, when the chronometer showed March 1ᵈ 3ʰ 46ᵐ 50ˢ M.T.G., obs. alt. sun's L.L. 15° 50′ 10″, the ship having made 42 miles on a true E. by N. ½ N. course in the interval, height of eye 15 feet: required the line of bearing when the first altitude was taken, and the position of the ship by Sumner's Method when the second altitude was observed, assuming latitudes 49° 15′ N. and 49° 45′ N.

Suppose the ship's position (by calculation) at the time of taking the second altitude to be as follows:—

$$B = \text{Lat. } 49° \text{ 15}' \text{ N.} \qquad \text{Long. } 1° \text{ 9}' \text{ E.}$$
$$B' = \text{Lat. } 49° \text{ 45}' \text{ N.} \qquad \text{Long. } 0° \text{ 39}\tfrac{1}{4}' \text{ E.}$$

8. 1882, February 3rd, A.M. at ship, being uncertain of my position, when the chronometer showed February 2ᵈ 16ʰ 33ᵐ 36ˢ M.T.G., obs. alt. sun's L.L. 20° 1' 23"; and again, P.M. on the same day, the chronometer showed February 2ᵈ 21ʰ 55ᵐ 54ˢ, obs. alt. sun's L.L. 11° 10' 30", the ship having made 40 miles on a true E.N.E. course in the interval, height of eye 19 feet: required the line of bearing when the first altitude was taken, and the position of the ship by Sumner's Method when the second altitude was observed, assuming the latitudes 47° 10' N. and 47° 40' N.

Suppose the ship's position (by calculation) at the time of taking the second altitude to be as follows :—

$$B = \text{Lat. } 47° \text{ 10}' \text{ N.} \qquad \text{Long. } 86° \text{ 14}\tfrac{3}{4}' \text{ E.}$$
$$B' = \text{Lat. } 47° \text{ 40}' \text{ N.} \qquad \text{Long. } 85° \text{ 37}\tfrac{1}{4}' \text{ E.}$$

9. 1882, January 24th, A.M. at ship, being uncertain of my position, when the chronometer showed January 24ᵈ 9ʰ 14ᵐ 44ˢ, obs. alt. sun's L.L. 9° 40' 10"; and again, P.M. same day, when the chronometer showed January 24ᵈ 13ʰ 14ᵐ 30ˢ M.T.G., obs. alt. sun's L.L. 18° 31' 15", the ship having made 23 miles on a true S.E. ¼ S. course in the interval, height of eye 22 feet: required the line of bearing when the first altitude was taken, and the position of the ship by Sumner's Method when the second altitude was observed, assuming the ship to be between the latitudes 51° 15' N. and 50° 45' N.

Suppose that by calculation the ship's position at the time of the second observation is as follows :—

$$B = \text{Lat. } 51° \text{ 15}' \text{ N.} \qquad \text{Long. } 178° \text{ 26}' \text{ E.}$$
$$B' = \text{Lat. } 50° \text{ 45}' \text{ N.} \qquad \text{Long. } 178° \text{ 40}\tfrac{1}{4}' \text{ W.}$$

10. 1882, March 21st, A.M. at ship, being uncertain of my position, when the chronometer showed March 20ᵈ 10ʰ 38ᵐ 58ˢ, obs. alt. sun's L.L. 22° 49' 50"; again, P.M. same day, when chronometer showed March 21ᵈ 1ʰ 18ᵐ 50ˢ M.T.G., obs. alt. sun's L.L. 32° 29' 50', the ship having made 29 miles on a true N. 52° W. course in the interval, height of eye 21 feet: required the line of bearing when the first altitude was taken, and the position of the ship by Sumner's Method when the second altitude was observed, assuming the ship to be between the latitudes 50° 10' N. and 50° 50' N.

Suppose that by calculation the ship's position at the time of the second observation is as follows :—

$$B = \text{Lat. } 50° \text{ 10}' \text{ N.} \qquad \text{Long. } 0° \text{ 26}\tfrac{1}{2}' \text{ E.}$$
$$B' = \text{Lat. } 50° \text{ 50}' \text{ N.} \qquad \text{Long. } 0° \text{ 48}\tfrac{1}{4}' \text{ W.}$$

11. 1882, March 15th, A.M. at ship, being uncertain of my position, when chronometer showed March 14ᵈ 19ʰ 53ᵐ 51ˢ M.T.G., obs. alt. sun's L.L. 14° 14'; and again, P.M. same day, when the chronometer showed March 15ᵈ 3ʰ 30ᵐ 20ˢ M.T.G., obs. alt. sun's L.L. 23° 17', the ship sailed N. by E. ¾ E. 21 miles in the interval, height of eye 16 feet: required the line of bearing when the first altitude was taken, and the position of the ship by Sumner's Method when second altitude was observed, assuming the ship to be between lats. 49° 0' N. and 49° 40' N.

Suppose the ship's position by calculation to be at the time of the second observation as follows :—

$$B = \text{Lat. } 49° \text{ 0}' \text{ N.} \qquad \text{Long. } 0° \text{ 40}\tfrac{1}{4}' \text{ W.}$$
$$B' = \text{Lat. } 49° \text{ 40}' \text{ N.} \qquad \text{Long. } 1° \text{ 22}' \text{ W.}$$

12. 1882, May 4th, A.M. at ship, and uncertain of my position, when a chronometer showed May 4ᵈ 1ʰ 25ᵐ 18ˢ G.M.T., obs. alt. sun's L.L. 50° 30' 10"; and again, P.M. same day, when the chronometer showed May 4ᵈ 5ʰ 5ᵐ 4ˢ, obs. alt. sun's L.L. 52° 50', the ship having made 32 miles on a true S. by W. ¾ W. course in the interval, height of eye 21 feet: required the line of bearing when the first altitude was taken, and the position of the ship by Sumner's Method when the second altitude was observed, assuming the ship to be between latitudes 47° 0' N. and 46° 20' N.

Suppose the position of the ship by calculation to be at the time of the second observation as follows:—

$\quad\quad\quad\quad$ B = Lat. 47° 0' N. $\quad\quad$ Long. 52° 36' W.
$\quad\quad\quad\quad$ B' = Lat. 46° 20' N. $\quad\quad$ Long. 51° 32½' W.

13. 1882, June 17th, A.M. at ship, and uncertain of my position, when chronometer showed June 16^d 18^h 24^m 3^s, obs. alt. sun's L.L. 51° 50' 40"; again, P.M. same day, when chronometer showed June 16^d 21^h 23^m 30^s, obs. alt. sun's L.L. 63° 34' 1", the ship having made 33 miles on a true S.E. ½ S. course, height of eye 20 feet: required the line of position when the first altitude was taken, and the position of the ship by Sumner's Method when the second altitude was observed, assuming latitudes 48° 50' N. and 48° 10' N.

For the second observation the positions as given by the Examiner would be

$\quad\quad\quad\quad$ B = In Lat. 48° 50' N., and Long. 47° 29½' E.
$\quad\quad\quad\quad$ B' = In Lat. 48° 10' N., and Long. 50° 13¾' E.

14. 1882, October 29th, A.M. at ship, and uncertain of my position, when a chronometer showed October 29^d 4^h 9^m 12^s G.M.T., obs. alt. sun's L.L. 28° 18'; again, P.M. same day, when chronometer showed Oct. 29^d 7^h 39^m 12^s, obs. alt. sun's L.L. 16° 40', the ship having sailed in the interval 15 miles on a true N. by E. course: required the line of position when the first altitude was taken, and the position of the ship by Sumner's Method when the second altitude was observed, assuming latitudes 47° 30' N. and 48° 0' N.

For the second observation the positions as given by the Examiner would be

$\quad\quad\quad\quad$ B = Lat. 47° 30' N. $\quad\quad$ Long. 73° 42¼' W.
$\quad\quad\quad\quad$ B' = Lat. 48° 0' N. $\quad\quad$ Long. 74° 25¾' W.

REMARKS ON THE DEGREE OF DEPENDENCE.

337. We have seen that the line of position is at right-angles to the direction of the object, hence the line of position itself can be obtained from *one* assumed latitude if the true bearing of the sun or celestial object observed be known. The tables of the Sun's True Bearing, or Time Azimuth Tables, by BURDWOOD, for the parallels 60° to 30°, and the tables of DAVIS for the parallels of 30° to the equator give the true bearing of any celestial object whose declination does not exceed 23°. The tables of LABROSSE may also be used for the same purpose. By means of these tables the computation of a Sumner is shortened by one-half.

338. When the altitude of the celestial object is low, the circle of equal altitudes is large, the small portion of it comprised between two assumed latitudes is very nearly a straight line; but when the altitude is high, the circle is small, a small portion of it may be much curved, and the direction of the two extremities very different; that is, the bearing of the land and the sun's azimuth may be sensibly different for different parts of the same portion.* An error in the assumed latitudes has therefore most effect when the altitude is high, and least when it is low, which last is always the preferable case.

---

* It will always be judicious to check the line of position, which is very easily done by means the Time Azimuth Tables just referred to. By entering the tables with each assumed latitude, the declination of the object, and its given hour-angle, the true bearing for each latitude is found; then at each end of the line of position, project the line perpendicular to the true bearing just found; the point of crossing of the two perpendiculars will show whether or not the line of position is good.

339. If the sun is on the prime vertical at both observations, the lines of position will run North or South, and there will be no intersection.

If the latitude and declination are equal nearly, the lines of position will not change their direction sufficiently to depend upon their intersection.

When the body is near the prime vertical, errors in the latitudes have the least effect upon the corresponding longitudes.

When the body is near the meridian, errors in the longitudes have their least effect upon corresponding latitudes.

The value of the method depends entirely on the latitude assumed being near the truth.

340. When the coast trends parallel to the line of position, the distance of the ship from the shore is ascertained, though her absolute position is uncertain. When the observation is near noon, the line of position lies nearly E. and W., and the bearing of land in this direction is ascertained; but when the sun is E. or W., the line of position lies nearly N. or S., and the position of the ship depends altogether upon the chronometer.

341. If there is an uncertainty in the altitude, draw on each side of the line of position lines parallel to it, and distant from it, the amount of the supposed uncertainty, and the position will be somewhere within this *belt* or linear space. Again, if a second altitude which gives another line of position is also doubtful, we project, as before, lines on each side this last line of position, and to the extent of the error of the altitude; we thus get a space (indicated by a quadrilateral) within which is the ship's position, and the area of the space is more circumscribed than either belt.

In the same manner, if there is an uncertainty in the Greenwich time, parallels may be drawn upon each side of the line of position equal to this uncertainty.

The error of the chronometer places the line of position too far East or West bodily, but does alter its direction.

VARIATION BY AN AZIMUTH.

In this problem the Error of the Compass is required by computing the true bearing of the sun, and taking the difference between the true bearing and the bearing by an Azimuth Compass.

342. **The Azimuth** of a heavenly body is the arc of the horizon intercepted between the cardinal point adjacent to the elevated pole, and the circle of altitude passing through the body, or it is the angle at the Zenith contained between the vertical circle passing through the elevated pole (the meridian) and the vertical circle passing through the object. Azimuth is usually reckoned from the north or south point, eastward and westward from $0°$ to $180°$.

343. **True Azimuth** is the bearing of an object from the *true* North or South point, and is the azimuth found by calculation from the observed altitude or hour-angle of the body. It is in general simply called *The "Azimuth,"* but it is thus qualified as the *True* Azimuth to distinguish it from the *Magnetic* Azimuth, which is the bearing of the object from the compass North or South point, and which is found by direct observation with an instrument carrying a magnetic needle. The difference between the true and magnetic azimuth gives the entire correction of the compass—variation and deviation combined.

344. Given the latitude, altitude, and declination of an object, to find the true azimuth.

RULE XCVIII.

$1°$ *Add together the polar distance, the latitude, and the altitude,* take half the sum, and take the difference between the half sum and the polar dist.*

Note.—When the latitude is 0, suppose it to be of same name to the declination when finding the polar distance.

$2°$. *Add together the log. sec. of latitude, the log. sec. of altitude (rejecting tens) the log. cosines of the half sum and remainder; the sum (rejecting tens) is log. sine square of true azimuth* (Table 69, RAPER). *Or half the sum of the four logs. is the log. sine of half of the true azimuth, which take out of the table* (Table 24, NORIE), *and double it; the result is the true azimuth.*

$3°$. *Reckon the true azimuth from S., when the latitude is N., but from the N. when the latitude is S.; towards E. when it is* A.M., *or when the altitude is increasing, but towards W. when it is* P.M., *or when the altitude is decreasing.*

(a) *When* latitude *is* $0°$, *if* declination *is* N., *reckon the azimuth from the South; if* declination *is* S., *reckon the azimuth from the North.*

(b) *When both latitude and declination are* $0°$, *the object moves on the prime vertical, or is* E. *while the altitude is* increasing *and* W. *while the altitude is* decreasing.

Note.—The logs. are taken out in these examples to the nearest second.

---

* The learner will observe that in this formula the pol. dist., lat., and alt., occur in the reverse order of that in Rule XCIV (finding hour-angle) in which the initials form the word *alp*. In finding the azimuth the initials form *pla*. The 2nd and 3rd terms take secants; the last two, cosines. By this arrangement the term which has to be taken for the half sum is always on the top.

EXAMPLES.

Ex. 1. Given the latitude 47° 46′ S.; declination 22° 27′ 22″ (or polar distance 67° 32′ 38″), true altitude 26° 44′ *decreasing* or (being P.M.)

Polar dist.	67° 32′ 38″		
Latitude	47 46 0	sec.	0·172533
Altitude	26 44 0	sec.	0·049095
Sum	142 2 38		
Half sum	71 1 19	cos.	9·512159
Half sum—p.d.	3 28 41	cos.	9·999200
		2)	19·732987
	47° 20′ 12″	sine	9·866493
	2		470
True az. N.	94 40 W.		12″ = 23

The sum of logs. (less 10 from index) 9·732987 being found in Table of log. sine squares (RAPER, Table 69) gives the corresponding arc, 94° 20′, whence true azimuth is N. 94° 40′ W. The *true* azimuth is here marked N. because the latitude is S., and W. because the altitude is *decreasing*, it being P.M.

Ex. 3. Latitude 28° 3′ N., declination 12° 39′ 50″ S., true altitude 25° 12′ 4″ + (A.M.)

Polar dist.	102° 39′ 50″		
Latitude	28 3 0	sec.	0·054267
Altitude	25 12 4	sec.	0·043438
Sum	155 54 54		
Half sum	77 57 27	cos.	9·319392
P. dist.—½ sum	24 42 23	cos.	9·958307
		2)	19·375404
Half azimuth	29° 9½′	sine	9·687702
	2		
True az.	S. 58 19 E.		

The sum of the four logs., rejecting 10 from the index, 9·375404, is the log. sine square of true azimuth; seek for log. in Table 69, RAPER, and the corresponding arc is 58° 19′, whence true azimuth is S. 58° 19′ E., the same as by NORIE's Tables.

Ex. 5. Latitude 34° 19′ S., declination 7° 5′ 27″ S., true altitude 40° 55′ 57″ + (P.M.)

Polar dist.	82° 54′ 33″		
Latitude	34 19 0	sec.	0·083054
Altitude	40 55 57	sec.	0·121776
Sum	158 9 30		
Half sum	79 4 45	cos.	9·277500
Remainder	3 49 48	cos.	9·999029
		2)	19·481359
	33° 23′ 40″	sine	9·740679
	2		
True az.	N. 66 47 20 E.		

Ex. 2. Latitude 37° 15′ N., declination 22° 22′ 58″ N., true altitude 39° 20′ 8″—(P.M.)

Polar dist.	67° 37′ 2″		
Latitude	37 15 0	sec.	0·099086
Altitude	39 20 8	sec.	0·111570
Sum	144 12 10		
Half sum	72 6 5	cos.	9·487610
Half sum—p.d.	4 29 3	cos.	9·998668
		2)	19·696934
	44° 51′ 58″	sine	9·848467
	2		
True az.	S. 89 44 W.		

The sum of logs. (less 10 in the index), viz., 9·696934 being found in the log. sine square (Table 69 RAPER), gives true azimuth as above, S. 89° 44′ W.

Ex. 4. Latitude 38° 46′ N., declination 7° 41′ 56″ S., true altitude 27° 16′ 8″—(P.M.)

Polar dist.	97° 41′ 56″		
Latitude	38 46 0	sec.	0·108071
Altitude	27 16 8	sec.	0·051164
Sum	163 44 4		
Half sum	81 52 2	cos.	9·150657
Remainder	15 49 54	cos.	9·983206
		2)	19·293098
	26° 18¼′	sine	9·646549
	2		
True az.	S. 52 36½ W.		

The sum of logs. (rejecting 10 from index) being found in the log. sine square (Table 69 RAPER) gives the corresponding arc 52° 36½′, whence the true azimuth is S. 52° 36½′ W.

Ex. 6. Lat. 0°, declination 15° 2′ 27″ N., true altitude 24° 12′ 10″—.

Polar dist.	105° 2′ 27″		
Latitude	0 0 0		
Altitude	24 12 10	sec.	0·039957
Sum	129 14 37		
Half sum	64 37 18	cos.	9·632045
Remainder	40 25 9	cos.	9·881568
		2)	19·553570
	36° 44′	sine	9·776785
	2		
	N. 73 28 W.		

Examples for Practice.

In each of the following examples it is required to find the true azimuth. (The sign + means A.M., and the sign — means P.M.)

	True altitude		Declination		Latitude
1.	7° 43′ 27″ +	,,	11° 28′ 32″ N.	,,	51° 10′ N.
2.	,, 28 30 53 +	,,	21 56 45 S.	,,	26 20 N.
3.	,, 12 50 46 —	,,	9 36 51 N.	,,	15 47 S.
4.	,, 29 41 59 +	,,	2 38 14 N.	,,	4 22 N.
5.	,, 7 15 55 —	,,	12 14 38 S.	,,	51 2 N.
6.	,, 13 47 28 —	,,	17 50 57 N.	,,	42 36 S.
7.	,, 45 30 0 —	,,	23 2 0 S.	,,	0 0
8.	,, 25 40 10 +	,,	0 0 0	,,	0 0
9.	,, 40 7 21 —	,,	17 4 3 S.	,,	33 51 S.

345. Given the true bearing and compass bearing, to find the error of the compass.

RULE XCIX.

1°. **To find the amount of the Error of the Compass.**—*Reckon the True and Magnetic Azimuths from the same point of the compass*—North or South.

Caution.—Be careful when taking the true bearing from 180°, *not to change* the *East* or *West* name; only the *North* or *South*.

(a) *If one of the azimuths be expressed from the* North *and the other from the* South, *take either of them from* 180°, *and it will then be* reckoned *from the* same *point as the other*.

(b) *If the bearing by compass be reckoned from* East *or* West, *towards* North *or* South, *take it from* 90°, *and reverse the position of the letters; or*, add 90°, *and it will then be expressed from the opposite point to that from which it is reckoned when taken from* 90°.

Example.

Ex. Suppose magnetic azimuth to be W. 18° 30′ N.; then subtract the magnetic azimuth from 90° thus:—

$$\begin{array}{c} 90° \ 0′ \\ \text{W. } 18 \ 30 \ \text{N.} \\ \hline \text{N. } 71 \ 30 \ \text{W.} \end{array}$$

The azimuth is thus reckoned from the North Pole.

Or add 90° to the magnetic azimuth, thus:—

$$\begin{array}{c} \text{W. } 18° \ 30′ \ \text{N.} \\ 90 \ \ 0 \\ \hline \text{S. } 108 \ 30 \ \text{W.} \end{array}$$

The azimuth is thus reckoned from the South Pole.

(c) *When the magnetic azimuth is either* East *or* West, *it is to be reckoned as* 90° *from* North *or* South, *according as the true azimuth is* North *or* South.

2°. Take the difference of the true and magnetic azimuths *when measured towards the same point of the compass*, East *or* West; *but when measured towards different points, i.e., when one is reckoned* towards East *and the other towards* West, *take the* sum; the result is the error of the compass or correction.

3°. **To name the Correction of Compass.**—*Let the observer look at the two azimuths (or bearings) from the centre of the compass—then if the* true *azimuth is to the right of the* magnetic *azimuth, the correction is* East; *but if the* true *azimuth is to the* left *of the* magnetic *azimuth, the error is* West.

EXAMPLES.

Ex. 1. Given true azimuth N. 44° 20′ E. and the sun's bearing by compass (or magnetic azimuth) N. 17° 10′ E.: required the error of compass.

True az.　N. 44° 20′ E.
Mag. az.　N. 17 10 E.
─────────
Error　　　27 10 E.

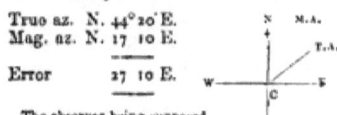

The observer being supposed looking from the centre of the compass in the direction of the magnetic azimuth, then the *true azimuth* lies to the *right* hand of the magnetic azimuth, whence the error of compass is to be marked *East*.

Ex. 3. Given true azimuth S. 69° W., magnetic azimuth S. 47° W.: required the error of compass.

True az.　S. 69° W.
Mag. az.　S. 47 W.
─────────
Error　　　22 E.

The observer being supposed looking from the centre of the compass in the direction of the magnetic azimuth C M, then the true azimuth, C T, lies to the *right* hand of the magnetic azimuth, whence the error of the compass is *East*.

Ex. 5. The true azimuth S. 62° 41′ E., and magnetic azimuth E.S.E.: required the error of compass.

True az. S. 62° 41′ E.
S. 6 pts. E. = Mag. az. S. 67 30 E.
─────────
Error of compass　4 49 E.

Here the error of compass is East, since the *true azimuth* is on the *right* of the magnetic azimuth, the observer looking from the centre of the compass in the direction of the magnetic azimuth.

Ex. 7. True azimuth N. 72° E., magnetic azimuth East.

True azimuth　　　　N. 72° E.
Mag. azimuth East = N. 90 E.
─────────
Error of compass　18 W.

Ex. 9. The true azimuth S. 90° 33′ E., and magnetic azimuth N. 81° 20′ E.: find the error of compass.

True azimuth　　S. 90° 33′ E.
　　　　　　　　　180　0
　　　　　　　　─────────
　　　　　　or　N. 89 27 E.
Mag. azimuth　　N. 81 20 E.
─────────
Error of compass　8 7 E.

The *true azimuth* being reckoned from S., while the magnetic azimuth is expressed from N., the true is subtracted from 180°, in order to reckon it from the same point as the magnetic azimuth, viz., from N.

Ex. 2. Given true azimuth S. 70° 57′ E. the magnetic azimuth S.E. by E. ¾ E.: required the error of compass.

Mag. az. S.E. by E. ¾ E. = S. 64° 41′ 15″ E.
True az. S. 70° 57′ 0″ E.
Mag. az. S. 64 41 15 E.
─────────
Error　　6 15 45 W.

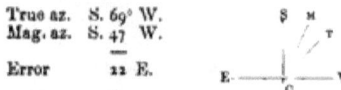

The error of compass is in this instance West, because when looking from the centre of the compass in the direction of the magnetic azimuth, the *true azimuth* is on the *left* hand of the magnetic.

Ex. 4. True azimuth N. 50° 12′ E., and the magnetic azimuth N. 61° 50′ E.: required the correction of compass.

True azimuth　　　N. 50° 12′ E.
Magnetic azimuth　N. 61 50 E.
─────────
Error of compass　11 38 W.

The error of compass is here West, because the *true azimuth* is to the *left* hand of the magnetic azimuth, the observer being supposed to look from the centre of the compass in the direction of the magnetic azimuth.

Ex. 6. The true azimuth S. 82° 50′ W., and magnetic azimuth W. 15° N.

　　　　　　　　　True az. S. 82° 50′ W.
W. 15° N. = Mag. az. S. 105　0 W.
─────────
Error of compass　22 10 W.

The error of compass is West, the *true azimuth* being to the *left of magnetic*, 90° is *added* to the compass bearing in order to reckon it from the same point as the true azimuth; thus, from S. to W. is 90°, and from W. to W. 15° N. is 15° more; hence magnetic azimuth is S. 105° W.

Ex. 8. The true azimuth is S. 76° W., and the magnetic azimuth West.

True azimuth　S. 76° 0′ W.
Mag. azimuth　S. 90　0 W.
─────────
Error of compass　14　0 W.

The magnetic azimuth West is reckoned as 90° from S., because the true azimuth is reckoned from S.

Ex. 10. The true azimuth N. 69° 39′ W., and magnetic azimuth S. 93° 30′ W.: find the error of compass.

True azimuth　　N. 69° 39′ W.
　　　　　　　　180　0
　　　　　　　　─────────
　　　　　　or　S. 110 21 W.
Mag. azimuth　　S. 93 30 W.
─────────
Error of compass　16 51 E.

The *true azimuth* is here taken from 180°, in order to reckon it from the same point as the magnetic azimuth.

Ex. 11. True azimuth S. 36° W., magnetic azimuth S. 9° E.

True azimuth   S. 36° W.
Mag. azimuth   S. 9° E.
                  ─────
Error of compass   45 E.

Ex. 12. True azimuth N. 68° W., magnetic azimuth N. 5° E.

True azimuth   N. 68° W.
Mag. azimuth   N. 5 E.
                  ─────
Error of compass   73 W.

Ex. 13. True azimuth N. 49° E., magnetic azimuth N. 3° W.

True azimuth   N. 49° E.
Mag. azimuth   N. 3 W.
                  ─────
Error of compass   52 E.

Ex. 14. True azimuth S. 50° E., magnetic azimuth S. 8° W.

True azimuth   S. 50° E.
Mag. azimuth   S. 8 W.
                  ─────
Error of compass   58 W.

RULE C.

1°. **To find Greenwich Date.**—*With ship time and longitude in time find the Greenwich date* (Rule LXXVIII, page 227).

2°. **To find Polar Distance.**—*Take from* page II, Nautical Almanac, *the sun's declination and reduce it to Greenwich date* (Rule LXXIX, page 230); *also take out sun's semi-diameter.*

If apparent time is given, use *Nautical Almanac*, page I.

3°. **To find True Altitude.**—*Correct observed altitude for index error, dip, refraction, parallax, and semi-diameter, and thus get the true altitude* (Rule LXXXIV, page 242).

4°. **To find True Azimuth.**—*Proceed according to* Rule XCVIII, page 304, *to find the true azimuth.*

5°. **To find Error of Compass.**—*Having found the true azimuth, proceed by* Rule XCIX, page 306, *to find the entire correction or error of the compass.*

6°. **To find Deviation.**—*Next place the* variation *below the* error of compass *and proceed as in the amplitude* (7° and 8° of Rule LXXXVI, page 245), *the result is the* deviation *for the position of the ship's head at the time of observation.*

EXAMPLES.

Ex. 1. 1882, May 19th, 3ʰ 7ᵐ 44ˢ P.M., mean time at ship, latitude 41° 53' N., longitude 60° 19' W., sun's bearing by compass S. 104° 40' W., observed altitude sun's L.L. 43° 56' 7", height of eye 18 feet, index correction 0': required the true azimuth and error of the compass; and supposing the variation to be 17° 10' W.: required the deviation of the compass for the position of the ship's head at the time of observation.

Ship date (M.T.) May   19ᵈ 3ʰ 7ᵐ44ˢ
Long. 60° 19' W. in time +   4   1 16
                          ─────────────
Green. date (M.T.) May   19   7   9 0

By *Raper:* Dip — 4' 10", refr. — 1' 1", par. + 6", semid. + 15' 50". True alt. 44° 6' 54".

Obs. alt. ☉'s L.L. 43° 56' 7"
Dip   —   4   4
                    ───────────
Corr. altitude   43 52   3
                  —   53
                    ───────────
                    43 51 10
Semi-diameter   +   15 50
                    ───────────
True altitude   44   7   0

Polar dist.	70° 7' 48"			Decl., page II, N.A.		H.D.
Latitude	41 53 0	sec.	0·128132	19th, noon 19° 48' 23" N. +		32"·03
Altitude	44 7 0	sec.	0·143922	Correction + 3 49		7'·15
Sum	156 7 48			Red. decl. 19 52 12		16015
				90 0 0		3203
Half sum	78 3 54	cos.	9·315554			22421
Remainder	7 56 6	cos.	9·995821	Polar dist. 70 7 48		6,0)22,9·0145
		2)	19·583429	Correction	+ 3' 49"	
Half azimuth	38° 14' 44"	sine	9·791714			
	2					

True azimuth S. 76 29 W.
Mag. azimuth S. 104 40 W.

Error of compass 28 11 W.
Variation   17 10 W.

Deviation   11 1 W.

The true azimuth being to the left of the magnetic, the *error* of compass is West, and the *deviation* is West, because the error is to the *left* of variation.

Ex. 2. 1882, September 2nd, mean time at ship $8^h 59^m$ A.M., latitude 39° 31' S., longitude 127° 45' W., sun's bearing by compass N. 29° 50' E., observed altitude sun's L.L. 26° 40' 37", height of eye 18 feet: required the true azimuth and error of the compass: and supposing the variation to be 9° 50' E.: required the deviation of the compass for the position of the ship's head at the time of observation.

Ship date (M.T.), Sept.	$1^d 20^h 59^m$		Obs. alt. ☉'s L.L.	26° 40' 37"	
Long. 127° 45' W.	+ 8 31		Dip (T. 4)	− 4 4	
Green. date (M.T.) Sept.	2 5 30		Corr. altitude	26 36 33	
				− 1 45	
By *Raper*: Dip − 4' 10", refr. − 1' 56",			Semi-diameter	26 34 48	
par. + 8", semid. + 15' 54". True alti-				+ 15 54	
tude 26° 50' 33".			True altitude	26 50 42	
Decl., 2nd, p. II, N.A.	7° 52' 13" N. (−)		Hourly diff., noon	54"·82	
Correction	− 5 2		$5^h 30^m =$	× 5·5	
Reduced declination	7 47 11 N.			6,0)30,1'·510	
	90 0 0			5' 3"	
Polar distance	97 47 11				
Polar dist.	97° 47' 11"		Half true azimuth	26° 3'	
Latitude	39 31 0	sec. 0·112698		2	
Altitude	26 50 42	sec. 0·049523			
			True azimuth	N. 52 6 E.	
Sum	164 8 53		Magnetic azimuth	N. 29 50 E.	
Half sum	82 4 26	cos. 9·139531	Error of compass	22 16 E.	
			Variation	9 50 E.	
Remainder	15 42 45	cos. 9·983461			
		2)19·285233	Deviation	12 26 E.	
Half azimuth	26° 3' 0"	sine 9·642616			

The true azimuth being to the *right* of magnetic the error is named E., and the error being to the *right* of the variation, the *deviation* is named E.

Ex. 3. 1882, July 5th, mean time at ship 6ʰ 55ᵐ 51ˢ P.M., latitude 50° 53′ N., longitude 119° 8′ E., sun's bearing by compass N. 69° 0′ W., observed altitude sun's L.L. 9° 40′, index correction + 3′ 50″, height of eye 18 feet, variation 4° 0′ W.

Ship date (M.T.), July	5ᵈ 6ʰ 55ᵐ 51ˢ		Obs. alt. ☉'s L.L.	9° 40′ 0″
Long. 119° 8′ E.	− 7 56 32		Index correction	+ 3 50
Green. date (M.T.), July	4 22 59 19			9 43 50
			Dip	− 4 4
Hourly diff.	14″·20			9 39 46
Time from noon 1ʰ	1		Corr. altitude	− 5 17
Correction	14″·20			9 34 29
	or + 14′		Semi-diameter	+ 15 46
Decl., noon, July 5th	22 47 19 N.		True altitude	9 50 15
Red. decl.	22 47 33 N.			
	90		By *Raper*: Index corr. + 3′ 50″, dip − 4′ 10″, refr. − 5′ 31″, par. + 8″, semid. + 15′ 46″. True alt. 9° 50′ 3″.	
Polar dist.	67 12 27			

Polar dist.	67° 12′ 27″			True azimuth	N. 65° 48′ W.
Latitude	50 53 0	sec.	0·200038	Magnetic azimuth	N. 69 0 W.
Altitude	9 50 15	sec.	0·006434		
				Error of compass	3 12 E.
Sum	127 55 42			Variation	4 0 W.
Half sum	63 57 51	cos.	9·642399	Deviation	7 12 E.
Remainder	3 14 36	cos.	9·999304		
			2)19·848175		
Half azimuth	57° 6′ 3″	sine	9·924087		
	2				
True azimuth S.	114 12 W.				
	180 0				
or, N.	65 48 W.				

The *true* and *magnetic* azimuths being reckoned from *opposite* points, the *true* is taken from 180, and the remainder reckoned from the *opposite* point, whence true azimuth is N. 65° 52′ W. The true azimuth being to the *right* of magnetic, the error of compass is East. The error and variation being of *opposite* names, their sum is *deviation*, and is named East, because the error is to the *right* of the variation, or takes the *same name* as the error.

Ex. 4. 1882, February 10th, at 8ʰ 2ᵐ A.M., mean time at ship, latitude 50° 48′ N., longitude 77° 30′ W., sun's bearing by compass S.E. by E. ¼ E., observed altitude sun's L.L. 7° 10′ 40″, index correction − 1′ 6″, height of eye 15 feet; required the true azimuth and error of compass; variation 14° 0′ W.

Ship date, M.T., Feb.	9ᵈ 20ʰ 2ᵐ		Obs. alt. ☉'s L.L.	7° 10′ 40″
Long. 77° 30′ W.	+ 5 10		Index correction	− 1 6
Green. date, M.T., Feb.	10 1 12			7 9 34
			Dip for 15 feet	− 3 42
	or, 1ʰ·2			7 5 52
Hourly diff., decl., noon	− 48″·89		Correction of altitude	− 7 6
Time from noon 1ʰ 12ᵐ	× 1·2			6 58 46
	58″·668		Semi-diameter	+ 16 14
Correction	− 59″		True altitude	7 15 0
Decl., 10th, noon	14 16 52 S.			
Red. declination	14 15 53 S.		By *Raper*: Index corr. − 1′ 4″, dip − 3′ 50″, refr. − 7′ 19″, par. + 9″, semid. + 16′ 14″, true alt. 7° 14′ 50″.	
	90			
Polar distance	104 15 53			

Variation by an Azimuth.   311

Polar dist.	104° 15' 58"		
Latitude	50 48 0	sec.	0·199263
Altitude	7 15 0	sec.	0·003486
Sum	162 18 53		
Half sum	81 9 26	cos.	9·186740
Remainder	23 6 27	cos.	9·963680
			2)19·353169
Half azimuth	28° 21' 6"	sine	9·671584

	28° 21'
	2
True azimuth	S. 56 42 E.
Mag. az. (S.E. by E. ¼ E.)	S. 59 4 E.
Error of compass	2 22 E.
Variation	14 0 W.
Deviation	16 22 E.

The error is East, the *true* azimuth being to the *right* of the *magnetic*, and the *deviation* is East, the error being to the *right* of variation.

Ex. 5. 1882, January 21st, at 10ʰ 14ᵐ A.M. apparent time at ship, lat. 39° 3' S., long. 96° 28' E., sun's bearing by compass E. 2° 30' S., observed altitude sun's U.L. 46° 15', index correction − 2' 43", height of eye 19 feet : required the true azimuth and error of compass ; variation 17° 0' W. : find the deviation.

Green. date, M.T., Jan   20ᵈ 15ʰ 48ᵐ 8ˢ

Time from noon, Jan. 21ᵈ   8 12

Hourly diff. 21st, noon   − 33"·67
8ʰ 12ᵐ =   × 8·2

6,0)27,6·094

Correction   + 4' 36"
Decl. 21st, noon, page I   19 51 24 S.

Red. decl.   19 56 0 S.

Polar distance   70 4 0

By *Norie*; Index corr. − 2ʰ 43ᵐ, dip − 4' 15", corr. of alt. − 0' 49", semid. − 16' 17", true alt. 45° 51' 0".

By *Raper*: Index corr. − 2' 43", dip − 4' 15", refr. − 56", par. + 6", semid. − 16' 17", true alt. 45° 50' 55".

Polar dist.	70° 4' 0"		
Latitude	39 3 0	sec.	0·109805
Altitude	45 51 0	sec.	0·157054
Sum	154 58 0		
Half sum	77 29 0	cos.	9·335906
Polar dist. − do.	7 25 0	cos.	9·996351
			2)19·599116
	39° 4' 25"	sine	9·799558
	2		

True az.   N. 78 9 0 E.

True azimuth	N. 78° 9' E.
Mag. azimuth	N. 92 30 E.
Error of compass	14 21 W.
Variation	17 0 W.
Deviation	2 39 E.

In this example the best way is to reckon the *magnetic* azimuth from the North, the same as the *true*; thus from N. to E. is 90°, and from E. to E. 2° 30' S., is 2° 30'; therefore the magnetic azimuth is N. 92° 30' E.

Ex. 6. 1882, June 1st, at 9ʰ 40ᵐ A.M. mean time at ship, latitude 60° N., longitude 40° 20' W., observed altitude sun's L.L. 44° 48' 50", index correction + 3' 17", height of eye 18 feet, sun's bearing by compass S. ½ W.: required the true azimuth and error of compass; variation 51° 15' W.

The Greenwich date is June 1ᵈ 0ʰ 21ᵐ 20ˢ. True altitude (NORIE) 45° 3' 0", hourly diff. of decl. 20"·15 × ⅓ʰ = 7", which, *added* to decl. June 1st at noon, viz., 22° 4' 57" N. *incr.*, gives the red. decl. 22° 5' 4" N., polar distance 67° 54' 56", sum of logs. 19·217516, true azimuth S. 47° 56' E.

True azimuth S. 47° 56' E.   *Add* when one East
S. ½ point W. = Mag. azimuth S.   5 37½ W.   and the other West.

Error of compass   53 33½ W. *true* azimuth *left* of *mag.*
Variation   51 15 W.

Deviation   2 18½ W. because *error* is left of variation.

Ex. 7. 1882, March 20th, at 8ʰ 50ᵐ A.M., mean time at ship, lat. 0°, long. 123° 30′ W., obs. alt. sun's L.L. 36° 24′, height of eye 15 feet, sun's bearing by compass East: required the true azimuth and error of compass; and suppose the variation to be 24° W., find the deviation.

Green. date (M.T.) March 20ᵈ 5ʰ 4ᵐ

or 5ʰ·07

Decl., page II, N.A.	H.D.
March 20th 0° 5′ 0″ S. (—)	59″·25
Corr. for 5ʰ·07 — 5 0	5·07
Red. decl. 0 0 0	6,0)30,0″·3975
	5ᵐ 0ˢ

Lat. 0°, and decl. 0°, and A.M., the sun's true bearing is East, and since the observed bearing is also East, the error of compass is 0° 0′, and the deviation is equal in amount to the variation with the name reversed; hence the deviation is 24° E. The example need not be calculated beyond the reduced declination.

Ex. 8. 1882, September 1st, at 8ʰ 0ᵐ A.M., mean time at ship, lat. 31° 22′ S., long. 33° W., obs. alt. sun's L.L. 15° 2′ 0″, height of eye 22 feet, sun's bearing by compass N. 70° 6′ E., variation 19° 15′ E.: required the deviation.

Green. date (M.T.) Aug. 31ᵈ 17ʰ 48ᵐ, decl. noon Sept. 1st = 8° 14′ 4″·7 N. *decr.*, H.D. 54″·49 × 6ʰ·2 (time wanting to noon Sept. 1st) gives corr. 5′ 37″·8 to be *added* to decl. at noon Sept. 1st, as decl. is *increasing*; whence red. decl. is 8° 19′ 42″ N. By *Norie*: ind. corr. — 2′ 30″, dip 4′ 30″, corr. alt. — 3′ 23″, and semid. + 15′ 54″, give true alt. 15° 7′ 31″. Sum of logs. = 19·518194; log. sine of ½ sum of logs. 9·759092 gives ½ azimuth 35° 2′ 48″; whence true azimuth N. 70° 6′ E.: the error is 0, and deviation 19° 40′ W.—being the same amount as the variation, but of the opposite name.

EXAMPLES FOR PRACTICE.

In each of the following examples it is required to find the true azimuth, also the error of compass and deviation for the position of the ship's head at the time of observation.

No.	Civil date. 1882.	M.T. ship.	Latitude.	Longitude.	Sun's bearing by compass.	Obs. alt. sun's L.L.	Ht. of eye.	Variation.
1	Jan. 24th	8ʰ 22ᵐ 35ˢ A.M.	26° 5′ S.	50° 53′ W.	E. by N.	38° 23′ 10″	15	4° 36′ E.
2	Feb. 28th	3 14 0 P.M.	38 46 N.	47 16 W.	N. 42° 30′ W.	26 57 14	17	11 30 E.
3	March 27th	4 6 40 P.M.	4 22 N.	53 7 E.	W. by N.	29 30 50	20	3 30 W.
4	April 3rd	6 20 0 P.M.	49 52 N.	109 58 E.	West	11 43 0	17	9 10 E.
5	May 27th	9 3 20 A.M.	55 0 N.	1 33 W.	S. 61 45 E.	43 8 51	18	15 45 W.
6	June 20th	6 10 0 P.M.	44 45 N.	11 26 W.	N. 43 20 W.	16 40 20	18	23 0 W.
7	July 31st	8 46 30 A.M.	38 18 N.	65 4 W.	S. 71 20 E.	43 24 58	18	8 50 W.
8	Aug. 23rd	5 24 58 A.M.	51 10 N.	135 10 W.	N. 66 20 E.	7 38 0	15	25 30 E.
9	Sept. 1st	3 47 50 P.M.	10 40 S.	138 42 E.	W. ¼ N.	30 4 10	15	3 30 W.
10	Nov. 25th	4 7 0 P.M.	39 58 S.	50 52 W.	N. ¼ W.	33 51 0	11	9 22 E.
11	Dec. 17th	9 10 30 A.M.	29 10 S.	26 53 W.	S. 84 20 E.	51 1 13	16	9 48 W.
12	July 3rd	8 26 50 A.M.	32 10 S.	62 0 E.	N. 62 0 E.	14 11 37	19	17 0 W.
13	Jan. 6th	5 2 14 P.M.	47 46 S.	11 11 E.	N. 81 40 W.	26 37 27	25	33 15 W.
14	April 20th	7 56 41 A.M.	27 20 S.	86 43 W.	N. 61 50 E.	18 44 55	30	14 30 W.
15	Jan. 29th	3 36 35 P.M.	42 26 N.	49 18 W.	W. 9 10 S.	13 38 40	16	20 30 W.
16	Feb. 1st	3 41 51 P.M.	33 51 S.	20 37 E.	N. 70 50 W.	39 56 10	18	20 40 W.
17	March 21th	9 5 50 A.M.	43 6 N.	51 2 W.	S. 44 50 E.	32 40 0	18	20 0 W.
18	Feb. 26th	2 48 0 P.M.	5 0 N.	167 0 E.	N. 118 34 W.	60 37 0	19	9 15 E.
19	June 21st	3 22 0 P.M.	66 40 N.	55 20 W.	N. ¾ E.	15 38 0	18	67 50 W.
20	Sept. 11th	7 0 0 A.M.	37 0 S.	19 0 W.	N. 44 50 E.	42 28 0	20	12 20 W.
21	April 25th	8 50 0 A.M.	5 25 N.	94 30 E.	N. 75 46 E.	12 55 0	13	2 30 E.
22	March 1st	0 35 0 P.M.	15 30 N.	65 0 E.	S. ¼ E.	66 14 0	12	0 0
23	1883, Jan. 1st	9 27 10 A.M.	0 0	0 25 E.	S.E.	45 10 50	17	21 0 W.

ON FINDING THE LATITUDE BY REDUCTION TO THE MERIDIAN.

346. The latitude of a place is most simply determined by observation of the meridian altitude of a known heavenly body. When such an observation cannot be obtained by reason of the state of the weather, the altitude of the body may often be obtained a little before or a little after its meridian passage. And if at the time of observing such an altitude near the meridian, the hour-angle of the body is known, we may find by computation very nearly the difference of altitude by which to reduce the observed to the Meridian altitude. The correction is called the "Reduction to the Meridian." This method, in point of simplicity, is little inferior to the meridian altitude, to which it is next in importance. The altitude may also be determined by a direct process, deduced from spherical trigonometry. The former is the method used in the following pages. The term "near the meridian" implies a meridian distance limited according to the latitude and declination, and also the degree of precision with which the time is known (see RAPER, Table 47).

RULE CI.

$1°$. **To find the Apparent Time at Ship.**—*To the time shown by the watch, expressed astronomically, apply the error of the watch for apparent time,*[*] *adding when the watch is slow (rejecting 24^h when the sum exceeds 24^h, and putting the day one forward), subtracting when the watch is fast (increasing the time shown by watch by 24^h, if necessary, and putting the day one back).*

$2°$. *Next turn into time the difference of longitude made since the error of the watch was determined,* adding *when the difference of longitude is* East, subtracting *when the difference of longitude is* West; *the result is apparent time at ship when the observation was made.*[†]

$3°$. **To find the Time from Noon.**—*If* P.M. *at ship the apparent time at ship is the time from noon; when it is* A.M. *(reckoning from the preceding noon) subtract apparent time at ship from 24^h; the remainder is the time from noon.*

---

[*] The error of chronometer for *apparent time* at place should be noted when the morning sights are taken for determining the longitude. This, with the diff. long. made in the interval between this last time and the time of observing the ex-meridian altitude, will give the apparent time at ship. If the ship has not changed her meridian since the time of morning sights, the result obtained by applying the error of chronometer is, of course, the apparent time at ship.

[†] The reason for this rule will appear on considering that if a watch is set to the time at any given meridian, it will be *slow* for any meridian to the *eastward*, but *fast* for any meridian to the *westward*, at the rate of 1^m for $15'$ diff. long., since the sun comes to the *easterly* meridian earlier, and to the *westerly* meridian later.

Examples.

Ex. 1. Suppose it is P.M. at ship, and the watch when *corrected* shows January 2^d 0^h 16^m 56^s (see Ex. 1 following): then the time from noon is 16^m 56^s past noon of the 2nd.

Ex. 2. Again, suppose it is A.M. at ship, and the watch when *corrected* indicates Feb. 5^d 23^h 37^m 16^s (see Ex. 2 following): then we have

$$\begin{array}{r} 24^h\ 0^m\ 0^s \\ 23\ 37\ 16 \\ \hline 22\ 44 \end{array}$$

In this instance it is 22^m 44^s before noon of the 6th.

4°. *To find Greenwich Apparent Time.*—*With apparent time at ship and longitude, find Greenwich date in apparent time* (Rule LXXVIII, page 227).

5°. *Take out of* Nautical Almanac, page I, *the declination, and reduce it to the Greenwich date* (Rule LXXIX, page 230).

6°. *Correct the observed altitude of sun's upper or lower limb, and so get the true altitude of sun's centre* (Rule LXXXIV, page 242).

Method I.

7°. *Take out log. rising of time from noon* (Table 29, NORIE), *log. cos. declination* (Table 25, NORIE), *and log. cos. of latitude* (Table 25, NORIE).

NOTE.—In using the natural sines and cosines to six places, it will be necessary to add 1 to the index of the log. rising, because, as given in the table, it is only adapted to five places of figures.

Caution.—In the use of the table of log. rising (XXIX, NORIE), care must be taken that the correct indices are used when the minutes of the time from noon are 1, 3, 10, or 32. It is necessary to notice that the indices in the table sometimes change in the column where they could not be inserted for want of room; this may, however, be easily known by observing that the first figure of the decimal part of the log. changes from 9 to 0.

Thus the log. rising of 0^h 1^m 0^s is 9·97860,
but the log. rising of 0^h 1^m 5^s is 0·04813.

The index, as given in the table, is in the form $\frac{9}{0}$, which means that it changes from 9 to 0 somewhere in that line. Similarly, opposite 10^m, the index is in the form $\frac{1}{2}$, and the numerator 1 is the index of the log. rising of 10^m 0^s, 10^m 5^s, 10^m 10^s and of 10^m 15^s, and changes to 2 somewhere between 10^m 15^s and 10^m 20^s.

8°. *Take the sum of these and find the natural number corresponding thereto.* (Table 24, NORIE).

9°. *To the natural number just found add the natural sine of the true altitude* (Table 26, NORIE); *the sum is natural cosine of meridian zenith distance, which take out of the table, and name it* North, *when the observer is* North *of the sun, or when the sun bears* South; *but call zenith distance* South *when the observer is* South *of sun, or when it bears* North. (See Rule LXXXV, 4°, page 244).

10°. *Apply the reduced declination to the zenith distance, taking their sum if they are of the same name, but their* difference *if of* contrary *names; the result, in either case, is the latitude of the same name as the* greater. (Rule LXXXV, 5°).

NOTE.—The foregoing Method (Method I) is only convenient when the computer is provided with a table of natural sines and cosines, as well as a table of log. versed sines, or the logarithmic value of 2 sine^2 $\frac{1}{2}$ t.

347. We may also compute directly the reduction of the observed altitude to the meridian altitude by the following:—

Method II.

1°. Add together the following logarithms:—

> *Constant log.* $5 \cdot 615455$; *(this is the log. of $\frac{2}{\sin 1''}$.)**
>
> *Log. cosine of latitude by account* (Table 25, NORIE).
>
> *Log. cosine of declination* (Table 25, NORIE).
>
> *Log. cosecant of meridian zenith distance as deduced from latitude by D.R. and declination* (Table 25, NORIE).
>
> *The log. of time from noon; (this is twice the log. sine of half the hour-angle).* (Table 31, NORIE, and 69, RAPER).

The sum of these logs. (rejecting tens from the index), will be the log. of the reduction in seconds (″). (Table 24, NORIE).

The zenith distance from latitude by D.R. is found as follows:—When the *latitude* and *declination* are both of the *same* name, take their *difference*; when *latitude* and *declination* have *different* names, take their *sum*: the result in either case will be zenith distance by D.R.

2°. Add *the reduction to the true altitude: the* result *is the* meridian altitude.†

3°. *Having the meridian altitude; find the* latitude *as by the method of meridian altitudes* (Rule LXXXV, page 244).

NOTE.—This Method (Method II) does not approximate so rapidly as the preceding (Method I), but the objection is of little weight when the observations are very near the meridian. On the other hand, it has the great advantage of not requiring the use of the Table of Natural Sines.

Method III.—(By Towson's Ex-Meridian Tables.)‡

1°. *Enter* Table I (TOWSON) *under nearest declination and find nearest hour-angle, against which stands Augmentation I, which* add *to declination, at the same time take out corresponding index number in the margin.*

2°. *Enter* Table II *under true altitude and opposite index number, find Augmentation II, which* add *to true altitude, and thence find latitude as in meridian altitude.*

---

* If we use the constant log. $0 \cdot 301030$ (this is log. of 2) instead of that given above, viz., $5 \cdot 615455$, the sum of logs. (rejecting tens from index), will be log. sine of reduction in minutes (′) and seconds (″). (Table 66, RAPER, or Table 25, NORIE). If we omit the constant altogether the sum of the other four logs. is the log. sine of half the reduction, in minutes (′) and seconds (″), which must be multiplied by 2 to get the reduction.

† This is only an approximate meridian altitude, in strictness a second reduction should be computed.

‡ At the Liverpool Local Marine Board Examinations the candidate is expected to solve this problem by means of TOWSON'S Ex-Meridian Tables.

EXAMPLES.

Ex. 1. 1882, January 2nd, P.M. at ship, latitude by account 52° 6′ S., longitude 71° 23′ W., observed altitude sun's L.L. North of observer 60° 20′ 30″, index correction + 2′ 58″, height of eye 20 feet, time by watch January 2d 0h 48m 22s, which was found to be 29m 16s *fast* on apparent time at ship, difference of longitude 32·4 miles to West: required the latitude by reduction to meridian.

Time by watch, Jan.	2d 0h 48m 22s	App. time at ship, Jan.	2d 0h 16m 56s
Watch *fast*	− 0 29 16	Long. in time	+ 4 45 32
	2 0 19 6	Green. date (A.T.), Jan.	2 5 2 28
Diff. long. $\frac{32\cdot 4 \times 4}{60}$	− 2 10	Decl., page I, N.A., January 2nd, at noon 22° 54′ 20″ S. *(decreasing)*.	
Time from noon, Jan.	2 0 16 56		
		H. diff., Jan. 2nd, noon	13″·80
Obs. alt. ☉'s L.L.	60° 20′ 30″	Greenwich date 5h 2m, or	× 5
Index correction	+ 2 58		
	60 23 28		6,0)69,00
Dip for 20 feet	− 4 17	Correction	− 1′ 9″
	60 19 11	Decl., Jan. 2nd, noon	22° 54′ 20″ S. *dec.*
Correction of alt.	− 0 28	Correction	− 1 9
	60 18 43	Red. decl.	22 53 11 S.
Semi-diameter	+ 16 18	By *Raper*: Index corr. + 2′ 58″, dip − 4′ 20″, refr. − 0′ 34″, par. + 4″, semid. + 16′ 18″, true alt. 60° 45′ 53″.	
True altitude	60 35 1		

Method I.

Time from noon	16m 56s	rising* 3·435880
Latitude	52° 6′	cos. 9·788370
Declination	22 53	cos. 0·964400
	1544	nat. no. 3·188650
		1544
T. alt. 60° 35′ 1″	nat. sine	871073
Z. dist. 29 14 9 S.	nat. cos.	872617
Decl. 22 53 11 S.	(next greater)	638
Lat. 52 7 20 S.		239)2100(9″

* The index of log. rising is *increased* by 1. See Note to 7°, page 314.

Method II.

Constant log.			5·61546
Lat. by D.R.	52° 6′ S.	cos.	9·78837
Declination	22 53 S.	cos.	9·96440
Mer. zen. dist.	29 13	cosec.	0·31148
Time from noon	16m 56s	log.	7·13486
	6,0)65,2	log.	2·81457
Reduction	+ 10′ 52″		
True altitude	60 35 1		
Meridian alt.	60 45 53		
	90		
Zenith distance	29 14 7 S.		
Declination	22 53 11 S.		
Latitude	52 7 18 S.		

By *Raper*: Lat. 52° 7′ 29″ S.

Method III.—By *Towson's Ex-Meridian Tables.*

☉'s red. declination	22° 53′ 11″ S.	True altitude	60° 35′ 1″ S.
Aug. Table 1, Index 30	+ 3 21	Aug. Table 2, Index 30	+ 14 4
Augmented declination	22 56 32 S.		60 49 5
		Meridian zen. dist.	29 10 55 S.
This is the method required of Candidates at Liverpool.		Declination	22 56 32 S.
		Latitude	52 7 27 S.

Reduction to Meridian. 517

Ex. 2. 1882, February 6th, A.M. at ship, lat. acct. 51° 58' N., long. 105° 41' W., obs. alt. sun's L.L. South of observer 22° 10' 30", index corr. + 56", height of eye 22 feet, time by watch 6ᵈ 0ʰ 4ᵐ 4ˢ, found to be 28ᵐ 47ˢ *fast* on app. time at ship, diff. of long. made to East 29·8 miles since error of watch on app. time at ship was determined: required the latitude by reduction to meridian.

Time by watch, Feb.	6ᵈ 0ʰ 4ᵐ 4ˢ	App. time at ship, Feb.	5ᵈ 23ʰ 37ᵐ 16ˢ	
Watch *fast*	− 28 47	Long. 105° 41' W.	+ 7 2 44	
Diff. long. $\frac{29'·8 \times 4}{60}$	5 23 35 17 + 1 59	Greenwich date, Feb.	6 6 40 0	
			or, 6ʰ·67	
App. time at ship, Feb.	5 23 37 16 24 0 0	Hourly diff. 6ʰ 40ᵐ =	− 46"·02 × 6·67	
Time from noon, Feb.	6 0 22 44			
Obs. alt. ☉'s L.L.	22° 10' 30"		32214 27612 27612	
Index correction	+ 56			
Dip	22 11 26 − 4 30		6,0)30,6·9534	
Corr. of alt.	{22 6 56 − 2 11		5' 7"	
	22 4 45	Declination, page I, N.A. Feb. 6th, noon, 15° 43' 7" S. *dcr.* Correction − 5 7		
Semi-diameter	+ 16 15			
True altitude	22 21 0	Red. decl.	15 38 0	

Method I.

		260
Time from noon	22ᵐ 44ˢ	rising 3·689030
Latitude	51° 50'	cos. 9·789665
Declination	15 28	cos. 9·983981
		2919 nat. no. 3·465276
		2919
True altitude	22° 21' 0"	nat. sine 380263
Zen. distance	67 28 9 N.	nat. cos. 383182
Declination	15 27 46 S.	
Latitude	52 0 23 N.	

The nat. sine being worked to six places of figures, 1 is added to index of log. rising.

Method II.

Constant log.		5·615455
Latitude	51° 58' N.	cos. 9·789665
Declination	15 28 S.	cos. 9·983981
Mer. zen. dist.	67 26	cosec. 0·034594
T. from noon	22ᵐ 44ˢ	log. 7·390540
	6,0)65,2	log. 2·814235
Reduction	+ 10' 52"	
True altitude	22° 21 0	
Mer. altitude	22 31 52	
Zenith distance	67 28 8 N.	
Declination	15 27 46 S.	
Latitude	52 0 22 N.	

Method III.—By Towson's Ex-Meridian Tables.

☉'s red. declination	15° 27' 46" S.	True altitude	22° 21' 0"	
Aug. Table 1, Index 57	+ 4 23	Aug. Table 2, Index 57	+ 6 35	
Augmented declination	15 32 9		22 27 35	
		Meridian zen. dist.	67 32 25 N.	
		Augmented declination	15 32 9 S.	
		Latitude	52 0 16 N.	

Ex. 3. 1882, August 7th, A.M. at ship, lat. acct. 40° 52′ N., long. 36° 47′ W., obs. alt. sun's L.L. South of observer 65° 1′, index corr. + 17″, eye 14 feet, time by watch 11ʰ 15ᵐ 46ˢ, found to be 26ᵐ 16ˢ *slow* of app. time at ship, the diff. of long. made to East was 17 miles after the error on app. time at ship was determined: required the latitude.

Time by watch, Aug.	6ᵈ 23ʰ 15ᵐ 46ˢ	App. time at ship, Aug.	6ᵈ 23ʰ 43ᵐ 10ˢ	
Watch *slow*	+ 26 16	Long. 36° 47′ W.	+ 2 27 8	
	6 23 42 2	Greenwich date, Aug.	7 2 10 18	
Diff. of long.	+ 1 8			
App. time at ship, Aug.	6 23 43 10	By *Raper*: True altitude	65° 13′ 3″	
	24 0 0	H.D., 7th noon,	42″·07	
Time from noon, Aug.	7 0 16 50	2ʰ 10ᵐ =	× 2·2	
			8414	
Obs. alt. ☉'s L.L.	65° 1′ 0″		8414	
Index correction	+ 17		6,0)9,2·554	
	65 1 17		1′ 33″	
Dip	− 3 36			
	64 57 41	Decl., page I, N.A.		
Corr. altitude	− 23	Aug. 7th	16° 23′ 47″ N. *decr.*	
	64 57 18	Correction	− 1 33	
Semi-diameter	+ 15 49	Red. decl.	16 22 14 N.	
True altitude	65 13 7			

Method I.

Time from noon 16ᵐ 50ˢ	rising	3·430750	
Latitude	40° 52′	cos.	9·878656
Declination	16 22	cos.	9·982035
		1956 nat. no.	3·291441
		nat. no.	1956
True altitude 65° 13′ 7″	nat. sine	907913	
Zen. distance 24 30 46 N.	nat. cos.	909869	
Declination	16 22 14 N.		
Latitude	40 53 0 N.		

Method II.

Constant log.			5·615455
Latitude acct.	40° 52′ N.	cos.	9·878656
Declination	16 22 N.	cos.	9·982035
Mer. zen. dist.	24 30	cosec.	0·382273
Time from noon 16ᵐ 50ˢ		log.	7·129720
	6,0)97,3	log.	2·988139
Reduction	+ 16′ 13″		
True altitude	65° 13 7		
Meridian alt.	65 29 20		
Mer. zen. dist.	24 30 40 N.		
Declination	16 22 14 N.		
Latitude	40 52 54 N.		
By *Raper*: Lat.	40 52 48 N.		

Method III.—By *Towson's* Ex-Meridian Tables.

☉'s red. declination	16° 22′ 14″ N.	True altitude	65° 13′ 7″
Aug. Table 1, Index 33	+ 2 30	Aug. Table 2, Index 33	+ 19 2
Augmented declination	16 24 44 N.	Augmented altitude	65 32 9
		Zenith distance	24 27 51 N.
		Augmented declination	16 24 44 N.
		Latitude	40 52 35 N.

Ex. 4. 1882, Sept. 23rd, P.M. at ship, lat. acct. 51° 2' N., long. 117° 8' E., obs. alt. sun's L.L. South of observer 38° 44' 20", index corr. + 1' 8', height of eye 21 feet, time by watch 50ᵐ 0ˢ (or 23ᵈ 0ʰ 50ᵐ), found to be 39ᵐ 2ˢ *fast* on app. time at ship, diff. of long. made to *West* was 8·2 miles after the error on app. time was determined: required the latitude.

Time by watch, Sept.	23ᵈ 0ʰ 50ᵐ 0ˢ		App. time at ship, Sept.		23ᵈ 0ʰ 10ᵐ 25ˢ
Watch *fast*	− 39 2		Long. 173° 53' E.		− 7 48 32
	23 0 10 58		Greenwich date, Sept.		22 16 21 53
Diff. of longitude	− 33		By *Raper*: " True altitude		38° 55' 55"
App. time at ship, Sept.	23 0 10 25		H.D.		− 58"·44
Time from noon, 23rd is	10 25		12ʰ 35ᵐ		× 16·37
Obs. alt. ☉'s L.L.	38° 44' 20"				40908
Index correction	+ 1 8				17532
	38 45 28				35064
Dip	− 4 23				5844
	38 41 5				6,0)95,6·6628
Corr. altitude	− 1 4				15' 56"·7
	38 40 1		Decl., page I, N.A.		
Semi-diameter	+ 15 59		Sept. 22nd 0° 15' 21" N. *decr*.		
			Correction − 15 57		
True altitude	38 56 0		Red. decl. 0 0 36 S.		

Method I.

Time from noon	10ᵐ 25ˢ	rising	3·01399
Latitude	51° 2'	cos.	9·798560
Declination	0 1	cos.	0·000000
		649 nat. no.	2·812550
		nat. no.	649
True altitude	38° 56'	nat. sine	628416
Mer. zen. dist.	51° 1' 8" N.	nat. cos.	629065
Declination	0 0 36 S.		
Latitude	51 0 32 N.		

In taking out log. rising for 10ᵐ 25ˢ, it will be noticed that the index given at the beginning of the line is ⅓, meaning that the index at the commencement of the line is 1, but that it changes somewhere along the line, which may easily be known by observing that when the first figure of the decimal part of the log. changes from 9 to 0, the index changes from 1 to 2.

Method II.

Constant log.			5·615455
Latitude acct.	51° 2' N.	cos.	9·798560
Declination	0 1	cos.	0·000000
Zen. dist. by D.R.	51 3	cosec.	0·109191
Time from noon	10ᵐ 25ˢ	log.	6·712960
		6,0)17,2 log.	2·236166
Reduction	+ 2' 52"		
True altitude	38 56 0		
Meridian altitude	38 58 52		
Mer. zen. dist.	51 1 8 N.		
Declination	0 0 36 S.		
Latitude	51 0 32 N.		

Method III.—By Towson's Ex-Meridian Tables.

☉'s red. declination	0° 0' 36" S.	True altitude	38° 56' 0"
Aug. Table 1, Index 13	+ 0	Aug. Table 2, Index 13	+ 2 47
Augmented declination	0 0 36 S.	Augmented altitude	38 58 47

As the decl. is less than any given in the head of Table I, augmentation is alone required. In this case enter Table I, under least declination, and with given hour-angle find corresponding Index number; with this and the altitude, augmentation II is determined as usual.

Zenith distance	51 1 13 N.
Augmented declination	0 0 36 S.
Latitude	51 0 37 N.

Ex. 5. 1882, May 5th, P.M. at ship, latitude account 5° 13' N., longitude 61° E., observed altitude sun's L.L. 78° 41' N., eye 17 feet, time by watch 5ʰ 1ᵐ 7ˢ, which was found *fast* 4ʰ 50ᵐ 57ˢ, difference of longitude made since, 20½ miles West.

App. time at ship, May 5ᵈ 0ʰ 8ᵐ48ˢ   Green. date, app. time, May 4ᵈ 20ʰ 4ᵐ 48ˢ.

Time from noon is   8 48

Hourly diff., 5th noon, 42″·66 × Green. time 3ʰ·92 = 167″·2272 ÷ 60 = 2' 47″, decl. noon 5th, 16° 18' 6' N. − 2' 47″ = red. decl. 16° 15' 19″ N.

By *Norie*: True altitude 78° 52' 47″.   By *Raper*: True altitude 78° 52' 36″.

Method I.

Time from noon	8ᵐ48ˢ	rising	2·867510
Latitude acct.	5° 13'	cos.	9·998197
Declination	16 15	cos.	9·982294
		705 log.	2·848000

		nat. no. 705
True altitude 78° 52' 47″		nat. sine 981227
M. Z. dist. 10° 54' 31″ S.		nat. cos. 981932
Declination 16 15 19 N.		
Latitude 5 20 48 N.		

This example cannot be solved by means of Towson's Ex-Meridian Tables, as the altitude exceeds the limits of the Tables.

Method II.

Constant log.			5·61546
Latitude D.R.	5° 13' N.	cos.	9·99820
Declination	16 15 N.	cos.	9·98229
Mer. zen. dist.	11 2	cosec.	0·71810
Time from noon	8ᵐ48ˢ	log.	6·56649
	6,0)76,0	log.	2·88054

Reduction	+ 12' 40″
True altitude	78° 52 47
Meridian altitude	79 5 27
Mer. zen. dist.	10 54 33 S.
Declination	16 15 19 N.
Latitude	5 20 46 N.

EXAMPLES FOR PRACTICE.

1. 1882, January 4th, A.M. at ship, latitude by account 34° 47' N., long. 27° 12' W., observed altitude sun's L.L. South of the observer was 32° 12' 10″, index correction + 4' 19″, height of eye 28 feet, time by watch 0ʰ 13ᵐ 24ˢ, which had been found to be 25ᵐ 35ˢ *fast* of apparent time at ship, difference of longitude made to *West* was 29'·2 after the error on apparent time at ship was determined: required the latitude.

2. 1882, February 28th, P.M. at ship, lat. acct. 43° 46' N., long. 12° 31' W., obs. alt. sun's L.L. 38° 1' 15″ S., index corr. − 5' 10″, eye 23 feet, time by watch 22ᵐ 3ˢ, which had been found to be 8ᵐ 14ˢ *fast* of app. time at ship, diff. of long. made to *East* was 14' after error on app. time at ship was found: required the latitude.

3. 1882, March 20th, A.M. at ship, lat. acct. 41° 24' S., long. 105° E., obs. alt. sun's L.L. 47° 46' N., index corr. + 26″, eye 22 feet, time by chron. 19ᵈ 16ʰ 58ᵐ 12ˢ, which had been found to be 6ʰ 34ᵐ 34ˢ *slow* on app. time at ship, diff. of long. made to *East* was 23' after the error on app. time at ship was determined: required the latitude.

4. 1882, April 21st, A.M. at ship, lat. acct. 39° 54' N., long. 6° 6' E., obs. alt. sun's L.L. 61ᵈ 26' 35″ S., index corr. + 1', eye 18 feet, time by watch 21ᵈ 0ʰ 8ᵐ 10ˢ, which had been found to be 27ᵐ 0ˢ *fast* on app. time at ship, diff. of long. made to *East* was 5' after the error on app. time at ship was determined.

5. 1882, May 29th, P.M. at ship, lat. acct. 37° 15' S., long. 107° W., obs. alt. sun's L.L. 30° 22' 30″ N., index corr. + 49″, eye 22 feet, time by watch 29ᵈ 7ʰ 9ᵐ 11ˢ, which had been found to be 6ʰ 36ᵐ 56ˢ *fast* on app. time at ship, diff. of long. made to *West* was 27' after the error on app. time at ship was determined.

6. 1882, June 19th, A.M. at ship, lat. acct. 44° 24' N., long. 14° 5' W., obs. alt. sun's L.L. 68° 37' 5″ South of observer, eye 18 feet, time by watch 11ʰ 40ᵐ 40ˢ, which was found to be 2ᵐ 2ˢ *slow* on app. time at ship, diff. of long. made to *East* was 32½' after the error on app. time at ship was determined.

7. 1882, July 16th, P.M. at ship, lat. acct. 0° 38' S., long. 2° E., obs. alt. sun's L.L. 67° 41' (zen. S.), eye 15 feet, time by watch 0ʰ 11ᵐ 9ˢ, found *fast* on app. time at ship 56ˢ, diff. of long. made since 1½' to *East*.

8. 1882, August 30th, P.M. at ship, lat. acct. 41° 5' N., long. 139° 15' E., obs. alt. sun's L.L. 57° 20' S., index corr. + 2' 21", eye 14 feet, time by watch 22ᵐ 22ˢ, found to be 18ˢ *slow* of app. time at ship, diff. of long. made to *West* was 34'.

9. 1882, September 9th, P.M. at ship, lat. acct. 9° 20' N., long. 178° 30' E., obs. alt. sun's L.L. 85° 19' (zen. N.), eye 10 feet, time by watch 11ʰ 59ᵐ 40ˢ, *slow* on app. time at ship 9ᵐ 21ˢ, diff. of long. made to *East* was 10½'.

10. 1882, October 11th, P.M. at ship, lat. acct. 45° 51' N., long. 85° 3' E., obs. alt. sun's L.L. 36° 38' 15" S., index corr. − 5' 15", eye 16 feet, time by watch 10ʰ 18ʰ 50ᵐ 10ˢ which was 5ʰ 40ᵐ 12ˢ *slow* on app. time, diff. of long. 33' W.

11. 1882, November 3rd, P.M. at ship, lat. acct. 32° S., long. 109° 39' E., obs. alt. sun's L.L. 71° 50' N., index corr. + 32", eye 18 feet, time by watch 2ᵈ 22ʰ 22ᵐ, which was found 2ʰ *slow*, diff. of long. 28°·7 *West*.

12. 1882, December 23rd, A.M. at ship, lat. acct. 47° 22' S., long. 27° 3' W., obs. alt. sun's L.L. 65° 10' 15" N., index corr. + 45", eye 12 feet, time by watch 11ʰ 29ᵐ 42ˢ, found to be 18ᵐ 40ˢ *slow*, diff. of long. was 36' *East*.

13. 1882, January 5th, P.M. at ship, lat. acct. 8° 50' N., long. 130° 14' W., obs. alt. sun's L.L. 58° 6' 10" S., eye 21 feet, time by watch 0ʰ 2ᵐ 40ˢ, found 13ᵐ 48ˢ *slow* on app. time, diff. of long. made since 16' *East*.

14. 1882, April 28th, A.M. at ship, lat. acct. 18° 36' S., long. 34° 12' W., obs. alt. sun's L.L. 56° 28' (zen. S.), index corr. + 1' 5", eye 21 feet, time by watch 11ʰ 49ᵐ 50ˢ, found *fast* 2ᵐ 17ˢ on app. time at ship, diff. of long. made since 17½' *West*.

15. 1882, March 20th, A.M. at ship, lat. acct. 19° S., long. 33° 33' E., obs. alt. sun's L.L. 70° 21' N., index corr. − 2' 10", eye 16 feet, time by watch 8ʰ 17ˢ, found *fast* on app. time at ship 26ᵐ 11ˢ, diff. of long. made since 14½' *East*.

16. 1882, April 12th, A.M. at ship, lat. acct. 0°, long. 164° 12' W., obs. alt. sun's L.L. 80° 30' N., index corr. − 5' 10", eye 21 feet, time by watch 12ᵈ 0ʰ 0ᵐ 2ˢ, *fast* on app. time at ship 10ᵐ 51ˢ, diff. of long. made to *East* 7½'.

17. 1882, September 16th, A.M. at ship, lat. acct. 42° 36' S., long. 137° 10' E., obs. alt. sun's L.L. 44° 6' N., index corr. + 2' 10", eye 19 feet, time by watch 16ᵈ 8ʰ 41ᵐ 43ˢ, which had been found to be 9ʰ 2ᵐ 47ˢ *fast* on app. time at ship, the diff. of long. made to *West* was 14' after the error on app. time at ship was determined.

18. 1882, March 16th, A.M. at ship, lat. acct. 37° 42' N., long. 61° 40' E., obs. alt. sun's L.L. 50° 0' 30" S., index corr. + 34", eye 15 feet, time by watch 10ʰ 53ᵐ 31ˢ, found *slow* on app. time at ship 1ʰ 3ᵐ 22ˢ, diff. of long. made since 18' *West*.

19. 1882, March 5th, P.M. at ship, lat. acct. 31° 35' N., long. 78° E., obs. alt. sun's L.L. 49° 53' 15" S., index corr. − 3' 15", eye 22 feet, time by watch 4ᵈ 19ʰ 2ᵐ 12ˢ, found to be 5ʰ 17ᵐ 12ˢ *slow*, diff. of long. was 10' E.

20. 1882, September 22nd, A.M. at ship, lat. acct. 45° 45' S., long. 111° 42' W., obs. alt. sun's L.L. 43° 50' N., index corr. − 5' 40", eye 18 feet, time by watch 22ᵈ 7ʰ 41ᵐ 10ˢ, found to be 8ʰ 4ᵐ 10ˢ *fast*, diff. of long. was 13'·5 *East*.

MERIDIAN ALTITUDE OF A FIXED STAR.

RULE CII.

1°. *Take from* Nautical Almanac *the star's declination.*

2°. *To the observed altitude apply the index error, as the sign attached directs.*

3°. Subtract *the dip answering to the height of eye* (Table 5, NORIE; Table 30, RAPER).

4°. Subtract *the refraction* (Table 4, NORIE; Table 31, RAPER), *and thus get the true altitude.*

5°. Subtract *the true altitude from* 90; *the remainder is zenith distance.*

6°. *Mark the zenith distance* N. *or* S., *according as the observer is* North *or* South *of the star.*

7°. *Underneath this last place the declination, and take their* sum *if they have the* same *names; but take their* difference *if they have* unlike *names: the result, in either case, will be the latitude.*

The declination of a fixed star changes so slowly that it may be taken out of the *Nautical Almanac* by inspection, without any practical error resulting; a Greenwich date, therefore, is clearly unnecessary.

8°. *When the zenith distance and declination are of the* same *name, the latitude is of* that name; *when the zenith distance and declination are of* different *names, the latitude takes the name of the* greater.

The stars are inserted in the *Nautical Almanac* in the order of their Right Ascension, from 0^h to 24^h; it will, therefore, very much facilitate the finding of the given star in the *Nautical Almanac*, to turn, in the first instance, to the three pages (290—293, *Nautical Almanac*, 1882, and seek the given star under the head "Mean Places of Stars" for January, and thence obtain the star's Right Ascension, which find at the top of one of the pages following 311 — 366, *Nautical Almanac*, 1882), which will give the star, and the declination will be found opposite the day in the side column which is *nearest* the given day. The degrees (°) and minutes (′) are placed at the top of the column (as annexed), and the seconds (″) are ranged below, for the sake of economizing space in the second column below the name of the star. If the seconds exceed 60″, only take the excess of 60″ and increase the minutes (′) at the top by 1. Thus, on May 10th, 1876, (see table annexed) the declination of α Andromedæ is 28° 22′ 49″ N., and on January 1st the declination is 28° 23′ 3″ N., 62″·8 being 1′ 3″, which being added to 28° 22′, which stands at the head of the column, gives the declination 28° 23′ 3″.

Date.	α Andromedæ	
	R.A.	Decl. N.
	$0^h\ 1^m$	28° 22′
Jan. 1	45"·17	62"·8
11	45·04	61·8
21	44·90	60·6
31	44·78	59·2
&c.	&c.	&c.
May 10	45·50	49·2
20	45·80	49·9
21	46·12	50·9
&c.	&c.	&c.

EXAMPLES.

Ex. 1. 1882, Dec. 29th, long. 140° W., the obs. mer. alt. of the star α Leonis *(Regulus)*, bearing South, was 52° 7′ 30″, index corr. — 27″, height of eye 15 feet: required the latitude.

Observed altitude of star	52° 7′ 30″ S.	
Index correction	— 27	
	52 7 3	
Dip 15 feet	— 3 42	
	52 3 21	
Refraction	— 44	
True altitude	52 2 37	
	90 0 0	
Zenith distance	37 57 23 N.	
Declination (N.A., p. 338)	12 32 6 N.	
Latitude	50 29 29 N.	

By *Raper:* Index corr. — 27″, dip — 3′ 50″, refr. — 46″, true alt. 52° 2′ 27″, lat. 50° 29′ 39″ N.

Ex. 2. 1882, March 12th, long. 10° E., the obs. mer. alt. of the star *Pollux*, bearing North, was 71° 59′ 10″, index corr. + 1′ 15″, height of eye 18 feet: required the latitude.

Observed altitude of star	71° 59′ 10″N.	
Index correction	+ 1 15	
	72 0 25	
Dip 18 feet	— 4 4	
	71 56 21	
Refraction	— 18	
True altitude	71 56 3	
	90 0 0	
Zenith distance	18 3 57 S.	
Declination (N.A., p. 334)	28 18 28 N.	
Latitude	10 14 31 N.	

By *Raper:* Index corr. + 1′ 15″, dip — 3′ 10″, refr. — 19″, true alt. 71° 55′ 56″, lat. 10° 14′ 24″ N.

Ex. 3. 1882, March 11th, long. 84° W., the obs. mer. alt. of the star α Argus *(Canopus)*, bearing South, was 37° 26', index corr. + 47", height of eye 16 feet.

Observed altitude of star	37° 26' 0" S.
Index correction	+ 47
	37 26 47
Dip 16 feet	− 3 50
	37 22 57
Refraction	− 1 15
True altitude	37 21 42
	90 0 0
Zenith distance	52 38 18 N.
Declination (N.A., p. 331)	52 38 18 N.
Latitude	0 0 0

By *Raper:* Index corr. + 1' 12", dip — 4' 0", refr. — 1' 16", true alt. 37° 21' 56", latitude 0° 0' 14" N.

Ex. 4. 1882, January 1st, long. 100° E., the obs. mer. alt. of the star α Canis Majoris *(Sirius)*, bearing South, was 59° 59' 50", index corr. + 4' 12", height of eye 24 feet.

Observed altitude of star	59° 59' 50" S.
Index correction	+ 4 12
	60 4 2
Dip 24 feet	− 4 42
	59 59 20
Refraction	− 33
True altitude	59 58 47
	90 0 0
Zenith distance	30 1 13 N.
Declination (N.A., p. 332)	16 33 26 S.
Latitude	13 27 47 N.

By *Raper:* Index corr. + 4' 12", dip — 4' 50", refr. — 34", true alt. 59° 58' 38", latitude 13° 28' 38" N.

EXAMPLES FOR PRACTICE.

In each of the following examples it is required to find the latitude:—

NO.	CIVIL DATE. 1882.	LONG.	STAR.	OBS. ALT.	CORR.	EYE.
1.	Nov. 7th,	90° W.	α Andromedæ	75° 10' 30" S.	+ 0' 27"	25 ft.
2.	Jan. 1st,	27 W.	α Aurigæ *(Capella)*	54 0 15 N.	− 1 45	18
3.	Aug. 19th,	84 E.	α Lyræ *(Vega)*	50 0 20 N.	0 0	22
4.	Dec. 22nd,	82 E.	α Persei	51 51 45 N.	+ 0 40	26
5.	April 11th,	142 W.	α Virginus *(Spica)*	63 14 30 S.	+ 3 47	22
6.	June 10th,	151 E.	α Eridani *(Achernar)*	40 10 25 S.	+ 0 55	24
7.	Dec. 27th,	91 W.	*(Algenib)*	78 16 45 S.	− 0 25	24
8.	Nov. 30th,	24 W.	α Arietis	68 23 0 N.	− 1 38	28
9.	Feb. 2nd,	76 E.	α Tauri *(Aldebaran)*	29 52 10 N.	+ 5 20	15
10.	June 1st,	97 E.	α¹ Crucis	75 10 30 S.	− 1 40	14
11.	May 22nd,	178 W.	α Hydræ	30 28 53 S.	− 7 38	11
12.	July 17th,	29 E.	α Cygni	20 13 50 N.	0 0	18
13.	Oct. 17th,	165 E.	α Aquilæ *(Altair)*	60 49 10 N.	+ 0 55	17
14.	March 2nd,	154 W.	α Canis Majoris *(Sirius)*	58 58 50 N.	+ 1 10	20
15.	April 3rd,	111 E.	α Boötis *(Arcturus)*	79 49 40 S.	− 2 5	25
16.	Aug. 7th,	40 W.	α Scorpii *(Antares)*	68 49 30 S.	− 1 54	21
17.	May 1st,	8 E.	α² Centauri	10 2 50 S.	− 0 45	20
18.	Oct. 29th,	5 W.	α Piscis Australis *(Fomalhaut)*	70 6 0 N.	+ 0 55	12

ORDINARY EXAMINATION.

EXAMINATION PAPER

To be used by all candidates when appearing for Examination for the first time only.

DEFINITIONS.

The Candidate is requested to write at least ten of the following definitions. The writing should be clear, and the spelling should not be disregarded.

1. The Equator is a great circle passing round the earth at an equal distance from the two poles.
2. The Poles are the extremities of the axis of the earth.
3. A Meridian is a great circle passing through both poles, perpendicular to the equator.
4. The Ecliptic is the great circle of the celestial sphere in which the sun appears to move in consequence of the earth's motion in its orbit.
5. The Tropics of Cancer and Capricorn are the parallels of latitude 23° 28' N. and S.
6. Latitude is that portion of the meridian which is contained between the equator and the given place, and is reckoned in degrees, minutes, and seconds.
7. Parallels of Latitude are small circles parallel to the equator.
8. Longitude is an arc of the equator between the "first meridian" and the meridian of the place.
9. The Visible Horizon is the circle bounding the spectator's view at sea.
10. The Sensible Horizon is the plane on which the spectator stands, produced to meet the celestial concave.
11. The Rational Horizon is an imaginary plane parallel to the sensible horizon, and passing through the centre of the earth.
12. Artificial Horizon and its use. It is a small shallow trough, containing quicksilver, or any other fluid, the surface of which affords a reflected image of a heavenly body. It is used for observing altitudes on shore.
13. True course of a ship is the angle which the ship's track makes with the meridian, or N. and S. line of the horizon.
14. Magnetic Course (correct magnetic) is the angle which the ship's track makes with the magnetic meridian.
15. Compass Course is the angle which the ship's track makes with the N. and S. line of the compass card.
16. Variation of the Compass is the angle which the magnetic needle, under the influence of terrestrial magnetism only, makes with the meridian.
17. Deviation of the Compass is the angle the compass needle makes with the (correct) magnetic meridian.
18. The Error of the Compass is the angle the compass needle makes with the true meridian, being the combined effect of the variation and deviation.
19. Leeway is the angle included between the direction of the ship's keel and the direction of the wake she leaves on the surface of the water.
20. Meridian Altitude of a celestial object is the angular height of that object above the horizon when it is on the meridian of the place of observation.
21. Azimuth of a celestial object is the arc of the horizon between the N. and S. points, and a vertical circle drawn through the object.

22. Amplitude is the arc of the horizon between the East point and the centre of the object when rising, or the West point when setting.

23. Declination of a celestial object is the arc of a circle of declination between the object and the equator.

24. Polar distance is an arc of a circle of declination between the body and the pole (complement of the declination).

25. Right Ascension of a body is an arc of the equator, or an angle at the pole intercepted between the meridian passing through the first point of Aries, and that over the object.

26. Dip is the angle through which the sea horizon is depressed in consequence of the elevation of the spectator above the surface of the earth.

27. Refraction is the correction to be applied to the place of a heavenly body as actually viewed through the atmosphere, which bends the rays of light which pass through it into a position more nearly vertical, and thus causes the apparent places of the heavenly bodies to be above the true place.

28. Parallax is a correction to reduce an altitude as observed from the surface of the earth, to what it would be if taken from the centre. It is the angle subtended at the object by that radius of the earth which is drawn to the place of observation.

29. Semi-diameter of a heavenly body is half the angle subtended by the diameter of the visible disc at the eye of the observer.

30. Moon's Augmented Semi-diameter is an increase of the moon's apparent dimension due to increase of altitude, because the Moon's distance from the spectator decreases as the altitude increases.

31. Observed Altitude is the angular distance of a heavenly body from the horizon, as observed with the sextant or other instrument.

32. Apparent Altitude is the altitude of a celestial body as seen from the surface of the earth; or, the observed altitude corrected for index error and dip.

33. True Altitude is the altitude of a celestial body as seen from the centre of the earth; that is, the apparent altitude corrected for refraction, semi-diameter, and parallax.

34. Zenith Distance is an arc of a circle of altitude between the body and the zenith (complement of the altitude).

35. Vertical circles are great circles passing through the zenith and nadir, perpendicular to the horizon. They are also called *Circles of Altitude*, because altitudes are measured on them; and *Circles of Azimuth*, as marking out all the points that have the same azimuth.

36. Prime vertical is a great circle passing through the zenith and nadir, and the East and West (true) points of the horizon.

37. Civil time is the time used in ordinary life to record events. It begins at midnight and ends at the following midnight, and its hours are reckoned through twice 12, from midnight to noon, denoted by A.M.; and then from noon to midnight, denoted by P.M.

38. Astronomical Time is the time used in all astronomical calculations: it begins at noon and ends at the following noon, its hours being reckoned from 0^h to 24^h.

39. Sidereal Time is the westerly hour-angle of the first point of Aries.

40. Mean Time is the hour-angle which the mean sun is westward of the meridian.

41. Apparent Time is the hour-angle of the apparent or true sun, always reckoning westward.

42. Equation of Time is an angle at the pole between a meridian over the true sun, and one over the mean sun.

43. Hour-angle of a Celestial Object is an angle at the pole included between the meridian of the observer and that over the object.

44. Complement of an Arc or Angle is that arc or angle which must be added to it to make a right-angle (90°).

45. Supplement of ditto is that angle which must be added to it to make two right-angles (180°).

EXAMINATION PAPER—No. I.
FOR SECOND MATE.

1. Multiply 7654 by 95, and 950 by 586, by common logarithms.
2. Divide 3654000 by 7308, and 35420 by 322, by common logarithms.
3.—

H.	Courses.	K. ½		Winds.	Leeway.	Deviation.	Remarks, &c.
					pts.		
1	W.S.W.	10	8	N.W.	½	11° W.	A point, lat. 37° 3′ N. long. 9° 0′ W., bearing by compass N.E. ¼ E., dist. 15 miles. Ship's head W.S.W. Deviation as per log.
2		11	4				
3		11	4				
4		11	4				
5	N.W. ½ N.	12	2	W.S.W.	¼	17° W.	
6		12	3				
7		12	3				
8		12	2				
9	N.N.W.	9	6	West.	¾	11° W.	
10		9	5				
11		9	5				
12		9	4				Variation 22° 30′ W.
1	N.W. by W.	7	8	S.W. by W.	1½	20° W.	
2		7	6				
3		7	4				
4		8	2				
5	S.W. ½ S.	9	3	S.S.E.	1	6° W.	
6		8	7				
7		9	3				A current set the ship S.W. by W. ¼ W. (correct magnetic) 8 miles from the time the departure was taken to the end of the day.
8		8	7				
9	W. ½ S.	10	3	S. by W.	½	14° W.	
10		10	2				
11		10	2				
12		10	3				

Correct the courses for deviation, variation, and leeway, and find the course and distance from the given point, and the latitude and longitude in by inspection.

4. 1882, January 1st, in longitude 102° 41′ W., the observed meridian altitude of the sun's L.L. was 59° 59′ 50″, bearing South, index error + 50″, height of eye 15 feet: required the latitude.

5. In latitude 37° N., the departure made good was 89·2 miles: required the difference of longitude by parallel sailing.

6. Required the course and distance from Toulon to Valencia, by Mercator's sailing.

    Lat. Toulon   43° 8′ N.   Long. Toulon   5° 56′ E.
    Lat. Valencia   39 29 N.   Long. Valencia   0 24 W.

ADDITIONAL FOR ONLY MATE.

7. 1880, January 14th: find the time of high water, A.M. and P.M., at Cherbourg, Christchurch, and Falmouth.

8. 1882, January 1st, at 8ʰ 4ᵐ A.M. apparent time at ship, in latitude 50° 32′ N., longitude 139° 51′ W., the sun's magnetic amplitude E. by S. ½ S.: required the true amplitude and error of the compass; and supposing the variation to be 23° 52′ E.: required the deviation of the compass for the position of the ship's head when the observation was taken.

9. 1881, January 29th, P.M. at ship, latitude 42° 26′ N., observed altitude sun's L.L. 13° 40′, index error — 1′ 14″, height of eye 16 feet, time by chronometer 29ᵈ 6ʰ 48ᵐ 40ˢ, which was *slow* 11ᵐ 22ˢ·3 for mean noon at Greenwich, December 1st, 1881, and on January 1st, 1882, was 8ᵐ 7ˢ *slow* for Greenwich mean noon: required the longitude by chronometer.

ADDITIONAL FOR FIRST MATE.

10. 1882, January 15th, mean time at ship 9ʰ 39ᵐ 44ˢ A.M., latitude 23° 39′ S., longitude 127° 52′ W., sun's magnetic azimuth S. 103° E., observed altitude sun's L.L. 55° 8′ 30″, index

error — 2' 30", height of eye 12 feet: required the true azimuth and error of the compass; and supposing the variation be 7° 50' E.: required the deviation of the compass for the position of the ship's head at the time the observation was taken.

11. 1882, January 17th, P.M. at ship, latitude by account 36° 2' N., longitude 149° 28' E., observed altitude sun's L.L. South of observer was 31° 54' 15", index error + 2' 18", height of eye 22 feet, time by watch 16ᵈ 23ʰ 59ᵐ, which had been found to be 20ᵐ 24ˢ *slow* on app. time at ship, the difference of longitude made to the *West* since the error of watch on app. time at ship was determined, was 39'·2: required the latitude by reduction to meridian.

12. 1882, January 16th, A.M. at ship, and uncertain of my position, when a chronometer showed January 15ᵈ 22ʰ 16ᵐ 1ˢ M.T.G., obs. alt. sun's L.L. 12° 52' 30"; again, P.M. at ship same day, when chronometer showed January 16ᵈ 2ʰ 47ᵐ 2ˢ, obs. alt. sun's L.L. 14° 1' 57", height of eye 17 feet, the ship having made 42 miles on a true N.N.W. course in the interval: required the line of position when the first altitude was observed, and the ship's position by Sumner's Method when the second altitude was taken, assuming the ship to be between the latitudes 48° 50' N. and 49° 30' N.

For the second observation the positions as given by the Examiner would be as follows:—

B = Lat. 48° 50' N.   Long. 5° 47½' W.
B' = Lat. 49° 30' N.   Long. 7° 28' W.

ADDITIONAL FOR MASTER ORDINARY.

13. 1882, January 24th, the observed meridian altitude of the star *a* Tauri *(Aldebaran)* was 52° 36', bearing South, index correction — 23", height of eye 20 feet: required the latitude.

DEVIATION OF THE COMPASS.

N.B.—*The Candidate is to answer correctly at least eight of such of the following questions as are marked with a cross by the Examiner. The Examiner will not mark less than twelve.*

1. What do you mean by Deviation of the Compass?

A. The deflection of the compass needle from the magnetic meridian caused by the attraction of the iron of the ship.

2. How do you determine the deviation (*a*) when in port (and *b*) when at sea?

A. By bringing the ship's head successively upon each of the thirty-two points of the Standard Compass, or on each alternate point, and then (*a*) by taking reciprocal simultaneous bearings; or by the observer on board taking the bearings of a distant object whose correct magnetic bearing is known, or of some conspicuous object in a line with figures on a dock wall. (*b*) At sea, by bearings of well-known conspicuous objects in a line on the coast, or by amplitudes and azimuths and the known variation at the place of the ship.

3. Having determined the deviation with the ship's head on the various points of the compass, how do you know when it is easterly and when westerly?

A. When the *correct* magnetic bearing of the distant object is to the *right* of the reading of the compass on board, the deviation is easterly, when to the *left*, westerly.

4. Why is it necessary, in order to ascertain the deviations, to bring the ship's head in more than one direction?

A. Because the deviation alters as the direction of the ship's head is changed.

5. For accuracy, what is the least number of points to which the ship's head should be brought?

A. Eight; although, if the deviations be known on the four quadrantal points, N.E., S.E., S.W., and N.W., with the aid of Napier's diagram a good deviation curve may be formed.

6. How would you find the deviation when sailing along a well known coast?

A. By taking the Standard Compass bearing of two well-defined objects in a line, as for instance, the bearings of two beacons, two lights, two points of land, not too near one another, and whose correct magnetic bearing is known, from the chart or otherwise; then the difference between the *correct* magnetic bearing and the compass bearing is the deviation for the direction of the ship's head when the bearing was taken.

7. In the following table give the correct magnetic bearing of the distant object, and thence the deviation.

Correct magnetic bearing.

Ship's Head by Standard Compass.	Bearing of Distant Object by Standard Compass.	Deviation Required.	Ship's Head by Standard Compass.	Bearing of Distant Object by Standard Compass.	Deviation Required.
North	N. 86° W.		South	S. 64° W.	
N.E.	S. 79 W.		S.W.	S. 72 W.	
East	S. 69 W.		West	S. 89 W.	
S.E.	S. 65 W.		N.W.	N. 80 W.	

8. With the deviation as above, give the courses you would steer by the Standard Compass to make the following courses correct magnetic.

Correct magnetic courses :— W.N.W.; N.N.E.; E.S.E.; S.S.W.

Compass courses :—

9. Supposing you have steered the following courses by the Standard Compass, find the correct magnetic courses made from the above deviation table.

Compass courses :— N.N.W.; E.N.E.; S.S.E.; W.S.W.

Magnetic courses :—

10. You have taken the following bearings of two distant objects by your Standard Compass as above, with the ship's head at N.W. by W., find the bearings, correct magnetic.

Compass bearings :— E. by S. ½ S. and N. ½ E.

Bearings, magnetic :—

11. Name some suitable objects by which you could readily obtain the deviation of the compass when sailing along the coasts of the English Channel.

The South Foreland lighthouses in one, bearing W. by N., *correct* magnetic; Beachy Head lighthouse just open of the cliffs to the eastward, bearing N.W. by W., *correct* magnetic; Portland lights in one N.N.W. ¾ W.; Prawl Point and Start lighthouse in one, and Lizard lights in one.

12. Do you expect the deviation to change; if so, state under what circumstances?

A. Yes, it changes rapidly for several months after the ship is launched, an alteration also takes place by change of magnetic latitude, and in ships running long upon one course and then changing the course, by the heeling of the ship, and by taking in a cargo of iron.

13. How often is it advisable to test the accuracy of your table of deviations?

A. Frequently, in a new vessel; when nearing land; and under the circumstances stated in last question.

14. State briefly what you have chiefly to guard against in selecting a position for the compass.

A. Elongated iron, especially if vertical, such as stanchions, davits, capstan, spindles, funnels, ventilating shafts, &c., and the compass should be as far removed as possible from transverse bulk heads.

15. The compasses of iron ships are more or less affected by what is termed the heeling error; on what courses does this error vanish, and on what courses is it the greatest?

A. It vanishes when the ship's head is East or West by compass, and is greatest when the ship's head by compass is North or South.

16. State to which side of the ship, in the majority of cases, is the North point of the compass drawn in the Northern hemisphere; and what effect has it on the assumed position of the ship when she is steering on Northerly, and also on Southerly courses?

A. The North point of the compass is drawn to the weather side in the majority of cases. The effect of this is to throw the ship to windward on Northerly courses, and to leeward on Southerly courses.

17. The effect being as you state, on what courses would you keep away, and on what courses would you keep closer to the wind, in order to make good a given compass course?

A. I would keep her away on either tack on Northerly courses, but on either tack on Southerly I would keep her close to the wind.

18. Does the same rule hold good in both hemispheres with regard to the heeling error?

A. No, ships which have a large heeling error to windward in Northern latitudes, will probably have as large a heeling error to leeward in high Southern latitudes; but it is recommended in order to determine it, that observations be made in every ship.

19. Your steering compass having a large error, how would you proceed to correct that compass by compensating magnets and soft iron in order to reduce the error within manageable limits?

A. Draw a line upon the deck, fore-and-aft, through the centre of the binnacle. Draw another line across the deck at right-angles to the former, through the same centre.

Bring the ship's head either North or South (correct magnetic), and placing a magnet on the deck athwartship, with its centre exactly on a fore-and-aft line drawn on the deck at some distance from the binnacle; move it gradually to or from the foot of the binnacle until the compass points correctly. If the compass needle deviates to the left, the north end of the magnet must be placed to the left, and conversely. Next swing the ship's head round to the East or West (correct magnetic), and steady her on one of these points, and place a magnet fore-and-aft, either on the port or starboard side of the binnacle, with its centre on the athwartship line drawn on the deck; move it to or from the foot of the binnacle until the compass points correctly.

Again: the semicircular deviation having thus been corrected, and the binnacle being properly fitted with two small brass boxes, one on each side of, and on a level with the compass; steady the ship's head on one of the quadrantal points, N.E., S.E., S.W., or N.W.: if there is any deviation fill one of the chain boxes with a quantity of small chain until the compass points correctly; if one chain box be not sufficient, fill the other. For greater certainty, swing the ship's head to each of the other quadrantal points.

EXAMINATION PAPER—No. II.
FOR SECOND MATE.

1. Multiply 50030 by 800, and 41375 by 48, by common logarithms.
2. Divide 9999·46 by 67·8, and 900009 by 9, by common logarithms.
3.—

H.	Courses.	K.	$\tfrac{1}{10}$	Winds.	Lee-way.	Deviation.	Remarks, &c.
					pts.		
1	S.E. by E.	13	2	North.	0	11° E.	A point of land, lat. 47°31' N., long. 52°33' W., bearing by compass W.S.W., dist. 18 miles. Ship's head S.E. by E. Deviation as per log.
2		13	3				
3		12	5				
4		13					
5	S.E.	11		E.N.E.	½	9° E.	
6		10	5				
7		10	4				
8		11	1				
9	E. by N.	8	8	S.E. by S.	1	17° E.	
10		9	4				
11		9	4				
12		9					
1	E.N.E.	6	8	S.E.	1½	15° E.	Variation 28° W.
2		6	7				
3		6	5				
4		7					
5	S.S.E.	5	8	East.	2	7° E.	
6		5	8				
7		6	4				A current set (correct magnetic) S. b E., 12 miles, from the time the departure was taken to the end of the day.
8		6					
9	S.E. by S.	7		E. by N.	1½	8° E.	
10		7	3				
11		7	4				
12		7	3				

4. 1882, February 1st, in longitude 78° 14′ E., the observed meridian altitude of sun's L.L. was 78° 4′ 10″, bearing South, index error + 55″, height of eye 12 feet: required the latitude.

5. In latitude 47° 30′ N., the departure made good was 115·5 miles: required the difference of longitude by parallel sailing.

6. Required the course and distance from St. Helena to Cape Horn, by calculation on Mercator's principle.

Latitude St. Helena 15° 55′ S.   Longitude St. Helena 5° 44′ W.
Latitude Cape Horn 55 59 S.   Longitude Cape Horn 67 16 W.

ADDITIONAL FOR ONLY MATE.

7. 1880, February 5th: find A.M. and P.M. tides at Filey Bay, Milford Haven, and Cromarty.

8. 1882, February 20th, at 6^h 9^m P.M., apparent time at ship, latitude 11° 58′ S., longitude 179° 42′ E., sun's magnetic amplitude S.W. by W. ¼ W.: required the true amplitude and error of compass; and supposing the variation to be 10° 20′ E.: required the deviation of the compass for the position of the ship's head at the time of observation.

9. 1882, February 10th, A.M. at ship, latitude 50° 48′ N., observed altitude sun's L.L. 9° 10′ 50″, index correction − 3′ 20″, height of eye 18 feet, time by chronometer February 9^d 9^h 59^m 25^s, which was 37^m 58^s·8 *fast* for mean noon at Greenwich, December 20th, 1881, and on January 10th, 1882, was 34^m 12^s *fast* for mean noon at Greenwich: required the longitude.

ADDITIONAL FOR FIRST MATE.

10. 1882, February 16th, mean time at ship 5^h 7^m 35^s A.M., latitude 51° 2′ N., longitude 140° 34′ W., sun's magnetic azimuth S. 36° 20′ E., observed altitude sun's L.L. 7° 16′ 40″, index correction − 6′ 10″, height of eye 15 feet: required the error of compass; and supposing the variation to be 25° W.: required the deviation for the position of the ship's head at the time of observation.

11. 1882, February 15th, A.M. at ship, latitude acct. 55° 59′ S., longitude 54° 18′ E., observed altitude sun's L.L. North of observer was 46° 22′ 10″, index correction − 1′ 50″, height of eye 19 feet, time by watch 0^m 5^s, which had been found to be 30^m *fast* on apparent time at ship, the difference of longitude made to the *East* was 16′·8: required the latitude.

12. 1882, February 3rd, A.M. at ship, and uncertain of my position, when a chronometer showed February 2^d 10^h 9^m 2^s M.T.G., obs. alt. sun's L.L. 19° 50′; again, P.M. at ship same day, when chronometer showed February 2^d 15^h 39^m 53^s, obs. alt. sun's L.L. 11° 5′, height of eye 18 feet, the ship having made 36 miles on a true E. ¾ N. course in the interval: required the line of bearing when the first altitude was taken, and the ship's position by Sumner's Method when the second altitude was observed, assuming the ship to be between the latitudes 49° 15′ N. and 49° 45′ N.

For the second observation the positions as given by the Examiner would be as follows:—

B = Lat. 47° 10′ N.   Long. 179° 42¼′ W.
B′ = Lat. 47 40 N.   Long. 179 40¼ E.

ADDITIONAL FOR MASTER ORDINARY.

13. 1882, February 12th, the observed meridian altitude of star Procyon, South of observer, was 77° 18′ 10″, index correction + 19″, height of eye 16 feet: required the latitude.

In the following table give the correct magnetic bearing of the distant object, and thence the deviation:—

Correct magnetic bearing.

Ship's Head by Standard Compass.	Bearing of Distant Object by Standard Compass.	Deviation Required.	Ship's Head by Standard Compass.	Bearing of Distant Object by Standard Compass.	Deviation Required.
North	S. 34° E.		South	S. 43° E.	
N.E.	S. 58 E.		S.W.	S. 31 E.	
East	S. 62 E.		West	S. 17 E.	
S.E.	S. 52 E.		N.W.	S. 15 E.	

With the deviation as above, give the courses you would steer by the Standard Compass to make the following courses correct magnetic.

Correct magnetic courses:—N.E. by E. ½ E.; W. ¼ S.; W. ¾ N.; E. by S. ½ S.

Compass courses:—

Supposing you have steered the following courses by the Standard Compass, find the correct magnetic courses made from the above deviation table.

Compass courses:—W. by N. ½ N.; N.E. ¾ N.; S.W. ¼ S.; S.E. by S.

Magnetic courses:—

You have taken the following bearings of two distant objects by your Standard Compass as above, with the ship's head at W. ½ S., find the bearings, correct magnetic.

Compass bearings:—N. 79° W., and S. 19° W.

Bearings, magnetic:—

EXAMINATION PAPER—No. III.

FOR SECOND MATE.

1. Multiply 84·8 by 62·8, and 3914 by 600, by common logarithms.

2. Divide 666·666 by 8·88, and 55700 by 2785, by common logarithms.

3.—

H.	Courses.	K.	$\frac{1}{10}$	Winds.	Lee-way.	Deviation.	Remarks, &c.
					pts.		
1	S.S.E.	5	6	East.	2¼	3° E.	A point of land in lat. 62° N. long. 150° E. bearing by compass W. by S. ¾ S., distance 17 miles. Ship's head S.S.E. Deviation as per log.
2		5	6				
3		5					
4		4	8				
5	S.S.W. ½ W.	4	7	West.	2¾	4° W.	
6		4	8				
7		5	2				
8		5	3				
9	W.S.W.	5		South.	2½	9° W.	
10		6					
11		6	5				
12		6	5				
1	W. ½ N.	6	6	N. by E.	0	11° W.	Variation 31° E.
2		7					
3		6	4				
4		6					
5	East.	4	6	S.S.E.	2½	10° E.	
6		5					
7		4	8				A current set the ship (correct magnetic) N.N.E., 21 miles, from the time the departure was taken to the end of the day.
8		4	6				
9	E.S.E.	4		S. by W.	0	9° E.	
10		4	5				
11		4	5				
12		5					

4. 1882, March 20th, longitude 173° 18′ W., the observed meridian altitude of sun's L.L. was 89° 37′, bearing North, index correction + 4′ 27″, height of eye 18 feet: required the latitude.

5. In latitude 34° 28′ S., the departure made good was 394·2 miles: required the difference of longitude by parallel sailing.

6. Required the course and distance from the Cape of Good Hope to Cape Frio.

Lat. Cape of Good Hope	34° 28′ S.	Long. Cape of Good Hope	18° 28′ E.
Lat. Cape Frio	23 0 S.	Long. Cape Frio	41 57 W.

ADDITIONAL FOR ONLY MATE.

7. 1880, March 14th: find the times of high water, A.M. and P.M. (by Admiralty Tables) at Quillebœuf, Havre, Pool, Yarmouth Roads, Lerwick, and Beaumaris.

8. 1882, March 6th, at $5^h\ 31^m\ 52^s$ P.M., apparent time at ship, in latitude $52°\ 12'$ N., longitude $138°\ 54'$ W., the sun's magnetic amplitude was W. by S. $\frac{1}{2}$ S.: required the error of compass, and supposing the variation to be $24°$ E.: required the deviation for the position of the ship's head at the time of observation.

9. 1882, March 31st, A.M. at ship, latitude $26°\ 9'$ N., observed altitude sun's L.L. $29°\ 10'\ 20''$, height of eye 26 feet, time by chronometer $31^d\ 0^h\ 4^m\ 50^s$, which was $5^m\ 58^s$ *fast* for mean noon at Greenwich, November 20th, 1881, and on December 31st, 1881, was $1^h\ 2^m\ 55^s\cdot8$ *fast* for mean time at Greenwich: required the longitude.

ADDITIONAL FOR FIRST MATE.

10. 1882, March 10th, mean time at ship $7^h\ 35^m\ 25^s$ A.M., latitude $42°\ 41'$ S., longitude $148°\ 5'$ E., sun's bearing by compass S. $108°\ 37'\ 30''$ E., observed altitude sun's L.L. $17°\ 57'\ 40''$, height of eye 19 feet: required the error of the compass; and supposing the variation to be $10°\ 50'$ E.: required the deviation for the position of the ship's head at the time of observation.

11. 1882, March 25th, P.M. at ship, latitude acct. $20°\ 1'$ N., longitude $89°\ 10'$ E., observed altitude sun's L.L. South of observer was $71°\ 9'$, height of eye 18 feet, time by watch $11^h\ 38^m\ 12^s$ (or $24^d\ 23^h\ 38^m\ 12^s$), which had been found to be $31^m\ 8^s$ *slow* on apparent time at ship, the difference of longitude made to *East* was $13\frac{1}{2}$ miles after the error on apparent time was determined: required the latitude by reduction to meridian.

12. 1882, March 27th, A.M. at ship, uncertain of my position, when a chronometer showed March $27^d\ 10^h\ 26^m\ 24^s$ M.T.G., obs. alt. sun's L.L. $36°\ 38'\ 50''$; again, P.M. same day, when chronometer showed March $27^d\ 16^h\ 37^m\ 10^s$, obs. alt. sun's L.L. $15°\ 15'\ 20''$, height of eye 22 feet, the ship having made 28 miles on a true S. $\frac{1}{2}$ E. course in the interval: required the line of bearing when the first altitude was taken, and the position of the ship by Sumner's Method when the second altitude was observed, assuming latitudes $51°\ 30'$ N. and $51°\ 0'$ N.

ADDITIONAL FOR MASTER ORDINARY.

13. 1882, March 19th, the observed meridian altitude of *a Bootis (Arcturus)* $36°\ 10'\ 20''$, bearing North, index correction $+\ 2'\ 42''$, height of eye 20 feet: required the latitude.

In the following table give the correct magnetic bearing of the distant object, and thence the deviation.

Correct magnetic bearing.

Ship's Head by Standard Compass.	Bearing of Distant Object by Standard Compass.	Deviation Required.	Ship's Head by Standard Compass.	Bearing of Distant Object by Standard Compass.	Deviation Required.
North	S. 67° E.		South	S. 72° E.	
N.E.	East.		S.W.	S. 46 E.	
East	N. 85 E.		West	S. 45 E.	
S.E.	N. 87 E.		N.W.	S. 52 E.	

With the deviation as above, give the courses you would steer by the Standard Compass to make the following courses correct magnetic.

Correct magnetic courses:— N.E. $\frac{1}{2}$ E.; S.W. by W.; W. by S. $\frac{3}{4}$ S.; S.S.E. $\frac{1}{4}$ E.

Compass courses:—

Supposing you have steered the following courses by the Standard Compass, find the correct magnetic courses made from the above deviation table.

Compass courses: S.E. by E. ¾ E.; N.E. by E. ½ E.; N.W. by W. ¾ W.; N.N.W.

Correct magnetic courses:—

You have taken the following bearings of two distant objects by your Standard Compass as above, with ship's head at N. 24° E., find the bearings, correct magnetic.

Compass bearings:—W. ¾ S. and E.N.E.

Bearings, magnetic:—

EXAMINATION PAPER—No. IV.

FOR SECOND MATE.

1. Multiply, by common logarithms, 456 by 28·9, and 145·544 by 500.

2. Divide, by common logarithms, 462927 by 142·8, and 8712360 by 80670.

3.—

H.	Courses.	K.	1/10	Winds.	Lee-way.	Deviation.	Remarks, &c.
1	S.W. ½ W.	14	5	S.E.	pts. 0	6° W.	A point, lat. 50° 12′ S. long. 179° 40′ W., bearing by compass N. ¾ W. dist. 19 miles. Ship's head S.W. ½ W. Deviation as per log.
2		14	2				
3		14	6				
4		14	7				
5	N. ¾ E.	4		E.N.E.	3¼	3° E.	
6		3	6				
7		3	6				
8		3	8				
9	S. by E. ½ E.	2	4	S.W. ½ W.	2¾	6° E.	
10		2	3				
11		2	3				
12		2					
1	W. by S.	12	2	S. by W.	¼	14° W.	Variation 14° E.
2		12	4				
3		12	6				
4		12	8				
5	E.N.E.	3		S.E.	2½	19° E.	
6		2	3				
7		3	4				A current set the ship (correct magnetic) S.W. ¼ W., 42 miles, from the time the departure was taken to the end of the day.
8		3	3				
9	S.S.W. ½ W.	5	6	S.E.	1¾	4° W.	
10		5	7				
11		5	3				
12		5	4				

4. 1882, April 1st, in longitude 87° 42′ W., observed meridian altitude sun's L.L. South of observer was 48° 42′ 30″, index correction + 1′ 42″, height of eye 18 feet: required the latitude.

5. In latitude 49° 57′ N., the departure made good was 149 miles: required the difference of longitude by parallel sailing.

6. Required the course and distance from A to B.

Latitude A	56° 35′ S.	Longitude A	2° 15′ E.
Latitude B	51 10 S.	Longitude B	3 10 W.

ADDITIONAL FOR ONLY MATE.

7. 1880, April 18th: required the times of high water, A.M. and P.M., at Ecrehous, Blakeney, Portree, Llanelly, Cardiff, and New Ross.

8. 1882, April 28th, at $5^h\ 14^m\ 2^s$ P.M. apparent time at ship, latitude 38° 19′ S., longitude 88° 48′ E., sun's magnetic amplitude N.W. by W., variation 19° 10′ W.: required the deviation.

9. 1882, April 15th, P.M. at ship, latitude 37° 49′ S., observed altitude sun's L.L. was 26° 27′ 30″, index correction — 49″, height of eye 13 feet, time by chronometer 14ᵈ 21ʰ 48ᵐ 17ˢ, which was 4ᵐ 51ˢ *fast* for Greenwich mean noon, January 22nd, and on February 3rd, was 2ᵐ 35ˢ·4 *fast* for mean noon at Greenwich: required the longitude.

ADDITIONAL FOR FIRST MATE.

10. 1882, April 17th, mean time at ship 2ʰ 49ᵐ 45ˢ P.M., latitude 39° 50′ N., longitude 1° 35′ E., sun's bearing by compass West, observed altitude sun's L.L. 42° 10′, index corr. — 45″, height of eye 14 feet: required the error of compass; and supposing the variation to be 19° 50′ W.: required the deviation of the compass for the position of the ship's head when the observation was taken.

11. 1882, April 19th, A.M. at ship, latitude account 46° 15′ N., longitude 178° 12′ E. observed altitude of sun's L.L. South of observer 54° 7′, index correction + 2′ 12″, height of eye 20 feet, time by watch 11ʰ 24ᵐ 22ˢ, or 18ᵈ 23ʰ 24ᵐ 22ˢ, which had been found to be 5ᵐ *slow* on apparent time at ship, the difference of longitude made to the *East* was 30 miles, after the error on apparent time was determined.

12. 1882, April 6th, P.M. at ship, and uncertain of my position, when a chronometer showed April 6ᵈ 1ʰ 17ᵐ 25ˢ M.T.G., obs. alt. sun's L.L. 45° 7′ 10″; again, P.M. at ship same day, when chronometer showed April 6ᵈ 4ʰ 23ᵐ 24ˢ, obs. alt. sun's L.L. 24° 52′ 10″, height of eye 18 feet, the ship having made 46 miles on a true N.E. course in the interval: required the line of bearing when the first altitude was taken, and the position of the ship by Sumner's Method when the second altitude was observed, assuming latitudes 50° 20′ N. and 50° 50′ N.

ADDITIONAL FOR MASTER ORDINARY.

13. 1882, April 12th, the observed meridian altitude of the star Spica, South of observer, was 20° 58′ 40″, index correction — 45″, height of eye 25 feet: required the latitude.

In the following table give the correct magnetic bearing of the distant object, and thence the deviation:—

Correct magnetic bearing.

Ship's Head by Standard Compass.	Bearing of Distant Object by Standard Compass.	Deviation Required.	Ship's Head by Standard Compass.	Bearing of Distant Object by Standard Compass.	Deviation Required.
North	S. 2° W.		South	S. 5° E.	
N.E.	S. 5 W.		S.W.	S. 24 E.	
East	S. 10 W.		West	S. 17 E.	
S.E.	S. 16 W.		N.W.	S. 3 E.	

With the deviation as above, give the courses you would steer by the Standard Compass to make the following courses correct magnetic.

Correct magnetic courses:—E. by N. ¼ N.; N.W. ¾ W.; S.W. by S. ½ S.; S.E. by S. ¼ S.
Compass courses:—

Supposing you have steered the following courses by the Standard Compass, find the correct magnetic courses made from the above deviation table.

Compass Courses:—N.W. by W. ½ W.; W. by S. ½ S.; E. by S. ¼ S.; N.E. ¾ E.
Correct magnetic courses:—

You have taken the following bearings of two distant objects by your Standard Compass as above, with the ship's head at S. 24° W., find the bearings, correct magnetic.

Compass bearings:—N. 84° W. and W.S.W.
Bearings, magnetic:—

EXAMINATION PAPER—No. V.

FOR SECOND MATE.

1. Multiply 767 by 89·8, and 10003 by 110, by common logarithms.
2. Divide 66889·2 by 99·7, and 3972096 by 144, by common logarithms.
3.—

H.	Courses.	K.	$\tfrac{1}{10}$	Winds.	Lee-way.	Deviation.	Remarks, &c.
					pts.		A point, lat. 64° 2′ S. long. 140° 21′ E., bearing by compass W. b S. ⅜ S, distance 23 miles. Ship's head S.E. ½ E. Deviation as per log.
1	S.E. ½ E.	14		S.W.	0	8° E.	
2		14					
3		14	4				
4		13	6				
5	E. by S. ¾ S.	4	4	N.E.	3	14° E.	
6		4	2				
7		4	2				
8		4	2				
9	W.N.W.	4		North.	2¾	19° W.	
10		3					
11		3					
12		3					
1	N.E. ¼ N.	9	2	N.N.W.	¼	8° E.	Variation 37° E.
2		9	2				
3		9	2				
4		9	4				
5	S.W. by S.	8	4	W.N.W.	1½	10° W.	
6		8	4				
7		7	2				A current set the ship (correct magnetic) N.E. ¼ E., 48 miles, from the time the departure was taken to the end of the day.
8		7					
9	N.N.W.	5		West.	2	11° W.	
10		4					
11		4					
12		4					

4. 1882, May 8th, in longitude 105° 17′ W., observed meridian altitude of sun's L.L., bearing North, was 76° 3′, index correction − 1′ 27″, height of eye 10 feet: required the latitude.

5. In latitude 3° 24′ N., the departure made good was 982 miles: required the difference of longitude by parallel sailing.

6. Required the course and distance from A to B, by calculation on Mercator's principle.

    Latitude A   39° 39′ N.   Longitude A   51° 51′ E.
    Latitude B   27 27 N.   Longitude B   33 33 E.

ADDITIONAL FOR ONLY MATE.

7. 1880, May 20th: find the times of high water A.M. and P.M. at Loch Ryan, Tarn Point, Berwick, St. Malo, and Dungeness.

8. 1882, May 21st, at $7^h\ 29^m$ A.M., apparent time at ship, latitude 45° 53′ S., longitude 50° 39′ E., sun's magnetic amplitude N.E. ¼ E.: required the true amplitude and error of the compass; and supposing the variation to be 31° 50′ E.: required the deviation of the compass for the position of the ship's head at the time of observation.

9. 1882, May 22nd, A.M. at ship, latitude 43° 25′ N., observed altitude sun's L.L. 32° 8′, index correction + 47″, height of eye 15 feet, time by chronometer $21^d\ 21^h\ 6^m\ 10^s$, which was *slow* $12^s·6$ for mean noon at Greenwich, February 24th, and on April 1st was $2^m\ 45^s\ \textit{fast}$ for mean noon at Greenwich: required the longitude.

ADDITIONAL FOR FIRST MATE.

10. 1882, May 25th, mean time at ship $5^h\ 29^m\ 47^s$ P.M., latitude 41° 58' N., longitude 96° 1' W., sun's bearing by compass N. 118° 30' W., observed altitude sun's L.L. 40° 40' 40", index correction + 2' 15", height of eye 12 feet: required the true azimuth and error of the compass; and supposing the variation is 10° 30' E.: required the deviation of the compass for the position of the ship's head at the time of observation.

11. 1882, May 10th, P.M. at ship, latitude account 28° 13' S., longitude 112° 15' W., observed altitude of sun's L.L., North of observer, was 43° 35' 20", index correction − 6' 12", height of eye 19 feet, time by watch $30^m\ 26^s$ (or $10^d\ 0^h\ 30^m\ 26^s$), which had been found to be $6^m\ 45^s$ *fast* on apparent time at ship, the difference of longitude made to the *East* was 26', after the error on apparent time was determined: required the latitude.

12. 1882, May 30th, A.M. at ship, and uncertain of my position, when a chronometer showed May $29^d\ 19^h\ 21^m\ 15^s$ M.T.G., obs. alt. sun's L.L. 23° 42' 10"; again, P.M. at ship same day, when chronometer showed May $30^d\ 2^h\ 7^m\ 5^s$, obs. alt. sun's L.L. 56° 45', height of eye 15 feet, the ship having made 49 miles on a true West course in the interval: required the line of bearing when the first altitude was taken, and the position of the ship by Sumner's Method when the second altitude was observed, assuming latitudes 50° 0' N. and 50° 30' N.

ADDITIONAL FOR MASTER ORDINARY.

13. 1882, May 10th, the observed meridian altitude of Spica, bearing North, was 70° 10' 25", index correction + 42", height of eye 22 feet: required the latitude.

In the following table give the correct magnetic bearing of the distant object and thence the deviation.

Correct magnetic bearing.

Ship's Head by Standard Compass.	Bearing of Distant Object by Standard Compass.	Deviation Required.	Ship's Head by Standard Compass.	Bearing of Distant Object by Standard Compass.	Deviation Required.
North	East.		South	N. 85° E.	
N.E.	S. 78 E.		S.W.	N. 63 E.	
East	S. 70 E.		West	N. 64 E.	
S.E.	S. 71 E.		N.W.	N. 74 E.	

With the deviation as above, give the courses you would steer by the Standard Compass to make the following courses correct magnetic.

Correct magnetic courses:—N.N.E. ¼ E.; N. 84° W.; S. 72° E.; S.W. by W. ½ W.
Compass courses:—

Supposing you have steered the following courses by the Standard Compass, find the correct magnetic courses made from the above deviation table.

Compass courses:—N.N.W. ½ W.; N. 64° E.; S.E. ¾ E.; S. 39° W.
Correct magnetic courses:—

You have taken the following bearings of two distant objects by your Standard Compass as above; with the ship's head at S. 87° E., find the bearings, correct magnetic.

Compass bearings:—S. 15° W., and N. 72° W.
Bearings, magnetic:—

EXAMINATION PAPER—No. VI.

FOR SECOND MATE.

1. Multiply 987 by 543, and 5900 by ·00071, by common logarithms.
2. Divide 50800 by 4·835, and 999999 by 10101, by common logarithms.

3.—

H.	Courses.	K.	₁/₁₀	Winds.	Lee-way.	Deviation.	Remarks, &c.
					pts.		A point, lat. 56° 12' N.,
1	E. by S.	4	3	S. by E.	3½	15' E.	long. 135° 40' W., bear-
2		4	4				ing by compass WSW
3		4	3				dist. 22½ miles. Ship's
4							head E. by S. Devia-
5	S.E. by E.	5	6	S. by W.	1¼	12° E.	tion as per log.
6		5	8				
7		5	6				
8		5					
9	S. by E.	5	7	S.W. by W.	1½	4° E.	
10		5	8				
11		6					
12		6	5				
1	E. ¾ S.	6	4	S. by E.	1¾	15° E.	Variation 15° E.
2		6	5				
3		6	4				
4		5	7				
5	S.W. ¼ W.	4	8	S. by E.	3½	7° W.	
6		4	6				
7		4	2				A current set the ship
8		4	4				(correct magnetic)
9	S. by E.	4	5	E. by S.	2¼	4° E.	N. by W ¼ W., 16 mls.,
10		4	5				from the time the de-
11		3	6				parture was taken to
12		4	4				the end of the day.

4. 1882, June 1st, in longitude 96° 17' E., the observed meridian altitude of sun's L.L. was 75° 38' 15", bearing North, index correction + 27", height of eye 26 feet: required the latitude.

5. In latitude 35° 54' S., the departure made good was 249 miles: required the difference of longitude.

6. Required the course and distance from A to B, by Mercator's Sailing.

    Lat. of A   3° 19' N.   Long. of A   71° 42' W.
    Lat. of B 33 2 S.   Long. of B 122 20 W.

ADDITIONAL FOR ONLY MATE.

7. 1880, June 19th: find the time of high water, A.M. and P.M., at Rotterdam, Heligoland, and Rio Janeiro, longitude 43° 12' W.

8. 1882, June 21st, at 9ʰ 16ᵐ P.M. apparent time at ship, in latitude 59° 51' N., longitude 64° 42' W., the sun's magnetic amplitude N. ¾ E.: required the true amplitude and error of the compass; and supposing the variation to be 52° 30' W.: required the deviation of the compass for the position of the ship's head when the observation was taken.

9. 1882, June 14th, P.M. at ship, latitude 2° 2' S., observed altitude sun's L.L. 28° 38', index error + 48", height of eye 12 feet, time by chronometer 14ᵈ 0ʰ 3ᵐ 18ˢ, which was 2ʰ 28ᵐ 19ˢ·7 *fast* for mean noon at Greenwich, April 1st, and on April 30th, was 1ʰ 24ᵐ 19ˢ *fast* for Greenwich mean noon: required the longitude.

ADDITIONAL FOR FIRST MATE.

10. 1882, June 8th, mean time at ship 7ʰ 50ᵐ A.M., latitude 15° 45' N., longitude 32° 33' W., sun's magnetic azimuth N. 70° E., observed altitude sun's L.L. 31° 10', index error − 1' 22", height of eye 18 feet: required the error of the compass, and supposing the variation be 14° 40' W.: required the deviation of the compass for the position of the ship's head at the time the observation was taken.

11. 1882, June 5th, P.M. at ship, latitude by account 61° 58' N., longitude 155° 21' E., observed altitude sun's L.L. South of observer was 49° 50' 30", index error + 2' 10", height of eye 21 feet, time by watch 11ʰ 48ᵐ 26ˢ, (or 4ᵈ 23ʰ 48ᵐ 26ˢ), which had been found to be 50ᵐ 10ˢ slow on app. time at ship, the difference of longitude made to the West was 17'·5, after the error on apparent time was determined.

12. 1882, June 30th, A.M. at ship, and uncertain of my position, when a chronometer showed June 29ᵈ 11ʰ 44ᵐ 8ˢ M.T.G., obs. alt. sun's L.L. 62° 47' 40"; again, P.M. at ship same day, when chronometer showed June 29ᵈ 16ʰ 42ᵐ 24ˢ, obs. alt. sun's L.L. 30° 3' 40", height of eye 21 feet, the ship having made 21 miles on a true N.E. by E. ½ E. course in the interval: required the line of position when the first altitude was observed, and the ship's position by Sumner's Method when the second altitude was taken, assuming the ship to be between the latitudes 49° 40' N. and 50° 0' N.

For the second observation the position as given by the Examiner would be as follows:—

$$B = \text{Lat. } 49° 40' \text{ N.} \qquad \text{Long. } 179° 44\tfrac{1}{2}' \text{ W.}$$
$$B' = \text{Lat. } 50° 0' \text{ N.} \qquad \text{Long. } 179° 44' \text{ W.}$$

ADDITIONAL FOR MASTER ORDINARY.

13. 1882, June 11th, the observed meridian altitude of the star Pollux, bearing North, was 48° 40' 24", index correction − 1' 32", height of eye 20 feet.

In the following table give the correct magnetic bearing of the distant object, and thence the deviation.

Correct magnetic bearing.

Ship's Head by Standard Compass.	Bearing of Distant Object by Standard Compass.	Deviation Required.	Ship's Head by Standard Compass.	Bearing of Distant Object by Standard Compass.	Deviation Required.
North	N. 89° W.		South	S. 67° W.	
N.E.	S. 79 W.		S.W.	S. 75 W.	
East	S. 62 W.		West	N. 83 W.	
S.E.	S. 58 W.		N.W.	N. 77 W.	

With the deviation as above, give the courses you would steer by the Standard Compass to make the following courses correct magnetic.

Correct magnetic courses:—N.W. by W.; S.W. by W. ½ W.; N.N.E.; S. by W.

Compass courses:—

Supposing you have steered the following courses by the Standard Compass, find the correct magnetic courses made from the above deviation table.

Compass courses:—N.W. by N.; W.S.W.; S. ½ W.; S.E. by E. ½ E.

Magnetic courses:—

You have taken the following bearings of two distant objects by your Standard Compass as above, with the ship's head at S.E. by S., find the bearings, correct magnetic.

Compass bearings:—N. ¾ W. and S. 73° W.

Bearings, magnetic:—

EXAMINATION PAPER—No. VII.

FOR SECOND MATE.

1. Multiply 483 by 28·7, and 98400 by 6·5, by common logarithms.
2. Divide 242880 by 704, and 309760 by 7040, by common logarithms.
3.—

H.	Courses.	K.	$\frac{1}{10}$	Winds.	Lee-way.	Deviation.	Remarks, &c.
					pts.		
1	S.E. by S.	6		S.W. by S.	1¼	31° W.	A point of land in lat. 51° 25′ N., long. 178° 56′ E., bearing by compass N.E. by N., dist. 17 miles. Ship's head S.E. by S. Deviation as per log.
2		5	6				
3		5	4				
4		5					
5	South.	5	5	W.S.W.	2¾	20° W.	
6		5	5				
7		6	5				
8		6	5				
9	West.	6	4	S.S.W.	¾	30° E.	
10		6	6				
11		6	5				
12		6	5				
1	S.S.E.	4	4	S.W.	1½	27½° W.	Variation 28° W.
2		4	4				
3		4	2				
4		4					
5	S.E. by E.	10		S. by W.	½	32½° W.	
6		10	6				
7		9	4				
8		10					
9	S.S.E.	12	4	East.	¼	20½° W.	A current set the ship N.W.bN. (correct magnetic), 9 miles, from the time the departure was taken to the end of the day.
10		12	6				
11		12	6				
12		13	4				

4. 1882, July 26th, in longitude 12° 19′ W., the observed meridian altitude of the sun's L.L. was 15° 41′, bearing North, index correction − 3′ 10″, height of eye 19 feet: required the latitude.

5. In latitude 25° 20′ S., the departure made good was 389 miles: required the difference of longitude by parallel sailing.

6. Required the course and distance from Start Point to St. Michael's.

  Lat. Start Point 50° 13′ N. Long. Start Point 3° 38′ W.
  Lat. St. Michael's 37 48 N. Long. St. Michael's 25 10 W.

ADDITIONAL FOR ONLY MATE.

7. 1880, July 18th: find the time of high water, A.M. and P.M., at Bayonne, Ile de Noirmoutier, Port Navalo, Belle Isle, and Bordeaux (by Admiralty Tables).

8. 1882, July 12th, at $5^h\ 9^m$ P.M. apparent time at ship, in latitude 29° 3′ S., longitude 21° 53′ W., the sun's magnetic amplitude was W. by N. ¾ N.: required the true amplitude and error of the compass; and supposing the variation to be 11° 20′ W.: required the deviation for the position of the ship's head when the observation was taken.

9. 1882, July 17th, P.M. at ship, latitude 31° 32′ S., observed altitude sun's L.L. 13° 23′ 10″, index correction $+$ 5″, height of eye 16 feet, time by chronometer July $16^d\ 22^h\ 3^m\ 49^s$, which was $9^m\ 17^s$ fast for mean noon at Greenwich, June 6th, and on June 14th was fast $8^m\ 32^s·6$ on mean time at Greenwich: required the longitude.

ADDITIONAL FOR FIRST MATE.

10. 1882, July 4th, mean time at ship 8^h 39^m 2^s A.M., latitude $38°$ $10'$ S., longitude $78°$ $35'$ W., sun's bearing by compass N. $19°$ $16'$ E., observed altitude sun's L.L. $12°$ $16'$ $10''$, index correction — $2'$ $38''$, height of eye 14 feet: required the true azimuth and error of the compass; and supposing the variation be $17°$ $20'$ E.: required the deviation of the compass for the position of the ship's head at the time the observation was taken.

11. 1882, July 31st, P.M. at ship, latitude by account $45°$ $5'$ S., longitude $83°$ $12'$ E., observed altitude sun's L.L. North of observer was $16°$ $15'$ $10''$, index corr. — $40''$, height of eye 19 feet, time by watch 11^h 50^m, (or 30^d 23^h 50^m), which had been found to be 36^m 16^s *slow* on apparent time at ship, the difference of longitude made to the *West* was 14 miles, after the error on apparent time was determined: required the latitude by reduction to meridian.

12. 1882, July 5th, P.M. at ship, and uncertain of my position, when a chronometer showed July 5^d 1^h 4^m 2^s G.M.T., obs. alt. sun's L.L. $61°$ $15'$; and again, P.M. at ship same day, when chronometer showed July 5^d 6^h 17^m 2^s, obs. alt. sun's L.L. $20°$ $18'$ $50''$, height of eye 24 feet, the ship having made 19 miles on a true W. by N. ¼ N. course in the interval: required the line of bearing when the first altitude was taken, and the position of the ship by Sumner's Method when the second altitude was observed, assuming latitudes $50°$ $40'$ N. and $51°$ $0'$ N.

For the second observation the positions as given by the Examiner would be as follows:—

$B =$ Lat. $50°$ $40'$ N.   Long. $8°$ $12'$ W.
$B' =$ Lat. 51 0 N.   Long. 8 0 W.

ADDITIONAL FOR MASTER ORDINARY.

13. 1882, July 6th, the observed meridian altitude of the star *a* Scorpii *(Antares)*, bearing North, $70°$ $10'$ $30''$, height of eye 21 feet: required the latitude.

In the following table give the correct magnetic bearing of the distant object, and thence the deviation:—

Correct magnetic bearing.

Ship's Head by Standard Compass.	Bearing of Distant Object by Standard Compass.	Deviation Required.	Ship's Head by Standard Compass.	Bearing of Distant Object by Standard Compass.	Deviation Required.
North	West.		South	N. 85° W.	
N.E.	S. 72° W.		S.W.	N. 78 W.	
East	S. 70 W.		West	N. 70 W.	
S.E.	S. 82 W.		N.W.	N. 73 W.	

With the deviation as above, give the courses you would steer by the Standard Compass to make the following courses correct magnetic.

Correct magnetic courses:— E. by N. ¾ N.; S.E. by E. ½ E.; S. by W. ½ W.; N. 1° E.
Compass courses:—

Supposing you have steered the following courses by the Standard Compass, find the correct magnetic courses made from the above deviation table.

Compass courses:— W. 1° S.; N. ¼ E.; N.E. by E. ¼ E.; S.E. ½ S.
Magnetic courses:—

You have taken the following bearings of two distant objects by your Standard Compass as above, with the ship's head at N.W. by W., find the bearings, correct magnetic.

Compass bearings:— E. ¼ S., and E. ½ N.
Bearings, magnetic:—

EXAMINATION PAPER—No. VIII.

FOR SECOND MATE.

1. Multiply 777 by 999, and 209·36 by 46, by common logarithms.
2. Divide 111111 by 234, and 1962820 by 10·04, by common logarithms.
3.—

H.	Courses.	K.	⅒	Winds.	Lee-way.	Deviation.	Remarks, &c.
					pts.		
1	S.S.E.	7	2	S.W.	1	6° E.	A point of land in lat. 0° 10' N., long. 173° 50' E., bearing by compass S.W., dist. 15 miles. Ship's head S.S.E. Deviation as per log.
2		6	4				
3		6	4				
4		6	4				
5	W.N.W.	7	3	S.W.	1½	18° W.	
6		7					
7		6	7				
8		6					
9	W. ½ N.	5	4	S.W. by S.	2	16° W.	
10		5	5				
11		5	6				
12		5	5				
1	S. by E. ¼ E.	6	3	S.W.	1½	5° E.	Variation 8° E.
2		6	3				
3		6	2				
4		6	2				
5	S.S.E.	5	2	S.W.	1¾	5° E.	
6		5	3				
7		5	5				A current set the ship S. by W. correct magnetic 18 miles, from the time the departure was taken to the end of the day.
8		6					
9	W. by N.	9	6	S.S.W.	0	17° W.	
10		10	4				
11		11	5				
12		11	5				

4. 1882, August 12th, longitude 92° 12' E., the observed meridian altitude of sun's L.L. bearing North, was 42° 42' 10", index correction − 2' 50", height of eye 17 feet: required the latitude.

5. In latitude 56° 11' S., the departure made good was 356 miles East: required the difference of longitude by parallel sailing.

6. Required the course and distance from A to B.

  Latitude A 47° 50' S.   Longitude 42° 16' E.
  Latitude B 40 49 S.   Longitude 46 15 E.

ADDITIONAL FOR ONLY MATE.

7. 1880, August 1st: find the times of high water, A.M. and P.M., at Ushant, Cadiz, Antwerp, and Penzance.

8. 1882, August 20th, at 5ʰ 16ᵐ P.M., apparent time at ship, in latitude 42° 5' S., longitude 88° 36' W., the sun's magnetic amplitude was W. by S., variation 19° 15' E.: required the deviation of the compass for the position of the ship's head at the time of observation.

9. 1882, August 7th, P.M. at ship, latitude 6° 4' N., observed altitude sun's L.L. 24° 5', index correction + 1' 30", height of eye 12 feet, time by chronometer August 6ᵈ 20ʰ 30ᵐ 36ˢ, which was 36ˢ·2 *fast* for Greenwich mean noon, July 14th, and on July 21st was 10ˢ *slow* for Greenwich mean time: required the longitude.

ADDITIONAL FOR FIRST MATE.

10. 1882, August 20th, mean time at ship 2^h 35^m 25^s P.M., latitude 52° 2' S., longitude 89° 26' E., sun's bearing by compass N.W. ¾ N., observed altitude sun's L.L. 17° 26', index correction $+$ 1' 45", height of eye 21 feet: required the error of compass; and supposing the variation to be 33° 50' E.: find the deviation of the compass for the position of the ship's head at the time of observation.

11. 1882, August 11th, A.M. at ship, latitude account 39° 3' S., longitude 157° 25' E., observed altitude of sun's L.L., North of observer, was 34° 37', height of eye 12 feet, time by watch 7^h 41^m 25^s (or 10^d 19^h 41^m 25^s), which had been found to be 3^h 41^m 8^s *slow* on apparent time at ship, the difference of longitude made to the *East* was 33', after the error on apparent time was determined.

12. 1882, August 10th, A.M. at ship, and uncertain of my position, when a chronometer showed August 9^d 21^h 8^m 37^s M.T.G., obs. alt. sun's L.L. 38° 17'; again, P.M. at ship same day, when chronometer showed August 10^d 3^h 15^m 57^s, obs. alt. sun's L.L. 40° 10' 15", height of eye 18 feet, the ship having made 21 miles on a true N.E. course in the interval: required the line of position when the first altitude was taken, and the ship's position by Sumner's Method when the second altitude was observed, assuming latitudes 49° 40' N. and 50° 0' N.

ADDITIONAL FOR MASTER ORDINARY.

13. 1882, August 10th, the observed meridian altitude of the star *a* Aquilæ *(Altair)*, bearing North, was 66° 51' 10", index correction $+$ 58", height of 13 feet: required the latitude.

In the following table give the correct magnetic bearing of the distant object, and thence the deviation:—

Correct magnetic bearing.

Ship's Head by Standard Compass.	Bearing of Distant Object by Standard Compass.	Deviation Required.	Ship's Head by Standard Compass.	Bearing of Distant Object by Standard Compass.	Deviation Required.
North	S. 12° E.		South	S. 4° E.	
N.E.	S. 10 E.		S.W.	S. 17 E.	
East	S. 6 W.		West	S. 20 E.	
S.E.	S. 9 W.		N.W.	S. 16 E.	

With the deviation as above, give the courses you would steer by the Standard Compass to make the following courses correct magnetic.

Correct magnetic courses:—E. by N. ½ N.; S.W. by W. ½ W.; N.N.W. ¾ W.; E.S.E.

Compass courses:—

Supposing you have steered the following courses by the Standard Compass, find the correct magnetic courses made, from the above deviation table.

Compass courses:—N.W. ½ N.; W. by S. ½ S.; N. ⅓ W.; W. ¾ S.

Magnetic courses:—

You have taken the following bearings of two distant objects by your Standard Compass as above; with the ship's head at W.N.W., find the bearings, correct magnetic.

Compass bearings:—S.S.E. and S.E. by S.

Bearings, magnetic:—

EXAMINATION PAPER—No. IX.
FOR SECOND MATE.

1. Multiply 247·55 by 56·72, and ·03948 by 0·1959, by common logarithms.
2. Divide 69·7565 by 97564, and 33248100 by 830000, by common logarithms.
3.—

H.	Courses.	K.	$\frac{1}{10}$	Winds.	Lee-way.	Deviation.	Remarks, &c.
1	S.E. ½ E.	10	4	S. by W. ½ W.	pts. ¾	15° E.	A point, lat. 54° 7' N. long. 0° 5' W., bearing by compass W. ¾ N., dist. 20 miles. Ship's head S.E. ½ E. Deviation as per log.
2		10	2				
3		10	2				
4		10	2				
5	S.E. by S.	12	4	S.W. by S.	½	5° E.	
6		12	6				
7		12	5				
8		12	5				
9	East.	11	2	N.N.E.	¾	21° E.	
10		10	6				
11		10	6				
12		10	6				Variation 22° W.
1	S.E. ½ S.	8	5	S.S.W. ½ W.	1	7° E.	
2		8	3				
3		8	2				
4		8					
5	S. by E. ½ E.	7	3	S.W. ½ W.	1¼	3½° E.	
6		6	8				
7		6	6				A current set the ship E. by S. ½ S. (correct magnetic) 32 miles, from the time the departure was taken to the end of the day.
8		6	3				
9	S.E. by E.	8	6	N.E. by E.	1	17° E.	
10		9	2				
11		9	2				
12		9					

4. 1882, September 23rd, in longitude 123° 45' E., observed meridian altitude of sun's L.L. bearing North, was 89° 49' 50", index error − 52", height of eye 26 feet: required the latitude.

5. In latitude 20° 15' S., the departure made good was 352 miles W.: required the difference of longitude by parallel sailing.

Latitude A 25° 39' N.   Longitude A 48° 19' W.
Latitude B 34 28 S.   Longitude B 18 28 E.

ADDITIONAL FOR ONLY MATE.

7. 1880, September 11th: find A.M. and P.M. tides at Alderney, Heligoland, Nieuport, Fowey, Hastings, and Dornock Road.

8. 1882, September 30th, at 5ʰ 45ᵐ P.M., apparent time at ship, latitude 52° 30' N., longitude 12° 10' W., sun's magnetic amplitude N.W. ¾ W.: required error of compass; and supposing the variation to be 30° 28' E.: required the deviation of the compass for the position of the ship's head at the time of observation.

9. 1882, September 1st, P.M. at ship, latitude 9° 9' N., observed altitude sun's L.L. 62° 13' 14", index correction + 15", height of eye 16 feet, time by chronometer August 31ᵈ 15ʰ 34ᵐ 28ˢ, which was 2ᵐ 10ˢ *slow* for mean noon at Greenwich, July 28th, and on August 12th was 1ᵐ 31ˢ *slow* on mean noon at Greenwich: required the longitude.

ADDITIONAL FOR FIRST MATE.

10. 1882, September 16th, mean time at ship $8^h\ 3^m\ 18^s$ A.M., latitude 4° 22' N., longitude 81° 39' W., sun's bearing by compass N. 93° 20' E., observed altitude sun's L.L. 29° 30' 30", index correction + 1' 22", height of eye 20 feet: required the true azimuth and error of compass; and supposing the variation is 8° 20' E.: required the deviation for the position of the ship's head at the time of observation.

11. 1882, September 23rd, A.M. at ship, latitude acct. 27° 32' S., longitude 168° 51' E. observed altitude sun's L.L., North of observer, was 61° 59' 40", index correction – 1' 50", height of eye 18 feet, time by watch $11^h\ 10^m\ 10^s$ (or $22^d\ 23^h\ 10^m\ 10^s$), which had been found to be $31^m\ 31^s$ *slow* on apparent time at ship, the difference of longitude made to the *East* was 24'·4, after the error on apparent time was determined : required the latitude.

12. 1882, September 9th, A.M. at ship, and uncertain of my position, when a chronometer showed September $8^d\ 16^h\ 23^m\ 50^s$ M.T.G., obs. alt. sun's L.L. 11° 2' 30"; again, A.M. at ship same day, when chronometer showed September $8^d\ 20^h\ 39^m\ 25^s$, obs. alt. sun's L.L. 45° 45' 30", height of eye 18 feet, the ship having made 19 miles on a true S. 76° E. course in the interval : required the line of position when the first altitude was observed, and the ship's position by Sumner's Method when the second altitude was taken, assuming the ship to be between the latitudes 47° 50' N. and 47° 10' N.

For the second observation the positions as given by the Examiner would be as follows :—

$$B = \text{Lat. } 47°\ 50'\ N. \qquad \text{Long. } 35°\ 58\tfrac{1}{2}'\ E.$$
$$B' = \text{Lat. } 47°\ 10'\ N. \qquad \text{Long. } 33°\ 27\tfrac{1}{4}'\ E.$$

ADDITIONAL FOR MASTER ORDINARY.

13. 1882, September 7th, the observed meridian altitude of star Arcturus was 86° 35' 50", bearing North, index correction – 1' 10", height of eye 12 feet: required the latitude.

In the following table give the correct magnetic bearing of the distant object, and thence the deviation :—

Correct magnetic bearing.

Ship's Head by Standard Compass.	Bearing of Distant Object by Standard Compass.	Deviation Required.	Ship's Head by Standard Compass.	Bearing of Distant Object by Standard Compass.	Deviation Required.
North	S. 86° W.		South	N. 88° W.	
N.E.	S. 68 W.		S.W.	N. 80 W.	
East	S. 66 W.		West	N. 72 W.	
S.E.	S. 79 W.		N.W.	N. 75 W.	

With the deviation as above, give the courses you would steer by the Standard Compass to make the following courses correct magnetic.

Correct magnetic courses :—W. by S. ¾ S.; N. ½ E.; E. ¾ N.; S.E. ¼ E.

Compass courses :—

Supposing you have steered the following courses by the Standard Compass, find the correct magnetic courses made from the above deviation table.

Compass courses :—North ; S.S.W. ¼ W.; E. by S. ¼ S.; N.E. ½ E.

Correct magnetic courses :—

You have taken the following bearings of two distant objects by your Standard Compass as above, with the ship's head at N.N.E. ¾ E., find the bearings, correct magnetic.

Compass bearings :—N. 79° E. and W. ¼ S.

Bearings, magnetic :—

EXAMINATION PAPER—No. X.

FOR SECOND MATE.

1. Multiply 560072 by 50, and 10·5526 by 317·145, by common logarithms.
2. Divide 8491·9 by 98·4, and 2064840 by 3800·62, by common logarithms.
3.—

H.	Courses.	K.	$\frac{1}{10}$	Winds.	Lee-way.	Deviation.	Remarks, &c.
					pts.		
1	W.S.W.	11		South.	½	11° W.	A point, Cape Farewell, in lat. 59° 49′ N., long. 43° 54′ W., bearing by compass N. ½ E., distance 36 miles. Ship's head W.S.W. Deviation as per log.
2		11					
3		10	4				
4		10	6				
5	West.	5		S.S.W.	1	16° W.	
6		5					
7		4	5				
8		4	5				
9	S.E.	13		S.S.W.	½	10° E.	
10		12	2				
11		12	4				
12		12	4				
1	S. by E.	6		S.W. by W.	2¼	4° E.	
2		5	5				Variation 70° W.
3		5					
4		4	5				
5	S.W. by S.	1	5	S.E. by S.	3¼	5° W.	
6		1	5				
7		1	5				A current set the ship (correct magnetic) S.S.E., 48 miles, from the time the departure was taken to the end of the day.
8		1	5				
9	S.W.	6		W.N.W.	1¼	6° W.	
10		5	6				
11		5	4				
12		·5					

4. 1882, October 20th, in longitude 150° 25′ W., observed meridian altitude of sun's L.L., bearing North, was 49° 58′ 50″, index correction + 1′ 10″, height of eye 19 feet: required the latitude.

5. In latitude 59° 36′ N., the departure made good was 52·9 miles East: required the difference of longitude by parallel sailing.

6. Required the course and distance from A to B, by calculation on Mercator's principle.

  Latitude A 9° 36′ S. Longitude A 2° 10′ W.
  Latitude B 7 16 S. Longitude B 1 24 W.

ADDITIONAL FOR ONLY MATE.

7. 1880, October 1st: find the times of high water A.M. and P.M. at Gibraltar, Ramsgate, Wick, and Berwick.

8. 1882, October 9th, at $5^h\ 51^m$ A.M., apparent time at ship, latitude 18° 45′ S., longitude 99° 18′ E., sun's magnetic amplitude E. ¼ N.: required the true amplitude and error of the compass; and supposing the variation to be 1° 50′ W.: required the deviation of the compass for the position of the ship's head at the time of observation.

9. 1882, October 30th, P.M. at ship, latitude 32° 45′ N., observed altitude sun's L.L. 28° 30′, index correction + 2° 30″, height of eye 18 feet, time by chronometer Oct. 30^d 11^h 56^m 43^s, which was *slow* $2^m\ 28^s$ for mean noon at Greenwich, October 1st, and on October 8th was $2^m\ 44^s·8$ *slow* for mean noon at Greenwich: required the longitude.

ADDITIONAL FOR FIRST MATE.

10. 1882, October 1st, mean time at ship $4^h\ 54^m$ P.M., latitude $17°\ 8'$ S., longitude $152°\ 33'$ E., sun's bearing by compass W. ¼ N., observed altitude sun's L.L. $13°\ 59'$, index correction — 22", height of eye 17 feet: required the true azimuth and error of the compass; and supposing the variation is $7°\ 40'$ E.: required the deviation of the compass for the position of the ship's head at the time of observation.

11. 1882, October 2nd, A.M. at ship, latitude account $38°\ 12'$ N., longitude $23°\ 34'$ W., observed altitude of sun's L.L., South of observer, was $47°\ 30'$, index correction — $1'\ 38"$, height of eye 17 feet, time by watch $1^h\ 50^m$ (or $2^d\ 1^h\ 50^m$), which had been found to be $2^h\ 10^m$ *fast* on apparent time at ship, the difference of longitude made to the *East* was $43'$, after the error on apparent time was determined: required the latitude.

12. 1882, October 1st, A.M. at ship, and uncertain of my position, when a chronometer showed October $1^d\ 0^h\ 8^m\ 56^s$ M.T.G., obs. alt. sun's L.L. $23°\ 19'\ 50"$; again, P.M. at ship same day, when chronometer showed October $1^d\ 5^h\ 53^m\ 52^s$, obs. alt. sun's L.L. $25°\ 30'$, height of eye 23 feet, the ship having made 29 miles on a true N.W. ¾ W. course in the interval: required the line of bearing when the first altitude was taken, and the position of the ship by Sumner's Method when the second altitude was observed, assuming latitudes $50°\ 30'$ N. and $51°\ 0'$ N.

For the second observation the positions as given by the Examiner would be as follows:—

$$B = \text{Lat. } 50°\ 30'\ N. \qquad \text{Long. } 49°\ 54\tfrac{3}{4}'\ W.$$
$$B' = \text{Lat. } 51\ 0\ N. \qquad \text{Long. } 50\ 40\ W.$$

ADDITIONAL FOR MASTER ORDINARY.

13. 1882, October 7th, the observed meridian altitude of *a Pegasi (Markab)* bearing South, was $54°\ 10'\ 15"$, height of eye 13 feet: required the latitude.

In the following table give the correct magnetic bearing of the distant object and thence the deviation.

Correct magnetic bearing.

Ship's Head by Standard Compass.	Bearing of Distant Object by Standard Compass.	Deviation Required.	Ship's Head by Standard Compass.	Bearing of Distant Object by Standard Compass.	Deviation Required.
North	S. 37° W.		South	S. 41° W.	
N.E.	S. 55 W.		S.W.	S. 20 W.	
East	S. 60 W.		West	S. 14 W.	
S.E.	S. 57 W.		N.W.	S. 19 W.	

With the deviation as above, give the courses you would steer by the Standard Compass to make the following courses correct magnetic.

Correct magnetic courses:—S.E.; N.E. ¼ E.; S. 10° W.; E. ¼ N.

Compass courses:—

Supposing you have steered the following courses by the Standard Compass, find the correct magnetic courses made from the above deviation table.

Compass courses:—S.S.E. ¾ E.; S. ¾ W.; E. by N. ¾ N.; N. ¼ W.

Correct magnetic courses:—

You have taken the following bearings of two distant objects by your Standard Compass as above; with the ship's head at S.E. by E. ½ E., find the bearings, correct magnetic.

Compass bearings:—E. by S. ½ S., and W.N.W.

Bearings, magnetic:—

EXAMINATION PAPER—No. XI.

FOR SECOND MATE.

1. Multiply 45·3 by 9·76, and 40·405 by 10·8, by common logarithms.
2. Divide 100·002 by 1·0012, and 829440 by 288, by common logarithms.
3.—

H.	Courses.	K.	₁/₁₀	Winds.	Leeway.	Deviation.	Remarks, &c.
					pts.		
1	N. by E.	4	2	E. by N.	2¼	3° E.	A point of land, lat. 52° N., long. 120° E., bearing by compass N. by E. ¼ E., dist. 16 miles. Ship's head N. by E. Deviation as per log.
2		3	8				
3		4	5				
4		4	5				
5	N.E. ¾ E.	4	5	N. by W.	3½	17° E.	
6		5					
7		5					
8		4	5				
9	W. ¾ N.	7	5	N. by W.	1¾	17° W.	
10		7	5				
11		8					
12		8					
1	E.S.E.	4	5	South.	3	13° E.	Variation 25° E.
2		4	5				
3		4					
4		4					
5	S.E. ½ S.	4	6	N. by E.	0	9° E.	
6		4	5				
7		4	8				A current set (correct magnetic) E.N.E., 22 miles, from the time the departure was taken to the end of the day.
8		5	1				
9	W. ¾ S.	6		N.W. by N.	3¼	14° W.	
10		6	3				
11		6	4				
12		6	3				

4. 1882, November 15th, in longitude 80° 11' E., the observed meridian altitude of sun's L.L. was 67° 44', bearing North, index error + 1' 38", height of eye 15 feet: required the latitude.

5. In latitude 40° 50' S., the departure made good was 149 miles *East*: required the difference of longitude by parallel sailing.

6. Required the course and distance from the ship's position to the Lizard, by calculation on Mercator's principle.

    Latitude of position   17° 50' N.   Longitude of position 76° 42' W.
    Latitude of Lizard   49 58 N.   Longitude of Lizard   5 12 W.

ADDITIONAL FOR ONLY MATE.

7. 1880, November 12th: find A.M. and P.M. tides at Newhaven, Torbay, Kilrush, and St. Nazaire.

8. 1882, November 10th, at $4^h\ 3^m\ 52^s$ A.M., apparent time at ship, latitude 58° 13' S., longitude 55° 47' E., sun's magnetic amplitude S. by E. ¼ E.: required the true amplitude and error of compass; and supposing the variation to be 16° 30' E.: required the deviation of the compass for the position of the ship's head at the time of observation.

9. 1882, November 30th, A.M. at ship, latitude 40° 40' S., observed altitude sun's L.L. 39° 30', index correction + 6' 24", height of eye 22 feet, time by chronometer, November $30^d\ 2^h\ 58^m\ 45^s$, which was $10^m\ 50^s$·4 *fast* for mean noon at Greenwich, October 13th, and on October 25th, was $10^m\ 36^s$ *fast* for mean noon at Greenwich: required the longitude.

ADDITIONAL FOR FIRST MATE.

10. 1882, November 15th, mean time at ship $2^h\ 46^m\ 43^s$ P.M., latitude $45°\ 31'$ S., longitude $119°\ 56'$ W., sun's magnetic azimuth S. $98\frac{1}{2}°$ W., observed altitude sun's L.L. $43°\ 45'$, index correction $-56'$, height of eye 20 feet: required the error of compass; and supposing the variation to be $7°\ 50'$ W.: required the deviation for the position of the ship's head at the time of observation.

11. 1882, November 13th, A.M. at ship, latitude acct. $50°\ 52'$ S., longitude $48°\ 52'$ W., observed altitude sun's L.L. was $56°$ N., index correction $+23'$, height of eye 19 feet, time by watch $4^m\ 34^s$ (or $13^d\ 0^h\ 4^m\ 34^s$), which had been found to be $43^m\ 24^s$ *fast* on apparent time at ship, the difference of longitude made to *West* was 9 miles after the error on apparent time was determined: required the latitude by reduction to meridian.

12. 1882, November 1st, A.M. at ship, and uncertain of my position, when a chronometer showed October $31^d\ 23^h\ 19^m\ 10^s$ M.T.G., obs. alt. sun's L.L. $19°\ 8'$; again, P.M. at ship same day, when chronometer showed November $1^d\ 4^h\ 33^m\ 29^s$, obs. alt. sun's L.L. $12°\ 38'$, height of eye 15 feet, the ship having made 12 miles on a true S.S.W. course in the interval: required the line of bearing when the first altitude was taken, and the ship's position by Sumner's Method when the second altitude was observed, assuming the ship to be between the latitudes $50°\ 50'$ N. and $50°\ 20'$ N.

For the second observation the positions as given by the Examiner would be as follows:—

$B =$ Lat. $50°\ 50'$ N. Long. $25°\ 20'$ W.
$B' =$ Lat. $50\ 20$ N. Long. $24\ 36$ W.

ADDITIONAL FOR MASTER ORDINARY.

13. 1882, November 7th, the observed meridian altitude of the star *a Piscis Australis (Fomalhaut)*, bearing North, was $59°\ 40'$, index correction $+1'\ 12''$, height of eye 23 feet: required the latitude.

In the following table give the correct magnetic bearing of the distant object, and thence the deviation:—

Correct magnetic bearing.

Ship's Head by Standard Compass.	Bearing of Distant Object by Standard Compass.	Deviation Required.	Ship's Head by Standard Compass.	Bearing of Distant Object by Standard Compass.	Deviation Required.
North	S. 88° W.		South	N. 88° W.	
N.E.	S. 70 W.		S.W.	N. 80 W.	
East	S. 68 W.		West	N. 72 W.	
S.E.	S. 80 W.		N.W.	N. 75 W.	

With the deviation as above, give the courses you would steer by the Standard Compass to make the following courses correct magnetic.

Correct magnetic courses:—W. by S. $\frac{1}{4}$ S.; N.W. by W. $\frac{1}{2}$ W.; E. by S. $\frac{1}{2}$ S.; S. $\frac{1}{4}$ E.

Compass courses:—

Supposing you have steered the following courses by the Standard Compass, find the correct magnetic courses made from the above deviation table.

Compass courses:—W. $\frac{1}{2}$ N.; N. 42° W.; S. 64° E.; N.E. $\frac{1}{2}$ N.

Magnetic courses:—

You have taken the following bearings of two distant objects by your Standard Compass as above, with the ship's head at W. by N. $\frac{1}{2}$ N., find the bearings, correct magnetic.

Compass bearings:—W. $\frac{1}{2}$ N., and S. 36° E.

Bearings, magnetic:—

EXAMINATION PAPER—No. XII.

FOR SECOND MATE.

1\. Multiply 758900 by 13·5, and 0·006994 by 0·33318, by common logarithms.

2\. Divide 999·43 by 67·832, and 1875000 by 15000, by common logarithms.

3\.—

H.	Courses.	K.	1/10	Winds.	Lee-way.	Deviation.	Remarks, &c.
					pts.		
1	E. by N. ¼ N.	3	5	N. ½ E.	2	11° E.	A point of land in lat. 50° N. long. 40° W. bearing by compass E.N.E. ¾ E., distance 16 miles. Ship's head E. by N. ¼ N. Deviation as per log.
2		3	3				
3		4					
4		4	2				
5	W.N.W.	6	5	North.	1½	10° W.	
6		6	2				
7		5	6				
8		4	7				
9	S.S.W. ½ W.	4	2	West.	2½	4° W.	
10		4					
11		3	6				
12		3	2				
1	N.N.W. ¼ W.	5	6	West.	1¾	5° W.	Variation 36½° W.
2		5	6				
3		6	4				
4		6	4				
5	S.E. ¾ E.	6	2	S.S.W.	1½	7° E.	
6		5	6				
7		5	2				
8		5					A current set the ship (correct magnetic) S.S.W., 6 miles, from the time the departure was taken to the end of the day.
9	S.W. ¾ W.	2	5	S. by E. ¼ E.	3¼	6° W.	
10		3	2				
11		3					
12		3	3				

4\. 1882, December 31st, longitude 123° 45′ W., observed meridian altitude of sun's L.L. was 67° 8′ 10″, bearing South, index correction + 9″, height of eye 13 feet: required the latitude.

5\. In latitude 60° N., the departure made good was 111 miles *East*: required the difference of longitude by parallel sailing.

6\. Required the course and distance from Port San Francisco to Cape Palliser, by Mercator's Sailing.

Lat. Port San Francisco	37° 48′ N.	Long. Port San Francisco	122° 24′ W.
Lat. Cape Palliser	41 38 S.	Long. Cape Palliser	175 21 E.

ADDITIONAL FOR ONLY MATE.

7\. 1880, December 28th: find the times of high water, A.M. and P.M., at Skull, Westport, Valentia, Limerick, Coleraine, and Tenby.

8\. 1882, December 28th, at $4^h\ 11^m\ 13^s$ A.M., apparent time at ship, latitude 46° 47′ S., longitude 179° 54′ W., sun's magnetic amplitude S.E. by E. ¾ E.: required error of compass; and supposing the variation to be 15° 30′ E.: required the deviation of the compass for the position of the ship's head at the time of observation.

9\. 1882, December 24th, A.M. at ship, latitude 33° 33′ S., observed altitude sun's L.L. 40° 40′ 40″, index correction + 2′ 20″, height of eye 19 feet, time by chronometer $8^h\ 7^m\ 37^s$ P.M., which was $6^m\ 8^s$ *slow* for Greenwich mean noon, October 31st, and on November 12th, was $7^m\ 16^s\!\cdot\!2$ *slow* for mean noon at Greenwich: required the longitude by chronometer.

ADDITIONAL FOR FIRST MATE.

10. 1882, December 27th, mean time at ship 8ʰ 0ᵐ 10ˢ A.M., latitude 15° 12′ N., longitude 130° W., sun's bearing by compass E. by S. ¾ S., observed altitude sun's L.L. 20° 15′, index correction + 2′ 5″, height of eye 16 feet: required the error of compass; and supposing the variation to be 7° 20′ E.: required the deviation of the compass for the position of the ship's head when the observation was taken.

11. 1882, December 4th, A.M. at ship, latitude account 51° 54′ S., longitude 30° 10′ W., observed altitude sun's L.L., North of observer, 59° 59′, index correction — 3′ 12″, height of eye 20 feet, time by watch 12ʰ 10ˢ (or 4ᵈ 0ʰ 12ᵐ 10ˢ), which had been found to be 42ᵐ 10ˢ *fast* on apparent time at ship, the difference of longitude made to the *West* was 10 miles after the error on apparent time was determined: required the latitude.

12. 1882, December 25th, A.M. at ship, and uncertain of my position, when a chronometer showed December 24ᵈ 22ʰ 11ᵐ 2ˢ M.T.G., obs. alt. sun's L.L. 9° 8′; again, P.M. at ship same day, when chronometer showed December 25ᵈ 1ʰ 39ᵐ, obs. alt. sun's L.L. 15° 3′, height of eye 15 feet, the ship having made 35 miles on a true S. 54° E. course in the interval: required the line of bearing when the first altitude was taken, and the position of the ship by Sumner's Method when the second altitude was observed, assuming latitudes 46° 30′ N. and 46° 0′ N.

ADDITIONAL FOR MASTER ORDINARY.

13. 1882, December 21st, the observed meridian altitude of star α Canis Minoris *(Procyon)* was 52° 51′ 50″, bearing North, index correction — 49″, height of eye 21 feet: required the latitude.

In the following table give the correct magnetic bearing of the distant object, and thence the deviation:—

Correct magnetic bearing.

Ship's Head by Standard Compass.	Bearing of Distant Object by Standard Compass.	Deviation Required.	Ship's Head by Standard Compass.	Bearing of Distant Object by Standard Compass.	Deviation Required.
North	S. 2° W.		South	S. 5° E.	
N.E.	S. 4 W.		S.W.	S. 24 E.	
East	S. 10 W.		West	S. 16 E.	
S.E.	S. 16 W.		N.W.	S. 3 E.	

With the deviation as above, give the courses you would steer by the Standard Compass to make the following courses correct magnetic.

Correct magnetic courses:—N. 78° E.; E.S.E.; S.W. by W. ½ W.; N. ¾ W.

Compass courses:—

Supposing you have steered the following courses by the Standard Compass, find the correct magnetic courses made from the above deviation table.

Compass Courses:—S.S.E.; W. by S. ½ S.; S. by W. ½ W.; S.E. ½ S.

Correct magnetic courses:—

You have taken the following bearings of two distant objects by your Standard Compass as above, with the ship's head at S.W. by S., find the bearings, correct magnetic.

Compass bearings:—W. by S. ½ S. and N.N.W.

Bearings, magnetic:—

EXAMINATION PAPER—No. XIII.

FOR SECOND MATE.

1. Multiply 198400 by 6·5, and 448000 by ·0000448, by common logarithms.

2. Divide 2208000 by 3450, and ·085375 by ·07425, by common logarithms.

3.—

H.	Courses.	K.	$\frac{1}{10}$	Winds.	Lee-way.	Deviation.	Remarks, &c.
					pts.		
1	S.S.W. ¼ W.	12	2	West.	¼	42° W.	A point, lat. 62° 18′ N. long. 83° 17′ E., bearing by compass N. by E. ¼ E., dist. 23 miles. Ship's head S.S.W. Deviation as per log.
2		12	6				
3		13	2				
4		13					
5	S.W. ¾ W.	11	5	S. by E.	½	8° W.	
6		11	4				
7		11	1				
8		11					
9	E. ¾ S.	5	4	S. by E.	1¾	15° E.	
10		5	6				
11		5	6				
12		5	4				
1	W.N.W.	4	4	North.	3	19° W.	Variation 42° E.
2		4	4				
3		4	2				
4		5					
5	N.W. ½ N.	10	6	S. by W.	0	16° 30′ W.	
6		10	2				
7		11	4				A current set the ship (correct magnetic) S.W. ¼ W., 52 miles, from the time the departure was taken to the end of the day.
8		11	8				
9	E. ¾ N.	3	2	N. by E.	3¼	17° 14′ E.	
10		3	2				
11		3	2				
12		2	4				

4. 1882, August 11th, in longitude 92° 12′ E., observed meridian altitude sun's L.L. was 42° 42′ 10″, zenith South of sun, index correction − 2′ 50″, height of eye 17 feet: required the latitude.

5. In latitude 80° the departure made good was 80 miles: required the difference of longitude by parallel sailing.

6. Required the course and distance from A to B, by Mercator's Sailing.

Latitude A	51° 30′ N.	Longitude A	3° 30′ 30″ W.
Latitude B	20 0 N.	Longitude B	33 4 56 W.

ADDITIONAL FOR ONLY MATE.

7. 1880, July 24th: required the times of high water, A.M. and P.M., at Point de Galle, long. 80° E., St. Nazaire, and Jersey.

8. 1882, October 28th, at 8^h 30^m A.M., apparent time at ship, in latitude 49° 40′ N., longitude 116° 12′ W., the sun's magnetic amplitude was E. 10° 40′ N.: required the error of compass, and supposing the variation to be 23° 50′ E.: required the deviation for the position of the ship's head at the time of observation.

9. 1882, April 18th, A.M. at ship, latitude 50° 48′ N., observed altitude sun's L.L. 38° 10′ 50″, index correction + 45″, height of eye 16 feet, time by chronometer 9^h 27^m 2^s, A.M. at Greenwich, which was 0^m 49^s·3 *slow* for mean noon at Greenwich, March 17th, and on April 1st was 1^m 58^s·7 *fast* for mean time at Greenwich: required the longitude.

ADDITIONAL FOR FIRST MATE.

10. 1882, March 9th, mean time at ship 8ʰ 11ᵐ 42ˢ A.M., latitude 29° 58′ S., longitude 57° 24′ E., observed altitude sun's L.L. 28° 23′ 15″, height of eye 16 feet, sun's azimuth E. 9° 40′ S.: required the error of compass; and supposing the variation to be 17° 10′ W.: required the deviation for the position of the ship's head at the time of observation.

11. 1882, July 28th, A.M. at ship, latitude account 38° 54′ N., longitude 39° W., observed altitude sun's L.L. 69° 10′ S., index corr. + 1′ 27″, height of eye 23 feet, time by watch 11ʰ 3ᵐ 15ˢ, *slow* on apparent time at ship 28ᵐ 45ˢ, the difference of longitude made to *East* was 32 miles after the error on apparent time was determined: required the latitude by reduction to meridian.

12. 1882, February 28th, A.M. at ship, and uncertain of my position, when a chronometer showed February 28ᵈ 9ʰ 4ᵐ 12ˢ M.T.G., obs. alt. sun's L.L. 27° 31′; again, P.M. at ship same day, when chronometer showed February 28ᵈ 12ʰ 25ᵐ 35ˢ, obs. alt. sun's L.L. 32° 40′, height of eye 20 feet, the ship having made 27 miles on a true N.E. ¼ E. course in the interval: required the line of bearing when the first altitude was taken, and the position of the ship by Sumner's Method when the second altitude was observed, assuming latitudes 47° 10′ N. and 47° 40′ N.

For the second observation the positions as given by the Examiner would be as follows:—

$$B = \text{Lat. } 47° 10′ \text{ N.} \qquad \text{Long. } 165° 23′ \text{ W.}$$
$$B' = \text{Lat. } 47° 40′ \text{ N.} \qquad \text{Long. } 167° 23′ \text{ W.}$$

ADDITIONAL FOR MASTER ORDINARY.

13. 1882, October 8th, the observed meridian altitude of *a* Gruis was 50° 0′, bearing South, index correction − 1′ 12″, height of eye 17 feet: required the latitude.

In the following table give the correct magnetic bearing of the distant object, and thence the deviation.

Correct magnetic bearing.

Ship's Head by Standard Compass.	Bearing of Distant Object by Standard Compass.	Deviation Required.	Ship's Head by Standard Compass.	Bearing of Distant Object by Standard Compass.	Deviation Required.
North	S. 25° W.		South	S. 1° W.	
N.E.	S. 21 W.		S.W.	S. 7 E.	
East	S. 21 W.		West	S. 6 W.	
S.E.	S. 16 W.		N.W.	S. 21 W.	

With the deviation as above, give the courses you would steer by the Standard Compass to make the following courses correct magnetic.

Correct magnetic courses:—S.W. by W.; E.N.E.; S. by W. ½ W.; N.N.E.
Compass courses:—

Supposing you have steered the following courses by the Standard Compass, find the correct magnetic courses made from the above deviation table.

Compass courses:—N.E. by E.; N.W. ½ N.; N. ½ E.; S. by E.
Correct magnetic courses:—

You have taken the following bearings of two distant objects by your Standard Compass as above, with ship's head at S.S.W. ½ W., find the bearings, correct magnetic.

Compass bearings:—N.E. by E. and S.W. by W.
Bearings, magnetic:—

EXAMINATION PAPER—No. XIV.

FOR SECOND MATE.

1. Multiply 173·4 by 1·734, and 6003004·1 by ·000273004, by common logarithms.

2. Divide 57634·1 by 276, and 471 by 964315, by common logarithms.

3.—

H.	Courses.	K.	$\frac{1}{10}$	Winds.	Lee-way.	Deviation.	Remarks, &c.
					pts.		
1	E. ¼ S.	10	4	S. by E. ½ E.	¼	27½° W.	A point of land in lat. 30° 16′ N., long. 179° 52′ E., bearing by compass N. by E. ½ E., dist. 15 miles. Ship's head E. ¼ S. Deviation as per log.
2		10	6				
3		10	6				
4		10	4				
5	E. by S. ¼ S.	8	6	S. by E.	1¾	34½° W.	
6		8	4				
7		7	8				
8		7	2				
9	W. by S. ½ S.	9	6	South.	2½	7° E.	
10		9	4				
11		9	5				
12		8	5				
1	S. ¾ W.	9	3	S.E. by E.	1¼	46° W.	Variation 12° 30′ E.
2		9	4				
3		9	3				
4		9					
5	N. ¾ W.	9	5	E.N.E.	¾	50° E.	
6		9	5				
7		9	6				
8		9	4				A current set the ship N. ¾ W. (correct magnetic), 17½ miles, from the time the departure was taken to the end of the day.
9	E. ½ S.	8	4	S.S.E.	2¼	28° W.	
10		7	6				
11		6	4				
12		5	6				

4. 1882, March 20th, in longitude 174° W., the observed meridian altitude of the sun's L.L. was 89° 56′ 10″, bearing North, index correction — 1′ 15″, height of eye 15 feet: required the latitude.

5. In latitude 71° 44′ N., the departure made good was 164 miles: required the difference of longitude by parallel sailing.

6. Required the course and distance from A to B, by calculation on Mercator's principle.

Lat. A	2° 49′ N.	Long. A	130° 9′ E.
Lat. B	0 20 S.	Long. B	82 16 W.

ADDITIONAL FOR ONLY MATE.

7. 1880, June 1st: find the time of high water, A.M. and P.M., at Plymouth Breakwater, Falmouth, Exmouth, and Flambro' Head.

8. 1882, December 6th, at 0^h 16^m A.M. apparent time at ship, latitude 67° 19′ S., longitude 19° 2′ E., the sun's magnetic amplitude was S. ¾ W.: required the true amplitude and error of the compass; and supposing the variation to be 24° 50′ W.: required the deviation for the position of the ship's head when the observation was taken.

9. 1882, September 1st, A.M. at ship, lat. 47° 48′ N., observed altitude sun's L.L. 39° 46′ 50″, index correction + 2′ 10″, height of eye 13 feet, time by chronometer August 31^d 10^h 17^m 20^s, which was 4^m 50^s *slow* for mean noon at Greenwich, May 1st, and on June 30th was *slow* 8^m 20^s on mean time at Greenwich: required the longitude.

ADDITIONAL FOR FIRST MATE.

10. 1882, March 20th, mean time at ship $9^h\ 35^m$ A.M., latitude 43° 18′ N., longitude 32° 25′ W., sun's bearing by compass S. ¼ E., observed altitude sun's L.L. 35° 2′ 50″, height of eye 12 feet: required the true azimuth and error of the compass; and supposing the variation be 27° 10′ W.: required the deviation of the compass for the position of the ship's head at the time the observation was taken.

11. 1882, September 23rd, A.M. at ship, latitude by acct. 28° 5′ S., longitude 170° 57′ E., observed altitude sun's L.L. North of observer was 61° 40′ height of eye 19 feet, time by watch September $22^d\ 23^h\ 12^m\ 4^s$, which had been found to be $36^m\ 29^s$ *slow* on apparent time at ship, the difference of longitude made to the *East* was 27·5 miles, after the error on apparent time was determined: required the latitude by reduction to meridian.

12. 1882, January 17th, A.M. at ship, and uncertain of my position, when a chronometer showed January $16^d\ 22^h\ 10^m$ G.M.T., obs. alt. sun's L.L. 11° 8′ 30″; and again, P.M. at ship same day, when chronometer showed January $17^d\ 2^h\ 29^m\ 24^s$, obs. alt. sun's L.L. 13° 59′ 30″, height of eye 18 feet, the ship having made 24 miles on a true N.W. course in the interval: required the line of bearing when the first altitude was taken, and the position of the ship by Sumner's Method when the second altitude was observed, assuming latitudes 50° 50′ N. and 51° 20′ N.

For the second observation the positions as given by the Examiner would be as follows:—

$$B = \text{Lat. } 50° 50' \text{ N.} \qquad \text{Long. } 5° 59\tfrac{3}{4}' \text{ W.}$$
$$B' = \text{Lat. } 51\ 20\ \text{ N.} \qquad \text{Long. } 7\ 34\ \text{ W.}$$

ADDITIONAL FOR MASTER ORDINARY.

13. 1882, December 21st, the observed meridian altitude of the star *a* Cygni Deneb, bearing North, 56° 18′ 10″, height of eye 15 feet: required the latitude.

In the following table give the correct magnetic bearing of the distant object, and thence the deviation:—

Correct magnetic bearing.

Ship's Head by Standard Compass.	Bearing of Distant Object by Standard Compass.	Deviation Required.	Ship's Head by Standard Compass.	Bearing of Distant Object by Standard Compass.	Deviation Required.
North	N. 77° W.		South	S. 62° W.	
N.E.	N. 88 W.		S.W.	S. 59 W.	
East	S. 85 W.		West	S. 77 W.	
S.E.	S. 78 W.		N.W.	N. 81 W.	

With the deviation as above, give the courses you would steer by the Standard Compass to make the following courses correct magnetic.

Correct magnetic courses:—N.E. by E.; S.W. by S.; E.S.E.

Compass courses:—

Supposing you have steered the following courses by the Standard Compass, find the correct magnetic courses made from the above deviation table.

Compass courses:—W. ½ S.; N.W. by N.; N.E. by N.

Magnetic courses:—

You have taken the following bearings of two distant objects by your Standard Compass as above, with the ship's head at N.W., find the bearings, correct magnetic.

Compass bearings:—North, and E.S.E.

Bearings, magnetic:—

EXAMINATION PAPER—No. XV.

FOR SECOND MATE.

1. Multiply 100001 by 8, and 160800 by 325, by common logarithms.

2. Divide 37·149 by 523·76, and 615 by ·03075, by common logarithms.

3.—

H.	Courses.	K.	$\frac{1}{10}$	Winds.	Lee-way.	Deviation.	Remarks, &c.
					pts.		
1	S. by E. ½ E.	3	3	S.W. ½ W.	2½	5° E.	A point, Cape of Good Hope, latitude 34° 28′ S., longitude 18° 28′ E., bearing by compass N. *b* W. ¾ W. dist. 21 miles. Ship's head S. by E. ½ E. Deviation as per log.
2		3	4				
3		3	6				
4		3	7				
5	N.W. ½ W.	2	4	S.W. by W.	3½	18° W.	
6		2	3				
7		2	3				
8		2					
9	N. by E. ½ E.	4	6	N.W. ½ W.	2¼	6° E.	
10		4	4				
11		4	7				
12		4	3				
1	S.W. *b* W. ½ W.	5	6	N.W. ½ W.	1¾	7° W.	Variation 25° W.
2		5	7				
3		5	4				
4		5	3				
5	W. by N. ½ N.	7	5	N. ½ W.	½	16° W.	
6		7	6				
7		7	2				A current set the ship (correct magnetic) E. by S. ½ S., 14 miles, from the time the departure was taken to the end of the day.
8		6	7				
9	N.E. ¼ E.	5	3	E.S.E.	2	16° E.	
10		4	4				
11		5	5				
12		5	4				

4. 1882, February 11th, in longitude 32° 20′ E., the observed meridian altitude of sun's L.L. was 30° 25′ 10″, observer North of sun, index correction — 3′ 15″, height of eye 12 feet: required the latitude.

5. In latitude 51° 10′, the departure made good was 64·3 miles: required the difference of longitude.

6. Required the course and distance from A to B, by Mercator's Sailing.

Lat. of A	43° 24′ S.	Long. of A	65° 39′ W.
Lat. of B	26 38 N.	Long. of B	15 8 E.

ADDITIONAL FOR ONLY MATE.

7. 1880, April 2nd: find times of high water at Cape Virgin, longitude 68° W., Waterford Harbour, and Banff.

8. 1882, March 31st, at 6^h 1^m 48^s A.M. apparent time at ship, in latitude 6° 31′ N., longitude 155° 10′ E., the sun's magnetic amplitude was E. 3° 51′ S.: required the true amplitude and error of the compass; and supposing the variation to be 6° E.: required the deviation of the compass for the position of the ship's head when the observation was taken.

9. 1881, May 27th, A.M. at ship, latitude 55° N., observed altitude sun's L.L. 43° 9′ 5″, index error — 14″, height of eye 14 feet, time by chronometer 9^h 13^m 12^s A.M., which was $48^s{\cdot}5$ *slow* for mean noon at Greenwich, April 9th, and on April 24th, was *fast* 25^s: required the longitude.

ADDITIONAL FOR FIRST MATE.

10. 1882, July 10th, $9^h\ 44^m$ A.M., mean time at ship, latitude 59° 56′ N., longitude 40° 20′ W., sun's magnetic azimuth S. ½ W., observed altitude sun's L.L. 44° 49′, height of eye 20 feet; required the error of the compass, and supposing the variation be 51° W.: required the deviation of the compass for the position of the ship's head at the time the observation was taken.

11. 1882, November 8th, P.M. at ship, latitude by account 33° 9′ N., longitude 89° 42′ E., observed altitude sun's L.L. South of observer was 40° 0′, index error − 6′ 12″, height of eye 19 feet, time by watch $8^h\ 20^m\ 20^s$, (or $7^d\ 20^h\ 20^m\ 20^s$), which had been found to be $4^h\ 8^m\ 12^s$ *slow* on app. time at ship, the difference of longitude made to the *East* was 32′·3, after the error on apparent time was determined: required the latitude by reduction to meridian.

12. 1882, April 25th, P.M. at ship, and uncertain of my position, when a chronometer showed April $25^d\ 5^h\ 13^m\ 20^s$ M.T.G., obs. alt. sun's L.L. 51° 3′ 42″; again, P.M. at ship same day, when chronometer showed April $25^d\ 9^h\ 12^m\ 4^s$, obs. alt. sun's L.L. 16° 20′ 3″, height of eye 18 feet, the ship having made 21 miles on a true N.E. ¼ N. course in the interval: required the line of position when the first altitude was observed, and the ship's position by Sumner's Method when the second altitude was taken, assuming the ship to be between the latitudes 48° 10′ N. and 48° 40′ N.

For the second observation the position as given by the Examiner would be as follows:—

$$B = \text{Lat. } 48°\ 30'\ N. \qquad \text{Long. } 58°\ 26'\ W.$$
$$B' = \text{Lat. } 49°\ 0'\ N. \qquad \text{Long. } 58°\ 24\tfrac{3}{4}'\ W.$$

ADDITIONAL FOR MASTER ORDINARY.

13. 1882, July 19th, the observed meridian altitude of the star *a* Pavonis, bearing South, was 32° 50′ 15″, index correction + 4′ 48″, height of eye 23 feet: required the latitude.

In the following table give the correct magnetic bearing of the distant object, and thence the deviation.

Correct magnetic bearing.

Ship's Head by Standard Compass.	Bearing of Distant Object by Standard Compass.	Deviation Required.	Ship's Head by Standard Compass.	Bearing of Distant Object by Standard Compass.	Deviation Required.
North	S. 44° E.		South	S. 11° W.	
N.E.	S. 56 E.		S.W.	S. 13 W.	
East	S. 39 E.		West	S. 4 W.	
S.E.	S. 12 E.		N.W.	S. 12 E.	

With the deviation as above, give the courses you would steer by the Standard Compass to make the following courses correct magnetic.

Correct magnetic courses:—N.N.W.; W.N.W.; S.W. by W.; W.S.W.

Compass courses:—

Supposing you have steered the following courses by the Standard Compass, find the correct magnetic courses made from the above deviation table.

Compass courses:—E.N.E.; S.S.E.; N.W. by W.; N.E. by E.

Magnetic courses:—

You have taken the following bearings of two distant objects by your Standard Compass as above, with the ship's head at N.E. ½ E., find the bearings, correct magnetic.

Compass bearings:—N. by W. and E. by N.

Bearings, magnetic:—

EXAMINATION PAPER—No. XVI.

FOR SECOND MATE.

1. Multiply 108·581 by 500, and 964·3204 by ·000690041, by common logarithms.

2. Divide 408848 by 202, and ·000694321 by ·000014798, by common logarithms.

3.—

H.	Courses.	K.	$\frac{1}{10}$	Winds.	Lee-way.	Deviation.	Remarks, &c.
					pts.		
1	W. by N. ¼ N.	8	9	N. by W.	2½	21° 10′ W.	A point of land in lat. 59° 16′ S., long. 179° 42′ W., bearing by compass S.E. ½ E., dist. 13 miles. Ship's head W. by N. ¼ N. Deviation as per log.
2		8	3				
3		8	4				
4		8	4				
5	W. ¾ N.	9	4	N. by W.	2¼	19° 50′ W.	
6		9	8				
7		8	6				
8		8	4				
9	W. ¼ N.	9	8	N. by W.	2¾	18° 15′ W.	
10		9	7				
11		9	6				
12		9	5				Variation 15° 40′ E.
1	S.S.W. ½ W.	8	8	W. ½ N.	3¼	4° 30′ W.	
2		8	2				
3		8	5				
4		8	6				
5	S.S.E. ¼ E.	3	6	S.W.	¾	27° 40′ E.	
6		3	2				
7		4	8				A current set the ship S. ½ E., correct magnetic 21 miles, from the time the departure was taken to the end of the day.
8		5	6				
9	W. ¾ S.	8	6	South.	2¼	19° 30′ W.	
10		8	2				
11		8	3				
12		8	4				

4. 1881, May 16th, longitude 45° 26′ W., the observed meridian altitude of sun's L.L. bearing North, was 86° 34′ 20″, index correction + 4′ 16″, height of eye 15 feet: required the latitude.

5. In latitude 44° 20′ S., the departure made good was 44·2 miles: required the difference of longitude by parallel sailing.

6. Required the course and distance from A to B, by calculation on Mercator's principle.

Latitude A	9° 2′ N.	Longitude	171° 19′ W.
Latitude B	1 2 S.	Longitude	83 17 E.

ADDITIONAL FOR ONLY MATE.

7. 1880, April 20th: find the times of high water, A.M. and P.M., at Aberyrach, Morlaix, Gibraltar, and St. Nazaire.

8. 1882, June 24th, at 11ʰ 5ᵐ P.M., apparent time at ship, in latitude 66° 31′ N., longitude 9° W., the sun's magnetic amplitude was N. ¾ E.: required the true amplitude and error of the compass; and supposing the variation to be 33° 30′ W.: required the deviation of the compass for the position of the ship's head at the time of observation.

9. 1882, June 15th, A.M. at ship, latitude 29° 10′ S., observed altitude sun's L.L. 20° 40′, index correction 0, height of eye 2? feet, time by chronometer June 14ᵈ 22ʰ 59ᵐ 20ˢ, which was 4ᵐ 35ˢ *fast* for Greenwich mean noon, March 20th, and on May 3rd was 1ᵐ 17ˢ *slow* for Greenwich mean time: required the longitude.

ADDITIONAL FOR FIRST MATE.

10. 1882, September 22nd, mean time at ship 0ʰ 42ᵐ P.M., latitude 49° 40′ S., longitude 146° 56′ W., sun's bearing by compass N. by E. ¼ E., observed altitude sun's L.L. 40° 18′, height of eye 14 feet: required the error of compass; and supposing the variation to be 12° 20′ E.: find the deviation of the compass for the position of the ship's head at the time of observation.

11. 1882, September 23rd, P.M. at ship, latitude account 50° 47′ N., longitude 169° 54′ E., observed altitude of sun's L.L., South of observer, was 88° 47′, height of eye 12 feet; time by watch September 23ᵈ 0ʰ 58ᵐ 14ˢ, which had been found to be 40ᵐ 43ˢ *fast* on apparent time at ship, the difference of longitude made to the *East* was 27′·5, after the error on apparent time was determined: required the latitude by reduction to the meridian.

12. 1882, August 28th, A.M. at ship, and uncertain of my position, when a chronometer showed August 27ᵈ 21ʰ 56ᵐ 4ˢ M.T.G., obs. alt. sun's L.L. 39° 21′ 10″; again, P.M. at ship same day, when chronometer showed August 28ᵈ 3ʰ 20ᵐ 48ˢ, obs. alt. sun's L.L. 35° 42′ 40″, height of eye 16 feet, the ship having made 36 miles on a true N. ¾ W. course in the interval: required the line of bearing when the first altitude was taken, and the ship's position by Sumner's Method when the second altitude was observed, assuming latitudes 48° 50′ N. and 49° 30′ N.

ADDITIONAL FOR MASTER ORDINARY.

13. 1882, March 31st, the observed meridian altitude of the star *a* Virgiuas *(Spica)*, bearing North, was 52° 14′, height of 19 feet: required the latitude.

In the following table give the correct magnetic bearing of the distant object, and thence the deviation:—

Correct magnetic bearing.

Ship's Head by Standard Compass.	Bearing of Distant Object by Standard Compass.	Deviation Required.	Ship's Head by Standard Compass.	Bearing of Distant Object by Standard Compass.	Deviation Required.
North	S. 35° W.		South	S. 10° E.	
N.E.	S. 13 W.		S.W.	S. 11 E.	
East	S. 6 W.		West	S. 14 W.	
S.E.	S. 3 W.		N.W.	S. 37 W.	

With the deviation as above, give the courses you would steer by the Standard Compass to make the following courses correct magnetic.

Correct magnetic courses:—East; W.S.W.; S.E. by S.

Compass courses:—

Supposing you have steered the following courses by the Standard Compass, find the correct magnetic courses made, from the above deviation table.

Compass courses:—N.W. by W. ½ W.; North; E.N.E.

Magnetic courses:—

You have taken the following bearings of two distant objects by your Standard Compass as above; with the ship's head at W.S.W., find the bearings, correct magnetic.

Compass bearings:—W. by S. and South.

Bearings, magnetic:—

EXAMINATION PAPER—No. XVII.

FOR SECOND MATE.

1. Multiply 37340 by 1200, and 53·62 by 0·4188, by common logarithms.
2. Divide 9145752 by 22·22, and 5·6949 by 53·058, by common logarithms.
3.—

H.	Courses.	K.	$\tfrac{1}{10}$	Winds.	Lee-way.	Deviation.	Remarks, &c.
1	N.E.	9		N.N.W.	pts. ¾	6½° E.	A point, lat. 37° 37′ N. long. 0° 41′ W., bearing by compass N.W. by W. ¼ W., dist. 25 miles. Ship's head N.E. Deviation as per log.
2		8	6				
3		9	2				
4		8	6				
5	E.N.E.	12	3	North.	¼	11° E.	
6		12	3				
7		11	4				
8		12					
9	N.N.W.	10	5	N.E.	½	6° W.	
10		11	1				
11		10	6				
12		10	8				
1	E.S.E.	6	6	N.E.	1½	9½° E.	Variation 19° W.
2		6	4				
3		6	5				
4		6	5				
5	N.N.E.	4	3	East.	2¼	4° E.	
6		4	8				
7		4	5				A current set the ship E. by S. (correct magnetic) 36 miles, from the time the departure was taken to the end of the day.
8		4	4				
9	S.E.	8	5	E.N.E.	1½	4° E.	
10		8	7				
11		7	4				
12		7	4				

4. 1882, November 21st, in longitude 70° 20′ E., observed meridian altitude of sun's L.L. bearing North, was 80° 20′, index error − 2′ 50″, height of eye 10 feet: required the latitude.

5. In latitude 35° 39′, the departure made good was 66 miles: required the difference of longitude by parallel sailing.

6. Required the course and distance from A to B, by Mercator's Sailing.

    Latitude A   6° 1′ N.   Longitude A   60° 14′ E.
    Latitude B   6 10 S.   Longitude B   39 15 E.

ADDITIONAL FOR ONLY MATE.

7. 1880, September 1st: find A.M. and P.M. tides at Lynn Deep, Ramsgate, and Antwerp, and also at Victoria River, longitude 130° E.

8. 1882, July 20th, at 7ʰ 0ᵐ P.M., apparent time at ship, latitude 43° 4′ S., longitude 179° 12′ W., sun's magnetic amplitude S.W. by W. ¼ W.: required error of compass; and supposing the variation to be 9° 40′ E.: required the deviation of the compass for the position of the ship's head at the time of observation.

9. 1882, June 5th, A.M. at ship, latitude 2° 5′ S., observed altitude sun's L.L. 28° 4′, index correction + 4′ 25″, height of eye 15 feet, time by chronometer June 4ᵈ 12ʰ 18ᵐ 42ˢ, which was 1ᵐ 4ˢ *fast* for mean noon at Greenwich, March 6th, and on March 24th was 0ᵐ 8ˢ *slow* on mean noon at Greenwich: required the longitude.

ADDITIONAL FOR FIRST MATE.

10. 1882, November 10th, mean time at ship $8^h 45^m 38^s$ A.M., latitude 50° 30′ N., longitude 86° 43′ E., sun's bearing by compass S. 49° 50′ E., observed altitude sun's L.L. 6° 7′ 10″, height of eye 15 feet: required the true azimuth and error of compass; and supposing the variation is 7° 20′ E.: required the deviation for the position of the ship's head at the time of observation.

11. 1882, January 8th, A.M. at ship, latitude acct. 35° 10′ S., longitude 55° 12′ W., observed altitude sun's L.L., North of observer, was 76° 44′, index correction $+$ 1′ 18″, height of eye 14 feet, time by watch $39^m 34^s$ (or $8^d 0^h 39^m 34^s$), which had been found to be $50^m 3^s$ *fast* on apparent time at ship, the difference of longitude made to the *East* was 21′, after the error on apparent time was determined: required the latitude.

12. 1882, June 14th, P.M. at ship, and uncertain of my position, when a chronometer showed June $14^d 7^h 18^m 50^s$ M.T.G., obs. alt. sun's L.L. 54° 12′ 50″; again, P.M. at ship same day, when chronometer showed June $14^d 12^h 1^m 13^s$, obs. alt. sun's L.L. 11° 15′ 10″, height of eye 21 feet, the ship having made 25 miles on a true S.W.½W. course in the interval: required the line of position when the first altitude was observed, and the ship's position by Sumner's Method when the second altitude was taken, assuming the ship to be between the latitudes 50° 10′ N. and 49° 40′ N.

ADDITIONAL FOR MASTER ORDINARY.

13. 1882, February 1st, longitude 50° W., observed meridian altitude of the star *a* Canis Majoris *(Sirius)* was 37° 50′ 20″, bearing South, index correction $+$ 1′ 4″, height of eye 19 feet: required the latitude.

In the following table give the correct magnetic bearing of the distant object, and thence the deviation:—

Correct magnetic bearing.

Ship's Head by Standard Compass.	Bearing of Distant Object by Standard Compass.	Deviation Required.	Ship's Head by Standard Compass.	Bearing of Distant Object by Standard Compass.	Deviation Required.
North	North.		South	N. 14° E.	
N.E.	N. 12° E.		S.W.	N. 5 E.	
East	N. 29 E.		West	N. 5 W.	
S.E.	N. 36 E.		N.W.	N. 5 W.	

With the deviation as above, give the courses you would steer by the Standard Compass to make the following courses correct magnetic.

Correct magnetic courses:—W.N.W.; W.S.W.; S.E. by E.; S.S.E.

Compass courses:—

Supposing you have steered the following courses by the Standard Compass, find the correct magnetic courses made from the above deviation table.

Compass courses:—E. by S. ½ S.; S. by E. ½ E.; N.W. by W.; W. ½ S.

Correct magnetic courses:—

You have taken the following bearings of two distant objects by your Standard Compass as above, with the ship's head at E. by S. ½ S., find the bearings, correct magnetic.

Compass bearings:—N.W. by W. ½ W. and S. by E. ½ E.

Bearings, magnetic:—

EXAMINATION PAPER—No. XVIII.
FOR SECOND MATE.

1. Multiply 5940 by 530, and ·00087214 by ·001963, by common logarithms.
2. Divide 9504000 by 98, and ·9649 by 35·0583, by common logarithms.
3.—

H.	Courses.	K.	$\frac{1}{10}$	Winds.	Leeway.	Deviation.	Remarks, &c.
					pts.		
1	W.N.W.	12	6	North.	$\frac{1}{4}$	10° W.	A point, lat. 36°27′S., long. 68° 37′ W., bearing by compass E. $\frac{3}{4}$ S., distance 25 miles. Ship's head W.N.W. Deviation as per log.
2		12	6				
3		12	8				
4		13					
5	S.W. by W.	10	6	N.W. by W.	$\frac{3}{4}$	7° W.	
6		10	4				
7		10	4				
8		10	6				
9	N. by E. ½ E.	7	3	N.W. ½ W.	1¼	2½° W.	
10		7	6				
11		7	8				
12		7	3				
1	N.W.	11	4	N.N.E.	½	8° W.	Variation 22½° E.
2		11	4				
3		11	8				
4		11	4				
5	S.W. ¼ W.	3	3	W.N.W.	3¼	7° W.	
6		2	8				
7		2	6				A current set the ship (correct magnetic) S.S.W. ¼ W., 32 miles, from the time the departure was taken to the end of the day.
8		2	3				
9	N.E. ½ E.	4	7	N. by W. ½ W.	2¾	8° E.	
10		4	4				
11		3	6				
12		3	3				

4. 1883, January 1st, in longitude 167° 54′ E., observed meridian altitude of sun's L.L., bearing North, was 83° 40′, index correction + 47″, height of eye 23 feet: required the latitude.

5. In latitude 60° 5′ S., longitude 179° 17′ W., a ship sails due West 96 miles: find the longitude in.

6. Required the course and distance from A to B, by Mercator's Sailing.

    Latitude A   8° 57′ N.   Longitude A   79° 31′ W.
    Latitude B   36 50 S.   Longitude B   174 49 E.

ADDITIONAL FOR ONLY MATE.

7. 1880, March 28th: find the times of high water, A.M. and P.M., at Gibraltar, Port Louis (Mauritius), long. 57½° E., and Halifax, long. 64° W.

8. 1881, November 4th, at 4^h 52^m 42^s A.M., apparent time at ship, latitude 46° 40′ S., longitude 8° 57′ W., sun's magnetic amplitude S.E. ½ S.: required the true amplitude and error of the compass; and supposing the variation to be 16° 30′ W.: required the deviation of the compass for the position of the ship's head at the time of observation.

9. 1882, September 1st, A.M. at ship, latitude 15° 31′ S., observed altitude sun's L.L. 15° 18′ 20″, index correction — 20″, height of eye 26 feet, time by chronometer, August 31^d 20^h 12^m 40^s, *slow* 1^m 30^s on April 15th, and on April 29th was 0^m 29^s *fast* for Greenwich mean time: required the longitude.

A A A

ADDITIONAL FOR FIRST MATE.

10. 1882, June 1st, mean time at ship 8^h 19^m A.M., latitude 21° 10′ N., longitude 61° 30′ E., observed altitude sun's L.L. 39° 10′, index correction — 15″, height of eye 18 feet: sun's magnetic azimuth E. ¾ N.: required the error of compass; and supposing the variation to be 0° 50′ E.: required the deviation of the compass for the position of the ship's head at the time of observation.

11. 1882, April 13th, A.M. at ship, latitude account 0°, longitude 147° 10′ E., observed altitude of sun's L.L., North of observer, was 80° 30′, index correction + 1′ 10″, height of eye 16 feet, time by watch 0^h 0^m 12^s, which had been found to be 11^m 1^s fast on apparent time at ship, the difference of longitude made to the East was 8½ miles, after the error on apparent time was determined: required the latitude.

12. 1882, April 1st, A.M. at ship, and uncertain of my position, when a chronometer showed March 31^d 22^h 22^m 13^s M.G., obs. alt. sun's L.L. 34° 52′ 40″; again, P.M. at ship same day, when chronometer showed April 1^d 3^h 39^m 22^s, obs. alt. sun's L.L. 31° 20′ 30″, height of eye 17 feet, the ship having made 19½ miles on a true S.E. ¾ S. course in the interval: required the line of bearing when the first altitude was taken, and the position of the ship by Sumner's Method when the second altitude was observed, assuming latitudes 51° 0′ N. and 50° 30′ N.

ADDITIONAL FOR MASTER ORDINARY.

13. 1882, May 10th, the observed meridian altitude of a^2 Centauri was 10° 4′ 15″, (zenith North), index correction — 2′ 10″, height of eye 20 feet: required the latitude.

14. At what time will the star a Aquilae *(Altair)* pass the meridian of the Land's End, on December 8th, 1882, and how far North or South of the zenith.

15. 1882, January 8th, at 2^h 18^m, what stars will be near the meridian of a place in long. 45° 20′ E.

In the following table give the correct magnetic bearing of the distant object and thence the deviation.

Correct magnetic bearing.

Ship's Head by Standard Compass.	Bearing of Distant Object by Standard Compass.	Deviation Required.	Ship's Head by Standard Compass.	Bearing of Distant Object by Standard Compass.	Deviation Required.
North	S. 27° E.		South	S. 12° W.	
N.E.	South.		S.W.	S. 21 E.	
East	S. 21 W.		West	S. 37 E.	
S.E.	S. 27 W.		N.W.	S. 39 E.	

With the deviation as above, give the courses you would steer by the Standard Compass to make the following courses correct magnetic.

Correct magnetic courses:—N. by E. ¾ E.; E. ½ N.; S.S.W. ¼ W.; N.W. ¼ N.

Compass courses:—

Supposing you have steered the following courses by the Standard Compass, find the correct magnetic courses made from the above deviation table.

Compass courses:—N.W. ¼ N.; S.S.E. ½ E.; E. by N. ½ N.; N.N.E. ¾ E.

Correct magnetic courses:—

You have taken the following bearings of two distant objects by your Standard Compass as above; with the ship's head at S.E. ½ S., find the bearings, correct magnetic.

Compass bearings:—N.E. ½ E., and N. by W. ½ W.

Bearings, magnetic:—

EXAMINATION PAPER—No. XIX.

FOR SECOND MATE.

1. Multiply 2410050 by 5, and 47 by 1·405, by common logarithms.
2. Divide ·00001400018 by ·00000700009, and 2004·64 by 34, by common logarithms.
3.—

H.	Courses.	K.	$\frac{1}{10}$	Winds.	Lee-way.	Deviation.	Remarks, &c.
					pts.		
1	S. by E. ½ E.	12		E. ¾ S.	¼	19½° E.	A point, lat. 46° 20′ S. long. 176° 44′ W., bearing by compass
2		11	6				
3		12	2				E. by N. ¼ N.,
4		12	2				dist. 13 miles. Ship's
5	S. by W.	9	2	S.E. by E.	¾	25° E.	head S. by E. ½ E.
6		9					Deviation as per log,
7		9	6				19½° E.
8		9	4				
9	S.S.W. ¼ W.	7	5	West.	1¼	28° E.	
10		7	5				
11		7	4				
12		7	6				
1	W. by S.	11	3	N.W. by N.	½	21° E.	Variation 15° E.
2		10	8				
3		10	8				
4		10	6				
5	West.	9	8	N.N.W.	1	20° E.	
6		9	6				
7		9	4				A current set the ship
8		9	6				(correct magnetic)
9	N.N.W.	9	3	West.	¾	15½° W.	N.W. ½ W., 36 miles,
10		9	4				from the time the departure was taken to
11		9	6				the end of the day.
12		9	5				

4. 1882, September 23rd, in longitude 119° 54′ E., observed meridian altitude sun's L.L. was 83° 46′, zenith South of sun, index correction − 5′ 30″, height of eye 18 feet: required the latitude.

5. In latitude 63° 54′ N., the departure made good was 63·5 miles: required the difference of longitude by parallel sailing.

6. Required the course and distance from A to B, by Mercator's Sailing.
    Latitude A   9° 59′ S.   Longitude A   140° 5′ W.
    Latitude B 10 0 N.   Longitude B   163 5 E.

ADDITIONAL FOR ONLY MATE.

7. 1880, November 30th: required the times of high water, A.M. and P.M., at Margate, Gibraltar, Bordeaux, Ilfracombe, Cromarty, and Batavia, long. 106° 48′ E.

8. 1882, June 12th, at 11^h 15^m P.M., apparent time at ship, in latitude 66° 24′ N., longitude 4° 54′ E., the sun's magnetic amplitude was N.N.E. ½ E.: required the error of compass, and supposing the variation to be 23° 10′ W.: required the deviation for the position of the ship's head at the time of observation.

9. 1882, June 14th, P.M. at ship, latitude 31° 10′ N., observed altitude sun's L.L. 48° 59′ 10″, index corr. + 1′, height of eye 18 feet, time by chronometer June 14^d 12^h 15^m 40^s, which was *fast* 4^m 30^s, January 2nd, for mean noon at Greenwich, and on March 13th was 9^m 16^s *fast* for mean time at Greenwich: required the longitude.

ADDITIONAL FOR FIRST MATE.

10. 1882, January 1st, mean time at ship 4^h 10^m P.M., latitude 31° 50′ S., longitude 176° 25′ E., observed altitude sun's L.L. 34° 50′, height of eye 16 feet, sun's azimuth W. ¼ N.: required the error of compass; and supposing the variation to be 20° E.: required the deviation for the position of the ship's head at the time of observation.

11. 1882, December 31st, A.M. at ship, latitude account 52° N., longitude 78° E., observed altitude sun's L.L. 14° 46′ S., height of eye 19 feet, time by watch 0^h 56^m, *fast* on app. time at ship 1^h 5^m 20^s, the difference of longitude made to *West* was 21·4 miles after the error on apparent time was determined: required the latitude by reduction to meridian.

12. 1882, January 24th, A.M. at ship, and uncertain of my position, when a chronometer showed January 23^d 21^h 19^m 50^s M.T.G., obs. alt. sun's L.L. 9° 34′ 50″; again, P.M. at ship same day, when chronometer showed January 24^d 1^h 19^m 12^s, obs. alt. sun's L.L. 18° 28′ 10″, height of eye 17 feet, the ship having made 23 miles on a true S.E. by E. course in the interval: required the line of bearing when the first altitude was taken, and the position of the ship by Sumner's Method when the second altitude was observed, assuming latitudes 51° 15′ N. and 50° 45′ N.

For the second observation the positions as given by the Examiner would be as follows:—

B = Lat. 51° 15′ N.   Long. 0° 18′ W.
B' = Lat. 50° 45′ N.   Long. 3° 37½′ W.

ADDITIONAL FOR MASTER ORDINARY.

13. 1882, April 11th, the observed meridian altitude of star Sirius was 61° 3′ 50″, bearing North, height of eye 16 feet: required the latitude.

In the following table give the correct magnetic bearing of the distant object, and thence the deviation.

Correct magnetic bearing.

Ship's Head by Standard Compass.	Bearing of Distant Object by Standard Compass.	Deviation Required.	Ship's Head by Standard Compass.	Bearing of Distant Object by Standard Compass.	Deviation Required.
North	N. 75° W.		South	N. 75° W.	
N.E.	West.		S.W.	N. 68 W.	
East	N. 88 W.		West	N. 57 W.	
S.E.	N. 80 W.		N.W.	N. 53 W.	

With the deviation as above, give the courses you would steer by the Standard Compass to make the following courses correct magnetic.

Correct magnetic courses:—North; E. by S.; S. 40° W.;

Compass courses:—

Supposing you have steered the following courses by the Standard Compass, find the correct magnetic courses made from the above deviation table.

Compass courses:—W. by S.; E.N.E.; N.W. by N.

Correct magnetic courses:—

You have taken the following bearings of two distant objects by your Standard Compass as above, with ship's head N.E. ¼ E., find the bearings, correct magnetic.

Compass bearings:—South and W. by N.

Bearings, magnetic:—

EXAMINATION PAPER—No. XX.

FOR SECOND MATE.

1. Multiply 30·24 by 12·5, and ·034632 by ·397302, by common logarithms.
2. Divide 8100900 by 900, and ·00005 by 2·5, by 25, and by ·0000025, by common logs.
3.—

H.	Courses.	K.	$\tfrac{1}{10}$	Winds.	Leeway.	Deviation.	Remarks, &c.
					pts.		
1	N.E. ¼ E.	13	2	N. by W. ½ W.	¼	26° E.	A point of land, lat. 50°25′ S., long. 179°40′ E bearing by compass N. by W. ¼ W., dist. 16 miles. Ship's head N.E. ¼ E. Deviation as per log.
2		12	9				
3		13	5				
4		13	4				
5	W.S.W.	3	6	N.W.	2¼	4½° W.	
6		4					
7		4					
8		3	4				
9	E. by N.	12	2	N. by E.	½	28° W.	
10		12	4				
11		12	6				
12		12	8				
1	N. by W. ½ W.	2	4	N.E. ½ E.	3	51° E.	Variation 14° E.
2		2	3				
3		2	3				
4		2					
5	S. ¾ W.	6	9	W. by S.	¼	50° W.	
6		6	8				
7		6	8				A current set (correct magnetic) E.N.E., 42 miles, from the time the departure was taken to the end of the day.
8		7	5				
9	E. by N. ½ N.	11	5	N. by E. ½ E.	¾	28° W.	
10		12	2				
11		11	6				
12		11	7				

4. 1882, September 23rd, in longitude 57° 45′ E., the observed meridian altitude of sun's L.L. was 84° 10′ 50″, bearing North, index error − 1′ 36″, height of eye 16 feet: required the latitude.

5. In latitude 52° S., longitude 0° 40′ W., a ship sails 136 miles due East: required the longitude in.

6. Required the course and distance from A to B, by calculation on Mercator's principle.

    Latitude of A   5° 21′ N.   Longitude of A 163° 1′ E.
    Latitude of B 36 50 S.   Longitude of B 73 6 W.

ADDITIONAL FOR ONLY MATE.

7. 1880, December 12th: find A.M. and P.M. tides at Aberdeen Bar, Penzance, King's Road (Bristol Channel), Southampton, and Ferrol.

8. 1882, November 5th, at 5ʰ 10ᵐ P.M., apparent time at ship, latitude 20° 45′ N., longitude 116° 45′ E., sun's magnetic amplitude S.W. ¾ W.: required the true amplitude and error of compass; and supposing the variation to be 1° E.: required the deviation of the compass for the position of the ship's head at the time of observation.

9. 1882, August 5th, A.M. at ship, latitude at noon 30° 30′ N., observed altitude sun's L.L. 35° 6′, height of eye 16 feet, time by chronometer 8ʰ 39ᵐ 22ˢ P.M., which was *fast* 29ᵐ 32ˢ·4 on Greenwich mean noon, July 8th, and on July 20th, was *fast* 30ᵐ 0ˢ on Greenwich mean noon; course till noon West (true) 48 miles: required the longitude in at noon.

ADDITIONAL FOR FIRST MATE.

10. 1882, August 13th, mean time at ship $9^h\ 5^m\ 20^s$ A.M., latitude 30° 46′ S., longitude 78° 50′ W., sun's magnetic azimuth N. 25° E., observed altitude sun's L.L. 27° 12′, index correction $+ 1'\ 45''$, height of eye 21 feet: required the true azimuth and error of compass; and supposing the variation to be 16° 20′ E.: required the deviation of the compass for the position of the ship's head at the time of observation.

11. 1882, June 12th, P.M. at ship, latitude account 15° 50′ S., longitude 72° 12′ E., observed altitude of sun's L.L. 50° 10′ 10″, zenith South of observer, index correction $- 5'\ 40''$, height of eye 26 feet, time by watch $28^m\ 40^s$ (or $12^d\ 0^h\ 28^m\ 40^s$), which had been found to be $4^m\ 44^s$ *slow* on apparent time at ship, the difference of longitude made to *West* was 16½ miles after the error on apparent time was determined: required the latitude.

12. 1882, November 3rd, A.M. at ship, and uncertain of my position, when a chronometer showed November $2^d\ 22^h\ 38^m\ 45^s$ M.T.G., obs. alt. sun's L.L. 18° 53′ 50″; again, P.M. at ship same day, when chronometer showed November $3^d\ 2^h\ 44^m\ 15^s$, obs. alt. sun's L.L. 16° 32′ 30″, height of eye 19 feet, the ship having made 15 miles on a true E. by S. ½ S. course in the interval: required the line of bearing when the first altitude was taken, and the ship's position by Sumner's Method when the second altitude was observed, assuming the ship to be between the latitudes 51° 45′ N. and 51° 15′ N.

For the second observation the positions as given by the Examiner would be as follows:—

B = Lat. 51° 45′ N.   Long. 10° 41′ W.
B′ = Lat. 51 15 N.   Long. 9 33 W.

ADDITIONAL FOR MASTER ORDINARY.

13. 1882, December 7th, the observed meridian altitude of the star α Arietis was 60° 29′ 50″, zenith North of star, index correction $- 2'\ 10''$, height of eye 18 feet: required the latitude.

In the following table give the correct magnetic bearing of the distant object, and thence the deviation:—

Correct magnetic bearing.

Ship's Head by Standard Compass.	Bearing of Distant Object by Standard Compass.	Deviation Required.	Ship's Head by Standard Compass.	Bearing of Distant Object by Standard Compass.	Deviation Required.
North	S. 51° W.		South	N. 30° W.	
N.E.	W. 6 S.		S.W.	W. 21 N.	
East	N. 57 W.		West	W. 8 S.	
S.E.	W. 64 N.		N.W.	S. 46 W.	

With the deviation as above, give the courses you would steer by the Standard Compass to make the following courses correct magnetic.

Correct magnetic courses:—W. ¾ N.; S.E. ½ E.; N. by E. ½ E.; S. by E. ¼ E.

Compass courses:—

Supposing you have steered the following courses by the Standard Compass, find the correct magnetic courses made from the above deviation table.

Compass courses:—E. ¼ N.; S. by W. ¼ W.; N.W. ¾ W.; S.W. by W. ½ W.

Magnetic courses:—

You have taken the following bearings of two distant objects by your Standard Compass as above, with the ship's head at S.E. ½ E., find the bearings, correct magnetic.

Compass bearings:—S. ½ W.; W. by N. ¼ N.; E. ¼ N.

Bearings, magnetic:—

EXAMINATION PAPER—No. XXI.

FOR SECOND MATE.

1. Multiply 7642 by 7429·5, and 0·00064 by 10·0004, by common logarithms.

2. Divide ·39765 by 25, and 1000000 by ·0000001, by common logarithms.

3.—

H.	Courses.	K.	$\frac{1}{10}$	Winds.	Lee-way. pts.	Deviation.	Remarks, &c.
1	N. by W.	6	4	W. by N.	$1\frac{1}{2}$	8° W.	A point of land in lat. 57° N. long. 40° W. bearing by compass N.E. *b* E. ½ E., distance 19 miles. Ship's head N. by W. Deviation as per log.
2		6	3				
3		5	6				
4		5	4				
5	S.S.W. ¾ W.	4	8	W. ½ N.	$2\frac{1}{4}$	5° W.	
6		4	2				
7		3	7				
8		3	6				
9	N.N.E. ¾ E.	4	2	N.W. ¾ N.	2	13° E.	
10		4	4				
11		4	5				
12		4	5				
1	W. by N.	3	4	N. by W.	$2\frac{1}{2}$	17½° W.	Variation 48° W.
2		3	4				
3		3	6				
4		3	7				
5	S.E. ¾ E.	9	5	S.S.W.	$\frac{1}{4}$	11° E.	
6		10	5				
7		11	2				
8		10	8				A current set the ship (correct magnetic) W.N.W. for the last 5 hours, 3 miles an hour.
9	S. ½ W.	3	2	W.S.W.	$2\frac{3}{4}$	2° W.	
10		3	3				
11		2	8				
12		2	7				

4. 1882, June 25th, in longitude 59° 15′ E., the observed meridian altitude of sun's U.L. was 60° 23′ 15″, bearing North, index correction + 2′ 21″, height of eye 30 feet: required the latitude.

5. A ship sailed due West 120 miles from Cape Roca, in latitude 38° 46′ N., and longitude 9° 30′ W.: required the longitude of the ship.

6. Required the compass course and distance from Cape East, New Zealand, to San Francisco. Variation 14° 20′ E., and deviation 5° 40′ E.

Lat. Cape East	37° 40′ S.	Long. Cape East	178° 36′ E.
Lat. San Francisco	37 48·5 N.	Long. San Francisco	122° 24′ W.

ADDITIONAL FOR ONLY MATE.

7. 1880, August 7th: find the times of high water, A.M. and P.M., at Hong Kong, long. 114° E., New York (Sandy Hook), long. 74° W., and Skull.

8. 1882, June 24th, at 6^h A.M., apparent time at ship, latitude 0° N., longitude 12° 3′ W., sun at setting bore by compass S.E. by E. ½ E., variation by chart was 21° 40′ W.: required the error of compass and the deviation.

9. 1882, September 22nd, A.M. at ship, on the Equator, observed altitude sun's U.L. 17° 20′ 40″, index correction — 1′ 18″, height of eye 20 feet, time by chronometer September $22^d\ 4^h\ 59^m\ 16^s$, which was 15^s *slow* for Greenwich mean noon, April 30th, and on June 1st was $10^s{\cdot}6$ *fast* for mean time at Greenwich: required the longitude.

ADDITIONAL FOR FIRST MATE.

10. 1882, March 21st, mean time at ship $3^h\ 15^m$ P.M., latitude $9°\ 7'$ S., longitude $159°\ 4'$ W., sun's bearing by compass W. ½ S., observed altitude sun's L.L. $42°\ 49'\ 45''$, index correction $- 3'\ 14''$, height of eye 21 feet, variation by chart $7°\ 50'$ E.: required the error of compass and deviation.

11. 1882, October 4th, A.M. at ship, latitude account $30°\ 24'$ S., longitude $140°\ 30'$ E., observed altitude sun's L.L., North of observer, was $63°\ 37'\ 10''$, index correction $- 1'\ 15''$, height of eye 21 feet, time by watch October $3^d\ 22^h\ 37^m\ 15^s$, which had been found to be $1^h\ 10^m\ 20^s$ *slow* on apparent time at ship, the difference of longitude made to the *East* was 23½ miles after the error on apparent time was determined: required the latitude.

12. 1882, January 16th, A.M. at ship, and uncertain of my position, when a chronometer showed January $16^d\ 7^h\ 7^m\ 58^s$ M.T.G., obs. alt. sun's L.L. $13°\ 10'\ 30''$; again, P.M. at ship same day, when chronometer showed January $16^d\ 11^h\ 22^m\ 8^s$, obs. alt. sun's L.L. $15°\ 17'\ 30''$, height of eye 17 feet, the ship having made 24 miles on a true N.E. by N. course in the interval: required the line of bearing when the first altitude was taken, and the position of the ship by Sumner's Method when the second altitude was observed, assuming latitudes $49°\ 0'$ N. and $49°\ 40'$ N.

ADDITIONAL FOR MASTER ORDINARY.

13. 1882, June 10th, longitude $25°$ W., the observed meridian altitude of star α Cassiopeæ, was $85°\ 0'\ 20''$, bearing South, index correction $+ 34''$, height of eye 18 feet: required the latitude.

In the following table give the correct magnetic bearing of the distant object, and thence the deviation:—

Correct magnetic bearing.

Ship's Head by Standard Compass.	Bearing of Distant Object by Standard Compass.	Deviation Required.	Ship's Head by Standard Compass.	Bearing of Distant Object by Standard Compass.	Deviation Required.
North	N. 78° E.		South	S. 89° E.	
N.E.	E. 22 S.		S.W.	N. 66 E.	
East	E. 34 S.		West	E. 23 N.	
S.E.	E. 29 S.		N.W.	N. 79 E.	

With the deviation as above, give the courses you would steer by the Standard Compass to make the following courses correct magnetic.

Correct magnetic courses:—N. ¾ E.; E. ¼ S.; S. ½ E.; W. ¼ S.

Compass courses:—

Supposing you have steered the following courses by the Standard Compass, find the correct magnetic courses made from the above deviation table.

Compass Courses:—N.E. ½ E.; S.S.E. ½ E.; W. ½ N.; N. by W. ¼ W.

Correct magnetic courses:—

You have taken the following bearings of two distant objects by your Standard Compass as above, with the ship's head at S.W. by W. ½ W., find the bearings, correct magnetic.

Compass bearings:—N.E. ½ N. and E. ½ N.

Bearings, magnetic:—

EXAMINATION PAPER—No. XXII.

FOR SECOND MATE.

1. Multiply 10003 by 110, and 12344 by 57, by common logarithms.

2. Divide 3972096 by 144, and 120500 by 120·5, by common logarithms.

3.—

H.	Courses.	K.	$\frac{1}{10}$	Winds.	Lee-way.	Deviation.	Remarks, &c.
					pts.		
1	W.N.W.	13	4	S.E.	0	43¼° E.	A point, lat. 51° 8½′ N. long. 1° 25′ E., bearing by compass S.W. ¾ S., dist. 25 miles. Ship's head W.N.W. Deviation as per log.
2		13					
3		13	6				
4		10					
5	W. ½ S.	9	6	S.S.W.	½	43° E.	
6		9	4				
7		9	5				
8		9	5				
9	West.	8	4	N.N.W.	¾	43½° E.	
10		8	4				
11		8	4				
12		8	6				
1	N. ¾ E.	9	5	N.W. by W.	½	17¾° E.	Variation 25° W.
2		9	4				
3		9	6				
4		9	4				
5	North.	10	5	W.N.W.	¼	19° E.	
6		10	6				
7		11					A current set the ship N.W. by W. ¾ W. (correct magnetic) 32 miles, from the time the departure was taken to the end of the day.
8		10	4				
9	N.W. ¼ W.	7	4	N. by E. ½ E.	1¼	44° E.	
10		7					
11		7	4				
12		7	2				

4. 1882, August 26th, in longitude 92° 3′ E., observed meridian altitude of sun's L.L., bearing North, was 35° 35′ 20″, index error + 2′ 17″, height of eye 12 feet: required the latitude.

5. In latitude 17° 15′ N., the departure made good was 171·5 miles: required the difference of longitude by parallel sailing.

6. Required the course and distance from A to B, by Mercator's Sailing.

Latitude A	55° 40′ N.	Longitude A	2° 25′ W.
Latitude B	50 25 N.	Longitude B	3 40 E.

ADDITIONAL FOR ONLY MATE.

7. 1880, January 8th: find A.M. and P.M. tides at Flambro' Head, Peterhead, Crinan, St. Ives, Cape Wrath, and Whitby.

8. 1882, September 23rd, at 6^h 0^m P.M., apparent time at ship, latitude 51° 3′ S., longitude 152° 17′ E., sun's magnetic amplitude W. 15° S.: required error of compass; and supposing the variation to be 20° 16′ E.: required the deviation of the compass for the position of the ship's head at the time of observation.

9. 1882, August 28th, P.M. at ship, latitude 5° S., observed altitude sun's L.L. 38°, index correction + 5′ 20″, height of eye 21 feet, time by chronometer August 27^d 22^h 20^m 30^s, which was 10^m 0^s *slow* for mean noon at Greenwich, February 19th, and on May 30th was 2^m 20^s *slow* on mean noon at Greenwich: required the longitude.

ADDITIONAL FOR FIRST MATE.

10. 1882, June 15th, mean time at ship 8ʰ 40ᵐ P.M., latitude 55° N., longitude 92° 17′ E., sun's bearing by compass N. ⅛ E., observed altitude sun's L.L. 9° 15′ 40″, height of eye 17 feet: required the true azimuth and error of compass; and supposing the variation is 50° 40′ W.: required the deviation for the position of the ship's head at the time of observation.

11. 1882, December 23rd, P.M. at ship, latitude acct. 42° 16′ N., longitude 4° 39′ W., observed altitude sun's U.L., South of observer, was 24° 14′ 10″, height of eye 11 feet, time by watch December 23ᵈ 0ʰ 50ᵐ 58ˢ, which had been found to be 19ᵐ 38ˢ *fast* on apparent time at ship, the difference of longitude made to the *West* was 21′·3, after the error on apparent time was determined: required the latitude.

12. 1882, August 9th, A.M. at ship, and uncertain of my position, when a chronometer showed August 8ᵈ 16ʰ 12ᵐ 1ˢ M.T.G., obs. alt. sun's L.L. 50° 30′ 10″; again, P.M. at ship same day, when chronometer showed August 8ᵈ 19ʰ 51ᵐ 47ˢ, obs. alt. sun's L.L. 52° 50′, height of eye 21 feet, the ship having made 32 miles on a true S. by W. ¾ W. course in the interval: required the line of position when the first altitude was observed, and the ship's position by Sumner's Method when the second altitude was taken, assuming the ship to be between the latitudes 47° 0′ N. and 46° 20′ N.

ADDITIONAL FOR MASTER ORDINARY.

13. 1882, September 11th, observed meridian altitude of the star a Cassiopeæ, bearing North, was 62° 24′ 50″, index correction − 7′ 30″, height of eye 19 feet: required the latitude.

In the following table give the correct magnetic bearing of the distant object, and thence the deviation:—

Correct magnetic bearing.

Ship's Head by Standard Compass.	Bearing of Distant Object by Standard Compass.	Deviation Required.	Ship's Head by Standard Compass.	Bearing of Distant Object by Standard Compass.	Deviation Required.
North	N. 44° W.		South	S. 17° W.	
N.E.	S. 45 W.		S.W.	South.	
East	S. 45 W.		West	S. 14 W.	
S.E.	S. 40 W.		N.W.	S. 35 W.	

With the deviation as above, give the courses you would steer by the Standard Compass to make the following courses correct magnetic.

Correct magnetic courses:—S.W.; N. 15° E.; E.S.E.

Compass courses:—

Supposing you have steered the following courses by the Standard Compass, find the correct magnetic courses made from the above deviation table.

Compass courses:—S.W.; West; E. by N.

Correct magnetic courses:—

You have taken the following bearings of two distant objects by your Standard Compass as above, with the ship's head at North, find the bearings, correct magnetic.

Compass bearings:—North and E. ½ N.

Bearings, magnetic:—

EXAMINATION PAPER—No. XXIII.
FOR SECOND MATE.

1. Multiply 496 by 735, and 80400 by 325, by common logarithms.
2. Divide 460860 by 153620, and 100000 by ·125, and by ·00125, by common logarithms.
3.—

H.	Courses.	K.	⅒	Winds.	Lee-way.	Deviation.	Remarks, &c.
					pts.		
1	W. ½ N.	8	2	N. by W.	¾	24° W.	A point, in latitude 62° 10′ N., longitude 178° 52′ W., bearing by compass E.N.E., dist. 12 miles. Ship's head W. ½ N. Deviation as per log.
2		7	8				
3		8	8				
4		9	2				
5	W. by S.	7	8	S. by W.	1¾	21° W.	
6		7	6				
7		7	6				
8		7					
9	S.W. by W.	7	6	S. by E.	1	16° W.	
10		7	8				
11		8	8				
12		7	8				
1	S.S.W.	8	8	S.E.	½	7° W.	Variation 20° W.
2		8	8				
3		9					
4		7	4				
5	S. by E.	7		E. by S.	2¼	3° E.	
6		8	4				
7		8	6				A current set the ship (correct magnetic) W. by S., 30 miles, from the time the departure was taken to the end of the day.
8		8					
9	S.S.E.	8	4	East.	1½	7° E.	
10		8	6				
11		9					
12		8					

4. 1882, September 1st, in longitude 97° 42′ E., the observed meridian altitude of sun's L.L. was 51° 4′ 50″, observer South of sun, index correction − 6′, height of eye 23 feet: required the latitude.

5. In latitude 2° 10′ N., the departure made good was 210 miles: required the difference of longitude by Parallel Sailing.

6. Required the course and distance from A to B, by Mercator's Sailing.

    Lat. of A 58° 30′ S.   Long. of A 147° 40′ W.
    Lat. of B 40 58 N.   Long. of B 176 59 E.

ADDITIONAL FOR ONLY MATE.

7. 1880, July 25th: find the times of high water A.M. and P.M. at Cardigan, Coleraine, Margate, Peterhead, St. Ives, New Ross, Beaumaris, and Nagasaki, longitude 129° 52′ E.

8. 1882, December 24th, at $10^h 51^m$ A.M. apparent time at ship, in latitude 64° 56′ N., longitude 35° 48′ W., the sun's magnetic amplitude was S. ½ W.: required the true amplitude and error of the compass; and supposing the variation to be 1° 20′ E.: required the deviation of the compass for the position of the ship's head when the observation was taken.

9. 1882, April 15th, A.M. at ship, latitude 48° 52′ N., observed altitude sun's L.L. 22° 18′, index error − 4′, height of eye 17 feet, time by chronometer April $14^d 22^h 30^m 42^s$, which was $0^m 4^s$ *slow* for mean noon at Greenwich, January 1st, and on January 12th, was *fast* $0^m 2^s$: required the longitude by chronometer.

ADDITIONAL FOR FIRST MATE.

10. 1882, September 23rd, 3ʰ 10ᵐ P.M., mean time at ship, latitude 40° N., longitude 130° 32' E., sun's magnetic azimuth W. ½ S., observed altitude sun's L.L. 32° 8' 20", height of eye 20 feet: required the true azimuth and error of the compass; and supposing the variation to be 20° W.: required the deviation of the compass for the position of the ship's head at the time the observation was taken.

11. 1882, September 22nd, P.M. at ship, latitude by account 40° 35' N., longitude 121° 34' W., observed altitude sun's L.L. South of observer was 49° 20' 15", height of eye 15 feet, time by watch September 21ᵈ 19ʰ 15ᵐ 10ˢ, which had been found to be 4ʰ 55ᵐ 24ˢ *slow* on apparent time at ship, the difference of longitude made to the *East* was 26' after the error on apparent time was determined: required the latitude by reduction to meridian.

12. 1882, November 26th, A.M. at ship, and uncertain of my position, when a chronometer showed November 26ᵈ 2ʰ 39ᵐ 15ˢ M.T.G., obs. alt. sun's L.L. 11° 20'; again, P.M. at ship same day, when chronometer showed November 26ᵈ 6ʰ 49ᵐ 49ˢ, obs. alt. sun's L.L. 16° 35' 10", height of eye 23 feet, the ship having made 19 miles on a true N.E. by E. ½ E. course in the interval: required the line of position when the first altitude was observed, and the ship's position by Sumner's Method when the second altitude was taken, assuming the ship to be between the latitudes 49° 0' N. and 49° 30' N.

ADDITIONAL FOR MASTER ORDINARY.

13. 1882, August 20th, the observed meridian altitude of the star β Centauri, bearing South, was 59° 47' 13", height of eye 25 feet: required the latitude.

In the following table give the correct magnetic bearing of the distant object, and thence the deviation.

Correct magnetic bearing.

Ship's Head by Standard Compass.	Bearing of Distant Object by Standard Compass.	Deviation Required.	Ship's Head by Standard Compass.	Bearing of Distant Object by Standard Compass.	Deviation Required.
North	N. 79° E.		South	N. 76° E.	
N.E.	N. 54 E.		S.W.	East.	
East	N. 49 E.		West	S. 82 E.	
S.E.	N. 56 E.		N.W.	East.	

With the deviation as above, give the courses you would steer by the Standard Compass to make the following courses correct magnetic.

Correct magnetic courses:—S.E. by E.; W. ½ S.; N. 70° E.

Compass courses:—

Supposing you have steered the following courses by the Standard Compass, find the correct magnetic courses made from the above deviation table.

Compass courses:—East; N.W. by W.; S.W.

Magnetic courses:—

You have taken the following bearings of two distant objects by your Standard Compass as above, with the ship's head at S.E. ½ E., find the bearings, correct magnetic.

Compass bearings:—N. ½ W. and W. ½ N.

Bearings, magnetic:—

EXAMINATION PAPER—No. XXIV.

FOR SECOND MATE.

1. Multiply 41375 by 240, and 217·71 by ·001, by common logarithms.

2. Divide 900009 by 100001, and 306250 by 1·75, by common logarithms.

3.—

H.	Courses.	K.	$\frac{1}{10}$	Winds.	Lee-way.	Deviation.	Remarks, &c.
					pts.		
1	E. ¼ S.	12	4	North.	0	16½° E.	A point of land in lat. 37° 45′ N., long. 15° 0′ E., bearing by compass W. ½ N., dist. 42 miles. Ship's head E. ¼ S. Deviation as per log.
2		13					
3		13	6				
4		14					
5	E. ½ S.	12		N.N.E. ½ E.	¼	15½° W.	
6		11	4				
7		11	6				
8		12					
9	S.S.E.	5	4	East.	2¼	11° E.	
10		5	5				
11		5	6				
12		5	5				
1	N.N.E.	3	4	East.	3¼	6° W.	Variation 14° W.
2		3					
3		3	6				
4		4					
5	E. ½ S.	9	5	N.N.E. ½ E.	½	15½° W.	
6		9	5				
7		9					
8		9					A current set the ship E.S.E. (correct magnetic), 52½ miles, from the time the departure was taken to the end of the day.
9	E.S.E.	8	4	N.E.	¾	9½° W.	
10		8	6				
11		9					
12		9					

4. 1882, December 1st, in longitude 67° 56′ E., the observed meridian altitude of the sun's L.L. was 18° 48′ 10″, bearing South, index correction — 3′ 6″, height of eye 18 feet: required the latitude.

5. In latitude 37° 45′ S., the departure made good was 208·9 miles: required the difference of longitude by parallel sailing.

6. Required the course and distance from A to B, by calculation on Mercator's principle.

Lat. A	40° 58′ S.	Long. A	176° 59′ E.
Lat. B	58 30 N.	Long. B	147 40 W.

ADDITIONAL FOR ONLY MATE.

7. 1880, August 1st: find the time of high water, A.M. and P.M., at Morlaix, Ile de Sein, Ferehous, Penzance, Limerick, Dunmanus Harbour, and Hobarton, in long. 147° 22′ E.

8. 1882, February 1st, at 11ʰ 18ᵐ P.M. apparent time at ship, latitude 72° 58′ S., longitude 89° W., the sun's magnetic amplitude was S. ¾ E.: required the true amplitude and error of the compass; and supposing the variation to be 36° 20′ E.: required the deviation for the position of the ship's head when the observation was taken.

9. 1882, August 31st, P.M. at ship, lat. 0° N., observed altitude sun's L.L. 45° 5′ 30″, index correction — 2′ 10″, height of eye 15 feet, time by chronometer August 31ᵈ 9ʰ 11ᵐ 28ˢ, which was 5ᵐ 20ˢ *fast* for mean noon at Greenwich, April 15th, and on June 16th was *fast* 2ᵐ 43ˢ on mean time at Greenwich: required the longitude.

ADDITIONAL FOR FIRST MATE.

10. 1882, March 20th, mean time at ship $0^h\ 45^m$ P.M., latitude 36° 18′ N., longitude 39° 52′ W., sun's bearing by compass S. ¼ E., observed altitude sun's L.L. 52° 59′, height of eye 14 feet: required the true azimuth and error of the compass; and supposing the variation be 24° 50′ W.: required the deviation of the compass for the position of the ship's head at the time the observation was taken.

11. 1882, March 21st, P.M. at ship, latitude by acct. 18° 50′ N., longitude 108° 47′ E., observed altitude sun's L.L. South of observer was 71° 6′ 50″, height of eye 18 feet, time by watch March $20^d\ 23^h\ 58^m\ 12^s$, which had been found to be $11^m\ 8^s$ *slow* on apparent time at ship, the difference of longitude made to the *East* was 18½ miles, after the error on apparent time was determined: required the latitude by reduction to meridian.

12. 1882, August 17th, A.M. at ship, and uncertain of my position, when a chronometer showed August $16^d\ 8^h\ 0^m\ 20^s$ G.M.T., obs. alt. sun's L.L. 28° 14′ 10″; again, P.M. at ship same day, when chronometer showed August $16^d\ 14^h\ 0^m\ 20^s$, obs. alt. sun's L.L. 46° 21′, height of eye 12 feet, the ship having made 20 miles on a true *East* course in the interval: required the line of bearing when the first altitude was taken, and the position of the ship by Sumner's Method when the second altitude was observed, assuming latitudes 49° 30′ N. and 50° 0′ N.

ADDITIONAL FOR MASTER ORDINARY.

13. 1882, March 31st, the observed meridian altitude of the star *a Pegasi (Markab)*, bearing North, 33° 20′ 50″, index correction + 1′ 20″, height of eye 20 feet: required the latitude.

In the following table give the correct magnetic bearing of the distant object, and thence the deviation:—

Correct magnetic bearing.

Ship's Head by Standard Compass.	Bearing of Distant Object by Standard Compass.	Deviation Required.	Ship's Head by Standard Compass.	Bearing of Distant Object by Standard Compass.	Deviation Required.
North	N. 22° W.		South	N. 7° W.	
N.E.	N. 13 W.		S.W.	N. 16 W.	
East	N. 3 E.		West	N. 26 W.	
S.E.	N. 11 E.		N.W.	N. 18 W.	

With the deviation as above, give the courses you would steer by the Standard Compass to make the following courses correct magnetic.

Correct magnetic courses:—E. by S.; West; North.

Compass courses:—

Supposing you have steered the following courses by the Standard Compass, find the correct magnetic courses made from the above deviation table.

Compass courses:—E. by S.; West; North.

Magnetic courses:—

You have taken the following bearings of two distant objects by your Standard Compass as above, with the ship's head at West, find the bearings, correct magnetic.

Compass bearings:—West, and N. 14° W.

Bearings, magnetic:—

EXAMINATION PAPER—No. XXV.
FOR SECOND MATE.

1. Multiply 6054 by 912, and 2070·5 by 62·0898, by common logarithms.
2. Divide 117·658 by 146·932, and 167·342 by ·002, by common logarithms.
3.—

H.	Courses.	K.	$\frac{1}{10}$	Winds.	Lee-way.	Deviation.	Remarks, &c.
					pts.		
1	S. ½ W.	4	5	W. by S.	2¼	50° W.	A point of land in lat. 62° 20′ N., long. 64° 40′ W., bearing by compass W. by N. ½ N., dist. 21 miles. Ship's head S. ½ W. Deviation as per log.
2		4	2				
3		4					
4		3	9				
5	S.W. ¾ W.	3	5	S. by E.	3½	9° W.	
6		3	4				
7		3	2				
8		3	3				
9	E. ¾ S.	5	4	S. by E.	1¾	15° W.	
10		5	3				
11		4	4				
12		4	2				
1	W.N.W.	3	6	North.	3	37° E.	Variation 59° W.
2		4	5				
3		5	3				
4		5	7				
5	N.W. ½ N.	10	2	E.N.E.	0	28½° E.	
6		11	4				
7		12	6				A current set the ship E. by S. ½ S., correct magnetic, 49 miles, from the time the departure was taken to the end of the day.
8		13	4				
9	E. ¾ N.	5	5	N. by E.	3¼	26½° W.	
10		5	4				
11		5	4				
12		5					

4. 1882, June 1st, longitude 44° 40′ E., the observed meridian altitude of sun's L.L. bearing North, was 72° 14′ 10″, index correction + 3′ 45″, height of eye 22 feet: required the latitude.

5. In latitude 32° 3′ S., longitude 179° 45′ W., a ship makes 54 miles West, then 80 miles North: what is the longitude in, also find the compass course and distance; variation 18° E.: 1st deviation 4° 5′ E.; 2nd deviation 3° 10′ W.

6. Required the course and distance from Cape Lopatka to Callao.
   Lat. Cape Lopatka 50° 33′ N.   Long. Cape Lopatka 156° 46′ E.
   Lat. Callao 12 4 S.   Long. Callao 77 14 W.

ADDITIONAL FOR ONLY MATE.

7. 1880, May 7th: find the times of high water, A.M. and P.M., at Aberdeen, Montrose, Wick, Fécamp, Hellevoetsluis.

8. 1882, December 21st, at 11ʰ 15ᵐ P.M., apparent time at ship, in latitude 66° 25′ S., longitude 93° 57′ W., the sun's magnetic amplitude was S. ¾ E.: required the true amplitude and error of the compass; and supposing the variation to be 28° 10′ E.: required the deviation of the compass for the position of the ship's head at the time of observation.

9. 1882, January 29th, P.M. at ship, latitude 28° 45′ N., observed altitude sun's L.L. 17° 46′ 30″, index correction — 3′ 25″, height of eye 16 feet, time by chronometer January 28ᵈ 16ʰ 31ᵐ 30ˢ, which was 1ᵐ 16·5ˢ *fast* for Greenwich mean noon, December 17th, 1881, and on January 1st, 1882, was 1ᵐ 3ˢ *slow* for Greenwich mean time; course since noon N.W. by W. (true), distance 20 miles: required the longitude at the time of observation, and also at noon.

ADDITIONAL FOR FIRST MATE.

10. 1882, July 10th, mean time at ship 3^h 14^m 2^s P.M., latitude $38°$ $2'$ S., longitude $140°$ $58'$ E., sun's bearing by compass N. $2°$ $15'$ E., observed altitude sun's U.L. $14°$ $56'$ $30''$, index correction $+$ $3'$ $30''$, height of eye 19 feet, variation by chart $6°$ $45'$ E.: required the deviation of the compass for the position of the ship's head at the time of observation.

11. 1882, November 29th, P.M. at ship, latitude account $6°$ $20'$ S., longitude $123°$ $25'$ E., observed altitude of sun's L.L. $74°$, index correction $+$ $4'$ $0''$, height of eye 19 feet, time by watch November 28^d 22^h 46^m, which had been found to be 1^h 27^m *slow* on apparent time at ship, the difference of longitude made to the *West* was $12·3$ miles after the error on apparent time was determined: required the latitude.

12. 1882, October 20th, P.M. at ship, and uncertain of my position, when a chronometer showed October 19^d 16^h 30^m 54^s M.T.G., obs. alt. sun's L.L. $29°$ $37'$ $15''$; again, P.M. at ship same day, when chronometer showed October 19^d 19^h 44^m 50^s, obs. alt. sun's L.L. $12°$ $21'$, height of eye 20 feet, the ship having made 24 miles on a true N.N.E. ½ E. course in the interval: required the line of bearing when the first altitude was taken, and the ship's position by Sumner's Method when the second altitude was observed, assuming latitudes $49°$ $20'$ N. and $49°$ $50'$ N.

ADDITIONAL FOR MASTER ORDINARY.

13. 1882, May 15th, the observed meridian altitude of the star β Orionis $52°$ $20'$ $30''$, zenith North of star, index correction $-$ $4'$ $10''$, height of eye 15 feet: required the latitude.

14. 1882, September 4th, what bright stars in the *Nautical Almanac* will pass the meridian of a place in longitude $54°$ $40'$ E., between the hours of seven and ten.

15. 1882, June 15th, observed meridian altitude of η Argus, under the South pole, was $47°$ $50'$ $30''$, index correction $+$ $3'$ $20''$, height of eye 20 feet: required the latitude.

In the following table give the correct magnetic bearing of the distant object, and thence the deviation:—

Correct magnetic bearing.

Ship's Head by Standard Compass.	Bearing of Distant Object by Standard Compass.	Deviation Required.	Ship's Head by Standard Compass.	Bearing of Distant Object by Standard Compass.	Deviation Required.
North	S. $23°$ E.		South	S. $6°$ W.	
N.E.	S. 11 E.		S.W.	S. 18 E.	
East	S. 5 W.		West	S. 21 E.	
S.E.	S. 20 W.		N.W.	S. 22 E.	

With the deviation as above, give the courses you would steer by the Standard Compass to make the following courses correct magnetic.

Correct magnetic courses:—S. ½ W.; E. by N.; S.E. by S.; W. by N.

Compass courses:—

Supposing you have steered the following courses by the Standard Compass, find the correct magnetic courses made, from the above deviation table.

Compass courses:—N.W. by N.; W.N.W.; S.E. by E.; N.N.E.

Magnetic courses:—

You have taken the following bearings of two distant objects by your Standard Compass as above; with the ship's head at S.E. by S., find the bearings, correct magnetic.

Compass bearings:—N. $84°$ W., and N.W. by W. ½ W.

Bearings, magnetic:—

EXAMINATION PAPER—No. XXVI.
FOR SECOND MATE.

1. Multiply 6893 by 11300, and ·0001468 by ·000395, by common logarithms.
2. Divide 7122 by 8·9596, and 268430 by ·00310, by common logarithms.
3.—

H.	Courses.	K.	$\tfrac{1}{10}$	Winds.	Lee-way.	Deviation.	Remarks, &c.
1	S.E. ¼ E.	13	7	N.E. by E. ½ E.	pts. ¼	11° E.	A point, lat. 59°49′ N. long. 44° 10′ W., bearing by compass N.W. ¼ W., distance 30 miles. Ship's head S.E. ¼ E. Deviation as per log.
2		14					
3		13	1				
4		13	3				
5	E. ½ S.	10	4	S. by E. ½ E.	¾	14° E.	
6		10	4				
7		10	5				
8		10	3				
9	N. by W. ½ W.	4		N.E. ½ E.	2¼	8½° W.	
10		3	6				
11		3	4				
12		3	4				
1	S.S.W. ½ W.	11	6	S.E. ½ S.	½	2½° W.	Variation 53½° W.
2		11	7				
3		11	8				
4		11	4				
5	N.E. ¼ N.	7	2	E. by S. ½ S.	1	14° E.	
6		7	3				A current set the ship (correct magnetic) S.E. ¼ E., 1·7 knots per hour, from the time the departure was taken to the end of the day.
7		7	4				
8		7	2				
9	S. ¼ E.	12	5	E.S.E.	¼	3° E.	
10		12					
11		12	3				
12		12	4				

4. 1882, October 1st, in longitude 84° 40′ E., observed meridian altitude of sun's U.L., zenith North, was 57° 20′ 30″, index correction — 3′ 36″, height of eye 17 feet: required the latitude.

4.* 1882, July 2nd, in longitude 45° 15′ E., observed meridian altitude of the sun's L.L. below the pole, was 10° 19′ 45″, index correction — 1′ 15″, height of eye 12 feet: required the latitude.

5. A ship from latitude 35° 30′ S., longitude 27° 28′ W., sailing due East (true) 301 miles: required the compass course steered, and what will be the longitude in, variation 1¾ point E., and deviation 8° 50′ E.

6. Required the course and distance from A to B, by Mercator's Sailing.

   Latitude A 10° 8′ S.   Longitude A 175° 18′ E.
   Latitude B 23 12 N.   Longitude B 141 15 E.
   Variation ½ point West, and deviation 7° 15′ West.

ADDITIONAL FOR ONLY MATE.

7. 1880, March 7th: find the times of high water, A.M. and P.M., at Santander, Arcachon, Scarborough, Holy Island, Montrose, and Angra Pequena (S.W. coast of Africa), long. 15° E.

8. 1882, April 25th, at 7ʰ 22ᵐ 8ˢ P.M., apparent time at ship, latitude 57° 18′ S., longitude 101° 50′ E., sun's setting by compass N. ¼ E., variation by chart 35° 50′ W.: required the error of the compass and deviation.

9. 1882, August 24th, A.M. at ship, latitude at noon 37° 59′ N., observed altitude sun's L.L. 37° 13′ 30″, index correction + 2′ 40″, height of eye 18 feet, time by chronometer, August 24ᵈ 6ʰ 13ᵐ 24ˢ, A.M. at Greenwich, which was 1ᵐ 5ˢ *fast* for mean noon at Greenwich, August 1st, and on August 10th was 0ᵐ 42ˢ *slow* for mean time at Greenwich, course since observation N.N.W., 22′·4 (true): required the longitude at noon.

ADDITIONAL FOR FIRST MATE.

10. 1881, November 1st, mean time at ship $8^h\ 40^m$ A.M., latitude $50°\ 21'$ N., longitude $23°\ 56'$ W., sun's bearing by compass S. ¼ W., observed altitude sun's L.L. $12°\ 19'$, index correction $-\ 3'\ 20''$, height of eye 21 feet: required the error of the compass; and supposing the variation to be $33°\ 20'$ W.: required the deviation of the compass for the position of the ship's head at the time of observation.

11. 1882, May 29th, A.M. at ship, latitude account $0°\ 31'$ S., longitude $150°\ 40'$ W., observed altitude of sun's L.L. $67°\ 41'$ N., index correction $+\ 1'$, height of eye 20 feet, time by watch May $29^d\ 3^h\ 32^m$, *fast* on apparent time at ship $3^h\ 38^m$, the difference of longitude made to *East* was 26·9 miles, after the error on apparent time was determined: required the latitude by reduction to meridian.

12. 1882, April 10th, P.M. at ship, and uncertain of my position, when a chronometer showed April $9^d\ 16^h\ 40^m\ 24^s$ M.T.G., obs. alt. sun's L.L. $45°\ 17'\ 15''$; again, P.M. at ship same day, when chronometer showed April $9^d\ 20^h\ 31^m\ 52^s$, obs. alt. sun's L.L. $18°\ 17'$, height of eye 18 feet, the ship having made 12 miles on a true W. ¼ N. course in the interval: required the line of bearing when the first altitude was taken, and the position of the ship by Sumner's Method when the second altitude was observed, assuming latitudes $51°\ 10'$ N. and $51°\ 40'$ N.

ADDITIONAL FOR MASTER ORDINARY.

13. 1882, June 17th, the longitude $98°$ W., observed meridian altitude of *a* Serpentis, zenith South of object, was $29°\ 0'\ 40''$, index correction $+\ 4'\ 20''$, height of eye 24 feet: required the latitude.

14. 1882, June 15th, at what time will *a* Serpentis pass the meridian of a place in latitude $37°$ N., and longitude $15°\ 30'$ E.; what distance N. or S. of the zenith?

15. 1882, May 18th, observed meridian altitude of η Draconis under the North Pole was $34°\ 56'\ 15''$, index correction $-\ 5'\ 45''$, height of eye 22 feet: required the latitude.

16. At the Cape of Good Hope the variation is about $28°$ W., if the sun at noon bears due North by compass, what is the deviation?

In the following table give the correct magnetic bearing of the distant object and thence the deviation.

Correct magnetic bearing.

Ship's Head by Standard Compass.	Bearing of Distant Object by Standard Compass.	Deviation Required.	Ship's Head by Standard Compass.	Bearing of Distant Object by Standard Compass.	Deviation Required.
North	S. 29° E.		South	S. 69° E.	
N.E.	S. 33 E.		S.W.	S. 64 E.	
East	S. 47 E.		West	S. 48 E.	
S.E.	S. 63 E.		N.W.	S. 38 E.	

With the deviation as above, give the courses you would steer by the Standard Compass to make the following courses correct magnetic.

Correct magnetic courses:—S.W. ½ W.; S.S.E. ¼ E.; W. by N. ¼ N.; N. ½ E.

Compass courses:—

Supposing you have steered the following courses by the Standard Compass, find the correct magnetic courses made from the above deviation table.

Compass courses:—N.E. by N. ½ N.; S.W. by W. ¾ W.; S. ¼ E.; S.W.

Correct magnetic courses:—

You have taken the following bearings of two distant objects by your Standard Compass as above; with the ship's head at N.N.E., find the bearings, correct magnetic.

Compass bearings:—S. by W., and W. by N. ½ N.

Bearings, magnetic:—

QUADRANT AND SEXTANT.

348. The **Quadrant and Sextant*** are reflecting astronomical instruments for measuring angles, and are the instruments chiefly in use for taking the observations required for the solution of a number of the most useful problems in navigation, such as to find the time, the latitude and longitude of a place. The **Quadrant** contains an arc of $45°$ in real extent, and measures a few degrees more than $90°$;† it is usually of wood, and the graduated arc, which is ivory, reads to minutes, and sometimes to $30''$. The **Sextant** is constructed on the same principles as the Quadrant; has a graduated limb of more than $60°$ in real extent; and furnishes the means of measuring the angle between two objects in whatever direction they may be placed, so that the angle does not exceed $140°$. The quadrant serves for common purposes at sea; but the sextant is used when considerable precision is required, as, for instance, in taking a lunar observation.

349. The form of a sextant, as at present in common use, consists of a single frame of brass, so constructed as to combine strength with lightness; and in others a double frame connected by pillars (see Fig., page 380). The graduated arc, inlaid in the brass, is usually of silver, sometimes of gold, or platinum. The explanation of the parts of a sextant, and of the adjustments of that instrument, will answer for the quadrant, since the parts and appendages are common to both.

350. The flat surface of the sextant is called the *plane of the sextant;* the circular part B C is the arc or limb, which is graduated from right to left from the zero point $0°$ to about $140°$, and each degree in the best instruments is again sub-divided into six equal parts of $10'$ each, while the vernier g, used in estimating the sub-divisions of the arc, shows $10''$. The divisions are also continued a short distance in the opposite direction on the other side of zero (O), towards C, forming what is termed the *arc of excess*, for the purpose of determining the index error in the manner that will be subsequently explained. The microscope M, and its reflector r, secured at the point d by a movable arm $d\,r$ to the index bar A E, may be adjusted to read off the divisions on the graduated limb and the vernier g. A E is the radius, or index bar, movable along the arc and round a centre, and having a dividing scale (called the *vernier*) close to the arc, by which the sub-divisions of the arc are read off. The index bar is secured to the arc B C by the intervention of a mill-headed clamp screw *s* at its back, which must be loosened when the index has to be moved any considerable distance, and when the contact nearly has been made

---

* The first *inventor* of the sextant (or quadrant) was Newton, among whose papers a description of such an instrument was found after his death, not, however, until after its re-invention by Thomas Godfray, of Philadelphia, in 1730, and perhaps by Hadley in 1731.

† This depends on the properties of light, which we cannot consider here. The principle of the sextant is this:—The angle between the first and last direction of a ray which has suffered two reflections in one plane is equal to twice the inclination of the reflecting surfaces to each other.

by hand, the screw is again to be fixed, and a tangent screw s' enables the index bar and the vernier* upon it to be moved by a small quantity along the limb, so as to render the contact of the objects observed more perfect than could be effected by moving the index solely by hand; the other extremity of the index bar has a silvered glass or reflector I fixed perpendicular to the plane of the instrument, with its face parallel to the length of the index bar, and directly over the centre; another glass b is fixed perpendicular to the

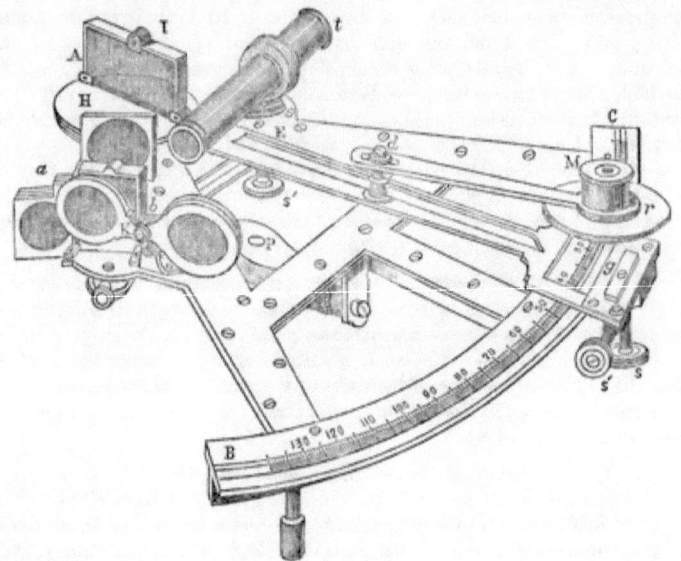

plane of the instrument frame H, and facing the index-glass, the lower half only is silvered (being a reflector), and the upper transparent; it is usually provided with screws, by which its position with respect to the plane of the sextant may be rectified; the plane of this glass, usually termed the horizon-glass, is made parallel to the plane of index-glass I, when the vernier g is adjusted to zero on the divided arc B C, or if not so made, the want of parallelism constitutes what is termed the index error of the instrument. The telescope t is carried by a ring fastened to a stem E, which can be raised or lowered by a mill-headed screw s'' at the back of the frame, for the purpose of so placing the field of the telescope that it may be bisected by the line on the horizon-glass, separating the silvered from the unsilvered part, whereby the brightness of the reflected object and that seen by direct vision may be made equal, and the quality of the observations improved; the ring and its elevating apparatus are technically known as an "up-and-down piece." It is usual to supply a direct and inverting telescope, of which the latter is to-

---

* VERNIER—so called after its inventor, PETER VERNIER, of France, who lived about 1630. By some it is called a *nonius*, after the Portuguese, NUNEN or NONIUS; but the invention of the latter (who died in 1577) was quite different.

be preferred, as possessing greater magnifying power, and thus showing a better contact of the images of the objects. Two wires parallel to each other, and to the plane of the instrument, are placed in the inverting telescope, within which limit the observation should be made. In the quadrant the telescope is omitted, and the eye is applied to a small circular orifice in a piece of brass, placed in the same position as the telescope in the drawing.

Dark glasses of different colours and shades are a necessary accompaniment to the sextant to enable the sun to be observed, and they are usually attached to a hinged joint at K. Four of these glasses or shades are placed at a, between the index and horizon-glasses, so as to admit of one or more of them being interposed between the index and horizon-glass, to moderate the light of any brilliant object seen by reflection. Three more such glasses, sometimes called back shades, are placed behind the horizon-glass at K, any one or more of which can also be turned down to moderate the intensity of the light before meeting the eye when observing a bright object, such as the sun. There is also a dark glass which can be placed at the eye-end t of the telescope, which method is preferable to the other, as no error in this is liable to be introduced in the passage of the rays from the index to the horizon-glass.*

When observing, the instrument is to be held with one hand by the handle P placed at the back of the frame, while the other hand moves the index.

351. **Reading off the Angle.**—The following brief directions for reading off will be more readily understood by the learner, if he place a sextant before him for reference and examination.

It will be seen that the arc (limb) is divided into degrees and parts of a degree, from 0° (zero) to about 140°; every 10th degree is numbered from 0° to 140°; the space between every 10° is divided into 10 equal parts by straight lines; consequently every part is 1°; every fifth line is made a little longer than the others, to represent every fifth degree; and (in the best instruments) every degree is sub-divided into six equal parts by lines shorter than those which represent the degrees; those short lines divide every degree into sixths of a degree, or 10′, every third line of these short ones being made a little longer to denote 30′. On any part of the arc, therefore, the first short stroke is 10′, the second is 20′, the third is 30′, the fourth is 40′, and the fifth is 50′. We will suppose it is an instrument of this kind before the learner. The *index*, up to which an arc is read off, is a line cut in a plate at the end of the movable radius, and is generally distinguished from the other lines on the plate by a diamond-shaped mark, resembling a spear head, and sometimes by O. Supposing this index to stand exactly at any of the long lines on the arc, that is, so that the two lines are in the same direction; in such a case the reading off is easily known, for it must be a certain number

---

* With respect to the dark glasses, when it is possible (as in observing altitudes of the sun in the mercurial horizon, &c.) to make the observation with a single dark glass on the eye-end of the telescope, without using any shade, this should always be done, for the error of this dark glass does not affect the contact at all, and the distortion caused by it is not magnified, whereas any fault in the dark shade between the index and horizon-glasses produces actual error in the observation, and the distortion is magnified subsequently by the telescope.

of divisions, of which the value is seen at once, the reading being degrees and *no minutes*; it may be 10°, 12°, 20°, 30°, &c.—any number: but if the index exactly coincides with a short stroke on the arc, in such a case the reading off must be a certain number of divisions and sub-divisions. Thus, if it coincide, for example, with the second line to the left of 40°, then the reading off will be 40° 20′, or if it coincide with the fifth stroke to the left of 30°, the reading will be 30° 50′, since each line on the arc represents 10′.

But suppose the index not to stand exactly at any line whatever on the arc, but somewhere between two, as in the above example, between the second and third line from 40°, suppose it appeared to be about half-way between the second and third lines (the learner may place it in that position). But as this is a rough and imperfect way of estimating the additional minutes and seconds beyond the second division from 40°, the exact value of this small space is known by means of a few divisions on the index plate to the left of the index, and called the **Vernier**. These divisions are made less than the arc divisions, so that the line on the plate immediately to the left of the index is somewhat nearer to the corresponding one on the arc than the small space to be determined. It is nearer thereto, as is manifest by difference of a division on the arc and one on the index plate. In like manner the second line, reckoning from the index, must be nearer to the corresponding line by two differences, the third by three, and so on. At length, therefore, there must be a coincidence of two lines, or nearly so; that is, they must appear to an eye placed directly over them to lie in the same direction, or nearly so. And since, upon the whole, the lines on the vernier have approached those upon the arc through the small part the index is in advance of 20′, this excess must be equal to as many times the difference of two divisions, as there are lines, reckoning from the index, before this coincidence takes place. Hence, if we know the value of a difference, we shall know the value of the small arc to be measured.

This difference is known as follows:—By examining the arc of the sextant before us, it will be seen that 60 divisions of the vernier just cover or coincide with 59 divisions on the arc, or the difference between a division on the arc and one on the vernier is $\frac{1}{60}$ of a division of the arc; if, therefore, a division on the arc is 10′, the difference will be $\frac{1}{60}$ of 10′, or 10″. Every sixth division of the vernier being distinguished by a figure denoting minutes, and the interval between each of these figures is divided into six parts of 10″ each.

Hence, to **read off on a Sextant**, we proceed thus:—First examine the divisions and sub-divisions on the arc, up to the line which stands before the index. We then move the microscope on the vernier and examine the numbered lines. If any one of these coincides in direction with the opposite one on the arc, the reading off to be added will be so many minutes; if not, we observe between which numbered lines the coincidence actually takes place, and then reckon the preceding minutes as numbered, and afterwards the sub-divisions of the vernier, as so many minutes or seconds. Let us now suppose the index to stand between the second and third divisions from 40°. In reading off, first 40° 20′ is noted on the arc, and then running the microscope farther on the arc, it is observed that a line on the vernier and an arc line are in the same direction, between the lines on the vernier marked 5 and 6. The farther reading off is therefore 5' and some seconds. On examining the interval between 5 and 6, which is divided into six equal parts, the fourth line to the left of 5 is found to be in the same direction with the opposite one on the arc. The remaining reading off is therefore 40". Hence the whole reading off is 40° 25' 40".

The sextant supposed under examination is marked to read off to the nearest 10"; some instruments are graduated to 15" or 30", &c., but the same method of reading off is to be followed as pointed out above.

Take a sextant cut to 15", then every degree is sub-divided into four equal parts by lines shorter than those which represent the degrees; these short strokes divide every degree into fourths of a degree, or 15': then, on any part of the arc the first short stroke is 15', the second 30', and the third 45'. On the vernier the short strokes from minute to minute are each 15". If 0 on the vernier is made to exactly coincide with a large stroke on the arc, the reading is degrees and *no minutes*; but if 0 on the vernier coincides with a short stroke, then the reading is so many divisions and sub-divisions: thus, for example, if it coincides with the third short stroke to the left of 30°, the reading is 30° 45', since each short stroke to the left of 30° represents 15'. Again, suppose 0 on the vernier to stand somewhere between the first and second strokes to the left of 36°, in the first place the reading would be 36° 15', but it must be something more, because the vernier indicates minutes between 15' and 30'. Next look along the vernier and see which stroke on it coincides with any stroke on the arc of the sextant: let us say it is the second short stroke to the left of tenth minute stroke, that is, 10' 30", then the reading will be 36° 15', and 10' 30" to add to it, making 36° 25' 30", and so on for any other indication.

Once more let us suppose the sextant to be graduated to 20', then on any part of the arc the first short stroke is 20', and the second 40'. If 0 on the vernier exactly coincides with a long mark on the arc, the reading must be a certain number of divisions, that is, degrees and *no minutes*; but if 0 on the vernier coincides with any short stroke or sub-division to the left of division, the reading is evidently degrees and minutes (twenties); then suppose it stands at the second short stroke to the left of 32°, the reading is 32° 40'.

Lastly, suppose 0 on the vernier to stand somewhere between the first and second short strokes to the left of 43°, in the first place the reading will be 43° 20'. Next look along the vernier and see which stroke on it coincides with any stroke on the arc of the sextant: let us suppose it is the second short stroke after the twelfth *long* stroke, then the reading will be 43° 20', and 12' 40" to add to it, making 43° 32' 40".

352. **To read off on the arc of excess.**—As has been observed before, the graduation of the arc of the sextant is usually continued to the right of 0, or zero, in which case we have to read off an arc divided from left to right by means of an index which is divided from right to left; this, however, is easily done if we remember that the line on the vernier marked 10' must be considered as the commencement of the divisions, 9' must be considered as 1', 8' as 2', 7' as 3', &c.; or else take the difference between the minutes and

seconds denoted by the vernier and 10'; thus, if the coincidence of lines on the arc and vernier is at 7' 20", we must read this as 2' 40"; if at 5' 40" we must read this as 4' 20", and so on. Similarly, for a sextant cut to 15", if O on the vernier stands between the second and third strokes to the right of O on the arc, and the seventh minute stroke of the vernier coincides with a stroke on the arc, then the reading on the arc of excess, that is *off*, will be 38', since the seventh stroke is 8 when reckoned from the left of the vernier.

The preceding remarks relate to the true *Quadrant* and *Sextant*, but instruments of the first kind are now not unfrequently graduated on the limb to 120°, and the second kind up to 160°; this arrangement is effected by placing the index-glass at an angle with the index-bar, and so fixing the horizon-glass that it shall be parallel with the index-glass, when 0° on the vernier coincides with 0° on the limb.

ADJUSTMENTS OF THE SEXTANT AND QUADRANT.

353. The *adjustments* of the sextant and quadrant are:—(1) *To set the index-glass and* (2) *the horizon-glass perpendicular to the plane of the instrument;* (3) *to adjust the line of collimation of the telescope,* i.e., *to set the axis of the telescope parallel to the plane of the instrument;* (4) *and to set the horizon-glass parallel with the index-glass, when* O *(zero) on the vernier coincides with* O *(zero) on the arc;* then, if the adjustments cannot be perfected, (5) *to find the index error* of the instrument:—

1*st*. **The index-glass, or central mirror, must be perpendicular to the plane of the instrument.**—Place the index near the middle of the arc. Hold the sextant with its face up, the index-glass being placed near the eye, and the limb turned from the observer. Look obliquely down the glass; then, if the part of the arc to the right, viewed by direct vision, and its image in the mirror, appear as one *continued arc of a circle*, the adjustment is perfect; if the reflection seems to *droop* from the arc itself, the glass leans *back*; if it *rises upward*, the glass leans *forward*. The position is rectified by screws at the back.

2*nd*. **The horizon-glass, or fixed mirror, must be perpendicular to the plane of the instrument.**—(a) *By the sea horizon*—Set O on the index to O on the arc. Hold the instrument with its *face up;* direct the sight to the horizon-glass, give the instrument a small nodding motion; then if the horizon, as seen through the transparent part of the horizon-glass, and its image, as seen in the silvered part, appear to be in a *continued straight line*, the adjustment is perfect.

For this method of (a) testing there must be no index error, which caution is unnecessary when (b) the sun is used.

(b) *By the sun*.—The instrument being held perpendicular, look at the sun (using the shades); sweep the index-glass along the limb, and if the reflected image pass exactly over the object itself, appearing neither to the right nor left of the object, then the horizon-glass is perpendicular to the plane of the instrument; if not, turn the adjusting screw, which in some instruments is a mill-headed one at the back of the instrument, while in others it is a small screw behind and near the upper part of the glass itself, which can be turned by placing a capstan-pin into the hole in the head of the screw.

3rd. **The line of collimation,* or in other words, the axis of the telescope, must be parallel to the plane of the instrument.†**—Turn the eye-piece of the telescope till two of the parallel wires in its focus appear parallel to the plane of the instrument; then select two objects, as the sun and moon, whose angular distance must not be *less* than from 100° to 120°, because an error is more easily discovered when the distance is great; bring the reflected image of the sun exactly in contact with the direct image of the moon, at the wire nearest the plane of the sextant, and fix the index; then, by altering a little the position of the instrument, make the object appear on the other wire; if the contact still remains perfect, no adjustment is required; if they separate, *slacken* the screw *furthest* from the instrument in the ring which holds the telescope, and tighten the other, and *vice versa* if they overlap.

4th. **The horizon-glass must be parallel to the index-glass.**—Set O on the index to O on the arc; screw the tube or telescope into its socket, and turn the screw at the back of the instrument till the line which separates the transparent and silvered parts of the horizon-glass appears in the middle of the tube or telescope. Hold the sextant vertically, that is, with its arc or limb downwards, and direct the sight through the tube or telescope to the horizon; then, if the reflected and true horizons do not coincide, turn the screw at the back and at the lower part of the horizon-glass till they are made to appear in the same straight line. Then will the horizon-glass be truly parallel to the index-glass.‡

354. *Def.*—**Index Error** of reflecting instruments, such as the sextant, is the difference between the zero point of the graduated limb, and where the zero point ought to be, as shown by the index when the index-glass is parallel to the horizon-glass.

5th. **To find the Index Correction.**—The two objects generally used to determine the index error are (a) the sea horizon, and (b) the sun.

(a) *By the horizon.*—Holding the instrument vertically, move the index till the horizon coincides with its image, and the distance of O on the index from O on the limb is the index error; *subtractive* when O on the index is to the *left*, and *additive* when it is to the *right* of O on the limb; but if O on the index stands exactly at O on the limb, there is *no* index error.

Ex. 1.—The horizon and its image being made to coincide, the reading is 2′ *on* the arc. Then 2′ is the *Index Correction* to be *subtracted* from every angle observed.

Ex. 2.—When the horizon and its image were made to coincide, the reading was 3′ 20″ *off* the arc; the index correction therefore was + 3′ 20″.

NOTE.—The reading may be on the arc proper, or on the arc of excess; it is on the latter when O on the vernier is to the right of O on the limb, and the reading is then said to be *off* the arc; when O on the vernier is to the left of O on the limb, the reading is said to be *on* the arc.

---

* The line of collimation, *i.e.*, the path of a visual ray passing through the centre of the object-glass, and the middle point between the cross wires.

† The error caused by the imperfection of this adjustment is called the "*Error of Collimation*," and the observed angle is always too great.

‡ Some sextants, as THROUGHTON's Pillar Sextants, are not provided with the means for making this adjustment, because it is not absolutely necessary. An allowance, called *Index Error*, being made for the want of parallelism of the two glasses when the zeroes coincide.

(2). *By measuring the sun's semi-diameter.*—Fitting the telescope and arranging the shades so that the reflected and direct images of the sun may be viewed clearly and seen of the same brightness, measure the sun's horizontal diameter, moving the index forward on the divisions until the images of the true and reflected suns touch at the edges; read off the measure which will be *on* the arc; then cause the images to change sides, by moving the index back; take the measure again and read off; this reading will (as is generally the case) be *off* the arc; half the difference of the two readings is the index correction.

When the reading *on* the arc is the *greater*, the correction is *subtractive*; when the *lesser*, *additive*.

EXAMPLES.

Ex. 1. On the arc — 33′ 10″
       Off + 30 50
             2)2 20
  Index corr. *sub.* 1 10

Ex. 2. On the arc — 30′ 20″
       Off + 33 30
             2)3 10
  Index corr. *add* 1 35

If both readings are *on* the arc, or both *off* the arc, half their sum is the index correction—*subtractive* when both *on*, *additive* when both *off* the arc.

Ex. 3. 1st reading on the arc — 65′ 30″
     2nd ,, ,,   — 1 40
                   2)67 10
  Index corr. *sub.* 33 35

Ex. 4. 1st reading off the arc + 1′ 30″
     2nd ,, ,,   + 66 50
                   2)68 20
  Index corr. *add*. 34 10

One-fourth of the sum of the two readings should be equal to the sun's semi-diameter in the *Nautical Almanac* for the day; but if both readings be *on* or both *off* the arc, one-fourth their *difference* should be the sun's semi-diameter.

Thus, suppose the observations, in Example 1, to be made on September 26th, 1882; here one-fourth of the sum of the two readings is 16′ 0″, agreeing with the semi-diameter as given in the *Nautical Almanac* for the given day.

This affords a test of the accuracy with which the observation has been made. But in order that the comparison may be a good criterion, we should measure the sun's *horizontal* diameter which is not sensibly affected by refraction.

Obs.—In order to obtain the index correction with the greatest precision, the mean of a number of measurements of the sun's diameter should be taken; also, the limb should be placed (by hand, before the tangent screw is used) alternately a little open and a little overlapping, so that in making the contact the tangent screw may be turned different ways.

EXAMPLES FOR PRACTICE.

Ex. 1. 1882, April 17th, the reading on the arc 29′ 40″, the reading off the arc 34′ 10″: required the index correction and semi-diameter.

Ex. 2. 1882, July 4th, the reading on 33′ 10″, off 29′ 50″: find index correction and semi-diameter.

Ex. 3. 1882, November 13th, on 4′ 40″, off 60′ 10″: find index correction and semi-diameter.

Ex. 4. 1882, July 10th, on 31′ 45″, off 31′ 30″: find index correction and semi-diameter.

Ex. 5. 1882, March 21st, off 1° 10′ 0″, off 6′ 40″: find index correction and semi-diameter.

Ex. 6. 1882, January 17th, on 67′ 40″, on 2′ 30″: find index correction and semi-diameter.

355. **The Prismatic Sextant.**—In the form of instrument just described, and which is all but universally employed, the angle measureable is limited to 140°; but we may perhaps add that PISTOR and MARTINS, of Berlin, have, by an ingenious modification of the horizon-glass (for which they substitute a prism), produced a sextant which will measure any angle up to 180°. This instrument is called the PRISMATIC SEXTANT.

The following shows the form of Examination Paper on the Adjustment of the Sextant.

EXAMINATION PAPER.

Exn. 9a.

Port of

Rotation No.

ADJUSTMENTS OF THE SEXTANT.

The applicant will answer in writing, on a sheet of paper which will be given him by the Examiner, all the following questions, numbering his answers with the numbers corresponding to the questions.

1.—What is the first adjustment of the sextant?

A.—To set the index-glass perpendicular to the plane of the sextant.

2.—How do you make that adjustment?

A.—Place the index near the middle of the arc, and look into the index-glass so that you can see both the arc and its reflection; if they be in one line, the glass is perpendicular, but if not continuous, they are brought so by the screws in the frame upon which the glass stands.

3.—What is the second adjustment?

A.—To set the horizon-glass perpendicular to the plane of the sextant.

4.—Describe how you make that adjustment?

A.—Place 0 of the vernier on 0 on the arc, hold the instrument obliquely, with its face upwards, and look from the sight vane at the horizon; if the reflected part and the direct portions of the horizon are in one line, this adjustment is perfect, but if not, they must be brought in line by gently moving a screw at the back (top) of the glass.

5.—What is the third adjustment?

A.—To set the index and horizon-glasses parallel when the index is at 0.

6.—How would you make the third adjustment?

A.—Place the index at 0, and holding the instrument vertically, look at the horizon; if the reflected and direct parts are in one line, this adjustment is perfect, but if they are not in one line, move a screw at the back of the horizon-glass until they are.

7.—In the absence of a screw how would you proceed?

A.—I would find the index correction, or as it is called, the index error.

8.—How would you find the index error by the horizon?

A.—Hold the instrument vertically, and, looking at the horizon, move the tangent screw until the horizon in both parts of the horizon-glass form one line; the reading is the index error.

9.—How is it to be applied?

A.—To be added when the reading is off the arc, and to subtract when the reading is on the arc.

10.—Place the index at the error of   minutes to be added, clamp it, and leave it.

The Examiner will see that it is correct. This is a reading off the arc, i.e., on the arc of excess.

11.—The Examiner will then place the zero of the vernier on the arc, not near any of the marked divisions, and the Candidate will read it.

In all cases the Candidate will name or otherwise point out the screws used in the various adjustments.

NOTE to 10 and 11.—When the Examiner is satisfied that the Candidate can read the arc of the sextant both on and off the arc, it will be sufficient to place his initials against 10 and 11 on the paper containing the answer.

The above completes the examination of Second and Only Mates.

In addition to the above, First Mates and Masters will be required to state in writing:—

12.—How do you find the index error by the sun?

A.—Place the index at about 30' on the arc, and holding the instrument vertically, look at the sun, two suns will be seen; bring their upper and lower limbs in exact contact, read off and mark down, then place the index at about 30 off the arc, or to the right of O, bring down the upper and lower limbs in contact as before, read off and mark down; half the difference of these two readings will be the index error.

13.—How is the same applied?

A.—To be added when the greater reading is off the arc, and subtracted when the greater reading is on the arc.

14.—What proof have you that these measurements or angles have been taken with tolerable accuracy?

A.—By adding the two readings together, and dividing the sum by 4; if the measurements are correct, the result should be nearly equal to the semi-diameter for the day, as given in the *Nautical Almanac*. If they do not so agree, repeat the observations until they do.

CURRENT.—SOUNDINGS.

TO FIND THE COURSE TO STEER IN ORDER TO MAKE GOOD ANY COURSE IN A KNOWN CURRENT, AND ALSO THE DISTANCE MADE GOOD.

356. Draw a line on a chart to represent the course to be made good; from the ship's place on the chart lay off a line in the direction of the set of the current, on which mark off from the ship's place the rate of the current per hour; then take in the compasses the distance the ship sails in an hour by log, and put one foot on the last-named mark, and from the point where the other foot reaches the first line draw a line to the mark on the line representing the direction of the current. The course to be steered is represented by the line last drawn, and the parallel ruler being placed to it, and moved to the centre of the compass on the chart, will give the course of the ship; and that portion of the first line drawn, intersected by the last line drawn, will be the distance the ship will make good per hour.

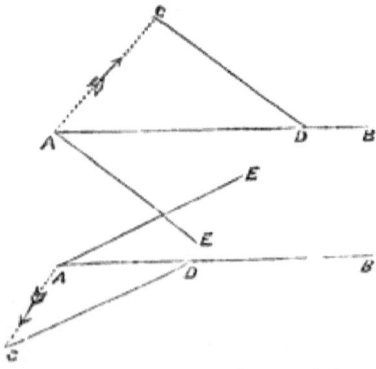

On a chart, suppose A to be the place of the ship, B the port of destination; also A C the set of the current, the rate per hour being taken from the scale of miles and laid off in the direction of the line. Take the distance sailed by the ship per hour from the scale of miles, and with one foot of the dividers at C, make an arc cutting A at D. Join C D, and move the parallel ruler from C D to A, drawing A E parallel to C D; then A E will be the direction of the ship's head; and the parallel ruler being moved to the centre of the compass on the chart, will give the course of the ship on the chart; and A D will be the distance the ship will make good.

SOUNDINGS.

357. In the open sea, the tide requires about six hours and a quarter to rise from low to high water, and an equal interval to fall from high to low water. If the rise or fall was an uniform quantity throughout, by simply

taking a proportionate part of the rise or fall due to the time of tide, we should at once obtain the quantity required to reduce the soundings to the low water of that day. But the water does not rise in equal proportions, the rise during the first and last hours being very small (about one-sixteenth of the whole range); in the second hour there is a considerable increase of rise; in the third and fourth hours a still greater increase of rise; and then the rise begins to take off in the same proportion as it increased.*

The correct amount for every half-hour, and for various ranges, is given in the "Tide Tables for the English and Irish Ports for 1880," (p. 98, Table B), published by the Hydrographic Office, Admiralty.†

358. As the soundings upon the chart are all referred to or measured downwards from the mean level of low water of *ordinary* spring tides,‡ casts of the lead taken at any other time of the tide, or any other day than full and change, will exceed the depth marked on the chart (except when it happens to be low water of *greatest* spring tides). It is necessary for the seaman to be able to calculate the difference between the actual depth obtained by means of his lead, and that marked on his chart, in order to the identification of his ship's place, more especially when the range of the tide is considerable, and the depth not great. Also, when about to enter a port in a vessel whose draught of water is nearly equal to the depth, it is necessary to find the height of the tide as exactly as circumstances will permit.

359. Two classes of questions may be proposed in reference to this subject —*firstly*, to find the depth of water at a given place and time; *secondly*, having obtained the actual depth by a cast of the lead, to find the sounding on the chart corresponding thereto, and thence to identify the ship's place. Both these classes of questions require us to know the *time* of high water and the *range* of the tide on the given day; and for this purpose almanacs are published. The most correct, and by far the most useful of all these, are the "Tide Tables" published by the Admiralty, and to which we have already referred. In this book are given the times of *high water* and the *height of the tide* for every day in the year, at each of the principal ports in Great Britain.

---

* The reader may obtain an idea of this law, sufficiently exact for practical purposes, in the following manner:—Describe a circle, and divide the circumference into six equal parts on each side, corresponding to the hours of the tide; then divide the diameter into proportional parts, corresponding to a given (assumed) range of tide. Connect the segments of the circle by straight lines drawn across the figure, when it will be perceived that they intersect the diameter at certain divisions of the range. These are the correct quantities respectively due to each hour's rise or fall of such a tide from low to high water, and *vice versa*. An examination of these quantities will show, that in the first hour of the tide the rise is equal to one-sixteenth of the whole range; at two hours from low or high water, the tide has risen or fallen *one-fourth* of the whole range; at three hours it has risen just *half* its range; at four hours it has risen *three-fourths* of the whole range; at five hours to within a *sixteenth* of the whole range. The above method, which is constructed on principles theoretically correct, will represent with sufficient exactness all that is necessary for practical purposes.

† Table XIX, RAPER, which the author, in 1847, computed for RAPER's work, also shows the space through which the surface of the water rises and falls at given intervals from high or low water.

‡ On most charts the soundings expressed are reduced to low water of *ordinary spring tides*; but in some charts, however, the soundings are reduced to the low water of extraordinary spring tides—such, for example, is the case on the chart of Liverpool, surveyed by Captain DENHAM, R.N., the soundings on which are reduced to a spring range of thirty feet, while the mean spring range for that place, as deduced from observations made for two years at the Tide Gauge, St. George's Pier, is 26 feet.

360. To find how much we must subtract from a cast of the lead, in order to a comparison with the soundings marked on the chart, proceed by

RULE CIII.

1°. *Open the* Admiralty Tide Tables *at the proper month; and in the column under the head of the place near your position, and opposite the day of the month,* take out the "*time*" *of high water in the morning or afternoon, as the case requires, and also from the adjoining column, under* "*height*," *take out the height of the tide*.

2°. *Next, underneath the time of high water place the time at ship, and take the* difference *and call it* "time from high water."

3°. *From the* height of tide *subtract the half mean Spring Range, which stands at the foot of the column.*

The remainder is the *half range* of the day.

4°. *Enter* Table B, page 98, Admiralty Tide Tables, *and* under *the time from high water, and* opposite *the half range for the given day, take out the correction corresponding thereto, observing whether it is to be* added *or* subtracted.

5°. Add or subtract *the correction, as directed, to the mean half Spring Range marked on the chart.*

The result is the excess *of the* sounding observed *above the* sounding recorded *on the chart, or is the height of the tide above zero.*

6°. Subtract *this last from the sounding shown by the* lead, *the* remainder *is the sounding shown by the* chart.

NOTE I. When it happens to be an *extraordinary* low ebb tide, the quantity given in Table B will be *greater* than the half mean spring range, and will be *subtractive*. In such cases, subtract the half mean spring range from the correction by Table B, and *add* the result to the soundings by lead; the sum will be the sounding on the chart.

EXAMPLES.

Ex. 1. 1880, September 16th, at 7^h 41^m P.M., a ship off Liverpool strikes soundings in 8 fathoms: required the corrected soundings to compare with the chart. (The half spring range by Captain DENHAM's chart is 15 feet.)

Admiralty Tide Tables (page 70); time of high water at Liverpool, September 16th, 1880 .. .. 9^h 41^m P.M.
Time of sounding 7 41

Time from high water 2 0

	ft.	in.
Height at Liverpool	26	7
Half mean spring range	13	0
Half-range of the day	13	7
In Table B, page 98, under 2^h, opposite $13\frac{1}{2}$ ft., stands *add*	6	9
Half spring range by chart	15	0
Correction $3\frac{1}{2}$ fathoms, or	21	9

Depth by lead .. .. 8 fathoms.
Correction .. .. $3\frac{1}{2}$,,

Showing the depth by comparison .. .. $4\frac{1}{2}$,,

Whence the depth to compare with the chart is only $4\frac{1}{2}$ fathoms instead of 8 fathoms.

Ex. 2. 1880, October 6th, at $7^h\ 23^m$ A.M., a vessel anchored off Weston-super-mare in $6\frac{1}{2}$ fathoms; at low water the vessel was "high and dry:" required the cause of this. (Half spring range by chart 23 feet.)

By Table: October 16th, the time of high water at Weston-super-mare	$7^h\ 47^m$ A.M.
Time of anchoring	7 23
Time before high water	0 24
	ft. in.
Height of tide by Tables	38 8
Half spring range	18 7
Half range	20 1
By Table B, 24^m and half range 20 feet 1 inch give *add*	19 6
By chart: half spring range	23 0
Correction to low water	42 6
Sounding $6\frac{1}{2}$ fathoms, or	39 0
	3 6

Water below the sounding; or, the ship is found to be 3 feet 6 inches dry at low water.

Ex. 3. 1880, March 2nd, at $4^h\ 18^m$ A.M., a vessel has to cross the Victoria Bar, Liverpool; it is required to know what water she will have over the bar. (Depth at low water springs on chart, 11 feet.)

By Table: March 7th, time of high water at Liverpool	$2^h\ 15^m$ A.M.
Time of crossing the bar	4 18
Time after high water	2 3
	ft. in.
Height of tide by Tables	25 7
Half spring range	13 9
Half range for the day	12 10
By Table B: $2^h\ 3^m$ and half range 12ft. 7in. *add*	6 5
Half spring range by chart	15 0
Correction	21 5
Depth on Bar at $2^h\ 3^m$ from high water, March 7th	11 0
By chart: depth on Victoria Bar at low water springs	32 5
or $5\frac{1}{2}$ fathoms, nearly.	

Ex. 4. 1880, August 21st, at $2^h\ 11^m$ P.M., off Weston-super-mare, sounded in $4\frac{1}{2}$ fathoms: required the soundings on the chart.

Time of high water, Weston-super-mare, August 21st	$7^h\ 43^m$ P.M.
Time of sounding	2 11
Time from high water	5 32
Height of tide, Weston-super-mare, August 21st	39ft. 2in.
Half mean spring range	18 7
Height above half tide	20 7
By Table B: $5^h\ 32^m$ and half range 20 ft. 7 in. *subt.*	19 10
Half spring range	18 7
Level of tide below zero	1 3
Soundings by lead $4\frac{1}{2}$ fathoms, or	27 0
Correction	$+$ 1 3
Soundings on chart	28 3
Or a little less than 5 fathoms.	

EXAMPLES FOR PRACTICE.

1. 1880, August 8th, at $9^h\ 37^m$ A.M.: required the depth of water on the "Four-fathom Ledge," off Weston-super-mare.
2. 1880, June 9th, at $2^h\ 28^m$ A.M.: off Brest, the depth of water by the lead was $10\frac{1}{2}$ fathoms: required the soundings on the chart.
3. 1880, August 18th, at $8^h\ 8^m$ P.M., sounded in the Victoria Channel, Liverpool, in 5 fathoms: required the soundings on the chart.
4. 1880, March 28th, at $7^h\ 55^m$ P.M., a vessel anchored off Weston-super-mare, in 6 fathoms: required the depth at low water (half spring range by chart 23 feet).
5. 1880, March 12th, at $5^h\ 41^m$ P.M.: required the height of the tide above mean low water of spring tides at Liverpool.
6. 1880, December 13th, at $8^h\ 10^m$ A.M.: going up the Firth of Forth, the lead showed 12 fathoms: required the soundings on the chart.

361. The following is the form of the Rule as used at the Liverpool Examinations:—

1°. *Take the* difference *between the time of high water,* full and change, *at Liverpool and full and change at ship, and take this* difference *from the time of high water on the given day at Liverpool;* the result is *time of high water at ship.*

2°. *Next find the time from high water when the "cast" was taken.*

3°. *Take 13 feet, the half mean spring range from the height of tide on given day at Liverpool.*

4°. *Apply a correction from* Table B *to the* half mean spring range, *as directed at the head of the Table;* the result is *the Reduction at Liverpool.*

Lastly.—Find the Reduction at ship (by proportion), thus:—

*As Spring Range at Liverpool
Is to Spring Range at Ship,
So is the Reduction at Liverpool
To the Reduction at Ship.* } The Reduction or Correction of Soundings *is to be taken from the Cast.*

Ex. 1880, September 22nd, at $1^h\ 57^m$ P.M. at ship, off Holyhead, sounded in 45 fathoms: required the corrected cast to compare with the chart.

Full and change at Liverpool $11^h\ 23^m$ } Page 152, Admiralty Tide
Full and change at Holyhead   10 11   }   Tables, 1880.

        Difference   — 1 12
Time high water, Liverpool, Sept. 19th 1 9 P.M., and Height of tide $25\frac{1}{2}$ ft.
                                                     Half mean spring range 13
Time high water at ship   11 57 A.M.,
Time of cast   1 57 P.M., Half range for day $12\frac{1}{2}$

Time of cast from high water   2 0 } give in Table B correction + 6
Half range $12\frac{1}{2}$ ft. } Half spring range   13
       By Proportion.
  ft.   ft.   ft.   Reduction at Liverpool   19
  26 : 19 :: 16
        16
       ——   fath. ft.
  26)304(12 feet   = 2 0
     26 (nearly)   45 0 cast taken.
     ——
     44   43 0 cast corrected.

NOTE.—In the above proportion, 26 is the spring range at Liverpool, 16 the spring range at ship, and 20 the reduction at Liverpool.

EEE

ON THE CHART.

361. A Chart is a map or plan of a sea or coast. It is constructed for the purpose of ascertaining the position of the ship with reference to the land, and of shaping a course to any place.

362. The use to be made of the chart in each case determines the method of projection, and the particulars to be inserted. (1) The chart may be required for coasting purposes, for the use of the pilot, &c., and then only a very small portion of the surface of the globe being represented at once, no practical error results from considering that surface a plane, and a "*plane chart*" is constructed in which the different headlands, lighthouses, &c., are laid down according to their bearings. The soundings on these charts are marked with great accuracy; the rocks, banks, and shoals, the channels, with their buoys, the local currents, and circumstances connected with the tides, are also noted. (2) Again, for long sea passages the seaman requires a chart on which his course may conveniently laid down. The track of a ship always steering the same course appears as a straight line (and can at once be drawn with a ruler) on the *Mercator's chart*. Hence the charts used in navigation are Mercator's charts. (3) When great circle sailing is practicable, and of advantage, a chart on the "*central projection*," or gnomic, exhibits the track as a straight line, and is therefore convenient.*

ON MERCATOR'S CHARTS.

(See Norie, pages 126—131; or Raper's "Practice of Navigation," pages 120—127, on this subject).

363. A chart used at sea for marking down a ship's track and for other purposes, exhibits the surface of the globe on a plane on which the meridians are drawn parallel to each other, and therefore the parts BH, CI, DK, &c. (fig. chap. def. nav.), arcs of parallels of latitude, are increased and become equal to the corresponding parts of the equator UV, VW, &c. Now, in order that every point of this plane may occupy the same relative position with respect to each other that the points corresponding to them do on the surface of the globe, the distance between any points, A and O, and A and F must be increased in the same proportion as the distance FO has been increased. The true difference of latitude, AO, is thus projected on the

---

* The method lately introduced by HUGH GODFRAY, Esq., M.A., St. John's College, Cambridge, deserves special mention, as its beauty and simplicity will ultimately lead to its general adoption. A chart on the central projection, as stated above, exhibits the great circle as a straight line, and thus it is seen at once, whether the track between two places is a practicable one; hence, also, we have by inspection the point of highest latitude. An accompanying diagram then gives the different courses, and distances to be run on each, in order to keep within ¼ of a point to the great circle. This chart and diagram is fully described in the *Transactions of the Cambridge Philosophical Society*, vol. X, part II, and is published by J. D. Potter, Poultry.

chart into what is called the *meridional difference* of latitude, and the departure BH + CI + DK, &c., into the difference of longitude, and the representation is called a Mercator's projection. It is evidently a true representation as to *form* of every particular small track, but varies greatly as to point of *scale* in its different regions, each portion being more and more enlarged as it lies farther from the equator, and thus giving an appearance of distortion.*

(1.) In charts generally, the upper part as the spectator holds it, is the North, the lower part South, and that towards his right hand the East, that towards the left West, as on the compass card.

In a case which sometimes happens when the upper part is not the North, the North part may be known by the North part of the compass.

(2.) On Mercator's chart the parallel lines from North to South (from top to bottom) are termed *meridians*, and they are all perpendicular to the equator; the meridians on the extreme *right* and *left* are the *graduated* meridians —so called from showing the divisions for degrees and minutes. The *latitude* is measured on the graduated meridians, and also the *distance*.

(3.) The parallel lines from West to East (from left to right) are called *parallels*, and they are all parallel to the equator, the parallels at the top and bottom are *graduated* to degrees and minutes—and longitude is measured on the graduated parallels. Distance cannot be taken from them.

(4.) The depth of water is denoted, as also in some places the quality of the bottom. The numerals or figures in harbours, bays, channels, &c., indicate *soundings reduced to low water ordinary spring tides*. The *Roman figures* indicate the time of high water at full and change of the moon. Thus: XI hrs. 34ᵐ F & C means that the time of high water is thirty-four minutes past eleven on days of full and new moon. The *anchors* on the chart denote

---

* It is plain from the principles of Mercator's projection, and from the diagram (page 100) which connects the enlarged meridian with the difference of longitude, that if a ship set out on any point on the globe, and sail on the same oblique rhumb towards the pole, it can reach it only after an infinite number of revolutions round it. For from any point to the pole, the projected meridian is infinite in length, and so, therefore, is the difference of longitude due to this advance in latitude upon an oblique course. Consequently, this latitude can be reached only after the ship has circulated round the pole an infinite number of times.

These endless revolutions, however, are all performed in a finite time, the entire track of the ship being of limited extent. This, however paradoxical it may appear, is necessarily true from the principles of plane sailing, which shows that any finite advance in latitude is always connected with a finite length of track, this length being $\dfrac{\text{diff. lat.}}{\text{cos. course.}}$

The apparent paradox of the infinite number of revolutions about the pole being performed in a finite time, becomes explicable when we consider that, whatever be the progressive rate of the ship along its undeviating course, the times of performing the successive revolutions continually diminish as the ship approaches the pole, both the extent of circuit and the time of tracing it tending to zero, the limit actually attained at the pole itself; hence there must ultimately be an infinite number of such circuits to occupy a finite time.

When the pole is reached the direction all along preserved may still be continued, and a descending path will be described similar to that just considered, and which will conduct the ship to the opposite pole, after an infinite number of revolutions round it, as in the former case. In receding from this pole the track described will at length unite with that at first traced, the point of junction being that from which the ship originally departed. But for the strict mathematical proof of these latter circumstances the student may consult Professor DAVIES' curious and instructive papers on *Spherical Co-ordinates* in the Edinburgh Transactions, vol. XII.

anchorage. The small *arrows* show the *direction of the set of the current*, the current going with the arrow.

(5.) Lines called *Compasses*, similar to those on the compass card, are drawn at convenient intervals on the chart. In charts of large seas, as the Atlantic, these compasses are generally drawn so that the line from the North to the South point corresponds with the true meridian; but in coasting charts the same line generally coincides with the *correct* magnetic meridian.

(6.) When the *true* course between two places is known, it must be remembered that *Westerly* variation is allowed to the right, and *Easterly* to the left hand of the true course in order to obtain the compass course.

(7.) In "*cross bearings*," both bearings must be corrected for the deviation due to the direction of the ship's head at the instant of making the observations.

(8.) With respect to the method of determining the ship's position by cross bearings, it may be observed that this is the most complete of all methods when the difference of bearings is near 90°; but if the difference is small—as, for example, less than 10° or 20°, or near 180°—the ship's position will be uncertain, because a small error in the bearing will cause a great error in the distance.—(RAPER, page 120, No. 367.)

EXAMINATION PAPER.

Exn. 9b.

Rotation No.

Port of

EXAMINATION IN CHART.

The applicant will be required to answer in writing, on a sheet of paper which will be given him by the Examiner, all the following questions according to the grade of Certificate required, numbering his answers with the numbers corresponding with those in the question paper.

1.—A strange chart being placed before you, what should be your special care to determine, before you answer any questions concerning it, or attempt to make use of it?
A.—Which is the North part of the chart.
NOTE.—If a foreign chart, note what meridian it is projected for.

2.—How do you ascertain that in our British charts?
A.—In our British charts there is always at least one compass, the true north point of which is designated by a star or other ornament.

3.—Describe how you would find the course by the chart between any two places, A and B.
A.—Lay the edge of a parallel ruler over the two given places, A and B, then taking care to preserve the direction, move one edge of the ruler until it comes over the centre of the nearest compass on the chart; the circumference of the compass cut by the edge of the ruler would show the course according to the direction the one place is from the other.

4.—Supposing there to be   points of   variation at the first named place, what would the course be magnetic? *the true course being*
In answering this question, merely write down the magnetic course corresponding to the true course given.

Rule.—To turn true course into a magnetic course, allow

        Easterly variation to the left hand,
        Westerly variation to the right hand, thus—

True course N.E. by N.,	with 2 points W. variation, gives	N.E. by E.	magnetic.
,, ,,	with 2 points E. ,,	N. by E.	,,
True course S.E. by S.,	with 2 points W. variation, gives	S. by E.	magnetic.
,, ,,	with 2 points E. ,,	S.E. by E.	,,
True course S.W. by S.,	with 2 points W. variation, gives	S.W. by W.	magnetic.
,, ,,	with 2 points E. ,,	S. by W.	,,
True course N.W. by N.,	with 2 points W. variation, gives	N. by W.	magnetic.
,, ,,	with 2 points E. ,,	N.W. by W.	,,

A.—   points of   variation should be allowed to the   and the magnetic course would be

5.—How would you measure the distance between those two or any other two places on the chart?

A.—With a pair of dividers measure half the distance on the chart between, then placing one leg of the dividers on the middle latitude on the graduated meridian, measure on each side of the same, and the number of degrees measured between those two extreme points, brought into miles, will be the distance required.

6.—Why would you measure it in that particular manner?

A.—Because on a Mercator's chart the degrees of latitude increase in length as the latitude increases.

The above comprises all the questions on the chart that are put to Mates and Only Mates.

In addition to the above, the Masters are required to answer:

7.—What do you understand those small numbers to indicate that you see placed about the chart?

A.—Depths of water in fathoms, or feet, as specified on the chart.

8.—At what time of the tide?

A.—At low water ordinary springs.

9.—What are the requisites you should know in order that you may compare the depths obtained by your lead-line on board with the depths marked on the chart?

A.—The time from high water "rise and fall," or as it is now called, the "mean spring range."

NOTE.—The rise of tide at springs and neaps, as well as the range, is given on a chart to facilitate finding the height of tide at different hours between high and low water.

10.—What do the Roman numerals indicate that are occasionally seen near the coast and in harbours?

A.—The time of high water at that place at full and change of the moon.

NOTE.—It is generally expressed thus,—H.W. at F. and C., VIII^h 32^m, that is, high water at full and change of the moon occurs at 8^h 32^m.

11.—How would you find the time of high water at any place, the Admiralty Tide Tables not being at hand, nor any other special tables available?

A.—To the time of high water at full and change add 49 minutes for every day that has elapsed since the full or change of moon, the sum will be the P.M. tide for the given day approximately; or, to the time of the moon's meridian passage, corrected for longitude, add the port establishment, the sum will be the P.M. tide required.

All the above questions should be answered, but this does not preclude the Examiner from putting any other questions of a practical character, or which the local circumstances of the port may require.

PRACTICAL EXAMINATION IN THE USE OF THE CHART.

In this problem the chart is that used for Sumner's Method, on which the compass is *true*: the amount of variation, E. and W., will be written within or near the compass: and the Candidate is supplied with a Table of Deviations; these deviations are given in two columns, one for the *ship's head correct magnetic*, and the other for the *ship's head by compass*.

The form of the question is as follows:—

FOR ALL GRADES WHERE THE CHART IS USED.

a.—Using Deviation Card No.   find the course to steer by compass from   to   , also the distance.

    Answer.—Compass Course

              Distance

              Variation

b.—With the Ship's head on the above-named Compass Course   bore by Compass   Distant   miles; find the Lat. and Long. of the Ship when it was on that bearing.

    Answer.—Latitude

              Longitude

c.—With the Ship's head as above   bore by Compass   also   bore   by the same Compass. Find the Ship's position.

    Answer.—Latitude

            Longitude

Signature

Date

On the different days of Examination, the positions in (Question a), and consequently the directions of the line from one place to the other, the various bearings of the letters (which are taken to be lighthouses or land marks) are changed. The form above, however, indicates the character of the question.

DEVIATION CARD—No. I.

SHIP'S HEAD WHEN BUILDING.

Ship's Head Correct Magnetic	Deviations.	Ship's Head By Compass	Deviations.
North	11½° W.	North	15° W.
N. by E.	9 W.	N. by E.	12 W.
N.N.E.	6 W.	N.N.E.	7½ W.
N.E. by N.	3 W.	N.E. by N.	4 W.
N.E.	0	N.E.	0
N.E. by E.	2½ E.	N.E. by E.	3 E.
E.N.E.	6 E.	E.N.E.	5½ E.
E. by N.	7 E.	E. by N.	8 E.
East	9 E.	East	10 E.
E. by S.	10 E.	E. by S.	11½ E.
E.S.E.	11½ E.	E.S.E.	12 E.
S.E. by E.	12½ E.	S.E. by E.	13 E.
S.E.	13 E.	S.E.	14 E.
S.E. by S.	14 E.	S.E. by S.	15 E.
S.S.E.	15 E.	S.S.E.	15 E.
S. by E.	15 E.	S. by E.	15 E.
South	15 E.	South	15 E.
S. by W.	15 E.	S. by W.	14 E.
S.S.W.	15 E.	S.S.W.	12 E.
S.W. by S.	12½ E.	S.W. by S.	10 E.
S.W.	10 E.	S.W.	7 E.
S.W. by W.	5 E.	S.W. by W.	4 E.
W.S.W.	1½ W.	W.S.W.	1 W.
W. by S.	9½ W.	W. by S.	6 W.
West	15 W.	West	10 W.
W. by N.	18 W.	W. by N.	14 W.
W.N.W.	20½ W.	W.N.W.	17 W.
N.W. by W.	20½ W.	N.W. by W.	19 W.
N.W.	20 W.	N.W.	21 W.
N.W. by N.	18½ W.	N.W. by N.	20½ W.
N.N.W.	16 W.	N.N.W.	20 W.
N. by W.	14 W.	N. by W.	18 W.

DEVIATION CARD—No. II.

SHIP'S HEAD WHEN BUILDING.

Ship's Head Correct Magnetic	Deviations.	Ship's Head By Compass	Deviations.
North	6½° E.	North	10° E.
N. by E.	10½ E.	N. by E.	18 E.
N.N.E.	16 E.	N.N.E.	25 E.
N.E. by N.	20 E.	N.E. by N.	29 E.
N.E.	24 E.	N.E.	32 E.
N.E. by E.	28½ E.	N.E. by E.	33 E.
E.N.E.	30 E.	E.N.E.	31 E.
E. by N.	32 E.	E. by N.	28½ E.
East	32 E.	East	24 E.
E. by S.	30 E.	E. by S.	18 E.
E.S.E.	25 E.	E.S.E.	10 E.
S.E. by E.	10½ E.	S.E. by E.	5 E.
S.E.	1 W.	S.E.	1 W.
S.E. by S.	4½ W.	S.E. by S.	4 W.
S.S.E.	7 W.	S.S.E.	6 W.
S. by E.	10 W.	S. by E.	8 W.
South	12 W.	South	10 W.
S. by W.	13½ W.	S. by W.	11 W.
S.S.W.	15 W.	S.S.W.	13 W.
S.W. by S.	18½ W.	S.W. by S.	15 W.
S.W.	21 W.	S.W.	17 W.
S.W. by W.	23 W.	S.W. by W.	19 W.
W.S.W.	22½ W.	W.S.W.	21 W.
W. by S.	22 W.	W. by S.	23 W.
West	20½ W.	West	23 W.
W. by N.	19 W.	W. by N.	22 W.
W.N.W.	17 W.	W.N.W.	20 W.
N.W. by W.	14 W.	N.W. by W.	18 W.
N.W.	10½ W.	N.W.	15 W.
N.W. by N.	6½ W.	N.W. by N.	10 W.
N.N.W.	2 W.	N.N.W.	3 W.
N. by W.	2½ E.	N. by W.	3½ E.

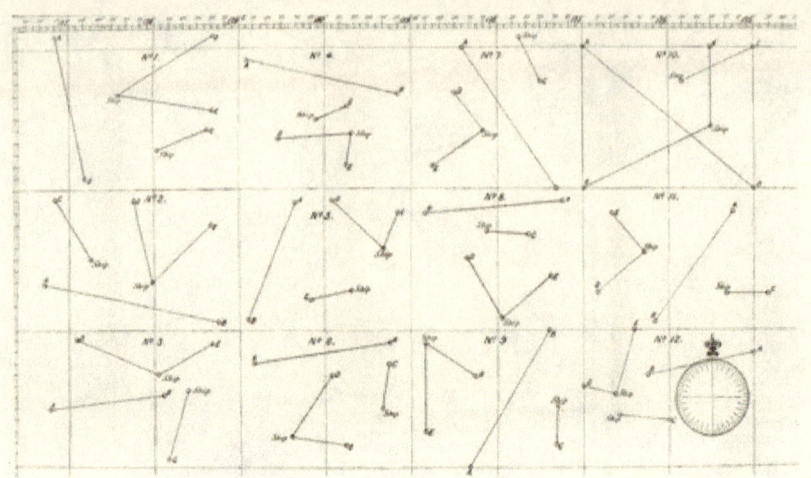

EXPLANATION OF QUESTIONS ON THE CHART.

Given **True** Course, to find Correct Magnetic and Compass Courses.

> Allow EAST variation to the *left*,
> WEST variation to the *right*; also,
> EAST deviation to the *left*,
> WEST deviation to the *right*.

Given **Compass** Bearings, to get True Bearings.

> Allow EAST variation to the *right*,
> WEST variation to the *left*.

EXAMPLE I. (See chart.)

(*a*) Using Deviation Card No. I: find the course to steer by compass from A to B, also the distance.

Lay an edge of a parallel rule over the positions A and B, then move the parallel rule (strictly preserving the direction) until an edge pass through the centre of the compass, then the point of the compass coinciding with the edge of the ruler shows the true course from A to B; suppose it to be S. by E., then proceed as follows:—

True course	S. by E. = S. 11° E. (L. of S.)
Variation, suppose	11 E. allow to *left*.
Correct magnetic course	S. 22 E. = S.S.E.
Deviation by card (left side)	15 E. allow to *left*.
Compass course	S. 37 E. = S.E. ¾ S.

The variation being 11° E., this allowed to the left makes correct magnetic course S. 22° E., or S.S.E.; then referring to Deviation Card No. I, page 399, under Ship's Head Correct Magnetic, the deviation is found opposite S.S.E. to be 15° E.; this must be allowed to the left on the correct magnetic course, making *compass* course S. 37° E., or S.E. ¾ S. Write down this compass' course, and also the distance from A to B, as measured on the graduated meridian.

> *Answer.*—Compass course S.E. ¾ S.
> Distance 61 miles.

Next, a position C is marked on the chart, and the Candidate is required to answer the following question:—

(*b*) With the ship's head on the above-named compass course, C bore by compass N. 44° E., distant 23 miles: find the latitude and longitude of the ship when it was on that bearing. (See No. 1, small chart).

Proceed as follows:—

Compass bearing	N. 44° E.
Deviation same as in *a*	15 E. allow to *right*.
Correct magnetic bearing	N. 59 E.
Variation same as in *a*	11 E. allow to *right*.
True bearing	N. 70 E.

Place the parallel rules on the compass in the corner of the chart, over N. 70° E. and S. 70° W., then work the parallels (strictly preserving the direction) till an edge pass over C. Draw a line trending S. 70° W. from C; also take off the distance from the graduated

F F F

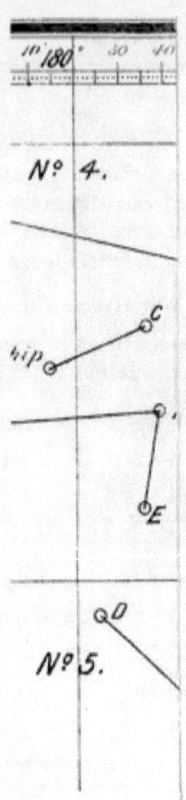

EXPLANATION OF QUESTIONS ON THE CHART.

Given **True** Course, to find Correct Magnetic and Compass Courses.

Allow **East** variation to the *left*,
West variation to the *right*; also,
East deviation to the *left*,
West deviation to the *right*.

Given **Compass** Bearings, to get True Bearings.

Allow **East** variation to the *right*,
West variation to the *left*.

EXAMPLE I. (See chart.)

(*a*) Using Deviation Card No. I: find the course to steer by compass from **A** to **B**, also the distance.

Lay an edge of a parallel rule over the positions A and B, then move the parallel rule (strictly preserving the direction) until an edge pass through the centre of the compass, then the point of the compass coinciding with the edge of the ruler shows the true course from A to B; suppose it to be S. by E., then proceed as follows:—

True course	S. by E. =	S. 11° E. (L. of S.)
Variation, suppose		11 E. allow to *left*.
Correct magnetic course		S. 22 E. = S.S.E.
Deviation by card (left side)		15 E. allow to *left*.
Compass course		S. 37 E. = S.E. ¾ S.

The variation being 11° E., this allowed to the left makes correct magnetic course S. 22° E., or S.S.E.; then referring to Deviation Card No. I, page 399, under Ship's Head Correct Magnetic, the deviation is found opposite S.S.E. to be 15° E.; this must be allowed to the left on the correct magnetic course, making *compass* course S. 37° E., or S.E. ¾ S. Write down this compass course, and also the distance from A to B, as measured on the graduated meridian.

Answer.—Compass course S.E. ¾ S.
Distance 61 miles.

Next, a position C is marked on the chart, and the Candidate is required to answer the following question:—

(*b*) With the ship's head on the above-named compass course, C bore by compass N. 44° E., distant 23 miles: find the latitude and longitude of the ship when it was on that bearing. (See No. 1, small chart.)

Proceed as follows:—

Compass bearing	N. 44° E.
Deviation same as in *a*	15 E. allow to *right*.
Correct magnetic bearing	N. 59 E.
Variation same as in *a*	11 E. allow to *right*.
True bearing	N. 70 E.

Place the parallel rules on the compass in the corner of the chart, over N. 70° E. and S. 70° W., then work the parallels (strictly preserving the direction) till an edge pass over C. Draw a line trending S. 70° W. from C; also take off the distance from the graduated

F F F

meridian, so that the middle latitude between C and the ship's position may bisect the distance contained between the points of the dividers; apply the distance thus found from C in the direction of the line drawn through it, and here is found the ship's position—latitude and longitude—which take from the proper scale and write down.

Answer.—Latitude 51° 16½' N.
Longitude 178° 2' E.

Finally, two positions D and E are marked on the chart, and the Candidate is required to answer the following questions. (See No. 1, small chart.)

(c) With the ship's head as above, D bore by compass N. 33° E., also E bore N. 74° E. by the same compass: find the ship's position.

In this question, the same corrections (variation 25° W. and deviation 11° E.) applied to the compass bearings of D and E, give their true bearings, viz., N. 59° E. and N. 80° E., thus—

	D	E
Compass bearings	N. 33° E.	N. 74° E.
Deviation same as in *a*, allow to *right*	15 E.	15 E.
Correct magnetic bearings	N. 48 E.	N. 89 E.
Variation same as in *a*, allow to *right*	11 E.	11 E.
True bearings	N. 59 E. = N.E. by E. ¼ E.	N. 100 E.
		180
		S. 80 E. = E. ⅞ N.

Proceed, as in question *b*, to draw a line S.W. by W. ¼ W. from D, and another line W. ⅞ S. from E; the point where these two lines intersect is the ship's position: find the latitude and longitude of this point, and write them down.

Answer.—Latitude 51° 40' N.
Longitude 177° 34½' E.

EXAMPLE II. (See chart).

(*a*) Using Deviation Card No. I: find the course to steer by compass from A to B, also the distance. (Suppose variation 29° E.)

True course	E. by S. = S. 79° E.
Variation, suppose	29 E. allow to *left*.
Exceeds 90°,	S. 108 E.
Subtract from	180
Correct magnetic course	N. 72 E. = E.N.E. ⅜ E.
Deviation by card	6½ E. allow to *left*.
Compass course	N. 65½ E. = N.E. by E. ¾ E.
Distance 77½ miles.	

Answer.—Compass course N.E. by E. ¾ E.
Distance 77½ miles.

(*b*) With the ship's head on the above-named compass course, C bore by compass N. 69½° W., distant 19 miles: find the latitude and longitude of the ship when it was on that bearing.

Compass bearing	N. 69½° W.
Deviation same as in *a*	6½ E. allow to *right*.
Correct magnetic bearing	N. 63 W. = N.W. by W. ⅞ W.
Variation same as in *a*	29 E. allow to *right*.
True bearing	N. 34 W. = N.W. by N.

Answer.—Latitude 50° 31' N.
Longitude 177° 15' E.

(*c*) With the ship's head as above, D bore by compass N. 46½° W., also E bore N. 3½° E. by the same compass: find the ship's position.

	D	E
Compass bearings	N. 46½° W.	N. 3½° E.
Deviation same as in *a*, allow to *right*	6½ E.	6½ E.
Correct magnetic bearing	N. 40 W.	N. 10 E.
Variation same as in *a*, allow to *right*	29 E.	29 E.
True bearings	N. 11 W. = N. by W.	N. 39 E. = N.E. ½ N.

Answer.—Latitude 50° 18½′ N.
Longitude 177° 59′ E.

EXAMPLE III. (See chart.)

(*a*) Using Deviation Card No. I: find the course to steer by compass from A to B, also the distance. Suppose variation 25° W.

True course from A to B is E. by N. =	N. 84° E.	
Variation	25 W.	allow to *right*
Exceeds 90	N. 109 E.	
Subtract from	180	
Correct magnetic course	S. 71 E. = E.S.E. ¾ E.	
Deviation by card	11 E.	allow to *left*.
Compass course	S. 82 E. = E. ¾ S.	

Answer.—Compass course E. ¾ S.
Distance 52 miles.

(*b*) With the ship's head on the above-named compass course, C bore by compass S.S.W. ¾ W., distant 31 miles.

Compass bearing S.S.W. ¾ W. =	S. 31° W.	
Deviation same as in *a*	11 E.	allow to *right*.
	S. 42 W.	
Variation same as in *a*	25 W.	allow to *left*.
True bearing of C	S. 17 W.	

Answer.—Latitude 49° 33′ N.
Longitude 178° 23′ E.

(*c*) With the ship's head as above, D bore by compass N.W. ¾ W., also E bore N.N.E. ¾ E. by the same compass: find the ship's position.

	D	E
Compass bearing	N.W. ¾ W. = N. 53½° W.	N.N.E. ¾ E. = N. 31° E.
Deviation same as in *a*, allow to *right*	11 E.	11 E.
Correct magnetic bearings	N. 42½ W.	N. 42 E.
Variation same as in *a*, allow to *left*	25 W.	25 W.
True bearings	N. 67½ W. = W.N.W.	N. 17 E. = N. *b* E. ½ E.

Answer.—Latitude 49° 40′ N.
Longitude 178° 3′ E.

EXAMPLE IV. (See chart).

(*a*) Using Deviation Card No. I: find the course to steer by compass from A to B, also the distance.

Answer.—Compass course S. 68½° E.
Distance 67 miles.
Variation 21° W.

(b) With the ship's head on the above-named compass course, C bore by compass E. by N. ½ N., distant 14 miles: find the latitude and longitude of the ship when it was on that bearing.

    Answer.—Latitude 51° 29′ N.
              Longitude 179° 55′ E.

(c) With the ship's head as above, D bore by compass N. 83½° W., also E bore S. 17° W. by the same compass: find the ship's position.

    Answer.—Latitude 51° 23′ N.
              Longitude 179° 42′ W.

EXAMPLE V. (See chart).

(a) Using Deviation Card No. I: find the course to steer by compass from A to B, also the distance.

    Answer.—Compass course S. 9° E.
               Distance 53 miles.

(b) With the ship's head on the above-named compass course C bore by compass S. 47° W., distant 18 miles: find the latitude and longitude of the ship when it was on that bearing.

    Answer.—Latitude 50° 17′ N.
              Longitude 179° 41′ W.

(c) With the ship's head as above, D bore by compass N. 80° W., also E bore N. 7° W. by the same compass: find the ship's position.

    Answer.—Latitude 50° 35′ N.
              Longitude 179° 19½′ W.

EXAMPLE VI. (See chart.)

(a) Using Deviation Card No. I.: find the course to steer by compass from A to B, also the distance.

    Answer.—Compass course N. 58° E.
               Distance 61 miles.
               Variation 20° E.

(b) With the ship's head on the above-named compass course, C bore by compass N. 15° W, distant 19 miles: find the latitude and longitude of the ship when it was on that bearing.

    Answer.—Latitude 49° 26′ N.
              Longitude 179° 19½′ W.

(c) With the ship's head as above, D bore by compass N. 10° E.; also E bore by the same compass E. by N. ½ N.: find the ship's position.

    Answer.—Latitude 49° 13′ N.
              Longitude 179° 37′ E.

EXAMPLE VII. (See chart).

(a) Using Deviation Card No. I: find the course to steer by compass from A to B, also the distance.

    Answer.—Compass course S. 66° E. (S.E. by E. ¼ E.)
              Distance 72 miles.
              Variation 19° E.

(b) With the ship's head on the above-named compass course, C bore by compass S. 57° E, distant 21 miles: find the latitude and longitude of the ship when it was on that bearing.

    Answer.—Latitude 52° 4′ N.
              Longitude 177° 44′ W.

(c) With the ship's head as above, D bore by compass N. 71° W., also E bore S. 22° W. by the same compass: find the ship's position.

   Answer.—Latitude 51° 24½' N.
       Longitude 178° 11' W.

EXAMPLE VIII. (See chart).

(a) Using Deviation Card No. I: find the course to steer by compass from A to B, also the distance.

   Answer.—Compass course S. 73° W. (W. by S. ½ S.)
       Distance 60 miles.
       Variation 14½° E.

(b) With the ship's head on the above-named compass course, C bore by compass N. 83½° E., distant 17 miles: find the latitude and longitude of the ship when it was on that bearing.

   Answer.—Latitude 50° 42' N.
       Longitude 178° 7' W.

(c) With the ship's head as above, D bore by compass S.E. ½ S., also E bore S.W. ½ S. by the same compass: find the ship's position.

   Answer.—Latitude 50° 5' N.
       Longitude 177° 58' W.

EXAMPLE IX. (See chart.)

(a) Using Deviation Card No. I: find the course to steer by compass from A to B, also the distance.

   Answer.—Compass course N. 46° E. (N.E. ¼ E.)
       Distance 69 miles.
       Variation 16° W.

(b) With the ship's head on the above-named compass course, C bore by compass S. 18° W., distant 17 miles: find the latitude and longitude of the ship when it was on that bearing.

   Answer.—Latitude 49° 27' N.
       Longitude 177° 19' W.

(c) With the ship's head as above, D bore by compass S. 43° E., also E bore S. 15° W. by the same compass; find the ship's position.

   Answer.—Latitude 49° 55' N.
       Longitude 178° 50' W.

EXAMPLE X. (See chart).

(a) Using Deviation Card No. I: find the course to steer by compass from A to B, also the distance.

   Answer.—Compass course N. 84½° E. (E. ½ N.)
       Distance 95 miles.
       Variation 34° E.

(b) With the ship's head on the above-named compass course, C bore by compass N.N.E., distant 35 miles; find the latitude and longitude of the ship when it was on that bearing.

   Answer.—Latitude 51° 45' N.
       Longitude 175° 51' W.

(c) With the ship's head as above, D bore by compass S.S.W., also E bore N. 43° W. by the same compass: find the ship's position.

   Answer.—Latitude 51° 26½' N.
       Longitude 175° 30' W.

EXAMPLE XI. (See chart).

(*a*) Using Deviation Card No. I: find the course to steer from A to B, also the distance. (See No. 11 in small chart). Variation 22° W.

Answer.—Compass course S. 56° W., or S.W. by W.
Distance 57 miles.

(*b*) With the ship's head on the above-named compass course, C bore by compass S. 71° E., distant 18 miles: find the latitude and longitude of the ship when it was on that bearing.

Answer.—Latitude 50° 15' N.
Longitude 175° 19' W.

(*c*) With the ship's head as above, D bore by compass W.S.W., also E bore N. by W. ¾ W. by the same compass: find the ship's position.

Answer.—Latitude 50° 33' N.
Longitude 176° 18' W.

EXAMPLE XII. (See chart).

(*a*) Using Deviation Card No. I: find the course to steer by compass from A to B, also the distance. (See No. 12 in chart). Variation 28° W.

Answer.—Compass course N. 54° W.
Distance 50 miles.

(*b*) With the ship's head on the above-named compass course, C bore by compass S. 35° E., distant 22 miles: find the latitude and longitude of the ship when it was on that bearing.

Answer.—Latitude 49° 23' N.
Longitude 176° 34' W.

(*c*) With the ship's head as above, D bore by compass N.N.E. ⅞ E., also E bore N. 64° E. by the same compass: find the ship's position.

Answer.—Latitude 49° 32' N.
Longitude 176° 37' W.

EXERCISES ON THE CHART.
FOR ONLY MATE, FIRST MATE, AND MASTER.
(No Deviation allowed).

North Sea.

1. Latitude 55° 5' N.
   Longitude 0 5 E.
   Required the course and distance to Hartlepool.

2. Latitude 57° 30' N.
   Longitude 0 40 E.
   Required the course and distance to Tynemouth Light.

3. Latitude 60° 21' N.
   Longitude 0 35 E.
   Required the course and distance to Udsire.

4. Latitude 57° 25' N.
   Longitude 7 25 E.
   Required the course and distance to the Naze of Norway.

5. Latitude 55° 40' N.
   Longitude 0 15 W.
   Required the compass course and the distance to St. Abb's Head Light.

6. Latitude 58° 25' N.
   Longitude 2 10 W.
   Required the compass course and the distance to Duncansby Head.

7. Required the true and magnetic bearing and distance between Whitby and the Naze of Norway.

8. Required the direct true and magnetic course and distance between Buchanness in Scotland to the entrance of the Texel.

9. A ship from Kinnaird's Head, in Scotland, sailed S.E. by E. (true) 186 miles: required the latitude and longitude she is come to, and the direct course and distance she must sail in order to arrive at Heligoland.

10. A ship from Heligoland sailed on a direct course between the North and West 197 miles, and spoke a ship which had run 170 miles on a direct course from Hartlepool: required the latitude and longitude of the place of meeting, also the course steered by each ship.

11.   Sunderland Light, bearing by compass S.W. ½ S.
     Coquet Island   ,, ,, N.W.

Required the latitude and longitude of ship; also the course and distance to Hartlepool Light.

12.   Buchanness Light, N. by W. ½ W., by compass.
     Girdleness Light, West.

Required the latitude and longitude of ship; also the course (by compass) and distance to the Staples.

13.   The Skerries, North, by compass.
     Sumburg Head, W. ¾ S. ,,

Required the latitude and longitude in; also the compass course and distance to Peterhead.

14.   Flambro' Head Light, S.W. by S., by compass.
     Whitby Lights,   N.W. by W. ¾ W. ,,

Required the latitude and longitude in; also the compass course and distance to Outer Dowsings.

15.   The Dudgeon Light, W. by N., by compass.
     Hasbro' Sand-end Light, S.S.W.   ,,

Required the latitude and longitude of ship; also the compass course and distance to Flambro' Head.

16. Scarbro' Light was observed to bear S.W. by compass, then sailed E.S.E. 11 miles, and the light then bore West: required the latitude and longitude of the ship at each station, and her distance from the light.

17. Coasting along shore, observed Tynemouth Light to bear W. by S. by compass; then sailed S. by W. 16 miles, and the light bore N.W. by N.: required the latitude and longitude of the ship, and her distance from the light.

English and Bristol Channels, and South Coast of Ireland.

1.   Latitude 50° 1' N.
    Longitude 2 4 W.
Required the compass course and distance to the Caskets.

2.   Latitude 48° 50' N.
    Longitude 5 50 W.
Required the compass course and distance to Ushant.

3.   Latitude 49° 30' N.
    Longitude 3 30 W.
Required the compass course and distance to the Start Point.

4.   Latitude 50° 10' N.
    Longitude 1 10 W.
Required the compass course and distance to St. Catherine's Light.

5.   Latitude 50° 30' N.
    Longitude 0 55 E.
Required the compass course and distance to Dungeness.

6.   Latitude 48° 55' N.
    Longitude 6 5 W.
Required the compass course and distance to the Lizard.

7.   Latitude 50° 50' N.
    Longitude 10 35 W.
Required the compass course and distance to the Fastnet Rock.

8.   Latitude 51° 52' N.
    Longitude 6 6 W.
Required the compass course and distance to the Tuskar Light.

9.   Latitude 50° 50' N.
    Longitude 7 20 W.
Required the compass course and distance to Old Head of Kinsale.

10.   Latitude 50° 30' N.
    Longitude 8 30 W.
Required the compass course and distance to Cape Clear.

11. Longships Light, bearing by compass E.N.E.
St. Agnes' Light,   „   „   N.N.W. ½ W.

Required the latitude and longitude in; also the compass course and distance to the Lizard.

12. Berry Head, bearing by compass N. ½ E.
Start Point   „   „   W. by N. ½ N.

Required the compass course and distance to Portland.

13. Bill of Portland, bearing by compass N.W. by W.
St. Alban's Head,   „   „   N.E. ½ E.

Required the latitude and longitude of ship, and the compass course and the distance to Start Point.

14. Longships Light, bearing by compass S.S.E.
Seven Stones Light „   „   W. by S.

Required the latitude and longitude of ship; also the compass course and the distance to Roches Point.

15. Tuskar Rock   N.E.   by compass.
Great Saltees Lightvessel N.W. ½ W.   „

Required the latitude and longitude of the ship; also the course (by compass) and distance to the Smalls.

16. Shipwash Light, bearing by compass W. by N.
Galloper   „   „   S.S.W.

Required the latitude and longitude in; also the compass course and distance to Corton Lightvessel.

17. Caldy Island Light, bearing by compass E.N.E.
Lundy Island Light, „   „   S. by E.

Required the latitude and longitude of ship; also the compass course and distance to the Smalls.

18. Lizard Lights, bearing by compass E. ¼ S.
Longships,   „   „   N. ¼ W.

Required the latitude and longitude of ship; also the compass course and distance to St. Agnes' Light.

19. Smalls Light,   bearing by compass N. ½ E.
St. Ann's (Milford Haven) „   „   E.S.E.

Required the latitude and longitude of ship; also the compass course and distance to Seal Rock (Lundy Island).

20. Dungeness, bearing by compass N.E. by E. ½ E.
Beachy Head „   „   N.W. ½ W.

Required the latitude and longitude of the ship; and her distance from each place.

21. A ship is bound to Boulogne, being 18 miles distant, and lying directly to windward, the wind being E. by N. (true). It is intended to reach her port on two boards, the first being on the port tack, and the ship can lie within six points of the wind; required the course and distance upon each tack.

ANSWERS.

NOTATION, *Pages* 16 *and* 17.

1. 100   2. 101   3. 110   4. 9009   5. 9090   6. 9909   7. 5074
8. 10700   9. 90090   10. 305000   11. 900900   12. 505550   13. 1003008
14. 5030049   15. 9900006   16. 58000009   17. 70302441   18. 222000035
19. 60406005   20. 800001033   21. 900900900   22. 700000007   23. 180000000
24. 500000000   25. 580245192   26. 707007077

NUMERATION, *Page* 18.

1. Forty-three.   2. Sixty.   3. Twelve.   4. Twenty-one.   5. One hundred.   6. One hundred and one.   7. One hundred and ten.   8. Five hundred.   9. Five hundred and five.   10. Five hundred and fifty.   11. One thousand.   12. Two thousand and twenty.   13. Three thousand three hundred and three.   14. Four thousand and four.   15. Seven thousand seven hundred and seven.   16. Eight thousand eight hundred and eighty.   17. Eighty-seven thousand and fifty-four.   18. Seventy thousand seven hundred and seven.   19. Sixty thousand eight hundred and eighty.   20. Ninety-nine thousand four hundred and four.   21. Nine hundred and three thousand seven hundred and fifty-six.   22. Two hundred and two thousand two hundred and two.   23. Four hundred thousand and four hundred.   24. Five hundred and fifty thousand five hundred and fifty.   25. One million and one.   26. Eight million and forty-seven thousand three hundred and twenty-eight.   27. Four million and ninety thousand and three hundred.   28. Five million two hundred and ten thousand and seven.   29. Six million and thirty thousand four hundred and five.   30. Nine million nine thousand and nine hundred.   31. Forty-one million forty-one thousand and fourteen.   32. Three million and six.   33. Twenty million eighty-four thousand two hundred and sixteen.   34. Five million one thousand eight hundred and sixty.   35. Eight million eighty thousand eight hundred and eight.   36. Fifty-five million seven hundred thousand and five.   37. Seventy-six million and fourteen thousand and fifty-nine.   38. Six million six thousand six hundred and six.   39. Fifty-six million seven hundred thousand five hundred and five.   40. One hundred and twenty million fifteen thousand and fifteen.   41. Two hundred and two million two hundred and two thousand and two hundred.   42. One hundred million one hundred thousand one hundred and one.   43. Two hundred and seventy-five million eight thousand and five.   44. One hundred million ten thousand and one.   45. Seventy-nine million thirty thousand one hundred and eighty-four.   46. Four hundred and eight million seventy-six thousand and thirty-two.   47. Four hundred and one million four hundred thousand and fifty-six.   48. Nine hundred and eight million five hundred thousand and sixty.

ADDITION, *Page* 20.

1.	1274170	7.	1648127	13.	3312667	19.	8518439
2.	1634607	8.	2067690	14.	3018498	20.	7498159
3.	1659291	9.	3329175	15.	2797285	21.	9560155
4.	2333431	10.	3724599	16.	3519772	22.	5621434
5.	3005313	11.	4483647	17.	9185198	23.	6524956
6.	1536206	12.	4105670	18.	7485613	24.	8238336

25. (1) 13788543   (2) 12844819   (3) 14661377   (4) 13937260   (5) 15878135
    (6) 10176138   (7) 10970368   (8) 13815798
26. 20566726566

SUBTRACTION, *Page* 23.

1.	621511	9.	681179	17.	2922930923	25.	4244103
2.	539540	10.	507871	18.	90889109901	26.	5460813
3.	1	11.	376099	19.	10020950993	27.	8026758
4.	9	12.	174386	20.	91089009099	28.	98599383
5.	676001	13.	107500	21.	238036793034	29.	983982
6.	554999	14.	222419	22.	5540058	30.	9985268
7.	480895	15.	157406	23.	5866974	31.	743187
8.	590098	16.	8409091	24.	6521913	32.	8457

33. 71880

MULTIPLICATION, Page 27.

1. 685295792
2. 1962965961
3. 1506172792
4. 1899328910
5. 550942443156
6. 45652143474
7. 3886950304
8. 5159176101
9. 9876543210
10. 9803614194
11. 7774239492
12. 11031283848

13. (2) 476949824 (3) 5432756589 (4) 30524788736 (5) 116442828125 (6) 347696421696 (7) 876760466309 (8) 1953586479104 (9) 3960479553381 (11) 13202276674301 (12) 22152570988544

Page 30.

1. 2684444024
2. 5629618680
3. 8918232255
4. 61286228934
5. 1599378c666
6. 999999999
7. 27349835014665
8. 32228449759163
9. 770930181732
10. 199999929143681
11. 22435037273423888
12. 355733452311336
13. 395130761574453
14. 66809446c288461
15. 83123729682c2684
16. 460937297776
17. 121932631112635269c
18. 28725564940c8787
19. 1614054474492415542
20. 53107710897987
21. 40155302248305278754132
22. 15232906283421580
23. 506034412716915c
24. 2961296118145871914806c56
25. 102030190786666880930302c1
26. 99940014998000149994c001
27. 99999995000000c04000
28. 99940014998c00149994c001

Page 31.

1. 7239334500
2. 10814314500
3. 24538039680000
4. 290675534724420
5. 229262443200000
6. 581643960
10. 12003400820050006000000
7. 2345457600
8. 2315085840
9. 35333670133890810

1. 10285980
2. 11838608
3. 38114062
4. 24335360
5. 20146968
6. 246649057c
7. 13085320545
8. 155726288c
9. 18532696320

DIVISION, Page 35.

1. 67896347-1
2. 194899128-2
3. 99836471
4. 59648952
5. 66779748-5
6. 39512348-1
7. 868427625-6
8. 274473675
9. 25409614-6
10. 100107478-9
11. 91261430-10
12. 4953087942-8

1. (2) 96168 (3) 8442-2 (4) 1502-2 (5) 393-4 (6) 131-5 (7) 52-2 (8) 23-3 (9) 11-5 (11) 3-5 (12) 2
2. (2) 109375 (3) 9602 (4) 1708-3 (5) 448 (6) 150 (7) 59-3 (8) 26-5 (9) 13-1 (11) 3-10 (12) 2-4
3. (2) 63856 (3) 5606 (4) 997-3 (5) 261-2 (6) 87-3 (7) 34-5 (8) 15-4 (9) 7-6 (11) 2-3 (12) 1-4
4. (2) 137828 (3) 12100 (4) 2153-2 (5) 564-2 (6) 189 (7) 74-6 (8) 33-5 (9) 16-5 (11) 4-10 (12) 2-11
5. (2) 31261 (3) 2744-1 (4) 483-1 (5) 128 (6) 42-5 (7) 17 (8) 7-5 (9) 3-6 (11) 1-1 (12) 0-8
6. (2) 146406 (3) 12853 (4) 2287-2 (5) 599-3 (6) 200-4 (7) 79-4 (8) 35-5 (9) 17-5 (11) 5-3 (12) 3-1
7. (2) 154907 (3) 13599-1 (4) 2420-1 (5) 634-2 (6) 212-2 (7) 84-1 (8) 37-6 (9) 18-5 (11) 5-6 (12) 3-3
8. (2) 73531 (3) 6455-1 (4) 1148-3 (5) 301 (6) 100-5 (7) 40 (8) 17-7 (9) 8-7 (11) 2-7 (12) 1-6

Page 37.

1. 75638-2
2. 34785-7
3. 542370
4. 220950-14
5. 127474-28
6. 100045-25
7. 461002-18
8. 725983
9. 706321-515
10. 8607936214-143
11. 576845
12. 245728
13. 567977-3852
14. 107356-2031
15. 100010-10
16. 999000-1000
17. 90000900c0-10000
18. 848585c-43614
19. 1068392-117002
20. 544023-195858
21. 70030401
22. 730956788-980154321
23. 63245553
24. 9007609600
25. 30452987817410c474
26. 4342944819-49016833

Page 39.

1. 463519673763533-5
2. 2720149043856034-10
3. 1582874324701-32
4. 95022741046776-8
5. 14964459409277-63
6. 4489339863279-30
7. 2720698019559-123
8. 34045491087172-1

1. 187157296759729-46
2. 9618695639112-25
3. 329218107-670
4. 40316322081100 56-69
5. 1039682583-834
6. 3396-5094687

MISCELLANEOUS EXAMPLES, Page 40.

1. 10004; 4004; 44004.   2. 474788.   3. 2808846363.
4. 7398981889800.   5. 9576108.   6. 100100100.
7. 87846125.   8. 99912350214.   9. 1000622528890200.
10. 2768884-85187.   11. 103080207.   12. 1202609.
13. 71625861494.   14. 128721301414200.   15. 607862510254-15696883.

16. The one is larger than the other by forty-nine thousand nine hundred and fifty, i.e., by 49950.   17. 60768396; of 129847 and 40068.   18. 847021, 36865365.   19. 6 and 3.
20. 324937594.   21. 300490090, sum; 275798734, difference; 3557338118051336, product. 5555656, sum; 3086522, difference; 5334673883465, product.   22. 372 tons.   23. 127 years.
24. 7852 times.   25. 34 ships.   26. 141.   27. 1002.   28. 65280.   29. 129115.
30. 146, after subtracting it 390 times.   31. 203.   32. 1666350, sum; 1639900, difference; 2186257 8125, product; 125, quotient.   33. 9843750, sum; 9687500, difference; 762939453125 product; 125, quotient.

NOTATION OF DECIMALS, Page 44.

1. ·3, ·03, ·003, and 3·3; also ·7, 11·7, ·33, and 1·015.
2. ·01, ·0021, ·0117, ·0000003, ·1, ·53, ·007, ·0011, and ·00137.
3. 30·1, 400·01, 53·00415, 50·000101, ·441, 33·1, and ·000000000501.
4. ·9178, 91·78, ·09178, ·0091, ·00009, 520·3, and ·90.
5. 3·0141, 6·72819, ·000672819, and 6728·19.
6. 7·06, 43·1143, 9·07823457, 1·000001, and 35·721341.
7. ·073, ·0197, ·000001, ·00261, and ·0001001.
8. 1·54, 24·079, 315·008005, ·00000011, and ·00903.
9. ·1, ·03, ·005, ·105, ·000002, ·000060, 41·08, 1000·001, 30·000006, ·00001, and ·00002375.
10. Two hundred and eighty-three thousandths; Five thousand three hundred and twenty-one ten thousandths; Seventy-four thousand eight hundred and ninety-five hundred thousandths; Eight hundred and twenty-one thousand and fifty-six millionths; Twenty-seven, and eight thousand three hundred and fifty-four ten thousandths; Thirty-four, and nine ten thousandths; Forty-three, and one one hundred and one thousand and seven millionths; Twenty-three, and seventy-five hundredths; Two, and three hundred and seventy-five thousandths; Two thousand three hundred and seventy-five ten thousandths; Two thousand three hundred and seventy-five hundred millionths.
11. Six tenths; Ninety-two hundredths; Five thousand four hundred and ninety-eight ten thousandths; Seven, and seven hundredths; Twenty-six, and four hundred and five thousandths; One millionth; Thirty-seven hundred thousandths; Eleven, and one hundred and one thousand one hundred and one millionth; Four hundred and forty thousand three hundred and eight ten millionths; Eighty-two thousand three hundred and forty-four hundred thousandths; Thirteen thousand two hundred and thirty-six hundred thousandths.
12. Nine, and four hundred and fifty-seven ten thousandths; Four thousand and four, and three hundred and forty-five ten millionths; Three, and four hundred thousandths; Five hundred and twenty-four millions six hundred and thirty-four, and eight thousand and thirty-four ten millionths; Three thousand seven hundred and five thousand millionths; Twenty-four thousand and fifty-six thousand millionths; Seven thousand and five, and six hundred and seventy-four thousand millionths; One hundred thousand, and one ten millionth; Ten, and one thousandth; Nine, and twenty-eight millionths; One, and six thousand and three ten millionths.

13. One, and one millionth; One million and one ten millionth; One hundred millionths; One, and thirteen thousand and four hundred thousandths; Nine, and two hundred and three thousand one hundred and sixty-seven millionths; Four, and three million eight thousand and four ten millionths; Twenty-seven, and four million six hundred and twenty-seven thousand three hundred and fifty ten millionths.

ADDITION OF DECIMALS, *Page* 45.

1. 745·0261; 2·919563
2. 886·9326; 1681·679
3. 1437·4179; 330·875521
4. 4009·0; 501·15998
5. 538·6422021
6. 140·1996; 1408·25559
7. 53·6769; 127·050340
8. 1·1111; 42·7162

9. 1·2345; 945·5993

SUBTRACTION OF DECIMALS, *Page* 47.

1. 3·431; 8·20001; ·0011; 8·000001
2. 39·8479194; 31·99968; 7·336606; 91·7423
3. ·01; 98·99999901; 9·999999; 995·710; 541·787
4. 64·0317753; 8·20001; 72·5193401
5. ·000099; ·000396; 31·99968; 24680·12377
6. 699·930; ·0000999
7. ·0378; ·062156
8. ·00510; 28999·908

MULTIPLICATION OF DECIMALS, *Page* 48.

1. 10·0; 10·0; 1190·0; 11·9; ·0119; ·00119
2. ·000000101; 3·06034; ·000000112; ·00210175
3. ·075460; 1·8019; 74·9265; ·0010488693·3696
4. ·0306002448; 470116914·4360; 536·660075952
5. 2·5067823; ·000011816009; ·00000006676542672
6. 47·83; 500·0; 75000·0
7. ·001301400; 1·5; ·00000072
8. 5·314410; 4·096; ·032016
9. ·00012343·21; ·000444080; 6138·36

DIVISION OF DECIMALS, *Page* 51.

1. 19·82421; 14·16015; 11·01345; 2·7533625; 1·12637557
2. ·2017386; ·1008693; ·0672462; ·025217325; ·0009456496
3. 134·88057; 790·9882353; 59·406396; 24·82661
4. 14·789983; 255·121; 1210·234426; ·02
5. ·8810891; 908·83768; ·1108754; ·0174532922
6. 711·855; 2280·28; 234508
7. ·03; 74·84; 43206·7; ·000007375; 83671000; ·000000000003; ·061096
8. 2681·081081; ·0000360074; ·0001; 6·578947; ·00862.
9. 15000·0; 5060000·0; ·008; ·0375313.
10. ·013; 4·57; ·008; 73939·39.
11. ·050005; 1250·0; ·0125; 60·2589.
12. ·0000000125; 125·0; 125·0; ·00004.
13. ·00000125; ·00001; 20200; 77485·93.
14. ·10; 10·0; ·001; ·001; ·1000; 10000; ·00001.
15. ·0093536; 7393·939; 39723·66; 241·6192; 100·0; ·60000000; 4000000; ·000006; 31000000.
16. ·0036; ·93; ·52306; ·0008; ·0000010076364.
17. ·0882352941; ·0172566371768; ·000099900099900099000999.

REDUCTION OF DECIMALS, *Page* 53.

1. ·4375; ·73; ·2142857; ·34375; ·1875; ·076923; ·0112; ·275.
2. ·5384615; ·6470588; ·6315789; ·185; ·7167235; ·3183098; ·4683544; ·0104895.

Page 55.

1. ·2833; ·4833; ·7; ·4166; ·8; ·9666
2. ·788260449735; ·05
3. 11·5925 days; 19·71649305 days; 15·74219164 days; 119·22170138 days
4. 8°·1875; 19°·67916; 104°·26875; 82°·325
5. ·046875 tons; ·61964185714 tons; ·7125 tons; ·51875 tons; 13·303125 tons; 1·65 tons; 17·9015625 tons; ·8669642 tons
6. ·316; ·435416; ·23125; ·796875; ·9917083; 1·697916; and ·476190

EXAMPLES FOR PRACTICE, Page 56.

1. 1 qr.; 10 lbs. 8 oz.; 10 oz.; 19 cwt. 2 qrs.; 15 cwt. 2 qrs. 4 lbs. 0 oz. 8·192 drs.; 4 cwt. 3 qrs. 27 lbs. 0 oz. 3·6864 drs.; 7 lbs. 8 oz. 15·36 drs.
2. $11^h 47^m 2\frac{2}{5}^s$; $2^d 9^h 37^m 26\frac{2}{3}^s$; $195^d 1^h 2^m 38\frac{4}{5}^s$; $12^h 44^m 2 \cdot 868$
3. 17° 44' 8·061"; 7° 51' 15"·3; 64° 22' 57"; 10° 52' 21"
4. 17 cub. ft. 29511·59847552 cub. in.
5. 106·434785 gallons

MISCELLANEOUS EXAMPLES, Pages 56—57.

1. ·0006628⅜
2. $\frac{3}{8}$
3. 29^d·53059027
4. 12·175 hours; ·0013
5. 1 N.M. = 1·15202 I.M.; 1 I.M. = ·86804 N.M.; 1·151515, and ·868421
6. ·997269560 day
7. 80·655 degrees, or 80° 39' 30"; 89·616 grades
8. 1 kilomètre = ·621382 mile; 1 mile = 1·609315 kilomètre
9. 3 ft. 3·42 in.
10. 34·002; 83100·000831
11. ·000038
12. ·0434027 seconds
14. 574·425
15. 43·039783 feet; 6·1575164 yards; 89·03894378 miles
16. 1·240346
17. 7 ft. 5⅜ in.
18. 277·2738 cub. in.
19. 3·883 in.
20. 62·321 lbs.; 35·943 cub. ft.
21. 227·205 tons; 8639·6805 tons
22. 500 sovereigns
23. ·001021; 20210; ·1902; 1902000000; 1902000
24. 7970 miles
25. ·00000033

CHARACTERISTICS OF LOGARITHMS, Page 61.

1. 2	6. 4	11. 5	16. 2
2. 0	7. 2	12. 3	17. 7
3. 2	8. 3	13. 0	18. 4
4. 1	9. 2	14. 0	19. 0
5. 0	10. 0	15. 1	20. 1

Page 62.

1. $\bar{1}$ or 8	6. $\bar{7}$ or 3	11. $\bar{1}$ or 9	16. $\bar{2}$ or 8
2. $\bar{1}$ or 9	7. $\bar{2}$ or 8	12. $\bar{3}$ or 7	17. $\bar{7}$ or 3
3. $\bar{4}$ or 6	8. $\bar{4}$ or 6	13. $\bar{7}$ or 3	18. $\bar{7}$ or 3
4. $\bar{1}$ or 9	9. $\bar{2}$ or 8	14. $\bar{5}$ or 5	19. $\bar{1}$ or 9
5. $\bar{3}$ or 7	10. $\bar{1}$ or 9	15. $\bar{4}$ or 6	20. $\bar{11}$ or 9

LOGARITHMS OF NATURAL NUMBERS, Page 65.

1. 0·698970
2. 0·954243
3. $\bar{3}$·954243 or 7·954243
4. $\bar{7}$·000000 or 8·000000
5. $\bar{7}$·000000 or 6·000000
6. 1·146128
7. 1·612784
8. $\bar{3}$·602060 or 7·602060
9. 0·380211
10. 1·380211
11. $\bar{1}$·380211 or 9·380211
12. $\bar{3}$·322219 or 7·322219
13. 1·973128
14. $\bar{7}$·698970 or 9·698970
15. $\bar{1}$·875061 or 9·875061
16. 0·397940
17. $\bar{1}$·397940 or 9·397940
18. $\bar{7}$·954243 or 8·954243
19. $\bar{3}$·959041 or 7·959041
20. 1·397940
21. $\bar{2}$·380211 or 8·380211
22. $\bar{3}$·544068 or 5·544068
23. $\bar{5}$·755875 or 5·755875
24. $\bar{4}$·698970 or 6·698970

1. 2·000000
2. 2·161368
3. 0·468347
4. 2·557507
5. 2·828015
6. 2·899820
7. 2·992111
8. 0·681241
9. 0·952308
10. 0·167317
11. $\bar{1}$·167317 or 9·167317
12. $\bar{1}$·954725 or 9·954725
13. 2·167317 or 8·167317
14. 2·627366
15. 5·651278 or 5·651278
16. 5·651278

Answers

Page 66.

1. 3·000000
2. 3·091315
3. 1·409087
4. 3·734960
5. 1·415974
6. 0·415974
7. 1̄·005180 or 8·005180
8. 1·977129
9. 0·890812
10. 2·994581
11. 1̄·835247 or 8·835247
12. 3̄·444669 or 7·444669

Page 67.

1. 4·585178
2. 2·585178
3. 4·091491
4. 2·734968
5. 4·823904
6. 3·965898
7. 2·639088
8. 1·895445
9. 0·343507
10. 1̄·894105 or 9·894105
11. 4·000000
12. 4̄·903120 or 6·903120
13. 5·301030
14. 2̄·749845 or 8·749845
15. 3̄·993714 or 7·993714
16. 5·808742
17. 3·052717
18. 1̄·999172 or 9·999172
19. 5·562474
20. 2·998755
21. 1·507732
22. 0·014001
23. 3·000003
24. 2·775555

NATURAL NUMBER OF LOGARITHMS, Pages 70—72.

1. 3
2. 9·4
3. 14·5
4. 6·49
5. 586
6. 2·48
7. 30·09
8. 7916
9. 345·6
10. 24·83
11. 7000
12. 10000000
13. 669000
14. 50000
15. 100000
16. 978·5
17. 34800
18. ·5547
19. ·3171
20. ·00000075
21. 4000000
22. ·00000007
23. 4029
24. 2784
25. ·0000009797
26. 80080000
27. ·04183
28. ·000000007968
29. ·0046
30. ·00071
31. 8199000
32. 738800

1. 853·52167
2. 4220·3
3. 71105·9
4. 23000·1
5. 53·133
6. 93·8689
7. 456780
8. 543210
9. 666660
10. 98765
11. 84321
12. 123456
13. 342·945
14. 5555·54
15. 678945·3
16. 260418
17. 69500·645
18. 12375·426
19. 1·7
20. 1651374
21. ·0096532
22. ·000290888
23. ·0174533
24. 2349632·4
25. ·0000017645
26. ·99727
27. ·7854
28. ·000856735
29. ·000036808

LOGARITHMS OF NATURAL NUMBERS, Page 72.

1. 0·903090
2. 1̄·000000
3. 0·690196
4. 1·579784
5. 2·579784
6. 2·000000
7. 6·000000
8. 1·390935
9. 0·588831
10. 2·954243
11. 1·802774
12. 3·805501
13. 1̄·165244
14. 0·588160
15. 3·829561
16. 2̄·942504
17. 1̄·539954
18. 0·034628
19. 1̄·096910
20. 3̄·954243
21. 2·926548
22. 1̄·964240
23. 2̄·953760
24. 4·000000
25. 4·681241
26. 3·958121
27. 4·763428
28. 2̄·554755
29. 4·651278
30. 7·651278
31. 1·972043
32. 4·722552
33. 4·698970
34. 5·845154
35. 5·421604
36. 5·606388
37. 5·699759
38. 1·686877
39. 1·970876
40. 2̄·515397
41. 7·000000
42. 2·792392
43. 4·477134
44. 4·000039
45. 5·774152
46. 7·947385
47. 2·458852
48. 3̄·551938
49. 4·932847
50. 7·816109

NATURAL NUMBERS OF LOGARITHMS, Page 73.

1. 204
2. 4753
3. 9
4. 50
5. 1
6. 100
7. 366·855
8. 3659
9. 418·557
10. 3·673
11. 6·004
12. 588·172
13. 594500
14. 264000
15. 1000
16. 2480000
17. 26·042
18. 15·438
19. ·09
20. ·0031
21. 50800
22. 2·606
23. ·1
24. ·009
25. ·052
26. 451070
27. 2·71828
28. 404007
29. 100000
30. ·0762
31. ·147
32. ·00000075
33. 1·00043
34. 8859000
35. ·0918504
36. 5·80693
37. ·763888
38. 4220·3
39. 53·1329
40. ·042404
41. ·0048553
42. 2·5152
43. 100591
44. ·000209675
45. 7·5

Answers.   415

MULTIPLICATION BY LOGARITHMS, *Page* 76.

1. $3.774517 = 5950$; $3.022429 = 1053$; $3.000000 = 1000$; $3.521269 = 3321$.
2. $2.971331 = 936.12$; $4.034147 = 10818$; $3.494850 = 3125$; $4.443232 = 27748$.
3. $1.009026 = 10.21$; $0.436878 = 2.7345$; $\bar{1}.818753 = .6588$; $1.575742 = 37.648$.
4. $5.425758 = 266537$; $4.532375 = 34070.2$; $2.639870 = 436.385$; $1.292881 = 19.62826$; $6.783260 = 6071000$.
5. $4.586678 = 38608$; $4.677607 = 47600$; $2.680225 = 478.878$; $5.237543 = 172800$; $5.786113 = 611101$.
6. $3.971387 = 9362.39$; $2.993736 = 985.68$; $4.659678 = 45675$; $5.749272 = 561400$; $3.723999 = 5296.6$.
7. $7.146212 = 14002718$; $6.445142 = 2787032$; $6.919110 = 8300615$; $7.498480 = 31512319$.
8. $3.100249 = 1259.64$; $2.511391 = 324.632$; $5.000000 = 100000$; $8.696466 = .0497125$.
9. $7.499467 = 31583985.5$; $3.782115 = 6055$; $3.842614 = 6960.08$; $4.327379 = 21250.98$.
10. $5.599311 = 397476.1$; $3.590806 = 3897.68$; $5.477728 = 300419.31$; $7.623683 = 42041942$.
11. $6.314887 = 2064842.8$; $4.808914 = 64404.2$; $2.552762 = 357.077$; $3.983651 = 9630.555$.
12. $3.394677 = 2481.28$; $4.312842 = 20551.4$; $0.123363 = 1.32850$; $3.519872 = 3310.34$.
13. $8.763323 = .000000057986$; $\bar{3}.778168 = .006000236$; $\bar{7}.740796 = .00000055055$; $\bar{1}.233799 = .171317$.
14. $\bar{5}.755418 = .00005694$; $0.622110 = 4.189$; $3.473368 = .00297418$; $4.147399 = 14041.03$.
15. $2.951043 = 893.394$; $3.524617 = 3346.7$; $3.398070 = .00250075$; $\bar{4}.783612 = .000607593$.
16. $2.000000 = .01$; $\bar{1}.000000 = .000000001$; $\bar{5}.050035 = .0000112211$; $5.000000 = 100000$.

DIVISION BY LOGARITHMS, *Page* 79.

1. $1.919078 = 83$; $2.778875 = 601$; $1.924279 = 84$; $2.096910 = 125$.
2. $2.986680 = 969.8$; $3.698970 = 5000$; $5.880170 = 758875.4$; $2.775257 = 596.015$.
3. $\bar{1}.809855 = .64544$; $3.644712 = 4412.78$; $3.477122 = 3000$; $1.822584 = 66.4637$.
4. $0.494768 = 3.1244$; $1.614317 = 41.145$; $1.428589 = 26.828$; $0.139814 = 1.3798$.
5. $0.472427 = 2.967762$; $1.785908 = .06108$; $2.385683 = .0243043$; $\bar{3}.571412 = .037274$; $\bar{3}.301030 = .00002$; $\bar{4}.301030 = .000002$; $1.301030 = 20$.
6. $3.886534 = 7700.76 +$; $\bar{1}.999234 = .998236$; $1.999489 = 99.8815$; $3.763429 = 5800$.
7. $1.569001 = 37.0684$; $\bar{1}.425115 = .266143$; $4.376859 = 23815.5$; $1.378403 = 23.9003$.
8. $1.995636 = 99$; $1.319142 = 20.8517$; $1.854294 = 71.498$; $0.793259 = 6.2124$.
9. $4.699490 = 5006$; $4.514747 = 32715$; $4.317415 = 20769$; $2.096910 = 125$.
10. $0.942505 = 8.76$; $2.525020 = 334.98$; $5.291146 = 195500$; $1.004364 = 10.1010$.
11. $1.832752 = .68038$; $4.903504 = .00080763$; $\bar{1}.030734 = .107333$; $2.509203 = 323$; $5.778151 = 600000$.
12. $1.940506 = 87.1978$; $\bar{2}.297735 = .0198488$; $\bar{1}.278331 = .189815$; $0.833525 = 6.81592$.
13. $\bar{1}.421422 = .265388$; $1.421422 = 26.5388$; $\bar{1}.421422 = .265388$; $3.421422 = 2653.88$; $7.301030 = 20000000$; $7.477122 = 30000000$.
14. $\bar{1}.057101 = .11405$; $2.537395 = 344.6631$; $\bar{3}.003528 = .0000100816$; $0.833525 = 6.816$.
15. $2.339894 = 218.723$; $2.934196 = 859.4$; $\bar{1}.004751 = .1011$; $7.505150 = 32000000$; $\bar{5}.028878 = .0000106875$.
16. $0 = 1$; $13 = 10000000000000$; $\bar{1} = .1$; $\bar{2} = .01$; $4 = 10000$.

NATURAL SINES AND COSINES, Page 83.
Natural Sines.

1. 570774
2. 867085
3. 947463
4. 370382
5. 723895
6. 974228
7. 800000
8. 997630

Natural Cosines, Page 83.

1. 969227
2. 328958
3. 167237
4. 995612
5. 688000
6. 868805
7. 782397
8. 989472

Arcs of Natural Sines, Page 84.

1. 63° 53' 47"
2. 21 44 21
3. 53° 7' 49"
4. 66 59 10
5. 26° 21' 34"
6. 11 45 52
7. 47° 48' 33"
8. 31 57 10
9. 48° 46' 34"
10. 73 44 23

Arc of Natural Cosines, Page 84.

1. 63° 19' 58"
2. 18 29 12
3. 43 21 52
4. 45° 24' 39"
5. 59 0 47
6. 23 54 9
7. 39° 42' 4"
8. 12 53 4
9. 35 8 32
10. 77° 33' 15"
11. 2 33 48
12. 53 7 48

LOG. SINES, TANGENTS, SECANTS, ETC., Page 87.

1. 9·202234
2. 10·185981
3. 9·989071
4. 9·883934
5. 10·039843
6. 10·135990
7. 9·774729
8. 9·895443
9. 10·308556
10. 9·275658
11. 10·144904
12. 9·907590

Pages 92—93.

NO.	SINE.	TANGENT.	SECANT.	COSINE.	COTANGENT.	COSECANT.
1.	9·079607	9·082763	10·003156	9·996844	10·917237	10·920393
2.	9·611999	9·651805	10·039806	9·960194	10·348195	10·388001
3.	9·787595	9·890004	10·102409	9·897591	10·109996	10·212405
4.	9·923122	10·185903	10·262781	9·737219	9·814097	10·076878
5.	9·246845	9·253720	10·006876	9·993124	10·746280	10·753155
6.	9·975130	10·457990	10·482859	9·517141	9·542010	10·024870
7.	8·504189	8·504410	0·000221	9·999779	11·495590	1·495811
8.	9·999580	11·356298	1·356719	8·643281	8·643702	0·000421
9.	8·246654	8·246721	0·000068	9·999932	11·753279	1·753346
10.	9·955206	10·319983	0·364777	9·635223	9·680017	0·044794
11.	9·938922	0·244180	0·305259	9·694741	9·755820	0·061078
12.	9·990916	10·684913	0·693987	9·306013	9·315087	0·009074
13.	8·668140	17. 8·361681	21. 10·348195	25. 8·024643	29. 10·714159	
14.	9·217118	18. 9·505271	22. 10·100598	26. 8·658227	30. 9·972464	
15.	8·504188	19. 9·297036	23. 10·203779	27. 8·305785		
16.	6·297326	20. 11·263695	24. 8·546002	28. 8·258261		

ARCS OF LOG. SINES, Page 95.

1. 33° 26' 48"
2. 57 30 53
3. 2 55 26
4. 2° 17' 7"
5. 19 15 35
6. 58 15 30
7. 18° 26' 6"
8. 39 7 15
9. 54 13 20
10. 0° 9' 50"
11. 87 38 20
12. 70 34 18
13. 52° 35' 30"
14. 4 1 28
15. 1 39 39

ARCS OF LOG. COSINES, Page 95.

1. 52° 13' 35"
2. 55 45 8
3. 89 13 8
4. 8° 6' 31"
5. 81 18 0
6. 79 11 16
7. 70° 47' 25"
8. 80 37 20
9. 88 40 54
10. 31° 9' 33"
11. 3 56 40
12. 88 54 16
13. 89° 38' 20"
14. 84 15 39
15. 84 4 38

ARCS OF LOG. SECANTS, Page 95.

1. 14° 23' 15"
2. 51 28 50
3. 3° 24' 0"
4. 26 33 0
5. 79° 39' 51"
6. 84 19 47
7. 18° 22' 17"
8. 88 41 42
9. 61° 4' 15"
10. 86 22 57

ARCS OF LOG. COSECANTS, Page 95.

1. 26° 43' 0"
2. 34 1 14
3. 5 40 16
4. 6° 5' 13"
5. 49 11 9
6. 2 4 7
7. 5° 43' 39"
8. 1 57 4
9. 3 54 45
10. 58° 15' 30"
11. 7 13 56
12. 4 46 56
13. 78° 22' 32"
14. 60 13 52
15. 2 56 20

ARCS OF LOG. TANGENT, Page 95.

1. 77° 0′ 23″
2. 45 0 24
3. 62 42 21
4. 81° 31′ 58″
5. 54 43 26
6. 5 13 23
7. 86° 58′ 16″
8. 1 8 7
9. 27 28 54
10. 23° 43′ 17″
11. 35 3 32
12. 87 0 46
13. 48° 58′ 24″
14. 2 40 10
15. 1 2 18

ARCS OF LOG. COTANGENTS, Page 95.

1. 61° 2′ 39″
2. 7 34 16
3. 27 16 43
4. 41° 1′ 35″
5. 11 0 41
6. 3 37 50
7. 88° 46′ 54″
8. 44 20 2
9. 86 32 24
10. 82° 49′ 23″
11. 8 30 34
12. 88 55 35
13. 88° 20′ 53″
14. 76 40 15
15. 0 29 16

MISCELLANEOUS EXAMPLES.

For practice in natural and logarithmic Sines, Tangents, and Secants, *Page 95.*

1. Nat. sine ·432651, its common log. is 9·636138, which is the log. sine required.
2. Nat. tang. 3, its common log. is 10·477121, which is the log. tang. required.
3. Log. 9·236713, its corresponding nat. no. is ·172470, the nat. cos. required.
4. The given log. tang. 9·850593, being subtracted from 20, gives 10·149407, the log. cotang.
5. The nat. sine of 68° 45′ 24″ is ·932050, the log. of which, or 9·969439, is its log. sine, which, being subtracted from 20, leaves 10·030561 for the log. cosec.
6. The log. sec. 11·024680 subtracted from 20, leaves 8·975320, the log. cosine, the nat. no. corresponding to which, or ·094476, is the nat. cosine sought.
7. 1. The quantity 9·450981 is found in the tables to be the log. cosine of the arc 73° 35′ 31″. 2. The nat. no. corresponding to the given log. is ·282476, which is the nat. cos. of 73° 35′ 31″, the arc A sought.
8. 1. The square of radius, or 1, divided by the nat. sec. 2·005263, gives ·498688, the nat. cosine of A, which is found in the tables of nat. cosines to correspond to 60° 5′ 12″, the value A. 2. The common log. of the nat. sec. 2·005263 is 0·302171, which is found to be the log. sec. of 60° 5′ 12″, the arc A sought.

DIFFERENCE OF LATITUDE, Page 107.

1. 203′ N.
2. 470 S.
3. 293′ S.
4. 330 N.
5. 795′ N.
6. 157 S.
7. 610′ S.
8. 459 N.
9. 94′ N.

MERIDIONAL DIFFERENCE OF LATITUDE, Page 107.

1. 97
2. 2426
3. 345
4. 1216
5. 932
6. 260

LATITUDE IN, Page 108.

1. 34° 2′ N.
2. 27 54 N.
3. 0° 8′ N.
4. 3 1 N.
5. 2° 48′ S.
6. 2 54 S.
7. 0° 20′ S.
8. Equator.
9. Equator.
10. 39° 14′ S.

MIDDLE LATITUDE, Page 109.

1. 17° 19′
2. 2° 10½′
3. 35° 37′
4. 61° 31½′
5. 53° 12½′
6. 64° 31′

DIFFERENCE OF LONGITUDE, Page 110.

1. 300′ E.
2. 507 E.
3. 716′ W.
4. 260 W.
5. 270′ E.
6. 422 W.
7. 368′ E.
8. 420 W.
9. 180′ W.
10. 412 E.
11. 2835′ E.
12. 1200 W.

LONGITUDE IN, Page 111.

1. 7° 38′ W.
2. 1° 18′ E.
3. 31° 4′ E.
4. 0° 30′ W.
5. 1 15 E.
6. 0 45 W.
7. 39 10 W.
8. 92 9 E.
9. 103 56 E.
10. 178 26 W.
11. 178 57 E.
12. 179 59 E.

LEEWAY—CORRECTED COURSES, *Page* 121.

1. S.W. ½ S.	2. S.S.W. ¼ W.	3. N. ¼ E.	4. N.E. ½ E.
5. E. by S.	6. N.W. by W.	7. W. ¼ S.	8. N.E. by E. ¾ E.

DEVIATION—TRUE COURSES *Page* 155.

1. N. 38° 59′ E.	8. S. 48° 55′ E.	15. S. 28° 57′ W.	22. S. 78° 10′ W.
2. N. 0 22 W.	9. N. 19 25½ E.	16. S. 4 38½ E.	23. S. 56 2½ W.
3. N. 20 58½ W.	10. S. 0 16 W.	17. S. 73 10½ W.	24. S. 4 34½ W.
4. S. 43 39½ W.	11. S. 84 7½ W.	18. N. 32 24½ E.	25. S. 63 11½ W.
5. S. 84 7½ W.	12. S. 36 12½ E.	19. N. 3 0 E.	26. S. 74 3½ E.
6. S. 24 5½ W.	13. S. 87 0 E.	20. N. 89 55 W.	27. S. 79 16 E.
7. S. 89 35 E.	14. N. 59 12½ W.	21. S. 9 33 E.	28. N. 77 46 W.

MAGNETIC BEARINGS OF OBJECTS, *Page* 158.

1. N. 78° 10′ E.	5. N. 84° 54′ E.	9. S. 86° 2′ W.
2. S. 72 32 E.	6. S. 11 53 W.	10. S. 2 39 E.
3. N. 5 14 E.	7. N. 8 37 E.	11. N. 72 33 W.
4. S. 10 16 E.	8. S. 89 13 W.	12. N. 3 6 W.

LEEWAY, VARIATION, and DEVIATION—TRUE COURSES, *Page* 161.

1. E. by S. ½ S.	11. S. 79° W.	21. S. 9° E.
2. N.W. by N.	12. S. 6 E.	22. N. 59 E.
3. N. by W.	13. S. 73 W.	23. N. 89 E.
4. N.N.W.	14. S. 37 W.	24. S. 60 W.
5. S.E. by E.	15. S. 57½ W.	25. N. 10 E.
6. S.E.	16. S. 77 E.	26. S. 60½ E.
7. S. 44° E.	17. N. 84 W.	27. S. 76 W.
8. N. 64 E.	18. S. 81 E.	28. N.E. ½ E.
9. N. 78 W.	19. S. 32 W.	29. N. ¾ W.
10. S. 85 E.	20. S. 36 E.	30. W. ¼ N.

NAPIER'S DIAGRAM.

Deviations, *Page* 166.

(*a*) Curve A.—10° E.; 19° E.; 24° W.; 18½° W.; 25° W.; 2° E.; 23½° E.; 17° W.; 24½° E.; 24° W.; 15° E.

(*c*) Curve C.—15° W.; 11½° W.; 0½° W.; 28½° E.; 10½° E.; 6° E.; 14½° W.; 28½° E.; 24½° W.; 2° E.

Correct Magnetic Courses, *Page* 168.

Curve A.—N. 66° W.; S. 87° E; S. 23° E.; S. 54° W.; S. 63½° W.; N. 68° E.; S. 31½° E.

Curve C.—N. 37° W.; N. 33½° E.; S. 75° E.; N. 62° W.; N. 58° W.; N. 20° E.; N. 80° E.; N. 58½° W.; N. 27½° W.; S. 76° W.

Compass Courses, *Page* 169.

Curve A.—N. 79° W.; N. 27° E.; S. 21½° W.; N. 9° E.; N. 41½° E.; N. 78° E.; S. 66° W.; N. 21½° E.; N. 86° W.

Curve C.—S. 21° W.; N. 78° E.; South; N. 35½° E.; S. 73° E.; S. 31° E.; S. 8° W.; N. 65½° E.; S. 18° W.

DIFFERENCE OF LATITUDE AND DEPARTURE, *Page* 186.

No.	Diff. lat.	D. p.	No.	Diff. lat.	Dep.
1.	27′·7 S.	11′·5 E.	8.	10′·8 S.	33′·3 W.
2.	9·4 S.	47·1 E.	9.	22·9 N.	8·8 W.
3.	100·8 S.	91·3 W.	10.	10·9 S.	23·3 W.
4.	12·3 N.	83·1 W.	11.	2·5 S.	14·5 W.
5.	28·8 S.	48·0 E.	12.	27·3 N.	13·9 W.
6.	142·7 S.	173·9 W.	13.	7·4 S.	42·2 W.
7.	44·5 N.	177·5 E.	14.	33·2 N.	10·8 W.

COURSES AND DISTANCES, *Page* 188.

NO.	COURSES	DIST.	NO.	COURSES	DIST.
1.	S. 19° E.	77'	6.	N. 44° E.	52½'
2.	N. 67 E.	186½	7.	S. 55 W.	93
3.	N. 66½ W.	161	8.	S. 6 W.	161½
4.	S. 21 E.	105½	9.	S. 2½ W.	173
5.	N. 30 W.	480	10.	N. 58 E.	310

TRAVERSE SAILING, *Page* 193.

NO.	D. LAT.	DEP.	LAT. IN.	COURSE.	DIST.
1.	95·2 S.	92·1 W.	51° 23' N.	S. 44° W.	132'
2.	20·0 S.	128·8 S.	53 52 N.	S. 81 W.	130
3.	375·6 S.	0·0	2 26 S.	S.	376
4.	0·0	76 8 E.	19 0 S.	E.	77
5.	75·1 S.	77·8 E.	0 15 S.	S. 46 E.	108
6.	80·4 S.	36·0 W.	0 10 S.	S. 24 W.	88
7.	31·0 S.	8·4 W.	46 41 N.	S. 15 W.	32
8.	24·7 S.	145·1 W.	34 36 N.	N. 80 W.	147
9.	55·7 N.	129·9 W.	35 39 S.	N. 23½ W.	139½
10.	150·3 S.	56·8 W.	0 44 S.	S. 21 W.	161

PARALLEL SAILING, *Page* 195.

1. 250·3' W.   2. 344·4' E.   3. 519·2' W.   4. 1480·0' W.
5. 512·5 E.   6. 612·0 W.   7. 113·8 E.   8. 408·0 E.

Page 196.

1. N. 64° 17' 30" W., distance 396·7 miles.   2. 893·4 miles.
4. S. 79° 8' 45" E., distance 96·5 miles.   5. 60° and 70° 32'.
6. 61·6 miles, or 1° 1'·6.   7. West, distance 864·1 miles.

MIDDLE LATITUDE SAILING, *Page* 199.

	D. lat.	Dep.	Lat. in	D. long.	Long. in
1.	113'·3	273'·5	27° 28' N.	305'	54° 55' W.
2.	,, 99·9	,, 187·0	,, 34 10 N.	,, 224	,, 29 8 W.
3.	,, 89·8	,, 189·8	,, 41 0 S.	,, 248	,, 70 12 E.
4.	,, 165·6	,, 223·3	,, 49 10 S.	,, 334	,, 175 58 W.
5.	,, 96·7	,, 318·7	,, 18 52 N.	,, 358	,, 175 12 E.
6.	,, 114·6	,, 122·9	,, 0 59 S.	,, 123	,, 27 47 W.

MERCATOR'S SAILING, *Page* 203.

NO.	D. LAT.	M. D. LAT.	D. LONG.	COURSE.	DIST.
1.	97 N.	125	131 E.	N. 46° 21' E.	141
2.	85 S.	130	76 E.	S. 30 19 E.	98
3.	230 S.	500	270 E.	S. 28 22 E.	261
4.	1232 N.	1760	4732 E.	N. 69 36 E.	3534
5.	115 N.	166	191 E.	N. 49 0 E.	175
6.	1107 N.	1230	1452 W.	N. 49 44 W.	1713
7.	779 S.	1080	1200 W.	S. 48 1 W.	1165
8.	1011 N.	1139	3868 W.	N. 73 21 W.	3528
9.	792 S.	794	1254 E.	S. 57 39½ E.	1480
10.	128 N.	233	725 W.	N. 72 11 W.	418
11.	731 S.	733	2459 E.	S. 73 24 E.	2559
12.	150 N.	274	354 E.	N. 52 16 E.	245
13.	4483 N.	4842	3313 E.	N. 34 23 E.	5432
14.	1860 S.	1884	412 E.	S. 12 20 E.	1904
15.	3355 N.	3516	7583 W.	N. 65 8 W.	7978
16.	180 N.	190	1140 W.	N. 80 32 W.	1094

DAY'S WORKS CORRECTED FOR LEEWAY, VARIATION, AND DEVIATION, Pages 217—222.

NOTE.—In the following key, the first line for each day's work is explained by the titles at the top of the page. The second line contains the True Courses. The third line contains the Diff. Lat. and Dep. corresponding to each course: their names are not given because these are easily seen from the courses in the second line.

	Courses.	Distance.	Diff. Lat.	Departure.	Lat. in.	Mid. Lat.	Diff. Long.	Long. in.
1	N. 63° E.	227'	102'·2 N.	202'·6 E.	36° 57' N.	36° 6'	251' E.	71° 19' W.
	N. 89°E.19' 0'·3	S. 43°E. 50' 19'·0	S. 71°E. 18' 36'·6 34'·1	N. 55°E. 42' 12'·4 55'·9	N. 3°W. 42' 24'·1 31'·4	N. 15°E. 42' 41'·9 2'·2	N. 82°E. 17' 40'·6 10'·9	N. 31°E. 42' 5'·1 36'·6 35°·2 34'·1
2	S. 77½° E.	99'	21'·5 S.	96'·6 E.	53° 4½' N.	53° 59'	164' E.	2° 39' E.
	S. 68° E. 17' 6'·4 15'·8 N. 3°W. 6' 6'·0 0'·3	N. 73°E. 28' 8'·2 26'·8	S. 75°E. 31' 8'·0 20'·9	N. 72°E. 20' 6'·2 19'·0	S. 99 W. 19' 18'·8 3'·0	S. 5° W. 13' 13'·0 1'·1	S. 20° E. 10' 9'·4 3'·4	N. 24°E. 15' 13'·7 6'·1
3	North.	86'·6	86'·6 N.		45° 14' N.			7° 51' W.
	N. 17°E. 11' 12'·4 3'·8	North 38° 38'·0	N. 27°E. 18' 16'·0 8'·2	N. 79°E. 25' 4'·8 24'·3	S. 83°W. 23' 2'·8 26'·8	S. 48°E. 23'·2 16'·9 15'·7	N. 48°W 27'·1 18'·1 20'·1	N. 36°W. 21' 17'·0 12'·3
4	N 44° E	14'	10'·5 N.	9'·8 E	46° 13' S	46° 6'	13' E	1° 56' W
	N 50° W 20' 12'·9 15'·3	S 71° W 26' 7'·6 24'·9	N 40° E 31' 26'·0 21'·9	S 54° E 22' 12'·9 17'·8	N 67° E 25' 9'·8 23'·0	N 67° E 26' 10'·2 23'·9	S 44° W 24' 17'·3 16'·7	S 62° W 22'·5 10'·6 19'·9
5	S. 45° W.	82'	58'·3 S	58'·1 W	34° 12' N	34° 41'	71' W	6° 47' W
	S 87° W 9' 0'·5 9'·0	S 29° W 38' 33'·2 18'·4	N 23° W 45' 41'·4 17'·6	N 77° W 31' 7'·6 33'·1	South 16' 16'·0	S 29° W 37'·5 32'·8 18'·2	N 71° E 24'·8 8'·1 23'·5	S 79° E 15' 2'·9 14'·7
6	N 3° W	28'·5	28'·5 N	1'·6 W	30° 27'·5 N	30° 13'	2' W	32° 52' E.
	S 3° E 15' 15'·0 0'·8	N 75° W 12' 3'·1 11'·6	S 63° E 13' 5'·4 11'·6	S 10° E 26' 24'·6 4'·3	N 8° E 20' 19'·8 2'·8	S 76° W 15' 3'·6 14'·0	N 11° W 27' 26'·5 5'·2	N 20° E 30' 28'·2 10'·3
7	S 50° W	87'	56'·5 S	66'·5 W	45° 16½' S	44° 28'	94' W	178° 23' W
	N 70° W 18' 6'·2 16'·9	N 20° E 38' 35'·7 13'·0	S 88° W 11' 1'·2 34'·0	S 85° W 38' 5'·8 47'·6	S 6° W 25' 24'·7 3'·9	S 2° W 38' 38'·0 1'·3	S 10° E 50' 47'·3 1'·3	N 26° E 18' 16'·2 7'·9
8	N 70° E	161'	55'·3 N	151'·2 E	63° 13' N	62° 45'	330' E	57° 47' W
	N 66° E 21' 8'·5 19'·2	N 71° E 50' 16'·3 47'·3	N 75° E 17' 4'·4 16'·4	N 35° E 34' 27'·9 19'·5	S 28° W 14' 12'·4 6'·6	S 74° E 21' 5'·8 20'·2	S 2° W 17' 17'·0 0'·6	N 47° E 40' 33'·4 35'·8
9	S 77° E	105'	24'·3 S	102'·5 E	59° 25' N	59° 37'	202'·7 E	40° 31' W
	S 12° E 14' 13'·7 2'·9	S 37° E 19' 15'·2 11'·4	N 12° W 25' 24'·5 4'·8	S 13° E 22' 21'·4 4'·9	N 41° E 33' 20'·8 25'·6	S 4° W 10' 19'·0 1'·3	S 72° E 25' 8'·1 23'·6	N 79° E 41' 7'·8 40'·2
10	N 31° W	129'	110'·1 N.	66'·7 W.	34° 20' S.	35° 15'	81½' W.	111° 31½' W
	N 50 W 14' 9'·0 10'·7	N 53° W 20'·5 12'·3 16'·4	N 33° W 17'·2 14'·4 9'·4	N 10° W 24' 23'·6 4'·2	N 10° W 22'·3 22'·0 3'·9	N 40° W 31'·2 23'·9 20'·1	N 54° W 20'·3 11'·9 16'·4	S 64° E 16' 7'·0 14'·4
11	N 52° W	135'	83'·1 N.	106'·9 W	54° 35' S	55° 17'	188' W	70° 24' W.
	N 76° W 17' 4'·1 16'·5	N 73° W 33'·6 9'·8 32'·1	N 12° W 23'·2 22'·7 4'·8	N 13° W 41'·6 32'·6 10'·4	S 68° W 47'·8 17'·9 44'·3	N 7° E 26'·5 29'·3 3'·6	S 60° E 37'·8 18'·9 32'·7	N 34° W 27' 22'·4 15'·1
12	N 86° W	247'	16'·5 N	246'·1 W	41° 32' S	44° 12'	343' W	177° 28' E.
	West 16' 16'·0	N 62° W 40' 18'·8 35'·3	N 63° W 37' 16'·8 33'·0	S 86° W 38' 2'·7 37'·9	S 83° W 31' 3'·8 30'·8	S 50° W 31' 19'·9 23'·7	N 80° W 42' 7'·3 41'·4	West 28' 28'·0

ASTRONOMICAL DATES, Page 224.

1. Jan. 1ᵈ 16ʰ 38ᵐ 9ˢ
2. Feb. 27ᵈ 8ʰ 12ᵐ 0ˢ
3. Aug. 14ᵈ 6ʰ 28ᵐ 40ˢ
4. Mar. 31 19 54 19
5. June 3 16 18 3
6. Aug. 31 20 10 52
7. Dec. 31 6 18 34
8. July 1 8 3 24
9. June 30 23 30 10
10. Oct. 1 0 10 12
11. 1871, Dec. 31 20 9 50
12. 1872, Dec. 31 12 44 12

CIVIL DATE, Page 224.

1. Jan. 11th, 4ʰ 31ᵐ 15ˢ A.M., Feb. 3rd, 11ʰ 28ᵐ 56ˢ P.M.
2. Oct. 15th, 3 17 13 A.M., Dec. 3rd, 5 16 12 P.M.
3. May 17th, 7 15 11 P.M., Mar. 14th, 11 15 7 A.M.
4. April 1st, 11 10 16 A.M., Mar. 21st, 7 24 12 P.M.
5. Sept. 1st, 8 10 54 P.M., Sept. 1st, 8 10 54 A.M.
6. 1872, Jan. 1st, 9 50 41 P.M., 1873, Jan. 1st, 10 48 56 A.M.

DEGREES INTO TIME, Page 225.

1. 1ʰ 15ᵐ 36ˢ   0ʰ 50ᵐ 43ˢ   9ʰ 9ᵐ 48ˢ   6ʰ 24ᵐ 43ˢ   5ʰ 57ᵐ 4ˢ
2. 4 30 48   5 5 22   0 5 40   9 22 8·7   4 37 56
3. 0 3 54·4   3 16 17·33   0 1 47·2   0 56 10   8 41 16
4. 0 36 56   10 52 11·2   0 2 29·6   0 9 12·8   11 21 0
5. 7 14 28   0 41 48·9   0 9 56   5 38 50   0 2 18
6. 0 0 54   3 24 40·8   10 27 28   11 55 19   0 2 46·8

TIME INTO DEGREES, Page 226.

1. 18° 18′ 0″   2. 58° 1′ 0″   3. 10° 33′ 0″   4. 168° 50′ 15″
5. 67 16 15   6. 147 24 30   7. 8 44 33   8. 25 15 24
9. 89 46 0   10. 124 16 30   11. 5 22 43·5   12. 175 16 40
13. 0 58 0   14. 2 29 0   15. 0 13 0   16. 5 10 15
17. 9 14 0   18. 75 12 45   19. 179 59 15   20. 0 28 0

GREENWICH DATES, Page 229.

1. Jan. 6ᵈ 8ʰ 8ᵐ 0ˢ   6. Oct. 31ᵈ 21ʰ 22ᵐ 10ˢ   11. Dec. 27ᵈ 23ʰ 19ᵐ 30ˢ
2. Feb. 12 22 4 19   7. Dec. 1 6 32 45   12. July 8 at noon
3. Jan. 31 7 29 28   8. June 30 16 36 52   13. Jan. 31 13 45 20
4. Mar. 15 8 8 6   9. Aug. 3 23 50 22   14. May 31 18 24 40
5. May 15 6 6 0   10. Sept. 1 6 24 11   15. Mar. 2 0 5 40
                       16. 1882. Dec. 31 14 3 20

SUN'S DECLINATION, Page 237.

1. 22° 35′ 17″ S.   2. 16° 40′ 4″·5 S.   3. 4° 9′ 28″·5 N.   4. 2° 7′ 15″·9 N.
5. 19 8 39 N.   6. 14 38 43 N.   7. 23 2 1 N.   8. 14 28 6 S.
9. 8 27 48 N.   10. 3 13 58 S.   11. 23 19 26 S.   12. 18 17 30 S.
13. 20 14 24 S.   14. 12 19 31 S.   15. 0 6 11 N.   16. 17 8 32 N.
17. 23 27 10 N.   18. 0 3 37 N.   19. 18 57 52″·5 N.   20. 0 2 6 N.
21. 3 3 57 S.   22. 23 27 13 S.   23. 23 2 59″·5 S.   24. 0 3 17 N.

EQUATION OF TIME, Page 241.

1. + 5ᵐ 48ˢ·3   2. + 14ᵐ 6ˢ·1   3. + 6ᵐ 18ˢ·9   4. 0ᵐ 0ˢ·0
5. − 3 45·0   6. + 0 2·7   7. + 5 46·5   8. + 0 5·8
9. − 5 55·8   10. − 11 49·3   11. + 0 3·2   12. − 0 1·6
13. − 3 52·8   14. 0 0·1   15. − 15 15·4   16. + 6 7·8
17. + 0 13·4   18. − 15 58·4   19. + 0 5·2   20. 0 0·0

TRUE ALTITUDES, Page 243.

1. 17° 52′ 42″   2. 48° 17′ 14″   3. 30° 2′ 9″   4. 76° 14′ 16″
5. 58 48 28   6. 14 56 49   7. 65 13 4   8. 85 22 51
9. 17 51 38   10. 67 22 16   11. 13 44 33   12. 11 45 28

MERIDIAN ALTITUDES, Page 249.

No.	Green. date	Red. decl.	True alt.	Latitude.	Raper's True alt.	Raper's Latitude.
1.	Jan. 10ᵈ 3ʰ19ᵐ24ˢ	21° 54′ 35″ S.	68° 57′ 22″	42° 57′ 13″ S.	68° 57′ 19″	42° 57′ 16″ S.
2.	Jan. 31 21 20 36	17 4 6 S.	72 58 6	0 2 12 S.	72 58 3	0 2 9 S.
3.	Mar. 7 18 0 48	4 54 12 S.	51 58 1	33 7 42 N.	51 57 51	33 7 57 N.
4.	April 28 11 1 32	14 20 59 N.	82 35 15	6 56 14 N.	82 35 10	6 56 9 N.
5.	May 1 21 51 48	15 24 8 N.	45 57 1	59 27 7 N.	45 56 55	59 27 13 N.
6.	June 10 19 48 12	23 5 36 N.	42 37 28	24 16 56 S.	42 37 24	24 17 0 S.
7.	July 20 10 26 32	20 34 7 N.	52 7 9	17 18 44 S.	52 6 57	17 18 56 S.
8.	Aug. 19 5 30 0	12 39 46 N.	57 50 46	44 49 0 N.	57 50 45	44 49 1 N.
9.	Sept. 22 12 54 0	0 2 47 S.	41 42 33	48 20 14 N.	41 42 23	48 20 24 N.
10.	Oct. 23 6 0 48	11 33 38 N.	54 51 19	23 35 3 N.	54 51 13	23 35 9 N.
11.	Nov. 14 18 39 16	18 29 19 S.	67 56 49	3 33 53 N.	67 56 40	3 34 1 N.
12.	Dec. 9 20 18 40	22 55 49 S.	26 4 46	40 59 25 N.	26 4 36	40 59 35 N.
13.	Sept. 20 19 59 56	0 42 26 N.	56 37 9	32 40 15 S.	56 37 5	32 40 19 S.
14.	Mar. 19 18 2 0	0 1 56 N.	61 58 10	28 3 46 N.	61 58 1	28 3 55 N.
15.	April 7 9 19 0	7 2 33 N.	90 13 41	6 48 52 N.	90 13 40	6 48 53 N.
16.	Sept. 22 15 45 0	0 0 0	83 52 20	6 7 40 N.	83 52 14	6 7 46 N.
17.	Nov. 2 16 56 0	15 2 14.5 S.	70 42 48	34 19 27 S.	70 42 44	34 19 31 S.
18.	Sept. 22 11 35 52	0 3 16 N.	71 34 38	18 28 38 N.	71 34 32	18 28 44 N.
19.	Feb. 12 0 32 48	13 36 37 S.	30 4 42	46 18 41 N.	30 4 36	46 18 47 N.
20.	Mar. 19 18 49 0	0 0 14 N.	77 7 26	12 52 20 S.	77 7 23	12 52 23 S.
21.	Dec. 31 15 37 52	23 1 34 S.	54 38 7	12 20 19 N.	54 38 6	12 21 20 N.
22.	Sept. 30 19 14 40	3 10 24 S.	81 37 15	11 33 9 S.	81 37 11	11 33 13 S.

AMPLITUDES, Page 258.

No.	Green. date	Red. decl.	True amp.	Sine.	Error of compass.	Deviation.
1.	Jan. 25ᵈ 15ʰ 47ᵐ 8ˢ	13° 27′ 46″ S.	E. 22° 56′ S.	9·500741	33° 19′ W.	11° 29′ W.
2.	Feb. 17 4 7 28	11 30 43 S.	W. 14 30 S.	9·398805	10 14½ E.	11 34½ E.
3.	March 29 2 20 20	3 20 20 N.	E. 3 53 N.	8·830360	20 23 N.	2 43 W.
4.	April 4 10 53 0	0 1 40 N.	W. 0 29 N.	9·053145	0 11 W.	6 59 W.
5.	Nov. 6 22 5 8	16 10 12 S.	E. 18 30 S.	9·504420	21 10 E.	7 29 E.
6.	May 25 16 44 0	22 6 23 N.	E. 35 11 N.	9·761035	43 42 W.	8 20 W.
7.	June 2 9 56 20	22 15 2 N.	W. 34 30 N.	9·791008	26 11½ W.	11 8½ E.
8.	July 14 2 15 42	21 39 31 N.	E. 21 57 N.	9·625221	22 52 E.	11 12 E.
9.	Aug. 27 3 18 44	9 38 22 N.	W. 13 42 N.	9·260637	25 52 W.	2 42 W.
10.	Sept. 7 21 37 0	5 43 46 N.	E. 6 13 N.	9·030286	6 13 W.	6 13 W.
11.	Oct. 1 5 29 51	3 20 21 S.	E. 5 31 S.	8·803389	12 50 E.	5 51 W.
12.	Sept. 22 6 1 12	0 0 29 S.	E. 0 17 N.	—	3 0 W.	14 6 W.
13.	Nov. 2 21 27 40	15 3 40 N.	W. 17 23 S.	9·475106	5 7 E.	7 57 E.
14.	Dec. 3 21 13 45	22 13 17 S.	W. 39 7 S.	9·770427	2 22 W.	18 22 W.
15.	March 23 4 50 24	0 0 0	0 0	—	0 0	15 0 W.
16.	Sept. 22 15 45 0	0 0 0	0 0	—	0 0	21 50 E.
17.	June 8 18 41 23	22 56 0 N.	E. 21 56 N.	—	17 18 W.	7 3 E.
18.	Feb. 25 20 10 28	8 41 15 S.	E. 14 40 S.	9·503481	43 41 W.	12 53 W.
19.	April 30 15 41 30	15 1 30 N.	W. 1 37 N.	0·406310	0 15 W.	10 15½ W.
20.	May 27 17 32 12	21 26 54 N.	W. 32 53 N.	·734931	21 35 E.	1 21 E.
21.	June 18 1 24 20	23 25 40 N.	E. 64 36 N.	9·955851	39 28 E.	14 28 E.
22.	March 6 6 5 20	5 29 7 S.	W. 6 26 S.	9·040030	23 18 W.	5 28 W.
23.	April 9 18 50 48	7 56 5 N.	W. 13 31 N.	—	22 0 E.	5 50 E.
24.	Dec. 13 11 35 19	23 12 25 S.	E. 52 1 S.	—	57 58½ W.	38 38½ W.

TIDES, Page 265.

1.	7ʰ19ᵐ A.M.	7ʰ58ᵐ P.M.	16.	8ʰ25ᵐ A.M.	8ʰ50ᵐ P.M.
2.	8 55 „	9 15 „	17.	11 59 „	No „
3.	11 53 „	No „	18.	11 12 „	11 41 „
4.	0 39 „	0 58 „	19.	No „	0 15 „
5.	11 42 „	No „	20.	11 58 „	No „
6.	10 32 „	10 57 „	21.	—	Noon
7.	11 52 „	No „	22.	No „	0 1 „
8.	11 34 „	11 59 „	23.	No „	0 18 „
9.	11 59 „	No „	24.	0 7 „	0 24 „
10.	—	Noon	25.	11 43 „	No „
11.	2 56 „	3 39 „	26.	11 50 „	No „
12.	9 58 „	10 26 „	27.	No „	0 36 „
13.	10 0 „	10 39 „	28.	9 54 „	10 29 „
14.	No „	0 35 „	29.	11 36 „	No „
15.	No „	Noon	30.	10 10 „	10 58 „

TIDES—FOREIGN PORTS, Page 267.

1.	Constant — 0ʰ47ᵐ	corr. for long. + 6ᵐ	3ʰ 5ᵐ A.M.	"	3ʰ 29ᵐ P.M.	
2.	Constant — 0 17	corr. for long. + 10	No	"	0 13	"
3.	Constant + 3 18	corr. for long. — 23	11 8	"	11 42	"
4.	Constant — 1 47	corr. for long. — 10	10 15	"	10 43	"
5.	Constant — 3 41	corr. for long. + 15	1 39	"	1 57	"
6.	Constant + 3 43	corr. for long. — 14	0 16	"	0 56	"
7.	Constant + 4 28	corr. for long. — 18	10 37	"	10 56	"
8.	Constant + 6 58	corr. for long. + 8	5 24	"	6 3	"
9.	Constant + 0 49	corr. for long. — 24	9 29	"	10 5	"
10.	Constant + 7 43	corr. for long. + 8	11 26	"	11 47	"
11.	Constant — 1 47	corr. for long. — 10	4 29	"	4 49	"
12.	Constant + 6 13	corr. for long. — 15	No	"	0 2	"

FINDING DAILY RATE, Page 270.

1.	22 days	3ˢ·8 gaining	5.	31 days	4ˢ·2 losing
2.	18 "	3·0 losing	6.	22 "	1·1 losing
3.	17 "	4·5 gaining	7.	15 "	6·4 losing
4.	15 "	11·2 gaining	8.	14 "	7·0 gaining

GREENWICH DATE, Page 273.

	Daily Rate.	Acc. Rate.	Green. Date.		Daily Rate.	Acc. Rate.	Green. Date.
1.	6ˢ·8 losing	2ᵐ45ˢ·5	Feb. 1ᵈ 7ʰ47ᵐ42ˢ·5	7.	4ˢ·8 gaining	7ᵐ5ˢ·7	Nov. 8ᵈ 16ʰ26ᵐ50ˢ·7
2.	7·1 gaining	4 2·7	April 28 4 21 37	8.	8·7 losing	6 57·6	Aug. 1 0 5 55·1
3.	3·22 losing	3 16·2	May 7 6 25 15	9.	1·0 losing	1 2·5	May 1 14 28 1·5
4.	2·2 gaining	2 38·1	June 25 20 55 30	10.	0·8 losing	0 32·8	Jan. 20 0 4 50
5.	5·3 gaining	6 25·9	Oct. 25 8 31 43	11.	1·0 losing	4 20·2	Sept. 27 10 32 18
6.	4·0 losing	5 22	Jan. 19 12 33 28	12.	2·6 gaining	2 2·8	April 16 5 33 4·2

HOUR-ANGLE, Page 276.

1.	4ʰ26ᵐ34ˢ	2.	2ʰ50ᵐ42ˢ	3.	4ʰ50ᵐ20ˢ	4.	4ʰ 6ᵐ56ˢ
5.	4 8 45	6.	2 33 42	7.	4 3 50·1	8.	4 29 56

CHRONOMETER, Pages 284—286.

1. Interval 7 days, rate 9ˢ·8 losing, interval 25ᵈ 19ʰ, accumulated rate + 4ᵐ 12ˢ·9, Green. date January 1ᵈ 19ʰ 32ᵐ 13ˢ·5, red. decl. 22° 55′ 22″ S., red. eq. T. add 4ᵐ 16ˢ·25, true alt. 49° 19′ 18″, sum of logs. 9·157889, hour-angle 2ʰ 58ᵐ 18ˢ, M.T. ship January 1ᵈ 21ʰ 5ᵐ 58ˢ. Longitude 23° 26′ 7ˢ·5 E.

By *Raper*: True alt. 49° 19′ 13″, sine sq. 9·157923, hour-angle 2ʰ 58ᵐ 19ˢ. *Longitude* 23° 25′ 52″·5 E.

2. Interval 7 days, rate 7ˢ·6 gaining, interval 19ᵈ 20ʰ, accumulated rate — 2ᵐ 30ˢ·7, Green. date February 18ᵈ 19ʰ 45ᵐ 57ˢ, red. decl. 11° 15′ 51″, eq. T. 14ᵐ 3ˢ·35 add, true alt. 21° 34′ 14″, sum of logs. 9·531239, hour-angle 4ʰ 45ᵐ 15ˢ, M.T. ship February 18ᵈ 19ʰ 28ᵐ 48ˢ. Longitude 4° 17′ 15″ E.

By *Raper*: True alt. 21° 34′ 7″, log. sine sq. 9·531261, hour-angle 4ʰ 45ᵐ 16ˢ. *Longitude* 4° 17′ 30″ E.

3. Interval 43 days, rate 4ˢ·0, interval 115ᵈ 23½ʰ, accumulated rate + 7ᵐ 43ˢ·8, red. decl. 3° 4′ 16″ N., red. eq. T. 5ᵐ 7ˢ·97 add, true alt. 30° 21′ 7″, sum of logs. 9·342946, hour-angle 3ʰ 43ᵐ 55ˢ, M.T. ship February 28ᵈ 3ʰ 49ᵐ 3ˢ. *Longitude* 65° 48′ 45″ E.

By *Raper*: True alt. 30° 21′ 0″, log. sine sq. 9·342975, hour-angle 3ʰ 43ᵐ 56ˢ. *Longitude* 65° 49′ E.

4. Interval 28ᵈ, rate 5ˢ·8, interval 25ᵈ 19½ʰ, accumulated rate — 2ᵐ 29ˢ·6, Green. date April 5ᵈ 19ʰ 13ᵐ 41ˢ, red. decl. 6° 26′ 44″ N., red. eq. T. add 2ᵐ 27ˢ·86, true alt. 16° 17′ 12″, sum of logs. 9·531803, hour-angle 4ʰ 45ᵐ 28ˢ, M.T. ship April 5ᵈ 19ʰ 17ᵐ 0ˢ. *Longitude* 0° 49′ 45″ E.

By *Raper*: True alt. 16° 16′ 58″, log. sine sq. 9·531873, hour-angle 4ʰ 45ᵐ 30ˢ. *Longitude* 0° 49′ 15″ E.

5. Interval 28^d, rate 2^s·5 losing, interval 108^d, accumulated rate + 4^m 30^s, Green. date May 19^d 0^h 29^m 10^s, red. decl. 19° 48' 39" N., red. eq. T. 3^m 45^s·81 subt. from A.T., true alt. 30° 41' 2", sum of logs. 9·340198, hour-angle 3^h 43^m 9^s, M.T. ship May 19^d 3^h 39^m 23^s. Longitude 47° 33' 15" E.

By *Raper*: True alt. 30° 40' 55", log. sine sq. 9·340227, hour-angle 3^h 43^m 10^s. Longitude 47° 33' 30" E.

6. Interval 22^d, rate 9^s·4 gaining, interval 33^d 18^h, accumulated rate − 5^m 17^s·3, Green. date June 14^d 17^h 56^m 42^s, red. decl. 23° 20' 14" N., red. eq. T. + 0^m 6^s, sum of logs. 9·277995, hour-angle 3^h 26^m 32^s, M.T. ship June 14^d 20^h 33^m 34^s. Longitude 39° 13' E.

By *Raper*: True alt. 39° 50' 18", log. sine sq. 9·278041, hour-angle 3^h 26^m 33^s. Longitude 39° 12' 45" E.

7. Interval 26^d, rate 4^s·7 gaining, interval 34^d, accumulated rate − 2^m 40^s, Green. date July 5^d 0^h 33^m 8^s, red. decl. 22° 47' 11" N., red. eq. T. + 4^m 15^s·9, true alt. 48° 47'·5, sum of logs. 9·166615, hour-angle 3^h 0^m 12^s. Longitude 52° 16' W.

By *Raper*: True alt. 48° 46' 55", log. sine sq. 9·166672, hour-angle 3^h 0^m 13^s. Longitude 52° 16' 15" W.

8. Interval 21^d, rate 2^s·6 gaining, interval 104^d 2^h, accumulated rate − 4^m 31^s, Green. date August 13^d 2^h 20^m 42^s, red. decl. 14° 36' 29" N., red. eq. T. 4^m 38^s·16 add, true alt. 27° 23' 29", sum of logs. 9·163652, hour-angle 2^h 59^m 33^s, M.T. ship August 12^d 21^h 5^m 5^s. Longitude 78° 54' 15" W.

By *Raper*: True alt. 27° 23' 25", log. sine sq. 9·163683, hour-angle 2^h 59^m 34^s. Longitude 78° 54' 30" W.

9. Interval 28^d, rate 0^s·7, interval 32^d 19^h, accumulated rate 22^s·4, Green. date August 31^d 19^h 12^m 18^s, red. decl. 8° 18' 26" N., red. eq. T. 0^m 4^s·24 subt., true alt. 44° 44' 40", sum of logs. 9·057066, hour-angle 2^h 37^m 54^s, M.T. ship September 1^d 2^h 37^m 50^s. Longitude 11° 23' E.

By *Raper*: True alt. 44° 44' 36", log. sine sq. 9·057044, hour-angle 2^h 37^m 53^s. Longitude 11° 22' 45" E.

10. Interval 36^d, rate 4^s·7 gaining, interval 97^d 8½^h, accumulated rate − 7^m 38^s, Green. date October 25^d 8^h 35^m 41^s, red. decl. 12° 17' 42" S., red. eq. T. 15^m 52^s·6 subt., true alt. 40° 31' 1", sum of logs. 9·012568, hour-angle 2^h 29^m 42^s, M.T. ship October 25^d 2^h 13^m 49^s. Longitude 95° 28' 15" W.

By *Raper*: True alt. 40° 30' 56", log. sine sq. 9·012622, hour-angle 2^h 29^m 43^s. Longitude 95° 28' W.

11. Interval 20^d, rate 6^s·7, interval 18^d 7^h, accumulated rate + 2^m 2^s·6, Green. date November 27^d 7^h 13^m 53^s, red. decl. 21° 13' 53" S., eq. T. 12^m 5^s subt., true alt. 34° 50' 6", sum of logs. 9·420025, hour-angle 4^h 6^m 51^s, M.T. ship November 26^d 19^h 41^m 4^s. Longitude 173° 12' 15" W.

By *Raper*: True alt. 34° 49' 57", log. sine sq. 9·420039, hour-angle 4^h 6^m 51^s. Longitude 173° 12' 15" W.

12. Interval 18^d, rate 9^s·3 losing, interval 148^d 18^h, accumulated rate + 23^m 3^s, Green. date December 24^d 18^h 37^m 42^s S, true alt. 10° 42' 56", sum of logs. 9·638730, hour-angle 5^h 30^m 14^s, M.T. ship December 24^d 18^h 30^m 0^s. Longitude 1° 55' 45" W.

By *Raper*: True alt. 10° 42' 44", log. sine sq. 9·638761, hour-angle 5^h 30^m 15^s. Longitude 1° 56' W.

13. Interval 31^d, rate 6^s·9 losing, interval 80^d 14^h, accumulated rate 9^m 16^s, Green. date January 1^d 14^h 0^m 38^s, red. decl. 22° 56' 41" S., red. eq. T. + 4^m 9^s·87, true alt. 39° 9' 31", sum of logs 9·362235, hour-angle 3^h 49^m 25^s. Longitude 151° 45' 45" W.

By *Raper*: True alt. 39° 9' 28", log. sine sq. 9·362253, hour-angle 3^h 49^m 25^s. Longitude 151° 45' 45" W.

14. Interval 31^d, rate $8^s\cdot3$ *gaining*, interval $71^d\ 22^h$, accumulated rate $9^m\ 56^s\cdot9$, Green. date February $10^d\ 21^h\ 33^m\ 26^s$, red. decl. $13°\ 59'\ 12'$ S., red. eq. T. $14^m\ 27^s\cdot6$ *add*, true alt. $12°\ 17'\ 54''$, sum of logs. $9\cdot176958$, hour-angle $3^h\ 2^m\ 29^s$. *Longitude* $5°\ 21'\ 45''$ W.

By *Raper* : True alt. $12°\ 17'\ 48''$, log. sine sq. $9\cdot177030$, hour-angle $3^h\ 2^m\ 30^s$. *Longitude* $5°\ 22'$ W.

15. Interval 34^d, rate $2^s\cdot5$ *losing*, interval 52 days, accumulated rate $+\ 2^m\ 10^s$, Green. date October $26^d\ 0^h\ 26^m\ 10^s$, red. decl. $12°\ 31'\ 15'$ S., red. eq. T. $15^m\ 56^s\cdot6$ *subt.*, true alt. $25°\ 10'\ 24''$, sum of logs. $9\cdot286497$, hour-angle $3^h\ 28^m\ 44^s$. *Longitude* $62°\ 42'\ 45''$ W.

By *Raper* : True alt. $25°\ 10'\ 13''$, log. sine sq. $9\cdot286565$, hour-angle $3^h\ 28^m\ 45^s$. *Longitude* $62°\ 43'$ W.

16. Interval 18^d, rate $5^s\cdot4$ *losing*, interval 17 days (nearly), accumulated rate $1^m\ 32^s$, Green. date February $5^d\ 23^h\ 59^m\ 40^s$, red. decl. $15°\ 33'\ 6'$ S., red. eq. T. $+\ 14^m\ 19^s\cdot4$, true alt. $21°\ 21'\ 7''$, sum of logs. $9\cdot466313$, hour-angle $4^h\ 21^m\ 59^s$. *Longitude* $69°\ 9'\ 30''$ E.

By *Raper* : True alt. $21°\ 20'\ 57'$, log. sine sq. $9\cdot466348$, hour-angle $4^h\ 22^m\ 0^s$. *Longitude* $69°\ 9'\ 45''$ E.

17. Interval 48^d, rate $3^s\cdot3$, interval $72^d\ 19^h$, accumulated rate $-\ 4^m\ 0^s$, Green. date April $30^d\ 18^h\ 54^m\ 3^s$, red. decl. $15°\ 3'\ 56'$ N., red. eq. T. $2^m\ 56^s\cdot6$ *subt.*, true alt. $28°\ 18'\ 45''$, sum of logs. $9\cdot460500$, hour-angle $4^h\ 20^m\ 1^s$. *Longitude* $140°\ 45'\ 15''$ E.

By *Raper* : True alt. $28°\ 18'\ 32''$, sine sq. $9\cdot460546$, hour-angle $4^h\ 20^m\ 2^s$. *Longitude* $140°\ 45'\ 30''$ E.

18. Interval 58^d, rate $1^s\cdot6$, interval $99^d\ 16^h$, accumulated rate $-\ 2^m\ 39^s\cdot4$, Green. date April $20^d\ 15^h\ 48^m\ 55^s\cdot8$, red. decl. $11°\ 48'\ 12'$ N., red. eq. T. *subt*. $1^m\ 18^s$, true alt. $31°\ 59'\ 23'$, sum of logs. $9\cdot360584$, hour-angle $3^h\ 48^m\ 56^s$. *Longitude* at sight $179°\ 40'\ 30''$ E., diff. *longitude* $+\ 29'\ 54''$ give *longitude* at noon $179°\ 49'\ 36''$ W.

By *Raper* : True alt. $31°\ 59'\ 19''$, sine sq. $9\cdot360594$, hour-angle $3^h\ 48^m\ 56^s$. *Longitude* at noon $179°\ 49'\ 36''$ W.

19. Interval 48^d, rate $1^s\cdot25$, interval $113^d\ 8^h$, accumulated rate $+\ 2^m\ 21^s\cdot7$, Green. date August $21^d\ 8^h\ 22^m\ 2^s$, red. decl. $11°\ 57'\ 40'$, red. eq. T. $+\ 2^m\ 53^s$, true alt. $34°\ 2'\ 1''$, sum of logs. $9\cdot330311$, hour-angle $3^h\ 40^m\ 25^s$. *Longitude* at sights $179°\ 53'\ 30''$ W. $+\ 29'\ 54''$ W. *Longitude* at noon $179°\ 36'\ 36''$ E.

By *Raper* : True alt. $34°\ 1'\ 51''$, log. sine sq. $9\cdot330352$, hour-angle $3^h\ 40^m\ 25^s$. *Longitude* $179°\ 36'\ 36''$ E.

20. Interval 27^d, rate $0^s\cdot3$ *gaining*, interval 20^d, accumulated rate $-\ 6^s$, Green. date March $20^d\ 1^h\ 32^m\ 46^s$, red. decl. $0°\ 3'\ 28''$ S., red. eq. T. $+\ 7^m\ 33^s$, true alt. $29°\ 1'\ 4''$, sum of logs. $9\cdot410710$, hour-angle $4^h\ 3^m\ 56^s$, M.T. ship March $19^d\ 20^h\ 3^m\ 37^s$. *Longitude* $82°\ 17'\ 15''$ W.

21. Interval 32^d, rate $0^s\cdot8$ *gaining*, interval $113^d\ 15\frac{1}{2}^h$, accumulated rate $-\ 1^m\ 31^s$, Green. date September $22^d\ 15^h\ 38^m\ 0^s$, red. decl. $0°$, red. eq. T. $-\ 7^m\ 47^s$, hour-angle $4^h\ 3^m\ 50^s$, M.T. ship September $22^d\ 19^h\ 48^m\ 23^s$. *Longitude* $62°\ 35'\ 45''$ E.

By *Raper* : True alt. $29°\ 2'\ 15''$, hour-angle $4^h\ 3^m\ 51^s$. *Longitude* $62°\ 35'\ 30''$ E.

SUMNER'S METHOD, *Page* 299.

1. 1st red. decl. $7°\ 42'\ 55''$ S ; 1st red. eq. T. $12^m\ 37^s\cdot47$ (*added* to A.T.) ; 1st true alt. $29°\ 57'\ 17''$; lat. $49°\ 10'$ gives hour-angle $1^h\ 27^m\ 53^s$; long. $178°\ 27'\ 30''$ E. (A). Lat. $49°\ 40'$ gives hour-angle $1^h\ 21^m\ 5^s$; long. $180°\ 9'\ 30''$ E., or $179°\ 50'\ 30''$ W. (A'). 2nd red. decl. $7°\ 38'\ 10''$ S.; 2nd red. eq. T. $12^m\ 35^s\cdot4$ (*added* to A.T.) ; 2nd true alt. $15°\ 55'\ 57''$; lat. $49°\ 10'$ gives hour-angle $3^h\ 38^m\ 34^s$; long. $180°\ 5'\ 0''$ E., or $179°\ 55'$ W. (B). Lat. $49°\ 40'$ N. gives hour-angle $3^h\ 36^m\ 34^s$; long. $179°\ 35'$ E. The line of bearing when the first altitude was taken trends E.N.E. and W.S.W. (nearly). The position of ship when the second altitude was observed was lat. $49°\ 25'$ N., long. $179°\ 49'$ E.

2. 1st red. decl. $19°\ 47'\ 20''$ N.; 1st red. eq. T. $-\ 3^m\ 46^s$; true alt. $49°\ 8'\ 2''$; lat. $48°\ 30'$ gives hour-angle $2^h\ 25^m\ 6^s$; M.T. ship May $18^d\ 21^h\ 31^m\ 8^s$; long. $7°\ 37\frac{1}{2}'$ W. (A). Lat. $49°\ 0'$ gives hour-angle $2^h\ 23^m\ 11^s$; M.T. ship May $18^d\ 21^h\ 33^m\ 3^s$; long. $7°\ 9'$ W. ; 2nd red. decl. $19°\ 50'\ 19''$ N.; red. eq. T. $3^m\ 45^s\cdot4$ *subt.* ; true alt. $40°\ 56'\ 2''$; lat. $48°\ 30'$ gives hour-angle $3^h\ 19^m\ 30^s$; long. $5°\ 18\frac{1}{4}'$ W. (B). Lat. $49°\ 0'$ gives hour-angle $3^h\ 18^m\ 51^s$; long.

5° 38′ W. The line of bearing when the first altitude was taken trends N.E. by N. (northerly) and S.W. by S. (southerly). The position of ship at the time of the second observation is lat. 49° 58½′ N., long. 6° 1½′ W.

3. 1st red. decl. 16° 37′ 15″ S.; 1st. red. eq. T. $16^m\ 7^s{\cdot}4$ *subt.*; 1st true alt. 19° 58′ 31″; lat. 48° 10′ N. gives hour-angle $1^h\ 59^m\ 6^s$; long. 11° 3¾′ W. (A). Lat. 48° 40′ N. gives hour-angle $1^h\ 53^m\ 49^s$; long. 9° 44½′ W. (A′). 2nd red. decl. 16° 41′ 6″ S.; 2nd red. eq. T. $16^m\ 6^s$ *subt.*; lat. 48° 10′ gives hour-angle $3^h\ 22^m\ 48^s$; long. 10° 33¾′ W. (B). Lat. 48° 40′ N. gives hour-angle $3^h\ 20^m\ 9^s$; long. 11° 13½′ W. (B′). The line of bearing at first observation trends N.E. by E. ¼ E. and S.W. by W. ¼ W. The ship's position at time of second observation is lat. 48° 5′ N., long. 10° 27′ W.

4. 1st red. decl. 16° 28′ 46″ S.; 1st red. eq. T. $14^m\ 4^s{\cdot}59$ *add*; 1st true alt. 19° 59′ 26″; lat. 48° 10′ gives hour-angle $2^h\ 0^m\ 28^s$; long. 0° 54¼′ W. (A). Lat. 48° 40′ gives hour-angle $1^h\ 55^m\ 17^s$; long. 0° 23′ E. (A′). 2nd true alt. 11° 0′ 53″; lat. 48° 10′ gives hour-angle $3^h\ 24^m\ 30^s$; long. 0° 21¼′ E. (B). Lat. 48° 40′ gives hour-angle $3^h\ 21^m\ 54^s$; long. 0° 17¾′ W. The line of bearing at first observation runs N.E. by E. ¼ E. and S.W. by W. ¼ W. The ship's position at second observation is lat. 48° 22½′ N., long. 0° 4½′ E.

5. 1st red. decl. 19° 10′ 24″ S.; 1st eq. T. $12^m\ 22^s$ *add*; 1st true alt. 9° 46′ 49″; lat. 51° 15′ gives hour-angle $2^h\ 55^m\ 36^s$; long. 15° 33¾′ W. (A). Lat. 50° 45′ gives hour-angle $2^h\ 59^m\ 8^s$; long. 16° 26¾′ W. (A′). 2nd red. decl. 19° 7′ 58″ S.; 2nd red. eq. T. $12^m\ 24^s{\cdot}77$ *add*; lat. 51° 15′ gives hour-angle $0^h\ 57^m\ 35^s$; long. 17° 16¼′ W. (B). Lat. 50° 45′ gives hour-angle $1^h\ 8^m\ 51^s$; long. 14° 27½′ W. (B′). The line of bearing at first observation trends N.E. ¾ E. (easterly) and S.W. ¾ W (westerly). The position of ship at second observation is lat. 50° 56′ N., long. 15° 26½′ W.

6. 1st red. decl. 2° 27′ 0″ N.; 1st red. eq. T. $5^m\ 36^s{\cdot}45$ *add*; 1st true alt. 36° 48′ 53″; lat. 50° 40′ gives hour-angle $1^h\ 36^m\ 22^s$; long. 180° 42′ 30″ E., or 179° 17½′ W. (A). Lat. 51° 10′ gives hour-angle $1^h\ 41^m\ 35^s$; long. 179° 24½′ E. (A′); 2nd red. decl. 2° 33′ 2″ N.; 2nd red. eq. T. $5^m\ 31^s{\cdot}7$ *add*; 2nd true alt. 15° 23′ 33″; hour-angle $4^h\ 32^m\ 38^s$; long. 180° 15′ E., or 179° 0′ W. (B). Lat. 51° 10′ gives hour-angle $4^h\ 33^m\ 32^s$; long. 181° 28½′ E., or 178° 31½′ W. The line of bearing at first observation trends N.E. by E. ¾ E. and S.W. by W. ¾ W. The position of the ship at the second observation is lat. 51° 12¾′ N., long. 178° 37½′ W.

7. 1st red. decl. 7° 31′ 34″ S.; 1st red. eq. T. $12^m\ 31^s{\cdot}5$ *add*; 1st true alt. 29° 55′ 26″; lat. 49° 15′; hour-angle $1^h\ 29^m\ 49^s$; long. 1° 23′ W. (A). Lat. 49° 45′ gives hour-angle $1^h\ 23^m\ 7^s$; long. 0° 17′ 30″ E. (A′). 2nd red. decl. 7° 36′ 50″; 2nd red. eq. T. $12^m\ 29^s$ *add*; lat. 49° 15′ gives hour-angle $3^h\ 38^m\ 57^s$; long. 1° 9′ E. (B). Lat. 49° 45′ gives hour-angle $3^h\ 36^m\ 58^s$; long. 0° 39½′ E. (B′). The line of bearing at first observation trends N.E. by E. ¾ E. (easterly) and S.W. by W. ¾ W. (westerly). The position of the ship at the second observation is lat. 49° 45′ N., long. 0° 39½′ E.

8. 1st red. decl. 16° 33′ 0″ S.; 1st red. eq. T. $14^m\ 3^s{\cdot}14$ *add*; 1st true alt. 20° 11′ 0″; lat. 47° 10′ gives hour-angle $2^h\ 7^m\ 10^s$; long. 83° 19′ 15″ E. (A). Lat. 47° 40′ gives hour-angle $2^h\ 2^m\ 25^s$; long. 84° 30′ 30″ E. (A′). 2nd red. decl. 16° 29′ 2″ S.; 2nd red. eq. T. $14^m\ 4^s{\cdot}49$; 2nd true alt. 11° 17′ 59″; lat. 47° 10′ gives hour-angle $3^h\ 26^m\ 49^s$; long. 86° 14′ 45″ E. (B). Lat. 47° 40′ gives hour-angle $3^h\ 24^m\ 19^s$; long. 85° 37′ 15″ E. (B′). The line of bearing at the first observation trends N.E. by E. and S.W. by W. The position of the ship at the second observation is lat. 47° 53½′ N., long. 85° 20′ E.

9. 1st red. decl. 19° 3′ 46″ S.; 1st red. eq. T. $12^m\ 28^s{\cdot}88$ *add*; 1st true alt. 9° 46′ 38″; lat. 51° 15′ N. gives hour-angle $2^h\ 56^m\ 30^s$; long. 180° 18¾′ E., or 179° 41¼′ W. (A). 2nd red. decl. 19° 1′ 20″ S.; 2nd eq. T. $12^m\ 31^s$; 2nd true alt. 18° 40′ 21″; lat. 50° 45′ gives hour-angle $2^h\ 59^m\ 11^s$; long. 179° 38½′ E. (A′) Lat. 51° 15′ N. gives hour-angle $0^h\ 55^m\ 43^s$; long. 180° 34′ W., or 178° 26′ E. (B). Lat. 50° 45′ gives hour-angle $1^h\ 7^m\ 18^s$; long. 178° 40¼′ W. The line of bearing at first observation trends N.E. ½ N. (northerly) and S.W. ½ S. (southerly). The position of the ship at the second observation is lat. 50° 53′ N., long. 179° 24′ W.

10. 1st red. decl. 0° 15′ 23″ N.; 1st red. eq. T. $7^m\ 18^s{\cdot}5$ *add*; 1st true alt. 22° 59′ 25″; lat. 50° 10′ gives hour-angle $3^h\ 31^m\ 16^s$; long. 0° 43¾′ W. (A). Lat. 50° 50′ gives hour-angle $3^h\ 28^m\ 48^s$; Long. 0° 6′ E. (A′). 2nd red. decl. 0° 20′ 58″ N.; 2nd red. eq. T. $7^m\ 14^s{\cdot}3$

add; 2nd true alt. 32° 40′ 10″; lat. 50° 10′ gives hour-angle 2h 13m 22s; long. 0° 26½′ E. (B). Lat. 50° 50′ gives hour-angle 2h 8m 22s; long. 0° 48½′ W. (B′). The line of bearing at the first observation trends N.E. ½ N. and S.W. ½ S. The position of the ship at the second observation is lat. 50° 14′ N., long. 0° 19½′ E.

11. 1st red. decl. 2° 7′ 36″ S.; 1st red. eq. T. + 9m 4s·4; 1st true alt. 14° 22′ 43″; lat. 49° 0′ gives hour-angle 4h 20m 17s; long. 1° 16′ W. (A). Lat. 49° 40′ gives hour-angle 4h 18m 43s; long. 0° 52½′ W. (A′). 2nd red. decl. 2° 0′ 6″ S.; 2nd eq. T. + 8m 59s·4; 2nd true alt. 23° 27′ 13″; lat. 49° 0′ gives hour-angle 3h 18m 40s; long. 0° 40¼′ W. (B). Lat. 49° 40′ gives hour-angle 3h 15m 53s; long. 1° 22′ W. (B′). The line of bearing at the first observation trends N. by E. ¾ E. and S. by W. ¾ W. The position of the ship at the second observation is lat. 49° 22½′ N., long. 1° 4′ W.

12. 1st red. decl. 16° 1′ 57″ N.; 1st red. eq. T. — 3m 23s·2; 1st true alt. 50° 40′ 55″; lat. 47° 0′ gives hour-angle 1h 57m 10s; long. 51° 27¾′ W. (A). Lat. 46° 20′ gives hour-angle 2h 0m 35s; long. 52° 19′ W. (A′). The line of bearing at the first observation trends N.E. ¾ N. and S.W. ¾ S. The position of the ship at the second observation is lat. 46° 28′ N., long. 51° 45½′ W.

13. 1st red. decl. 23° 23′ 23″ N.; 1st red. eq. T. + 0m 32s; 1st true alt. 52° 1′ 30″; lat. 48° 50′ gives hour-angle 2h 23m 35s; long. 48° 13½′ E. (A). Lat. 48° 10′ gives hour-angle 2h 25m 45s; long. 47° 41′ E. (A′). 2nd red. decl. 23° 23′ 36″ N.; 2nd red. eq. T. + 0m 33s·7; 2nd true alt. 63° 45′ 6″. Lat. 48° 50′ gives hour-angle 0h 32m 41s; long. 47° 29½′ E. (B). Lat. 48° 10′ gives hour-angle 0h 43m 41s; long. 50° 13¾′ E. (B′). The line of bearing at the first observation trends N.E. by N. ¼ N. (northerly) and S.W. by S. ¼ S. (southerly). The position of the ship at the second observation is lat. 48° 30½′ N., long. 48° 49½′ E.

14. 1st red. decl. 13° 34′ 51″ S.; 1st red. eq. T. *subt.* 16m 11s·69; 1st true alt. 28° 28′ 14″; lat. 47° 30′ gives hour-angle 0h 33m 11s; long. 74° 38¾′ E. (A). Assumed lat. 47° 50′ (*not* 48°, which would give the sum more than 180°), this gives hour-angle 0h 16m 55s; long. 70° 34¾′ E. (A′). Red. decl. 13° 37′ 45; eq. T. 16m 13s·6 *subt.*; lat. 47° 30′ gives hour-angle 3h 0m 37s; long. 73° 42¼′ W. (B). Lat. 48° 0′ gives hour-angle 2h 57m 43s; long. 74° 25¾′ W. (B′). The line of bearing at the first observation trends E. ½ N. (northerly) and W. ½ S. (southerly). The position of the ship at the second observation is lat. 47° 46′ N., long. 74° 12′ W.

N.B.—It is scarcely necessary to warn the intelligent navigator against making an assumption of latitude that shall render the computation of the hour-angle impossible: the sum of the altitude, latitude, and polar distance, can never exceed 180°.

TRUE AZIMUTHS, *Page* 306.

1. S. 98° 43′ 30″ E.	4. S. 90° 32′ 54″ E.	7. S. 56° 4′ 0″ W.
2. S. 41 58 18 E.	5. S. 60 9 36 W.	8. East.
3. N. 75 58 4 W.	6. N. 49 7 14 W.	9. N. 84 5 2 W.

AZIMUTHS, *Page* 312.

No.	Green. date.	Red. decl.	True alt.	Sum of logs.	True azimuth.	Error.	Deviation.
1.	Jan. 23d 23h 46m 7s	19° 0′ 30″ S	38° 34′ 32″	19·731159	N 91° 24′ 44″ E	15° 39′ 41″ E	11° 3′ 44″ E
2.	Feb. 28 9 43 4	7 44 1 S	27 7 43	10·211153	S 51 41 38 W	10 8 38 E	1 21 22 W
3.	Mar. 27 0 34 12	2 40 49 N	20 41 3	19·703187	S 90 35 56 W	10 39 4 W	7 9 4 W
4.	April 2 19 0 8	5 18 4 N	11 50 42	19·651722	S 74 5 20 W	5 54 42 W	15 4 40 W
5.	May 25 21 0 32	21 18 32 N	43 19 41	19·918155	S 61 33 50 E	0 11 10 E	15 56 10 E
6.	June 20 6 55 44	23 27 4 N	16 49 1	19·808282	S 106 37 52 W	30 2 8 W	7 2 8 W
7.	July 31 1 6 45	18 14 9 N	43 35 48	19·601312	S 78 22 56 E	3 2 58 W	5 47 2 E
8.	Aug. 23 2 57 38	11 22 44 N	7 43 28	19·759124	S 48 32 37 E	15 7 24 E	10 22 36 W
9.	Aug. 31 18 33 2	8 10 2 N	39 14 51	19·940260	N 73 43 34 W	7 50 11 E	11 29 11 E
10.	Nov. 25 7 30 28	20 51 43 S	34 2 21	19·966021	N 80 41 22 W	2 39 7 W	11 52 7 W
11.	Dec. 16 22 58 2	23 22 28 S	51 13 0	19·712137	N 91 45 50 E	3 54 19 W	5 52 59 E
12.	July 21 16 18 50	22 59 26 N	14 19 37	19·250071	N 50 26 36 E	11 33 24 W	5 26 36 E
13.	Jan. 6 2 40 30	22 27 49 S	26 40 51	19·712700	N 91 38 0 W	0 57 0 W	23 17 0 E
14.	April 23 1 43 33	13 16 17 N	18 52 57	19·130408	N 63 15 44 E	1 25 44 E	15 55 41 E
15.	Jan. 20 6 53 47	17 47 45 S	13 47 28	19·240075	S 43 21 0 W	31 26 0 W	10 56 0 W
16.	Feb. 1 2 22 23	17 0 41 S	40 7 21	19·651011	N 83 59 54 W	13 0 54 W	16 39 6 E
17.	Mar. 26 0 20 58	2 17 17 N	32 59 30	19·362660	S 57 23 0 E	12 31 0 W	13 27 0 E
18.	Feb. 25 15 40 0	8 46 8 S	60 48 32	19·222946	S 61 56 30 W	0 39 38 E	8 44 30 W
19.	June 21 7 3 20	23 27 10 N	15 45 28	19·8417 6	S 112 51 32 W	75 31 43 W	7 41 43 W
20.	Sept. 10 20 16 0	4 31 48 N	42 38 43	18·930079	N 33 57 20 E	10 52 31 W	1 27 26 E
21.	April 23 14 32 0	13 8 50 N	43 6 34	19·784877	S 102 48 0 E	1 26 0 E	1 4 0 W
22.	Feb. 28 20 15 0	7 34 0 S	66 26 23	18·043319	S 12 4 2 W	24 52 47 E	14 52 47 E
23.	Dec. 31 21 25 30	23 1 23 S	45 22 20	19·891183	N 123 49 48 E	11 10 12 W	9 49 48 E

REDUCTION TO MERIDIAN, Pages 320—321.

No.	Green. date.	Time noon.	Red. decl.	True. alt.	Nat. no.	Latitude.
1.	Jan. 4ᵈ 1ʰ31ᵐ40ˢ	14ᵐ 8ˢ	22° 41′ 58″ S.	32° 26′ 19″	1442	34° 45′ 50″ N.
2.	Feb. 28 1 4 49	14 45	7 51 58 S.	38 6 34	1481	43 55 0 N.
3.	Mar. 19 16 34 18	25 42	0 12 13 S.	47 57 15	4711	41 50 41 S.
4.	April 20 23 17 6	18 30	11 54 31 N.	61 39 0	2444	39 57 44 N.
5.	May 29 7 38 27	30 27	21 41 51 N.	30 33 8	6517	37 18 52 S.
6.	June 19 0 41 12	15 8	23 26 19 N.	68 48 28	1427	44 24 13 N.
7.	July 16 0 2 19	10 19	21 21 22 N.	67 52 44	942	0 37 17 S.
8.	Aug. 29 15 2 44	20 24	9 5 25 N.	57 34 6	2945	41 12 21 N.
9.	Sept. 8 12 15 43	9 43	5 27 4 N.	85 30 34	883	9 14 31 N.
10.	Oct. 10 18 47 58	28 10	7 0 24 S.	36 44 4	5215	45 53 56 N.
11.	Nov. 2 17 1 29	20 5	15 2 18 S.	72 2 22	3143	32 24 10 S.
12.	Dec. 22 1 38 58	9 14	23 27 4 S.	65 23 36	504	47 59 18 S.
13.	Jan. 5 8 58 28	17 32	22 33 10 S.	58 17 34	2669	8 51 45 N.
14.	April 28 2 3 11	13 37	14 13 56 N.	56 40 4	1610	18 55 50 S.
15.	Mar. 19 21 28 52	16 56	0 7 22 S.	70 30 47	2580	19 9 42 S.
16.	April 12 10 46 29	10 19	8 55 58 N.	80 36 17	1001	0 6 13 S.
17.	Sept. 15 14 29 20	22 0	2 44 18 N.	44 19 4	3385	42 40 20 S.
18.	Mar. 15 19 49 1	4 19	1 43 52 S.	50 12 46	140	38 2 57 N.
19.	Mar. 4 19 8 4	20 4	6 2 56 N.	50 0 56	3174	33 39 6 N.
20.	Sept. 22 7 4 42	22 6	0 8 27 N.	43 55 22	3242	45 40 41 S.

METHOD II.

No.	Reduction.	True alt. (Norie)	True alt. (Raper)	Latitude (Norie)	Latitude (Raper)
1.	+ 5′ 52″	32° 26′ 19″	32° 26′ 12″	34° 45′ 51″ N.	34° 45′ 58″ N.
2.	+ 6 30	38 6 34	38 16 27	43 54 56 N.	43 55 5 N.
3.	+ 24 36	47 57 15	47 57 18	41 50 22 N.	41 50 19 S.
4.	+ 17 54	61 39 0	61 38 59	39 57 37 N.	39 57 38 N.
5.	+ 26 10	30 33 8	30 33 5	37 18 51 S.	37 18 54 S.
6.	+ 13 44	68 48 28	68 48 21	44 24 7 N.	44 24 14 N.
7.	+ 8 40	67 52 44	67 52 35	0 37 14 S.	0 37 23 S.
8.	+ 19 8	57 34 6	57 34 3	41 12 11 N.	42 12 14 N.
9.	+ 44 48	Exceeds limits.			
10.	+ 22 30	36 44 4	36 43 52	45 53 2 S.	45 53 14 S.
11.	+ 37 2	72 2 22	72 2 16	32 22 54 S.	31 23 0 S.
12.	+ 4 16	65 23 36	65 22 35	47 59 12 S.	47 59 13 S.
13.	+ 17 38	58 17 34	58 17 31	8 51 38 N.	8 51 41 N.
14.	+ 10 14	56 40 4	56 37 51	18 55 46 S.	18 55 49 S.
15.	+ 27 24	70 30 47	70 30 37	19 9 11 S.	19 9 21 S.
16.	+ 22 10	80 36 17	80 36 16	0 5 35 S.	0 5 36 S.
17.	+ 16 22	44 19 4	44 18 59	42 40 16 S.	42 40 21 S.
18.	+ 0 46	50 12 46	50 12 37	38 2 36 N.	38 2 45 N.
19.	+ 17 6	50 0 56	50 18 2	33 39 8 S.	33 39 2 N.
20.	+ 15 31	43 55 22	44 10 45	45 40 40 S.	45 40 48 S.

BY TOWSON:

No.	Aug. I.	Aug. decl.	Index.	Aug. II.	Latitude.
1.	+ 2′ 20″	22° 44′ 18″ S.	21	+ 3′ 22″	34° 46′ 1″ N.
2.	+ 0 58	7 52 56 S.	26	+ 5 23	43 55 7 N.
3.	+ 0 11	0 12 24 S.	68	+ 24 11	41 50 58 S.
4.	+ 2 19	11 56 50 N.	41	+ 20 6	39 57 54 N.
5.	+ 10 31	21 52 22 N.	77	+ 15 38	37 18 52 S.
6.	+ 2 44	23 29 3 N.	24	+ 16 37	44 23 58 N.
7.	+ 1 8	21 22 30 N.	11	+ 7 12	0 37 34 S.
8.	+ 2 6	9 7 31 N.	50	+ 20 52	41 12 33 N.
9.	This hour-angle exceeds the limits of the Table.				
10.	+ 3 10	7 3 34 S.	76	+ 19 45	45 52 37 N.
11.	+ 3 19	15 5 37 S.	47	+ 38 56	32 24 19 S.
12.	+ 1 1	23 28 5 S.	9	+ 5 15	47 59 14 S.
13.	+ 3 34	22 36 44 S.	32	+ 13 43	8 51 59 N.
14.	+ 1 27	14 15 23 N.	22	+ 8 48	18 55 46 S.
15.	+ 0 0	0 7 22 S.	36	+ 27 12	19 9 23 S.
16.	This hour-angle exceeds the limits of the Table.				
17.	+ 0 42	2 45 0 N.	57	+ 15 31	42 40 25 S.
18.	+ 0 1	1 43 53 S.	2	+ 0 38	38 2 43 N.
19.	+ 1 23	6 4 19 S.	49½	+ 15 37	33 39 8 N.
20.	+ 0 0	0 8 27 N.	57	+ 15 18	45 40 53 S.

MERIDIAN ALTITUDE OF STAR, Page 323.

No.	Declination.	NORIE. True alt.	Latitude.	RAPER. True alt.	Latitude.
1.	28° 26' 54" N.	75° 5' 55" S.	43° 20' 59" N.	75° 5' 46" S.	43° 21' 8" N.
2.	45 52 36 N.	53 53 45 N.	9 46 21 S.	53 53 38 N.	9 46 14 N.
3.	38 40 50 N.	49 55 2 N.	1 24 8 S.	49 55 1 N.	1 24 9 S.
4.	49 26 44 N.	51 46 49 N.	11 13 33 N.	51 46 39 N.	11 13 23 N.
5.	10 32 59 S.	63 13 18 S.	16 13 43 N.	63 13 17 S.	16 13 44 N.
6.	57 49 47 S.	40 5 30 S.	7 55 17 S.	40 5 21 S.	7 55 8 S.
7.	14 32 9 N.	78 11 26 S.	26 20 43 N.	78 11 18 S.	26 20 51 N.
8.	22 54 38 N.	68 15 54 N.	1 10 32 N.	68 15 49 N.	1 10 27 N.
9.	16 16 14 N.	29 52 9 N.	43 51 37 S.	29 51 59 N.	43 51 47 S.
10.	62 27 12 S.	75 4 59 S.	47 32 11 S.	75 4 54 S.	47 32 6 S.
11.	8 9 13 S.	30 16 28 S.	51 34 19 N.	30 16 20 S.	51 34 27 N.
12.	44 51 47 N.	20 7 12 N.	25 1 1 S.	20 7 2 N.	25 1 11 S.
13.	8 33 52 N.	60 45 36 N.	20 40 32 S.	60 45 28 N.	20 40 40 S.
14.	16 33 37 S.	58 55 9 N.	47 38 28 S.	58 55 5 N.	47 38 32 S.
15.	19 47 35 N.	79 42 38 S.	30 4 57 N.	79 42 29 S.	30 5 6 N.
16.	26 10 13 S.	68 42 51 S.	2 53 4 S.	68 42 48 S.	4 53 1 S.
17.	60 21 10 S.	9 52 32 S.	19 46 18 N.	9 52 24 S.	19 46 26 N.
18.	30 14 31 S.	70 3 15 N.	50 11 16 S.	70 3 14 N.	50 11 17 S.

EXAMINATION PAPER—No. I, Pages 326—327.

1. Log. 5·861612 = nat. no. 727130. Log. 5·745622 = nat. no. 556700.

2. Log. 2·698971 = nat. no. 500. By *Raper*: log. 2·698970 = 500; log. 2·041393 = nat. no. 110.

3. True Courses.—S. 14° W., 15' dep. course; S. 28° W., 45'; N. 76° W., 49'; N. 48° W., 38'; N. 85° W., 31'; S. 22° W., 36'; S. 54° W., 41'; S. 42° W., 8' current course. *Diff. lat.* 77'·7 S., *dep.* 183'·4 W.; *course* S. 67° W.; *dist.* 199'. Lat. in 35° 45' N.; *mid. lat.* 36° 24'; *diff. long.* 228'; *long.* in 12° 48' W.

4. Green. date, Jan. 1ᵈ 6ʰ 50ᵐ 44ˢ; red. decl. 22° 58' 11" S.; true alt 60° 12' 47"; latitude 6° 49' 2" N.

By *Raper*: True alt. 60° 12' 39". Latitude 6° 49' 10" N.

5. Log. of diff. long. 2·048016 = diff. long. 111'·7.

6. Diff. lat. 219' S.; mer. diff. lat. 292'; diff. long. 380' W.; log. tang. of course 10·114401; course S. 52° 27' 38" W.; log. of distance 2·555608; *distance* 359'·4.

7. Standard, Brest   constant + 4ʰ 2ᵐ; 9ʰ 19ᵐ A.M., and 9ʰ 39ᵐ P.M.
   „ Portsmouth   „   − 2 41 ; 10 36 A.M., and 10 57 P.M.
   „ Devonport   „   − 0 46 ; 6 27 A.M., and 6 47 P.M.

8. Green. date, January 1ᵈ 5ʰ 23ᵐ 24ˢ; red. decl. 22° 58' 30" S.; log. sine true amp. 9·788227. True amp. E. 37° 53' 7" S. *Error of compass* 21° 0' 37" E. *Deviation* 2° 51' 23" W.

9. Interval 31ᵈ, rate 6ˢ·3 gains. Interval 28ᵈ 7ʰ, acc. rate 2ᵐ 58ˢ·2, Green. date, Jan. 29ᵈ 6ʰ 53ᵐ 49ˢ, red. decl. 17° 47' 45" S., true alt. 13° 47' 28", hour-angle 3ʰ 23' 1", red. eq. time *add* 13ᵐ 27ˢ·5, mean time at ship 29ᵈ 3ʰ 36ᵐ 29ˢ. Longitude 49° 20' W.

Raper: True alt. 13° 47' 12"; hour-angle 3ʰ 23ᵐ 3ˢ. Longitude 49° 19' 30" W.

10. Green. date, Jan. 15ᵈ 6ʰ 11ᵐ 12ˢ, red. decl. 21° 2' 27" S., true alt. 55° 18' 24", sum of logs. 19·722644, true azimuth N. 93° 12' 42" E. *Error of compass* 16° 12' 42" E. *Deviation* 8° 22' 42" E.

By *Raper*: True alt. 55° 18' 22", sine sq. of azimuth 9·722645, true azimuth N. 93° 12' 42" E. *Error of compass* 16° 12' 42" E. *Deviation* 8° 22' 42" E.

11. Time from noon 16ᵐ 47ˢ, Green. date, January 16ᵈ 14ʰ 18ᵐ 55ˢ, red. decl. 20° 46' 59" S., true alt. 33° 7' 1", nat. no. 2026, nat. cos. mer. zen. dist. 548376, mer. zen. dist. 56° 44' 40" N. Latitude 35° 57' 41" N.

METHOD II.—*Raper*: True alt. 33° 6' 58", reduction + 8' 19". Latitude 35° 57' 43" N.

Towson: Aug. I, 3' 5", aug. decl. 20° 50' 4", index 30. Latitude 35° 57' 45" N. Aug. II, 5' 10". Latitude 35° 57' 45" N.

12. 1st Red. decl. 20° 54' 49" S.; eq. T. + 10ᵐ 2ˢ·3; true alt. *(Norie)* 13° 0' 53"; alt. 13° 0' 53"; lat. 48° 50'; polar dist. 110° 54' 49"; hour-angle 2ʰ 26ᵐ 13ˢ, long. 8° 3' W. (A). Alt. 13° 0' 53"; lat. 49° 30'; polar dist. 110° 54' 49"; hour-angle 2ʰ 20ᵐ 12ˢ, *Longitude* 6° 32¾' W. (A'). The line of bearing at the first observation trends N.E. ¾ E. (easterly) and S.W. ¾ W. (westerly). The position of the ship at the second observation is lat. 49° 42½' N., long. 7° 57' W.

13. Star's decl. 16° 16' 14" N., true alt. 52° 30' 36". *Latitude* 53° 45' 38" N.
Raper: True alt. 52° 30' 32". *Latitude* 53° 45' 42" N.
The Curve.—Correct magnetic bearing S. 79° W.
Deviations.—15° W.; 0°; 10° E.; 14° E.; 15° E.; 7° E.; 10° W.; 21° W.
Compass courses.—N. 47° W.; N. 27½° E.; S. 79° E.; S. 8½° W.
Magnetic courses.—N. 41½° W.; N. 73° E.; S. 7° E.; S. 67° W.
Bearings, magnetic.—N. 84° E.; N. 14° W.

EXAMINATION PAPER—No. II, *Pages* 329—331.

1. 7·602321 = 40024074. Log. 6·297979 = nat. no. 1986000.
2. 2·168747 = 147·485. Log. 5·000004 = nat. no. 100001.
3. **True Courses.**—N. 51° E., 18' dep. course; S. 73° E., 52'; S. 58° E., 43'; N. 57° E., 36'; N. 38° E., 27'; S. 21° E., 24'; S. 40° E., 29'; S. 39° E., 12', current course. *Diff. lat.* 39'·7 S.; *dep.* 181'·8 E.; *course* S. 78° E., 186'. *Lat. in* 46° 51' N.; *mid. lat.* 47° 11'; *diff. long. (mid. lat.)* 267½', or 4° 27½' E. *Longitude in* 48° 5½' W.
4. Green. date, Jan. 31ᵈ 18ʰ 47ᵐ 4ˢ, or long. in time 5ʰ 12ᵐ 56ˢ; red. decl. 17° 5' 56" S.; true alt. 78° 17' 52". *Latitude* 5° 23' 48" S.
By *Raper:* True alt. 78° 17' 51". *Latitude* 5° 23' 47" S.
5. Log. of diff. long. 2·232899 = diff. long. 171'·0.
6. Diff. lat. 2404' S.; mer. diff. lat. 3104'; diff. long. 3692' W.; tang. course 10·075340; *course* S. 49° 56' 42" W.; log. of distance 3·572370, *distance* 3736'.
7. Standard, Filey Bay, constant + 0ʰ 58ᵐ; 10ʰ 49ᵐ A.M., and 11ʰ 32ᵐ P.M.
   " Milford Haven, " — 0 16 ; no A.M., and 0 7 P.M.
   " Cromarty, " — 2 21 ; 6 34 A.M, and 7 18 P.M.
8. Green. date, February 19ᵈ 18ʰ 10ᵐ 12ˢ; red. decl. 10° 55' 36" S.; sine 9·287272; true amp. W. 11° 10' 21" S. *Error of compass* 19° 45' 54" E. *Deviation* 9° 25' 54" E.
9. Interval 21ᵈ; rate 10ˢ·8 *losing*; Interval 30ᵈ 9½ʰ; acc. rate 5ᵐ 28ˢ; Green. date, Feb. 9ᵈ 9ʰ 30ᵐ 41ˢ; red. decl. 14° 28' 38" S.; true alt. 9° 14' 3"; red eq. time *add* 14ᵐ 27ˢ; log. 9·323467; hour-angle 3ʰ 38ᵐ 31ˢ. *Longitude* 166° 18' 30" E.
Raper: True alt. 9° 13' 52"; sine sq. 9·323548; hour-angle 3ʰ 38ᵐ 34ˢ. *Long.* 166° 18' E.
10. Green. date, Feb. 16ᵈ 5ʰ 29ᵐ 51ˢ; red. decl. 12° 10' 43" S.; true alt. 7° 15' 55"; sum of logs. 19·401578; true azimuth S. 60° 16' 40" E. *Error* 23° 56' 40" W. *Deviation* 1° 3' 20" E.
Raper: True alt. 7° 15' 43"; sine sq. 9·401643 = S. 60° 16' 58" E. *Deviation* 1° 3' 2" E.
11. Time from noon 28ᵐ 48ˢ; Green. date, Feb. 14ᵈ 19ʰ 54ᵐ; red. decl. 12° 39' 34" S.; true alt. 46° 31' 34" N.; nat. no. 4305; nat. cos. 729994; zen. dist. 43° 7' 7" S. *Latitude* 55° 46' 41" S.
METHOD II.—*Raper:* True alt. 46° 31' 28"; reduction + 21' 34". *Latitude* 55° 46' 32" S.
Towson: Aug. I, 5' 46"; index 76; aug. decl. 12° 45' 20"; Aug. II, 27' 28". *Latitude* 55° 46' 18" S.

12. 1st red. decl. 16° 37' 43" S.; red. eq. T. + 14ᵐ 1ˢ·5; true alt. 19° 59' 43"; lat. 47° 10' gives hour-angle 2ʰ 8ᵐ 26ˢ; long. 179° 8½' E. (A). Lat. 47° 40' gives hour-angle 2ʰ 3ᵐ 44ˢ; long. 180° 18½' E. or 179° 41½' W. (A'). 2nd red. decl. 16° 35' 40" S.; eq. T. + 14ᵐ 3ˢ; .true alt. 11° 12' 34"; lat. 47° 10' gives hour-angle 3ʰ 27ᵐ 0ˢ; long. 180° 17½' E. or 179° 42½' W. (B). Lat. 47° 40' gives hour-angle 3ʰ 24ᵐ 31ˢ; long. 179° 40½' E. The line of bearing at first observation lies N.E. by E. (easterly), and S.W. by W. (westerly). The position of ship at the time of second observation is lat. 47° 18½' N.; long. 179° 53' W.

13. Star's decl. 5° 31' 20" N.; true alt. 77° 14' 26". *Latitude* 18° 16' 54" N.
Raper: True alt. 77° 13' 38". *Latitude* 18° 17' 42" N.
The Curve.—Correct magnetic bearing S. 39° E.
Deviations.—5° W.; 19° E.; 23° E.; 13° E.; 4° E.; 8° W.; 22° W.; 24° W.
Compass courses.—N. 43° E.; N. 68° W.; N.W. by W.; N. 83° E.
Magnetic courses.—S. 82° W.; N. 52° E.; S. 35° W.; S.S.E.
Bearings, magnetic.—S. 80½° W.; S. 1½° E.

EXAMINATION PAPER—No. III, *Pages* 331—333.

1. 3·726356 = 5325·44; 6·370772 = 2348400.
2. 1·875495 = 75·075; 2·301030 = 200.
3. **True Courses.**—S. 70° E., 17' dep. course; S. 37° W., 21'; S. 24° W., 20'; N. 62° W., 24'; N. 64° W., 26'; S. 77° E., 19'; S. 28° E, 18'; N. 54° E., 21' current course. *Diff. lat.* 26'·1 S.; *dep.* 5'·3 W.; *course* S. 11° W.; *dist.* 27'. *Lat. in* 61° 34' N.; *diff. long.* 11' W. *Long. in* 149° 49' E.
4. Green. date, March 20ᵈ 11ʰ 33ᵐ 12ˢ, or long. in time 11ʰ 33ᵐ 12ˢ; red. decl. 0° 6' 32" S.; true alt. 89° 53' 28". *Latitude* 0.
Raper: True alt. 89° 53' 22". *Latitude* 0° 0' 6" S.
5. Log. of diff. long. 2·679550 = *diff. long.* 478"·1.
6. Diff. lat. 688' N.; mer. diff. lat. 786'; diff. long. 3625' W.; tang. 10·663885; *course* N. 77° 45' 58" W.; log. of *distance* = 3·511449; *distance* 3247'.
7. Standard, Brest   constant + 6ʰ19ᵐ; 11ʰ51ᵐA.M., and no P.M.
   ,,   Brest   ,,   + 5 31 ; 11 3 A.M., and 11ʰ19ᵐP.M.
   ,,   Portsmouth   ,,   − 2 31 ; 11 2 A.M., and 11 18 P.M.
   ,,   Harwich   ,,   − 2 51 ; 11 11 A.M., and 11 27 P.M.
   ,,   Thurso   ,,   + 2 2 ; midnight, and 0 17 P.M.
   ,,   Liverpool   ,,   − 0 55 ; no A.M., and 0 15 P.M.
8. Green. date, March 6ᵈ 14ʰ 47ᵐ 28ˢ; red. decl. 5° 20' 41" S.; sine amp. 9·181777; true amp. W. 8° 44' 30" S.; *Correction* S° S' E.; *deviation* 15° 52' W.
9. Interval 41ᵈ; rate 5ˢ·8 *gaining*; interval 89ᵈ 23ʰ; acc. rate 8ᵐ 41ˢ·7; Green. date March 30ᵈ 22ʰ 53ᵐ 12ˢ·5; red. decl. 4° 12' 31" N.; true alt. 29° 19' 56"; hour-angle 3ʰ 57' 4"; red. eq. time *a* ld 4ᵐ 13ˢ; mean time ship March 30ᵈ 20ʰ 7ᵐ 9ˢ. *Longitude* 41° 30' 52"·5 W.
Raper: True alt. 29° 19' 45"; hour-angle 3ʰ 57ᵐ 5ˢ. *Longitude* 41° 31' 7"·5 W.
10. Green. date, March 9ᵈ 9ʰ 43ᵐ 5ˢ; red. decl. 4° 15' 37" S.; true alt. 18° 6' 51"; sum of logs. 19·604572, true azimuth N. 78° 44' 4" E. *Correction* 7° 21' 34" E.; *deviation* 3° 28' 26" W.
11. Time from noon 10ᵐ 14ˢ; Green. date March 24ᵈ 18ʰ 13ᵐ 34ˢ; red. decl. 1° 47' 40" N.; true alt. 71° 20' 43"; nat. no. 937; nat. cos. mer. zen. dist. 948400 = 18° 29' 10" N. *Latitude* 20° 16' 44" N.
METHOD II.—*Raper:* True alt. 71° 20' 37" + 10' 17". *Latitude* 20° 16' 46" N.
Towson: Aug. I, + 0' 7"; index no. 13. Aug. II, + 10' 14". *Latitude* 20° 16' 50" N.
12. 1st red. decl. 2° 50' 28"·5 N.; 1st eq. T. + 5ᵐ 18ˢ·0; true alt. 36° 49' 14"; lat. 51° 30' gives hour-angle 1ʰ 42ᵐ 34ˢ; long. 180° 55' W. or 179° 5' E. (A); lat. 51° 0' gives hour-angle 1ʰ 47ᵐ 24ˢ; long. 182° 7½' W. or 177° 52½' E. (A). 2nd red. d. l. 2° 56' 31" N.; and red. eq. T. + 5ᵐ 13ˢ; true alt. 15° 23' 34"; lat. 51° 30' gives hour-angle 4ʰ 35ᵐ 1ˢ; long. 179° 14' W.; lat. 51° 0' gives hour-angle 4ʰ 35ᵐ 53ˢ; long. 179° 1' W. The line of bearing trends N.E. by E. and S.W. by W.; the position of ship at the time of second observation, lat. 51° 39' N., long. 179° 18¾' W.
13. Arcturus' decl. 19° 47' 33" N.; true alt. 36° 7' 27". *Latitude* 34° 5' 0" S.
Raper: True alt. 36° 7' 22". *Latitude* 34° 5' 5" S.
The Curve.—Correct magnetic bearing S. 70° E.
Deviations.—3° W.; 20° E.; 15° E.; 23° E.; 2° E.; 24° W.; 25° W.; 18° W.
Compass courses.—N. 35° E.; S. 82° W.; N. 85½° W.; S. 49° E.
Magnetic courses.—S.E. by S.; N. 84½° E.; N. 86° W.; N.W. by N.
Bearings, magnetic.—N. 87½° W.; N. 78½° E.

EXAMINATION PAPER—No. IV, *Pages* 333—334.

1. Log. $5'119863 = 13178'4$ Log. $4'861964 = 7'1772$.
2. Log. $3'510784 = 3241'78$ nearly. Log. $2'033424 = 108$.
3. **True Courses.**—S., 19' dep. course; S. 59° W., 58'; N. 11° W., 15'; S. 28° E., 9'; S. 82° W., 50'; N. 72° E., 12'; S. 58° W., 22'; S. 62° W., 42', current course. *Diff. lat.* 76'·8 S.; *d. p.* 142'·3 W.; *course* S. 62° W.; *dist.* 162'. *Lat. in* 51° 29' S.; *diff. long.* 225' W. *Longitude in* 183° 25' W., or 176° 35' E.
4. Green. date, April $1^d\ 5^h\ 50^m\ 48^s$; red. decl. 4° 42' 29" N.; true alt. 48° 55' 26". *Lat.* 45° 47' 3" N.

Raper: True alt. 48° 55' 19". *Latitude* 45° 47' 11" N.

5. Log. of diff. long. $2'364667 =$ *Diff. long.* 231'·6.
6. Diff. lat. 325' N.; mer. diff. lat. 552'; diff. long. 325' W.; tang. course $9'769944$; *course* N. 30° 29' 17" W.; *log. dist.* $2'576509$; *dist.* 377'·2.
7.

Standard, Brest	$+ 2^h 45^m$	no A.M.,	and $0^h 18^m$ P.M.
" Hull	$+ 0\ 1$	no A.M.,	and 0 20 P.M.
" Thurso	$- 1\ 56$	$0^h\ 4^m$ A.M.,	and 0 46 P.M.
" Pembroke	$+ 0\ 4$	11 42 A.M.,	and no P.M.
" Weston-super-mare	$+ 0\ 2$	no A.M.,	and 0 14 P.M.
" Waterford	$+ 0\ 44$	11 43 A.M.,	and no P.M.

8. Green. date, April $27^d\ 23^h\ 18^m\ 50^s$; red. decl. 14° 11' 48" N.; log. sine $9'494965$; true amplitude W. 18° 11' 54" N. *Error* 15° 32' 6" W.; *deviation* 3° 37' 54" E.
9. Interval 12 days; rate $11^m\cdot 3$ *losing*; interval $70^d\ 22^h$; acc. rate $13^m\ 21^{s}\cdot 2$; Green. date, April $14^d\ 21^h\ 59^m\ 3^s$; re l. decl. 9° 48' 0" N.; red. eq. time *subt.*, $0^m\ 0^s$; true alt. 26° 37' 25"; hour-angle $2^h\ 59^m\ 9^s$. *Longitude* 75° 1' 30" E.

Raper: True alt. 26° 37' 20"; sine sq. $9'161814$; hour-angle $2^h\ 59^m\ 9^s$. *Long.* 75° 1' 30" E.

10. Green. date, April $17^d\ 2^h\ 43^m\ 25^s$; red. decl. 10° 34' 40" N.; true alt. 42° 20' 39"; sum of logs. $19'449652$; true azimuth S. 64° 6' 18" W. *Error* 25° 53' 42" W.; *deviation* 6° 3' 42" W.

Raper: True alt. 42° 20' 34"; sine sq. $9'449680$; true azimuth S. 64° 6' 15" W. *Error of compass* 25° 53' 42" W.; *deviation* 6° 3' 42" W.

11. Time from noon $28^m\ 38^s$; Green. date, April $18^d\ 11^h\ 38^m\ 34^s$; red. decl. 11° 3' 23" N.; true alt. 54° 20' 16"; nat. no. 5291; nat. cos. mer. zen. dist. $817759 = 35° 8' 20"$ N. *Lat.* 46° 11' 43" N.

METHOD II.—*Raper:* True alt. 54° 20' 12"; $+ 31'\ 33"$. *Latitude* 46° 11' 38" N.

Towson: Aug. I, $+ 5'\ 10"$; index no. 76. Aug. II, $+ 36'\ 28"$. *Latitude* 46° 11' 49" N.

12. 1st red. decl. 6° 32' 26" N.; eq. T. $+ 2^m\ 24^s$; true alt. 45° 18' 15"; lat. 50° 20' gives hour-angle $0^h\ 42^m\ 45^s$; long. 7° 34' W. (A); lat. 50° 50' gives hour-angle $0^h\ 28^m\ 45^s$; long. 11° 34' W. (A'). 2nd red. decl. 6° 35' 22" N.; eq. T. $+ 2^m\ 21^s\cdot 6$; true alt. 25° 2' 12"; lat. 50° 20' gives hour-angle $3^h\ 52^m\ 30^s$; long. 7° 8' W.; lat. 50° 50' gives hour-angle $3^h\ 51^m\ 15^s$; long. 7° 26½' W. The line of bearing at first observation trends E. by S. and W. by N.; the position of ship at the time of second observation is lat. 50° 59' N., long. 7° 32' W.

13. Spica's decl. 10° 32' 59" S. *Latitude* 58° 36' 21" N.

Raper: True alt. 20° 50' 30". *Latitude* 58° 36' 31" N.

The Curve.—Correct magnetic bearing S. 2° E.

Deviations.—4° W.; 7° W.; 11° W.; 18° W.; 3° E.; 22° E.; 15° E.; 1° E.

Compass courses.—N. 88° E.; N. 59° W.; S. 14° W.; S. 19° E.

Magnetic courses.—N. 56° W.; N. 87½° W.; East; N.E.

Bearings, magnetic.—N. 67° W.; S. 84½° W.

EXAMINATION PAPER—No. V, *Page* 335—336.

1. $4'838071 =$ nat. no. $68876'5$.
2. $2'826661 =$ nat. no. $670'905$.

3. **True Courses.**—S. 63° E., 23' dep. course; S. 6° E., 56'; S. 14° W., 17'; N. 80° W., 13'; E., 37'; S. 44° W., 31'; N. 26° E., 17'; N. 85° E., 48' current course. Diff. lat. 82'·4 S., dep. 80'·6 E.; course S. 44° E., dist. 115'. Lat. in 65° 24' S., mid. lat. 64° 43', diff. long. 189' E. Long. in 178° 30' W.

4. Green. date, May 3d 7h 1m 8s; red. decl. 17° 12' 48" N.; true alt. 76° 14' 11". Lat. 3° 26' 59" N.
Raper: True alt. 76° 14' 3". *Latitude* 3° 26' 50" N.

5. Log. of diff. long. 2·992876 = *Diff. long.* 983'·7.

6. Diff. lat. 732' S.; mer. diff. lat. 881'; diff. long. 1098' W.; log. tang. 10·095626; course S. 51° 15' 27" W.; *log. of distance* 3·068060; *distance* 1169'·6.

7. Standard, Greenock constant — 0h 56m; 6h 58m A.M., and 7h 28m P.M.
„ Liverpool „ — 0 1 ; 7 21 A.M., and 7 51 P.M.
„ Sunderland „ — 1 4 ; 10 38 A.M., and 11 5 P.M.
„ Brest „ + 2 18 ; 1 59 A.M., and 2 30 P.M.
„ Dover „ — 0 27 ; 6 28 A.M., and 6 59 P.M.

8. Green. date, May 20d 16h 6m 24s; red. decl. 20° 9' 18" N.; log. sine 9·694581; true amp. E. 29° 40' 5" N. *Error of compass* 12° 31' 10" E.; *deviation* 19° 18' 50" W.

9. Interval 36d; rate 4s·9 gaining; interval 50d 21h; acc. rate — 4m 9s; Green. date, May 21d 20h 59m 16s; red. decl. 20° 23' 48" N.; red. eq. T. subt. 3m 35s·6; true alt. 32° 19' 31"; sum of logs. 9·452117; hour-angle 4h 17m 13s; mean time at ship May 21d 19h 39m 11s. *Longitude* 20° 1' 15" W.
Raper: True alt. 32° 19' 21"; hour-angle 4h 17m 14s. *Longitude* 20° 1' 30" W.

10. Green. date, May 25d 9h 53m 51s; red. decl. 21° 3' 26" N.; true alt. 40° 54' 26"; sum of logs. 19·603544; true azimuth S. 81° 57' 40" W. or N. 98° 2' 20" W. *Error* 20° 27' 40" E.; *deviation* 9° 57' 40" E.
Raper: True alt. 40° 54' 22"; sine sq. 9·633554; true azimuth S. 81° 57' 44" W. or N. 98° 2' 16" W. *Error* 20° 27' 40" E.; *deviation* 8° 57' 40" E.

11. Time from noon 25m 25s; Green. date 10d 7h 54m 25s; red. decl. 17° 45' 6" N.; true alt. 43° 39' 55"; nat no. 5144; mer. zen. dist. 45° 55' 34" S. *Latitude* 28° 10' 28' S.
METHOD II.—*Raper:* True alt. 43° 39' 49"; + 24' 39". *Latitude* 28° 10' 26" S.
Towson: Aug. I, + 6' 16'; index 63. Aug. II, + 18' 17". *Latitude* 28° 10' 26" S.

12. 1st red. decl. 21° 46' 17" N.; 1st red. eq. T. — 2m 46s·3; 1st true alt. 23° 52' 16"; lat. 50° 0' gives hour-angle 5h 13m 24s; long. 9° 21¼' W. (A). Lat. 50° 30' gives hour-angle 5h 13m 43s; long. 9° 26' W. (A'). 2nd red. decl. 21° 48' 46" N.; 2nd red. eq. T. subt. 2m 44s; and true alt. 56° 56' 35". Lat. 50° 0' gives hour-angle 1h 27m 50s; long. 10° 29½' W. (B). Lat. 50° 30' gives hour-angle 1h 25m 54s; long. 10° 29¾' W. (B'). The line of bearing at the first observation trends N. ½ W. and S. ½ E. The position of the ship at the second observation is lat. 50° 5' N., long. 10° 40½' W.

13. Star's decl. 10° 33' 0" S. *Latitude* 30° 16' 44" S.
Raper: True alt. 70° 6' 16'. *Latitude* 30° 26' 44" S.

The Curve.—Correct magnetic bearing N. 87° E.
Deviations.—3° W.; 15° W.; 23° W.; 22° W.; 2° E.; 24° E.; 23° E.; 13° E.
Compass courses.—N. 45½° E.; S. 71½° W.; S. 49° E.; S. 39½° W.
Magnetic courses.—N. 20½° W.; N. 46° E.; S. 76½° E.; S. 61° W.
Bearings, magnetic.—S. 8° E.; S. 85° W.

EXAMINATION PAPER—No. VI, *Pages* 336—338.

1. 0·622110 = nat. no. 4·189. (The product).

2. 4·021468 = nat. no. 10506·77. (The quotient).

3. **True Courses.**—S. 73° E., 23 dep. course; S. 75° E., 17'; S. 33° E., 22'; S. 1° W., 24'; S. 61° E., 25'; N. 69° W., 18'; S. 43° W., 17'; N. 11° E., 16' current course. Diff. lat. 55'·9 S., dep. 46'·6 E.; course S. 40° E., dist. 73'. Lat. in 55° 16' N.; diff. long. (Mercator) 82'·9 E.; long. in 134° 17' W. Diff. long. (by mid. lat.) 83' E.; long. in 179° 57' W.

K K K

4. Green. date, May 31^d 17^h 34^m 52^s; red. decl. $22°$ $6'$ $44''$ N.; true alt. $75°$ $49'$ $26''$. Latitude $7°$ $56'$ $9''$ N.

5. Log. of diff. long. $2·487692 =$ *Diff. long.* $307'·4$.

6. Diff. lat. $2181'8$.; mer. diff. lat. $2301'$; diff. long. $3038''$ W.; tangent course $10·120671$; course S. $52°$ $51'$ $34''$ W.; *log. distance* $3·880471$; *distance* $3612'·3$.

7. Standard, Dover $+ 4^h 33^m$; $11^h 46^m$ A.M., no P.M.
   „ Harwich $- 0$ 33 ; 7 39 A.M., 8^h 12^m P.M.
   „ Brest $- 0$ 47 ; corr. for long. $+ 7^m$; 3^h 9^m A.M., 3^h 28^m P.M.

8. Green. date, June 21^d 13^h 34^m 48^s; red. decl. $23°$ $27'$ $11''$ N.; sine of true amplitude $9·898948$; true amplitude W. $52°$ $24'$ $39''$ N. *Error* $46°$ $1'$ $36''$ W.; *deviation* $6°$ $28'$ $24''$ E.

9. Interval 29^d, daily rate $8^s·3$ *losing*, interval 44^d 22^h, accumulated rate 6^m 13^s, Green. date June 13^d 21^h 45^m 12^s, red. decl. $23°$ $16'$ $38''$ N., true alt. $28°$ $49'$ $40''$, red. eq. time 0^s 5^s *subt.*, hour-angle 3^h 49^m 7^s, M.T. ship June 13^d 27^h 49^m 2^s. *Longitude* $90°$ $57'$ $30''$ E.

By *Raper*: True alt. $31°$ $18'$ $45''$, log. sine sq. $9·813214$, true azimuth S. $107°$ $30'$ $42''$ E. *Deviation* $17°$ $9'$ $18''$ E.

10. Green. date, June 7^d 22^h 0^m 12^s; red. decl. $22°$ $51'$ $39''$ N.; true alt. $31°$ $18'$ $54''$; logs. $19·813215$; true azimuth S. $107°$ $30'$ $42''$ E. *Error* $2°$ $29'$ $18''$ E.; *deviation* $17°$ $9'$ $18''$ E.

By *Raper*: True alt. $28°$ $49'$ $36''$; hour-angle 3^h 49^m 8^s. *Longitude* $90°$ $57'$ $45''$ E.

11. Time from noon 37^m 26^s; Green. date, June 4^d 14^h 16^m 2^s; red. decl. $22°$ $31'$ $36''$ N.; true alt. $50°$ $3'$ $22''$; nat. no. 5778; mer. zen. dist. $39°$ $25'$ $32''$ N. *Latitude* $61°$ $57'$ $8''$ N.

Method II. $+ 31'$ $18''$. True alt. *(Norie)* $50°$ $3'$ $22''$; *latitude* $61°$ $56'$ $58''$ N. True alt. *(Raper)* $50°$ $3'$ $18''$; *latitude* $61°$ $57'$ $2''$ N.

Towson: Beyond the limits of the Table.

12. 1st red. decl. $23°$ $12'$ $34''$ N.; 1st red. eq. T. $+ 3^m$ $14^s·3$; 1st true alt. $62°$ $58'$ $35''$; lat. $49°$ $40'$ gives hour-angle 0^h 28^m 3^s; long. $177°$ $45\frac{3}{4}'$ E. (A). Lat. $50°$ $0'$ gives hour-angle 0^h 18^m 6^s; long. $180°$ $15'$ E. or $179°$ $45'$ W. (A'). 2nd red. decl. $23°$ $11'$ $49''$ N.; 2nd eq. T. $+ 3^m$ 17^s, and true alt. $30°$ $13'$ $27''$; lat. $49°$ $40'$ gives hour-angle 4^h 40^m 9^s; long. $180°$ $15\frac{1}{2}'$ E. or $179°$ $44\frac{1}{2}'$ W. (B). Lat. $50°$ $0'$ gives hour-angle 4^h 40^m 1^s; long. $180°$ $16'$ E. or $179°$ $44'$ W. (B'). The line of bearing at the first observation trends E. by N. (northerly) and W. by S. (southerly). The position of the ship at the second observation is lat. $50°$ $6'$ N., long. $179°$ $44'$ W.

13. Star's decl. $28°$ $18'$ $29''$ N.; true alt. *(Norie)* $48°$ $33'$ $45''$. *Latitude* $13°$ $7'$ $46''$ S.
Raper: True alt. $48°$ $33'$ $40''$. *Latitude* $13°$ $7'$ $51''$ S.

The Curve.—Correct magnetic bearing S. $79°$ W.

Deviations.—$12°$ W.; $0°$; $17°$ E.; $21°$ E.; $12°$ E.; $4°$ E.; $18°$ W.; $24°$ W.

Compass courses.—N.W. by N.; S. $70°$ W.; N. $27\frac{1}{2}°$ E.; S. $1\frac{1}{2}°$ E.

Magnetic courses.—N.W. by W.; S. $61°$ W.; S. $17°$ W.; S. $40\frac{1}{2}°$ E.

Correct magnetic bearing.—N. $11°$ E.; N. $87\frac{1}{2}°$ W.

EXAMINATION PAPER—No. VII, *Pages* 339—340.

1. $4·141829 = 13862·1$; $5·805908 = 639600$.

2. $2·537819 = 345·0$; $1·643452 = 44·0$.

3. True Courses.—S. $25°$ E., $17'$; N. $73°$ E., $21'$; S. $79°$ E., $24'$; N. $80°$ W., $26'$; N. $85°$ E., $17'$; N. $58°$ E. $40'$; S. $68°$ E., $51'$; N. $62°$ W., $9'$ current course. *Diff. lat.* $1'·3$ S., *dep.* $116'·4$ E., *course* S. $87\frac{1}{2}°$ E., *dist.* $116'$. *Lat. in* $51°$ $24'$ N., *mid. lat.* $51°$ $24'$, *long.* $187°$ E. *Long. in* $182°$ $3'$ E., or $177°$ $57'$ W.

4. Green. date, July 26^d 0^h 49^m 16^s; red. decl. $19°$ $24'$ $44''$ N.; true alt. $15°$ $46'$ $12''$. Latitude $54°$ $49'$ $4''$ S.

Raper: True alt. $15°$ $46'$ $3''$. *Latitude* $54°$ $49'$ $13''$ S.

5. Log. of diff. long. $2·633861 =$ *diff. long.* $430'·4$ nearly.

6. Diff. lat. $745'8$.; mer. diff. lat. $1042'$; diff. long. $1292'$ W.; log. tang. $10·093395$; course S. $51°$ $6'$ $50''$ W.; *log. distance* $3·074353$, *distance* $1186''·7$.

7. Standard, Brest constant — $0^h 2^m$; no A.M., $0^h 10^m$ P.M.
 ,, Brest ,, — 0 45 ; $11^h 27^m$ A.M., no P.M.
 ,, Brest ,, — 0 5 ; no A.M., 0 7 P.M.
 ,, Brest ,, — 0 29 ; 11 43 A.M., no P.M.
 ,, Brest ,, + 3 3 ; 9 9 A.M., 9 48 P.M.

8. Green. date, July $12^d 6^h 36^m 32^s$; red. decl. $21° 55' 38'' $ N.; sine amp. 9·630599; true amp. W. $25° 17' 17'' $ N. *Correction* $5° 36' 2'' $ E.; *deviation* $16° 56' 2'' $ E.

9. Interval 8^d, rate 5^s·55 *losing*, interval $32^d 22^h$, accumulated rate + $3^m 2^s$·7, Green. date July $16^d 21^h 58^m 19^s$, red. decl. $21° 12' 14''$ N., true alt. $13° 31' 23''$, red. eq. T. $5^m 51^s$ *add*; hour-angle $3^h 5^m 21^s$, M.T. ship $16^d 27^h 57^m$·12^s. *Longitude* $89° 43' 15''$ E.
Raper: True alt. $13° 31' 7''$; hour-angle $3^h 51^m 23^s$. *Longitude* $89° 43' 45''$ E.

10. Green. date, July $4^d 1^h 53^m 21^s$; red. decl. $22° 52' 22''$ N.; true alt. $12° 21' 31''$; sum of logs. 19·106389; true azimuth N. $47° 17' 12''$ E. *Error* $28° 1' 12''$ E.; *deviation* $10° 39' 12''$ E.
Raper: True alt. $12° 21' 21''$; sine sq. 9·206465; true azimuth N. $47° 17' 30''$ E. *Error* $28° 1' 30''$ E.; *deviation* $10° 41' 30''$ E.

11. Time from noon $25^m 20^s$; Green. date, July $30^d 18^h 52^m 32^s$; red. decl. $18° 17' 56''$ N.; true alt. $26° 24' 19''$; nat. no. 4092; mer. zen. dist. $63° 19' 57''$ S. *Latitude* $45° 2' 0''$ S.
Method II. + $15' 44''$. *Latitude* $45° 2' 0''$ S.
Raper: True alt. $26° 24' 13''$. *Latitude* $45° 2' 7''$ S.
Towson: Aug. I, + $6' 17''$; index 63. Aug. II, + $9' 27''$. *Latitude* $45° 2' 0''$ S.

12. 1st red. decl. $22° 47' 3''$ N.; 1st rod. eq. T. + $4^h 16^m 11^s$; 1st true alt. $61° 25' 37''$; lat. $50° 40'$ N. gives hour-angle $0^h 32^m 2^s$; long. $6° 56'$ W. (A). Lat. $51° 0'$ N. gives hour-angle $0^h 23^m 11^s$; long. $9° 8\frac{3}{4}'$ W. (A'). 2nd red. decl. $22° 45' 50''$ N.; 2nd red. eq. T. + $4^m 18^s·36$; 2nd true alt. $20° 27' 28''$; lat. $50° 40'$ gives hour-angle $5^h 40^m 17^s$; long. $8° 6\frac{1}{2}'$ W. (B). Lat. $51° 0'$ N. gives hour-angle $5^h 40^m 42^s$; long. $8° 0\frac{1}{2}'$ W. (B'). The line of bearing at first observation trends W. by N. (northerly) and E. by S. (southerly). The ship's position at time of second observation is lat. $50° 51'$ N., long. $8° 3\frac{1}{2}'$ W.

13. Star's decl. $26° 10' 13''$ S.; true alt. $70° 5' 46''$. *Latitude* $46° 4' 27''$ S.
Raper: True alt. $70° 5' 44''$. *Latitude* $46° 4' 29''$ S.

The Curve.—Correct magnetic bearing N. $89°$ W.
Deviations.—$1°$ E.; $19°$ E.; $21°$ E.; $9°$ E.; $4°$ W.; $11°$ W.; $19°$ W.; $16°$ W.
Compass courses.—N. $51°$ E.; S. $81\frac{1}{2}°$ E.; S. $23\frac{1}{2}°$ W.; North.
Magnetic courses.—S. $70°$ W.; N. $6°$ E.; N. $80\frac{1}{2}°$ E.; S. $33°$ E.
Bearings, magnetic.—N. $75°$ E.; N. $66\frac{1}{2}°$ E.

EXAMINATION PAPER—No. VIII, *Pages* 341—342.

1. 5·889986 = 776221·4; 3·983651 = 9630·555.
2. 2·676541 = 474·833; 5·291146 = 195500.
3. True Courses.—N. $59°$ E., $15'$ dep. course; S. $20°$ E., $26'$; N. $61°$ W., $27'$; N. $70°$ W., $22'$; S. $24°$ E., $25'$; S. $29°$ E., $22'$; N. $88°$ W., $43'$; S. $19°$ W., $18'$ current course. *Diff. lat.* $53'$·6 S.; *dep.* $50'$·5 W.; *course* S. $43\frac{1}{2}°$ W.; *dist.* $73\frac{1}{2}'$. *Latitude in* $0°$ $44'$ S.; *diff. long.* $50\frac{1}{2}'$ W. *Longitude in* $172°$ $59\frac{1}{2}'$ E.
4. Green. date, Aug. $11^d 17^h 51^m 12^s$; red. decl. $15° 1' 3''$ N.; true alt. $42° 50' 18''$. *Lat.* $32° 8' 39''$ S.
Raper: True alt. $42° 50' 6''$. *Latitude* $32° 8' 51''$ S.
5. Log. of diff. long. 2·805956 = *Diff. long.* 639'·7.
6. Diff. lat. $421'$ N.; mer. diff. lat. $590'$; diff. long. $249'$ E.; tangent 9·625347; *course* N. $22° 52' 53''$ E.; *log. dist.* 2·659876; *distance* $457'$ nearly.
7. Standard, Brest — $0^h 15^m$; no A.M. and $0^h 5^m$ P.M.
 ,, Brest — 1 51 ; $10^h 29^m$ A.M. and 11 5 P.M.
 ,, Dover + 5 13 ; no A.M. and 0 7 P.M.
 ,, Devonport — 1 13 ; no A.M. and 0 16 P.M.

8. Green. date, Aug. 20^d 11^h 10^m 24^s; red. decl. 12° 15' 18" N.; sine 9˙456381; true amp. W. 16° 37' 8" N. *Correction* 27° 52' 8' E.; *deviation* 8° 37' 8" E.

9. Interval 7^d, rate 6^s˙6 *losing*, interval 16^d 20$\frac{1}{2}$^{h}, accumulated rate $+$ 1^m 51^s, Green. date, Aug. 6^d 20^h 32^m 37^s, red. decl. 16° 26' 16" N., true alt. 24° 17' 1", red. eq. T. 5^m 33^s *additive*, hour-angle 4^h 25^m 44^s, M.T. ship 6^d 28^h 31^m 17^s. *Longitude* 119° 40' E.

Raper: True alt. 24° 16' 57"; hour-angle 4^h 25^m 44^s. *Longitude* 119° 40' E.

10. Green. date, Aug. 19^d 20^h 37^m 41^s; red. decl. 12° 27' 22" N.; true alt. 17° 36' 21"; sum of logs. 19˙052161; true azimuth N. 39° 14' 32" W. *Error* 2° 40' 47" W.; *deviation* 36° 30' 47" W.

Raper: True alt. 17° 36' 14"; sine sq. logs. 9˙052148 $=$ N. 39°14' 47" W. *Error* 2° 4' 12" W.; *deviation* 36° 31' 2" W.

11. Time from noon 35^m 15^s; Green. date August 10^d 12^h 55^m 5^s; red. decl. 15° 22' 39' N.; true alt. 34° 48' 15"; nat. no. 8840; mer. zen. dist. 54° 34' 32" S *Latitude* 39° 11' 53" S.

METHOD II.—*Reduction* $+$ 37° 22'. *Latitude* 39° 11' 44" S.

Raper: True alt. 34° 48' 11". *Latitude* 39° 11' 48" S.

Towson: Aug. I, $+$ 10' 26"; index 90. Aug. II, $+$ 26' 35". *Latitude* 39° 12' 5" S.

12. 1st red. decl. 15° 34' 16" N.; 1st eq. T. 5^m 9^s˙23 *add*; 1st true alt. 38° 27' 40"; lat. 49° 40' gives hour-angle 3^h 11^m 55^s; long. 3° 50$\frac{3}{4}$' W. (A). Lat. 50° 0' gives hour-angle 3^h 10^m 59^s; long. 3° 36$\frac{1}{2}$' W. (A'). 2nd red. decl. 15° 29' 47" N.; 2nd red. eq. T. 5^m 7^s *add*; 2nd true alt. 40° 20' 59"; lat. 49° 40' gives hour-angle 2^h 58^m 34^s; long. 3° 4' W. (B). Lat. 50° 0' gives hour-angle 2^h 57^m 29^s; long. 4° 20$\frac{1}{2}$' W. (B'). The line of bearing at first observation trends N.N.E. (easterly) and S.S.W. (westerly). The position of ship at second observation is lat. 49° 47$\frac{1}{2}$' N., long. 3° 33' W.

13. Star's decl. 8° 33' 49" N.; true alt. 66° 48' 17". *Latitude* 14° 37' 54" S. *(Norie).*
    "   "   "   66 48 13   "   14 37 58 S. *(Raper).*

The Curve.—Correct magnetic bearing S. 8° E.

Deviations.—4° E.; 2° E.; 14° W.; 17° W.; 4° W.; 9° E.; 12° E.; 8° E.

Compass courses.—N. 85$\frac{1}{2}$° E.; S. 51$\frac{1}{2}$° W.; N. 38° W.; S. 50° E.

Magnetic courses.—N. 32° W.; S. 85° W.; N. 1° W.; N. 86$\frac{1}{2}$° W.

Bearings, magnetic.—S. 12° E.; S. 23$\frac{1}{2}$° E.

EXAMINATION PAPER—No. IX, *Pages* 343—344.

1. 4˙147399 $=$ 14041˙03. 7˙888411 $=$ ˙007734.

2. 6˙854294 $=$ ˙0071498. 1˙602689 $=$ 40˙058.

3. True Courses.—S. 89° E., 20' dep. course; S. 66° E., 41'; S. 56° E., 50'; S. 83° E., 43'; S. 66° E., 33'; S. 49° E., 27'; S. 50° E., 36'; N. 85° E., 32' current course. *Diff. lat.* 101'˙6 S.; *dep.* 251''˙7 E.; *course* S. 68° E.; *dist.* 271$\frac{1}{2}$'. *Latitude in* 52° 25' N.; *diff. long.* 420'˙9 E. *Longitude in* 6° 56' E.

4. Green. date, Sept. 22^d 15^h 45^m, (*long.* 8° 15'); red. decl. 0° 0' 0"; true alt. 90° 0' 5". *Latitude* 0° 0' 5" N.

Raper: True alt. 89° 59' 57". *Latitude* 0° 0' 3" S.

5. Log. of diff. long. 2˙019277 $=$ *diff. long.* 104'˙5.

6. Diff. lat. 3607' S.; mer. diff. lat. 3798'; diff. long. 4007' E.; tang. 10˙023264; *course* S. 46° 32' E.; *log. distance* 3˙719604; *distance* 5243'.

7. Standard, Brest   — 1^{h}51^m; no   A.M., and   0^{h}16^{m}P.M.
   "   Sunderland — 1 14; no   A.M., and   0 22 P.M.
   "   Leith   — 1 17; 11^{h}20^{m}A.M., and 11 44 P.M.
   "   Greenock $+$ 4 47; 2 8 A.M., and   2 27 P.M.

8. Green. date, September 30^d 6^h 33^m 40^s; red. decl. 2° 58' 4" S.; sine 8˙929667; true amp. W. 4° 52' 44" S. *Error* 41° 26' 19" W.; *deviation* 10° 58' 29" W.

9. Interval 15^d, rate 2^s˙6 *gaining*, interval 19^d 15$\frac{1}{2}$^{h}, accumulated rate — 0^m 51^s, Green. date August 31^d 15^h 35^m 8^s, red. decl. 8° 21' 41" N., red. eq. T. 0^m 1^s *subt.*, true alt. 62° 25' 7", hour-angle 1^h 51^m 36^s, M.T. ship August 31^d 25^h 51^m 23^s. *Long.* 154° 6' 45" E.

Raper: True alt. 62° 24' 56"; hour-angle 1^h 51^m 37^s. *Longitude* 154° 7' 0" E.

10. Green. date, September $16^d\,1^h\,29^m\,54^s$; rod. decl. $2°\,33'\,37''$ N.; true alt. $29°\,41'\,59''$; sum of logs. 19·702438; true azimuth S. $90°\,27'\,32''$ E. Error $3'\,47'\,32''$ W. Deviation $12°\,7'\,32''$ W.

Raper: True alt. $29°\,41'\,53''$; sine sq. 9·702440; true azimuth S. $90°\,27'\,34''$ E. Error $3°\,47'\,34''$ W.; deviation $12°\,7'\,34''$ W.

11. Time from noon $16^m\,41^s$, Green. date, September $22^d\,12^h\,27^m\,55^s$, red. decl. $0°\,3'\,13''$ S., true alt. $62°\,9'\,19''$, nat. no. 2348, mer. zen. dist. $27°\,33'\,19''$ S. Latitude $27°\,30'\,36''$ S.

METHOD II. $+\,17'\,26''$. Latitude $27°\,30'\,2''$ S.

Raper: True alt. $62°\,9'\,11''$. Latitude $27°\,30'\,11''$.

Towson: Aug. I, $+$ 0, index 35. Aug. II, $+\,17'\,29''$. Latitude $27°\,29'\,59''$ S.

12. 1st red. decl. $5°\,23'\,1''$ N.; 1st red. eq. T. $-\,2^m\,46^s$; 1st true alt. $11°\,9'\,42''$; lat. $47°\,50'$ gives hour-angle $5^h\,17^m\,13^s$; long. $34°\,2\frac{3}{4}'$ E. (A). Lat. $47°\,10'$ gives hour-angle $5^h\,17^m\,30^s$; long. $33°\,58\frac{1}{2}'$ E. (A'); 2nd red. decl. $5°\,18'\,59''$ N.; 2nd rod. eq. T. $-\,2^m\,43''58^s$; 2nd true alt. $45°\,56'\,32''$; lat. $47°\,50'$ hour-angle $0^h\,53^m\,57^s$; long. $35°\,58\frac{1}{2}'$ E. (B). Lat. $47°\,10'$ gives hour-angle $1^h\,4^m\,2^s$; long. $33°\,27\frac{1}{2}'$ E. (B'). The line of bearing at first observation trends N. $\frac{3}{4}$ E. and S. $\frac{1}{4}$ W. The position of the ship at the second observation is lat. $47°\,26'$ N., long. $34°\,28'$ W.

13. Star's decl. $19°\,47'\,49''$ N., true alt. $86°\,31'\,18''$. Latitude $16°\,19'\,7''$ N. (*Norie*).
         „   „ $86\,31\,17$ „ $16\,19\,5$ N. (*Raper*).

The Curve.—Correct magnetic bearing S. $88°$ W.

Deviations.—$2°$ E., $20°$ E., $22°$ E., $9°$ E., $4°$ W., $12°$ W., $20°$ W., $17°$ W.

Compass courses.—West, N. $2°$ E., N. $59\frac{1}{2}°$ E., S. $61\frac{1}{2}°$ E.

Magnetic courses.—N. $2°$ E., S. $17°$ W., S. $57\frac{1}{2}°$ E., N. $72°$ E.

Bearings, magnetic.—S. $84\frac{1}{2}°$ E., N. $76\frac{1}{2}°$ W.

EXAMINATION PAPER—No. X, Pages 345—346.

1. $7·417213 = 28003548$; $3·524617 = 3346·7$.

2. $1·936010 = 86·30$; $2·735031 = 543·288$.

3. **True Courses.**—S. $75°$ E., $36'$ dep. course; S. $8°$ E., $43'$; S. $15°$ W., $19'$; N. $69°$ E., $50'$; N. $77°$ E., $21'$; S. $5°$ E., $6'$; S. $38°$ W., $3'$; S. $51°$ E., $22'$; N. $88°$ E., $48'$ current course. Diff. lat. $65'·8$ S.; dep. $168'·7$ E.; course S. $68\frac{3}{4}°$ E.; dist. $181'$. Lat. in $58°\,43'$ N.; mid. lat. $59°\,16'$; diff. long. $330'$ E. Long. in $38°\,24'$ W.

4. Green. date, Oct. $20^d\,10^h\,1^m\,40^s$, red. decl. $10°\,33'\,31''$ S., true alt. $50°\,11'\,14''$. Lat. $50°\,22'\,17''$ S.

Raper: True alt. $50°\,11'\,9''$. Latitude $50°\,22'\,22''$ S.

5. Log. of diff. long. 2·019277 $=$ Diff. long. $104'·5$.

6. Diff. lat. $140'$ N., mer. diff. lat. $142'$, diff. long. $46'$ E., tang. course 9·510470; course S. $17°\,57'$ E.; log. dist. 2·167799; dist. $147'·1$.

7. Standard, Brest   — $2^h\,0^m$; $11^h\,54^m$ A.M. and no P.M.
   „   London   — $2\,19$; $9\,44$ A.M. and $10\,\,6$ P.M.
   „   Leith   — $2\,55$; $9\,34$ A.M. and $9\,54$ P.M.
   „   Sunderland — $1\,\,4$; no A.M. and $0\,22$ P.M.

8. Green. date, Oct. $8^d\,11^h\,13^m\,48^s$, red. decl. $6°\,7'\,40''$ S., true amp. E. $6°\,28'\,21''$ S. Error $9°\,17'\,6''$ E.; deviation $11°\,7'\,6''$ E.

9. Interval 7^d, rate $2^s·4$ *losing*, interval $22^d\,12^h$, accumulated rate $0^m\,54^s$, Green. date, October $30^d\,12^h\,0^m\,22^s$, red. decl. $14°\,1'\,0''$ S., true alt. $28°\,42'\,58''$, hour-angle $2^h\,45^m\,51^s$, red. eq. T. $16^m\,16^s$ subt., M.T. ship Oct. $30^d\,2^h\,29^m\,35^s$. Longitude $142°\,41'\,45''$ W.

Raper: True alt. $28°\,42'\,50''$; hour-angle $2^h\,45^m\,52^s$. Longitude $142°\,41'\,30''$ W.

10. Green. date, Sept. $30^d\,18^h\,43^m\,48^s$; red. decl. $3°\,10'\,4''$ S.; true alt. $14°\,7'\,4''$; sum of logs. 19·691122; true azimuth N. $88°\,58'\,26''$ W. Error $4'\,35'\,56''$ W.; deviation $12°\,15'\,56''$ W.

Raper: True alt. $14°\,6'\,51''$; sine sq. 9·691132; true azimuth N. $88°\,58'\,30''$ W. Deviation $12°\,16'\,0''$ W.

11. Time from noon 17ᵐ 8ˢ; Green. date, Oct. 2ᵈ 1ʰ 17ᵐ 8ˢ; red. decl. 3° 39′ 31″ S.; true alt. 47° 39′ 40″; nat. no. 2190; mer. zen. dist. 42° 9′ 9″ N. Lat. 38° 29′ 31″ N.
Method II.—Red. + 11′ 15″: lat. 38° 29′ 31″ N. (Norie). 38° 29′ 38″ N. (Raper).
Towson: Aug. I, + 35′ index 36. Aug. II, + 10′ 24″. Latitude 38° 29′ 49″ N.

12. 1st red. decl. 3° 15′ 18″ S.; 1st red. eq. T. — 10ᵐ 21ˢ·4; 1st true alt. 23° 29′ 11″; lat. 50° 30′; hour-angle 3ʰ 3ᵐ 24ˢ; long. 50° 40½′ W. (A). Lat. 51° 0′ gives hour-angle 3ʰ 0ᵐ 50ˢ; long. 50° 1¾′ W. (A′). 2nd red. decl. 3° 20′ 55″ S.; 2nd red. eq. T. — 10ᵐ 26ˢ; true alt. 25° 39′ 34″; lat. 50° 30′ gives hour-angle 2ʰ 44ᵐ 39ˢ; long. 49° 54¾′ W. (B). Lat. 51° 0′ gives hour-angle 2ʰ 41ᵐ 30ˢ; long. 50° 40′ W. (B′). The line of bearing at first observation trends N.E. ⅓ N. and S.W. ⅓ S. The position of the ship at the second observation is lat. 51° 7′ N., long. 50° 51′ W.

13. Star's decl. 14° 34′ 46″ N., true alt. 54° 6′ 7″. Latitude 50° 28′ 39″ N. (Norie).
    "   "   "   "   54 6 3   "   50 28 43 N. (Raper).
The Curve.—Correct magnetic bearing S. 38° W.
Deviations.—1° E., 17° W., 22° W., 19° W., 3° W., 18° E., 24° E., 19° E.
Compass courses.—S.S.E. ¾ E., E. by N. ¼ N., S. 9° W., S. 72° E.
Magnetic courses.—S. 45½° E., S. 10° W., N. 49° E., North.
Bearings, magnetic.—N. 85½° W., N. 89° W.

EXAMINATION PAPER—No. XI, Pages 347—348.

1. 2·645548 = 442·128; 5·639859 = 436374.
2. 1·999489 = 99·88; 3·459392 = 2880.
3. **True Courses.**—S. 42° W., 16′ dep. course; N. 14° E., 17′; S. 45° E., 19′; S. 87° W., 31′; S. 63° E., 17′; S. 5° E., 19′; S. 56° W., 25′; S. 88° E., 22′ current course. Diff. lat. 51′·8 S.; dep. 6′·1 W.; course S. 7° W.; dist. 52′. Lat. in 51° 8′ N.; diff. long. 10′. Long. in 119° 50′ E.
4. Green. date, November 14ᵈ 18ʰ 39ᵐ 16ˢ; red. decl. 18° 29′ 19″ S.; true alt. 67° 57′ 49″. Latitude 40° 31′ 30″ S.
5. Log. of diff. long. 2·294311 = diff. long. 196′·9.
6. Diff. lat. 1928′ N.; mer. diff. lat. 2383′; diff. long. 4290′ E.; tang. 10·255333; course N. 60° 56′ 56″ E.; log. of distance = 3·598938; distance 3971′.
7. Standard, Dover   constant + 0ʰ39ᵐ;   7ʰ42ᵐA.M., and   8ʰ14ᵐP.M.
   "   Devonport   "   — 0 17 ; 1 6 A.M., and   1 40 P.M.
   "   Galway   "   + 0 7 ;   0 53 A.M., and   1 21 P.M.
   "   Brest   "   — 0 7 ;   no A.M., and   0 13 P.M.
8. Green. date, Nov. 9ᵈ 11ʰ 20ᵐ 44ˢ; red. decl. 17° 4′ 14″ S.; true amp. E. 33° 52′ 16″ S.; Correction 42° 3′ 59″ W.; deviation 58° 33′ 59″ W.
9. Interval 12ᵈ, rate 1ˢ·2 losing, interval 36ᵈ 3ʰ, accumulated rate 43ˢ·2, Green. date Nov. 30ᵈ 2ʰ 48ᵐ 52ˢ, red. decl. 21° 42′ 33″ S, red. eq. T. 11ᵐ 5ˢ·5 subt., true alt. 39° 47′ 8″, hour-angle 3ʰ 42ᵐ·7ˢ, M.T. ship Nov. 29ᵈ 20ʰ 6ᵐ 47ˢ. Longitude 100° 31′ 15″ W.
Raper: True alt. 39° 47′ 6″; hour-angle 3ʰ 42ᵐ 7ˢ. Longitude 100° 31′ 7″·5 W.
10. Green. date, Nov. 15ᵈ 10ʰ 46ᵐ 27ˢ; red. decl. 18° 39′ 44″; true alt. 43° 55′ 7″; sum of logs. 19·514224; true azimuth N. 69° 43′ 40″ W., or S. 110° 2′ 26″ W. Error 11° 46′ 20″ E. Deviation 19° 36′ 20″ E.
Raper: True alt. 43° 55′ 2″; sine sq. 9·514239; true azimuth N. 69° 43′ 43″ W. Deviation 19° 36′ 17″ E.
11. Time from noon 39ᵐ 26ˢ; Green. date, Nov. 13ᵈ 2ʰ 36ᵐ 2ˢ; red. decl. 18° 3′ 15″ S.; true alt. 56° 11′ 50″; nat. no. 8861; mer. zen. dist. 32° 52′ 46″ S. Latitude 50° 56′ 1″ S.
Method II.—Red. + 56′ 13″, true alt. 56° 11′ 50″, latitude 50° 55′ 12″ S., Norie. True alt. 56° 11′ 46″, latitude 50° 55′ 16″ S., Raper.
Towson: Hour-angle exceeds limits of Table.

12. 1st red. decl. 14° 29' 29" S.; 1st red. eq. T. — 16ᵐ 18ˢ·53; 1st true alt. 19° 17' 53"; lat. 50° 50' gives hour-angle 2ʰ 3ᵐ 45ˢ; long. 24° 48½' W. (A). Lat. 50° 20' gives hour-angle 2ʰ 8ᵐ 42ˢ; long. 26° 2¼' W. (A'). 2nd red. decl. 14° 33' 41" S.; 2nd red. eq. T. — 16ᵐ 19ˢ; and true alt. 12° 46' 17"; lat. 50° 50' gives hour-angle 3ʰ 8ᵐ 28ˢ; long. 25° 20' W. (B). Lat. 50° 20' gives hour-angle 3ʰ 11ᵐ 24ˢ; long. 24° 36' W. (B'). The line of bearing at the first observation trends N.E. by E. (easterly) and S.W. by W. (westerly). The position of the ship at the second observation is lat. 50° 36½' N., long. 25° 0½' W.

13. Star's decl. 30° 14' 33" S., true alt. 59° 36' 2", *latitude* 60° 38' 23" S., *Norie.* True alt. 59° 35' 58", *latitude* 60° 38' 35" S., *Raper.*

The Curve.—Correct magnetic bearing S. 89° W.

Deviations.—1° E.; 19° E.; 21° E.; 9° E.; 3° W.; 11° W.; 19° W.; 16° W.

Compass courses.—N. 85° W.; N. 45½° W.; N. 86° E.; South.

Magnetic courses.—S. 77° W.; N. 58° W.; S. 49° E.; N. 56° E.

Bearings, magnetic.—S. 76° W.; S. 56° E.

EXAMINATION PAPER—No. XII, *Pages* 349—350.

1. 6·010519 = 1024516·5; 3·367405 = ·00233026.
2. 1·168317 = 14·7339; 2·096910 = 125.
3. **True Courses.**—S. 45° W., 16' dep. course; N. 73° E., 15'; S. 49° W., 23'; S. 43° E., 15'; N. 47° W., 24'; N. 80° E., 22'; S. 53° W., 12'; S. 14° E., 6', current course. *Diff. lat.* 25'·8 S.; *dep.* 8'·2 W.; *course* S. 18° W.; *dist.* 27'. Lat. *in* 49° 34' N.; *diff. long.* 13' W. *Longitude in* 40° 13' W.
4. Green. date, Dec. 31ᵈ 8ʰ 15ᵐ; red. decl. 23° 4' 1" S.; true alt. 67° 20' 49"; *latitude* 0° 24' 50" S.

By *Raper:* True alt. 67° 20' 46'. *Latitude* 0° 24' 47" S.

5. Log. of diff. long. 2·346353 = *diff. long.* 222'.
6. Diff. lat. 4766' S.; mer. diff. lat. 5205'; diff. long. 3735' W.; tang. course 9·855870; course S. 35° 39' 45" W.; log. of distance 3·768350; *distance* 5866'.
7. Standard, Queenstown, constant — 0ʰ 59ᵐ; 0ʰ 6ᵐ A.M., and 0ʰ 43ᵐ P.M.
   " Sligo " — 0 21; 1 30 A.M., and 2 3 P.M.
   " Queenstown " — 1 19; no A.M., and 0 23 P.M.
   " Galway " + 1 35; 2 38 A.M., and 3 10 P.M.
   " Londonderry " — 1 37; 3 6 A.M., and 3 32 P.M.
   " Pembroke " — 0 30; 1 37 A.M., and 2 15 P.M.

8. Green. date, Dec. 28ᵈ 4ʰ 10ᵐ 49ˢ; red. decl. 23° 15' 10" S.; true amp. E. 35° 12½' S. *Correction* 9° 53½' E.; *deviation* 5° 36½' W.

9. Interval 12ᵈ, rate 5ˢ·7 *losing*; interval 42ᵈ 8ʰ, acc. rate 4ᵐ 0ˢ·5. Green. date, Dec. 24ᵈ 8ʰ 18ᵐ 44ˢ, red. decl. 23° 24' 46" S., red. eq. time add 0ᵐ 1ˢ, true alt. 40° 54' 8", hour-angle 3° 41' 17". *Longitude* 180° 0' or W.

10. Green. date, Dec. 27ᵈ 4ʰ 40ᵐ 40ˢ, red. decl. 23° 19' 18" S., true alt. 20° 27' 7", sum of logs. 19·362516, true azimuth S. 57° 22' 24" E. Error *of compass* 12° 56' 21" E. *Deviation* 5° 36' 21" E.

Raper: True alt. 20° 26' 54"; log. sine sq. 9·362629 = S. 57° 22' 53" E. *Error of compass* 12° 55' 52" E. *Deviation* 5° 35' 52" E.

11. Time from noon 30ᵐ 40ˢ, Green. date, Dec. 4ᵈ 1ʰ 30ᵐ, red. decl. 22° 17' 12" S., true alt. 60° 7' 18", nat. no. 5104, zen. dist. 29° 17' 12" S. *Latitude* 51° 34' 24" S.

METHOD II. + 35' 31", true alt. 60° 7' 18", *latitude* 51° 34' 23" S., *Norie.* True alt. 60° 7' 14", *latitude* 51° 34' 28" S., *Raper.*

Towson: Exceeds limits of Table.

12. 1st Green. date, Dec. 24ᵈ 22ʰ 11ᵐ 21ˢ; decl. 23° 24' 15" S.; eq. T. + 0ᵐ 18ˢ·51; true alt. 9° 14' 59"; lat. 46° 30' gives hour-angle 2ʰ 58ᵐ 53ˢ; long. 17° 24' W. (A). lat. 46° 0' gives hour-angle 3ʰ 2ᵐ 11ˢ; long. 18° 13½' W. (A'). 2nd Green. date, Dec. 25ᵈ 3ʰ 15ᵐ (not 1ʰ 39ᵐ); decl. 23° 23' 54" S.; eq. T. + 0ᵐ 22ˢ·78; true alt. 15° 12' 14"; lat. 46° 30' gives

hour-angle $1^h 57^m 39^s$; long. $19° 14\frac{1}{2}'$ W. (B); lat. $46° 0'$ gives hour-angle $2^h 3^m 0^s$; long. $17° 54\frac{1}{4}'$ W. (B'). The line of bearing at first observation trends N.E. by E. (easterly) and S.W. by W. (westerly). The position of ship at the time of second observation is latitude $45° 50'$ N., longitude $17° 28'$ W.

13. Star's decl. $5° 31' 15''$ N., true alt. $52° 45' 55''$, *latitude* $31° 42' 50''$ S., *Norie*. True alt. $52° 45' 52''$, *latitude* $31° 42' 53''$ S., *Raper*.

The Curve.—Correct magnetic bearing S. $2°$ E.

Deviations.—$4°$ W.; $6°$ W.; $12°$ W.; $18°$ W.; $3°$ E.; $21°$ E.; $14°$ E.; $1°$ E.

Compass courses.—East; S. $50°$ E.; S. $40°$ W.; N. $4°$ W.

Magnetic courses.—S. $33°$ E.; N. $88\frac{1}{2}°$ W.; S. $30°$ W.; S.E. by E.

Bearings, magnetic.—N. $86\frac{1}{2}°$ W.; N. $2°$ W.

EXAMINATION PAPER—No. XIII, *Pages* 351—352.

1. $6·110455 = 1289598$. $1·302556 = 20·0704$.
2. $2·806180 = 640$. $0·060635 = 1·14983$.
3. **True Courses.**—S. $14°$ W., $23'$ dep. course; S. $23°$ W., $51'$; N. $87°$ W., $45'$; S. $44°$ E., $22'$; N. $78°$ W., $18'$; N. $14°$ W., $44'$; S. $3°$ E., $12'$; West $52'$, current course. *Diff. lat.* $48'·2$ S.; *dep.* $134'·7$ W.; *course* S. $70\frac{1}{2}°$ W., *dist.* $143'$. *Lat. in* $61° 30'$ N.; *diff. long.* $286'$ W. *Longitude in* $78° 31'$ E.
4. Green. date, Aug. $10^d 17^h 51^m 12^s$; red. decl. $15° 18' 58''$ N.; true alt. $42° 50' 17''$. *Latitude* $31° 50' 45''$ S.

   By *Raper:* True alt. $42° 50' 7''$. *Latitude* $31° 50' 55''$ S.
5. Log. of diff. long. $2·663420 = $ *diff. long.* $460'·7$.
6. Diff. lat. $1890'$ S.; mer. diff. lat. $2392'$; diff. long. $1774'·43$ W.; *course* S. $36° 34'$ W.; *distance* $2353'$.
7. Standard, Brest — $1^h 47^m$; corr. for long. — 9^m; $3^h 21^m$ A.M., and $3^h 42^m$ P.M.
     ,,   Brest — $0\ \ 7$; $5\ 10$ A.M., and $5\ 30$ P.M.
     ,,   Brest $+ 2\ 42$; $7\ 59$ A.M, and $8\ 19$ P.M.
8. Green. date, October $28^d 4^h 14^m 48^s$; red. decl. $13° 14' 46''$ S.; true amp. E. $20° 44'$ S. *Error of compass* $31° 24'$ E. *Deviation* $7° 34'$ E.
9. Interval 15^d; rate $11^{s·2}$ *gaining*; Interval $16^d 21\frac{1}{2}^h$; acc. rate $3^m 9^s$; Green. date, April $17^d 21^h 21^m 54^s$; red. decl. $10° 50' 59''$ N.; true alt. $38° 22' 36''$; red eq. T. *subt.* $0^m 42^s$; log. $9·069755$; hour-angle $2^h 40^m 19^s$. *Longitude* $0° 43' 45''$ W.
10. Green. date, March $8^d 16^h 22^m 6^s$; red. decl. $4° 32' 35''$ S.; true alt. $28° 33' 55''$; sum of logs. $19·596707$, true azimuth N. $77° 53' 22''$ E. *Error* $21° 46' 38''$ W.; *deviation* $4° 36' 38''$ W.
11. Time from noon $25^m 52^s$; Green. date July $28^d 2^h 10^m 8^s$; red. decl. $18° 56' 42''$ N.; true alt. $69° 22' 19''$; nat. no. 4683; mer. zen. dist. $19° 51' 6''$ N. *Latitude* $38° 47' 48''$ N.

    METHOD II. $+ 46° 59''$; true alt. $69° 22' 19''$. *Latitude* $38° 41' 25''$ N. *Norie*.

    Towson: Exceeds limits of Table.
12. 1st red. decl. $7° 44' 36''$ S.; 1st red. eq. T. $+ 12^m 38^s·4$; 1st true alt. $27° 41' 12''$; lat. $47° 10'$ gives hour-angle $2^h 12^m 59^s$; long. $166°8\frac{1}{4}'$ W. (A). Lat. $47° 40'$ gives hour-angle $2^h 9^m 5^s$; long. $165° 9\frac{3}{4}'$ W. (A'). 2nd red. decl. $7° 41' 25''·5$; oq. T. $+ 12^m 36^s$; true alt. $32° 50' 32''$; lat. $47° 10'$ gives hour-angle $1^h 12^m 21^s$; long. $165° 23'$ W. (B). Lat. $47° 40'$ gives hour-angle $1^h 4^m 21^s$; long. $167° 23'$ W. (B'). The line of bearing at the first observation trends N.E. $\frac{3}{4}$ E. and S.W. $\frac{3}{4}$ W. The position of the ship at the second observation is lat. $47° 18\frac{1}{2}'$ N., long. $165° 57\frac{1}{2}'$ W.
13. Star's decl. $47° 31' 39''$ S.; true alt. $49° 54'\ 3''$. *Latitude* $7° 25' 42''$ S. *Norie*.
       ,,   ,,   ,, $49\ 53\ 54$,, $7\ 25\ 33$ S. *Raper*.

The Curve.—Correct magnetic bearing S. $13°$ W.

Deviations.—$12°$ W.; $8°$ W.; $8°$ W.; $3°$ W.; $12°$ E.; $20°$ E.; $7°$ E.; $8°$ W.

Compass courses.—S. $36°$ W.; N. $75°$ E.; S. $4°$ W.; N. $31°$ E.

Magnetic courses.—N. $49°$ E.; N. $47°$ W.; N. $5°$ W.; S. $2\frac{1}{2}°$ E.

Bearings, magnetic.—N. $75°$ E.; S. $75°$ W.

Answers. 441

EXAMINATION PAPER—No. XIV, *Pages* 353—354.

1. 2·478098 = nat. no. 300·675 ; 4·214537 = 16388·38.
2. 2·319771 = nat. no. 208·82 ; 6·688798 = ·0004884.
3. **True Courses.**—S. 1° E., 15′ dep. course; N. 75° E., 42′ ; N. 62° E., 32′ ; N. 59° W., 37′ ; S. 11° E., 37′ ; N. 51° E., 38′ ; N. 55° E., 28′ ; N. 10° E., 17′·5 current course. *Diff. lat.* 50″·9 N.; *dep.* 100″·0 E.; *course* N. 63° E.; *dist.* 112′. *Lat. in* 31° 7′ N.; *mid. lat.* 30° 41′ ; *diff. long.* 116′·3; *long. in* 181° 48′ E., or 178° 12′ W.
4. Green. date, March 20^d 11^h 36^m; red. decl. 0° 6′ 35″ N.; true alt. 90° 7′ 18″. Lat. 0° 13′ 53″ N.
   Raper : True alt. 90° 7′ 10″. *Latitude* 0° 13′ 45″ N.
5. Log. of diff. long. 2·718690 = *Diff. long.* 523″·2.
6. Diff. lat. 189′ S.; mer. diff. lat. 189′; diff. long. 147° 35′ E. = 8855′; tang. course 11·670727 ; *course* S. 88° 46′ 38″ E.; *log. dist.* 3·947270 ; *dist.* 8857′.
7. Standard, Devonport — 0^h 6^m ; 11^h 53^m A.M.,   and   no P.M.
   „ Devonport — 0 46 ; 11 13 A.M.,   and 11^h 47^m P.M.
   „ Devonport + 0 38 ; 0 4 A.M.,   and 0 37 P.M.
   „ Hull   — 1 59 ; 11 18 A.M.,   and 11 47 P.M.
8. Green. date, Dec. 5^d 10^h 59^m 52^s; red. decl. 22° 27′ 55″ S.; sine amp. 9·996025 ; true amp. E. 82° 15′ 36″ S. *Error* 16° 10′ 39″ W.; *deviation* 8° 39′ 21″ E.
9. Interval 60^d ; rate 3^s·5 *losing*; interval 62^d 10½^h; acc. rate + 3^m 38^s·5 ; Green. date, Aug. 31^d 10^h 29^m 18^s·5 ; rel. decl. 8° 26′ 20″ N.; eq. time + 0^m 2^s·5 ; true alt. 40° 0′ 24″ ; hour-angle 2^h 25^m 59^s; mean time ship Aug. 31^d 21^h 34^m 3^s·5. *Longitude* 166° 11′ 15″ E.
   Raper : True alt. 40° 0′ 20″ ; hour-angle 2^h 25^m 59^s. *Longitude* 166° 11′ 15″ E.
10. Green. date, March 19^d 23^h 44^m 40^s, red. decl. 0° 5′ 15″ S., true alt. 35° 14′ 22″, sum of logs. 19·219729 ; true azimuth S. 48° 3′ 54″ E. *Error* 45° 15′ 9″ W.; *deviation* 18° 5′ 9″ W.
11. Time from noon 9^m 37^s; Green. date Sept. 22^d 12^h 26^m 35^s; red. decl. 0° 3′ 13″ N.; true alt. 61° 51′ 21″; nat. no. 776 ; mer. zen. dist. 28° 2′ 58″ S. *Latitude* 27° 59′ 45″ S.
   METHOD II. + 5′ 40″, true alt. 61° 51′ 21″, *latitude* 27° 59′ 46″ S., *Norie.* True alt. 61° 51′ 16″, *latitude* 27° 59′ 51″ S., *Raper.*
   Towson : Aug. I, 0 ; Aug. II, + 5′ 27″. *Latitude* 27° 59′ 59″ S.
12. 1st red. decl. 20° 43′ 9″ S. ; 1st. red. eq. T. + 10^m 22^s·56 ; 1st true alt. 11° 16′ 8″; lat. 50° 50′ N. gives hour-angle 2^h 29^m 20^s; long. 7° 14¼′ W. (A). Lat. 51° 20′ N. gives hour-angle 2^h 24^m 45^s; long. 6° 11′ W. (A′). 2nd red. decl. 20° 40′ 59″·6 S.; 2nd red. eq. T. + 10^m 26^s; 2nd true alt. 14° 8′ 7″; lat. 50° 50′ gives hour-angle 1^h 54^m 59^s; long. 5° 59¼′ W. (B). Lat. 51° 20′ N. gives hour-angle 1^h 48^m 42^s; long. 7° 34′ W. (B′). The line of bearing at first observation trends N.E. ¾ E. and S.W. ¾ W. The ship's position at time of second observation is lat. 51° 17′ N., long. 7° 22′ W.
13. Decl. (α Cygni), Deneb, 44° 52′ 3″; true alt. 56° 13′ 50″. *Latitude* 11° 5′ 53″ N.
    Raper : True alt. 56° 13′ 41″. *Latitude* 11° 5′ 44″ N.
    The Curve.—Correct magnetic bearing S. 82° W.
    Deviations.—21° W.; 10° W.; 3° W.; 4° E.; 20° E.; 23° E.; 5° E.; 17° W.
    Compass courses.—N. 64½° E.; S. 10½° W.; S. 68½° E.
    Magnetic courses.—N. 87½° W.; N. 54° W.; N. 22° E.
    Bearings, magnetic.—N. 17° W.; S. 84½° E.

EXAMINATION PAPER—No. XV, *Pages* 355—356.

1. 7·718169 = 52160000. 5·903094 = 800007·4.
2. 4·301030 = 20000. 8·850814 = ·0709274.
3. **True Courses.**—S. 40° E., 21′ dep. course: S. 65° E., 14′; N. 54° W., 9′; N. 23° E., 18′; S. 10° W., 22′; S. 60° W., 29′; N. 16° E., 20′·6; N. 79° E., 14′ current course. *Diff. lat.* 13′·8 S., *dep.* 16′·4 E.: *course* S. 49½° E., *dist.* 21½′. *Lat. in* 34° 42′ S., *diff. long.* 20′ E. *Long. in* 18° 48′ E.

L L L

4. Green. date, February 10ᵈ 21ʰ 50ᵐ 40ˢ: red. decl. 13° 58′ 46″ S.: true alt. 30° 33′ 20″. *Latitude* 45° 27′ 54″ N.
Raper: True alt. 30° 33′ 17″. *Latitude* 45° 27′ 57″ N.
5. Log. of diff. long. 2·010904 = *Diff. long.* 102′·5.
6. Diff. lat. 4202′ N.: mer. diff. lat. 4555′: diff. long. 4847′ E.: tang. course 10·026985: *course* N. 46° 46′ 44″ E.: *log. dist.* 3·787,882: *distance* 6136′.
7. Standard, Brest, constant $+ 4^h 43^m$; corr. for long. $+ 10^m$; 1ʰ 6ᵐ A.M., 1ʰ41ᵐ P.M.
    „   Waterford $+ 0\ 46$;   9 56 A.M., 10 23 P.M.
    „   Leith $- 1\ 49$;   5 28 A.M., 6 7 P.M.
8. Green. date, March 30ᵈ 7ʰ 41ᵐ 8ˢ: red. decl. 3° 57′ 55″ N.: sine 8·842619: true amp. E. 3° 59′ 28″ N. Error 7° 50′ 28″ W.: *deviation* 13° 50′ 28″ W.
9. Interval 15ᵈ, rate 4ˢ·9 *gaining*, interval 32ᵈ 21ʰ, accumulated rate $- 2^m 41^s$, Green. date May 26ᵈ 21ʰ 10ᵐ 6ˢ, red. decl. 21° 18′ 36″ N., red. eq. T. 3ᵐ 8ˢ *subt.*, true alt. 43° 20′ 9″, hour-angle 2ʰ 53ᵐ 24ˢ. *Longitude* 1° 39′ 30″ W.
10. Green. date, July 10ᵈ 0ʰ 25ᵐ 20ˢ, red. decl. 22° 13′ 56″ N., true alt. 44° 59′ 38″, sum of logs. 19·231552, true azimuth S. 48° 46′ E. *Error* 54° 23½′ W.; *deviation* 3° 23½′ W.
11. Time from noon 30ᵐ 41ˢ: Green. date, Nov. 7ᵈ 18ʰ 31ᵐ 53ˢ: red. decl. 16° 34′ 11″ S.: true alt. 40° 4′ 47″; mer. zen. dist. 49° 22′ 53″ N. *Latitude* 32° 48′ 42″ N.
METHOD II. $+ 32′ 22″$. True alt. *(Norie)* 40° 4′ 47″; *latitude* 32° 48′ 40″ N.
Towson: Aug. I, $+ 8′ 30″$; index 81; Aug. II, $+ 24′ 10″$. *Latitude* 32° 40′ 21″ N.
12. 1st red. decl. 13° 19′ 6″ N.; 1st red. eq. T. $- 2^m 10^s$; 1st true alt. 51° 14′ 52″; lat. 48° 30′ gives hour-angle 1ʰ 18ᵐ 31ˢ; long. 59° 14½′ W. (A). Lat. 49° 0′ gives hour-angle 1ʰ 13ᵐ 20ˢ; long. 60° 32½′ W. (A′). 2nd red. decl. 13° 22′ 20″ N.; 2nd eq. T. $- 1^m 12^s$; 2nd true alt. 16° 28′ 49″; lat. 48° 30′ gives hour-angle 5ʰ 20ᵐ 32ˢ; long. 58° 26′ E. (B). Lat. 49° 0′ gives hour-angle 5ʰ 20ᵐ 37ˢ; long. 58° 24½′ W. (B′). The line of bearing at the first observation trends N.W. by W. ⅓ W. and S.E. by E. ⅓ E. The position of the ship at the second observation is lat. 48° 35½′ N., long. 58° 26′ W.

13. Star's decl. 57° 6′ 22″ S.; true alt. 32° 48′ 59″. *Latitude* 0° 4′ 39″ N.
Raper: True alt. 32° 48′ 53″. *Latitude* 0° 4′ 45″ N.
The Curve.—Correct magnetic bearing S. 17° E.
Deviations.—27° E.; 39° E.; 22° E.; 5° W.; 29° W.; 27° W.; 23° W.; 5° W.
Compass courses.—N. 28° W.; N. 58° W.; S. 79½° W.; S. 89° W.
Magnetic courses.—S. 80° E.; S. 42° E.; N. 66° W.; S. 87° E.
Correct magnetic bearing.—N. 28° E.; S. 62° E.

EXAMINATION PAPER—No. XVI, *Pages* 357—358.

1. 4·734724 = 54290·5; 9·823096 = ·66542.
2. 3·306211 = 2024; 1·671357 = 46·92.
3. **True Courses.**—N. 56° W., 13′ dep. course; S. 70° W., 34′; S. 69° W., 36′·2; S. 59° W., 38′·6; S. 3° W., 34′·1; S. 10° W., 17′·2; N. 77° W., 33′·5; S. 10° W., 21′ current course. *Diff. lat.* 101′·4 S.; *dep.* 150′·9 W.; *course* S. 56° W.; *dist.* 181¼′. *Lat. in* 60° 57′ S.; *diff. long.* 302′. *Long. in* 184° 44′ W., or 175° 16′ E.
4. Green. date, May 16ᵈ 3ʰ 1ᵐ 44ˢ, red. decl. 19° 10′ 10″ N., true alt. 86° 50′ 42″. *Lat.* 16° 0′ 52″ N.
5. Log. of diff. long. 1·790942 = *Diff. long.* 61·79.
6. Diff. lat. 604′ S., mer. diff. lat. 606′, diff. long. 6324′ E., log. tang. 11·018519; *course* S. 84° 31′ 35″ E.; *log. dist.* 3·801545; *dist.* 6332′.
7. Standard, Brest, constant $+ 0^h 27^m$;   no A.M.   and   0ʰ34ᵐ P.M.
    „   Brest   „ $+ 1\ 6$;   0ʰ38ᵐ A.M. and 1 13 P.M.
    „   Brest   „ $- 2\ 0$;   10 7 A.M. and 10 38 P.M.
    „   Brest   „ $- 0\ 7$;   no A.M.   and   noon.

8. Green. date, June 24^d 11^h 41^m, red. decl. $23°$ $14'$ $47"$ N., sine amp. 9·998771, true amp. W. $85°$ $41'$ $29"$ N. Error $12°$ $44'$ $46"$ W.; deviation $20°$ $45'$ $14"$ E.

9. Interval 44^d, rate 8^s losing, interval 42^d 23^h, accumulated rate $+ 5^m$ $43^{s}\cdot 7$, Green. date June 14^d 23^h 6^m 21^s, red. decl. $23°$ $19'$ $30"$ N., red. eq. T. $+ 0^m$ $8^{s}\cdot 76$, true alt. $20°$ $48'$ $38"$, hour-angle 3^h 7^m 27^s, M.T. ship June 14^d 20^h 52^m 42^s. Longitude $33°$ $24'$ $45"$ W.

10. Green. date, Sept. 23^d 10^h 29^m 44^s; red. decl. $0°$ $18'$ $24"$ S.; true alt. $40°$ $29'$ $23"$; sum of logs. 17·416341; true azimuth N. $5°$ $55'$ $22"$ W. Error $20°$ $0'$ $7"$ W.; deviation $32°$ $20'$ $7"$ W.

11. Time from noon 19^m 21^s, Green. date, September 22^d 12^h 59^m 45^s, red. decl. $0°$ $2'$ $41"$ N., true alt. $38°$ $53'$ $36"$, nat. no. 2253, mer. zen. dist. $50°$ $51'$ $26"$ N. Latitude $50°$ $54'$ $7"$ N.

METHOD II.—Red. $+ 10'$ $0"$: lat. $50°$ $54'$ $5"$ N. (Norie).

Towson: Aug. I, 0, index 46. Aug. II, $+ 9'$ $50"$. Latitude $50°$ $54'$ $15"$ N.

12. 1st red. decl. $9°$ $41'$ $59"$ N.; 1st red. eq. T. $+ 1^m$ $6^{s}\cdot 1$; 1st true alt. $39°$ $32'$ $10"$; lat. $48°$ $50'$ gives hour-angle 2^h 32^m 56^s; long. $6°$ $58\frac{1}{2}'$ W. (A). Lat. $49°$ $30'$ gives hour-angle 2^h 29^m 40^s; long. $6°$ $9\frac{1}{2}'$ W. (A'). 2nd red. decl. $9°$ $37'$ $11"$ N.; 2nd red. eq. T. $+ 1^m$ $2^{s}\cdot 6$; true alt. $35°$ $53'$ $31"$; lat. $48°$ $50'$ gives hour-angle 2^h 59^m 14^s; long. $5°$ $7\frac{3}{4}'$ W. (B). Lat. $49°$ $30'$ gives hour-angle 2^h 56^m 44^s; long. $5'$ $4\frac{1}{2}'$ W. (B'). The line of bearing at first observation trends N.E. $\frac{1}{2}$ N. (northerly) and S.W. $\frac{1}{2}$ S. (southerly). The position of the ship at the second observation is lat. $50°$ $6'$ N., long. $6'$ $17\frac{1}{2}'$ W.

13. Star's decl. $10°$ $32'$ $59"$ S.; true alt. $52°$ $9'$ $5"$. Latitude $48°$ $23'$ $54"$ S.

The Curve.—Correct magnetic bearing S. $11°$ W.

Deviations.—$24°$ W., $2°$ W., $5°$ E., $8°$ E., $21°$ E., $22°$ E., $3°$ W., $26°$ W.

Compass courses.—N. $85°$ E., S. $46°$ W., S. $42\frac{1}{2}°$ E.

Magnetic courses.—N. $81°$ W., N. $23°$ W., N. $70°$ E.

Bearings, magnetic.—(Deviation for ship's head W.S.W. $= 10\frac{1}{2}°$ E.) S. $8\frac{1}{2}°$ W., S. $10\frac{1}{2}°$ W.

EXAMINATION PAPER—No. XVII, Pages 359—360.

1. 7·651355 = 4480793$8\cdot 1$; 1·351334 = 22·4561.

2. 5·614476 = 411600·94; 9·030734 = ·107333.

3. **True Courses.**—S. $72°$ E., $25'$ dep. course; N. $41°$ E., $35'\cdot 4$; N. $62°$ E., $48'$; N. $53°$ W., $43'$; S. $60°$ E. $26'$; N. $18°$ W., $18'$; S. $46°$ E., $32'$; N. $82°$ E., $36'$ current course. Diff. lat. $54'\cdot 3$ N., dep. $130'\cdot 6$ E.; course N. $67\frac{1}{2}°$ E., dist. $141\frac{1}{4}'$. Lat. in $38°$ $31'$ N., diff. long. $166'$ E. Long. in $2°$ $5'$ E.

4. Green. date, Nov. 20^d 19^h 18^m 40^s; red. decl. $19°$ $55'$ $38"$ S.; true alt. $80°$ $28'$ $59"$. Lat. $29°$ $26'$ $39"$ S.

Raper: True alt. $80°$ $28'$ $55"$ N. Latitude $29°$ $26'$ $43"$ S.

5. Log. of diff. long. 1·909671 = Diff. long. $81\cdot 22$.

6. Diff. lat. $731'$ S.; mer. diff. lat. $733'$; diff. long. $1259'$ W.; log. tang. 10·234922; course S. $59°$ $47\frac{1}{2}'$ W.; log. of distance 3·162224; distance $1453'$.

7. Standard, Hull   constant $- 0^h 29^m$; $3^h 32^m$ A.M., and 4^h 0^m P.M.
   ,, London   ,, $- 2$ 19;   9 30 A.M., and 9 57 P.M.
   ,, Dover   ,, $+ 5$ 13;   1 24 A.M., and 1 53 P.M.
   ,, Brest   ,, $+ 3$ 28;   corr. for long.—16^m; $4^h 37^m$ A.M., & $5^h 2^m$ P.M.

8. Green. date, July 20^d 18^h 56^m 48^s; red. decl. $20°$ $30'$ $1"$ N.; true amp. W. $28°$ $38'$ $41"$ N. Error of compass $59°$ $34'$ $56"$ E.; deviation $49°$ $55'$ E.

9. Interval 18^d; rate $4^{s}\cdot 0$ losing; interval 72^d 12^h; Green. date, June 4^d 12^h 33^m 40^s; red. decl. $22°$ $31'$ $9"$ N.; true alt. $28°$ $18'$ $52"$; red. eq. T. subt. 1^m 54^s; hour-angle 3^h 52^m 17^s; mean time at ship June 4^d 20^h 5^m 49^s. Longitude $113°$ $2'$ $15"$ E.

10. Green. date, Nov. 9^d 14^h 58^m 46^s; red. decl. $17°$ $6'$ $15"$ S.; true alt. $6°$ $11'$ $26"$; sum of logs. 19·304632; true azimuth S. $53°$ $22'$ $6"$ E. Error $3°$ $32'$ $6"$ W.; deviation $10°$ $52'$ $6"$ W.

11. Time from noon 9^m 5^s; Green. date Jan. 8^d 3^h 31^m 43^s; red. decl. $22°$ $11'$ $54"$ S.; true alt. $76°$ $57'$ $49"$; nat no. 594; mer. zen. dist. $12°$ $55'$ $5"$ S. Latitude $35°$ $4'$ $59"$ S.

METHOD II. $+ 9'$ $6"$. Latitude $35°$ $4'$ $59"$ S., Norie.

Towson: Exceeds limits of Table.

12. 1st red. decl. 23° 17′ 47″ N.; 1st red. eq. T. 0ᵐ 0ˢ·3; 1st true alt. 54° 23′ 38″; lat. 50° 10′ gives hour-angle 2ʰ 0ᵐ 8ˢ; long. 79° 40½′ W. (A). Lat. 49° 40′ gives hour-angle 2ʰ 2ᵐ 30ˢ; long. 79° 5′ W. (A'). 2nd red. decl. 23° 18′ 21″ N.; 2nd red. eq. T. + 0ᵐ 2ˢ·8; 2nd true alt. 11° 22′ 0″. Lat. 50° 10′ gives hour-angle 6ʰ 41ᵐ 49ˢ; long. 79° 50¾′ W. (B). Lat. 49° 40′ gives hour-angle 6ʰ 40ᵐ 30ˢ; long. 80° 10′ W. (B'). The line of bearing at the first observation trends N.W. ½ N. (northerly) and S.E. ½ S. (southerly). The position of the ship at the second observation is lat. 49° 48½′ N., long. 80° 4′ W.

13. Star's decl. 16° 33′ 33″ S.; true alt. 37° 45′ 59″. *Latitude* 35° 40′ 28″ N., *Norie.*
Raper: True alt. 37° 45′ 54″. *Latitude* 35° 40′ 33″ N.
The Curve.—Correct magnetic bearing N. 12° E.
Deviations.—12° E.; 0°; 17° W.; 24° W.; 12° W.; 7° E.; 17° E.; 17° E.
Compass courses.—N. 85° W.; S. 57° W.; S. 34° E.; S. 8° E.
Magnetic courses.—N. 86° E.; S.E. by S.; N. 38½° W.; N. 79° W.
Bearings, magnetic.—(Deviation for ship's head E. by S. ½ S. = 20½° W.); N. 82½° W.; S. 37½° E.

EXAMINATION PAPER—No. XVIII, *Pages* 361—362.

1. Log. of product 6·498062 = product 3148195·6. Log. 6·233506, product ·00001712.
2. Log. of quotient 4·986680 = quotient 96979·6. Log. 2·439691, product ·027523.
3. True Courses.—N. 69° W., 25′ dep. course; N. 58° W., 51′; S. 63° W., 42′; N. 51° E., 30′; N. 36° W., 46′; S. 27° W., 11′; S. 68° E., 16; S. 51° W., 32′ current course. *Diff. lat.* 37·0 N., *dep.* 122′·8 W., *course* N. 73° W., *dist.* 128′. *Lat. in* 36° 50′ S., *mid. lat.* 36° 8′, *diff. long.* 152′ W. *Long. in* 71° 9′ W.
4. Green. date, 1882, December 31ᵈ 12ʰ 48ᵐ 24ˢ; red. decl. 23° 3′ 9″ S.; true alt. 83° 52′ 24″. *Latitude* 16° 53′ 33″ S.
Raper: True alt. 82° 52′ 19″. *Latitude* 16° 55′ 28″ S.
5. Diff. long. 192′·5 W. *Long. in* 182° 29′·5 W., or 177° 30½′ E.
6. Diff. lat. 2747′ S.; mer. diff. lat. 2919′; diff. long. 6340′ W.; tang. course 10·336855; course S. 65° 16′ 41″ W.; *log. distance* 3·817459, *distance* 6568.
7. Standard, Brest constant — 2ʰ 0ᵐ; 2ʰ39ᵐA.M., 2ʰ59ᵐP.M.
„ Brest „ — 3 17 ; corr. for long. — 8ᵐ 1 14 A.M., 1 34 P.M.
„ Brest „ + 4 2 ; „ + 9ᵐ 8 50 A.M., 9 10 P.M.
8. Green. date, Nov 3ᵈ 17ʰ 28ᵐ 30ˢ; red. decl. 15° 21′ 20″ S.; true amp. E. 22° 41′ 24″ S. *Error* 27° 56′ 6″ W.; *deviation* 11° 26′ 6″ W.
9. Interval 14ᵈ; rate 8ˢ·5 *gaining*; interval 124ᵈ 20ʰ; acc. rate 17ᵐ 41ˢ; Green. date August 31ᵈ 19ʰ 54ᵐ 30ˢ; red. decl. 8° 17′ 48″ N.; eq. T. — 0ᵐ 5ˢ; true alt. 15° 25′ 44″; sum of logs. 9·531791 ; hour-angle 4ʰ 45ᵐ 41ˢ. *Longitude* 10° 0′ 45″ W.
10. Green. date, May 31ᵈ 16ʰ 13ᵐ; red. decl. 22° 2′ 19″ N.; true alt. 39° 20′ 26″; sum of logs. 19·779199; true azimuth S. 101° 41′ 8″ E. *Error* 3° 14′ 53″ W.; *deviation* 4° 6′ 9″ W.
11. Green. date, April 12ᵈ 14ʰ 1ᵐ 5ˢ; time from noon 10ᵐ 15ˢ; red. decl. 8° 57′ 37″ N.; true alt. 80° 43′ 11″; nat. no. 988; mer. zen. dist. 8° 55′ 23″ S. *Latitude* 0° 2′ 14″ N.
METHOD II. + 21′ 48″. *Latitude* 0° 2′ 36″ N., *Norie.*
Towson: The altitude exceeds the limits of the Table.
12. 1st red. decl. 4° 35′ 13″ N.; 1st red. eq. T. + 3ᵐ 55ˢ·3; 1st true alt. 35° 3′ 31″; lat. 51° 0′ gives hour-angle 2ʰ 21ᵐ 1ˢ; long. 9° 49¾′ W. (A). Lat. 50° 30′ gives hour-angle 2ʰ 24ᵐ 9ˢ; long. 10° 36¼′ W. (A'). 2nd red. decl. 4° 40′ 16″ N.; 2nd red. eq. T. + 3ᵐ 51ˢ; 2nd true alt. 31° 31′ 9″; lat. 51° 0′ gives hour-angle 2ʰ 51ᵐ 35ˢ; long. 10° 59′ W. (B). Lat. 50° 30′ gives hour-angle 2ʰ 53ᵐ 57ˢ; long. 10° 23½′ W. (B'). The line of bearing at the first observation trends N.E. and S.W. The position of the ship at the second observation is lat. 50° 19′ N., long. 10° 10′ W.
13. Star's decl. 60° 21′ 13″ S., true alt. 9° 52′ 32″. *Latitude* 19° 46′ 15″ N.
The Curve.—Correct magnetic bearing S. 8° E.
Deviations.—19° E.; 8° W.; 29° W.; 35° W.; 20° W.; 13° E.; 29° E.; 31° E.
Compass courses.—N. 2° E.; S. 60½° E.; S. 29° W.; N. 81½° W.
Magnetic courses.—N. 11° W.; S. 56½° E.; N. 50° E.; N. 34° E.
Bearings, magnetic.—N. 15½° E.; N. 49° W.

EXAMINATION PAPER—No. XIX, *Pages* 363—364.

1. 7·080996 = 12030250; 1·819774 = 66·035.
2. 0·301030 = 2'; 1·770558 = 58·96.
3. *True Courses.*—N. 70° W., 13° dep. course; S. 20° W., 48'; S. 60° W., 37'·2; S. 54° W., 30'; N. 71° W., 43'·5; N. 66° W., 38'·4; N. 15° W., 37'·8; N. 36° W., 36' current course. Diff. lat. 18'·5 N., dep. 192'·2 W.; course N. 84½° W., dist. 193½'. Lat. in 46° 1½' S.; diff. long. 4° 37½' W. Long. in 181° 21½' W., or 178° 38½' E.
4. Green. date, Sept. 22ᵈ 16ʰ 0ᵐ 24ˢ; red. decl. 0° 0' 15" S.; true alt. 83° 52' 20". *Lat.* 6° 7' 25" N.
   Raper: True alt. 83° 52' 14". *Latitude* 6° 7' 31" N.
5. Log. of diff. long. 2·159381 = *Diff. long.* 144'·3.
6. Diff. lat. 1199' N.; mer. diff. lat. 1205'; diff. long. 3410' W.; tangent course 10·451767; course N. 70° 32' 17" W.; log. distance 3·556139; *distance* 3599'.
7. Standard, London   constant = 2ʰ 13ᵐ;   9ʰ 53ᵐ A.M., 10ʰ 17ᵐ P.M.
   ,, Brest   ,,   = 2 0 ;   no A.M., 0 4 P.M.
   ,, Brest   ,,   + 3 3 ;   4 43 A.M., 5 7 P.M.
   ,, Weston-super-mare,,   − 1 12 ;   3 21 A.M., 3 50 P.M.
   ,, Leith   ,,   − 2 21 ; 10 16 A.M., 10 40 P.M.
   ,, Brest   ,,   + 6 13 ; corr. for long. − 16ᵐ; 7ʰ 37ᵐ A.M., and 8ʰ 1ᵐ P.M.
8. Green. date, June 12ᵈ 10ʰ 55ᵐ 24ˢ; red. decl. 23° 11' 57" N.; sine of true amplitude 9·992978; true amplitude W. 79° 44' N. *Error* 38° 23½' W.; *deviation* 15° 13½' W.
9. Interval 70ᵈ, daily rate (4"·08 =) 4"·1 *gaining*, interval 93ᵈ 12ʰ, acc rate + 6ᵐ 23"·3, red. decl. 23° 18' 21" N., red. eq. time + 0ᵐ 2"·81, true alt. 49° 11' 9", hour-angle 3ʰ 1ᵐ 29ˢ, M.T. ship June 14ᵈ 3ʰ 1ᵐ 32ˢ. *Longitude* 134° 37' 15" W.
10. Green. date, Dec 31ᵈ 16ʰ 24ᵐ 20ˢ; red. decl. 23° 1' 15" S.; true alt. 35° 1' 13"; logs. 19·750904; true azimuth N. 97° 17' 52" W. *Error* 12° 55' 22" W.; *deviation* 32° 55' 22" W.
11. Time from noon 10ᵐ 46ˢ; Green. date, Dec. 30ᵈ 18ʰ 37ᵐ 14ˢ; red. decl. 23° 6' 34" S.; nat. no. 625; nat. cos. 257950; mer. zen. dist. 75° 3' 6" N. *Latitude* 51° 56' 32" N.
    METHOD II. Red. + 2' 14". *Latitude* 51° 56' 31" N., *Norie*.
    Towson: Aug. I, + 1' 21"; index 12. Aug. II, + 50". *Latitude* 51° 36' 34" N.
12. 1st red. decl. 19° 10' 59" S.; 1st red. eq. T. + 12ᵐ 22ˢ; true alt. 9° 41' 48"; lat. 51° 15' N. gives hour-angle 2ʰ 56ᵐ 19ˢ; long. 0° 56¾' W. (A). Lat. 50° 45' gives hour-angle 2ʰ 59ᵐ 49ˢ; long. 1° 49¼' W. (A'). 2nd red. decl. 19° 8' 34" S.; red. eq. T. + 12ᵐ 24ˢ; true alt. 18° 37' 49"; long. 3° 17½' W. (B). Lat. 50° 45' gives hour-angle 1ʰ 5ᵐ 36ˢ; long. 0° 18' W. (B'). The line of bearing when the first altitude was taken trends N.E. ¼ E. and S.W. ¼ W. The position of ship at the time of the second observation is lat. 51° 19½' N., long. 0° 15½' E.
13. Star's decl. 10° 33' 38" S., true alt. 60° 59' 28". *Latitude* 45° 34' 10" S.
    The Curve.—Correct magnetic bearing N. 73° W.
    Deviations.—0°; 17° E.; 15° E.; 7° E.; 2° E.; 5° W.; 16° W.; 20° W.
    Compass courses.—North; N. 85½° E.; S.W.
    Magnetic courses.—S. 65° W.; N. 84½° E.; N. 52° W.
    Bearings, magnetic.—S. 19° W., and N. 60° W.

EXAMINATION PAPER—No. XX, *Pages* 365—366.

1. Log. 2·577492 = product 378; 2·138598 = ·013759.
2. Log. 3·954292 = quotient 9001·0; 5̄·301030 = ·00002; 6̄·301030 = ·000002; and 1·301030 = 20.
3. *True Courses.*—S. 26° W., 16' dep. course; S. 87° E., 53'; S. 52° W., 15'; N. 70° E., 50'; N. 14° E., 9'; S. 30° E., 28'; N. 68° E., 47'; N. 82° E., 42' current course. *Diff. lat.* 1'·4 S.; *dep.* 182'·5 E.; *course* S. 89½° E.; *dist.* 182½'. *Latitude in* 50° 26' S.; *diff. long.* 286'·5 E. *Longitude in* 184° 26½' E., or 175° 33½' W.

4. Green. date, Sept. 22^d 20^h 9^m; red. decl. 0° 4′ 17″ S.; true alt. 84° 21′ 18″. *Latitude* 5° 42′ 59″ S.
Raper: True alt. 84° 21′ 7″. *Latitude* 5° 43′ 10″ S.
5. *Diff. long.* 220′·9 E., or 3° 41′ E. *Long. in* 3° 1′ E.
6. Diff. lat. 2531′ S.; mer. diff. lat. 2701′; diff. long. 7433′ E.; tang. 10·439639; *course* S. 70° 1′ 48″ E.; *distance* 7411′.
7. Standard, Leith   constant — 1^{h}17^m;   9^{h}58^m A.M., and 10^{h}26^m P.M.
    „   Devonport   „   — 1 13 ;   0 30 A.M., and   1 3 P.M.
    „   Weston-super-mare„   „   + 0 19 ;   3 5 A.M., and   3 40 P.M.
    „   Portsmouth   „   — 1 11 ;   6 47 A.M., and   7 20 P.M.
    „   Brest   „   — 0 47 ; 11 54 A.M., and no P.M.

8. Green. date, November 4^d 21^h 23^m; red. decl. 15° 42′ 43″ S.; sine 9·461777; true amp. W. 16° 50′ S. *Error* 19° 44′ E.; *deviation* 18° 44′ E.
9. Green. date Aug. 5^d 8^h 8^m 44^s, red. decl. 16° 51′ 26″ N., red. eq. T. + 5^m 43^s, true alt. 35° 16′ 45″, hour-angle 3^h 54^m 9^s. *Longitude at sight* 179° 17′ 30″ W., *diff. longitude since sight* 55′ 42″. *Longitude at noon* 180° 13′ 12″ W., or 179° 46′ 48″ E.
10. Green. date, Aug. 13^d 2^h 20^m 40^s; red. decl. 14° 36′ 29″ N.; true alt. 27° 23′ 29″; sum of logs. 19·256412; true azimuth N. 50° 16′ 44″ E. *Error* 25° 16′ 44″ E. *Deviation* 8° 56′ 44″ E.
11. Green. date, June 11^d 19^h 43^m 30^s; time from noon 32^m 18^s; red. decl. 23° 9′ 36″ N.; true alt. 50° 14′ 43″; nat. no. 8770; mer. zen. dist. 38° 57′ 47″ S. *Lat.* 15° 48′ 11″ S.
METHOD II. + 47′ 55″. *Latitude* 15° 47′ 46″ S.
Towson: Aug. I, + 11′ 16′, index 81. Aug. II, + 34′ 38″. *Latitude* 15° 46′ 41″ S.
12. 1st red. decl. 15° 6′ 55″ S.; 1st red. eq. T. — 16^m 19^s·3; 1st true alt. 19° 3′ 11″; lat. 51° 45′ gives hour-angle 1^h 49^m 6^s; long. 11° 2½′ W. (A). Lat. 51° 15′ gives hour-angle 1^h 54^m 59^s; long. 12° 30¾′ W. (A′); 2nd red. decl. 15° 10′ 6″ S.; and red. eq. T. — 16^m 19″·13; 2nd true alt. 16° 41′ 27″; lat. 51° 45′ hour-angle 2^h 17^m 50^s; long. 10° 41′ W. (B). Lat. 51° 15′ gives hour-angle 2^h 22^m 22^s; long. 9° 33′ W. (B′). The line of bearing at first observation trends N.E. by E. ½E. and S. W. by W. ½ W. The position of the ship at the second observation is lat. 51° 42′ N., long. 10° 34′ W.
13. Star's decl. 22° 54′ 38″ N., true alt. 60° 23′ 4″. *Latitude* 52° 31′ 34″ N. *(Norie).*
    „   „   60 22 57   „   52 31 41 N. *(Raper).*
The Curve.—Correct magnetic bearing S. 100° W. or N. 80° W.
Deviations.—49° E., 16° E., 23° W., 54° W., 50° W., 11° W., 18° E., 54° E.
Compass courses.—S. 85° W., South, N. 39° W., S. 16½° W.
Magnetic courses.—N. 65° E., S. 27½° E., N. 5° W., S. 62° W.
Bearings, magnetic.—S. 46° E., S. 52° W., N. 35° E.

EXAMINATION PAPER—No. XXI, *Pages* 367—368.

1. 7·754166 = 56776104; 3·806197 = ·00640025.
2. 8·201561 = ·015906; 13·000000 = 10000000000000.
3. **True Courses.**—S. 6° W., 19′ dep. course; N. 50° W., 23′·7; S. 47° E., 16′·3; N. 18° E., 17′·6; S. 8° W., 14′·1; N. 87° E., 42′; S. 75° E., 12′; S. 65° W., 15′, current course. *Diff. lat.* 19′·3 S.; *dep.* 35′ E.; *course* S. 61° E.; *dist.* 40′ E. *Lat. in* 56° 41′ N.; *diff. long.* 64′ E. *Longitude in* 38° 56′ W.
4. Green. date, June 24^d 20^h 3^m; red. decl. 23° 24′ 13″ N.; true alt. 60° 4′ 7″; *latitude* 6° 31′ 40″ S.
By *Raper:* True alt. 60° 4′ 1″ N. *Latitude* 6° 31′ 46″ S.
5. Diff. long. 153′·9 W., or 2° 34′ W. *Longitude in* 12° 4′ W.
6. Diff. lat. 4728′·5 N.; mer. diff. lat. 4896′·5; diff. long. 3540′ E.; log. tang. 9·859117; *course* N. 35° 52′ E. *Compass course* N. 15° 52′ E.; *distance* 5588′.

7. Standard, Brest, constant $+ 6^h 28^m$; corr. for long. $- 14^m$; $10^h 37^m$ A.M. & $10^h 54^m$ P.M.
,, Brest ,, $+ 3\ 42$; ,, $+ 9$: $8\ 14$ A.M. & $8\ 31$ P.M.
,, Queenstown, constant $- 0^m 59^s$; $4^h 41^m$ A.M. and $4^h 58^m$ P.M.

8. Green. date, June $23^d 18^h 48^m 12^s$; red. decl. $23° 25' 38''$ N.; true amp. $23° 25' 38''$ N. *Error of compass* $54° 22'$ W.; *deviation* $32° 42'$ W.

9. Interval 32^d, rate $0^{s \cdot}8$ *gaining*, interval $113^d 5^h$, accumulated rate $- 1^m 30^{s \cdot}5$, Green. date, Sept. $22^d 4^h 57^m 35^s$, red. decl. $0° 10' 24''$ N., true alt. $16° 56' 12''$, red. eq. T. $- 7^m 24^s$, hour-angle $4^h 52^m 15^s$. *Longitude* $149° 18' 30''$ W.

10. Green. date, March $21^d 13^h 51^m 16^s$; red. decl. $0° 32' 22''$ N.; true alt. $42° 57' 18''$; sum of logs. $19·621987$; true azimuth N. $80° 39' 4''$ W. *Error* $14° 58\frac{1}{2}'$ E.; *deviation* $7° 8\frac{1}{2}'$ E.

11. Time from noon $10^m 51^s$; Green. date October $3^d 14^h 27^m 9^s$; red. decl. $4° 15' 31''$ S.; true alt. $63° 47' 10''$; nat. no. 963; mer. zen. dist. $26° 5' 19''$ S. *Latitude* $30° 20' 50''$ S.
METHOD II. $+ 7' 31''$. *Latitude* $30° 20' 50''$ S.
Towson: Aug. I, $+ 0' 16''$; index $15''$. Aug. II, $+ 8' 6''$. *Latitude* $30° 20' 54''$ S.

12. 1st red. decl. $20° 20' 33''$ S.; 1st eq. T. $+ 10^m 10^s$; 1st true alt. $13° 18' 58''$; lat. $49° 0'$ gives hour-angle $2^h 22^m 12^s$; long. $140° 0'$ W. (A). Lat. $49° 40'$ gives hour-angle $2^h 15^m 57^s$; long. $138° 26\frac{1}{2}'$ W. (A'). 2nd red. decl. $20° 48' 31''$ S.; 2nd red. eq. T. $+ 10^m 13^{s \cdot}63$; lat. $49° 0'$ gives hour-angle $1^h 57^m 32^s$; long. $138° 35\frac{1}{2}'$ W. (B). Lat. $49° 40'$ gives hour-angle $1^h 49^m 39^s$; long. $140° 33\frac{3}{4}'$ W. (B'). The line of bearing at first observation trends N.E. by E. and S.W. by W. The position of ship at second observation is lat. $49° 21'$ N., long. $139° 37\frac{1}{2}'$ W.

13. Star's decl. $55° 53' 21''$ N.; true alt. $84° 56' 45''$. *Latitude* $60° 56' 36''$ N.
The Curve.—Correct magnetic bearing N. $95°$ E., or S. $85°$ E.
Deviations.—$7°$ W.; $17°$ W.; $29°$ W.; $24°$ W.; $4°$ E.; $29°$ E.; $28°$ E.; $16°$ E.
Compass courses.—N. $18\frac{1}{2}°$ E.; S. $62°$ E.; S. $5\frac{1}{2}°$ E.; S. $58°$ W.
Magnetic courses.—N. $32\frac{1}{2}°$ E.; S. $40°$ E.; S. $58\frac{1}{2}°$ W.; N. $15°$ W.
Bearings, magnetic.—(Deviation for ship's head S.W. by W. $\frac{1}{2}$ W. $= 30°$ E.), N. $69\frac{1}{2}°$ E.; S. $65\frac{1}{2}°$ E.

EXAMINATION PAPER—No. XXII, *Pages 369—370.*

1. $6·041523 = 1100330·8$; $5·847331 = 703608·06$.
2. $4·410657 = 27584$; $3·000000 = 1000$.
3. True Courses.—N. $55°$ E., $25'$ dep. course; N. $49°$ W., $50'$; N. $72°$ W., $38'$; N. $80°$ W., $33'·8$; N. $6°$ E., $37'·9$; N. $3°$ W., $42'·5$; N. $43°$ W., $29'$; West, $32'$ current course. *Diff. lat.* $166'·0$ N.; *dep.* $136'·7$ W.; *course* N. $39\frac{1}{2}°$ W.; *dist.* $215\frac{1}{2}'$. *Latitude in* $53° 54\frac{1}{2}'$ N.; *mid. lat.* $52° 31'$ N.; *diff. long.* $224'·6$. *Longitude in* $2° 20'$ W.
4. Green. date, Aug. $25^d 17^h 51^m 48^s$; red. decl. $10° 27' 36''$ N.; true alt. $35° 48' 58''$. Lat. $43° 43' 26'$ S.
5. Log. of diff. long. $2·254252 =$ diff. long. $179'·6$.
6. Diff. lat. $315'$ S.; mer. diff. lat. $524'$; diff. long. $365'$; tang. course $9·842962$; course S. $34° 51' 35''$ E.; log. of distance $2·584204$; *distance* $383'·9$.
7. Standard, Hull, constant $- 1^h 59^m$; $0^h 9^m$ A.M., and $0^h 45^m$ P.M.
,, Leith ,, $- 1\ 43$; $9\ 0$ A.M., and $9\ 36$ P.M.
,, Greenock ,, $+ 4\ 41$; no A.M., and $0\ 27$ P.M.
,, Weston-super-mare ,, $- 2\ 10$; no A.M., and $0\ 39$ P.M.
,, Thurso ,, $- 0\ 58$; $3\ 45$ A.M., and $4\ 23$ P.M.
,, Sunderland ,, $+ 0\ 23$; no A.M., and noon.
8. Green. date, Sept. $22^d 19^h 50^m 52^s$; red. decl. $0° 3' 59''$ S.; sine $7·267198$; true amp. W. $0° 6' 22''$ S. *Error* $14° 53' 38''$ E.; *deviation* $5° 22' 22''$ W.
9. Interval 10^d, rate $4^{s \cdot}6$ *gaining*; interval $89^d 22\frac{1}{2}^h$, acc. rate $6^m 54^s$. Green. date, Aug. $27^d 22^h 15^m 56^s$, red. decl. $9° 41' 40''$ N., red. eq. time $+ 1^m 6^{s \cdot}3$, true alt. $38° 15' 44''$, hour-angle $3^h 19^m 10^s$, M.T. ship Aug. $28^d 3^h 20^m 16^s$. *Longitude* $76° 5' 0''$ E.

10. Green. date, June $15^d\ 2^h\ 30^m\ 52^s$, red. decl. $23°\ 19'\ 52''$ N., true alt. $9°\ 21'\ 57''$, sum of logs. 19·864591, true azimuth S. 117° 39' 46" W. *Error of compass* $65°\ 9'$ W. *Deviation* 14° 29' W.

11. Time from noon $29^m\ 55^s$, Green. date, Dec. $23^d\ 0^h\ 48^m\ 31^s$, red. decl. $23°\ 26'\ 33''$ S., true alt. $23°\ 52'\ 44''$, nat. no. 5776, mer. zen. dist. $65°\ 45'\ 51''$ N. *Latitude* $42°\ 18'\ 58''$ N.
METHOD II. Red. $+ 21'\ 46''$: *latitude* $42°\ 18'\ 57''$ N.
Towson: Aug. I, $+ 10'\ 31''$: index 73. Aug. II, $+ 10'\ 51''$: *latitude* $42°\ 19'\ 21''$ S.

12. 1st red. decl. $15°\ 55'\ 17''$ N.; eq. T. $+ 5^m\ 19^s·4$; true alt. $50°\ 40'\ 55''$; lat. $47°\ 0'$ gives hour-angle $1^h\ 56^m\ 26^s$; long. $89°\ 13'$ E. (A); lat. $46°\ 20'$ gives hour-angle $1^h\ 59^m\ 52^s$; long. $88°\ 21\tfrac{1}{4}'$ E. (A'). 2nd red. decl. $15°\ 52'\ 38''$ N.; eq. T. $+ 5^m\ 18^s$; true alt. $53°\ 0'\ 48''$; lat. $47°\ 0'$ gives hour-angle $1^h\ 36^m\ 30^s$; long. $87°\ 30\tfrac{1}{2}'$ E. (B); lat. $46°\ 20'$ gives hour-angle $1^h\ 40^m\ 50^s$; long. $88°\ 35\tfrac{1}{2}'$ E. (B'). The line of bearing at first observation trends N.E. by E. $\tfrac{1}{4}$ E. and S.W. by W. $\tfrac{1}{4}$ W. The position of ship at the time of second observation is latitude $46°\ 1'$ N., longitude $87°\ 50\tfrac{1}{2}'$ E.

13. Star's decl. $55°\ 53'\ 41''$ N., true alt. $62°\ 12'\ 39''$, *latitude* $28°\ 6'\ 20''$ N.
The Curve.—Correct magnetic bearing S. 30° W.
Deviations.—14° W.; 15° W.; 15° W.; 10° W.; 15° E.; 30° E.; 16° E.; 5° W.
Compass courses.—S. $20\tfrac{1}{2}°$ W.; N. 30° E.; S. 55° E.
Magnetic courses.—S. 75° W.; N. 74° W.; N. 62° E.
Bearings, magnetic.—N. 14° W.; N. 70° E.

EXAMINATION PAPER—No. XXIII, *Pages* 371—372.

1. 5·561769 = 364560 nearly. 7·417139 = 26130000.
2. 0·477120 = 3. 5·903090 = 800000. 7·903090 = 80000000.
3. True Courses.—S. 24° W., 12' dep. course; S. 43° W., 34'; S. 57° W., 30'; S. 32° W., 32'; S. 1° W., 34'; S. 3° E., 32'; S. 19° E., 34'; S. 59° W. 30', current course. *Diff. lat.* 192'·9 S.; *dep.* 83'·8 W.; *course* S. $23\tfrac{1}{2}°$ W., *dist.* $210\tfrac{1}{2}'$. *Lat. in* 58° 57' N.; *mid. lat.* 60° 33'; *diff. long.* 170'·4 W. *Longitude in* 181° 42' W., or 178° 18' E.
4. Green. date, Aug. $31^d\ 17^h\ 29^m\ 12^s$; red. decl. $8°\ 19'\ 59''$ N.; true alt. 51° 9' 27". *Latitude* 30° 30' 34" S.
Raper: True alt. 51° 9' 22". *Latitude* 30° 30' 39" S.
5. Log. of diff. long. 2·322530 = *diff. long.* 210'·2.
6. Diff. lat. 5968' N.; mer. diff. lat. 7050'; diff. long. 2121' W.; tang. 9·478352; *course* N. 16° 44' 38" W.; *log. distance* 3·794644: *distance* 6232'.

7. | | | | |
    |---|---|---|---|
    | Standard, Holyhead | constant | — 3^h10^m : | 9^h12^m A.M., 9^h34^m P.M. |
    | ,, Londonderry | ,, | — 1 37 : | 8 30 A.M., 8 49 P.M. |
    | ,, London | ,, | — 2 13 : | 1 55 A.M., 2 17 P.M. |
    | ,, Weston-super-mare | ,, | — 2 10 : | 6 55 A.M., 7 14 P.M. |
    | ,, Waterford | ,, | + 0 44 : | 8 21 A.M., 8 40 P.M. |
    | ,, Liverpool | ,, | — 0 55 : | 0 19 A.M., 0 39 P.M. |
    | ,, Brest | ,, | + 3 28 : | corr. long.—$16^m,9^h\ 9^m$ A.M., $9^h\ 29^m$ P.M. |

8. Green. date, Dec. $24^d\ 1^h\ 14^m\ 12^s$; red. decl 23° 25' 31" S.; sine 9·972365 : true amp. E. 69° 46' 35". *Error of compass* 25° 50' 55" W. *Deviation* 27° 10' 55" W.
9. Interval 11^d; rate $0^m·55$ *gaining*; Interval $92^d\ 22^h$; acc. rate $0^m\ 51^s$; Green. date, April $14^d\ 22^h\ 29^m\ 49^s$; red. decl. 9° 48' 27" N.; true alt. 22° 23' 50"; red eq. time 0; hour-angle $4^h\ 28^m\ 12^s$·5; M.T. ship April $14^d\ 19^h\ 31^m\ 47^s$·5. *Longitude* 44° 30' 22"·5 W.
10. Green. date, Sept. $22^d\ 18^h\ 27^m\ 52^s$; red. decl. 0° 2' 46" S.; true alt. 32° 18' 38"; sum of logs. 19·369294, true azimuth S. 57° 51' 54" W. *Error* 26° 30' 36" W.; *deviation* 6° 30' 36" W.
11. Time from noon $12^m\ 18^s$; Green. date Sept. $22^d\ 8^h\ 18^m\ 34^s$; red. decl. 0° 7' 16" N.; true alt. 49° 31' 49"; nat. no. 1094; nat. cos. 761843; mer. zen. dist. 40° 22' 21" N. *Latitude* 40° 29' 38" N.
METHOD II. Red. $+ 5'\ 48''$. *Latitude* 40° 29' 39" N.
Towson: Aug. I, 0 : index 19. Aug. II. $+ 5'\ 51''$. *Latitude* 40° 29' 36" N.

12. 1st red. decl. 21° 0′ 47″ S.; 1st red. eq. T. — $12^m 28^s \cdot 3$; 1st true alt. $11^\circ 27′ 7″$; lat. 49° 0′ gives hour-angle $2^h 39^m 59^s$; long. 82° 55½′ W. (A). Lat. 49° 30′ gives hour-angle $2^h 35^m 58^s$; long. 81° 55¼′ W. (A'). 2nd red. decl. 21° 2′ 45″ S.; eq. T. — $12^m 25^s$; true alt. 16° 43′ 47″; lat. 49° 0′ gives hour-angle $1^h 36^m 26^s$; long. 81° 27′ W. (B). Lat. 49° 30′ gives hour-angle $1^h 29^m 4^s$; long. 83° 17½′ W. (B'). The line of bearing at the first observation trends N.E. ¾ E. and S.W. ¾ W. The position of the ship at the second observation is lat. 49° 19′ N., long. 82° 19′ W.

13. Decl. β Centauri 59° 48′ 36″ S.; true alt. 59° 41′ 53″. *Latitude* 29° 30′ 29″ S.

The Curve.—Correct magnetic bearing N. 74° E.

Deviations.—5° W.; 20° E.; 25° E.; 18° E.; 2° W.; 16° W.; 24° W.; 16° W.

Compass courses.—S. 81½° E.; N. 72° W.; N. 48½° E.

Magnetic courses.—S. 65° E.; N. 76° W.; S. 29° W.

Bearings, magnetic.—N. 14° E.; N. 65° W.

EXAMINATION PAPER—No. XXIV, *Pages* 373—374.

1. $6 \cdot 996948 = 9919977$; $9 \cdot 337878 = \cdot 21771$.

2. $0 \cdot 954242 = 9$; $5 \cdot 243038 = 175000$.

3. **True Courses.**—S. 82° E., 42′ dep. course; S. 85° E., 53′; N. 69° E., 47′; South, 22′; N. 34° W., 14′; N. 72° E., 37′; S. 83° E., 35′; S. 82° E., 52′·5 current course. *Diff. lat.* 4′·1 S.; *dep.* 252′·3 E.; *course* S. 89° E.; *dist.* 252′. Lat. in 37° 41′ N.; *diff. long.* 319′ E. Long. in 20° 19′ E.

4. Green. date, November $30^d 19^h 28^m 16^s$; red. decl. 21° 49′ 6″ S.; true alt. 18° 54′ 37″. *Latitude* 49° 16′ 17″ N.

5. Log. of diff. long. $2 \cdot 421932 = $ *diff. long.* 264″·2.

6. Diff. lat. 5968′ N.; mer. diff. lat. 7050′; diff. long. 2121′ E.; tang. $9 \cdot 478352$; *course* N. 16° 44′ 38″ E.; log. of *distance* $= 3 \cdot 794664$; *distance* 6232′.

7. Standard, Brest   constant $+ 1^h 6^m$; $0^h 45^m$ A.M., and $1^h 26^m$ P.M.
   ,,   Brest   ,, $- 0 26$; 11 54 A.M., and   no   P.M.
   ,,   Brest   ,, $+ 2 45$; 2 24 A.M., and   3 5 P.M.
   ,,   Devonport   ,, $- 1 13$; no A.M., and 0 16 P.M.
   ,,   Galway   ,, $+ 1 35$; 2 16 A.M., and 2 52 P.M.
   ,,   Queenstown   ,, $- 1 4$; no A.M., and 0 14 P.M.
   ,,   Brest   ,, $+ 4 28$; corr. for long. — 19^m; $3^h 48^m$ A.M., $4^h 29^m$ P.M.

8. Green. date, February $1^d 17^h 14^m$; red. decl. 16° 49′ 50″; sine $9 \cdot 994951$; true amp. W. 81° 17′ S. *Error* 17° 9′ E.; *deviation* 19° 11′ W.

9. Interval 62^d, rate $2^s \cdot 5$ *losing*, interval $76^d 9^h$, accumulated rate $+ 3^m 11^s$, Green. date August $31^d 9^h 11^m 56^s$, red. decl. 8° 27′ 30″ N., red. eq. T. $+ 0^m 3^s \cdot 60$, true alt. 45° 14′ 40″, hour-angle $2^h 56^m 28^s$, M.T. ship Aug. $31^d 2^h 56^m 32^s$. Longitude 93° 51′ 0″ W.

10. Green. date, March $20^d 3^h 24^m 28^s$; red. decl. 0° 1′ 39″ S.; true alt. 53° 10′ 51″; sum of logs. 17·948571; true azimuth S. 10° 48′ 58″ W. *Error* 13° 37′ 43″ E.: *deviation* 38° 27′ 43″ E.

11. Time from noon $10^m 34^s$; Green. date, March $20^d 16^h 55^m 26^s$; red. decl. 0° 11′ 49″ N.; true alt. 71° 18′ 34″; nat. no. 1024; mer. zen. dist. 18° 30′ 38″ N. *Latitude* 18° 42′ 25″ N.

12. 1st red. decl. 13° 35′ 56″ N.; 1st eq. T. $+ 4^m 1^s$; true alt. 28° 25′ 3″; lat. 49° 30′ gives hour-angle $4^h 7^m 41^s$; long. 179° 0′ E. (A); lat. 50° 0′ gives hour-angle $4^h 6^m 59^s$; long. 179° 10½′ E. (A'). 2nd red. decl. 13° 31′ 10″ N.; 2nd red. eq. T. $+ 3^m 58^s \cdot 2$; true alt. 46° 32′ 43″; lat. 49° 30′ gives hour-angle $1^h 59^m 3^s$; long. 180° 40¼′ E., or 179° 19¾′ W. (B). Lat. 50° 0′ gives hour-angle $1^h 55^m 50^s$; long. 179° 52′ E. (B') The line of bearing trends N. by E. and S. by W.; the position of ship at the time of second observation is lat. 50° 3½′ N., long. 179° 44′ E.

13. Murkab's decl. 14° 34′ 16″ N.: true alt. 33° 16′ 27″. *Latitude* 42° 9′ 17″ S. *Norie.*
    ,,   ,,   ,,   33 16 31   ,,   42 9 13 S. *Raper.*

M M M

The Curve.—Correct magnetic bearing N. 12° W.
Deviations.—10° E.: 10° E.: 15° W.: 23° W.: 5° W.: 14° E.: 14° E.: 6° E.
Compass courses.—S. 56° E.: S. 75° W.: N. 9° W.
Magnetic courses.—N. 83° E.: N. 76° W.: N. 10° E.
Bearings, magnetic.—N. 76° W.: North.

EXAMINATION PAPER—No. XXV, Pages 375—376.

1. $6{\cdot}742037 = 5521240{\cdot}5$. $5{\cdot}109096 = 128557$.
2. $1{\cdot}903505 = {\cdot}800767$. $7{\cdot}922575 = 83670961{\cdot}5$.
3. True Courses.—N. 2° W., 21° dep. course; N. 51° E., 16·6; S. 25° W., 13·4; N. 5° W., 19·3; S. 57° W., 19·1; N. 70° W., 48'; N. 33° E., 21·3; N. 48° E., 49' current course. Diff. lat. 95·2 N.; dep. 7·9; course N. 6° W.; dist. 96'. Lat. in 63° 55' N.; mid. lat. 63° 7'; diff. long. 18' W. Long. in 64° 57½' W.
4. Green. date, May 31ᵈ 21ʰ 1ᵐ 20ˢ, red. decl. 22° 3' 56" N., true alt. 72° 28' 57". Lat. 4° 32' 53" N. Raper: True alt. 72° 28' 57". Latitude 4° 32' 53" N.
5. Log. of diff. long. 1·804211 = diff. long. 63°·71. Long. in 179° 11' 17" E.
6. Diff. lat. 3757' S., mer. diff. lat. 4255', diff. long. 7560' E., log. tangent 10·249622, course S. 60° 38' E., distance 7661'.
7. Standard, Leith — 1ʰ 17ᵐ : 11ʰ 59ᵐ A.M., and no P.M.
    „ Leith — 0 52 : 0 3 A.M., and 0ʰ 24ᵐ P.M.
    „ Leith — 2 55 : 10 21 A.M., and 10 40 P.M.
    „ Brest + 6 57 : 9 18 A.M., and 9 37 P.M.
    „ Dover + 3 18 : 0 35 A.M., and 0 57 P.M.
8. Green. date, Dec. 21ᵈ 17ʰ 30ᵐ 48ˢ, red. decl. 23° 27' 5" S., true amp. W. 84° 5' S. Error of compass 14° 21' E. Deviation 13° 49' W.
9. Green. date, Jan. 28ᵈ 16ʰ 36ᵐ 50ˢ, red. decl. 17° 57' 24" S., red. eq. time + 13ᵐ 22ˢ, true alt. 17° 51' 42", D. lat. made since noon 11' 6" added to lat. at noon (28° 45') gives lat. at sights 28° 56' 6" N., hour-angle 3ʰ 47ᵐ 6ˢ. Long. at sight 170° 54' 30" E., diff. long. since noon 19' W. Subt. from long. at sights as the ship was farther East at noon. Long. at noon 171° 13' 27" E.
10. Green. date, July 9ᵈ 17ʰ 50ᵐ 10ˢ, red. decl. 22° 15' 58" N., true alt. 14° 36' 40", sum of logs. 19·174552, true azimuth N. 45° 29' 16' W. Error of compass 47° 44' 16" W. Deviation 54° 29' 16" W.
11. Green. date, Nov. 28ᵈ 15ʰ 58ᵐ 31ˢ; time from noon 12ᵐ 11ˢ; red. decl 21° 28' 8" S.; true alt. 74° 15' 51"; nat. no. 1306; mer. zen. dist. 15° 27' 27" N.; latitude 6° 0' 41" S.

Method II. + 17' 12". Latitude 6° 0' 11" S.

Towson: Hour-angle exceeds the limits of Table.

12. 1st red. decl. 10° 18' 1" S.; 1st red. eq. T. — 15ᵐ 4ˢ·85; 1st true alt. 29° 47' 31" lat. 49° 20' gives hour-angle 0ʰ 37ᵐ 44ˢ; long. 117° 50'] E. (A). Lat. 49° 50' gives hour-angle 0ʰ 13ᵐ 41ˢ; long. 111° 55½' E. (A') 2nd red. decl. 10° 20' 55" S., 2nd eq. T. — 15ᵐ 6ˢ·2, 2nd true alt. 12° 28' 41". Lat. 49° 20' gives hour-angle 3ʰ 46ᵐ 39ˢ; long. 116° 40¾' E. (B). Lat. 49° 50' gives hour-angle 3ʰ 44ᵐ 39ˢ; long. 116° 10¾' E. (B') The line of bearing at first observation trends E. ⅜ S. (southerly) and W. ⅜ N. (northerly). The position of the ship at the second observation is lat. 49° 52' N., long. 116° 9' E.
13. Star's decl. 8° 20' 29" S.; true alt. 52° 11' 54". Latitude 29° 27' 37" N. Norie. Raper: True alt. 52° 11' 45". Latitude 29° 27' 46" N.

The Curve.—Correct magnetic bearing S. 8° E.
Deviations.—15° E.; 3° E.; 13° W.; 28° W.; 14° W.; 10° E.; 13° E.; 14° E.
Compass courses.—S. by W.; S. 87° E.; S. 13° E.; S. 85½° W.
Magnetic courses.—N. 20° W.; N. 54° W.; S. 82° E.; N. 31° E.
Bearings, magnetic.—S. 68° W.; West.

EXAMINATION PAPER—No. XXVI, Pages 377—378.

1. $7{\cdot}891486 = 77890714$. $\bar{3}{\cdot}763323 = {\cdot}000000057986$.
2. $2{\cdot}900314 = 794{\cdot}904$. $7{\cdot}950265 = 89179388$.

3. **True Courses.**—East, 30′ dep. course; S. 88° E., 54′·2; N. 48° E., 41′·6; S. 76° W., 14′·4; S. 22° E., 46′·5; N. 9° W., 29′·1; S. 51° E., 49′·2; N. 79° E., 40′·8 current course. Diff. lat. 15′·2 S.; dep. 192′·2 E.: course S. 85½° E.: dist. 193′. Latitude in 59° 34′ N.: diff. long. 380½′ E.: longitude in 37° 49½′ W.

4. Green. date, Sept. 30^d 18^h 21^m 20^s; red. decl. 3^h 9^m 32^s S.: true alt. 56° 56′ 24″. Lat. 29° 54′ 4″ N.

Raper: True alt. 56° 56′ 15″ N., latitude 29° 54′ 13″ N.

4*. Green. date, July 2^d 8^h 59^m: red. decl. 23° 0′ 51″ N.: true alt. 10° 25′ 59″. Latitude 78° 25′ 58″ N.

When the sun is observed below the pole (at midnight), instead of subtracting the true alt. from 90° add 90° to it; the lat. will be of the same name as the declination.

5. Diff. long. 369′·7 E., or 6° 10′ E.: long. in 21° 18′ W. Compass course N. 61° 29′ E., or N.E. by E. ¼ E., nearly.

6. Diff. lat. 2000′ N.: mer. diff. lat. 2043′: diff. long. 2043′ W.: log. tang. 10·000000; true course N. 45° W. Compass course N. 32° 7½′ W. Dist. 2828′·4.

7.
Standard, Brest	constant	— 0^h 17^m;	0^h 25^m A.M., and	1^h 1^m P.M.
„ Brest	„	+ 0 50	1 32 A.M., and	2 8 P.M.
„ Sunderland	„	+ 0 49	0 46 A.M., and	1 26 P.M.
„ Sunderland	„	— 0 52	11 45 A.M., and	no P.M.
„ Leith	„	— 0 52	10 52 A.M., and	11 26 P.M.
„ Brest	„	— 1 17	corr. for long. — 1^m; no A.M., and noon.

8. Green. date, April 25^d 0^h 34^m 48^s; red. decl. 13° 15′ 18″ N.; true amp. W. 25° 6′ N. Error 67° 42′ W; deviation 31° 52′ W.

9. Green. date, Aug. 23^d 18^h 16^m 50^s; red. decl. 11° 8′ 40″ N.; red. eq. time + 2^m 16^s; true alt. 37° 26′ 49″; lat. at sight 37° 38′ 18″ N.; hour-angle 3^h 23^m 36^s. Long. at sight 35° 27′ 30″ E.; diff. long. 10′ 50″ W.; long. at noon 35° 16′ 40″ E.

10. Green. date, Oct. 31^d 22^h 15^m 44^s; red. decl. 14° 28′ 39″ S.; true alt. 12° 23′ 17″; sum of logs. 19·222304; true azimuth S. 48° 13′ 42″ E. Error 51° 2′ 21″ W.; deviation 17° 42′ 21″ W.

11. Green. date, May 29^d 9^h 58^m 28^s; time from noon 4^m 12^s; red. decl. 21° 42′ 46″ N.; true alt. 67° 53′ 11″; nat. no. 160; mer. zen. dist. 22° 5′ 21″ S. Latitude 0° 22′ 35″ S.

METHOD II. + 1′ 25″. Latitude 0° 22′ 38″ S.

Towson: Aug. I, + 0′ 12″; index 2. Aug. II, + 1′ 18″. Latitude 0° 22′ 33″ S.

12. 1st red. decl. 7° 54′ 3″ N.; 1st eq. T. + 1^m 22^s·11; 1st true alt. 45° 28′ 20″; lat. 51° 10′ gives hour-angle 0^h 50^m 56^s; long. 122° 58½′ E. (A). Lat. 51° 40′ gives hour-angle 0^h 39^m 51^s; long. 120° 12½′ E. (A′) 2nd red. decl. 7° 57′ 37″ N.; 2nd eq. T + 1^m 19^s; 2nd true alt. 18° 26′ 12″; lat 51° 10′ gives hour-angle 4^h 41^m 35^s; long. 122° 45½′ E. (B). Lat. 51° 40′ gives hour-angle 4^h 40^m 59^s; long. 122° 36½′ (B′). The line of bearing at the first observation trends E. by S. ¼ S. (southerly) and W. by N. ¼ N. (northerly). The position of the ship at the second observation is lat 51° 10′ N., long. 122° 45½′ E.

13. Star's decl. 6° 48′ 49″ N.; true alt. 28° 58′ 36″. Latitude 54° 12′ 35″ S.

The Curve.—Correct magnetic bearing S. 49° E.

Deviations.—20° W.; 16° W.; 2° W.; 14° E.; 20° E.; 15° E.; 1° W.; 11° W.

Compass courses.—S. 33° W.; S. 43° E.; N. 69° W.; N. 24° E.

Magnetic courses.—N. 9° E.; S. 74° W.; S. 16° W.; S. 60° W.

Bearings, magnetic.—S. 7½° E.; S. 88° W.

INDEX ERROR, Page 386.

1.	+ 2′ 15″	Semid. 15′ 57″	2.	— 1′ 40″	Semid. 15′ 45″
3.	+ 27 45	„ 16 12	4.	+ 0 52	„ 16 49
5.	+ 38 20	„ 15 50	6.	+ 35 5	„ 16 17·5

EXERCISES ON THE CHART.

FOR ONLY MATE, FIRST MATE, AND MASTER.

North Sea, Pages 406—407.

1.	Course W. $\frac{1}{2}$ S.	Dist. 49′	2.	Course S.W. $\frac{1}{4}$ W.	Dist. 163′
3.	„ S.E. $\frac{1}{4}$ S.	„ 149	4.	„ N. $\frac{1}{2}$ E.	„ 35
5.	„ N.W. by W.	„ 66	6.	„ N.W. $\frac{1}{2}$ N.	„ 30
7.	True course N.E. $\frac{1}{2}$ E. / Mag. do., E. by N. $\frac{1}{4}$ N.	329	8.	True course S.E. $\frac{1}{2}$ S. / Mag. do., S. by E. $\frac{1}{2}$ E.	351

9. The ship is in latitude 55° 59′ N., longitude 2° 40′ E., and must sail S.E. $\frac{1}{2}$ S. (mag.) 209 miles.

10. The place of meeting was lat. 56° 5′ N., long. 3° 11′ E.: the course steered by the ship from Heligoland was N.W. $\frac{3}{4}$ W. (true), and by the ship from Hartlepool was N.E. by E. $\frac{1}{2}$ E.

11.	Lat. 55° 15$\frac{1}{2}$′ N.	Long. 1° 11′ W.	Course S.S.W.	Distance 34′
12.	„ 57 16 N.	„ 1 28 W.	„ S.S.W. $\frac{1}{4}$ W.	„ 96
13.	„ 60 4 N.	„ 0 24 W.	„ S.W. $\frac{1}{2}$ S. nearly	„ 159
14.	„ 54 27 N.	„ 0 3 E.	„ S. $\frac{3}{4}$ E.	„ 69
15.	„ 53 20 N.	„ 1 36 E.	„ N.N.W. $\frac{3}{4}$ W.	„ 75$\frac{1}{2}$

16. 1st Station—Lat. 54° 24′ N., long. 0° 20′ W., distance 6 miles.
 2nd „ „ 54 24 N., „ 0 1 W., „ 14 „

17. 1st „ „ 55 8 N., „ 1 8 W., „ 12 „
 2nd „ „ 54 53 N., „ 1 1 W., „ 16 „

English and Bristol Channels, and South Coast of Ireland, Pages 407—408.

1.	Course S.W. by W.	Dist. 21′	2.	Course S.S.E. $\frac{1}{2}$ E.	Dist. 37′
3.	„ N. by E. $\frac{1}{2}$ E.	„ 44$\frac{1}{2}$	4.	„ N. by E. nearly	„ 25
5.	„ N.N.E. $\frac{1}{4}$ E.	„ 25$\frac{1}{2}$	6.	„ N.E. $\frac{3}{4}$ E.	„ 72$\frac{1}{2}$
7.	„ E. by N. $\frac{1}{4}$ N.	„ 50	8.	„ N. by E. $\frac{1}{4}$ E.	„ 21
9.	„ N. by W. $\frac{3}{4}$ W.	„ 65	10.	„ N. $\frac{3}{4}$ W.	„ 67

11.	Lat. 49° 48′ N.	Long. 6° 7$\frac{1}{2}$′ W.	Course E. by S.	Distance 37′
12.	„	„	„ E. $\frac{1}{4}$ N.	„ 40
13.	„ 50 28 N.	„ 2 9 W.	„	„ ..
14.	„	„	„ N. by W. $\frac{1}{2}$ W.	„ 134
15.	„ 52 2 N.	„ 6 19 W.	„ S.S.E. $\frac{1}{2}$ E.	„ 30
16.	„ 52 4 N.	„ 1 58 E.	„ N. $\frac{3}{4}$ E.	„ 27
17.	„ 51 25 N.	„ 4 59 W.	„ N.N.W. $\frac{1}{2}$ W.	„ 32
18.	„ 49 51$\frac{1}{2}$N.	„ 5 35 W.	„ N.W. by W. $\frac{1}{2}$ W.	„ 30
19.	„ 51 41 N.	„ 5 39 W.	„ S.E. by S. $\frac{1}{2}$ S.	„ 46$\frac{1}{2}$
20.	„ 50 39$\frac{1}{2}$N.	„ 0 35 E.	Dist. off Dungeness Light, 21 miles. / „ off Beachy Head Lt., 15 „	

21. 1st Course—S.E. by S. (true), or S. by E. (mag.), distance 24 miles.
 2nd „ N. by E. (true), or N.E. by N. (mag.), distance 24 miles.

SOUNDINGS.

DEPTHS, &c., *Page* 393.

1. Time from high water 1^h 31^m: half-range for day 18 feet 1 inch: Table B + 12 feet 9 inches: 12 feet 9 inches + 18 feet 1 inch + 24 feet (4 fathoms). Depth of water required 55 feet 7 inches, or 9$\frac{1}{4}$ fathoms.

2. Time *before* high water 1^h 58^m: half-range for day 7 feet 6 inches: Table B + 34 feet 9 inches. Corr. to low water 13 feet 3 inches. Depth 51 feet 3 inches, or 8$\frac{1}{2}$ fathoms.

4. Time from high water 0^h 15^m: half-range for day 20 feet 11 inches: Table B + 20 feet 7 inches. Corr to low water 43 feet 7 inches. Water below sounding 7 feet 7 inches, or the ship is found to be 7 feet 7 inches dry at low water.

6. Time from high water 3^h 59^m: half-range for day 5 feet 5 inches: Table B − 2 feet 9 inches. Sounding by chart 6 feet 1 inch, or 10 fathoms (better).

www.ingramcontent.com/pod-product-compliance
Lightning Source LLC
Chambersburg PA
CBHW051856300426
44117CB00006B/424